Essentials of Process Control

McGraw-Hill Chemical Engineering Series

Building the Literature of a Profession

Fifteen prominent chemical engineers first met in New York more than 60 years ago to plan a continuing literature for their rapidly growing profession. From industry came such pioneer practitioners as Leo H. Baekeland, Arthur D. Little, Charles L. Reese, John V. N. Dorr, M. C. Whitaker, and R. S. McBride. From the universities came such eminent educators as William H. Walker, Alfred H. White, D. D. Jackson, J. H. James, Warren K. Lewis, and Harry A. Curtis. H. C. Parmelee, then editor of *Chemical and Metallurgical Engineering,* served as chairman and was joined subsequently by S. D. Kirkpatrick as consulting editor.

After several meetings, this committee submitted its report to the McGraw-Hill Book Company in September 1925. In the report were detailed specifications for a correlated series of more than a dozen texts and reference books which have since become the McGraw-Hill Series in Chemical Engineering and which became the cornerstone of the chemical engineering curriculum.

From this beginning there has evolved a series of texts surpassing by far the scope and longevity envisioned by the founding Editorial Board. The McGraw-Hill Series in Chemical Engineering stands as a unique historical record of the development of chemical engineering education and practice. In the series one finds the milestones of the subject's evolution: industrial chemistry, stoichiometry, unit operations and processes, thermodynamics, kinetics, and transfer operations.

Chemical engineering is a dynamic profession, and its literature continues to evolve. McGraw-Hill, with its editor B. J. Clark and its consulting editors, remains committed to a publishing policy that will serve, and indeed lead, the needs of the chemical engineering profession during the years to come.

The Series

Bailey and Ollis: *Biochemical Engineering Fundamentals*
Bennett and Myers: *Momentum, Heat, and Mass Transfer*
Brodkey and Hershey: *Transport Phenomena: A Unified Approach*
Carberry: *Chemical and Catalytic Reaction Engineering*
Constantinides: *Applied Numerical Methods with Personal Computers*
Coughanowr: *Process Systems Analysis and Control*
de Nevers: *Air Pollution Control Engineering*
de Nevers: *Fluid Mechanics for Chemical Engineers*
Douglas: *Conceptual Design of Chemical Processes*
Edgar and Himmelblau: *Optimization of Chemical Processes*
Gates, Katzer, and Schuit: *Chemistry of Catalytic Processes*
Holland: *Fundamentals of Multicomponent Distillation*
Katz and Lee: *Natural Gas Engineering: Production and Storage*
King: *Separation Processes*
Lee: *Fundamentals of Microelectronics Processing*
Luyben: *Process Modeling, Simulation, and Control for Chemical Engineers*
Luyben and Luyben: *Essentials of Process Control*
McCabe, Smith, and Harriott: *Unit Operations of Chemical Engineering*
Marlin: *Process Control: Designing Processes and Control Systems for Dynamic Performance*
Middlemann and Hochberg: *Process Engineering Analysis in Semiconductor Device Fabrication*
Perry and Chilton (Editors): *Perry's Chemical Engineers' Handbook*
Peters: *Elementary Chemical Engineering*
Peters and Timmerhaus: *Plant Design and Economics for Chemical Engineers*
Reid, Prausnitz, and Poling: *Properties of Gases and Liquids*
Smith: *Chemical Engineering Kinetics*
Smith and Van Ness: *Introduction to Chemical Engineering Thermodynamics*
Treybal: *Mass Transfer Operations*
Valle-Reistra: *Project Evaluation in the Chemical Process Industries*
Wentz: *Hazardous Waste Management*

Essentials of Process Control

Michael L. Luyben

Du Pont Central Research and Development
Experimental Station

William L. Luyben

Department of Chemical Engineering
Lehigh University

THE McGRAW-HILL COMPANIES, INC.

New York St. Louis San Francisco Auckland Bogotá Caracas Lisbon
London Madrid Mexico City Milan Montreal New Delhi
San Juan Singapore Sydney Tokyo Toronto

McGraw-Hill

A Division of The ***McGraw-Hill*** *Companies*

ESSENTIALS OF PROCESS CONTROL

This book is printed on acid-free paper.

1 2 3 4 5 6 7 8 9 0 DOC DOC 0 9 0 8 7 6

ISBN 0-07-039172-6

This book was set in Times Roman by Publication Services, Inc.
The editors were B. J. Clark and John M. Morriss;
the production supervisor was Denise L. Puryear.
The jacket was designed by Wanda Kossak.
R. R. Donnelley & Sons Company was printer and binder.

Library of Congress Cataloging-in-Publication Data

Luyben, Michael L., (date)
Essentials of Process control / Michael L. Luyben, William L. Luyben
p. cm.
Includes index.
ISBN 0-07-039172-6. — ISBN 0-07-039173-4
1. Chemical process control. I. Luyben, William L. II. Title.
TP155.75.L89 1997
660'.2815—dc20 96-8642

ABOUT THE AUTHORS

William L. Luyben has devoted over 40 years to his profession as teacher, researcher, author, and practicing engineer. Dr. Luyben received his B.S. in Chemical Engineering from Pennsylvania State University in 1955. He then worked for Exxon for five years at the Bayway Refinery and at the Abadan Refinery (Iran) in plant technical service and petroleum processing design. After earning his Ph.D. from the University of Delaware in 1963, Dr. Luyben worked for the Engineering Department of Du Pont in process dynamics and control of chemical plants.

Dr. Luyben has taught at Lehigh University since 1967 and has participated in the development of several innovative undergraduate courses, from the introductory course in mass and energy balance through the capstone senior design course and an interdisciplinary controls laboratory. He has directed the theses of more than 40 graduate students and has authored or coauthored six textbooks and over 130 technical papers. Dr. Luyben is an active consultant for industry in the area of process control. He was the recipient of the Eckman Education Award in 1975 and the Instrumentation Technology Award in 1969 from the Instrument Society of America.

William L. Luyben is currently a Professor of Chemical Engineering at Lehigh University.

Michael L. Luyben received his B.S. in Chemical Engineering (1987) and B.S. in Chemistry (1988) from Lehigh University. While a student, he worked during several summers in industry, including two summers with Du Pont and one summer with Bayer in Germany. After completing his Ph.D. in Chemical Engineering at Princeton University in 1993, working with Professor Chris Floudas, he joined the Process Control and Modeling Group in the Central Research and Development Department of Du Pont. His work has focused on the dynamic modeling and control of chemical and polymer plants. He has worked on plant improvement studies and on the design of new facilities. Luyben has authored a number of papers on plantwide control and on the interaction of process design and process control.

Michael L. Luyben is currently a research Engineer with Du Pont's Central Research and Development Department.

To Janet Nichol Luyben—mother, wife, friend, loving grandmother, avid gardener, community volunteer, softball queen extraordinaire—for 34 years of love, care, level-headed financial advice, and many pieces of Grandmother Lester's apple pie.

CONTENTS

PART 4 Multivariable Processes

PART 5 Sampled-Data Systems

PART 6 Identification

PREFACE

The field of process control has grown rapidly since its inception in the 1950s. Direct evidence of this growth in the body of knowledge is easily found by comparing the lengths of the textbooks written over this time period. The first process control book (Cealgske, 1956) was a modest 230 pages. The popular Coughanowr and Koppel (1965) text was 490 pages. The senior author's first edition (1973) was 560 pages. The text by Seborg et al. (1989) was 710 pages. The recently published text by Ogunnaike and Ray (1994) runs 1250 pages!

It seems obvious to us that more material has been developed than can be taught in a typical one-semester undergraduate course in process control. Therefore, a short and concise textbook is needed that presents only the essential aspects of process control that every chemical engineering undergraduate ought to know. The purpose of this book is to fulfill this need.

Our intended audience is junior and senior undergraduate chemical engineering students. The book is meant to provide the fundamental concepts and the practical tools needed by all chemical engineers, regardless of the particular area they eventually enter. Since many advanced control topics are not included, those students who want to specialize in control can go further by referring to more comprehensive texts, such as Ogunnaike and Ray (1994).

The mathematics of the subject are minimized, and more emphasis is placed on examples that illustrate principles and concepts of great practical importance. Simulation programs (in FORTRAN) for a number of example processes are used to generate dynamic results. Plotting and analysis are accomplished using computer-aided software (MATLAB).

One of the unique features of this book involves our coverage of two increasingly important areas in process design and process control. The first is the interaction between steady-state design and control. The second is plantwide control with particular emphasis on the selection of control structures for an entire multi-unit process. Other books have not dealt with these areas in any quantitative way. Because we feel that these subjects are central to the missions of process design engineers and process control engineers, we devote two chapters to them.

We have injected some examples and problems that illustrate the interdisciplinary nature of the control field. Most control groups in industry utilize the talents of engineers from many disciplines: chemical, mechanical, and electrical. All engineering fields use the same mathematics for dynamics and control. Designing control systems for chemical reactors and distillation columns in chemical engineering has direct parallels with designing control systems for F-16 fighters, 747 jumbo jets, Ferrari sports cars, or garbage trucks. We illustrate this in several places in the text.

This book is intended to be a learning tool. We try to educate our readers, not impress them with elegant mathematics or language. Therefore, we hope you find the book readable, clear, and (most important) useful.

When you have completed your study of this book, you will have covered the essential areas of process control. What ideas should you take away from this study and apply toward the practice of chemical engineering (whether or not you specialize as a control engineer)?

1. The most important lesson to remember is that our focus as engineers must be on the process. We must understand its operation, objectives, constraints, and uncertainties. No amount of detailed modeling, mathematical manipulation, or supercomputer exercise will overcome our ignorance if we ignore the true subject of our work. We need to think of Process control with a capital *P* and a small *c*.
2. A steady-state analysis, although essential, is typically not sufficient to operate a chemical process satisfactorily. We must also understand something about the dynamic behavior of the individual units and the process as a whole. At a minimum, we need to know what characteristics (deadtimes, transport rates, and capacitances) govern the dynamic response of the system.
3. It is always best to utilize the simplest control system that will achieve the desired objectives. Sophistication and elegance on paper do not necessarily translate into effective performance in the plant. Careful attention must be paid to the practical consequences of any proposed control strategy. Our control systems must ensure safe and stable operation, they must be robust to changes in operating conditions and process variables, and they must work reliably.
4. Finally, we must recognize that the design of a process fundamentally determines how it will respond dynamically and how it can be controlled. Considerations of controllability need to be incorporated into the process design. Sometimes the solution to a control problem does not have anything to do with the control system but requires some modification to the process itself.

If we keep these ideas in mind, then we can apply the basic principles of process control to solve engineering problems.

Michael L. Luyben
William L. Luyben

Essentials of Process Control

CHAPTER 1

Introduction

As the field of process control has matured over the last 30 years, it has become one of the core areas in chemical engineering along with thermodynamics, heat transfer, mass transfer, fluid mechanics, and reactor kinetics. Any chemical engineering graduate should have some knowledge not only of these traditional areas but also of the fundamentals of process control. For those of us who have been part of this period of development, the attainment of parity with the traditional areas has been long overdue.

The literature in process control is enormous: over a dozen textbooks and thousands of papers have been published during the last three decades. This body of knowledge has become so large that it is impossible to cover it all at the undergraduate level. Therefore, we present in this book only those topics we feel are essential for gaining an understanding of the basic principles of process control.

One of the important themes that we emphasize is the need for control engineers to understand the process—its operation, constraints, design, and objectives. The way the plant is designed has a large impact on how it should be controlled and what level of control performance can be obtained. As the mechanical engineers say, you can't make a garbage truck drive like a Ferrari!

We present in the following section three simple examples that illustrate the importance of dynamic response; show the structure of a single-input, single-output conventional control system; and illustrate a typical plantwide control system. Throughout the rest of the book, many more real-life examples and problems are presented. All of these are drawn from close to 50 years of collective experience of the authors in solving practical control problems in the chemical and petroleum industries.

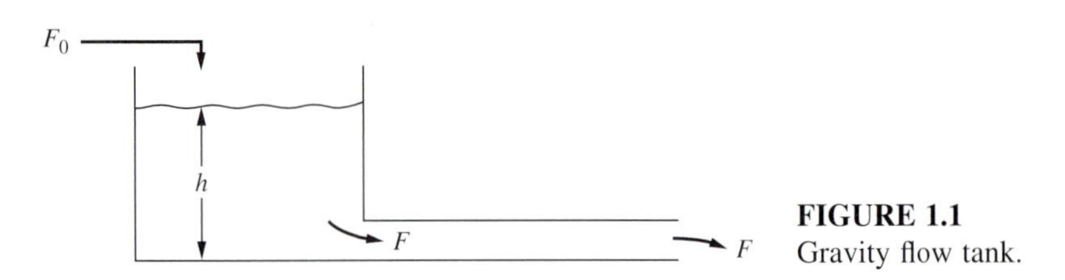

FIGURE 1.1
Gravity flow tank.

1.1 EXAMPLES OF PROCESS DYNAMICS AND CONTROL

EXAMPLE 1.1. Figure 1.1 shows a tank into which an incompressible (constant-density) liquid is pumped at a variable rate F_0 (gal/min). This inflow rate can vary with time because of changes in operations upstream. The height of liquid in the vertical cylindrical tank is h (ft). The flow rate out of the tank is F (gal/min).

Now F_0, h, and F will all vary with time and are therefore functions of time t. Consequently, we use the notation $F_{0(t)}$, $h_{(t)}$, and $F_{(t)}$. Liquid leaves the base of the tank via a long horizontal pipe and discharges into the top of another tank. Both tanks are open to the atmosphere.

Let us look first at the steady-state conditions. By "steady state" we mean the conditions when nothing is changing with time or when time has become very large. Mathematically this corresponds to having all time derivatives equal to zero or allowing time to approach infinity. At steady state the flow rate out of the tank must equal the flow rate into the tank: $\overline{F}_0 = \overline{F}$. In this book we denote the steady-state value of a variable by an overscore or bar.

For a given $\overline{F}$, the height of liquid in the tank at steady state $\overline{h}$ is a constant, and a larger flow rate requires a higher liquid level. The liquid height provides just enough hydraulic pressure head at the inlet of the pipe to overcome the frictional pressure losses of the liquid flowing down the pipe.

The steady-state design of the tank involves the selection of the height and diameter of the tank and the diameter of the exit pipe. For a given pipe diameter, the tank height must be large enough to prevent the tank from overflowing at the maximum expected flow rate. Thus, the design involves an engineering trade-off, i.e., an economic balance between the cost of a taller tank and the cost of a bigger-diameter pipe. A larger pipe diameter requires a lower liquid height, as illustrated in Fig. 1.2. A conservative design engineer would probably include a 20 to 30 percent over-design factor in the tank height to permit future capacity increases.

Safety and environmental reviews would probably recommend the installation of a high-level alarm and/or an interlock (a device to shut off the feed if the level gets too high) to guarantee that the tank could never overfill. The tragic accidents at Three Mile Island, Chernobyl, and Bhopal illustrate the need for well-designed and well-instrumented plants.

Now that we have considered the traditional steady-state design aspects of this fluid flow system, we are ready to examine its dynamics. What happens dynamically if we change F_0, and how will $h_{(t)}$ and $F_{(t)}$ vary with time? Obviously, F eventually has to end up at the new value of F_0. We can easily determine from the steady-state design curve of Fig. 1.2 where h will be at the new steady state. But what dynamic paths or time trajectories will $h_{(t)}$ and $F_{(t)}$ take to get to their new steady states? Fig. 1.3 shows two possible transient responses (curves 1 or 2) for F and h. Curves 1 show gradual increases

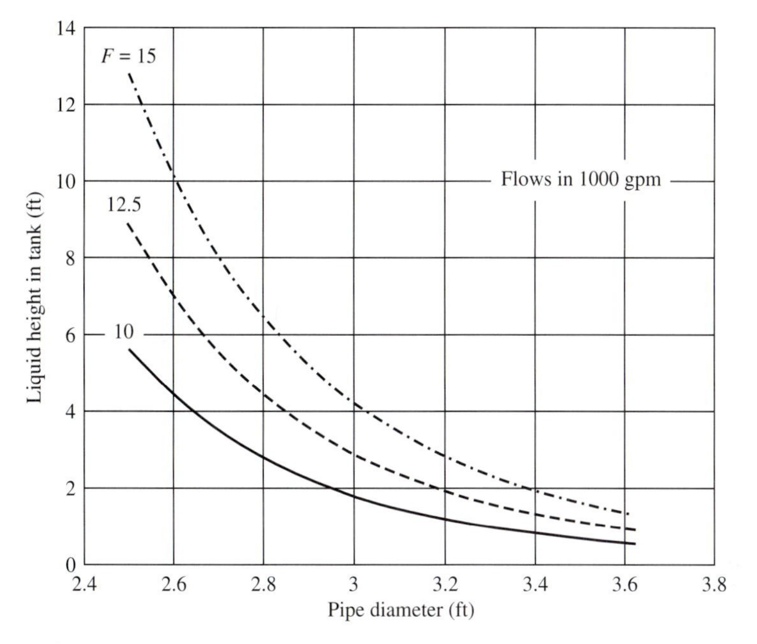

FIGURE 1.2
Gravity flow tank.

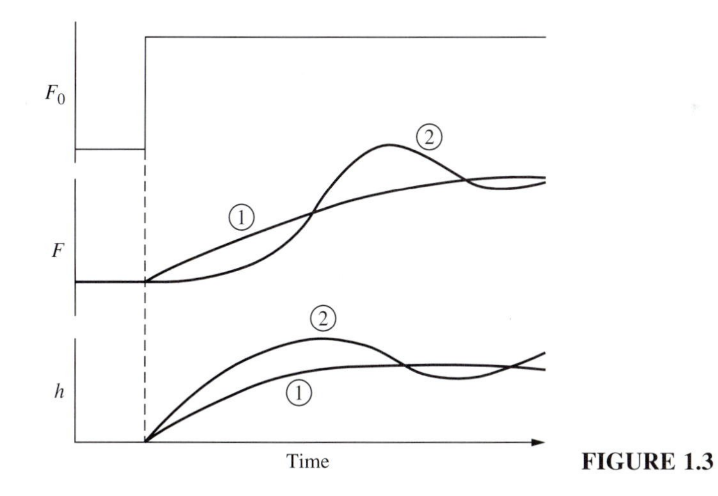

FIGURE 1.3

in h and F to their new steady-state values. Curves 2, however, show the liquid height rising above ("overshooting") its final steady-state value before settling out at the new liquid level. Clearly, if the peak of the overshoot in h were above the top of the tank, we would be in trouble.

Our steady-state design calculations tell us nothing about the dynamic response of the system. They tell us where we start and where we end but not how we get there. This kind of information is revealed by a study of the system's dynamics. ■

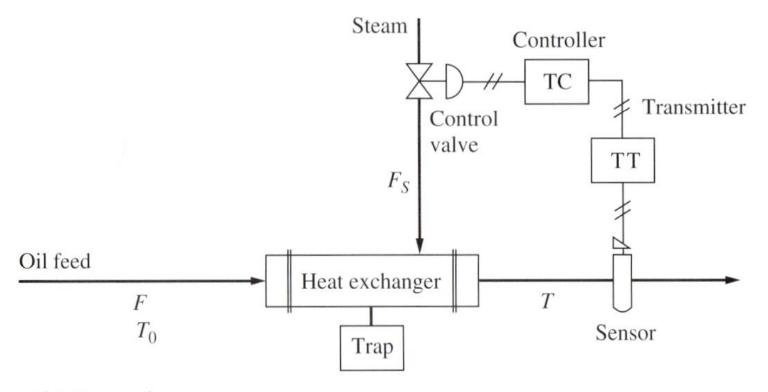

FIGURE 1.4

EXAMPLE 1.2. Consider the heat exchanger sketched in Fig. 1.4. An oil stream passes through the tube side of a tube-in-shell heat exchanger and is heated by condensing steam on the shell side. The steam condensate leaves through a steam trap (a device that permits only liquid to pass through it, thus preventing "blow-through" of the steam vapor). We want to control the temperature of the oil leaving the heat exchanger. To do this, a thermocouple is inserted in a thermowell in the exit oil pipe. The thermocouple wires are connected to a "temperature transmitter," an electronic device that converts the millivolt thermocouple output to a 4- to 20-mA "control signal." This current signal is sent to a temperature controller, an electronic, digital, or pneumatic device that compares the desired temperature (the "setpoint") with the actual temperature and sends out a signal to a control valve. The temperature controller opens the steam valve a little if the temperature is too low and closes the valve a little if the temperature is too high.

We consider all the components of this temperature control loop in more detail later in this book. For now we need only appreciate the fact that the automatic control of some variable in a process requires the installation of a sensor, a transmitter, a controller, and a final control element (usually a control valve). A major component of this book involves learning how to decide what type of controller should be used and how it should be "tuned," i.e., how the adjustable tuning parameters in the controller should be set so that we do a good control job. ■

EXAMPLE 1.3. Our third example illustrates a typical control scheme for a simplified version of an entire chemical plant. Figure 1.5 gives a sketch of the process configuration and its control system. Two liquid feeds are pumped into a reactor, in which they react to form products. The reaction is exothermic, and therefore heat must be removed from the reactor. This is accomplished by adding cooling water to a jacket surrounding the reactor. The reactor effluent is pumped through a preheater into a distillation column that splits it into two product streams.

Traditional steady-state design procedures are used to specify the various pieces of equipment in the plant:

Fluid mechanics: pump heads, rates, and power; piping sizes; column tray layout and sizing; heat-exchanger tube and shell side baffling and sizing

Heat transfer: reactor heat removal; preheater, reboiler, and condenser heat transfer areas; temperature levels of steam and cooling water

Chemical kinetics: reactor size and operating conditions (temperature, pressure, catalyst, etc.)

Thermodynamics and mass transfer: operating pressure, number of plates and reflux ratio in the distillation column; temperature profile in the column; equilibrium conditions in the reactor

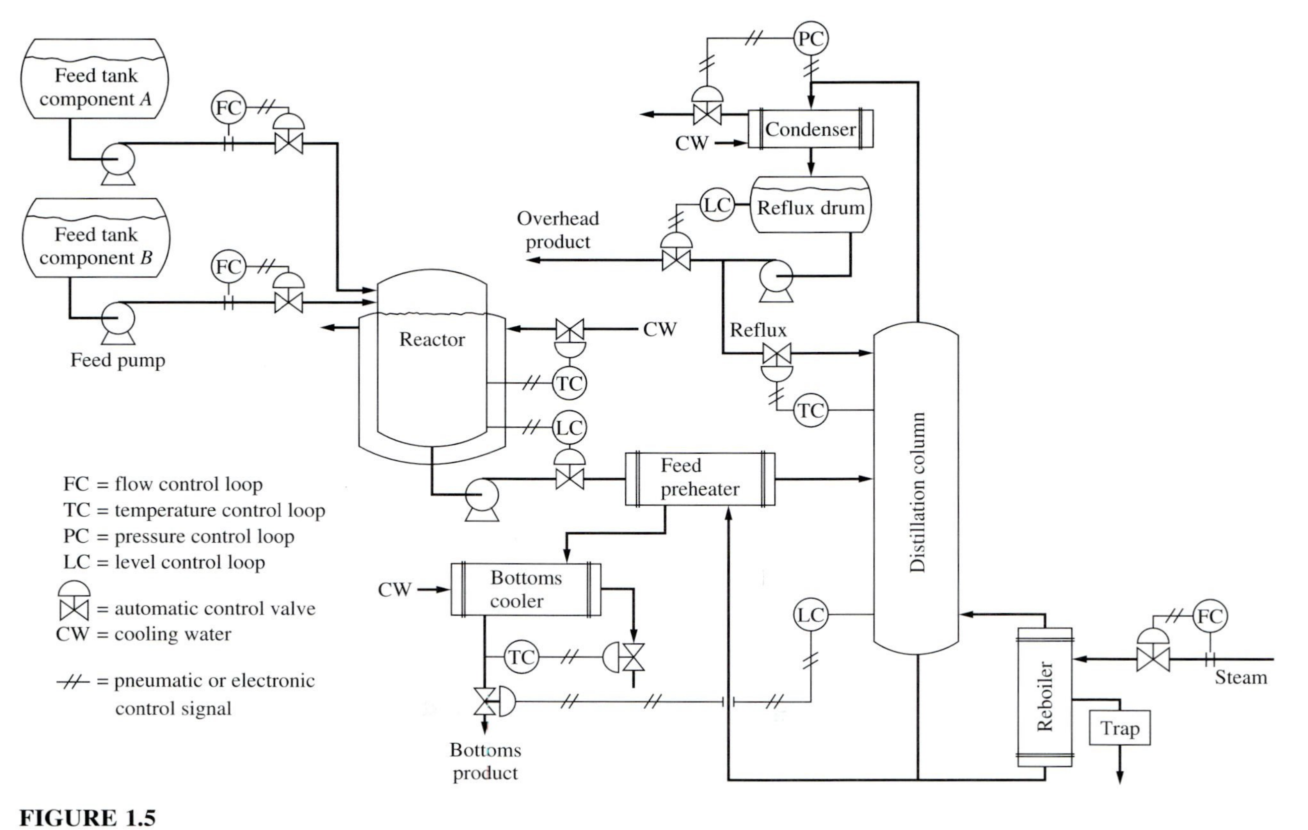

FIGURE 1.5
Typical chemical plant and control system.

But what procedure do we use to decide how to control this plant? We spend most of our time in this book exploring this important design and operating problem. Our studies of process control are aimed at understanding the dynamics of processes and control systems so that we can develop and design plants that operate more efficiently and safely, produce higher-quality products, are more easily controlled, and are more environmentally friendly.

For now let us merely say that the control system shown in Fig. 1.5 is a typical conventional system. It is about the minimum that would be needed to run this plant automatically without constant operator attention. Even in this simple plant, with a minimum of instrumentation, 10 control loops are required. We will find that most chemical engineering processes are multivariable. The key to any successful control system is understanding how the process works. ■

1.2 SOME IMPORTANT SIMULATION RESULTS

In the preceding section we discussed qualitatively some concepts of dynamics and control. Now we want to be more quantitative and look at two numerical examples of dynamic systems. The first involves level control in a series of tanks. The second involves temperature control in a three-tank process. These processes are simple, but their dynamic response is rich enough that we can observe some very important behavior.

1.2.1 Proportional and Proportional-Integral Level Control

The process sketched in Fig. 1.6 consists of two vertical cylindrical tanks with a level controller on each tank. The feed stream to the first tank comes from an upstream unit. The liquid level in each of the tanks is controlled by manipulating the flow rate of liquid pumped from the corresponding tank. The level signal from the level transmitter on each tank is sent to a level controller. The output signal from each controller goes to a control valve that sets the outflow rate.

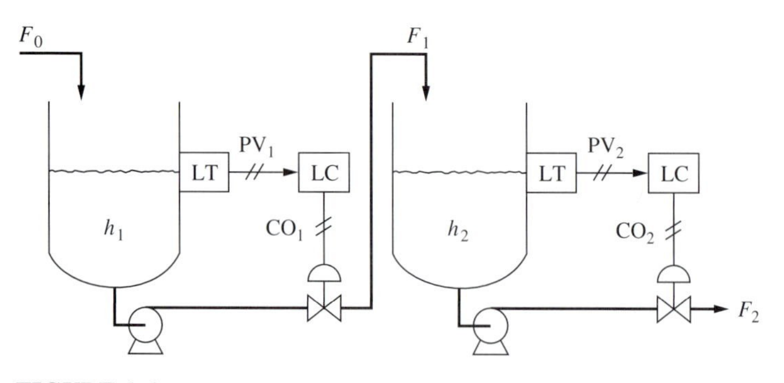

FIGURE 1.6
Level control.

A dynamic model of this process contains two ordinary differential equations, which arise from the total mass balance on each of the tanks. We assume constant density.

$$A_1 \frac{dh_1}{dt} = F_0 - F_1 \tag{1.1}$$

$$A_2 \frac{dh_2}{dt} = F_1 - F_2 \tag{1.2}$$

where A_n = cross-sectional area of nth tank
h_n = liquid height in nth tank
F_n = volumetric flow rate of liquid from nth tank
F_0 = volumetric flow rate of feed to process

The flow rates F_1 and F_2 are set by the controller output signals CO_1 and CO_2 from the two level controllers. The process variable signals PV_1 and PV_2 from the two level transmitters depend on the two liquid levels h_1 and h_2. In this example we express these PV and CO signals as fractions of the full-scale range of the signals. The signals from transmitters and controllers are voltage, current, or pressure signals, which vary over standard ranges (0 to 10 V, 4 to 20 mA, or 3 to 15 psig).

$$F_1 = CO_1 F_1^{max} \tag{1.3}$$

$$F_2 = CO_2 F_2^{max} \tag{1.4}$$

where F_n^{max} = flow rate when the control valve is wide open.

$$PV_1 = h_1/h_{1,span} \tag{1.5}$$

$$PV_2 = h_2/h_{2,span} \tag{1.6}$$

where $h_{n,span}$ = the "span" of the level transmitter, i.e., the difference between the maximum and minimum liquid levels measured in the tank. Numerical values of all parameters and the values of the variables at the initial steady-state conditions are given in Table 1.1. The FORTRAN program used to simulate the dynamics of the process is given in Table 1.2. For more background on dynamic modeling and simulation methods, refer to W. L. Luyben, *Process Modeling, Simulation and Control for Chemical Engineers,* 2d ed. (1990), McGraw-Hill, New York.

TABLE 1.1
Values of parameters and steady-state variables

Diameter of tank = 10 ft
Cross-sectional area of tank = 78.54 ft^2
Span of level transmitters = 20 ft
Maximum flow rate through control valves = 200 ft^3/min
Steady-state flow rates = 100 ft^3/min
Steady-state levels = 10 ft
Bias value of level controllers = Bias = 0.5 fraction of full scale
Setpoint signals of controller = SP = 0.5 fraction of full scale
Steady-state value of controller outputs = CO = 0.5 fraction of full scale
Steady-state value of level transmitter outputs = PV = 0.5 fraction of full scale

TABLE 1.2
FORTRAN simulation program for PI level control

```
c Program "level.f"
c P and PI control of two tank levels in series
      dimension h(2),f(2),co(2),reset(4),erint(2),e(2),sp(2)
      real pv(2),kc(4)
      dimension dh(2)
      dimension tp(2000),h1p(4,2000),h2p(4,2000)
      dimension f1p(4,2000),f2p(4,2000)
      open(7,file='lev1.dat')
      open(8,file='lev2.dat')
      open(9,file='lev3.dat')
      open(10,file='lev4.dat')

      data delta,tstop/.1,200./
      data sp/2*0.5/
      data area,fss/78.54,100./
      data kc/0.5,1.5,0.628,0.314/
      data reset/0.,0.,5.,5./
      data dtprint,dtplot/5.,.2/
c Disturbance is +10% fo
      fo=100.*1.1
c Make four runs with different controller settings
      do 1000 nc=1,4
      time=0.
      tprint=0.
      tplot=0.
      np=0
      do 10 ntank=1,2
      h(ntank)=10.
      erint(ntank)=0.
   10 f(ntank)=100.
c controller calculations to get flow rates
c All control signals (pv, sp, and co) are in fractions of full scale
  100 do 20 ntank=1,2
      pv(ntank)=h(ntank)/20.
      if(pv(ntank).gt.1.)pv(ntank)=1.
      if(pv(ntank).lt.0.)pv(ntank)=0.
      e(ntank)=sp(ntank)-pv(ntank)
      co(ntank)=0.5-kc(nc)*e(ntank)
      if(reset(nc).gt.0.)
     +co(ntank)=0.5-kc(nc)*(e(ntank)+erint(ntank)/reset(nc))
      if(co(ntank).gt.1.)co(ntank)=1.
      if(co(ntank).lt.0.)co(ntank)=0.
      f(ntank)=co(ntank)*fss*2.
   20 continue
```

TABLE 1.2 (CONTINUED)
FORTRAN simulation program for PI level control

```
c print and store for plotting
      if(time.lt.tprint)go to 30
      write(6,21)time,h,f
   21 format(' t=',f6.2,' h=',2f7.2,' f=',2f7.2)
      tprint=tprint+dtprint
   30 if(time.lt.tplot)go to 40
      np=np+1
      tp(np)=time
      h1p(nc,np)=h(1)
      h2p(nc,np)=h(2)
      f1p(nc,np)=f(1)
      f2p(nc,np)=f(2)
      tplot=tplot+dtplot
   40 continue
c evaluate all derivatives
      dh(1)=(fo-f(1))/area
      do 50 ntank=2,2
   50 dh(ntank)=(f(ntank-1)-f(ntank))/area
c integrate a la Euler
      time=time+delta
      do 60 ntank=1,2
      h(ntank)=h(ntank)+dh(ntank)*delta
   60 erint(ntank)=erint(ntank)+e(ntank)*delta
      if(time.lt.tstop)go to 100
 1000 continue
c store data for plotting using MATLAB
      do 110 j=1,np
      write(7,111)tp(j),h1p(1,j),h2p(1,j),f1p(1,j),f2p(1,j)
      write(8,111)h1p(2,j),h2p(2,j),f1p(2,j),f2p(2,j)
      write(9,111)h1p(3,j),h2p(3,j),f1p(3,j),f2p(3,j)
  110 write(10,111)h1p(4,j),h2p(4,j),f1p(4,j),f2p(4,j)
  111 format(7(1x,f7.3))
      stop
      end
```

Two types of controllers are studied in this example. The first is a "proportional" controller, in which the CO signal varies in direct proportion to the change in the PV signal.

$$CO_1 = \text{Bias}_1 - K_{c1}(SP_1 - PV_1) \tag{1.7}$$

$$CO_2 = \text{Bias}_2 - K_{c2}(SP_2 - PV_2) \tag{1.8}$$

where Bias_n = a constant (the value of CO when PV is equal to SP)
K_{cn} = controller gain
SP_n = setpoint of the controller, i.e., the desired value of PV

Note that if the liquid level goes up, PV goes up, CO goes up, and F increases. This is the correct response of the level controller to an increase in level.

Figure 1.7 shows the dynamic responses of the two liquid levels and the two outflow rates when a 10 percent increase in the feed flow rate to the process occurs

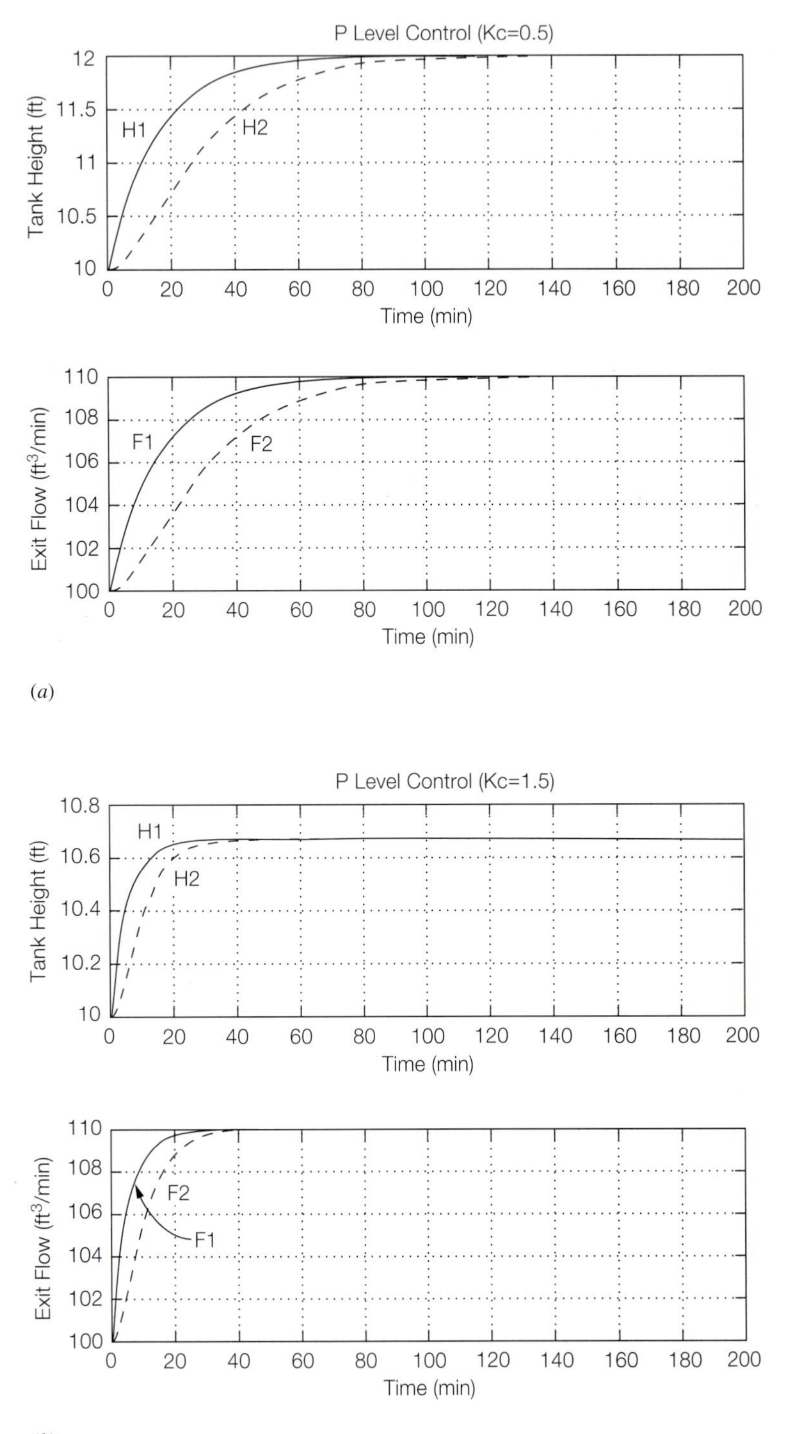

(*b*)

FIGURE 1.7

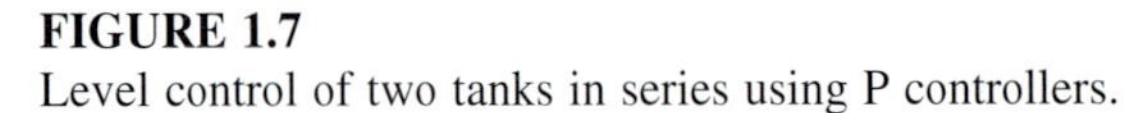
Level control of two tanks in series using P controllers.

at time equal zero. The commercial software package MATLAB is used to plot the results, and Table 1.3 gives the "m-file" used. If you do not already have some exposure to programming in MATLAB, you can pick up the essentials pretty quickly. A general reference for MATLAB programming is *The Student Edition of MATLAB*, 1992, The Math Works Inc., Prentice Hall, Englewood Cliffs, NJ. Two books that discuss the use of the MATLAB Control Toolbox to analyze dynamic systems

TABLE 1.3
MATLAB program to plot data

```
% matlab program "level.m" to plot results for P and PI controller
load lev1.dat
load lev2.dat
load lev3.dat
load lev4.dat

t=lev1(:,1);
h11=lev1(:,2);
h12=lev1(:,3);
f11=lev1(:,4);
f12=lev1(:,5);

h21=lev2(:,1);
h22=lev2(:,2);
f21=lev2(:,3);
f22=lev2(:,4);

h31=lev3(:,1);
h32=lev3(:,2);
f31=lev3(:,3);
f32=lev3(:,4);

h41=lev4(:,1);
h42=lev4(:,2);
f41=lev4(:,3);
f42=lev4(:,4);

clf
subplot(211)
plot(t,h11,'-',t,h12,'--')
title('P Level Control (Kc=0.5)')
ylabel('Tank Height (ft)')
xlabel('Time (min)')
legend('H1','H2')
grid
subplot(212)
plot(t,f11,'-',t,f12,'--')
ylabel('Exit Flow (ft3/min)')
xlabel('Time (min)')
legend('F1','F2')
grid
pause
print -dps plevel.ps
```

TABLE 1.3 (CONTINUED)
MATLAB program to plot data

```
clf
subplot(211)
plot(t,h21,'−',t,h22,'−−')
title('P Level Control (Kc=1.5)')
ylabel('Tank Height (ft)')
xlabel('Time (min)')
legend('H1','H2')
grid
subplot(212)
plot(t,f21,'−',t,f22,'−−')
ylabel('Exit Flow (ft3/min)')
xlabel('Time (min)')
legend('F1','F2')
grid
pause
print −dps −append plevel

clf
subplot(211)
plot(t,h31,'−',t,h32,'−−')
title('PI Level Control, Reset=5, Kc=0.628')
ylabel('Tank Height (ft)')
xlabel('Time (min)')
legend('H1','H2')
grid
subplot(212)
plot(t,f31,'−',t,f32,'−−')
ylabel('Exit Flow (ft3/min)')
xlabel('Time (min)')
legend('F1','F2')
grid
pause
print −dps −append plevel

clf
subplot(211)
plot(t,h41,'−',t,h42,'−−')
title('PI Level Control, Reset=5, Kc=0.314')
ylabel('Tank Height (ft)')
xlabel('Time (min)')
legend('H1','H2')
grid
subplot(212)
plot(t,f41,'−',t,f42,'−−')
ylabel('Exit Flow (ft3/min)')
xlabel('Time (min)')
legend('F1','F2')
grid
pause
print −dps −append plevel
```

are *MATLAB Tools for Control System Analysis and Design* by B. C. Kuo and D. C. Hanselman, 1994, Prentice Hall, Englewood Cliffs, NJ; and *Using MATLAB to Analyze and Design Control Systems* by N. E. Leonard and W. S. Levine, 1992, Benjamin-Cummings, New York.

In Fig. 1.7*a* the controller gain K_c is 0.5 and in Fig. 1.7*b* it is 1.5. The changes in both liquid levels and outflow rates are gradual, but the dynamic changes occur more quickly when a higher gain is used. The inflow rate has increased to 110 ft^3/min, so the flow rates from both tanks eventually climb up to 110 ft^3/min, and at this point the levels in both tanks stop changing.

You should note a very important point: the levels do not return to their original steady-state values of 10 ft. For a controller gain of 0.5, tank levels increase to 12 ft and stay there. For a gain of 1.5, they increase to about 10.7 ft. So at the new steady-state conditions, the value of SP is not equal to PV in Eqs. (1.7) and (1.8). We call this "steady-state error" or "offset." This example illustrates that a proportional controller does not give zero steady-state error. For the control of levels in surge tanks we normally are not concerned about holding a constant level, so offset is not a problem. But for many control loops, we do want to drive the PV back to the SP value. This is accomplished by adding "integral" or "reset" action to the controller.

The second type of level controller is a "proportional-integral" (PI) controller, in which the CO signal varies with both the PV signal and time integral of "error" (the difference between the SP and PV signals).

$$\text{CO}_n = \text{Bias}_n - \text{K}_{cn}\left[(\text{SP}_n - \text{PV}_n) + \frac{1}{\tau_I}\int(\text{SP}_n - \text{PV}_n)\,dt\right] \tag{1.9}$$

where τ_I = integral time or reset time (with units of minutes). The addition of the integral term forces the SP and PV signals to become equal at steady state because if the (SP − PV) term is not zero, the CO continues to change because of the integral action. Figure 1.8 demonstrates this for the same change in the feed rate. The outflow rates start and end at the same values as found with P control, but the liquid levels are returned to their SP values.

However, there is a price to be paid for this elimination of steady-state error. When P control was used, the flow rates simply increased to their new steady-state values. With PI control these flow rates increase above their final steady-state values for a period of time. This occurs because the only way that the level can be lowered back to its desired value is to have the instantaneous flow rate out of the tank be larger than the flow rate into the tank.

Figure 1.8*a* shows that the maximum instantaneous value of the flow rate F_1 is 114 ft^3/min and the maximum for F_2 is 119 ft^3/min. Remember that the initial flow rate is 100 and the final flow rate is 110, giving a change of 10 ft^3/min. The peak flow rate from the second tank of 119 corresponds to a change of 19 ft^3/min, which is an overshoot of almost 100 percent. Thus the use of PI control results in an amplification of the flow rate disturbances to the system, and this amplification becomes larger as we add more tanks in series. Figure 1.9 illustrates the difference between P and PI control in another way. We impose on the process a "noisy" disturbance—the feed flow rate F_0 into the process is changing in a random way—and compare the responses of the levels and flow rates in the system for P and PI control. The filtering of flow rate disturbances by P control is clearly demonstrated. If constant level

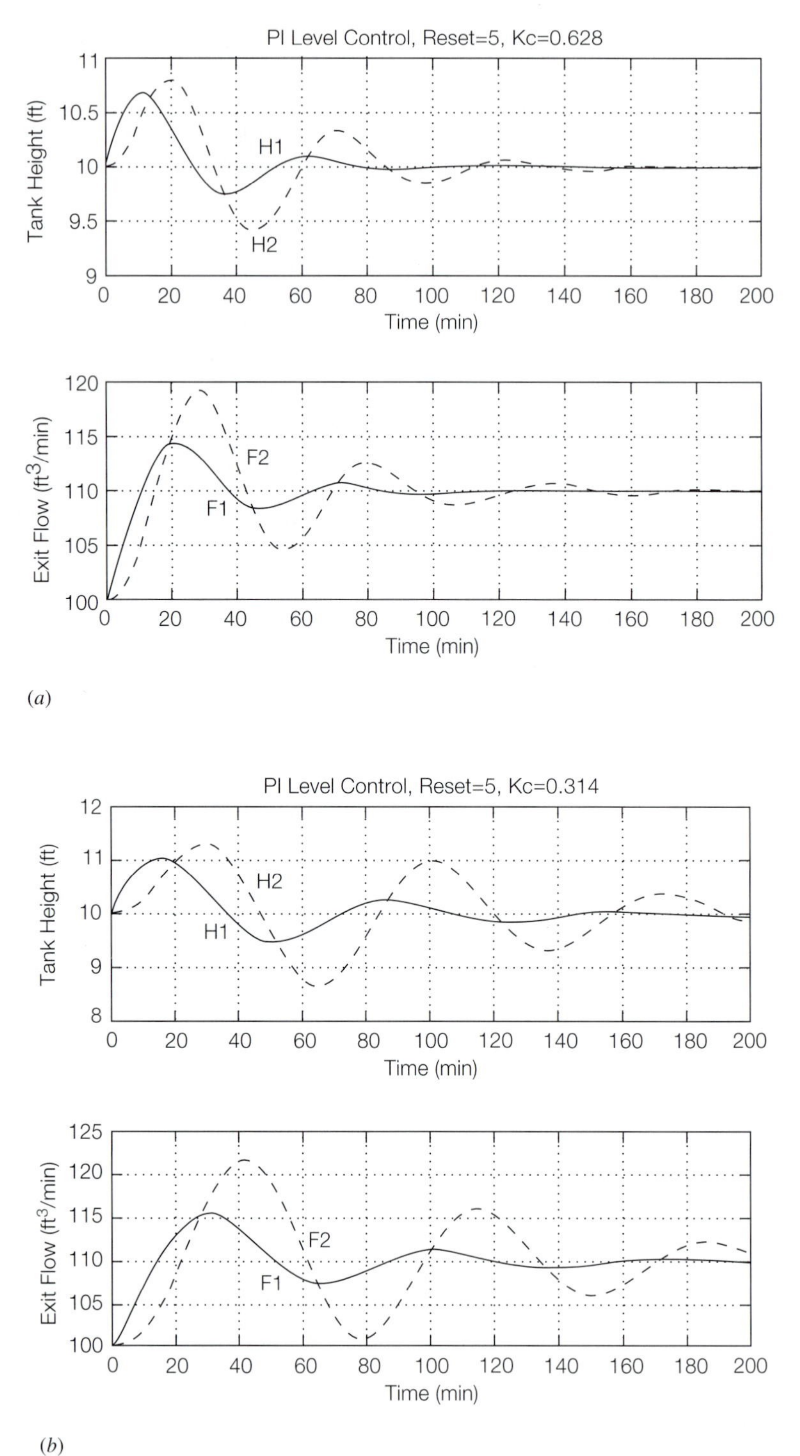

FIGURE 1.8
Level control using PI controllers.

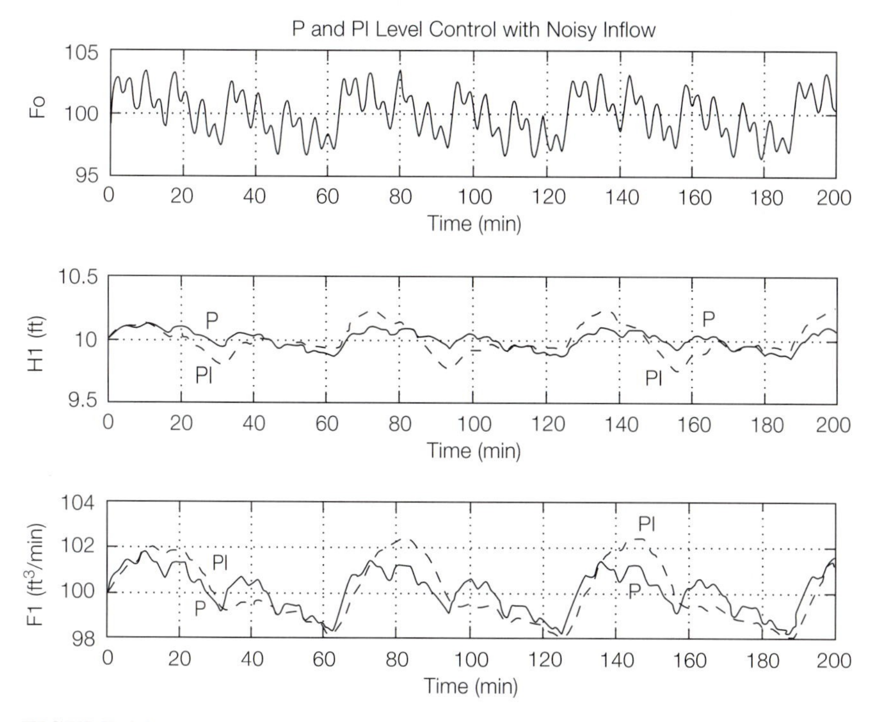

FIGURE 1.9

control is desired and flow rate variations are less important (e.g., level control in a chemical reactor in which we want to keep holdup constant), a PI controller does a better job.

You might suspect that we could change the controller tuning and reduce this amplification. Reducing controller gain usually slows down the dynamics of the system and produces less oscillatory response. However, this is not what happens for this process, as the results in Fig. 1.8*b* illustrate. The controller gain is cut in half, resulting in a slower response. But the peak flow rate from the second tank becomes larger (122 ft^3/min). We analyze this process quantitatively in Chapter 8 and explain mathematically why we observe these simulation results.

These results illustrate the importance of the selection of the type of controller and the control objectives. The simulation results have important implications for the plantwide control problem (multi-units connected in a complex flowsheet). They suggest that most level controllers should be proportional, not proportional-integral, to obtain smoothing (filtering or attenuation) of flow rate changes throughout a process.

1.2.2 Temperature Control of a Three-Tank Process

As a second simple example that demonstrates some very important and far-reaching principles, let us consider the control of temperature in a single tank and then in a series of two and three tanks. The dynamic model for three heated tanks in series is given in Eqs. (1.10) to (1.12). Constant holdup and constant physical properties are assumed. The FORTRAN program used to simulate the process is given in Appendix A.

The temperature in one of the tanks (T_1, T_2, or T_3) is controlled by a proportional temperature controller that manipulates the heat input Q_1 to the first tank. The disturbance is a drop in inlet feed temperature T_0 from 90°F to 70°F at time 0 hours. Three different values of controller gain ($K_c = 2, 4$, and 8) are used.

$$V_1 c_p \rho \frac{dT_1}{dt} = F c_p \rho (T_0 - T_1) + Q_1 \tag{1.10}$$

$$V_2 c_p \rho \frac{dT_2}{dt} = F c_p \rho (T_1 - T_2) \tag{1.11}$$

$$V_3 c_p \rho \frac{dT_3}{dt} = F c_p \rho (T_2 - T_3) \tag{1.12}$$

where V_n = tank volume in nth tank = 100 ft^3
c_p = heat capacity of process fluid = 0.75 Btu/lb °F
ρ = density of process fluid = 50 lb/ft^3
F = flow rate = 1000 ft^3/hr

Control signals of 4 to 20 mA are used. The range of the temperature transmitter is 50 to 200°F.

$$\text{PV} = 4 + \tfrac{16}{200}(T_{\text{control}} - 50) \tag{1.13}$$

where we will consider three cases for T_{control} (T_1, T_2, and T_3).

The control valve can pass enough steam to transfer 10×10^6 Btu/hr of heat into the first tank.

$$Q_1 = \frac{\text{CO} - 4}{16}(10 \times 10^6) \tag{1.14}$$

The proportional controller equation is

$$\text{CO} = 7.6 + K_c(\text{SP} - \text{PV}) \tag{1.15}$$

with setpoint SP = 12 mA.

The values of variables at the initial steady state are

$$T_0 = 90°\text{F} \qquad T_1 = T_2 = T_3 = 150°\text{F} \qquad \text{PV} = 12 \text{ mA}$$

A. Control of T_1

Figure 1.10 gives the temperature in the first tank and the heat input for three values of controller gain K_c. As gain increases, the dynamics of the system get faster and there is less steady-state offset: the final steady-state value of T_1 is closer to 150°F. The dynamic responses all show gradual asymptotic trajectories to their final values. There is no overshoot and no oscillation.

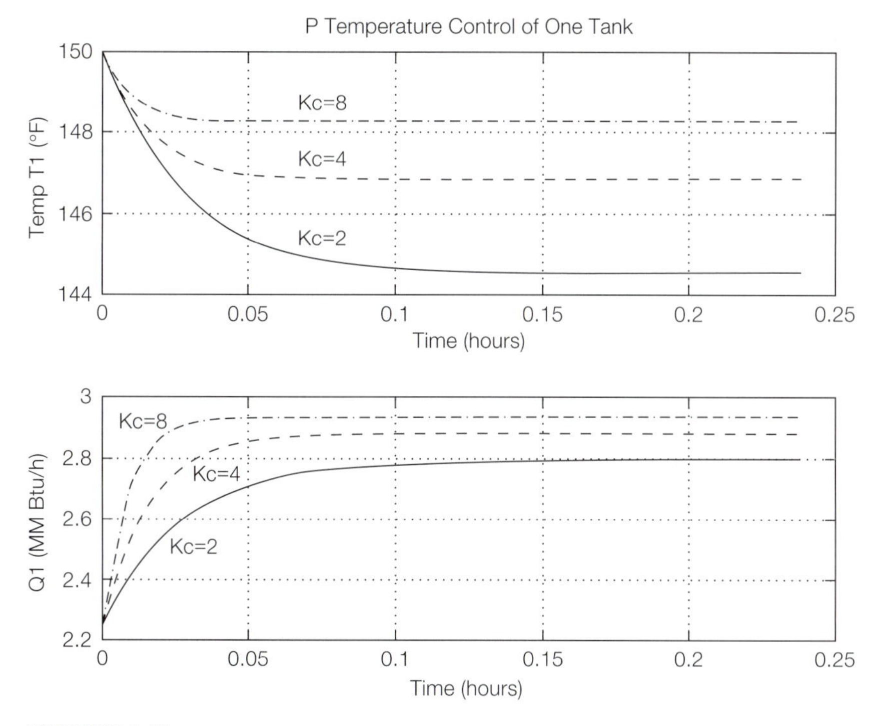

FIGURE 1.10

B. Control of T_2

Figure 1.11 shows what happens when we switch from controlling the temperature in the first tank to controlling the temperature in the second tank, T_2. The disturbance is the same, and the three controller gains are the same. Now we begin to see some overshooting and oscillatory responses for the larger values of controller gain.

C. Control of T_3

Figure 1.12 gives results when we control the temperature in the third tank, T_3. For a controller gain of $K_c = 2$, the system is only slightly oscillatory and the system settles out at a new steady state. The oscillations become larger for $K_c = 4$, and it takes longer for the system to settle out. However, for $K_c = 8$ the amplitude of the oscillations continues to grow. This system is "unstable."

D. Control of T_1 with deadtime

"Deadtime" is a term that we use to describe the situation where there is a delay between the input and the output of a system. A common chemical engineering example is the turbulent flow of fluid through a pipe. Let us assume that the flow is essentially plug flow with a residence time of D minutes. If the temperature of the stream entering the pipe changes, the temperature of the stream leaving the pipe will not change for D minutes. This is called a deadtime of D minutes.

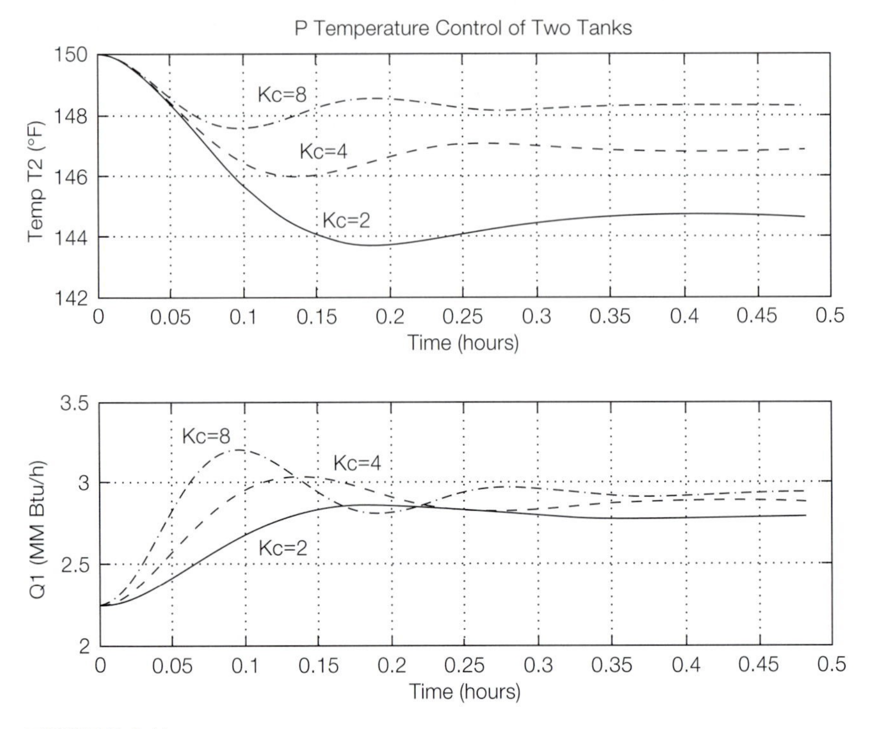

FIGURE 1.11

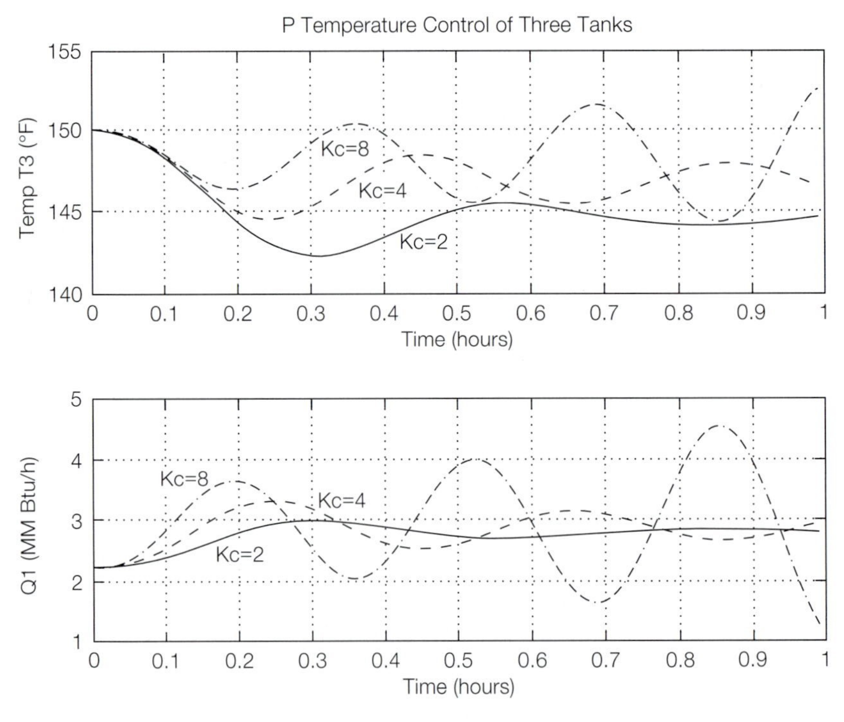

FIGURE 1.12

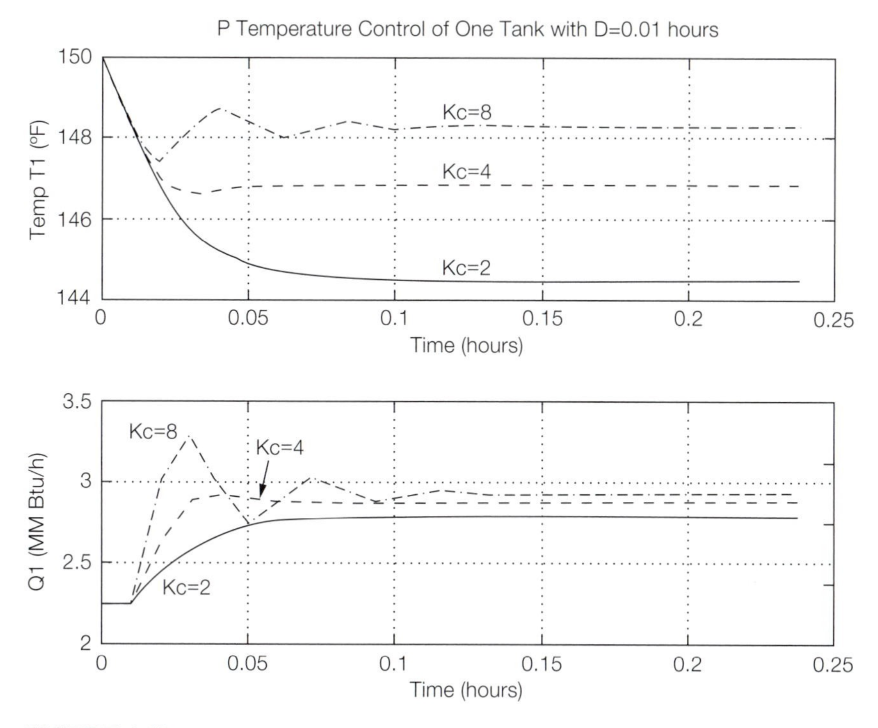

FIGURE 1.13

Suppose we have a deadtime of $D = 0.01$ hr in the measurement of the temperature T_1. The temperature controller sees a PV signal that is delayed by D hours. The effect this has on control is illustrated in Fig. 1.13. Notice that Q_1 does not change for 0.01 hr because of the deadtime. Comparing Figs. 1.10 and 1.13 shows clearly that the dynamic performance with deadtime is worse. Higher gains now give oscillatory behavior. The FORTRAN program used for the deadtime simulation is given in Appendix A.

These simulation results illustrate some profoundly important principles:

1. As controller gain is increased, the response of the process becomes faster but more oscillatory. This suggests that there is an inherent engineering trade-off between speed of response and oscillatory behavior. The terminology used in process control is the trade-off between "performance" and "robustness." A fast process response (a small time constant) is good performance. A less oscillatory response (a higher damping coefficient) is good robustness; i.e., the process is not close to the situation where the oscillations will continue to grow. We illustrate this trade-off in several other situations later in this book.
2. As more tanks are added to the system, the control becomes more difficult. Controlling T_1 with Q_1 (we call this a first-order system) is easy: the system is never oscillatory. Controlling T_2 with Q_1 (a second-order system) gives some oscillatory behavior, but controller gains have to be quite large before the oscillation

becomes a problem. However, controlling T_3 with Q_1 (a third-order system) is more difficult since high controller gains can lead to instability. Thus, as the order of the system is increased, the dynamic performance becomes worse. This, of course, suggests that we should avoid high-order systems in our plant designs and control system structures.

3. The addition of deadtime in a control loop degrades dynamic performance.

Later in this book we explain quantitatively and mathematically the results observed in the examples considered above.

1.3 GENERAL CONCEPTS AND TERMINOLOGY

It may be useful at this point to define some very broad and general concepts and some of the terminology used in the field.

1. *Dynamics:* Time-dependent behavior of a process. The behavior with no controllers in the system is called the *openloop* response. The dynamic behavior with controllers included with the process is called the *closedloop* response.
2. *Variables:*
   a. *Manipulated variables:* Typically flow rates of streams entering or leaving a process that we can change to control the plant.
   b. *Controlled variables:* Flow rates, compositions, temperatures, levels, and pressures in the process that we will try to control, either trying to hold them as constant as possible or trying to make them follow some desired time trajectory.
   c. *Uncontrolled variables:* Variables in the process that are not controlled.
   d. *Load disturbances:* Flow rates, temperatures, or compositions of streams entering (but sometimes leaving) the process. We are not free to manipulate them. They are set by upstream or downstream parts of the plant. The control system must be able to keep the plant under control despite the effects of these disturbances.
3. *Feedback control:* The traditional way to control a process is to measure the variable that is to be controlled, compare its value with the desired value (the setpoint to the controller), and feed the difference (the error) into a feedback controller that changes a manipulated variable to drive the controlled variable back to the desired value. Information is thus "fed back" from the controlled variable to a manipulated variable. Action is taken after a change occurs in the process.
4. *Feedforward control:* The basic idea is to take action before a disturbance reaches the process. As shown in Fig. 1.14, the disturbance is detected as it enters the process and an appropriate change is made in the manipulated variable such that the controlled variable is held constant. Thus, we begin to take corrective action as soon as a disturbance entering the system is detected instead of waiting

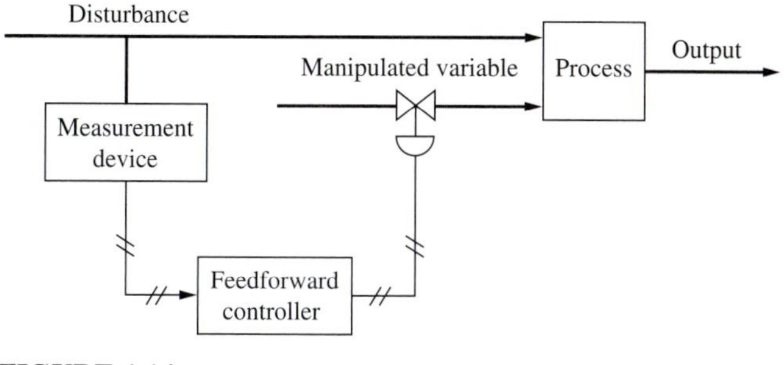

FIGURE 1.14
Feedforward control.

(as we do with feedback control) for the disturbance to propagate all the way through the process before a correction is made.

5. *Stability:* A process is said to be *unstable* if its output becomes larger and larger (either positively or negatively) as time increases. Examples are shown in Fig. 1.15. No real system actually does this, of course, because some constraint will be met; for example, a control valve will completely shut or completely open, or a safety valve will "pop." A linear process is right at the limit of stability if it oscillates, even when undisturbed, and the amplitude of the oscillations does not decay.

Most processes are *openloop stable,* i.e., stable with no controllers on the system. One important and very interesting exception that we will study in some detail is the exothermic chemical reactor, which can be openloop unstable. All real processes can be made *closedloop unstable* (unstable when a feedback controller is in the system) if the controller gain is made large enough. Thus, stability is of vital concern in feedback control systems.

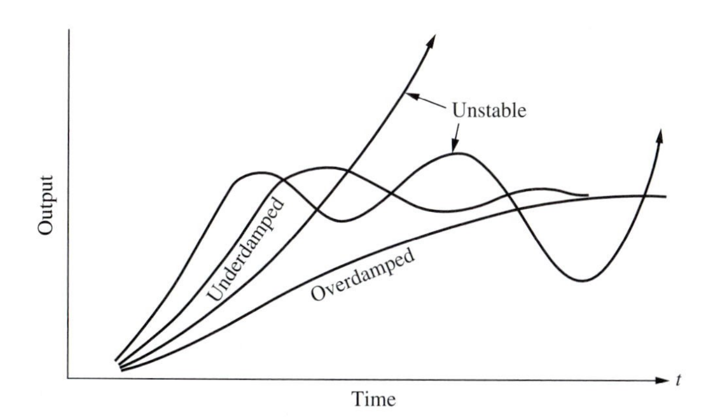

FIGURE 1.15
Stability.

1.4 LAWS, LANGUAGES, AND LEVELS OF PROCESS CONTROL

1.4.1 Process Control Laws

Several fundamental laws have been developed in the process control field as a result of many years of experience. Some of these may sound similar to some of the laws attributed to Parkinson, but the process control laws are not intended to be humorous.

First Law: The best control system is the simplest one that will do the job.

Complex and elegant control systems look great on paper but soon end up on "manual" (taken out of service) in an industrial environment. Bigger is definitely not better in control systems design.

Second Law: You must understand the process before you can control it.

No degree of sophistication in the control system (from adaptive control, to expert systems, to Kalman filters, to nonlinear model predictive control) will work if you do not know how your process works. Many people have tried to use complex controllers to overcome ignorance about the process fundamentals, and they have failed! Learn how the process works before you start designing its control system.

Third Law: Liquid levels must always be controlled.

The structure of the control systems must guarantee that the liquid levels in tanks, column base, reflux drums, etc. are maintained between their maximum and minimum values. A common error is to develop a control structure in which tank levels are not controlled and to depend on the operator of the plant to control tank levels manually. This increases the workload on the operator and results in poor plant performance because of inconsistencies among various operators concerning what should be done under various conditions. Having an automatic, fixed inventory control structure produces smoother, more consistent plant operation. The only exception to this law occurs in recycle systems, where the level in a recycle surge drum is typically not controlled, but floats up and down with recycle circulation rate.

1.4.2 Languages of Process Control

As you will see, several different approaches are used in this book to analyze the dynamics of systems. Direct solution of the differential equations to give functions of time is a "time domain" technique. The use of Laplace transforms to characterize the dynamics of systems is a "Laplace domain" technique. Frequency response methods provide another approach to the problem.

All of these methods are useful because each has its advantages and disadvantages. They yield exactly the same results when applied to the same problem but provide different perspectives. These various approaches are similar to the use of different languages by people around the world. A table in English is described by the word "table." In Russian a table is described by the word "СТОЛ." In Chinese a table is "桌子." In German it is "der Tisch." But in any language a table is still a table.

In the study of process dynamics and control we will use several languages.

English: time domain (differential equations, yielding exponential time function solutions)
Russian: Laplace domain (transfer functions)
Chinese: frequency domain (frequency response Bode and Nyquist plots)
Greek: state variables (matrix methods apply to differential equations)
German: z domain (sampled-data systems)

You will find that the languages are not difficult to learn because the vocabulary required is quite limited: only 8 to 10 "words" must be learned in each language. Thus, it is fairly easy to translate back and forth between the languages.

We will use "English" to solve some simple problems. We will find that more complex problems are easier to understand and solve using "Russian." As problems get even more complex and realistic, the use of "Chinese" is required. So we study in this book a number of very useful and practical process control languages.

We chose the five languages listed above simply because we have had some exposure to all of them over the years. Let us assure you that no political or nationalistic motives are involved. If you prefer French, Spanish, Italian, Japanese, and Swahili, please feel free to make the appropriate substitutions! Our purpose in using the language metaphor is to try to break some of the psychological barriers that students have to such things as Laplace transforms and frequency response. It is a pedagogical gimmick that we have used for over two decades and have found to be very effective with students.

1.4.3 Levels of Process Control

There are four levels of process control. Moving up these levels increases the importance, the economic impact, and the opportunities for process control engineers to make significant contributions.

The lowest level is controller tuning, i.e., determining the values of controller tuning constants that give the best control. The next level is algorithms—deciding what type of controller to use (P, PI, PID, multivariable, model predictive, etc.).

The third level is control system structure—determining what to control, what to manipulate, and how to match one controlled variable with one manipulated variable (called "pairing"). The selection of the control structure for a plant is a vitally important function. A good choice of structure makes it easy to select an appropriate algorithm and to tune. No matter what algorithm or tuning is used, it is very unlikely that a poor structure can be made to give effective control.

The top level is process design—developing a process flowsheet and using design parameters that produce an easily controllable plant. The steady-state economically optimal plant may be much more difficult to control than an alternative plant that is perhaps only slightly more expensive to build and operate. At this level, the economic impact of a good process control engineer can be enormous, potentially resulting in the difference between a profitable process and an economic disaster. Several cases have been reported where the process was so inoperable that it had to be shut down and the equipment sold to the junk man. Chapters 5 and 6 discuss this vitally important aspect in more detail.

1.5 CONCLUSION

In this chapter we have attempted to convey three basic notions:

1. The dynamic response of a process is important and must be considered in the process design.
2. The process itself places inherent restrictions on the achievable dynamic performance that no amount of controller complexity and elegance can overcome.
3. The choices of the control system structure, the type of controller, and the tuning of the controller are all important engineering decisions.

PART ONE

Time Domain Dynamics and Control

In this section we study the time-dependent behavior of some chemical engineering systems, both openloop (without control) and closedloop (with controllers included). Systems are described by differential equations and solutions are given in terms of time-dependent functions. Thus, our language for this part of the book will be "English." In the next part we will learn a little "Russian" so that we can work in the Laplace domain, where the notation is simpler than "English." In Part Three we will study some "Chinese" because of its ability to easily handle much more complex systems.

Most chemical engineering systems are modeled by equations that are quite complex and nonlinear. In the remaining parts of this book only systems described by *linear* ordinary differential equations will be considered (linearity is defined in Chapter 2). The reason coverage is limited to linear systems is that practically all the analytical mathematical techniques currently available are applicable only to linear equations.

Since most chemical engineering systems are nonlinear, studying methods that are limited to linear systems might initially appear to be a waste of time. However, linear techniques are of great practical importance, particularly for continuous processes, because the nonlinear equations describing most systems can be linearized around some steady-state operating condition. The resulting linear equations adequately describe the dynamic response of the system in some region around the steady-state conditions. The size of the region over which the linear model is valid varies with the degree of nonlinearity of the process and the magnitude of the disturbances. In many processes the linear model can be successfully used to study dynamics and, more important, to design controllers.

Complex systems can usually be broken down into a number of simple elements. We must understand the dynamics of these simple systems before we tackle the more complex ones. We start out looking at some simple uncontrolled processes in Chapter 2. We examine the openloop dynamics or the response of the system with no feedback controllers to a disturbance starting from some initial condition.

In Chapters 3 and 4 we look at closedloop systems. Instrumentation hardware, controller types and performance, controller tuning, and various types of control system structures are discussed.

CHAPTER 2

Time Domain Dynamics

Studying the dynamics of systems in the time domain involves the direct solution of differential equations. Computer simulations are general in the sense that they can give solutions to very complex nonlinear problems. However, they are also very specific in the sense that they provide a solution to only the particular numerical case fed into the computer.

The classical analytical techniques discussed in this chapter are limited to linear ordinary differential equations. But they yield general analytical solutions that apply for any values of parameters, initial conditions, and forcing functions.

We start by briefly classifying and defining types of systems and types of disturbances. Then we learn how to linearize nonlinear equations. It is assumed that you have had a course in differential equations, but we review some of the most useful solution techniques for simple ordinary differential equations.

The important lesson of this chapter is that the dynamic response of a *linear* process is a sum of exponentials in time, such as $e^{s_k t}$. The s_k terms multiplying time are the *roots of the characteristic equation* or the *eigenvalues* of the system. They determine whether the process responds quickly or slowly, whether it is oscillatory, and whether it is *stable.*

2.1 CLASSIFICATION AND DEFINITION

Processes and their dynamics can be classified in several ways:

1. Number of independent variables
   a. *Lumped:* if time is the only independent variable; described by ordinary differential equations
   b. *Distributed:* if time and spatial independent variables are required; described by partial differential equations

2. Linearity
   a. *Linear:* if all functions in the equations are linear (see Section 2.2)
   b. *Nonlinear:* if not linear

3. Stability
   a. *Stable:* if "self-regulatory" so that variables converge to some steady state when disturbed
   b. *Unstable:* if variables go to infinity (mathematically)

      Most processes are openloop stable. However, the exothermic irreversible chemical reactor is a notable example of a process that can be openloop unstable.

      All real processes can be made closedloop unstable (unstable with a feedback controller in service), and therefore one of the principal objectives in feedback controller design is to avoid closedloop instability.

4. Order: If a system is described by one ordinary differential equation with derivatives of order N, the system is called Nth order.

$$a_N \frac{d^N x}{dt^N} + a_{N-1} \frac{d^{N-1} x}{dt^{N-1}} + \cdots + a_1 \frac{dx}{dt} + a_0 x = f_{(t)} \tag{2.1}$$

where a_i are constants and $f_{(t)}$ is the forcing function or disturbance. Two very important special cases are for $N = 1$ and $N = 2$:

First-order:

$$a_1 \frac{dx}{dt} + a_0 x = f_{(t)} \tag{2.2}$$

Second-order:

$$a_2 \frac{d^2 x}{dt^2} + a_1 \frac{dx}{dt} + a_0 x = f_{(t)} \tag{2.3}$$

The "standard" forms that we will usually employ for these are

First-order:

$$\tau \frac{dx}{dt} + x = f_{(t)} \tag{2.4}$$

Second-order:

$$\tau^2 \frac{d^2 x}{dt^2} + 2\tau\zeta \frac{dx}{dt} + x = f_{(t)} \tag{2.5}$$

where τ = process time constant (either openloop or closedloop)
ζ = damping coefficient (either openloop or closedloop)

One of the most important parameters that we will use in the remaining sections of this book is the damping coefficient of the closedloop system. We typically tune a controller to give a closedloop system that has some specified damping coefficient.

Disturbances can also be classified and defined in several ways.

1. Shape (see Fig. 2.1)
   a. Step: Step disturbances are functions that change instantaneously from one level to another and are thereafter constant. If the size of the step is equal to unity, the disturbance is called the *unit step function* $u_{n(t)}$, defined as

$$\begin{aligned} u_{n(t)} &= 1 \quad \text{for } t > 0 \\ u_{n(t)} &= 0 \quad \text{for } t \leq 0 \end{aligned} \tag{2.6}$$

   The response of a system to a step disturbance is called the *step response* or the *transient response.*

   b. Pulse: A pulse is a function of arbitrary shape (but usually rectangular or triangular) that begins and ends at the same level. A rectangular pulse is simply the sum of one positive step function made at time zero and one negative step function made D minutes later. D is the length of the pulse.

$$\text{Rectangular pulse of height 1 and width } D = u_{n(t)} - u_{n(t-D)} \tag{2.7}$$

   c. Impulse: The impulse is defined as the Dirac delta function, an infinitely high pulse whose width is zero and whose area is unity. This kind of disturbance is, of course, a pure mathematical fiction, but we will find it a useful tool.
   d. Ramp: Ramp inputs are functions that change linearly with time.

$$\text{Ramp function} = Kt \tag{2.8}$$

   where K is a constant. The classic example is the change in the setpoint to an anti-aircraft gun as the airplane sweeps across the sky. Chemical engineering examples include batch reactor temperature or pressure setpoint changes with time.

   e. Sinusoid: Pure periodic sine and cosine inputs seldom occur in real chemical engineering systems. However, the response of systems to this kind of forcing function (called the *frequency response* of the system) is of great practical importance, as we show in our "Chinese" lessons (Part Three) and in multivariable processes (Part Four).

2. Location of disturbance in feedback loop: Let us now consider a process with a feedback controller in service. This closedloop system can experience disturbances at two different spots in the feedback loop: load disturbances and setpoint disturbances.

   Most disturbances in chemical engineering systems are load disturbances, such as changes in throughput, feed composition, supply steam pressure, and cooling water temperature. The feedback controller's function when a load disturbance occurs is to return the controlled variable to its setpoint by suitable changes in the manipulated variable. The closedloop response to a load disturbance is called the *regulator response* or the *closedloop load response.*

   Setpoint changes can also be made, particularly in batch processes or in changing from one operating condition to another in a continuous process. These setpoint changes also act as disturbances to the closedloop system. The function of the feedback controller is to drive the controlled variable to match the new

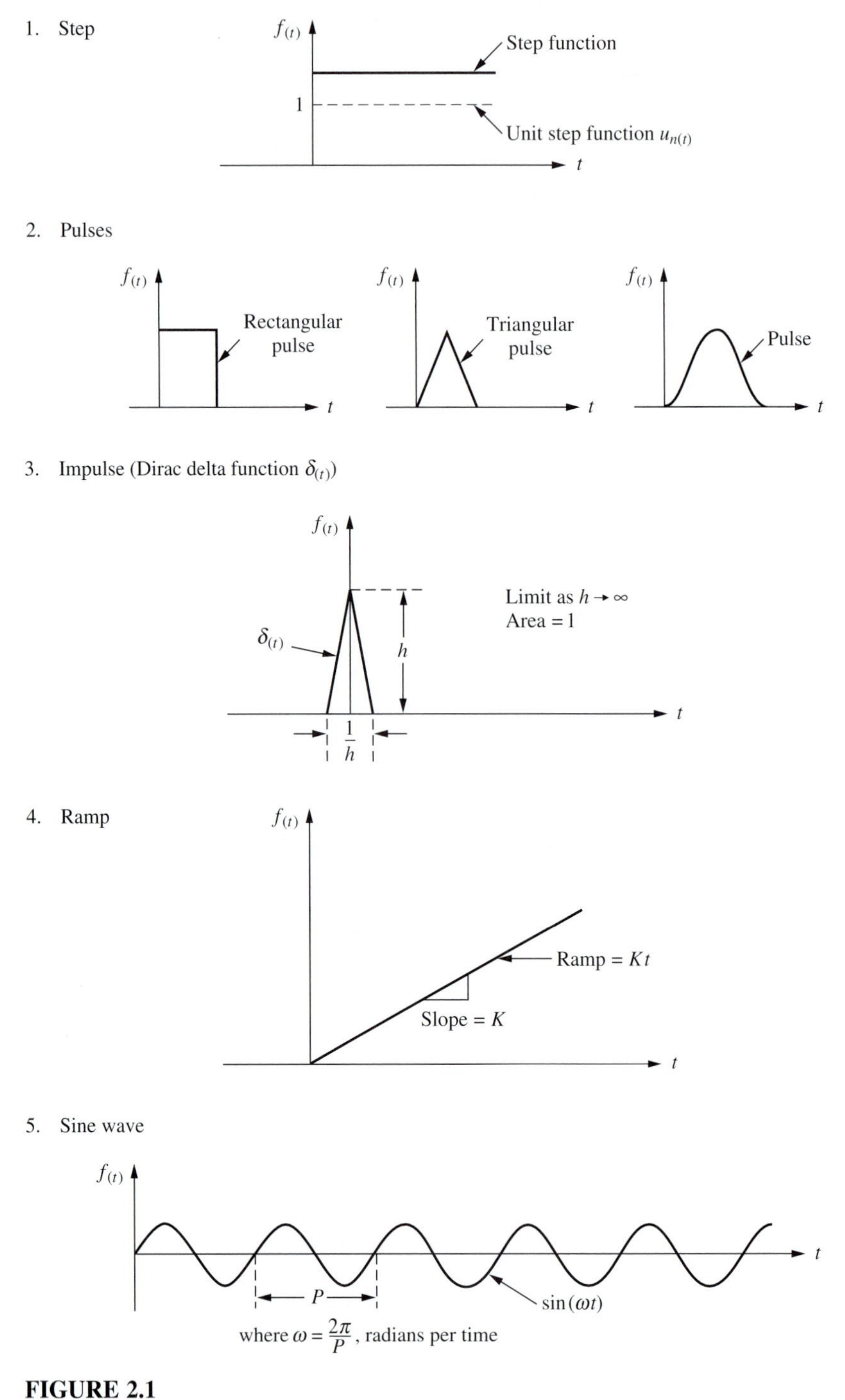

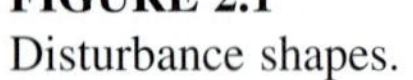

FIGURE 2.1
Disturbance shapes.

setpoint. The closedloop response to a setpoint disturbance is called the *servo response* (from the early applications of feedback control in mechanical servomechanism tracking systems).

2.2 LINEARIZATION AND PERTURBATION VARIABLES

2.2.1 Linearization

As mentioned earlier, we must convert the rigorous nonlinear differential equations describing a chemical system into linear differential equations so that we can use the powerful linear mathematical techniques.

The first question to be answered is, just what is a linear differential equation? Basically, it is one that contains variables only to the first power in any one term of the equation. If square roots, squares, exponentials, products of variables, etc. appear in the equation, it is nonlinear.

Linear example:

$$a_1 \frac{dx}{dt} + a_0 x = f_{(t)} \tag{2.9}$$

where a_0 and a_1 are constants or functions of time only, not of dependent variables or their derivatives.

Nonlinear examples:

$$a_1 \frac{dx}{dt} + a_0 x^{0.5} = f_{(t)} \tag{2.10}$$

$$a_1 \frac{dx}{dt} + a_0 (x)^2 = f_{(t)} \tag{2.11}$$

$$a_1 \frac{dx}{dt} + a_0 e^x = f_{(t)} \tag{2.12}$$

$$a_1 \frac{dx_1}{dt} + a_0 x_{1(t)} x_{2(t)} = f_{(t)} \tag{2.13}$$

where x_1 and x_2 are both dependent variables.

Mathematically, a linear differential equation is one for which the following two properties hold:

1. If $x_{(t)}$ is a solution, then $cx_{(t)}$ is also a solution, where c is a constant.
2. If x_1 is a solution and x_2 is also a solution, then $x_1 + x_2$ is a solution.

Linearization is quite straightforward. All we do is take the nonlinear functions, expand them in Taylor series around the steady-state operating level, and neglect all terms after the first partial derivatives.

Let us assume we have a nonlinear function f of the process variables x_1 and x_2: $f_{(x_1, x_2)}$. For example, x_1 could be mole fraction or temperature or flow rate. We

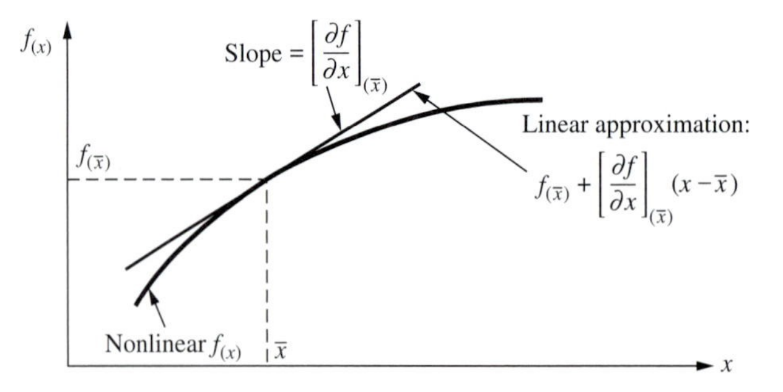

FIGURE 2.2
Linearization.

will denote the steady-state values of these variables by using an overscore:

$$\bar{x}_1 \equiv \text{steady-state value of } x_1$$
$$\bar{x}_2 \equiv \text{steady-state value of } x_2$$

Now we expand the function $f_{(x_1,x_2)}$ around its steady-state value $f_{(\bar{x}_1,\bar{x}_2)}$.

$$\begin{aligned} f_{(x_1,x_2)} &= f_{(\bar{x}_1,\bar{x}_2)} + \left(\frac{\partial f}{\partial x_1}\right)_{(\bar{x}_1,\bar{x}_2)} (x_1 - \bar{x}_1) + \left(\frac{\partial f}{\partial x_2}\right)_{(\bar{x}_1,\bar{x}_2)} (x_2 - \bar{x}_2) \\ &\quad + \left(\frac{\partial^2 f}{\partial x_1^2}\right)_{(\bar{x}_1,\bar{x}_2)} \frac{(x_1 - \bar{x}_1)^2}{2!} + \cdots \end{aligned} \tag{2.14}$$

Linearization consists of truncating the series after the first partial derivatives.

$$f_{(x_1,x_2)} \simeq f_{(\bar{x}_1,\bar{x}_2)} + \left(\frac{\partial f}{\partial x_1}\right)_{(\bar{x}_1,\bar{x}_2)} (x_1 - \bar{x}_1) + \left(\frac{\partial f}{\partial x_2}\right)_{(\bar{x}_1,\bar{x}_2)} (x_2 - \bar{x}_2) \tag{2.15}$$

We are approximating the real function by a linear function. The process is sketched graphically in Fig. 2.2 for a function of a single variable. The method is best illustrated by some common examples.

EXAMPLE 2.1. Consider the square-root dependence of flow out of a tank on the liquid height in the tank.

$$F_{(h)} = K\sqrt{h} \tag{2.16}$$

The Taylor series expansion around the steady-state value of h, which is $\bar{h}$ in our nomenclature, is

$$\begin{aligned} F_{(h)} &= F_{(\bar{h})} + \left(\frac{\partial F}{\partial h}\right)_{(\bar{h})} (h - \bar{h}) + \left(\frac{\partial^2 F}{\partial h^2}\right)_{(\bar{h})} \frac{(h - \bar{h})^2}{2!} + \cdots \\ &\simeq F_{(\bar{h})} + \left(\tfrac{1}{2} K h^{-1/2}\right)_{(\bar{h})} (h - \bar{h}) \\ &= K\sqrt{\bar{h}} + \frac{K}{2\sqrt{\bar{h}}}(h - \bar{h}) \end{aligned} \tag{2.17}$$

■

EXAMPLE 2.2. The Arrhenius temperature dependence of the specific reaction rate k is a highly nonlinear function that is linearized as follows:

$$k_{(T)} = \alpha e^{-E/RT} \tag{2.18}$$

$$\simeq k_{(\overline{T})} + \left(\frac{\partial k}{\partial T}\right)_{(\overline{T})} (T - \overline{T})$$

$$= \overline{k} + \frac{E\overline{k}}{R\overline{T}^2}(T - \overline{T}) \tag{2.19}$$

where $\overline{k} \equiv k_{(\overline{T})}$ ■

EXAMPLE 2.3. The product of two dependent variables is a nonlinear function of the two variables:

$$f_{(C_A,F)} = C_A F \tag{2.20}$$

Linearizing:

$$f_{(C_A,F)} \simeq f_{(\overline{C}_A,\overline{F})} + \left(\frac{\partial f}{\partial C_A}\right)_{(\overline{C}_A,\overline{F})} (C_A - \overline{C}_A) + \left(\frac{\partial f}{\partial F}\right)_{(\overline{C}_A,\overline{F})} (F - \overline{F}) \tag{2.21}$$

$$C_{A(t)}F_{(t)} \simeq \overline{C}_A\overline{F} + \overline{F}(C_{A(t)} - \overline{C}_A) + \overline{C}_A(F_{(t)} - \overline{F}) \tag{2.22}$$

Notice that the linearization process converts the nonlinear function (the product of two dependent variables) into a linear function containing two terms. ■

EXAMPLE 2.4. Consider the nonlinear ordinary differential equation (ODE) for a gravity-flow tank, which is derived from a momentum balance around the exit pipe.

$$\frac{dv}{dt} = \left(\frac{g}{L}\right)h - \left(\frac{K_F g_c}{\rho A_p}\right)v^2 \tag{2.23}$$

where
v = velocity of liquid in the pipe
h = liquid height in the tank
L = length of pipe
K_F = friction factor constant
ρ = density
A_p = cross-sectional area of pipe
g = gravitational force
g_c = gravitational constant

Linearizing the v^2 term gives

$$v^2 = \overline{v}^2 + (2\overline{v})(v - \overline{v}) \tag{2.24}$$

Thus Eq. (2.23) becomes

$$\frac{dv}{dt} = \left(\frac{g}{L}\right)h - \left(\frac{2\overline{v}K_F g_c}{\rho A_p}\right)v + \left(\frac{\overline{v}^2 K_F g_c}{\rho A_p}\right) \tag{2.25}$$

This ODE is now linear. The terms in the parentheses are constants; they depend, of course, on the steady state around which the system is linearized. ■

EXAMPLE 2.5. The component balance equation for an irreversible nth-order, non-isothermal reaction occurring in a constant-volume, variable-throughput continuous stirred-tank reactor (CSTR) is

$$V\frac{dC_A}{dt} = F_0C_{A0} - FC_A - V(C_A)^n\alpha e^{-E/RT} \tag{2.26}$$

Linearization gives

$$\begin{aligned} V\frac{dC_A}{dt} &= [\overline{F}_0\overline{C}_{A0} + \overline{F}_0(C_{A0} - \overline{C}_{A0}) + \overline{C}_{A0}(F_0 - \overline{F}_0)] \\ &\quad - [\overline{F}\,\overline{C}_A + \overline{F}(C_A - \overline{C}_A) + \overline{C}_A(F - \overline{F})] \\ &\quad - V[\overline{k}\,\overline{C}_A^n + n\overline{k}\,\overline{C}_A^{n-1}(C_A - \overline{C}_A) + \frac{\overline{k}\,\overline{C}_A^n E}{R\overline{T}^2}(T - \overline{T})] \end{aligned} \tag{2.27}$$

■

So far we have looked at examples where all the nonlinearity is in the derivative terms, i.e., the right-hand sides of the ODE. Quite often the model of a system will give an ODE that contains nonlinear terms inside the time derivative itself. For example, suppose the model of a nonlinear system is

$$\frac{d(h^3)}{dt} = K\sqrt{h} \tag{2.28}$$

The correct procedure for linearizing this type of equation is to rearrange it so that all the nonlinear functions appear only on the right-hand side of the ODE, and then linearize in the normal way. For the example given in Eq. (2.28), we differentiate the h^3 term to get

$$3h^2\frac{dh}{dt} = K\sqrt{h} \tag{2.29}$$

Then rearrangement gives

$$\frac{dh}{dt} = \frac{K}{3}(h)^{-1.5} \tag{2.30}$$

Now we are ready to linearize.

$$\frac{dh}{dt} = \frac{K}{3}(\overline{h})^{-1.5} + \frac{\partial}{\partial h}\left(\frac{K}{3}(h)^{-1.5}\right)_{(\overline{h})}(h - \overline{h}) \tag{2.31}$$

$$= \frac{K}{3}(\overline{h})^{-1.5} + \left(\frac{-1.5K}{3}(\overline{h})^{-2.5}\right)(h - \overline{h}) \tag{2.32}$$

$$= \frac{5K}{6}(\overline{h})^{-1.5} + \left(\frac{-K}{2}(\overline{h})^{-2.5}\right)h \tag{2.33}$$

This is a linear ODE with constant coefficients:

$$\frac{dh}{dt} = a_0 + a_1h \tag{2.34}$$

2.2.2 Perturbation Variables

For practically all the linear dynamics and control studies in the rest of the book, it is useful to look at the changes of variables away from steady-state values instead of the absolute variables themselves. Why this is useful will become apparent in the following discussion.

Since the total variables are functions of time, $x_{(t)}$, their departures from the steady-state values $\overline{x}$ will also be functions of time, as sketched in Fig. 2.3. These departures from steady state are called *perturbations, perturbation variables,* or *deviation variables.* We use, for the present, the symbol $x^P_{(t)}$. Thus, the perturbation in x is defined as

$$x^P_{(t)} \equiv x_{(t)} - \overline{x} \tag{2.35}$$

The equations describing the linear system can now be expressed in terms of these perturbation variables. When this is done, two very useful results occur:

1. The constant terms in the ordinary differential equation drop out.
2. The initial conditions for the perturbation variables are all equal to zero if the starting point is the steady-state operating condition around which the equations have been linearized.

Both of these results greatly simplify the linearized equations. For example, if the perturbations in velocity and liquid height are used in Eq. (2.25), we get

$$\frac{d(\overline{v} + v^P_{(t)})}{dt} = \left(\frac{g}{L}\right)(\overline{h} + h^P_{(t)}) - \left(\frac{2\overline{v}K_F g_c}{\rho A_p}\right)(\overline{v} + v^P_{(t)}) + \left(\frac{\overline{v}^2 K_F g_c}{\rho A_p}\right) \tag{2.36}$$

Since $\overline{v}$ is a constant,

$$\frac{dv^P_{(t)}}{dt} = \left(\frac{g}{L}\right)h^P_{(t)} - \left(\frac{2\overline{v}K_F g_c}{\rho A_p}\right)v^P_{(t)} + \left(\frac{g\overline{h}}{L} - \frac{\overline{v}^2 K_F g_c}{\rho A_p}\right) \tag{2.37}$$

Now consider Eq. (2.23) under steady-state conditions. At steady state v will be equal to $\overline{v}$, a constant, and h will be equal to $\overline{h}$, another constant.

$$\frac{d\overline{v}}{dt} = 0 = \left(\frac{g}{L}\right)\overline{h} - \left(\frac{K_F g_c}{\rho A_p}\right)\overline{v}^2 \tag{2.38}$$

Therefore the last term in Eq. (2.37) is equal to zero. We end up with a linear ordinary differential equation with constant coefficients in terms of perturbation variables.

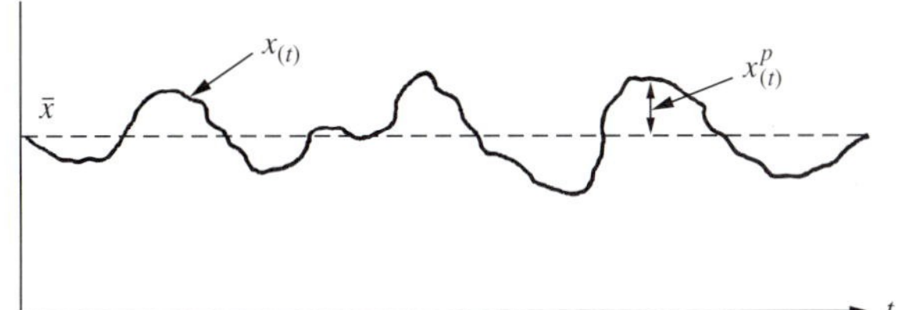

FIGURE 2.3
Perturbation variables.

$$\frac{dv_{(t)}^p}{dt} = \left(\frac{g}{L}\right)h_{(t)}^p - \left(\frac{2\bar{v}K_F g_c}{\rho A_p}\right)v_{(t)}^p \tag{2.39}$$

In a similar way Eq. (2.27) can be written in terms of perturbations in C_A, C_{A0}, F_0, F, and T.

$$\begin{aligned} V\frac{d(\overline{C}_A + C_A^p)}{dt} &= (\overline{F}_0)C_{A0}^p + (\overline{C}_{A0})F_0^p - (\overline{F})C_A^p - (\overline{C}_A)F^p \\ &\quad - (Vn\bar{k}\,\overline{C}_A^{n-1})C_A^p + \left(\frac{V\bar{k}\,\overline{C}_A^n E}{R\overline{T}^2}\right)T^p \\ &\quad + [\overline{F}_0\overline{C}_{A0} - \overline{F}\,\overline{C}_A - V\bar{k}\,\overline{C}_A^n] \end{aligned} \tag{2.40}$$

Application of Eq. (2.26) under steady-state conditions shows that the last term in Eq. (2.40) is just equal to zero. So we end up with a simple linear ODE in terms of perturbation variables.

$$\begin{aligned} V\frac{dC_A^p}{dt} &= (\overline{F}_0)C_{A0}^p + (\overline{C}_{A0})F_0^p - (\overline{F})C_A^p - (\overline{C}_A)F^p \\ &\quad - (Vn\bar{k}\,\overline{C}_A^{n-1})C_A^p + \left(\frac{V\bar{k}\,\overline{C}_A^n E}{R\overline{T}^2}\right)T^p \end{aligned} \tag{2.41}$$

Since we use perturbation variables most of the time, we often do not bother to write the superscript p. It is understood that whenever we write the linearized equations for the system, all variables are perturbation variables. Thus, Eqs. (2.39) and (2.41) can be written

$$\frac{dv}{dt} = \left(\frac{g}{L}\right)h - \left(\frac{2\bar{v}K_F g_c}{\rho A_p}\right)v \tag{2.42}$$

$$\begin{aligned} V\frac{dC_A}{dt} &= (\overline{F}_0)C_{A0} + (\overline{C}_{A0})F_0 - (\overline{F})C_A - (\overline{C}_A)F \\ &\quad - (Vn\bar{k}\,\overline{C}_A^{n-1})C_A + \left(\frac{V\bar{k}\,\overline{C}_A^n E}{R\overline{T}^2}\right)T \end{aligned} \tag{2.43}$$

Note that the initial conditions of all these perturbation variables are zero since all variables start at the initial steady-state values. This will simplify things significantly when we use Laplace transforms in Part Two.

2.3 RESPONSES OF SIMPLE LINEAR SYSTEMS

2.3.1 First-Order Linear Ordinary Differential Equation

Consider the general first-order linear ODE

$$\frac{dx}{dt} + P_{(t)}x = Q_{(t)} \tag{2.44}$$

with a given value of x known at a fixed point in time: $x_{(t_0)} = x_0$. Usually this is an initial condition where $t_0 = 0$.

Multiply both sides of Eq. (2.44) by the integrating factor $\exp(\int P\,dt)$.

$$\frac{dx}{dt}\exp\left(\int P\,dt\right) + P_{(t)}x\exp\left(\int P\,dt\right) = Q_{(t)}\exp\left(\int P\,dt\right)$$

Combining the two terms on the left-hand side of the equation gives

$$\frac{d}{dt}\left[x\exp\left(\int P\,dt\right)\right] = Q_{(t)}\exp\left(\int P\,dt\right)$$

Integrating yields

$$x\exp\left(\int P\,dt\right) = \int\left[Q_{(t)}\exp\left(\int P\,dt\right)\right]dt + c_1$$

where c_1 is a constant of integration and can be evaluated by using the boundary or initial condition. Therefore, the general solution of Eq. (2.44) is

$$x = \exp\left(-\int P\,dt\right)\left\{\int\left[Q_{(t)}\exp\left(\int P\,dt\right)\right]dt + c_1\right\} \tag{2.45}$$

EXAMPLE 2.6. An isothermal, constant-holdup, constant-throughput CSTR with a first-order irreversible reaction is described by a component continuity equation that is a first-order linear ODE:

$$\frac{dC_A}{dt} + \left(\frac{F}{V} + k\right)C_A = \left(\frac{F}{V}\right)C_{A0} \tag{2.46}$$

Let the concentrations C_{A0} and C_A be total values, not perturbations, for the present. The reactant concentration in the tank is initially zero.

Initial condition:

$$C_{A(0)} = 0$$

At time zero a step change in feed concentration is made from zero to a constant value $\overline{C}_{A0}$.

Forcing function:

$$C_{A0(t)} = \overline{C}_{A0}$$

Comparing Eqs. (2.44) and (2.46),

$$x = C_A \qquad P = \frac{F}{V} + k \qquad Q = \frac{F\overline{C}_{A0}}{V}$$

Therefore,

$$\exp\left(\int P\,dt\right) = e^{(F/V+k)t}$$

$$\int\left[Q_{(t)}\exp\left(\int P\,dt\right)\right]dt = \int\left(\frac{F\overline{C}_{A0}}{V}\right)e^{(F/V+k)t}\,dt$$

$$= \left(\frac{F\overline{C}_{A0}}{V}\right)\left(\frac{1}{F/V + k}\right)e^{(F/V+k)t} + c_1$$

The solution to Eq. (2.46) is, according to Eq. (2.45),

$$\begin{aligned} C_{A(t)} &= e^{-(F/V+k)t}\left\{\left(\frac{F\overline{C}_{A0}}{V}\right)\left(\frac{1}{F/V+k}\right)e^{(F/V+k)t}+c_1\right\} \\ &= \frac{F\overline{C}_{A0}}{F+kV}+c_1e^{-(F/V+k)t} \end{aligned} \tag{2.47}$$

The initial condition is now used to find the value of c_1.

$$C_{A(0)} = 0 = \frac{F\overline{C}_{A0}}{F+kV}+c_1(1)$$

Therefore, the time-dependent response of C_A to the step disturbance in feed concentration is

$$C_{A(t)} = \frac{\overline{C}_{A0}}{1+k\tau}\left[1-e^{-(1/\tau+k)t}\right] \tag{2.48}$$

where $\tau \equiv V/F$ and is the residence time of the vessel. The response is sketched in Fig. 2.4 and is the classical first-order exponential rise to the new steady state.

The first thing you should always do when you get a solution is check if it is consistent with the initial conditions and if it is reasonable physically. At $t = 0$, Eq. (2.48) becomes

$$C_{A(t=0)} = \frac{\overline{C}_{A0}}{1+k\tau}[1-1] = 0$$

so the initial condition is satisfied.

Does the solution make sense from a steady-state point of view? The new steady-state value of C_A that is approached asymptotically by the exponential function can be found from either the solution [Eq. (2.48)], letting time go to infinity, or from the original ODE [Eq. (2.46)], setting the time derivative dC_A/dt equal to zero. Either method predicts that at the final steady state

$$C_{A(t\to\infty)} \equiv \overline{C}_A = \frac{\overline{C}_{A0}}{1+k\tau} \tag{2.49}$$

Is this reasonable? It says that the consumption of reactant will be greater (the ratio of $\overline{C}_A$ to $\overline{C}_{A0}$ will be smaller) the bigger k and τ are. This certainly makes good chemical engineering sense. If k is zero (i.e., no reaction), the final steady-state value of $\overline{C}_A$ will be equal to the feed concentration $\overline{C}_{A0}$, as it should be. Note that $C_{A(t)}$ would not be dynamically equal to $\overline{C}_{A0}$; it would start at 0 and rise asymptotically to its final steady-

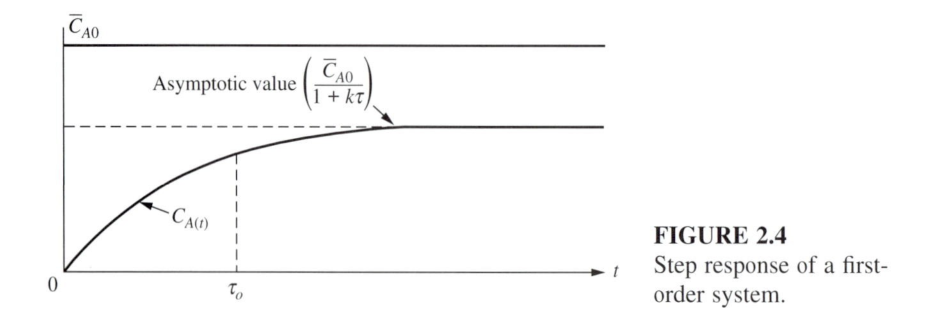

FIGURE 2.4
Step response of a first-order system.

state value. Thus, the predictions of the solution seem to check with the real physical world.

The ratio of the change in the steady-state value of the output divided by the magnitude of the step change made in the input is called the *steady-state gain* of the process K_p.

$$K_p \equiv \frac{\overline{C}_A}{\overline{C}_{A0}} = \frac{1}{1 + k\tau} \tag{2.50}$$

These steady-state gains will be extremely important in our dynamic studies and in controller design.

Does the solution make sense dynamically? The rate of rise will be determined by the magnitude of the $(k + 1/\tau)$ term in the exponential. The bigger this term, the faster the exponential term will decay to zero as time increases. The smaller this term, the slower the decay will be. Therefore, the dynamics are set by $(k + 1/\tau)$.

The reciprocal of this term is called the *process openloop time constant,* and we use the symbol τ_o. The bigger the time constant, the slower the dynamic response will be. The solution [Eq. (2.48)] predicts that a small value of k or a big value of τ will give a large process time constant. Again, this makes good physical sense. If there is no reaction, the time constant is just equal to the residence time $\tau = V/F$.

Before we leave this example, let us put Eq. (2.46) in the standard form

$$\tau_o \frac{dC_A}{dt} + C_A = K_p C_{A0} \tag{2.51}$$

This is the form in which we want to look at many systems of this type. Dividing by the term $(k + 1/\tau)$ does the trick.

$$\frac{1}{k + 1/\tau}\frac{dC_A}{dt} + C_A = \frac{1/\tau}{k + 1/\tau}C_{A0} = \frac{1}{k\tau + 1}C_{A0} \tag{2.52}$$

$$\tau_o = \frac{1}{k + 1/\tau} = \text{process openloop time constant with units of time}$$

$$K_p = \frac{1}{k\tau + 1} = \text{process steady-state gain with units of concentration in product stream divided by concentration in feed stream}$$

Then the solution [Eq. (2.48)] becomes

$$C_{A(t)} = \overline{C}_{A0} K_p (1 - e^{-t/\tau_o}) \tag{2.53}$$

In this example we have used total variables. If we convert Eq.(2.46) into perturbation variables, we get

$$\frac{d(\overline{C}_A + C_A^p)}{dt} + \left(\frac{F}{V} + k\right)(\overline{C}_A + C_A^p) = \left(\frac{F}{V}\right)(\overline{C}_{A0} + C_{A0}^p)$$

$$\frac{dC_A^p}{dt} + \left(\frac{F}{V} + k\right)C_A^p = \left(\frac{F}{V}\right)C_{A0}^p - \left\{\left(\frac{F}{V} + k\right)\overline{C}_A - \left(\frac{F}{V}\right)\overline{C}_{A0}\right\} \tag{2.54}$$

The last term in this equation is zero. Therefore, Eqs. (2.54) and (2.46) are identical, except one is in terms of total variables and the other is in terms of perturbations. Whenever the original ODE is already linear, either total or perturbation variables can be used. Initial conditions will, of course, differ by the steady-state values of all variables. ■

EXAMPLE 2.7. Suppose the feed concentration in the CSTR system considered above is ramped up with time:

$$C_{A0(t)} = Kt \tag{2.55}$$

where K is a constant. C_A is initially zero.

Rearranging Eq. (2.51) gives

$$\frac{dC_A}{dt} + \frac{1}{\tau_o} C_A = \frac{K_p K}{\tau_o} t \tag{2.56}$$

The solution, according to Eq. (2.45), is

$$C_{A(t)} = \exp\left(-\int \frac{1}{\tau_o} dt\right)\left\{\int \left[\frac{K_p K}{\tau_o} t \exp\left(\int \frac{1}{\tau_o} dt\right)\right] dt + c_1\right\}$$

$$= e^{-t/\tau_o}\left(\frac{K_p K}{\tau_o}\int t e^{t/\tau_o} dt + c_1\right) \tag{2.57}$$

The integral in Eq. (2.57) can be looked up in mathematics tables or found by integrating by parts. Let

$$u = t \quad \text{and} \quad dv = e^{t/\tau_o}\, dt$$

Then

$$du = dt \quad \text{and} \quad v = \tau_o e^{t/\tau_o}$$

Since $\int u\, dv = uv - \int v\, du$,

$$\int t e^{t/\tau_o}\, dt = \tau_o t e^{t/\tau_o} - \int \tau_o e^{t/\tau_o}\, dt$$
$$= \tau_o t e^{t/\tau_o} - (\tau_o)^2 e^{t/\tau_o} \tag{2.58}$$

Therefore Eq. (2.57) becomes

$$C_{A(t)} = K_p K(t - \tau_o) + c_1 e^{-t/\tau_o} \tag{2.59}$$

Using the initial condition to find c_1,

$$C_{A(0)} = 0 = K_p K(-\tau_o) + c_1 \tag{2.60}$$

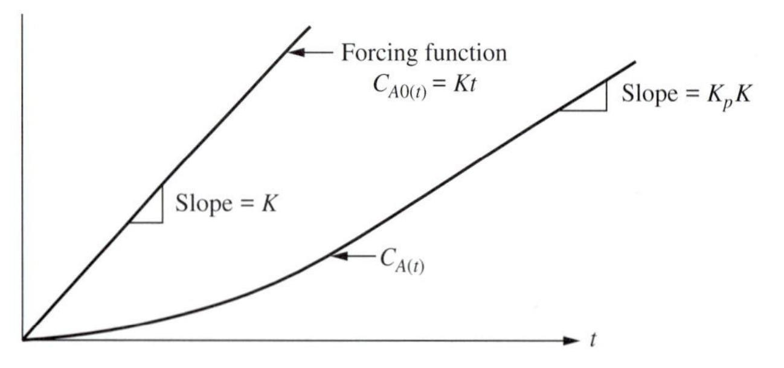

FIGURE 2.5
Ramp response of a first-order system.

The final solution is

$$C_{A(t)} = K_p K \tau_o \left(\frac{t}{\tau_o} - 1 + e^{-t/\tau_o} \right) \tag{2.61}$$

The ramp response is sketched in Fig. 2.5. ■

It is frequently useful to be able to determine the time constant of a first-order system from experimental step response data. This is easy to do. When time is equal to τ_o in Eq. (2.53), the term $(1 - e^{-t/\tau_o})$ becomes $(1 - e^{-1}) = 0.623$. This means that the output variable has undergone 62.3 percent of the total change it is going to make. Thus, the time constant of a first-order system is simply the time it takes the step response to reach 62.3 percent of its new final steady-state value.

2.3.2 Second-Order Linear ODEs with Constant Coefficients

The first-order system considered in the previous section yields well-behaved exponential responses. Second-order systems can be much more exciting since they can give an oscillatory or *underdamped* response.

The first-order linear equation [Eq. (2.44)] could have a time-variable coefficient; that is, $P_{(t)}$ could be a function of time. We consider only linear second-order ODEs that have constant coefficients (τ_o and ζ are constants).

$$\tau_o^2 \frac{d^2x}{dt^2} + 2\zeta\tau_o \frac{dx}{dt} + x = m_{(t)} \tag{2.62}$$

Analytical methods are available for linear ODEs with variable coefficients, but their solutions are usually messy infinite series, and we do not consider them here.

The solution of a second-order ODE can be deduced from the solution of a first-order ODE. Equation (2.45) can be broken up into two parts:

$$x_{(t)} = \left\{ c_1 \exp\left(-\int P\,dt \right) \right\} + \left\{ \exp\left(-\int P\,dt \right) \int \left[Q_{(t)} \exp\left(\int P\,dt \right) \right] dt \right\} \equiv x_c + x_p \tag{2.63}$$

The variable x_c is called the *complementary solution.* It is the function that satisfies the original ODE with the forcing function $Q_{(t)}$ set equal to zero (called the *homogeneous* differential equation):

$$\frac{dx}{dt} + P_{(t)}x = 0 \tag{2.64}$$

The variable x_p is called the *particular* solution. It is the function that satisfies the original ODE with a specified $Q_{(t)}$. One of the most useful properties of linear ODEs is that the total solution is the sum of the complementary solution and the particular solution.

Now we are ready to extend the preceding ideas to the second-order ODE of Eq. (2.62). First we obtain the complementary solution x_c by solving the homogeneous equation

$$\tau_o^2 \frac{d^2x}{dt^2} + 2\zeta\tau_o \frac{dx}{dt} + x = 0 \tag{2.65}$$

Then we solve for the particular solution x_p and add the two to obtain the entire solution.

A. Complementary solution

Since the complementary solution of the first-order ODE is an exponential, it is reasonable to guess that the complementary solution of the second-order ODE is also of exponential form. Let us guess that

$$x_c = ce^{st} \tag{2.66}$$

where c and s are constants. Differentiating x_c with respect to time gives

$$\frac{dx_c}{dt} = cse^{st} \quad \text{and} \quad \frac{d^2x_c}{dt^2} = cs^2e^{st}$$

Now we substitute the guessed solution and its derivatives into Eq. (2.65) to find the values of s that satisfy the assumed form [Eq. (2.66)].

$$\tau_o^2(cs^2e^{st}) + 2\zeta\tau_o(cse^{st}) + (ce^{st}) = 0$$

$$\boxed{\tau_o^2 s^2 + 2\zeta\tau_o s + 1 = 0} \tag{2.67}$$

This equation, called the *characteristic equation,* contains the system's most important dynamic features. The values of s that satisfy Eq. (2.67) are called the *roots* of the characteristic equation (they are also called the *eigenvalues* of the system). Their values, as we will shortly show, dictate if the system is fast or slow, stable or unstable, overdamped or underdamped. Dynamic analysis and controller design consist of finding the values of the roots of the characteristic equation of the system and changing their values to obtain the desired response. Much of this book is devoted to looking at roots of characteristic equations. They represent an extremely important concept that you should fully understand.

Using the general solution for a quadratic equation, we can solve Eq. (2.67) for its two roots

$$s = \frac{-2\zeta\tau_o \pm \sqrt{(2\zeta\tau_o)^2 - 4\tau_o^2}}{2\tau_o^2} = -\frac{\zeta}{\tau_o} \pm \frac{\sqrt{\zeta^2 - 1}}{\tau_o} \tag{2.68}$$

Two values of s satisfy Eq. (2.67). There are two exponentials of the form given in Eq. (2.66) that are solutions to the original homogeneous ODE [Eq. (2.65)]. The sum of these solutions is also a solution since the ODE is linear. Therefore, the complementary solution is (for $s_1 \neq s_2$)

$$x_c = c_1e^{s_1t} + c_2e^{s_2t} \tag{2.69}$$

where c_1 and c_2 are constants. The two roots s_1 and s_2 are

$$s_1 = -\frac{\zeta}{\tau_o} + \frac{\sqrt{\zeta^2 - 1}}{\tau_o} \tag{2.70}$$

$$s_2 = -\frac{\zeta}{\tau_o} - \frac{\sqrt{\zeta^2 - 1}}{\tau_o} \tag{2.71}$$

The shape of the solution curve depends strongly on the values of the physical parameter ζ, called the damping coefficient. Let us now look at three possibilities.

$\zeta > 1$ (overdamped system). If the damping coefficient is greater than unity, the quantity inside the square root is positive. Then s_1 and s_2 will both be real numbers, and they will be different (*distinct roots*).

EXAMPLE 2.8. Consider the ODE

$$\begin{aligned} \frac{d^2x}{dt^2} + 5\frac{dx}{dt} + 6x &= 0 \\ \left(\frac{1}{\sqrt{6}}\right)^2 \frac{d^2x}{dt^2} + 2\left(\frac{1}{\sqrt{6}}\right)\left(\frac{5}{2\sqrt{6}}\right)\frac{dx}{dt} + x &= 0 \end{aligned} \tag{2.72}$$

Its characteristic equation can be written in several forms:

$$s^2 + 5s + 6 = 0 \tag{2.73}$$

$$(s + 3)(s + 2) = 0 \tag{2.74}$$

$$\left(\frac{1}{\sqrt{6}}\right)^2 s^2 + 2\left(\frac{1}{\sqrt{6}}\right)\left(\frac{5}{2\sqrt{6}}\right)s + 1 = 0 \tag{2.75}$$

All three are completely equivalent. The time constant and the damping coefficient for the system are

$$\tau_o = \frac{1}{\sqrt{6}} \qquad \zeta = \frac{5}{2\sqrt{6}}$$

The roots of the characteristic equation are obvious from Eq. (2.74), but the use of Eq. (2.68) gives

$$\begin{aligned} s &= -\frac{\zeta}{\tau_o} \pm \frac{\sqrt{\zeta^2 - 1}}{\tau_o} = -\tfrac{5}{2} \pm \tfrac{1}{2} \\ s_1 &= -2 \\ s_2 &= -3 \end{aligned}$$

The two roots are real, and the complementary solution is

$$x_c = c_1 e^{-2t} + c_2 e^{-3t} \tag{2.76}$$

The values of the constant c_1 and c_2 depend on the initial conditions. ■

$\zeta = 1$ (critically damped system). If the damping coefficient is equal to unity, the term inside the square root of Eq. (2.68) is zero. There is only one value of s that satisfies the characteristic equation.

$$s = -\frac{1}{\tau_o} \tag{2.77}$$

The two roots are the same and are called *repeated* roots. This is clearly seen if a value of $\zeta = 1$ is substituted into the characteristic equation [Eq. (2.67)]:

$$\tau_o^2 s^2 + 2\tau_o s + 1 = 0 = (\tau_o s + 1)(\tau_o s + 1) \tag{2.78}$$

The complementary solution with a repeated root is

$$x_c = (c_1 + c_2 t)e^{st} = (c_1 + c_2 t)e^{-t/\tau_o} \tag{2.79}$$

This is easily proved by substituting it into Eq. (2.65) with ζ set equal to unity.

EXAMPLE 2.9. If two CSTRs like the one considered in Example 2.6 are run in series, two first-order ODEs describe the system:

$$\frac{dC_{A1}}{dt} + \left(\frac{1}{\tau_1} + k_1\right)C_{A1} = \left(\frac{1}{\tau_1}\right)C_{A0} \tag{2.80}$$

$$\frac{dC_{A2}}{dt} + \left(\frac{1}{\tau_2} + k_2\right)C_{A2} = \left(\frac{1}{\tau_2}\right)C_{A1} \tag{2.81}$$

Differentiating the second equation with respect to time and eliminating C_{A1} give a second-order ODE:

$$\frac{d^2C_{A2}}{dt^2} + \left(\frac{1}{\tau_1} + k_1 + \frac{1}{\tau_2} + k_2\right)\frac{dC_{A2}}{dt} + \left(\frac{1}{\tau_1} + k_1\right)\left(\frac{1}{\tau_2} + k_2\right)C_{A2} = \left(\frac{1}{\tau_1\tau_2}\right)C_{A0} \tag{2.82}$$

If temperatures and holdups are the same in both tanks, the specific reaction rates k and holdup times τ will be the same:

$$k_1 = k_2 \equiv k \qquad \tau_1 = \tau_2 \equiv \tau$$

The characteristic equation is

$$\begin{gathered} s^2 + 2\left(\frac{1}{\tau} + k\right)s + \left(\frac{1}{\tau} + k\right)^2 = 0 \\ \left(s + \frac{1}{\tau} + k\right)\left(s + \frac{1}{\tau} + k\right) = 0 \end{gathered} \tag{2.83}$$

The damping coefficient is unity and there is a real, repeated root:

$$s = -\left(\frac{1}{\tau} + k\right)$$

The complementary solution is

$$(C_{A2})_c = (c_1 + c_2 t)e^{-(k+1/\tau)t} \tag{2.84}$$

■

$\zeta < 1$ (underdamped system). Things begin to get interesting when the damping coefficient is less than unity. Now the term inside the square root in Eq. (2.68) is negative, giving an imaginary number in the roots.

$$s = -\frac{\zeta}{\tau_o} \pm \frac{\sqrt{\zeta^2 - 1}}{\tau_o} = -\frac{\zeta}{\tau_o} \pm i\frac{\sqrt{1-\zeta^2}}{\tau_o} \tag{2.85}$$

The roots are complex numbers with real and imaginary parts.

$$s_1 = -\frac{\zeta}{\tau_o} + i\frac{\sqrt{1-\zeta^2}}{\tau_o} \tag{2.86}$$

$$s_2 = -\frac{\zeta}{\tau_o} - i\frac{\sqrt{1-\zeta^2}}{\tau_o} \tag{2.87}$$

To be more specific, they are *complex conjugates* since they have the same real parts and their imaginary parts differ only in sign. The complementary solution is

$$\begin{aligned} x_c &= c_1 e^{s_1 t} + c_2 e^{s_2 t} \\ &= c_1 \exp\left\{\left(-\frac{\zeta}{\tau_o} + i\frac{\sqrt{1-\zeta^2}}{\tau_o}\right)t\right\} + c_2 \exp\left\{\left(-\frac{\zeta}{\tau_o} - i\frac{\sqrt{1-\zeta^2}}{\tau_o}\right)t\right\} \\ &= e^{-\zeta t/\tau_o}\left\{c_1 \exp\left(+i\frac{\sqrt{1-\zeta^2}}{\tau_o}t\right) + c_2 \exp\left(-i\frac{\sqrt{1-\zeta^2}}{\tau_o}t\right)\right\} \end{aligned} \tag{2.88}$$

Now we use the relationships

$$e^{ix} = \cos x + i \sin x \tag{2.89}$$

$$\cos(-x) = \cos x \tag{2.90}$$

$$\sin(-x) = -\sin x \tag{2.91}$$

Substituting into Eq. (2.88) gives

$$\begin{aligned} x_c &= e^{-\zeta t/\tau_o}\left(c_1\left\{\cos\left(\frac{\sqrt{1-\zeta^2}}{\tau_o}t\right) + i\sin\left(\frac{\sqrt{1-\zeta^2}}{\tau_o}t\right)\right\}\right. \\ &\quad \left. + c_2\left\{\cos\left(\frac{\sqrt{1-\zeta^2}}{\tau_o}t\right) - i\sin\left(\frac{\sqrt{1-\zeta^2}}{\tau_o}t\right)\right\}\right) \\ &= e^{-\zeta t/\tau_o}\left\{(c_1 + c_2)\cos\left(\frac{\sqrt{1-\zeta^2}}{\tau_o}t\right) + i(c_1 - c_2)\sin\left(\frac{\sqrt{1-\zeta^2}}{\tau_o}t\right)\right\} \end{aligned} \tag{2.92}$$

The complementary solution consists of oscillating sinusoidal terms multiplied by an exponential. Thus, the solution is oscillatory or underdamped for $\zeta < 1$. Note that as long as the damping coefficient is positive ($\zeta > 0$), the exponential term will decay to zero as time goes to infinity. Therefore, the amplitude of the oscillations decreases to zero, as sketched in Fig. 2.6.

If we are describing a real physical system, the solution x_c must be a real quantity and the terms with the constants in Eq. (2.92) must all be real. So the term $c_1 + c_2$ and the term $i(c_1 - c_2)$ must both be real. This can be true only if c_1 and c_2 are complex conjugates, as proved next.

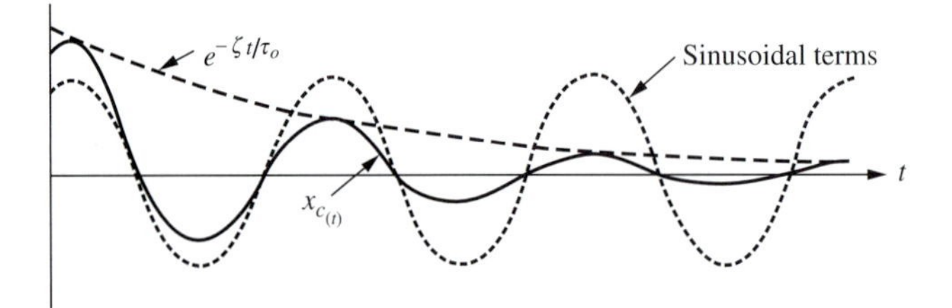

FIGURE 2.6
Complementary solution for $\zeta < 1$.

Let z be a complex number and $\bar{z}$ be its complex conjugate.

$$z = x + iy \quad \text{and} \quad \bar{z} = x - iy$$

Now look at the sum and the difference:

$$z + \bar{z} = (x + iy) + (x - iy) = 2x \qquad \text{a real number}$$
$$z - \bar{z} = (x + iy) - (x - iy) = 2yi \qquad \text{a pure imaginary number}$$
$$i(z - \bar{z}) = -2y \qquad \text{a real number}$$

So we have shown that to get real numbers for both $c_1 + c_2$ and $i(c_1 - c_2)$, the numbers c_1 and c_2 must be a complex conjugate pair. Let $c_1 = c^R + ic^I$ and $c_2 = c^R - ic^I$. Then the complementary solution becomes

$$x_{c(t)} = e^{-\zeta t/\tau_o}\left\{(2c^R)\cos\left(\frac{\sqrt{1-\zeta^2}}{\tau_o}t\right) - (2c^I)\sin\left(\frac{\sqrt{1-\zeta^2}}{\tau_o}t\right)\right\} \tag{2.93}$$

EXAMPLE 2.10. Consider the ODE

$$\frac{d^2x}{dt^2} + \frac{dx}{dt} + x = 0$$

Writing this in the standard form,

$$(1)^2\frac{d^2x}{dt^2} + 2(1)(0.5)\frac{dx}{dt} + x = 0$$

We see that the time constant $\tau_o = 1$ and the damping coefficient $\zeta = 0.5$. The characteristic equation is

$$s^2 + s + 1 = 0$$

Its roots are

$$\begin{aligned} s &= -\frac{\zeta}{\tau_o} \pm i\frac{\sqrt{1-\zeta^2}}{\tau_o} \\ &= -\tfrac{1}{2} \pm i\sqrt{1-(\tfrac{1}{2})^2} = -\tfrac{1}{2} \pm i\frac{\sqrt{3}}{2} \end{aligned} \tag{2.94}$$

The complementary solution is

$$x_{c(t)} = e^{-t/2}\left\{2c^R \cos\left(\frac{\sqrt{3}}{2}t\right) - 2c^I \sin\left(\frac{\sqrt{3}}{2}t\right)\right\} \tag{2.95}$$

■

$\zeta = 0$ (undamped system). The complementary solution is the same as Eq. (2.93) with the exponential term equal to unity. There is no decay of the sine and cosine terms, and therefore the system oscillates forever.

This result is obvious if we go back to Eq. (2.65) and set $\zeta = 0$.

$$\tau_o^2 \frac{d^2x}{dt^2} + x = 0 \tag{2.96}$$

You might remember from physics that this is the differential equation that describes a harmonic oscillator. The solution is a sine wave with a frequency of $1/\tau_o$. We discuss these kinds of functions in detail in Part Three, when we begin our "Chinese" lessons covering the frequency domain.

$\zeta < 0$ (unstable system). If the damping coefficient is negative, the exponential term increases without bound as time becomes large. Thus, the system is unstable.

This situation is extremely important because it shows the limit of stability of a second-order system. The roots of the characteristic equation are

$$s = -\frac{\zeta}{\tau_o} \pm i\frac{\sqrt{1-\zeta^2}}{\tau_o}$$

If the real part of the root of the characteristic equation $(-\zeta/\tau_o)$ is a positive number, the system is unstable. So the stability requirement is:

A system is stable if the real parts of all the roots of the characteristic equation are negative.

We use this result extensively throughout the rest of the book since it is the foundation upon which almost all controller designs are based.

B. Particular solution

Up to this point we have found only the complementary solution of the homogeneous equation

$$\tau_o^2 \frac{d^2x}{dt^2} + 2\zeta\tau_o \frac{dx}{dt} + x = 0$$

This corresponds to the solution for the unforced or undisturbed system. Now we must find the particular solutions for some specific forcing functions $m_{(t)}$. Then the total solution will be the sum of the complementary and particular solutions.

Several methods exist for finding particular solutions. Laplace transform methods are probably the most convenient, and we use them in Part Two. Here we present the *method of undetermined coefficients*. It consists of assuming a particular solution with the same form as the forcing function. The method is illustrated in the following examples.

EXAMPLE 2.11. The overdamped system of Example 2.8 is forced with a unit step function.

$$\frac{d^2x}{dt^2} + 5\frac{dx}{dt} + 6x = 1 \tag{2.97}$$

Initial conditions are

$$x_{(0)} = 0 \quad \text{and} \quad \left(\frac{dx}{dt}\right)_{(0)} = 0$$

The forcing function is a constant, so we assume that the particular solution is also a constant: $x_p = c_3$. Substituting into Eq. (2.97) gives

$$0 + 5(0) + 6c_3 = 1 \Rightarrow c_3 = \tfrac{1}{6} \tag{2.98}$$

Now the total solution is [using the complementary solution given in Eq. (2.76)]

$$x = x_c + x_p = c_1 e^{-2t} + c_2 e^{-3t} + \tfrac{1}{6} \tag{2.99}$$

The constants are evaluated from the initial conditions, using the total solution. A common mistake is to evaluate them using only the complementary solution.

$$x_{(0)} = 0 = c_1 + c_2 + \tfrac{1}{6}$$

$$\left(\frac{dx}{dt}\right)_{(0)} = 0 = (-2c_1 e^{-2t} - 3c_2 e^{-3t})_{(t=0)} = -2c_1 - 3c_2 = 0$$

Therefore

$$c_1 = -\tfrac{1}{2} \quad \text{and} \quad c_2 = \tfrac{1}{3}$$

The final total solution for the constant forcing function is

$$x_{(t)} = -\tfrac{1}{2}e^{-2t} + \tfrac{1}{3}e^{-3t} + \tfrac{1}{6} \tag{2.100}$$

■

EXAMPLE 2.12. A general underdamped second-order system is forced by a unit step function:

$$\tau_o^2 \frac{d^2x}{dt^2} + 2\zeta\tau_o\frac{dx}{dt} + x = 1 \tag{2.101}$$

Initial conditions are

$$x_{(0)} = 0 \quad \text{and} \quad \left(\frac{dx}{dt}\right)_{(0)} = 0$$

Since the forcing function is a constant, the particular solution is assumed to be a constant, giving $x_p = 1$. The total solution is the sum of the particular and complementary solutions [see Eq. (2.93)].

$$x_{(t)} = 1 + e^{-\zeta t/\tau_o}\left\{(2c^R)\cos\left(\frac{\sqrt{1-\zeta^2}}{\tau_o}t\right) - (2c^I)\sin\left(\frac{\sqrt{1-\zeta^2}}{\tau_o}t\right)\right\} \tag{2.102}$$

Using the initial conditions to evaluate constants,

$$x_{(0)} = 0 = 1 + [2c^R(1) - 2c^I(0)]$$

$$\frac{dx}{dt} = -\frac{\zeta}{\tau_o} e^{-\zeta t/\tau_o} \left\{ 2c^R \cos\left(\frac{\sqrt{1-\zeta^2}}{\tau_o} t\right) - 2c^I \sin\left(\frac{\sqrt{1-\zeta^2}}{\tau_o} t\right) \right\}$$

$$+ e^{-\zeta t/\tau_o} \left\{ -2c^R \frac{\sqrt{1-\zeta^2}}{\tau_o} \sin\left(\frac{\sqrt{1-\zeta^2}}{\tau_o} t\right) - 2c^I \frac{\sqrt{1-\zeta^2}}{\tau_o} \cos\left(\frac{\sqrt{1-\zeta^2}}{\tau_o} t\right) \right\}$$

$$\left(\frac{dx}{dt}\right)_{(0)} = 0 = -\frac{\zeta}{\tau_o}(2c^R) + \left(-2c^I \frac{\sqrt{1-\zeta^2}}{\tau_o}\right)$$

Solving for the constants gives

$$2c^R = -1 \quad \text{and} \quad 2c^I = \frac{\zeta}{\sqrt{1-\zeta^2}}$$

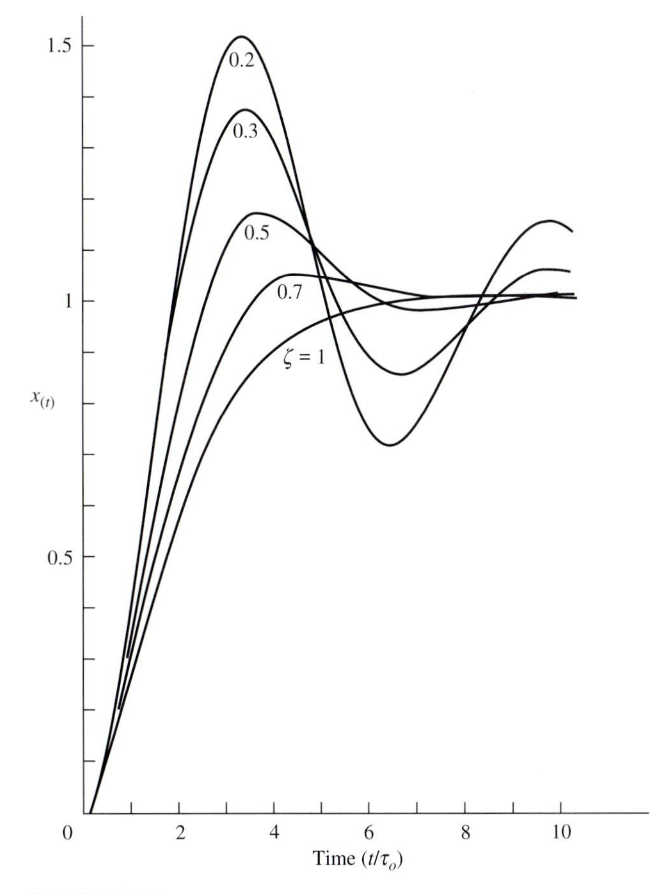

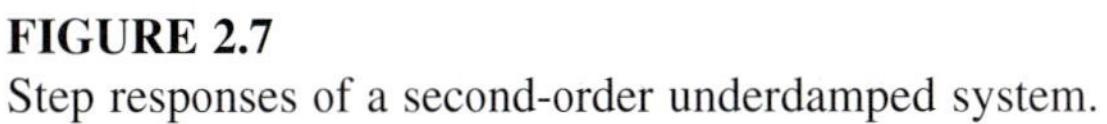

FIGURE 2.7
Step responses of a second-order underdamped system.

The total solution is

$$x_{(t)} = 1 - e^{-\zeta t/\tau_o}\left\{\cos\left(\frac{\sqrt{1-\zeta^2}}{\tau_o}t\right) + \frac{\zeta}{\sqrt{1-\zeta^2}}\sin\left(\frac{\sqrt{1-\zeta^2}}{\tau_o}t\right)\right\} \tag{2.103}$$

This step response is sketched in Fig. 2.7 for several values of the damping coefficient. Note that the amount the solution overshoots the final steady-state value increases as the damping coefficient decreases. The system also becomes more oscillatory. In Chapter 3 we tune feedback controllers so that we get a reasonable amount of overshoot by selecting a damping coefficient in the 0.3 to 0.5 range. ■

It is frequently useful to be able to calculate damping coefficients and time constants for second-order systems from experimental step response data. Problem 2.7 gives some very useful relationships between these parameters and the shape of the response curve. There is a simple relationship between the "peak overshoot ratio" and the damping coefficient, allowing the time constant to be calculated from the "rise time" and the damping coefficient. Refer to Problem 2.7 for the definitions of these terms.

EXAMPLE 2.13 The overdamped system of Example 2.8 is now forced with a ramp input:

$$\frac{d^2x}{dt^2} + 5\frac{dx}{dt} + 6x = t \tag{2.104}$$

Since the forcing function is the first term of a polynomial in t, we will assume that the particular solution is also a polynomial in t.

$$x_p = b_0 + b_1 t + b_2 t^2 + b_3 t^3 + \cdots \tag{2.105}$$

where the b_i are constants to be determined. Differentiating Eq. (2.105) twice gives

$$\frac{dx_p}{dt} = b_1 + 2b_2 t + 3b_3 t^2 + \cdots$$

$$\frac{d^2x_p}{dt^2} = 2b_2 + 6b_3 t + \cdots$$

Substituting into Eq. (2.104) gives

$$(2b_2 + 6b_3 t + \cdots) + 5(b_1 + 2b_2 t + 3b_3 t^2 + \cdots) + 6(b_0 + b_1 t + b_2 t^2 + b_3 t^3 + \cdots) = t$$

Now we rearrange the above expression to group together all terms with equal powers of t.

$$\cdots + t^3(6b_3 + \cdots) + t^2(6b_2 + 15b_3 + \cdots) + t(6b_3 + 10b_2 + 6b_1) + (2b_2 + 5b_1 + 6b_0) = t$$

Equating like powers of t on the left-hand and right-hand sides of this equation gives the simultaneous equations

$$\begin{aligned} 6b_3 + \cdots &= 0 \\ 6b_2 + 15b_3 + \cdots &= 0 \\ 6b_3 + 10b_2 + 6b_1 &= 1 \\ 2b_2 + 5b_1 + 6b_0 &= 0 \end{aligned}$$

Solving simultaneously gives

$$b_0 = -\tfrac{5}{36} \qquad b_1 = \tfrac{1}{6} \qquad b_2 = b_3 = \ldots = 0$$

The particular solution is

$$x_p = -\tfrac{5}{36} + \tfrac{1}{6}t \tag{2.106}$$

The total solution is

$$x_{(t)} = -\tfrac{5}{36} + \tfrac{1}{6}t + c_1 e^{-2t} + c_2 e^{-3t} \tag{2.107}$$

If the initial conditions are

$$x_{(0)} = 0 \quad \text{and} \quad \left(\frac{dx}{dt}\right)_{(0)} = 0$$

the constants c_1 and c_2 can be evaluated:

$$x_{(0)} = 0 = -\tfrac{5}{36} + c_1 + c_2$$

$$\left(\frac{dx}{dt}\right)_{(0)} = 0 = \tfrac{1}{6} - 2c_1 - 3c_2$$

Solving simultaneously gives

$$c_1 = \tfrac{1}{4} \quad \text{and} \quad c_2 = -\tfrac{1}{9} \tag{2.108}$$

And the final solution is

$$x_{(t)} = -\tfrac{5}{36} + \tfrac{1}{6}t + \tfrac{1}{4}e^{-2t} - \tfrac{1}{9}e^{-3t} \tag{2.109}$$

■

2.3.3 *N*th-Order Linear ODEs with Constant Coefficients

The results obtained in the last two sections for simple first- and second-order systems can now be generalized to higher-order systems. Consider the Nth-order ODE

$$a_N \frac{d^N x}{dt^N} + a_{N-1} \frac{d^{N-1} x}{dt^{N-1}} + \cdots + a_1 \frac{dx}{dt} + a_0 x = m_{(t)} \tag{2.110}$$

The solution of this equation is the sum of a particular solution x_p and a complementary solution x_c. The complementary solution is the sum of N exponential terms. The characteristic equation is an Nth-order polynomial:

$$a_N s^N + a_{N-1} s^{N-1} + \cdots + a_1 s + a_0 = 0 \tag{2.111}$$

There are N roots $s_k (k = 1, \ldots, N)$ of the characteristic equation, some of which may be repeated (twice or more). Factoring Eq. (2.111) gives

$$(s - s_1)(s - s_2)(s - s_3) \cdots (s - s_{N-1})(s - s_N) = 0 \tag{2.112}$$

where the s_k are the roots (or *zeros*) of the polynomial. The complementary solution is (for all distinct roots, i.e., no repeated roots)

$$x_{c(t)} = c_1 e^{s_1 t} + c_2 e^{s_2 t} + \cdots + c_N e^{s_N t}$$

And therefore the total solution is

$$x_{(t)} = x_{p(t)} + \sum_{k=1}^{N} c_k e^{s_k t} \tag{2.113}$$

The roots of the characteristic equation can be real or complex. But if they are complex, they must appear in complex conjugate pairs. The reason for this is illustrated for a second-order system with the characteristic equation

$$s^2 + a_1 s + a_0 = 0 \tag{2.114}$$

Let the two roots be s_1 and s_2.

$$\begin{aligned} (s - s_1)(s - s_2) &= 0 \\ s^2 + (-s_1 - s_2)s + s_1 s_2 &= 0 \end{aligned} \tag{2.115}$$

The coefficients a_0 and a_1 can then be expressed in terms of the roots:

$$a_0 = s_1 s_2 \quad \text{and} \quad a_1 = -(s_1 + s_2) \tag{2.116}$$

If Eq. (2.114) is the characteristic equation for a real physical system, the coefficients a_0 and a_1 must be real numbers. These are the coefficients that multiply the derivatives in the Nth-order differential equation. So they cannot be imaginary.

If the roots s_1 and s_2 are both real numbers, Eq. (2.116) shows that a_0 and a_1 are certainly both real. If the roots s_1 and s_2 are complex, the coefficients a_0 and a_1 must still be real and must also satisfy Eq. (2.116). Complex conjugates are the only complex numbers that give real numbers when they are multiplied *and* when they are added together. To illustrate this, let z be a complex number: $z = x + iy$. Let $\bar{z}$ be the complex conjugate of z: $\bar{z} = x - iy$. Now $z\bar{z} = x^2 + y^2$ (a real number), and $z + \bar{z} = 2x$ (a real number). Therefore, the roots s_1 and s_2 must be a complex conjugate pair if they are complex. This is exactly what we found in Eq. (2.85) in the previous section.

For a third-order system with three roots s_1, s_2, and s_3, the roots could all be real: $s_1 = \alpha_1$, $s_2 = \alpha_2$, and $s_3 = \alpha_3$. Or there could be one real root and two complex conjugate roots:

$$s_1 = \alpha_1 \tag{2.117}$$

$$s_2 = \alpha_2 + i\omega_2 \tag{2.118}$$

$$s_3 = \alpha_2 - i\omega_2 \tag{2.119}$$

where α_k = real part of $s_k \equiv \text{Re}[s_k]$
ω_k = imaginary part of $s_k \equiv \text{Im}[s_k]$

These are the only two possibilities. We cannot have three complex roots.

The complementary solution would be either (for distinct roots)

$$x_c = c_1 e^{s_1 t} + c_2 e^{s_2 t} + c_3 e^{s_3 t} \tag{2.120}$$

or

$$x_c = c_1 e^{\alpha_1 t} + e^{\alpha_2 t}[(c_2 + c_3)\cos(\omega_2 t) + i(c_2 - c_3)\sin(\omega_2 t)] \tag{2.121}$$

where the constants c_2 and c_3 must also be complex conjugates in the latter equation, as discussed in the previous section.

If some of the roots are repeated (not distinct), the complementary solution contains exponential terms that are multiplied by various powers of t. For example, if α_1 is a repeated root of order 2, the characteristic equation would be

$$(s - \alpha_1)^2(s - s_3)(s - s_4)\cdots(s - s_N) = 0$$

and the resulting complementary solution is

$$x_c = (c_1 + c_2 t)e^{\alpha_1 t} + \sum_{k=3}^{N} c_k e^{s_k t} \tag{2.122}$$

If α_1 is a repeated root of order 3, the characteristic equation would be

$$(s - \alpha_1)^3(s - s_4)\cdots(s - s_N) = 0$$

and the resulting complementary solution is

$$x_c = (c_1 + c_2 t + c_3 t^2)e^{\alpha_1 t} + \sum_{k=4}^{N} c_k e^{s_k t} \tag{2.123}$$

The stability of the system is dictated by the values of the real parts of the roots. The system is stable if the real parts of *all* roots are negative, since the exponential terms go to zero as time goes to infinity. If the real part of *any* one of the roots is positive, the system is unstable.

The roots of the characteristic equation can be very conveniently plotted in a two-dimensional figure (Fig. 2.8) called the "s plane." The ordinate is the imaginary

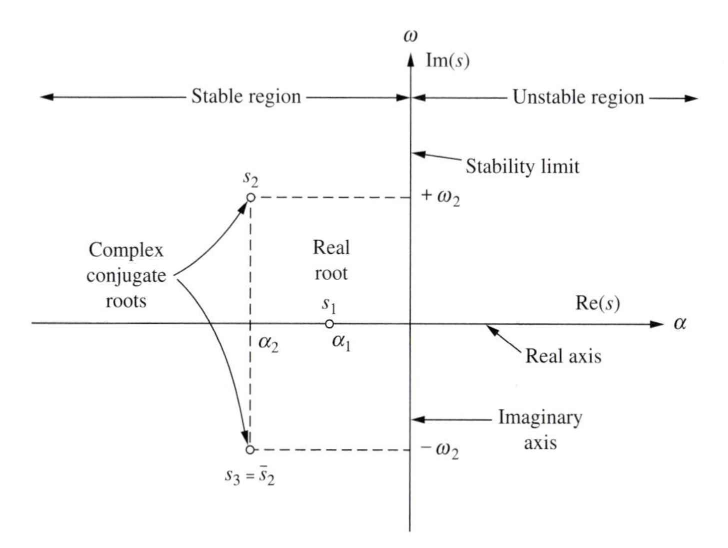

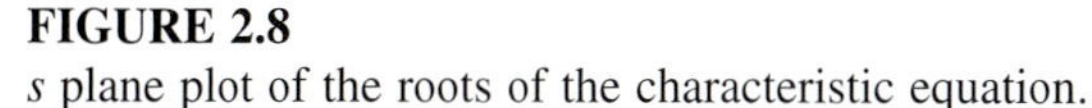

FIGURE 2.8
s plane plot of the roots of the characteristic equation.

part ω of the root s, and the abscissa is the real part α of the root s. The roots of Eqs. (2.117) to (2.119) are shown in Fig. 2.8. We will use these s-plane plots extensively in Part Two.

The stability criterion for an Nth-order system is:

The system is stable if all the roots of its characteristic equation lie in the left half of the s plane.

2.4 SOLUTION USING MATLAB

In the previous section we solved linear ordinary differential equations analytically, obtaining general solutions in terms of the parameters in the equations. Numerical methods can also be used to obtain solutions, using a computer. In Chapter 1 we looked at the dynamic responses of several processes by using numerical integration methods (Euler integration—see Table 1.2).

Solutions of linear ODEs can also be found using the software tool MATLAB. To demonstrate this, let us consider the three-heated-tank process studied in Chapter 1. The process is described by three linear ODEs [Eqs. (1.10), (1.11), and (1.12)]. If flow rate F, volume V (assuming equal volumes in the three tanks), and physical properties ρ and c_p are all constants, these three equations are linear and can be converted into perturbation variables by inspection.

$$\frac{dT_1}{dt} = \frac{F}{V}(T_0 - T_1) + \frac{1}{V\rho c_p}Q_1 \tag{2.124}$$

$$\frac{dT_2}{dt} = \frac{F}{V}(T_1 - T_2) \tag{2.125}$$

$$\frac{dT_3}{dt} = \frac{F}{V}(T_2 - T_3) \tag{2.126}$$

To solve these equations in MATLAB we put them into "state variable" form (this subject is discussed more fully in Chapter 12).

$$\frac{d\underline{x}}{dt} = \underline{\underline{A}}\,\underline{x} + \underline{\underline{B}}\,\underline{u} \tag{2.127}$$

$$\underline{y} = \underline{\underline{C}}\,\underline{x} + \underline{\underline{D}}\,\underline{u} \tag{2.128}$$

where $\underline{x}$ = vector of the three temperatures T_1, T_2, and T_3
$\underline{u}$ = vector of the two inputs T_0 and Q_1
$\underline{y}$ = vector of measured variables (in our case just the scalar quantity T_3)
$\underline{\underline{A}}$, $\underline{\underline{B}}$, $\underline{\underline{C}}$, and $\underline{\underline{D}}$ are matrices of constants

$$\underline{x} = \begin{bmatrix} T_1 \\ T_2 \\ T_3 \end{bmatrix} \qquad \underline{u} = \begin{bmatrix} T_0 \\ Q_1 \end{bmatrix}$$

$$\underline{\underline{A}} = \begin{bmatrix} -\frac{F}{V} & 0 & 0 \\ \frac{F}{V} & -\frac{F}{V} & 0 \\ 0 & \frac{F}{V} & -\frac{F}{V} \end{bmatrix} \tag{2.129}$$

$$\underline{\underline{B}} = \begin{bmatrix} \frac{F}{V} & \frac{1}{V\rho c_p} \\ 0 & 0 \\ 0 & 0 \end{bmatrix} \tag{2.130}$$

$$\underline{C} = \begin{bmatrix} 0 \\ 0 \\ 1 \end{bmatrix} \qquad \underline{D} = \begin{bmatrix} 0 & 0 \end{bmatrix} \tag{2.131}$$

Table 2.1 gives a MATLAB program that calculates the step response of the openloop process for a step change of $-20°F$ in the inlet temperature T_0. The numerical values of parameters are the same as those used in Chapter 1. The four matrices of constants are first defined. The time vector is defined, starting at zero and going to 1.5 hours at increments of 0.005 hours:

t=[0:0.005:1.5];

Then the *step* command is used to calculate the response of y (T_3) to a unit step input in the first input (T_0) by specifying *iu=1*.

[y,x]=step(a,b,c,d,iu,t); (2.132)

The variable y is the output T_3, and the variables x are the three state variables: T_1, T_2, and T_3. Figure 2.9 gives the response of T_3 for a 20°F decrease in T_0.

The above steps calculate the openloop response of the system with the Q_1 input fixed. To calculate the closedloop response with a P controller manipulating Q_1 to control T_3, we substitute for Q_1 in Eq. (2.124):

$$Q_1 = -(K_c G_V G_T) T_3 \tag{2.133}$$

where K_c = controller gain

G_V = valve gain = 10×10^6 Btu/hr/16 mA in the numerical example from Chapter 1

G_T = transmitter gain = 16 mA/200°F in the numerical example

Remember that all variables are perturbation variables and there is no change in the setpoint. The u input vector is now just a scalar: $u = T_0$. The four matrices for the closedloop system are:

$$\underline{\underline{A}} = \begin{bmatrix} -\frac{F}{V} & 0 & -\frac{K_c G_V G_T}{V\rho c_p} \\ \frac{F}{V} & -\frac{F}{V} & 0 \\ 0 & \frac{F}{V} & -\frac{F}{V} \end{bmatrix} \tag{2.134}$$

TABLE 2.1
MATLAB program—Openloop

```
% Program "tempstateol.m" uses Matlab to calculate openloop step responses
%   to change in To for three-heated-tank process
%
% Using state-space formulation for openloop
%
% Third-order system
%
% Openloop A matrix
a=[-10  0  0
    10  -10  0
    0  10  -10];
b=[10  1/3750
        0 0
        0 0];
c=[0  0  1];
d=[0  0];
%

% Define time vector (from 0 to 1 hours)
t=([0:0.005:1.5]);
% "iu"=1  is inlet temperature disturbance
iu=1;
% Use "step" function to get time responses for unit step in T0
[y,x]=step(a,b,c,d,iu,t);
%
% y=T3 for unit step (T0=1)
%
clf
plot(t,-20*y)
title('3 Heated Tanks; Openloop -20 Step Disturbance in T0')
xlabel('Time (hours)')
ylabel('Changes in T3 (degrees)')
grid
pause
print -dps pfig29.ps
```

$$\underline{B} = \begin{bmatrix} \frac{F}{V} \\ 0 \\ 0 \end{bmatrix} \tag{2.135}$$

$$\underline{C} = \begin{bmatrix} 0 \\ 0 \\ 1 \end{bmatrix} \qquad \underline{D} = 0 \tag{2.136}$$

Table 2.2 gives a MATLAB program that calculates the closedloop response of T_3 for two values of controller gain: $K_c = 4$ and 8. Figure 2.10 gives results, which are exactly the same as those in Chapter 1.

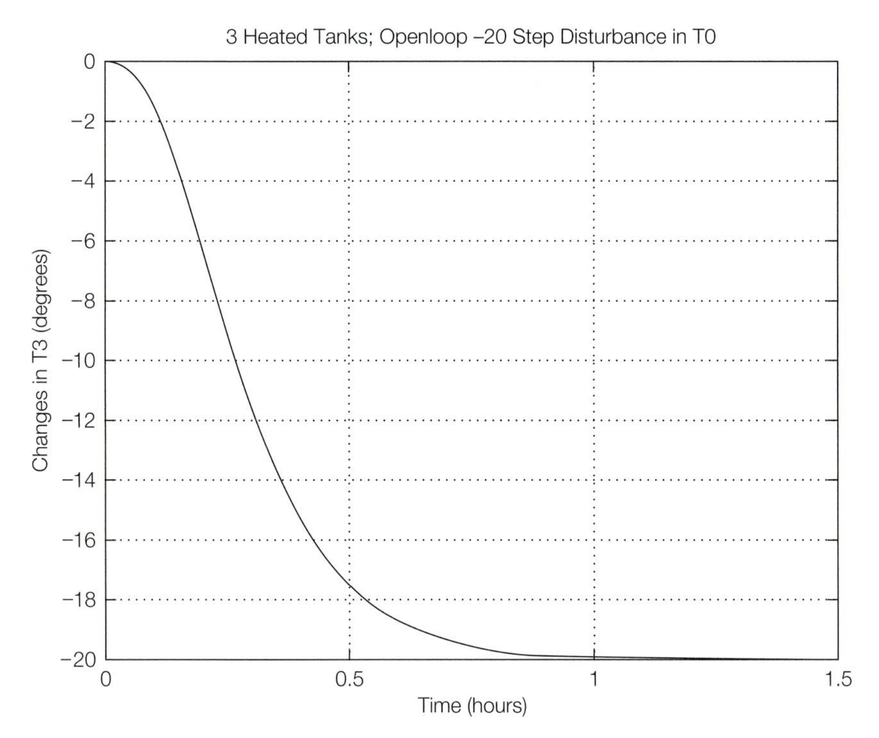

FIGURE 2.9

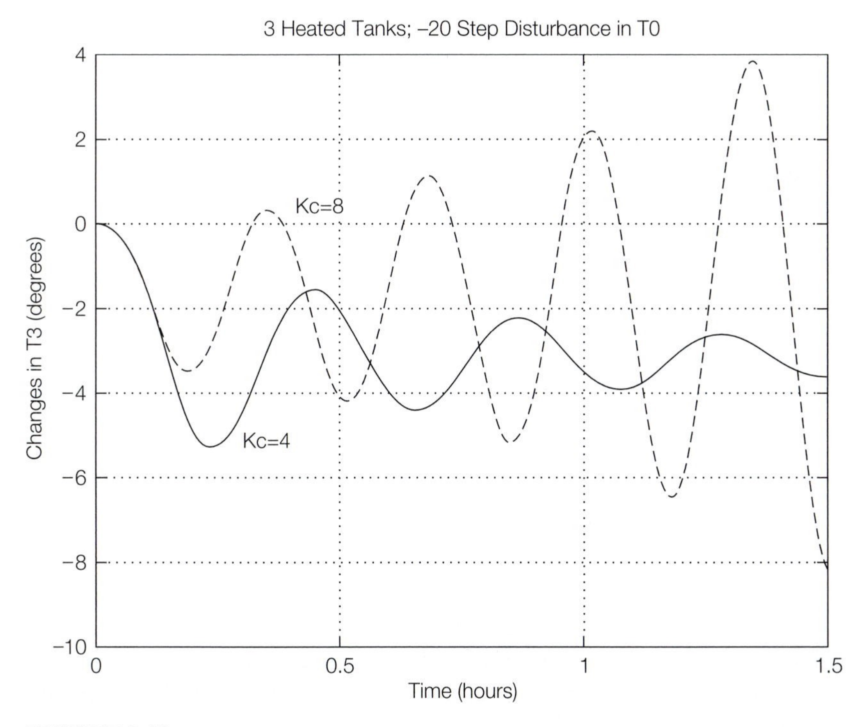

FIGURE 2.10

TABLE 2.2
MATLAB program—Closedloop

```
% Program "tempstatecl.m" uses MATLAB to calculate closedloop step responses
%   to change in To for three-heated-tank process
%
% Using state-space formulation
%
%
% Use "step" function to get time response
% Define time vector (from 0 to 1 hours)
t=([0:0.005:1.5]);
% "iu"=1 is inlet temperature disturbance
%
% Closedloop
%
kc=4;
a13=kc*10e6/3750/200;
acl=[-10  0  -a13
       10  -10  0
       0  10  -10];
bcl=[10  0  0];
dcl=0;
ccl=[0  0  1];
[ycl1,xcl1]=step(acl,bcl,ccl,dcl,iu,t);
%
kc=8;
a13=kc*10e6/3750/200;
acl=[-10  0  -a13
       10  -10  0
       0  10  -10];
[ycl2,xcl2]=step(acl,bcl,ccl,dcl,iu,t);

clf
plot(t,-20*ycl1,'-',t,-20*ycl2,'--')
title('3 Heated Tanks; -20 Step Disturbance in T0')
xlabel('Time (hours)')
ylabel('Changes in T3 (degrees)')
legend ('Kc=4', 'Kc=8')
grid
pause
print -dps pfig210.ps
```

2.5 CONCLUSION

The important concept contained in this chapter is that the dynamic response of a *linear* process is a sum of exponentials in time such as $e^{s_k t}$. The s_k terms multiplying time are the *roots of the characteristic equation* or the *eigenvalues* of the system. They determine whether the process responds quickly or slowly, whether it is oscillatory, and whether it is *stable.*

The values of s_k are either real or complex conjugate pairs. If these roots are complex, the dynamic response of the system contains some oscillatory components. If the real parts of all the roots are negative, the system is stable. If the real part of *any* of the roots is positive, the system is unstable. It only takes one bad apple to spoil the barrel!

PROBLEMS

2.1. Linearize the following nonlinear functions:

$$(a)\ f_{(x)} = y_{(x)} = \frac{\alpha x}{1 + (\alpha - 1)x} \quad \text{where } \alpha \text{ is a constant}$$

$$(b)\ f_{(T)} = P^S_{(T)} = e^{A/T+B} \quad \text{where } A \text{ and } B \text{ are constants}$$

$$(c)\ f_{(v)} = U_{(v)} = K(v)^{0.8} \quad \text{where } K \text{ is a constant}$$

$$(d)\ f_{(h)} = L_{(h)} = K(h)^{3/2} \quad \text{where } K \text{ is a constant}$$

2.2. A fluid of constant density ρ is pumped into a cone-shaped tank of total volume $H\pi R^2/3$. The flow out of the bottom of the tank is proportional to the square root of the height h of liquid in the tank. Derive the nonlinear ordinary differential equation describing this system. Linearize the ODE.

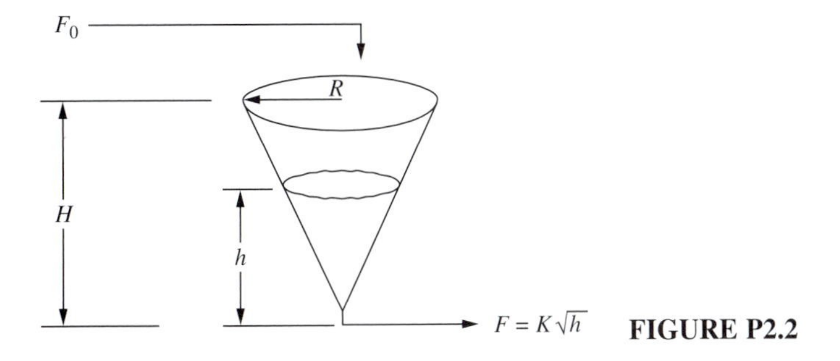

FIGURE P2.2

2.3. Solve the ODEs:

$$(a)\ \frac{d^2x}{dt^2} + 5\frac{dx}{dt} + 4x = 2 \qquad x_{(0)} = 0, \left(\frac{dx}{dt}\right)_{(0)} = 1$$

$$(b)\ \frac{d^2x}{dt^2} + 2\frac{dx}{dt} + 2x = 1 \qquad x_{(0)} = 2, \left(\frac{dx}{dt}\right)_{(0)} = 0$$

2.4. The gravity-flow tank discussed in Chapter 1 is described by two nonlinear ODEs:

$$A_T\frac{dh}{dt} = F_0 - F$$

$$\frac{dv}{dt} = \frac{g}{L}h - \frac{K_F g_c}{\rho A_p}v^2$$

Linearize these two ODEs and show that the linearized system is a second-order system. Solve for the damping coefficient and the time constant in terms of the parameters of the system.

2.5. Solve the second-order ODE describing the steady-state flow of an incompressible, Newtonian liquid through a pipe:

$$\frac{d}{dr}\left(r\frac{d\bar{v}_z}{dr}\right) = \left(\frac{\Delta P g_c}{\mu L}\right)r$$

What are the boundary conditions?

2.6. A feedback controller is added to the CSTR of Example 2.6. The inlet concentration C_{A0} is now changed by the controller to hold C_A near its setpoint value C_A^{set}.

$$C_{A0} = C_{AM} + C_{AD}$$

where C_{AD} is a disturbance composition. The controller has proportional and integral action:

$$C_{AM} = \overline{C}_{AM} + K_c\left(E + \frac{1}{\tau_I}\int E\,dt\right)$$

where K_c and τ_I are constants.

$$\overline{C}_{AM} = \text{steady-state value of } C_{AM}$$
$$E = C_A^{\text{set}} - C_A$$

Derive the second-order equation describing the closedloop process in terms of perturbation variables. Show that the damping coefficient is

$$\zeta = \frac{1 + k\tau + K_c}{2\sqrt{K_c\tau/\tau_I}}$$

What value of K_c will give critical damping? At what value of K_c will the system become unstable?

2.7. Consider the second-order underdamped system

$$\tau_o^2\frac{d^2x}{dt^2} + 2\tau_o\zeta\frac{dx}{dt} + x = K_p m_{(t)}$$

where K_p is the process steady-state gain and $m_{(t)}$ is the forcing function. The unit step response of such a system can be characterized by rise time t_R, peak time t_P, settling time t_S, and peak overshoot ratio POR. The values of t_R and t_P are defined in the sketch below. The value of t_S is the time it takes the exponential portion of the response to decay to a given fraction F of the final steady-state value of x, x_{SS}. The POR is defined as

$$\text{POR} \equiv \frac{x_{(t_P)} - x_{SS}}{x_{SS}}$$

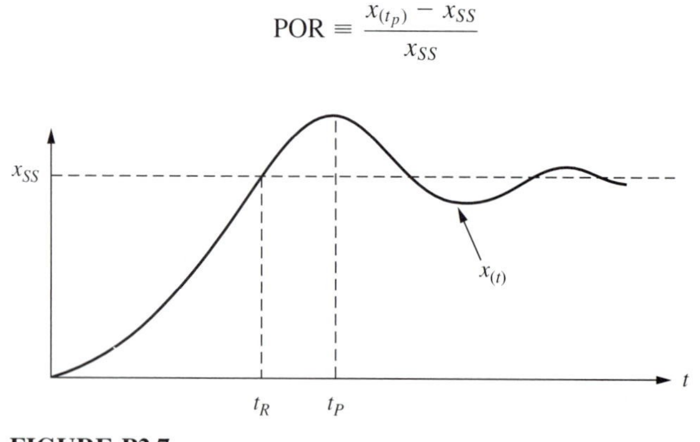

FIGURE P2.7

Show that

$$(a)\ \frac{x_{(t)}}{x_{SS}} = 1 - \frac{e^{-\zeta t/\tau_o}}{\sqrt{1-\zeta^2}} \sin\left(\frac{\sqrt{1-\zeta^2}}{\tau_o} t + \phi\right) \quad \text{where } \phi \equiv \cos^{-1}\zeta$$

$$(b)\ \frac{t_R}{\tau_o} = \frac{\pi - \phi}{\sin\phi}$$

$$(c)\ \frac{t_S}{\tau_o} = \frac{\ln[1/(F \sin\phi)]}{\cos\phi}$$

$$(d)\ \text{POR} = e^{-\pi \cot\phi}$$

2.8. (*a*) Linearize the following two ODEs, which describe a nonisothermal CSTR with constant volume. The input variables are T_0, T_J, C_{A0}, and F.

$$V\frac{dC_A}{dt} = F(C_{A0} - C_A) - VkC_A$$

$$V\rho C_p \frac{dT}{dt} = FC_p\rho(T_0 - T) - \lambda VkC_A - UA(T - T_J)$$

where $k = \alpha e^{-E/RT}$

(*b*) Convert to perturbation variables and arrange in the form

$$\frac{dC_A}{dt} = a_{11}C_A + a_{12}T + a_{13}C_{A0} + a_{14}T_0 + a_{15}F + a_{16}T_J$$

$$\frac{dT}{dt} = a_{21}C_A + a_{22}T + a_{23}C_{A0} + a_{24}T_0 + a_{25}F + a_{26}T_J$$

(*c*) Combine the two linear ODEs above into one second-order ODE and find the roots of the characteristic equation in terms of the a_{ij} coefficients.

2.9. The flow rate F of a manipulated stream through a control valve with equal-percentage trim is given by the following equation:

$$F = C_v \alpha^{x-1}$$

where F is the flow in gallons per minute and C_v and α are constants set by the valve size and type. The control valve stem position x (fraction of wide open) is set by the output signal CO of an analog electronic feedback controller whose signal range is 4 to 20 mA. The valve cannot be moved instantaneously. It is approximately a first-order system:

$$\tau_V \frac{dx}{dt} + x = \frac{\text{CO} - 4}{16}$$

The effect of the flow rate of the manipulated variable on the process temperature T is given by

$$\tau_o \frac{dT}{dt} + T = K_p F$$

Derive one linear ordinary differential equation that gives the dynamic dependence of process temperature on controller output signal CO.

2.10 To ensure an adequate supply for the upcoming set-to with the Hatfields, Grandpa McCoy has begun to process a new batch of his famous Liquid Lightning moonshine. He begins by pumping the mash at a constant rate F_0 into an empty tank. In this tank the ethanol undergoes a first-order reaction to form a product that is the source of the high potency of McCoy's Liquid Lightning. Assuming that the concentration of ethanol in the feed, C_0, is constant and that the operation is isothermal, derive the equations that describe how the concentration C of ethanol in the tank and the volume V of liquid in the tank vary with time. Assume perfect mixing and constant density.

Solve the ODE to show that the concentration C in Grandpa McCoy's batch of Liquid Lightning is

$$C_{(t)} = \frac{C_0(1 - e^{-kt})}{kt}$$

2.11. Suicide Sam slipped his 2000 lb_m hotrod into neutral as he came over the crest of a mountain at 55 mph. In front of him the constant downgrade dropped 2000 feet in 5 miles, and the local acceleration of gravity was 31.0 ft/sec^2.

Sam maintained a constant 55-mph speed by riding his brakes until they heated up to 600°F and burned up. The brakes weighed 40 lb_m and had a heat capacity of 0.1 Btu/lb_m °F. At the crest of the hill they were at 60°F.

Heat was lost from the brakes to the air, as the brakes heated up, at a rate proportional to the temperature difference between the brake temperature and the air temperature. The proportionality constant was 30 Btu/hr °F.

Assume that the car was frictionless and encountered negligible air resistance.

(*a*) At what distance down the hill did Sam's brakes burn up?

(*b*) What speed did his car attain by the time it reached the bottom of the hill?

2.12. A farmer fills his silo with chopped corn. The entire corn plant (leaves, stem, and ear) is cut up into small pieces and blown into the top of the cylindrical silo at a rate W_0. This is similar to a fed-batch chemical reaction system.

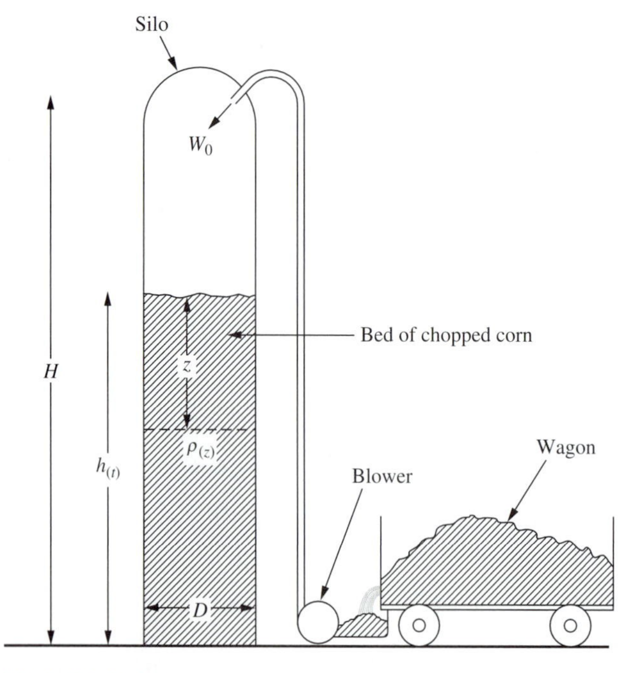

FIGURE P2.12

The diameter of the silo is D and its height is H. The density of the chopped corn in the silo varies with the depth of the bed. The density ρ at a point that has z feet of material above it is

$$\rho_{(z)} = \rho_0 + \beta z$$

where ρ_0 and β are constants.

(a) Write the equations that describe the system, and show how the height of the bed $h_{(t)}$ varies as a function of time.

(b) What is the total weight of corn fodder that can be stored in the silo?

2.13. Two consecutive, first-order reactions take place in a perfectly mixed, isothermal batch reactor.

$$A \xrightarrow{k_1} B \xrightarrow{k_2} C$$

Assuming constant density, solve analytically for the dynamic changes in the concentrations of components A and B in the situation where $k_1 = k_2$. The initial concentration of A at the beginning of the batch cycle is C_{A0}. There is initially no component B or C in the reactor.

What is the maximum concentration of component B that can be produced, and at what point in time does it occur?

2.14. The same reactions considered in Problem 2.13 are now carried out in a single, perfectly mixed, isothermal continuous reactor. Flow rates, volume, and densities are constant.

(a) Derive a mathematical model describing the system.

(b) Solve for the dynamic change in the concentration of component A, C_A, if the concentration of A in the feed stream is constant at C_{A0} and the initial concentrations of A, B, and C at time zero are $C_{A(0)} = C_{A0}$ and $C_{B(0)} = C_{C(0)} = 0$.

(c) In the situation where $k_1 = k_2$, find the value of holdup time ($\tau = V/F$) that maximizes the steady-state ratio of C_B/C_{A0}. Compare this ratio with the maximum found in Problem 2.13.

2.15. The same consecutive reactions considered in Problem 2.13 are now carried out in *two* perfectly mixed continuous reactors. Flow rates and densities are constant. The volumes of the two tanks (V) are the same and constant. The reactors operate at the same constant temperature.

(a) Derive a mathematical model describing the system.

(b) If $k_1 = k_2$, find the value of the holdup time ($\tau = V/F$) that maximizes the steady-state ratio of the concentration of component B in the product to the concentration of reactant A in the feed.

2.16. A vertical, cylindrical tank is filled with well water at 65°F. The tank is insulated at the top and bottom but is exposed on its vertical sides to cold 10°F night air. The diameter of the tank is 2 feet and its height is 3 feet. The overall heat transfer coefficient is 20 Btu/hr °F ft^2. Neglect the metal wall of the tank and assume that the water in the tank is perfectly mixed.

(a) Calculate how many minutes it will be until the first crystal of ice is formed.

(b) How long will it take to completely freeze the water in the tank? The heat of fusion of water is 144 Btu/lb_m.

2.17. An isothermal, first-order, liquid-phase, reversible reaction is carried out in a constant-volume, perfectly mixed continuous reactor.

$$\mathrm{A} \underset{k_2}{\overset{k_1}{\rightleftarrows}} \mathrm{B}$$

The concentration of product B is zero in the feed and is C_B in the reactor. The feed rate is F.

(*a*) Derive a mathematical model describing the dynamic behavior of the system.

(*b*) Derive the steady-state relationship between C_A and C_{A0}. Show that the conversion of A and the yield of B decrease as k_2 increases.

(*c*) Assuming that the reactor is at this steady-state concentration and that a step change is made in C_{A0} to $(C_{A0} + \Delta C_{A0})$, find the analytical solution that gives the dynamic response of $C_{A(t)}$.

2.18. An isothermal, first-order, liquid-phase, irreversible reaction is conducted in a constant volume batch reactor.

$$\mathrm{A} \xrightarrow{k} \mathrm{B}$$

The initial concentration of reactant A at the beginning of the batch is C_{A0}. The specific reaction rate k decreases with time because of catalyst degradation:

$$k = k_0 e^{-\beta t}$$

(*a*) Solve for $C_{A(t)}$.

(*b*) Show that in the limit as $\beta \rightarrow 0$, $C_{A(t)} = C_{A0} e^{-k_0 t}$.

(*c*) Show that in the limit as $\beta \rightarrow \infty$, $C_{A(t)} = C_{A0}$.

2.19. There are 3460 pounds of water in the jacket of a reactor that are initially at 145°F. At time zero, 70°F cooling water is added to the jacket at a constant rate of 416 pounds per minute. The holdup of water in the jacket is constant since the jacket is completely filled with water, and excess water is removed from the system on pressure control as cold water is added. Water in the jacket can be assumed to be perfectly mixed.

(*a*) How many minutes does it take the jacket water to reach 99°F if no heat is transferred into the jacket?

(*b*) Suppose a constant 362,000 Btu/hr of heat is transferred into the jacket from the reactor, starting at time zero when the jacket is at 145°F. How long will it take the jacket water to reach 99°F if the cold water addition rate is constant at 416 pounds per minute?

2.20. Hay dries, after being cut, at a rate that is proportional to the amount of moisture it contains. During a hot (90°F) July summer day, this proportionality constant is 0.30 hr^{-1}. Hay cannot be baled until it has dried down to no more than 5 wt% moisture. Higher moisture levels will cause heating and mold formation, making the hay unsuitable for horses.

The effective drying hours are from 11:00 A.M. to 5:00 P.M.. If hay cannot be baled by 5:00 P.M., it must stay in the field overnight and picks up moisture from the dew. It picks up 25 percent of the moisture that it lost during the previous day.

If the hay is cut at 11:00 A.M. Monday morning and contains 40 wt% moisture at the moment of cutting, when can it be baled?

2.21. Process liquid is continuously fed into a perfectly mixed tank in which it is heated by a steam coil. Feed rate F is 50,000 $\mathrm{lb_m}$/hr of material with a constant density ρ of

50 lb_m/ft^3 and heat capacity C_p of 0.5 Btu/lb_m °F. Holdup in the tank V is constant at 4000 lb_m. Inlet feed temperature T_0 is 80°F.

Steam is added at a rate S lb_m/hr that heats the process liquid up to temperature T. At the initial steady state, T is 190°F. The latent heat of vaporization λ_s of the steam is 900 Btu/lb_m.

(*a*) Derive a mathematical model of the system, and prove that process temperature is described dynamically by the ODE

$$\tau \frac{dT}{dt} + T = K_1 T_0 + K_2 S$$

where $\tau = V/F$
$K_1 = 1$
$K_2 = \lambda_s / C_p F$

(b) Solve for the steady-state value of steam flow $\overline{S}$.

(c) Suppose a proportional feedback controller is used to adjust steam flow rate,

$$S = \overline{S} + K_c(190 - T)$$

Solve analytically for the dynamic change in $T_{(t)}$ for a step change in inlet feed temperature from 80°F down to 50°F. What will the final values of T and S be at the new steady state for a K_c of 100 lb_m/hr/°F?

2.22. Use MATLAB to solve for the openloop and closedloop responses of the two-heated-tank process using a proportional temperature controller with K_c values of 0, 2, 4, and 8; T_2 is controlled by Q_1.

2.23. Use MATLAB to solve for the openloop and closedloop responses of the two-heated-tank process using a PI temperature controller with $\tau_I = 0.1$ hr and K_c values of 0, 2, 4, and 8.

2.24. A reversible reaction occurs in an isothermal CSTR.

$$A + B \underset{k_2}{\overset{k_1}{\rightleftarrows}} C + D$$

The reactor holdup V_R (moles) and the flow rates into and out of the reactor F (mol/hr) are constant. The concentrations in the reactor are z_j (mole fraction component j). The reaction rates depend on the reactor concentrations to the first power. The reactor feed stream concentration is z_{oj}.

(*a*) Write the dynamic component balance for reactant A.

(*b*) Linearize this nonlinear ODE and convert to perturbation variables.

2.25. A first-order reaction $A \xrightarrow{k} B$ occurs in an isothermal CSTR. Fresh feed at a flow rate F_0 (mol/hr) and composition z_0 (mole fraction A) is fed into the reactor along with a recycle stream. The reactor holdup is V_R (moles). The reactor effluent has composition z (mole fraction A) and flow rate F (mol/hr). It is fed into a flash drum in which a vapor stream is removed and recycled back to the reactor at a flow rate R (mol/hr) and composition y_R (mole fraction A).

The liquid from the drum is the product stream with flow rate P (mol/hr) and composition x_P (mole fraction A). The liquid and vapor streams are in phase equilibrium: $y_R = K x_P$, where K is a constant. The vapor holdup in the flash drum is negligible. The liquid holdup is M_D.

(*a*) Write the steady-state equations describing this system. If F_0, z_0, x_P, K, k, and V_R are all specified, solve for F, z, R, and y_R.
(*b*) Write the dynamic equations describing this system.

2.26. A vertical cylindrical tank, 10 feet in diameter and 20 feet tall, is partially filled with *pure* liquid methylchloride. The vapor phase in the tank is pure methylchloride vapor. The temperature in the tank is 100°F. The vapor pressure of methylchloride at 100°F is 125 psia. The liquid height in the tank is 2 feet.

A safety valve suddenly opens, releasing vapor from the top of the tank at a flow rate F (lb/min), which is proportional to the pressure difference between the tank and the atmosphere:

$$F = K(P - P_{\text{atm}})$$

where $K = 0.544$ lb/min/psi. Assume the gas is ideal and that the temperature of the contents of the tank remains *constant* at 100°F. The molecular weight of methylchloride is 50, and the density of liquid methylchloride at 100°F is 45 lb/ft^3.

Solve analytically for the dynamic changes in liquid level $h_{(t)}$, tank pressure $P_{(t)}$, and vapor flow rate $F_{(t)}$ from the tank.

2.27. A vertical cylindrical tank, 0.5 feet in diameter and 1 foot tall, is partially filled with *pure* liquid water. The vapor phase in the tank is pure water vapor. The temperature in the tank is 80°F. The vapor pressure of water at 80°F is 0.5067 psia. The liquid height in the tank is 2.737 in.

A small hole suddenly develops at the bottom of the tank. The flow rate of material out of the hole is proportional to the pressure difference between the pressure at the hole and the atmosphere.

$$F = K(P_{\text{hole}} - P_{\text{atm}})$$

where $K = 0.544$ lb/min/psi. Assume the gas is ideal and that the temperature of the contents of the tank remains *constant* at 80°F. The density of liquid water at 80°F is 62.23 lb/ft^3.

Solve analytically for the dynamic changes in liquid level $h_{(t)}$, tank pressure $P_{(t)}$, and flow rate $F_{(t)}$ from the tank.

2.28. A milk tank on a dairy farm is equipped with a refrigeration compressor that removes q Btu/min of heat from the warm milk. The insulated, perfectly mixed tank is initially filled with V_0 (ft^3) of warm milk (99.5°F). The compressor is then turned on and begins to chill the milk. At the same time, fresh warm (99.5°F) milk is continuously added at a constant rate F (ft^3/min) through a pipeline from the milking parlor. The total volume after all cows have been milked is V_T (ft^3).

Derive the equation describing how the temperature T of milk in the tank varies with time. Solve for $T_{(t)}$. What is the temperature of the milk at the end of the milking? How long does it take to chill the milk down to 35°F? Parameter values are $F = 1$ ft^3/min, $\rho = 62.3$ lb$_m$/ft^3, $C_p = 1$ Btu/lb$_m$ °F, $V_0 = 5$ ft^3, $V_T = 100$ ft^3, $q = 300$ Btu/min.

2.29. Calculate the $\underline{\underline{A}}$ and $\underline{\underline{B}}$ matrices for the state-variable representation of the closedloop two-heated-tank process when a PI temperature controller is used to control T_2 by manipulating Q_1.

CHAPTER 3

Conventional Control Systems and Hardware

In this chapter we study control equipment, controller performance, controller tuning, and general control system design concepts. Questions explored include: How do we decide what kind of control valve to use? What type of sensor can be used, and what are some of the pitfalls we should be aware of that can give faulty signals? What type of controller should we select for a given application? How do we "tune" a controller?

First we look briefly at some of the control hardware that is currently used in process control systems: transmitters, control valves, controllers, etc. Then we discuss the performance of conventional controllers and present empirical tuning techniques. Finally, we talk about some important design concepts and heuristics that are useful in specifying the structure of a control system for a process.

3.1 CONTROL INSTRUMENTATION

Some familiarity with control hardware and software is required before we can discuss selection and tuning. We are not concerned with the details of how the various mechanical, pneumatic, hydraulic, electronic, and computing devices are constructed. These nitty-gritty details can be obtained from the instrumentation and process control computer vendors. Nor are we concerned with specific details of programming a distributed control system (DCS). These details vary from vendor to vendor. We need to know only how they basically work and what they are supposed to do. Pictures of some typical hardware are given in Appendix B.

There has been a real revolution in instrumentation hardware during the last several decades. Twenty years ago most control hardware was mechanical and pneumatic (using instrument air pressure to drive gadgets and for control signals). Tubing had to be run back and forth between the process equipment and the central location (called the "control room") where all the controllers and recorders were installed. Signals were recorded on strip-chart paper recorders.

Today most new plants use DCS hardware—microprocessors that serve several control loops simultaneously. Information is displayed on CRTs (cathode ray tubes). Most signals are still transmitted in analog electronic form (usually current signals), but the use of digital data highways and networks is increasing. These systems provide much more computing power and permit mathematical models of the process to be run on-line (while the process is operating).

Despite all these changes in hardware, the basic concepts of control system structure and control algorithms (types of controllers) remain essentially the same as they were 30 years ago. It is now easier to implement control structures; we just reprogram a computer. But the process control engineer's job is the same: to come up with a control system that will give good, stable, robust performance.

As we preliminarily discussed in Chapter 1, the basic feedback control loop consists of a sensor to detect the process variable, a transmitter to convert the sensor signal into an equivalent "signal" (an air-pressure signal in pneumatic systems or a current signal in analog electronic systems), a controller that compares this process signal with a desired setpoint value and produces an appropriate controller output signal, and a final control element that changes the manipulated variable based on the controller output signal. Usually the final control element is an air-operated control valve that opens and closes to change the flow rate of the manipulated stream. See Fig. 3.1.

The sensor, transmitter, and control valve are physically located on the process equipment ("in the field"). The controller is usually located on a panel or in a computer in a control room that is some distance from the process equipment. Wires connect the two locations, carrying current signals from transmitters to the controller and from the controller to the final control element.

The control hardware used in chemical and petroleum plants is either analog (pneumatic or electronic) or digital. The analog systems use air-pressure signals (3 to 15 psig) or current/voltage signals (4 to 20 mA, 10 to 50 mA, or 0 to 10 V DC). They are powered by instrument air supplies (25 psig air) or 24 V DC electrical power. Pneumatic systems send air-pressure signals through small tubing. Analog electronic systems use wires.

Since most valves are still actuated by air pressure, current signals are usually converted to an air pressure. An "*I/P*" (current to pressure) transducer is used to convert 4 to 20 mA signals to 3 to 15 psig signals.

Also located in the control room is the manual/automatic switching hardware (or software). During start-up or under abnormal conditions, the plant operator may want to be able to set the position of the control valve instead of having the controller position it. A switch is usually provided on the control panel or in the computer system, as sketched in Fig. 3.2. In the "manual" position the operator can stroke the valve by changing a knob (a pressure regulator in a pneumatic system or a potentiometer in an analog electronic system). In the "automatic" position the controller output goes directly to the valve.

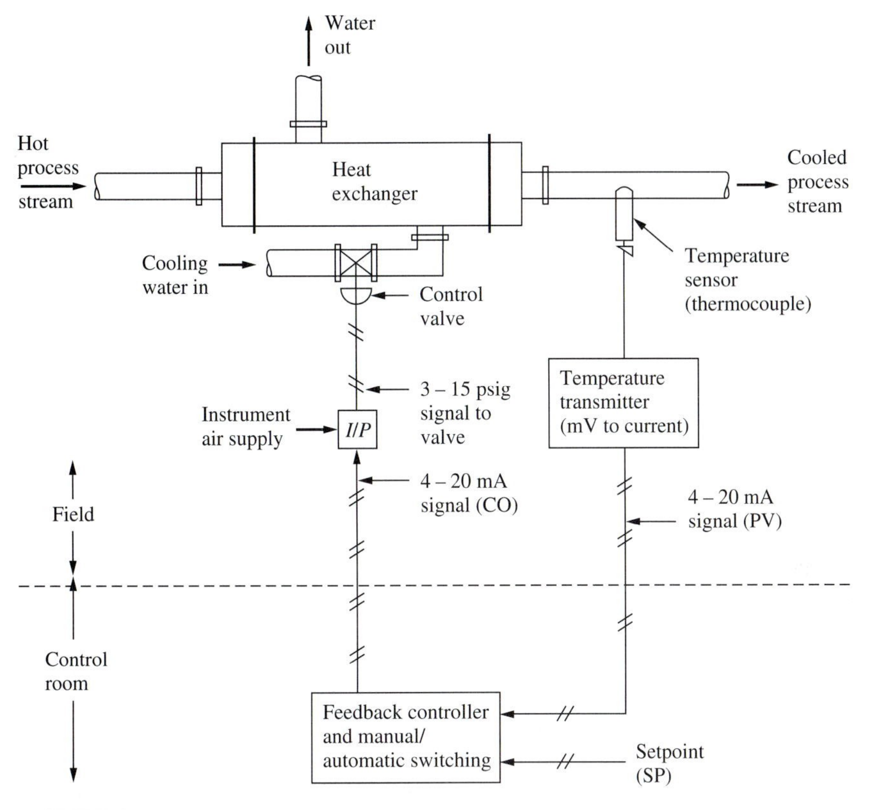

FIGURE 3.1
Feedback control loop.

Each controller must do the following:

1. Indicate the value of the controlled variable—the signal from the transmitter (PV).
2. Indicate the value of the signal being sent to the valve—the controller output (CO).
3. Indicate the setpoint signal (SP).
4. Have a manual/automatic/cascade switch.
5. Have a knob to set the setpoint when the controller is on automatic.
6. Have a knob to set the signal to the valve when the controller is on manual.

All controllers, whether 30-year-old pneumatic controllers or modern distributed microprocessor-based controllers, have these features.

3.1.1 Sensors

Let's start from the beginning of the control loop at the sensor. Instruments for on-line measurement of many properties have been developed. The most important variables

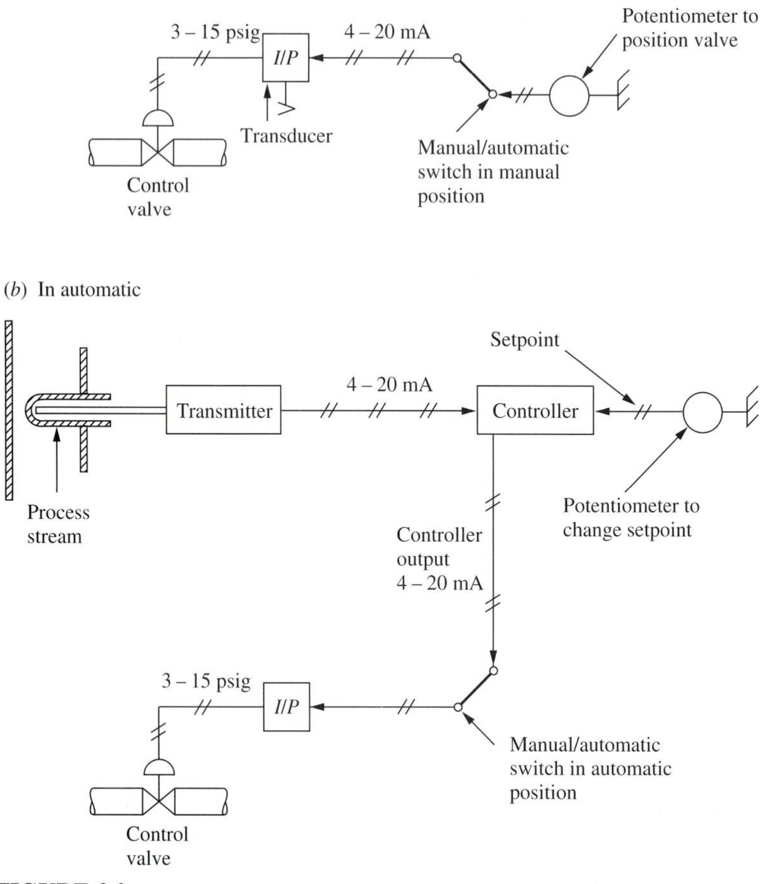

FIGURE 3.2
Manual/automatic switching.

are flow rate, temperature, pressure, and level. Devices for measuring other properties, such as pH, density, viscosity, infrared and ultraviolet absorption, and refractive index are available. Direct measurement of chemical composition by means of on-line gas chromatographs is quite widespread. These instruments, however, pose interesting control problems because of their intermittent operation (a composition signal is generated only every few minutes). We study the analysis of these discontinuous, "sampled-data" systems in Part Five.

We briefly discuss here some of the common sensing elements. Details of their operation, construction, relative merits, and costs are given in several handbooks, such as *Instrument Engineers' Handbook* by B. G. Liptak, Chilton, Radnor, PA, 1970; and *Measurement Fundamentals* by R. L. Moore, Instrument Society of America, Research Triangle Park, NC, 1982.

A. Flow

Orifice plates are by far the most common type of flow rate sensor. The pressure drop across the orifice varies with the square of the flow in turbulent flow, so measuring the differential pressure gives a signal that can be related to flow rate. Normally, orifice plates are designed to give pressure drops in the range of 20 to 200 inches of water. Turbine meters are also widely used. They are more expensive but give more accurate flow measurement. Other types of flow meters include sonic flow meters, magnetic flow meters, rotameters, vortex-shedding devices, and pitot tubes. In gas recycle systems, where the pressure drop through the flow meter can mean a significant amount of compressor work, low-pressure drop flow meters, such as the last two mentioned above, are used.

When a flow sensor is installed for accurate accounting measurements of the absolute flow rate, many precautions must be taken, such as providing a long section of straight pipe before the orifice plate. For control purposes, however, one may not need to know the absolute value of the flow, but only the changes in flow rate. Therefore, pressure drops over pieces of equipment, around elbows, or over sections of pipe can sometimes be used to get a rough indication of flow rate changes.

The signals from flow rate measurements are usually noisy, which means they fluctuate around the actual value because of the turbulent flow. These signals often need to be filtered (passed through an electronic device to smooth out the signal) before being sent to the controller.

B. Temperature

Thermocouples are the most commonly used temperature-sensing devices. They are typically inserted into a thermowell, which is welded into the wall of a vessel or pipe. The two dissimilar wires produce a millivolt signal that varies with the "hot junction" temperature. Iron-constantan thermocouples are commonly used over the 0 to 1300°F temperature range.

Filled-bulb temperature sensors are also widely used. An inert gas is enclosed in a constant-volume system. Changes in process temperature cause the pressure exerted by the gas to change. Resistance thermometers are used where accurate temperature or differential temperature measurement is required. They use the principle that the electrical resistance of wire changes with temperature.

The dynamic response of most sensors is usually much faster than the dynamics of the process itself. Temperature sensors are a notable and sometimes troublesome exception. The time constant of a thermocouple and a heavy thermowell can be 30 seconds or more. If the thermowell is coated with polymer or other goo, the response time can be several minutes. This can significantly degrade control performance.

C. Pressure and differential pressure

Bourdon tubes, bellows, and diaphragms are used to sense pressure and differential pressure. For example, in a mechanical system the process pressure force is balanced by the movement of a spring. The position of the spring can be related to the process pressure.

D. Level

Liquid levels are detected in a variety of ways. The three most common are:

1. Following the position of a float that is lighter than the fluid (as in a bathroom toilet).

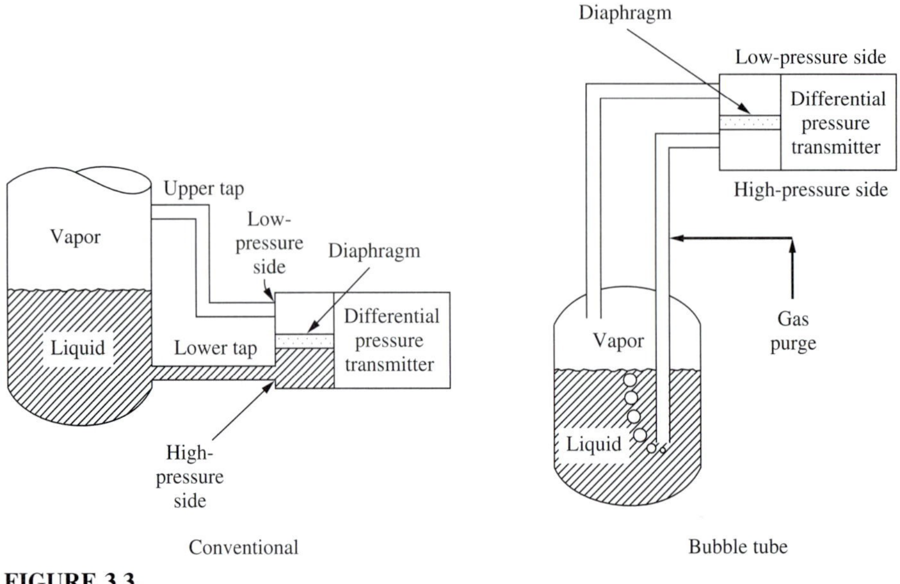

FIGURE 3.3
Differential-pressure level measurement.

2. Measuring the apparent weight of a heavy cylinder as it is buoyed up more or less by the liquid (these are called *displacement meters*).
3. Measuring the difference in static pressure between two fixed elevations, one in the vapor above the liquid and the other under the liquid surface. As sketched in Fig. 3.3, the differential pressure between the two level taps is directly related to the liquid level in the vessel.

In the last scheme the process liquid and vapor are normally piped directly to the differential-pressure measuring device (ΔP transmitter). Some care has to be taken to account for or to prevent condensation of vapor in the connecting line (called the "impulse line") from the top level tap. If the line fills up with liquid, the differential pressure will be zero even though the liquid level is all the way up to or above the top level tap, leading you to think that the level is low. If safety problems can occur because of a high level, a second level sensor should be used to *independently* detect high level. Keeping the vapor impulse line hot or purging it with a small vapor flow can sometimes keep it from filling with liquid. Purging it with a small liquid flow also works because you know that the line is always filled with liquid, so the "zero" (the ΔP at which the transmitter puts out its 4-mA signal) can be adjusted appropriately to indicate the correct level.

Because of plugging or corrosion problems, it is sometimes necessary to keep the process fluid out of the ΔP transmitter. This is accomplished by mechanical diaphragm seals or by purges (introducing a small amount of liquid or gas into the connecting lines, which flows back into the process).

If it is difficult to provide a level tap in the base of the vessel (for mechanical design reasons, for example, in a glass-lined or jacketed vessel), a bubble tube can be suspended from the top of the vessel down under the liquid surface, as shown in

Fig. 3.3. A small gas purge through the tube gives a pressure on the high-pressure side of the ΔP transmitter that is the same as the static pressure at the base of the bubble tube. This type of level measurement can give incorrect level readings when the pressure in the vessel is increasing rapidly because the liquid can back up in the dip-tube if the gas purge flow rate is not large enough to compensate for the pressure increase.

For very hard-to-handle process fluids, nuclear radiation gauges are used to detect interfaces and levels.

As you can tell from the preceding discussion, it is very easy to be fooled by a differential-pressure measurement of level. We have been bitten many times by these problems and highly recommend redundant sensors and judicious skepticism about the validity of instrument readings. Remember the Fourth Law of Process Control: "Never believe the instruments."

3.1.2 Transmitters

The transmitter serves as the interface between the process and its control system. The job of the transmitter is to convert the sensor signal (millivolts, mechanical movement, pressure differential, etc.) into a control signal (4 to 20 mA, for example).

Consider the pressure transmitter shown in Fig. 3.4*a*. Let us assume that this particular transmitter is set up so that its output current signal varies from 4 to 20 mA

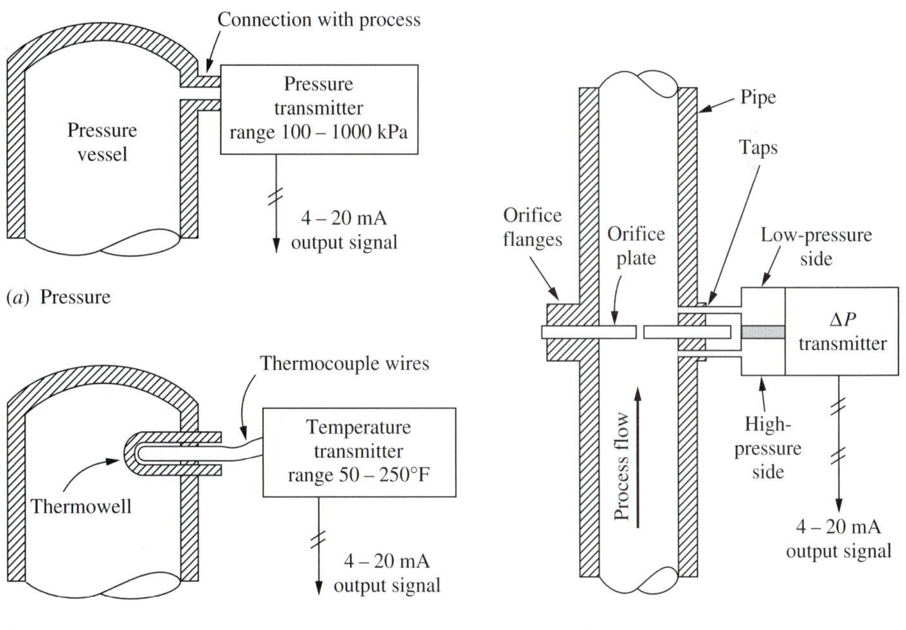

FIGURE 3.4
Typical transmitters. (*a*) Pressure. (*b*) Temperature. (*c*) Flow (orifice plate).

as the process pressure in the vessel varies from 100 to 1000 kPa gauge. This is called the *range* of the transmitter. The *span* of the transmitter is 900 kPa. The *zero* of the transmitter is 100 kPa gauge. The transmitter has two adjustment knobs that can be changed to modify the span or the zero. Thus, if we shifted the zero up to 200 kPa gauge, the *range* of the transmitter would now be 200 to 1100 kPa gauge while its span would still be 900 kPa.

The dynamic response of most transmitters is usually much faster than the process and the control valves. Consequently, we can normally consider the transmitter as a simple "gain" (a step change in the input to the transmitter gives an instantaneous step change in the output). The gain of the pressure transmitter considered above would be

$$\frac{20 \text{ mA} - 4 \text{ mA}}{1000 \text{ kPa} - 100 \text{ kPa}} = \frac{16 \text{ mA}}{900 \text{ kPa}} \tag{3.1}$$

Thus, the transmitter is just a "transducer" that converts the process variable to an equivalent control signal.

Figure 3.4*b* shows a temperature transmitter that accepts thermocouple input signals and is set up so that its current output goes from 4 to 20 mA as the process temperature varies from 50 to 250°F. The range of the temperature transmitter is 50 to 250°F, its span is 200°F, and its zero is 50°F. The gain of the temperature transmitter is

$$\frac{20 \text{ mA} - 4 \text{ mA}}{250°\text{F} - 50°\text{F}} = \frac{16 \text{ mA}}{200°\text{F}} \tag{3.2}$$

As noted in the previous section, the dynamics of the thermowell-thermocouple sensor are often not negligible and should be included in the dynamic analysis.

Figure 3.4*c* shows a ΔP transmitter used with an orifice plate as a flow transmitter. The pressure drop over the orifice plate (the sensor) is converted to a control signal. Suppose the orifice plate is sized to give a pressure drop of 100 in H_2O at a process flow rate of 2000 kg/hr. The ΔP transmitter converts inches of H_2O into milliamperes, and its gain is 16 mA/100 in H_2O. However, we really want flow rate, not orifice plate pressure drop. Since ΔP is proportional to the square of the flow rate, there is a nonlinear relationship between flow rate F and the transmitter output signal:

$$\text{PV} = 4 + 16\left(\frac{F}{2000}\right)^2 \tag{3.3}$$

where PV = transmitter output signal, mA
F = flow rate, kg/hr

Equation (3.3) comes from the square-root relationship between velocity and pressure drop. Dropping the flow by a factor of 2 cuts the ΔP signal by a factor of 4. For system analysis we usually linearize Eq. (3.3) around the steady-state value of flow rate, $\overline{F}$.

$$\text{PV} = \frac{32\overline{F}}{(F_{\text{max}})^2}F \tag{3.4}$$

where PV and F = perturbations from steady state
$\overline{F}$ = steady-state flow rate, kg/hr
$F_{\max}$ = maximum full-scale flow rate = 2000 kg/hr in this example

3.1.3 Control Valves

The interface with the process at the other end of the control loop is made by the final control element. In a vast majority of chemical engineering processes the final control element is an automatic control valve that throttles the flow of a manipulated variable. In mechanical engineering systems the final control element is a hydraulic actuator or an electric servo motor.

Most control valves consist of a plug on the end of a stem. The plug opens or closes an orifice opening as the stem is raised or lowered. As sketched in Fig. 3.5, the stem is attached to a diaphragm that is driven by changing air pressure above the diaphragm. The force of the air pressure is opposed by a spring.

There are several aspects of control valves: their *action, characteristics,* and *size.*

A. Action

Valves are designed to fail either in the completely open or the completely shut position. Which action is appropriate depends on the effect of the manipulated variable on the safety of the process. For example, if the valve is handling steam or fuel, we want the flow to be cut off in an emergency (valve to fail shut). If the valve is handling cooling water to a reactor, we want the flow to go to a maximum in an emergency (valve to fail wide open).

The valve shown in Fig. 3.5 is closed when the stem is completely down and wide open when the stem is at the top of its stroke. Since increasing air pressure closes the valve, this is an "air-to-close" (AC) valve. If the air-pressure signal should drop to zero because of some failure (for example, if the instrument-air supply line were cut or if it plugged with ice during a cold winter night), this valve would fail wide open since the spring would push the valve open. The valve can be made "air-to-open" (AO) by reversing the action of the plug to close the opening in the up position or by reversing the locations of the spring and air pressure (with the air pressure under the diaphragm). Thus, there are AO and AC valves, and the decision about which to use depends on whether we want the valve to fail shut or wide open.

B. Size

The flow rate through a control valve depends on the size of the valve, the pressure drop over the valve, the stem position, and the fluid properties. The design equation for liquids (nonflashing) is

$$F = C_v f(x) \sqrt{\frac{\Delta P_v}{\text{sp gr}}} \tag{3.5}$$

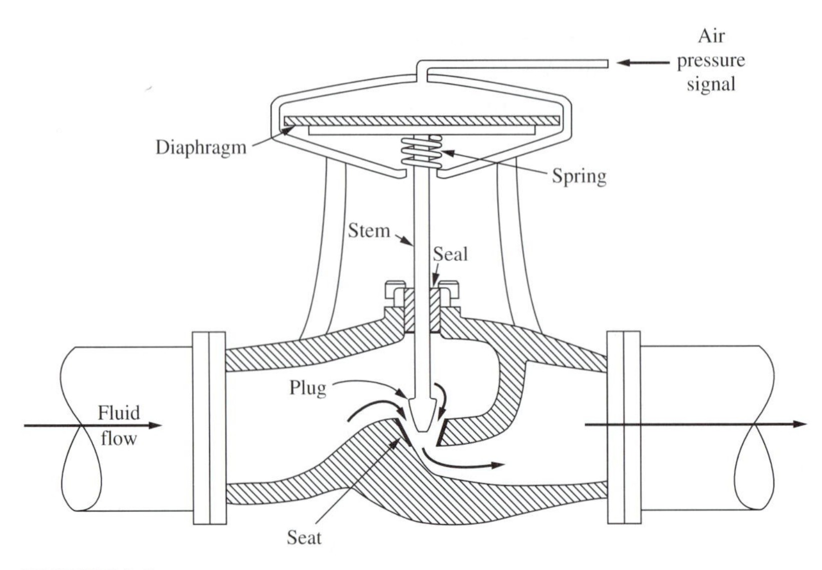

FIGURE 3.5
Typical air-operated control valve.

where F = flow rate, gpm
C_v = valve size coefficient
x = valve stem position (fraction of wide open)
$f_{(x)}$ = fraction of the total flow area of the valve (the curve of $f_{(x)}$ versus x is called the "inherent characteristics" of the valve, discussed later)
sp gr = specific gravity (relative to water)
ΔP_v = pressure drop over the valve, psi

More detailed equations are available in publications of the control valve manufacturers [for example, the *Masonielan Handbook for Control Valve Sizing,* 6th ed. (1977), Dresser Industries] that handle flows of gases, flashing liquids, and critical flows with either English or SI units.

Sizing of control valves is one of the more controversial subjects in process control. The sizing of control valves is a good example of the engineering trade-off that must be made in designing a plant. Consider the process sketched in Fig. 3.6. Sup-

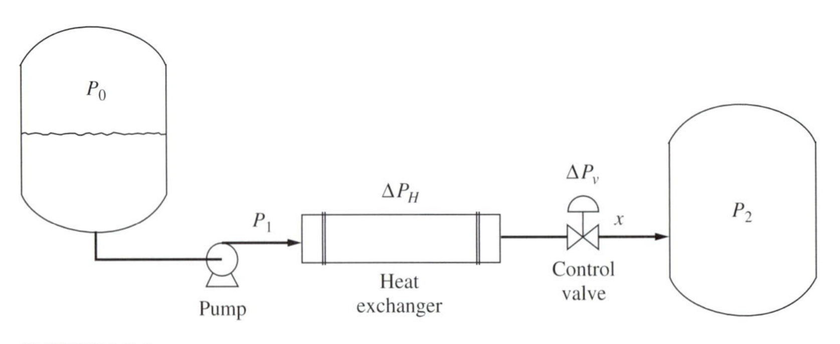

FIGURE 3.6

pose the flow rate at design conditions is 100 gpm, the pressure in the feed tank is atmospheric, the pressure drop over the heat exchanger (ΔP_H) at the design flow rate is 40 psi, and the pressure in the final tank, P_2, is 150 psig. Let us assume that we want the control valve half open ($f_{(x)} = 0.5$) at the design flow and that the specific gravity of the liquid is 1.

The process engineer has to size both the centrifugal pump and the control valve. The bigger the control valve, the less pressure drop it requires. This results in a pump with a lower pressure head and lower energy costs because less power is consumed by the pump motor. Without considering control, the process engineer wants to design a system that has a low pressure drop across the control valve. From a steady-state standpoint, this makes perfect sense.

However, the control engineer wants to take a large pressure drop over the valve. Why? Basically, it is a question of "rangeability": the larger the pressure drop, the larger the changes that can be made in the flow rate (in *both* directions—increase and decrease). Let's examine two different designs to show why it is desirable from a dynamic viewpoint to have more pressure drop over the control valve.

In case 1 we size the valve so that it takes a 20-psi pressure drop at design flow when it is half open. This means that the pump must produce a differential head of $150 + 40 + 20 = 210$ psi at design. In case 2 we will size the valve so that it takes an 80-psi pressure drop at design. Now a higher head pump will be needed: $150 + 40 + 80 = 270$ psi.

Using Eq. (3.5), both control valves can be sized.

Case 1:

$$F = C_v f_{(x)} \sqrt{\frac{\Delta P_v}{\text{sp gr}}}$$

$$100 = C_{v1}(0.5)\sqrt{20} \quad \Rightarrow \quad C_{v1} = 44.72 \quad \text{when the design valve pressure drop is 20 psi}$$

Case 2:

$$100 = C_{v2}(0.5)\sqrt{80} \quad \Rightarrow \quad C_{v2} = 22.36 \quad \text{when the design valve pressure drop is 80 psi}$$

Naturally, the control valve in case 2 is smaller than that in case 1.

Now let's see what happens in the two cases when we open the control valve all the way: $f_{(x)} = 1$. Certainly the flow rate will increase, but how much? From a control point of view, we may want to be able to increase the flow substantially. Let's call this unknown flow $F_{\max}$.

The higher flow rate will increase the pressure drop over the heat exchanger as the square of the flow rate.

$$\Delta P_H = 40\left(\frac{F_{\max}}{F_{\text{des}}}\right)^2 = 40\left(\frac{F_{\max}}{100}\right)^2 \tag{3.6}$$

where F_{des} = design flow. The higher flow rate might also reduce the head that the centrifugal pump produces if we are out on the pump curve where head is dropping rapidly with throughput. For simplicity, let us assume that the pump curve is flat. This means that the total pressure drop across the heat exchanger and the control valve is constant. Therefore, the pressure drop over the control valve must decrease as the pressure drop over the heat exchanger increases.

$$\Delta P_v = \Delta P_{\text{Total}} - \Delta P_H \tag{3.7}$$

Plugging in the numbers for the two cases yields the following results.

Case 1 (20-psi design):

$$\Delta P_{\text{Total}} = 60 \text{ psi} \qquad C_{v1} = 44.72$$

$$F_{\max} = (44.72)(1.0)\sqrt{60 - 40\left(\frac{F_{\max}}{100}\right)^2} \tag{3.8}$$

This equation can be solved for $F_{\max} = 115$ gpm. The maximum flow through the valve is only 15 percent more than design if a 20-psi pressure drop over the valve is used at design flow rate.

Case 2 (80-psi design):

$$\Delta P_{\text{Total}} = 120 \text{ psi} \qquad C_v = 22.36$$

$$F_{\max} = (22.36)(1.0)\sqrt{120 - 40\left(\frac{F_{\max}}{100}\right)^2} \tag{3.9}$$

Solving for $F_{\max}$ yields 141 gpm. So the maximum flow through this valve, which has been designed for a higher pressure drop, is over 40 percent more than design.

We can see from the results above that the valve designed for the larger pressure drop can produce larger flow rate increases at its maximum capacity.

Now let's see what happens when we want to reduce the flow. Control valves don't work too well when they are less than about 10 percent open. They can become mechanically unstable, shutting off completely and then popping partially open. The resulting fluctuations in flow are undesirable. Therefore, we want to design for a minimum valve opening of 10 percent. Let's see what the minimum flow rates will be in the two cases when the two valves are pinched down so that $f_{(x)} = 0.1$.

In this case the lower flow rate will mean a decrease in the pressure drop over the heat exchanger and therefore an *increase* in the pressure drop over the control valve.

Case 1 (20-psi design):

$$F_{\min} = (0.1)(44.72)\sqrt{60 - 40\left(\frac{F_{\min}}{100}\right)^2} \tag{3.10}$$

Solving gives $F_{\min} = 33.3$ gpm.

Case 2 (80-psi design):

$$F_{\min} = (0.1)(22.36)\sqrt{120 - 40\left(\frac{F_{\min}}{100}\right)^2} \tag{3.11}$$

This $F_{\min}$ is 24.2 gpm.

These results show that the minimum flow rate is lower for the valve designed for a larger pressure drop. So not only can we increase the flow more, but we also can reduce it more. Thus the *turndown ratio* (the ratio of F_{max} to F_{min}) of the big ΔP valve is larger.

$$\text{Turndown ratio for 20-psi design valve} = \frac{115}{33.3} = 3.46$$

$$\text{Turndown ratio for 80-psi design valve} = \frac{141}{24.2} = 5.83$$

We have demonstrated why the control engineer wants more pressure drop over the valve.

So how do we resolve this conflict between the process engineer wanting low pressure drop and the control engineer wanting large pressure drop?

A commonly used heuristic recommends that the pressure drop over the control valve at design should be 50 percent of the total pressure drop through the system. Although widely used, this procedure makes little sense to us. In some situations it is very important to be able to increase the flow rate above the design conditions (for example, the cooling water to an exothermic reactor may have to be doubled or tripled to handle dynamic upsets). In other cases this is not as important (for example, the feed flow rate to a unit).

A logical design procedure is based on designing the control valve and the pump so that both a specified maximum flow rate and a specified minimum flow rate can be achieved. The design flow conditions are used only to get the pressure drop over the heat exchanger (or fixed-resistance part of the process).

The designer must specify the maximum flow rate that is required under the worst conditions *and* the minimum flow rate that is required. Then the valve flow equations for the maximum and minimum conditions give two equations and two unknowns: the pressure head of the centrifugal pump ΔP_P and the control valve size C_v.

EXAMPLE 3.1. Suppose we want to design a control valve for admitting cooling water to a cooling coil in an exothermic chemical reactor. The normal flow rate is 50 gpm. To prevent reactor runaways, the valve must be able to provide three times the design flow rate. Because the sales forecast could be overly optimistic, a minimum flow rate of 50 percent of the design flow rate must be achievable. The pressure drop through the cooling coil is 10 psi at the design flow rate of 50 gpm. The cooling water is pumped from an atmospheric tank. The water leaving the coil runs into a pipe in which the pressure is constant at 2 psig. Size the control valve and the pump.

The pressure drop through the coil depends on the flow rate F:

$$\Delta P_c = 10\left(\frac{F}{50}\right)^2 \tag{3.12}$$

The pressure drop over the control valve is the total pressure drop available (which we don't know yet) minus the pressure drop over the coil.

$$\Delta P_v = \Delta P_T - 10\left(\frac{F}{50}\right)^2 \tag{3.13}$$

Now we write one equation for the maximum flow conditions and one for the minimum. At the maximum conditions:

$$150 = C_v(1.0)\sqrt{\Delta P_T - 10\left(\frac{150}{50}\right)^2} \tag{3.14}$$

At the minimum conditions:

$$25 = C_v(0.1)\sqrt{\Delta P_T - 10\left(\frac{25}{50}\right)^2} \tag{3.15}$$

Solving simultaneously for the two unknowns yields the control valve size ($C_v = 21.3$) and the pump head ($\Delta P_p = \Delta P_T + 2 = 139.2 + 2 = 141.2$ psi).

At the design conditions (50 gpm), the valve fraction open (f_{des}) will be given by

$$50 = 21.3 f_{\text{des}}\sqrt{139.2 - 10} \Rightarrow f_{\text{des}} = 0.206 \tag{3.16}$$

■

The control valve and pump sizing procedure proposed above is not without its limitations. The two design equations for the maximum and minimum conditions in general terms are:

$$F_{\max} = C_v\sqrt{\Delta P_T - (\Delta P_H)_{\text{des}}\left(\frac{F_{\max}}{F_{\text{des}}}\right)^2} \tag{3.17}$$

$$F_{\min} = f_{\min} C_v\sqrt{\Delta P_T - (\Delta P_H)_{\text{des}}\left(\frac{F_{\min}}{F_{\text{des}}}\right)^2} \tag{3.18}$$

where ΔP_T = total pressure drop through the system at design flow rates
$(\Delta P_H)_{\text{des}}$ = pressure drop through the fixed resistances in the system at design flow
$f_{\min}$ = minimum valve opening
F_{des} = flow rate at design

A flat pump curve is assumed in the above derivation. Solving these two equations for ΔP_T gives:

$$\frac{\Delta P_T}{(\Delta P_H)_{\text{des}}} = \frac{\left\{\dfrac{(F_{\max})^2 - (F_{\min})^2}{(F_{\text{des}})^2}\right\}}{1 - \left(\dfrac{f_{\min} F_{\max}}{F_{\min}}\right)^2} \tag{3.19}$$

It is clear from Eq. (3.19) that as the second term in the denominator approaches unity, the required pressure drop goes to infinity! So there is a limit to the achievable rangeability of a system.

Let us define this term as the rangeability index of the system, $\mathcal{R}$.

$$\mathcal{R} \equiv \frac{f_{\min} F_{\max}}{F_{\min}} \tag{3.20}$$

The parameters on the right side of Eq. (3.20) *must* be chosen such that $\mathcal{R}$ is less than unity.

This can be illustrated using the numbers from Example 3.1. If the minimum flow rate is reduced from 50 percent of design (where ΔP_T was 139.2 psi) to 40 percent, the new ΔP_T becomes 202 psi. If F_{min} is reduced further to 35 percent of design, ΔP_T is 335 psi. In the limit as F_{min} goes to 30 percent of design, the rangeability index becomes

$$\mathcal{R} \equiv \frac{f_{min} F_{max}}{F_{min}} = \frac{(0.1)(150)}{15} = 1$$

and the total pressure drop available goes to infinity.

The value of f_{min} can be reduced below 0.1 if a large turndown ratio is required. This is accomplished by using two control valves in parallel, one large and one small, in a split-range configuration. The small valve opens first, and then the large valve opens as the signal to the two valves changes over its full range.

C. Characteristics

By changing the shape of the plug and the seat in the valve, different relationships between stem position and flow area can be attained. The common flow *characteristics* used are *linear-trim* valves and *equal-percentage-trim* valves, as shown in Fig. 3.7. The term "equal percentage" comes from the slope of the $f_{(x)}$ curve being a constant fraction of f.

If constant pressure drop over the valve is assumed and if the stem position is 50 percent open, a linear-trim valve gives 50 percent of the maximum flow and an equal-percentage-trim valve gives only 15 percent of the maximum flow. The

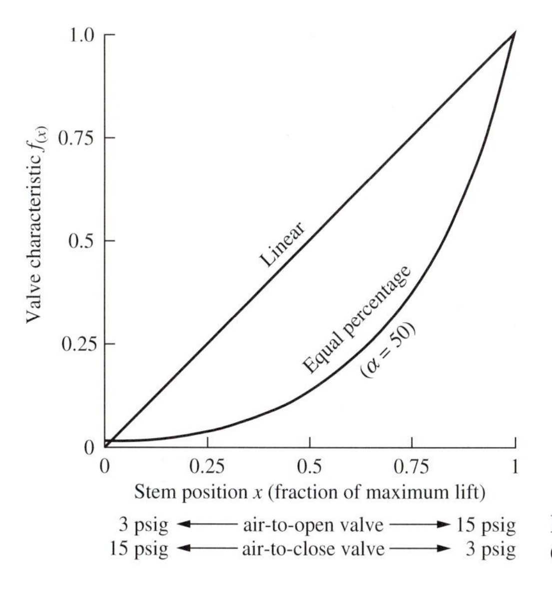

FIGURE 3.7
Control valve characteristics.

equations for these valves are

Linear:

$$f_{(x)} = x \tag{3.21}$$

Equal percentage:

$$f_{(x)} = \alpha^{x-1} \tag{3.22}$$

where α is a constant (20 to 50) that depends on the valve design. A value of 50 is used in Fig. 3.7.

The basic reason for using different control valve trims is to keep the stability of the control loop fairly constant over a wide range of flows. Linear-trim valves are used, for example, when the pressure drop over the control valve is fairly constant and a linear relationship exists between the controlled variable and the flow rate of the manipulated variable. Consider the flow of steam from a constant-pressure supply header. The steam flows into the shell side of a heat exchanger. A process liquid stream flows through the tube side and is heated by the steam. There is a linear relationship between the process outlet temperature and steam flow (with constant process flow rate and inlet temperature) since every pound of steam provides a certain amount of heat.

Equal-percentage-trim valves are often used when the pressure drop available over the control valve is not constant. This occurs when there are other pieces of equipment in the system that act as fixed resistances. The pressure drops over these parts of the process vary as the square of the flow rate. We saw this in the examples discussing control valve sizing.

At low flow rates, most of the pressure drop is taken over the control valve since the pressure drop over the rest of the process equipment is low. At high flow rates, the pressure drop over the control valve is low. In this situation the equal-percentage trim tends to give a more linear relationship between flow and control valve position than does linear trim. Figure 3.8 shows the *installed characteristics* of linear and equal-

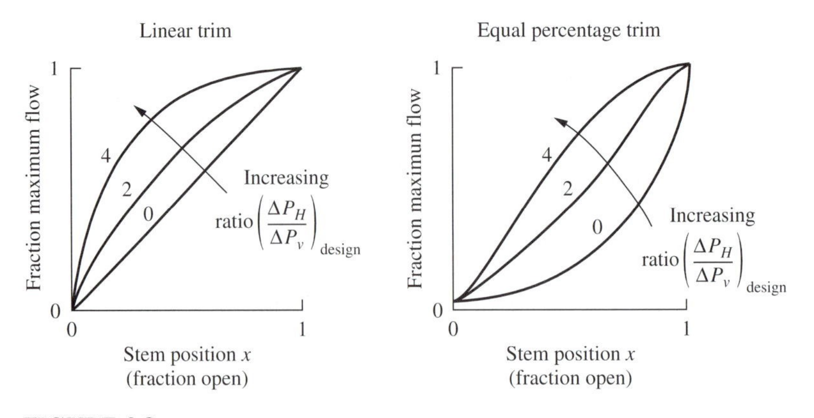

FIGURE 3.8
Control-valve performance in a system ("installed characteristics").

percentage valves for different ratios of the fixed resistance pressure drop (ΔP_H for the heat exchanger example) to the pressure drop over the control valve at design conditions. The larger this ratio, the more nonlinear are the installed characteristics of a linear valve.

The *inherent characteristics* are those that relate flow to valve position in the situation where the pressure drop over the control valve is constant. These are the $(\Delta P_H/\Delta P_v) = 0$ curves in Fig. 3.8. *Installed characteristics* are those that result from the variation in the pressure drop over the valve.

In conventional valves the air-pressure signal to the diaphragm comes from an I/P transducer in analog electronic systems. Control valves sometimes can stick, particularly large valves or valves in dirty service. A sticky valve can cause a control loop to oscillate; the controller output signal changes, but the valve position doesn't until the pressure force gets high enough to move the valve. Then, of course, the valve moves too far and the controller must reverse the direction of change of its output, and the same thing occurs in the other direction. So the loop will fluctuate around its setpoint even with no other disturbances.

This problem can be cured by using a "valve positioner." These devices are built into control valves and are little feedback controllers that sense the actual position of the stem, compare it to the desired position as given by the signal from the controller, and adjust the air pressure on the diaphragm to drive the stem to its correct position. Valve positioners can also be used to make valves open and close over various ranges (split-range valves).

Control valves are usually fairly fast compared with the process. With large valves (greater than 4 inches) it may take 20 to 40 seconds for the valve to move full stroke.

3.1.4 Analog and Digital Controllers

The part of the control loop with which we spend most of our time in this book is the controller. The job of the controller is to compare the process signal from the transmitter with the setpoint signal and to send out an appropriate signal to the control valve. We will go into more detail about the performance of the controller in Section 3.2. In this section we describe what kinds of action standard commercial controllers take when they detect a difference between the desired value of the controlled variable (the setpoint) and the actual value.

Analog controllers use continuous electronic or pneumatic signals. The controllers see transmitter signals continuously, and control valves are changed continuously. Digital computer controllers are discontinuous in operation, looking at a number of loops sequentially. Each individual loop is polled every sampling period. The analog signals from transmitters must be sent through analog-to-digital (A/D) converters to get the information into the computer in a form that it can use. After the computer performs its calculations in some control algorithm, it sends out a signal that must pass through a digital-to-analog (D/A) converter and a "hold" that sends a continuous signal to the control valve. We study these sampled-data systems in detail in Chapters 14 and 15.

Three basic types of controllers are commonly used for continuous feedback control. The details of construction of the analog devices and the programming of the digital devices vary from one manufacturer to the next, but their basic functions are essentially the same.

A. Proportional action

A proportional-only feedback controller changes its output signal, CO, in direct proportion to the error signal E, which is the difference between the setpoint signal SP and the process variable signal PV coming from the transmitter.

$$\text{CO} = \text{Bias} \pm \text{K}_\text{c}(\text{SP} - \text{PV}) \tag{3.23}$$

The Bias signal is a constant and is the value of the controller output when there is no error. K_c is called the *controller gain.* The larger the gain, the more the controller output will change for a given error. For example, if the gain is 1, an error of 10 percent of scale (1.6 mA in an analog electronic 4–20 mA system) will change the controller output by 10 percent of scale. Figure 3.9*a* sketches the action of a proportional controller for given error signals E.

Some instrument manufacturers use an alternative term, *proportional band* (PB), instead of gain. The two are related by

$$\text{PB} = \frac{100}{K_c} \tag{3.24}$$

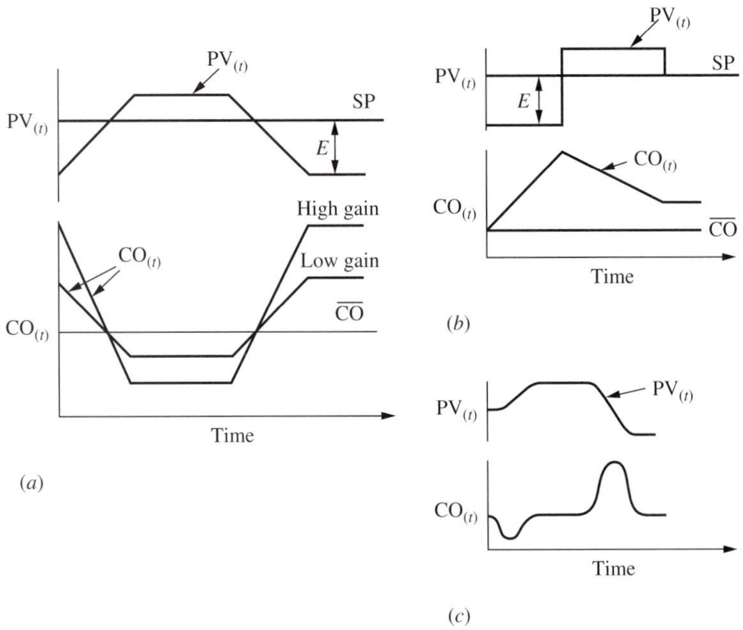

FIGURE 3.9
Action of a feedback controller. (*a*) Proportional. (*b*) Integral. (*c*) Ideal derivative.

The higher or "wider" the proportional band, the lower the gain, and vice versa. The term "proportional band" refers to the range over which the error must change to drive the controller output over its full range. Thus, a wide PB is a low gain, and a narrow PB is a high gain.

The gain on the controller can be made either positive or negative by setting a switch in an analog controller or by specifying the desired sign in a digital controller. A positive gain results in the controller output decreasing when the process variable increases. This "increase-decrease" action is called a "reverse-acting" controller. For a negative gain, the controller output increases when the process variable increases, and this is called a "direct-acting" controller. The correct sign depends on the action of the transmitter (which is usually direct), the action of the valve (air-to-open or air-to-close), and the effect of the manipulated variable on the controlled variable. Each loop should be examined closely to make sure the controller gives the correct action.

For example, suppose we are controlling the process outlet temperature of a heat exchanger as sketched in Fig. 3.10. A control valve on the steam to the shell side of the heat exchanger is manipulated by a temperature controller. To decide what action the controller should have we first look at the valve. Since this valve puts steam into the process, we would want it to fail shut. Therefore, we choose an air-to-open (AO) control valve.

Next we look at the temperature transmitter. It is direct-acting (when the process temperature goes up, the transmitter output signal, PV, goes up). Now if PV

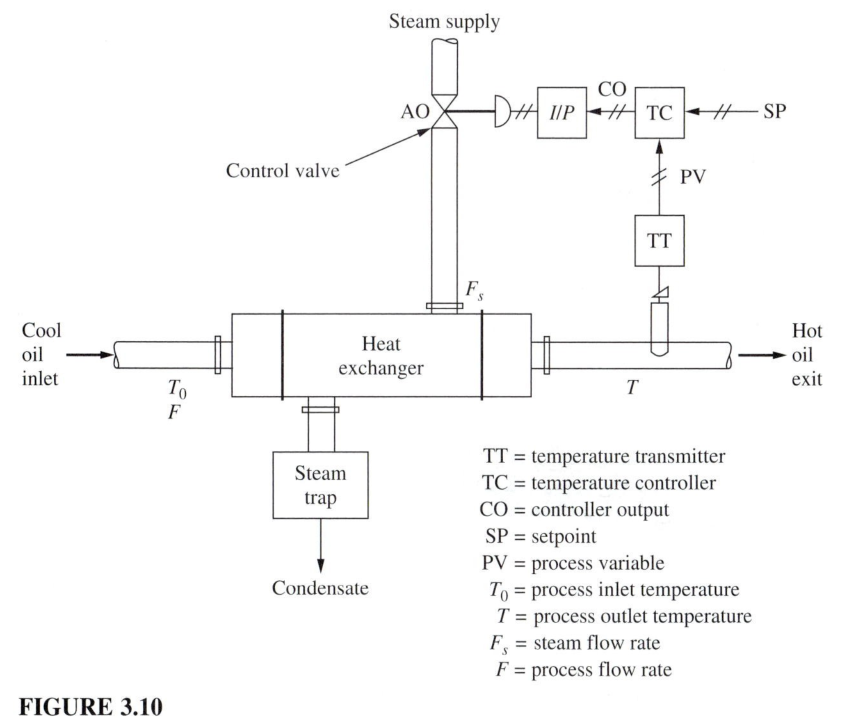

FIGURE 3.10
Heat exchanger.

increases, we want to have less steam. This means that the controller output must decrease since the valve is AO. Thus, the controller must be reverse-acting and have a positive gain.

If we were cooling instead of heating, we would want the coolant flow to increase when the temperature increased. But the controller action would still be reverse because the control valve would be an air-to-close valve, since we want it to fail wide open.

As a final example, suppose we are controlling the base level in a distillation column with the bottoms product flow rate. The valve would be AO because we want it to fail shut (and prevent the loss of base level in an emergency). The level transmitter signal increases if the level increases. If the level goes up, we want the bottoms flow rate to increase. Therefore, the base level controller should be "increase-increase" (direct-acting).

B. Integral action (reset)

Proportional action moves the control valve in direct proportion to the magnitude of the error. Integral action moves the control valve based on the time integral of the error, as sketched in Fig. 3.9*b*.

$$\text{CO} = \text{Bias} + \frac{1}{\tau_I} \int E_{(t)}\, dt \tag{3.25}$$

where τ_I is the *integral* time or the *reset* time with units of minutes.

If there is no error, the controller output does not move. As the error goes positive or negative, the integral of the error drives the controller output either up or down, depending on the action (reverse or direct) of the controller.

Most controllers are calibrated in *minutes* (or *minutes/repeat,* a term that comes from the test of putting into the controller a fixed error and seeing how long it takes the integral action to ramp up the controller output to produce the same change that a proportional controller would make when its gain is 1; the integral *repeats* the action of the proportional controller).

The basic purpose of integral action is to drive the process back to its setpoint when it has been disturbed. A proportional controller will *not* usually return the controlled variable to the setpoint when a load or setpoint disturbance occurs. This permanent error (SP − PV) is called *steady-state error* or *offset.* Integral action reduces the offset to zero.

Integral action usually degrades the dynamic response of a control loop. We demonstrate this quantitatively in Chapter 8. It makes the control loop more oscillatory and moves it toward instability. But integral action is usually needed if it is desirable to have zero offset. This is another example of an engineering trade-off that must be made between dynamic and steady-state performance.

C. Derivative action

The purpose of derivative action (also called *rate* or *preact*) is to anticipate where the process is heading by looking at the time rate of change of the controlled variable (its derivative). If we were able to take the derivative of the error signal (which we cannot do perfectly, as we explain more fully in Chapter 8), we would have ideal derivative action.

$$\text{CO} = \text{Bias} + \tau_D \frac{dE}{dt} \tag{3.26}$$

where τ_D is the *derivative* time (minutes).

In theory, derivative action should always improve dynamic response, and it does in many loops. In others, however, the problem of noisy signals (fluctuating process variable signals) makes the use of derivative action undesirable.

D. Commercial controllers

The three actions just described are used individually or combined in commercial controllers. Probably 60 percent of all controllers are PI (proportional-integral), 20 percent are PID (proportional-integral-derivative), and 20 percent are P-only (proportional). We discuss the reasons for selecting one type over another in Section 3.2.

3.1.5 Computing and Logic Devices

A wide variety of computations and logical operations can be performed on control signals. Electronic devices are used in analog systems, and computer software is used in DCS systems. For example, adders, multipliers, dividers, low selectors, high selectors, high limiters, low limiters, and square-root extractors can all be incorporated in the control loop. They are widely used in *ratio* control, in *computed variable* control, in *feedforward* control, and in *override* control. These are discussed in the next chapter.

In addition to the basic control loops, all processes have instrumentation that (1) sounds alarms to alert the operator to any abnormal or unsafe condition, and (2) shuts down the process if unsafe conditions are detected or equipment fails. For example, if a compressor motor overloads and the electrical control system shuts down the motor, the rest of the process will usually have to be shut down immediately. This type of instrumentation is called an "interlock." It either shuts a control valve completely or drives the control valve wide open. Other examples of conditions that can "interlock" a process down include electrical power failures, failure of a feed or reflux pump, detection of high pressure or temperature in a vessel, and indication of high or low liquid level in a tank or column base. Interlocks are usually achieved by pressure, mechanical, or electrical switches. They can be included in the computer software in a computer control system, but they are usually "hard-wired" for reliability and redundancy.

3.2 PERFORMANCE OF FEEDBACK CONTROLLERS

3.2.1 Specifications for Closedloop Response

There are a number of criteria by which the desired performance of a closedloop system can be specified in the time domain. For example, we could specify that the closedloop system be critically damped so that there is no overshoot or oscillation.

We must then select the type of controller and set its tuning constants so that, when coupled with the process, it gives the desired closedloop response. Naturally, the control specification must be physically attainable. We cannot make a Boeing 747 jumbo jet behave like an F-15 fighter. We cannot violate constraints on the manipulated variable (the control valve can only go wide open or completely shut), and we cannot require a physically unrealizable controller (more about the mathematics of this in Chapter 8).

There are a number of time-domain specifications. A few of the most frequently used dynamic specifications follow (see also Problem 2.7). The traditional test input signal is a step change in setpoint.

1. Closedloop damping coefficient (as discussed in Chapter 2)
2. Overshoot: the magnitude by which the controlled variable swings past the setpoint
3. Rise time (speed of response): the time it takes the process to come up to the new setpoint
4. Decay ratio: the ratio of maximum amplitudes of successive oscillations
5. Settling time: the time it takes the amplitude of the oscillations to decay to some fraction of the change in setpoint
6. Integral of the squared error: ISE $= \int_0^\infty (E_{(t)})^2 \, dt$

Notice that the first five of these assume an underdamped closedloop system, i.e., one that has some oscillatory nature.

Many years of experience have led to our personal preference of designing for a closedloop damping coefficient of 0.3 to 0.5. As we see throughout the rest of this book, this criterion is easy to use and reliable. A criterion such as ISE can be used for any type of disturbance, setpoint or load. Some "experts" (remember, an "expert" is one who is seldom in doubt but *often* in error) recommend different tuning parameters for the two types of disturbances. This makes little sense to us. What you want is a reasonable compromise between performance (tight control, small closedloop time constants) and robustness (not too sensitive to changes in process parameters). This compromise is achieved by using a closedloop damping coefficient of 0.3 to 0.5 since it keeps the real parts of the roots of the closedloop characteristic equation a reasonable distance from the imaginary axis, the point where the system becomes unstable (see Chapter 2). The closedloop damping coefficient specification is independent of the type of input disturbance.

The steady-state error is another time-domain specification. It is not a dynamic specification, but it is an important performance criterion. In many loops (but not all) a steady-state error of zero is desired, i.e., the value of the controlled variable should eventually level out at the setpoint.

3.2.2 Load Performance

The job of most control loops in a chemical process is one of regulation or load rejection, i.e., holding the controlled variable at its setpoint in the face of load disturbances. Let us look at the effects of load changes when the standard types of controllers are used.

We use a simple heat exchanger process (Fig. 3.10) in which an oil stream is heated with steam. The process outlet temperature T is controlled by manipulating the steam flow rate F_s to the shell side of the heat exchanger. The oil flow rate F and the inlet oil temperature T_0 are load disturbances. The signal from the temperature transmitter (TT) is the process variable signal, PV. The setpoint signal is SP. The output signal, CO, from the temperature controller (TC) goes through an I/P transducer to the steam control valve. The valve is AO because we want it to fail closed.

A. On/off control

The simplest controller would be an on/off controller like the thermostat in your home heating system. The manipulated variable is either at maximum flow or at zero flow. The on/off controller is a proportional controller with a very high gain and gives "bang-bang" control action. This type of control is seldom used in a continuous process because of the cycling nature of the response, surging flows, and wear on control valves.

In the heat exchanger example the controlled variable T cycles as shown in Fig. 3.11*a*. When a load disturbance in inlet temperature (a step decrease in T_0) occurs, both the period and the average value of the controlled variable T change. You have observed this in your heating system. When the outside temperature is colder, the furnace runs longer and more frequently, and the room temperature is lower on average. This is one of the reasons you feel colder inside on a cold day than on a warm day for the same setting of the thermostat. The system is really unstable in the classic linear sense. The nonlinear bounds or constraints on the manipulated variable (control valve position) keep it in a "limit cycle."

B. Proportional controller

The output of a proportional controller changes only if the error signal changes. Since a load change requires a new control valve position, the controller must end up with a new error signal. This means that a proportional controller usually gives a *steady-state error* or *offset.* This is an inherent limitation of P controllers and is why integral action is usually added. Our introductory simulation example in Chapter 1 illustrated this point.

As shown in Fig. 3.11*b* for the heat exchanger example, a decrease in process inlet temperature T_0 requires more steam. Therefore, the error must increase to open the steam valve more. The magnitude of the offset depends on the size of the load disturbance and on the controller gain. The bigger the gain, the smaller the offset. As the gain is made bigger, however, the process becomes underdamped, and eventually, at still higher gains, the loop will go unstable, acting like an on/off controller.

Steady-state error is not always undesirable. In many level control loops the absolute level is unimportant as long as the tank does not run dry or overflow. Thus, a proportional controller is often the best type for level control. We discuss this in more detail in Section 3.3.

C. Proportional-integral (PI) controller

Most control loops use PI controllers. The integral action eliminates steady-state error in T (see Fig. 3.11*c*). The smaller the integral time τ_I, the faster the error is

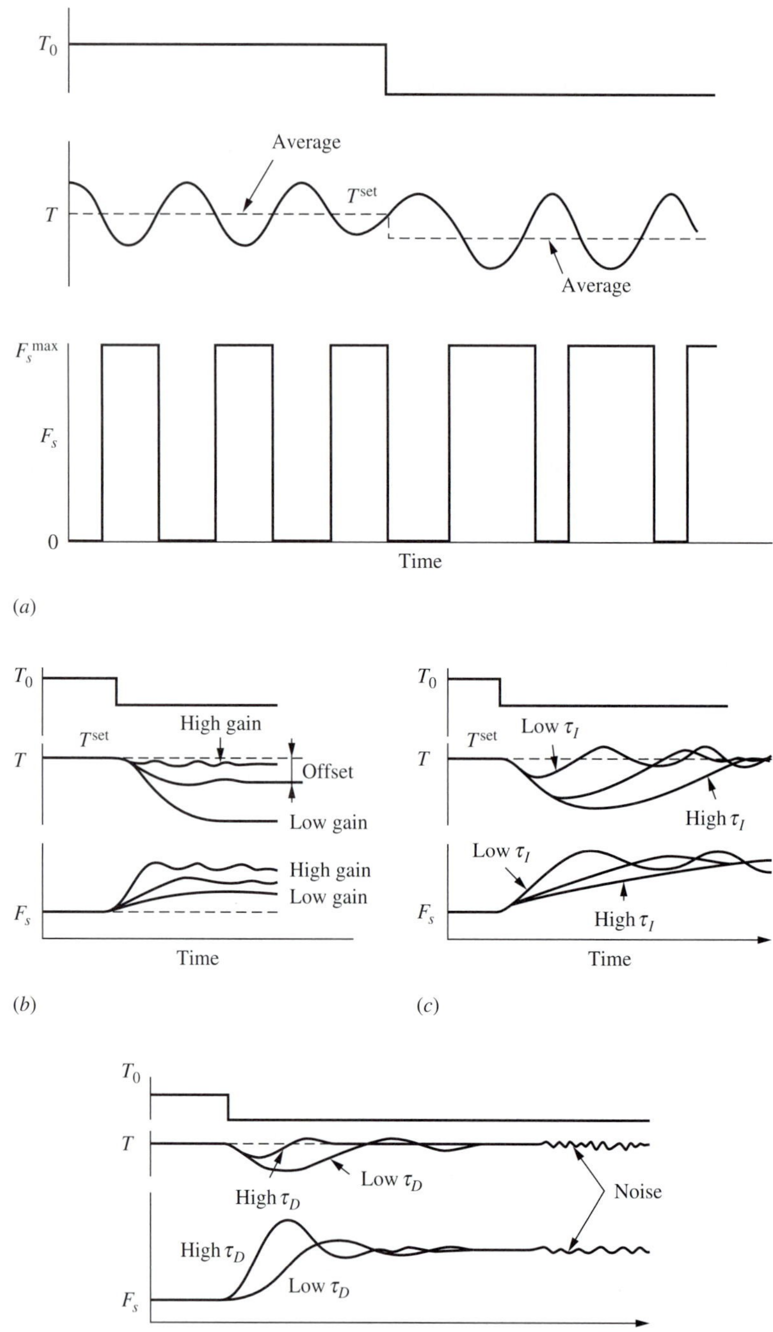

FIGURE 3.11
Controller load performance. (*a*) On/off controller. (*b*) Proportional (P). (*c*) Proportional-integral (PI). (*d*) Proportional-integral-derivative (PID).

reduced. But the system becomes more underdamped as τ_I is reduced. If it is made too small, the loop becomes unstable.

D. Proportional-integral-derivative (PID) controller

PID controllers are used in loops where signals are not noisy and where tight dynamic response is important. The derivative action helps to compensate for lags in the loop. Temperature controllers in reactors are usually PID. The controller senses the rate of movement away from the setpoint and starts moving the control valve earlier than with only PI action (see Fig. 3.11*d*).

(*a*) Derivative on error

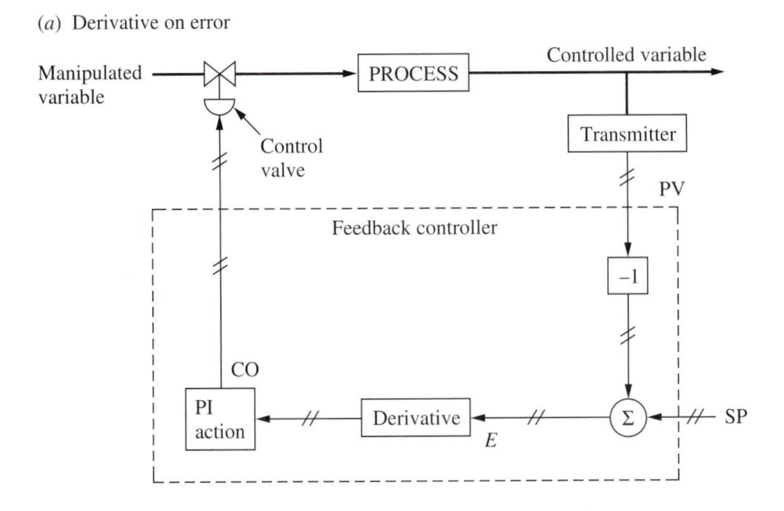

(*b*) Derivative on process variable

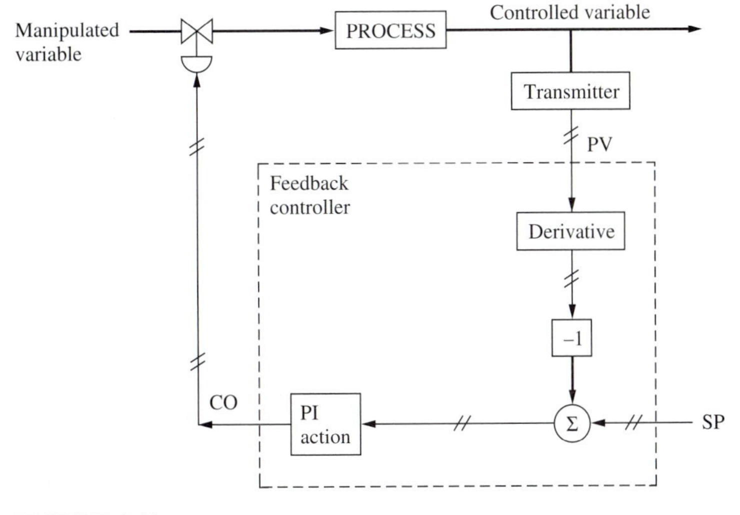

FIGURE 3.12
Derivative action.

Derivative action can be used on either the error signal (SP − PV) or just the process variable (PV). If it is on the error signal, step changes in setpoint will produce large bumps in the control valve. Therefore, in most process control applications, the derivative action is applied only to the PV signal as it enters the controller. The P and I action is then applied to the difference between the setpoint and the output signal from the derivative unit (see Fig. 3.12).

3.3 CONTROLLER TUNING

There are a variety of feedback controller tuning methods. Probably 80 percent of all loops are tuned experimentally by an instrument mechanic, and 75 percent of the time the mechanic can guess approximately what the settings will be by drawing on experience with similar loops. We discuss a few of the time-domain methods below. In subsequent chapters we present other techniques for finding controller constants in the Laplace and frequency domains.

3.3.1 Rules of Thumb

The common types of control loops are level, flow, temperature, and pressure. The type of controller and the settings used for any one type are sometimes pretty much the same from one application to another. For example, most flow control loops use PI controllers with wide proportional band and fast integral action.

Some heuristics are given next, but they are not to be taken as gospel. They merely indicate common practice, and they work in most applications.

A. Flow loops

PI controllers are used in most flow loops. A wide proportional band setting (PB = 150) or low gain is used to reduce the effect of the noisy flow signal due to flow turbulence. A low value of integral or reset time (τ_I = 0.1 minutes/repeat) is used to get fast, snappy setpoint tracking. The dynamics of the process are usually very fast. The sensor sees the change in flow almost immediately. The control valve dynamics are the slowest element in the loop, so a small reset time can be used.

There is one notable exception to fast PI flow control: flow control of condensate-throttled reboilers. As sketched in Fig. 3.13, the flow rate of vapor to a reboiler is sometimes controlled by manipulating the liquid condensate valve. Since the vapor flow depends on the rate of condensation, vapor flow can be varied only by changing the area for heat transfer in the reboiler. This is accomplished by raising or lowering the liquid level in this "flooded" reboiler. Changing the liquid level takes some time. Typical time constants are 3 to 6 minutes. Therefore, this flow control loop would have much different controller tuning constants than suggested in the rule of thumb cited above. Some derivative action may even be used in the loop to give faster flow control.

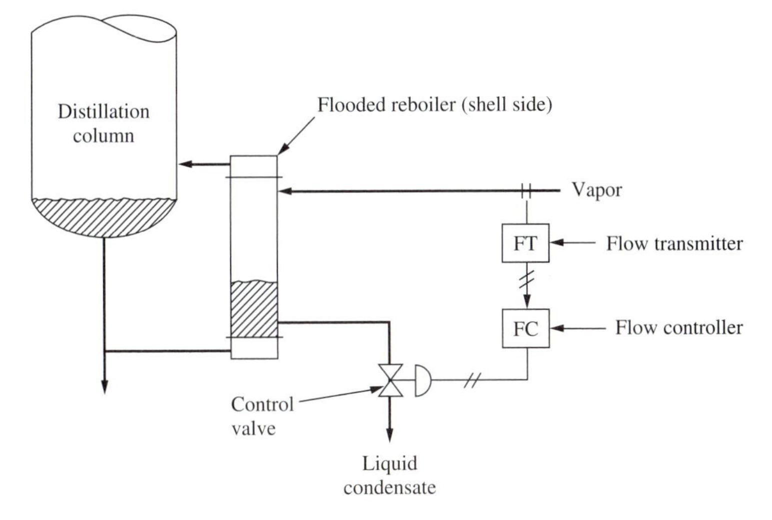

FIGURE 3.13
Condensate-throttle flow control.

B. Level loops

Most liquid levels represent material inventory used as surge capacity. In these cases it is relatively unimportant where the level is, as long as it is between some maximum and minimum values. Therefore, proportional controllers are often used on level loops to give smooth changes in flow rates and to filter out fluctuations in flow rates to downstream units. We demonstrated this important concept in one of our simulation examples in Chapter 1.

One of the most common errors in laying out a control structure for a plant with multiple units in series is the use of PI level controllers. If P controllers are used, the process flows rise or fall slowly down the train of units with no overshoot of flow rates. Liquid levels rise if flows increase and fall if flows decrease. Levels are *not* maintained at setpoints.

If PI level controllers are used, the integral action forces the level back to its setpoint. In fact, if the level controller is doing a "perfect" job, the level is held right at its setpoint. This means that any change in the flow rate into the surge tank will immediately change the flow rate out of the tank. This defeats the purpose of buffering. We might as well not even use a tank; just run the inlet pipe right into the outlet pipe! Thus, this is an example of where tight control is *not* desirable. We want the flow rate out of the tank to increase gradually when the inflow increases so that downstream units are not upset.

Suppose the flow rate F_0 increases to the first tank in Fig. 3.14. The level h_1 in the first tank will start to increase. The level controller will start to increase F_1. When F_1 has increased to the point that it is equal to F_0, the level will stop changing since the tank is just an integrator. Now, if we use a P level controller, nothing else will happen. The level will remain at the higher level, and the entering and exiting flows will be equal.

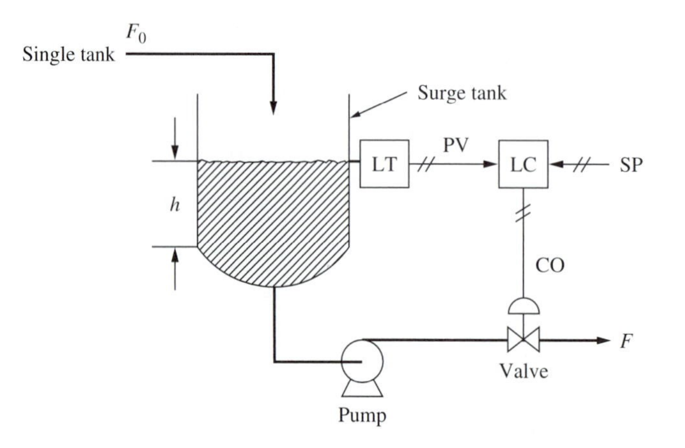

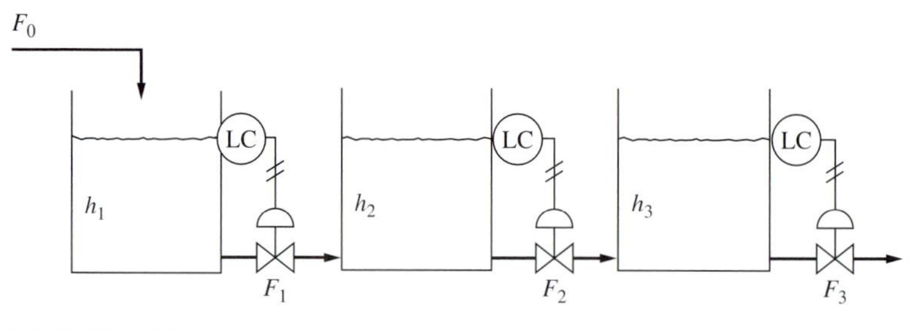

FIGURE 3.14
P versus PI level control.

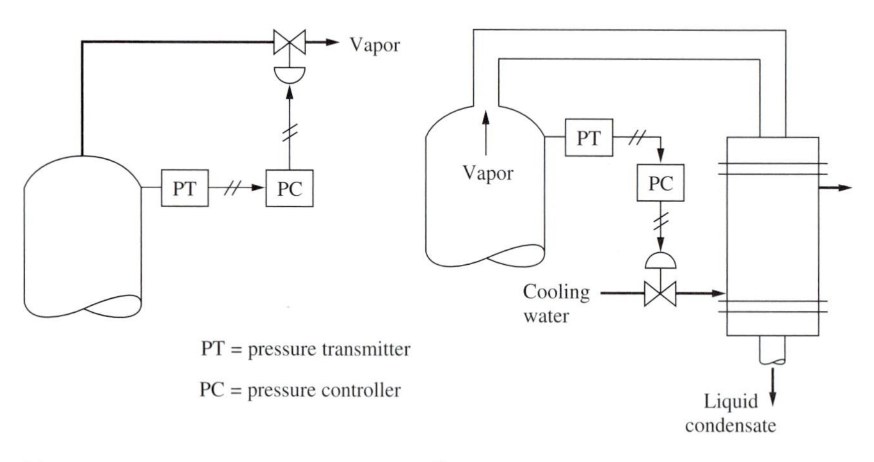

FIGURE 3.15
Pressure control. (*a*) Fast pressure loop; (*b*) slow pressure loop.

If, however, we use a PI level controller, the controller will continue to increase the outflow *beyond* the value of the inflow to drive the level back down to its setpoint. So an inherent problem with PI level controllers is that they amplify flow rate changes of this type. The change in the flow rate out of the tank is actually larger (for a period of time) than the change in the flow rate into the tank. This amplification gets worse as it works its way down through the series of units. What started out at the beginning as a small disturbance can result in large fluctuations by the time it reaches the last unit in the train.

There are, of course, many situations where it is desirable to control level tightly, for example, in a reactor where control of residence time is important.

The tuning of proportional level controllers is a trivial job. For example, we could set the bias value at 50 percent of full scale, the setpoint at 50 percent of full scale, and the proportional band at 50. This means that the control valve will be half open when the tank is half full, wide open when the tank is 75 percent full, and completely shut when the tank is 25 percent full. Changing the proportional band to 100 would mean that the tank would be completely full to have the valve wide open and completely empty to have the valve shut.

C. Pressure loops

Pressure loops vary from very tight, fast loops (almost like flow control) to slow averaging loops (almost like level control). An example of a fast pressure loop is the case of a valve throttling the flow of vapor from a vessel, as shown in Fig. 3.15*a*. The valve has a direct handle on pressure, and tight control can be achieved. An example of a slower pressure loop is shown in Fig. 3.15*b*. Pressure is held by throttling the water flow to a condenser. The water changes the ΔT driving force for condensation in the condenser. Therefore, the heat transfer dynamics and the lag of the water flowing through the shell side of the condenser are introduced into the pressure control loop.

D. Temperature loops

Temperature control loops are usually moderately slow because of the sensor lags and the process heat transfer lags. PID controllers are often used. Proportional band settings are fairly low, depending on temperature transmitter spans and control valve sizes. The reset time is of the same order as the process time constant; i.e., the faster the process, the smaller τ_I can be set. Derivative time is set something like one-fourth the process time constant, depending on the noise in the transmitter signal. We quantify these tuning numbers later in the book.

3.3.2 On-Line Trial and Error

To tune a controller on-line, a good instrument mechanic follows a procedure something like the following:

1. With the controller on manual, take all the integral and derivative action out of the controller; i.e., set τ_I at maximum minutes/repeat and τ_D at minimum minutes.

2. Set the K_c at a low value, perhaps 0.2.
3. Put the controller in automatic.
4. Make a small setpoint or load change and observe the response of the controlled variable. The gain is low, so the response will be sluggish.
5. Increase K_c by a factor of 2 and make another small change in setpoint or load.
6. Keep increasing K_c, repeating step 5 until the loop becomes very underdamped and oscillatory. The gain at which this occurs is called the *ultimate gain.*
7. Back off K_c to half this ultimate value.
8. Now start bringing in integral action by reducing τ_I by factors of 2, making small disturbances at each value of τ_I to see the effect.
9. Find the value of τ_I that makes the loop very underdamped, and set τ_I at twice this value.
10. Start bringing in derivative action by increasing τ_D. Load changes should be used to disturb the system, and the derivative should act on the process variable signal. Find the value of τ_D that gives the tightest control without amplifying the noise in the process variable signal.
11. Increase K_c again by steps of 10 percent until the desired specification on damping coefficient or overshoot is satisfied.

It should be noted that there are some loops for which these procedures do *not* work. Systems that exhibit "conditional stability" are the classic example. These processes are unstable at high values of controller gain *and* at low values of controller gain, but are stable over some intermediate range of gains. We discuss some of these in Chapter 9.

3.3.3 Ziegler-Nichols Method

The Ziegler-Nichols (ZN) controller settings (J. G. Ziegler and N. B. Nichols, *Trans. ASME 64:* 759, 1942) are pseudo-standards in the control field. They are easy to find and to use and give reasonable performance on some loops. The ZN settings are benchmarks against which the performance of other controller settings are compared in many studies. They are often used as first guesses, but they tend to be too underdamped for most process control applications. Some on-line tuning can improve control significantly. But the ZN settings are useful as a place to start.

The ZN method consists of first finding the ultimate gain K_u, the value of gain at which the loop is at the limit of stability with a proportional-only feedback controller. The period of the resulting oscillation is called the *ultimate period,* P_u (minutes per cycle). The ZN settings are then calculated from K_u and P_u by the formulas given in Table 3.1 for the three types of controllers. Notice that a lower gain is used when integration is included in the controller (PI) and that the addition of derivative permits a higher gain and faster reset.

As an example, let us consider the three-heated-tank process simulated in Chapter 1. As we prove in Chapter 8, the ultimate gain of this process is 6 and its ultimate period is 0.363 hours. The ZN settings for this system are given in Table 3.2. The response of the closedloop system to a step load disturbance in T_0 is shown in Fig. 3.16 with P and PI controllers using the ZN settings.

TABLE 3.1
Ziegler-Nichols and TLC settings

	P	PI	PID
	Ziegler-Nichols		
K_c	$\frac{K_u}{2}$	$\frac{K_u}{2.2}$	$\frac{K_u}{1.7}$
τ_I	—	$\frac{P_u}{1.2}$	$\frac{P_u}{2}$
τ_D	—	—	$\frac{P_u}{8}$
	Tyreus-Luyben		
K_c	—	$\frac{K_u}{3.2}$	$\frac{K_u}{2.2}$
τ_I	—	$2.2P_u$	$2.2P_u$
τ_D	—	—	$\frac{P_u}{6.3}$

TABLE 3.2
ZN and TLC settings for three-heated-tank process

Ultimate values: $K_u = 6$ $\omega_u = 17.3$ rad/hr $P_u = 0.363$ hr

	P	PI	PID
	Ziegler-Nichols		
K_c	3	2.73	3.53
τ_I (hr)	—	0.302	0.181
τ_D (hr)	—	—	0.0454
	Tyreus-Luyben		
K_c	—	1.88	—
τ_I (hr)	—	0.8	—

These results show some interesting things:

1. There is a steady-state error in the controlled variable T_3 when a P controller is used. This offset results because there is no integral term to drive the error to zero.
2. The ZN settings result in a fairly underdamped system: the responses show significant oscillation. The closedloop damping coefficient of this system is about 0.1 to 0.2.

In many process control applications, it is undesirable to have this kind of response. It is too snappy and calls for rapid and large changes in the manipulated variable. For example, in the control of a tray temperature in a distillation column, we want tight temperature control, but we do not want rapid or large changes in the heat input to the reboiler because the column may flood during one of the transients.

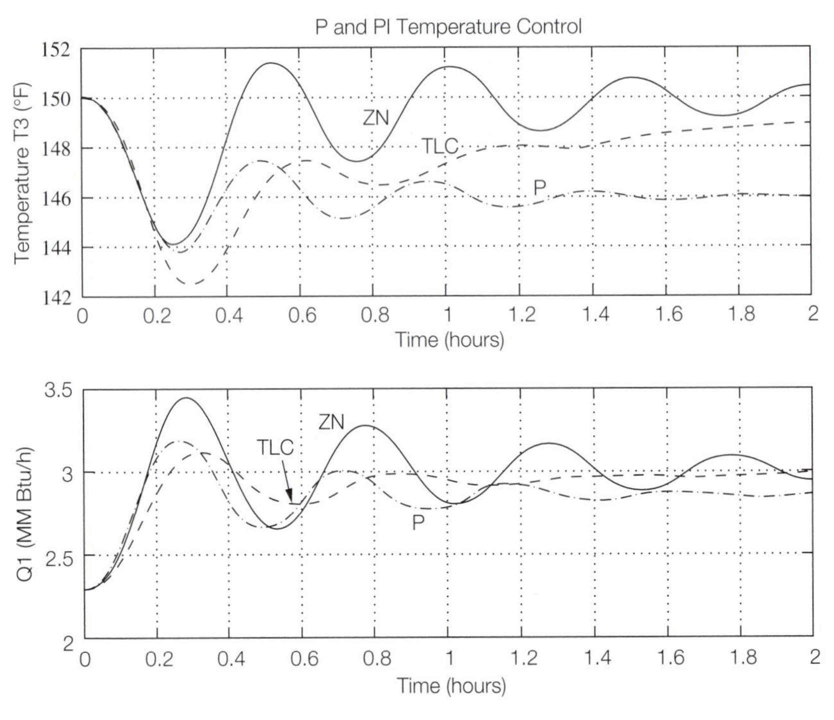

FIGURE 3.16

If the surge in vapor rate is too rapid, we could even mechanically damage the trays in the column. So we must sacrifice some performance (tight control) for smoother and less violent changes in the manipulated variable.

3.3.4 Tyreus-Luyben Method

The Tyreus-Luyben method procedure is quite similar to the Ziegler-Nichols method but gives more conservative settings (higher closedloop damping coefficient) and is more suitable for chemical process control applications. Our processes are more like 747 jumbo jets. We want to treat them with "tender loving care" (hence, we call these controller settings TLC settings) and will gladly sacrifice performance for robustness. If we were designing the flight controls for an F-16 jet fighter, we would probably design for two situations: an "attack" mode (when we are ready to engage in a dogfight with a MIG) and a "landing" mode (when we are trying to land on the deck of an aircraft carrier). The attack mode setting could be Ziegler-Nichols (small damping coefficient, small time constant), and the landing mode settings could be Tyreus-Luyben (big damping coefficient, large time constant).

The method (B. D. Tyreus and W. L. Luyben, *I&EC Research 31:* 2625, 1992) uses the ultimate gain K_u and the ultimate frequency ω_u. The formulas for PI and PID controllers are given in Table 3.1, and the TLC settings for the three-heated-tank process are given in Table 3.2. The performance of these settings is shown in Fig. 3.16.

The use of PID controllers in process control is limited, primarily because of problems with noisy signals. The derivative action amplifies this noise and gives poor performance in some applications. There are some reactor temperature control loops that can (and sometimes must) use derivative action, but these are usually in situations where the temperature signal is not noisy (filtering is sometimes required).

3.4 CONCLUSIONS

We have discussed concepts and ideas in this chapter that are of great practical importance. You need to have some appreciation of the hardware and software of process control. You need to know what conventional P and PI controllers can and cannot do. You need to be able to tune these controllers. In this chapter we have presented some of the basics. In the next three chapters we discuss more advanced topics.

PROBLEMS

3.1. (*a*) Calculate the gain of an orifice plate and differential-pressure transmitter for flow rates from 10 percent to 90 percent of full scale.
(*b*) Calculate the gain of linear and equal-percentage valves over the same range, assuming constant pressure drop over the valve.
(*c*) Calculate the total loop gain of the valve and the sensor-transmitter system over this range.

3.2. The temperature of a CSTR is controlled by an electronic (4 to 20 mA) feedback control system containing (1) a 100 to 200°F temperature transmitter, (2) a PI controller with integral time set at 3 minutes and proportional band at 25, and (3) a control valve with linear trim, air-to-open action, and $C_v = 4$ through which cooling water flows. The pressure drop across the valve is a constant 25 psi. If the steady-state controller output is 12 mA, how much cooling water is going through the valve? If a sudden disturbance increases reactor temperature by 5°F, what will be the immediate effect on the controller output signal and the water flow rate?

3.3. Simulate the three-CSTR system on a digital computer with an on/off feedback controller. Assume that the manipulated variable C_{AM} is limited to ± 1 mole of A/ft^3 around the steady-state value. Find the period of oscillation and the average value of C_{A3} for values of the load variable C_{AD} of 0.6 and 1.

3.4. Two ways to control the outlet temperature of a heat exchanger cooler are sketched on the following page. Comment on the relative merits of these two systems from the standpoints of both control and heat exchanger design.

3.5. Specify the following items for the bypass cooler system of Problem 3.4:
(*a*) The action of the valves (AO or AC) and kind of trim.
(*b*) The action and type of controller.

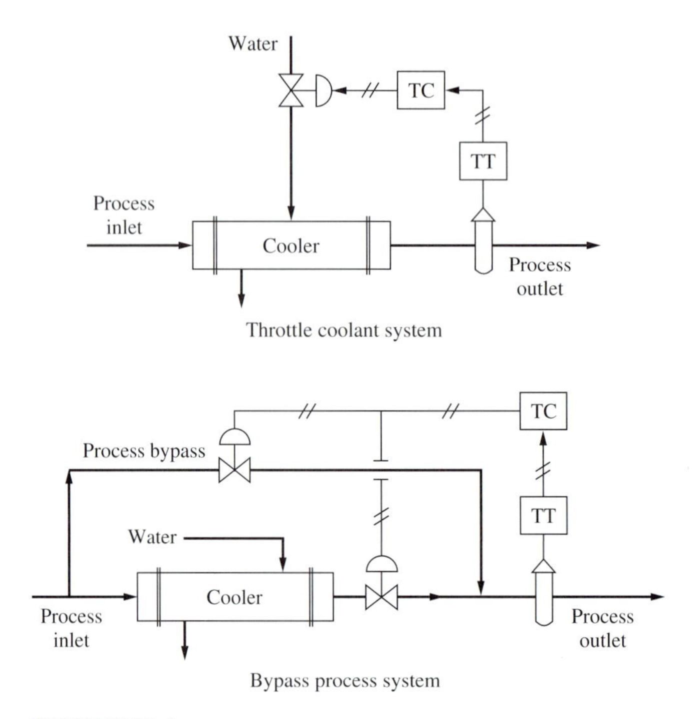

FIGURE P3.4

3.6. Assume that the bypass cooling system of Problem 3.4 is designed so that the total process flow of 50,000 lb_m/hr (heat capacity of 0.5 Btu/lb_m °F) is split under normal conditions with 25 percent going around the bypass and 75 percent going through the cooler. Process inlet and outlet temperatures under these conditions are 250 and 150°F. Inlet and outlet water temperatures are 80 and 120°F. Process side pressure drop through the exchanger is 10 psi. The control valves have linear trim and are designed to be half open at design rates with a 10 psi drop over the valve in series with the cooler. Liquid density is constant at 62.3 lb_m/ft^3.

What will the valve positions be if the total process flow is reduced to 25 percent of design and the process outlet temperature is held at 150°F?

3.7. A liquid (sp gr = 1) is to be pumped through a heat exchanger and a control valve at a design rate of 200 gpm. The exchanger pressure drop is 30 psi at design throughput. Make plots of flow rate versus valve position x for linear and equal-percentage ($\alpha = 50$) control valves. Both valves are set at $f_{(x)} = 0.5$ at design rate. The total pressure drop over the entire system is constant. The pressure drop over the control valve at design rate is:

(*a*) 10 psi
(*b*) 30 psi
(*c*) 120 psi

3.8. Process designers sometimes like to use "dephlegmators" or partial condensers mounted directly in the top of the distillation column when the overhead product is taken off as a vapor. They are particularly popular for corrosive, toxic, or hard-to-handle chemicals

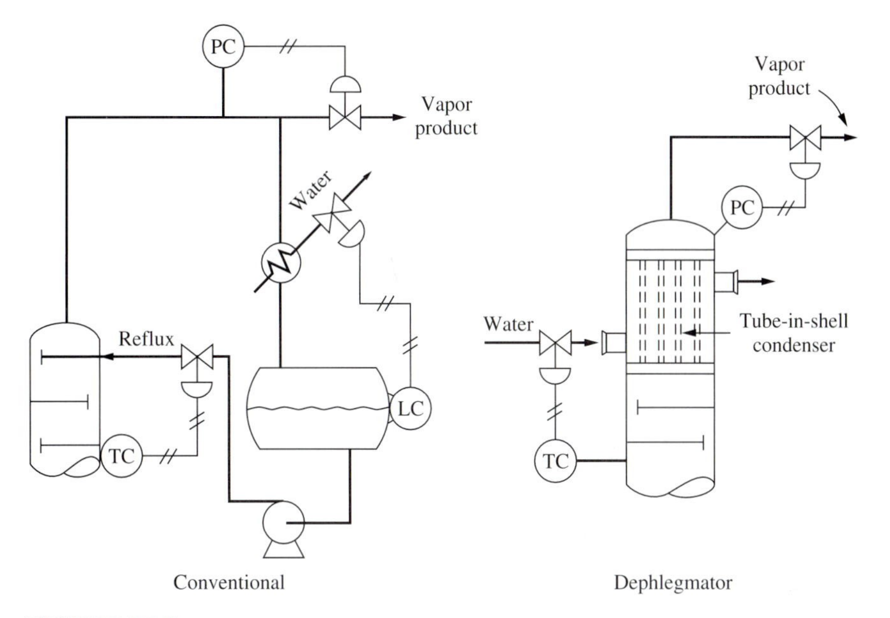

FIGURE P3.8

since they eliminate a separate condenser shell, a reflux drum, and a reflux pump. Comment on the relative controllability of the two process systems sketched above.

3.9. Compare quantitatively by digital simulation the dynamic performance of the three coolers sketched on the next page with countercurrent flow, cocurrent flow, and circulating water systems. Assume the tube and shell sides can each be represented by four perfectly mixed lumps. Process design conditions are:

Flow rate $= 50{,}000\ \text{lb}_m/\text{hr}$
Inlet temperature $= 250°\text{F}$
Outlet temperature $= 130°\text{F}$
Heat capacity $= 0.5\ \text{Btu/lb}_m\ °\text{F}$

Cooling-water design conditions are:

A. Countercurrent:
  Inlet temperature $= 80°\text{F}$
  Outlet temperature $= 130°\text{F}$
B. Cocurrent:
  Inlet temperature $= 80°\text{F}$
  Outlet temperature $= 125°\text{F}$
C. Circulating system:
  Inlet temperature to cooler $= 120°\text{F}$
  Outlet temperature from cooler $= 125°\text{F}$
  Makeup water temperature to system $= 80°\text{F}$

Neglect the tube and shell metal. Tune PI controllers experimentally for each system. Find the outlet temperature deviations for a 25 percent step increase in process flow rate.

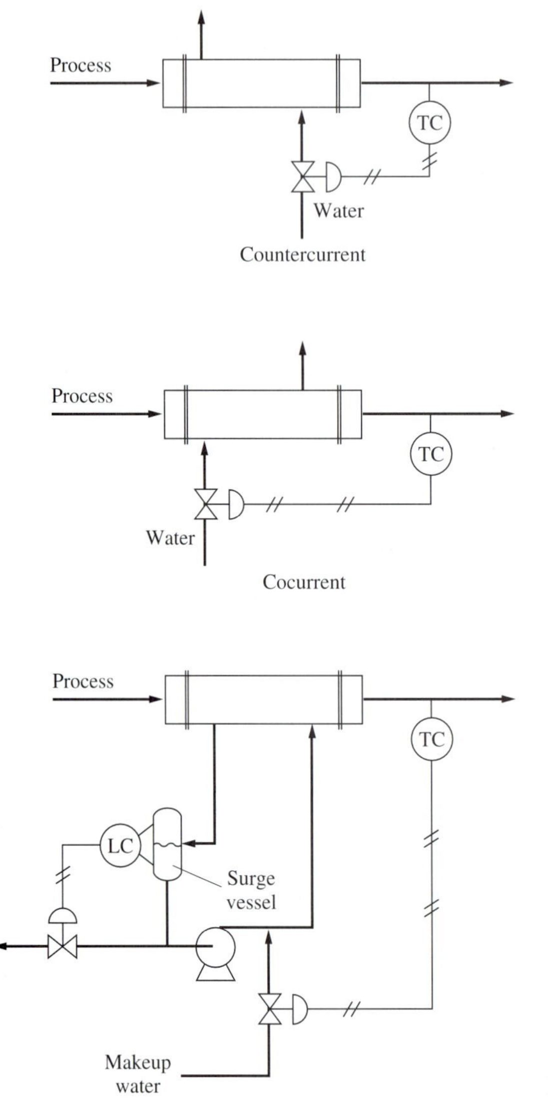

FIGURE P3.9

3.10. The overhead vapor from a depropanizer distillation column is totally condensed in a water-cooled condenser at 120°F and 227 psig. The vapor is 95 mol% propane and 5 mol% isobutane. Its design flow rate is 25,500 lb_m/hr, and average latent heat of vaporization is 125 Btu/lb_m.

Cooling water inlet and outlet temperatures are 80°F and 105°F, respectively. The condenser heat transfer area is 1000 ft^2. The cooling water pressure drop through the condenser at design rate is 5 psi. A linear-trim control valve is installed in the cooling water line. The pressure drop over the valve is 30 psi at design with the valve half open.

The process pressure is measured by an electronic (4–20 mA) pressure transmitter whose range is 100–300 psig. An analog electronic proportional controller with a gain of 3 is used to control process pressure by manipulating cooling water flow. The

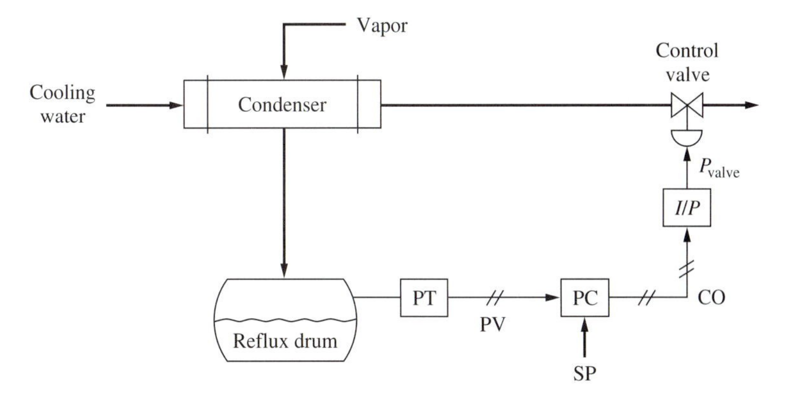

FIGURE P3.10

electronic signal from the controller (CO) is converted to a pneumatic signal in the I/P transducer.

(*a*) Calculate the cooling water flow rate (gpm) at design conditions.
(*b*) Calculate the size coefficient (C_v) of the control valve.
(*c*) Specify the action of the control valve and the controller.
(*d*) What are the values of the signals PV, CO, SP, and P_{valve} at design conditions?
(*e*) Suppose the process pressure jumps 10 psi. How much will the cooling water flow rate increase? Give values for PV, CO, and P_{valve} at this higher pressure. Assume that the total pressure drop over the condenser and control valve is constant.

3.11. A circulating chilled-water system is used to cool an oil stream from 90 to 70°F in a tube-in-shell heat exchanger. The temperature of the chilled water entering the process heat exchanger is maintained constant at 50°F by pumping the chilled water through a refrigerated cooler located upstream of the process heat exchanger.

The design chilled-water rate for normal conditions is 1000 gpm, with chilled water leaving the process heat exchanger at 60°F. Chilled-water pressure drop through the process heat exchanger is 15 psi at 1000 gpm. Chilled-water pressure drop through the refrigerated cooler is 15 psi at 1000 gpm. The heat transfer area of the process heat exchanger is 1143 ft^2.

The temperature transmitter on the process oil stream leaving the heat exchanger has a range of 50 to 150°F. The range of the orifice differential-pressure flow transmitter on the chilled water is 0 to 1500 gpm. All instrumentation is electronic (4 to 20 mA). Assume the chilled-water pump is centrifugal with a flat pump curve.

(*a*) Design the chilled-water control valve so that it is 25 percent open at the 1000 gpm design rate and can pass a maximum flow of 1500 gpm. Assume linear trim is used.
(*b*) Give values of the signals from the temperature transmitter, temperature controller, and chilled-water flow transmitter when the chilled-water flow is 1000 gpm.
(*c*) What is the pressure drop over the chilled-water valve when it is wide open?
(*d*) What are the pressure drop and fraction open of the chilled-water control valve when the chilled-water flow rate is reduced to 500 gpm? What is the chilled-water flow transmitter output at this rate?
(*e*) If electric power costs 2.5 cents/kilowatt-hour, what are the annual pumping costs for the chilled-water pump at the design 1000 gpm rate? What horsepower motor is required to drive the chilled-water pump? (1 hp $=$ 550 ft-lb$_f$/sec $=$ 746 W.)

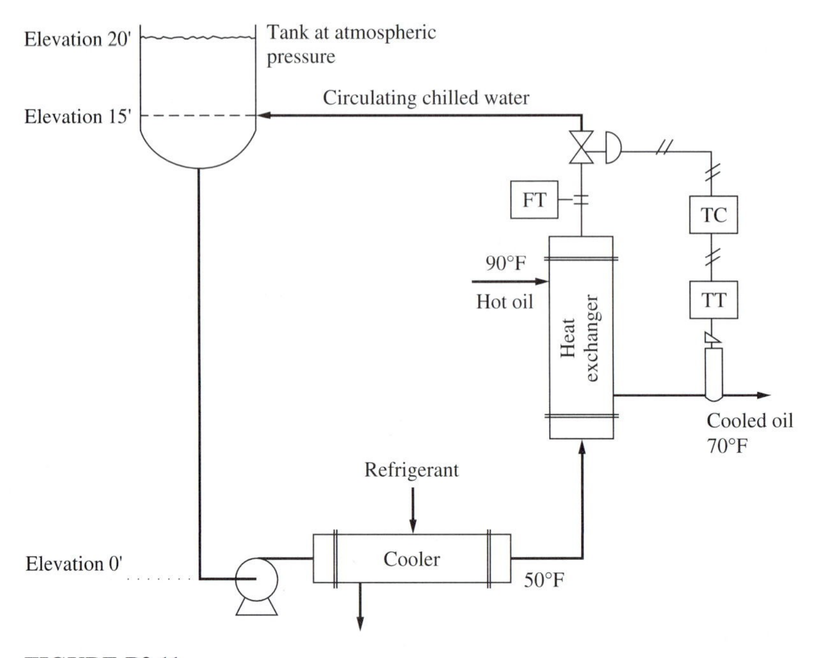

FIGURE P3.11

3.12. Tray 4 temperature on the Lehigh distillation column is controlled by a pneumatic PI controller with a 2-minute reset time and a 50 percent proportional band. Temperature controller output (CO_T) adjusts the setpoint of a steam flow controller (reset time 0.1 minutes and proportional band 100 percent). Column base level is controlled by a pneumatic proportional-only controller that sets the bottoms product withdrawal rate.

Transmitter ranges are:

Tray 4 temperature	60–120°C
Steam flow	0–4.2 lb_m/min (orifice/ΔP transmitter)
Bottoms flow	0–1 gpm (orifice/ΔP transmitter)
Base level	0–20 in H_2O

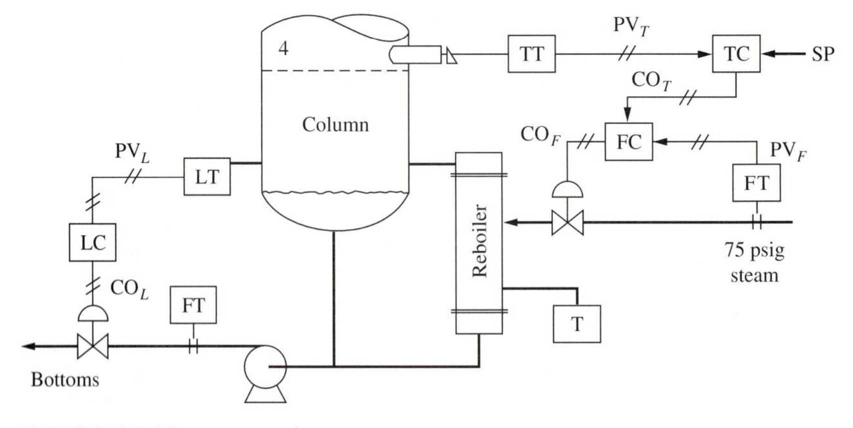

FIGURE P3.12

Steady-state operating conditions are:

Tray 4 temperature	83°C
Base level	55% full
Steam flow	3.5 lb_m/min
Bottoms flow	0.6 gpm

Pressure drop over the control valve on the bottoms product is constant at 30 psi. This control valve has linear trim and a C_v of 0.5. The formula for steam flow through a control valve (when the upstream pressure P_s in psia is greater than twice the downstream pressure) is

$$W = \frac{3C_v}{2} P_s X$$

where W = steam flow rate (lb_m/hr)
C_v = 4
X = valve fraction open (linear trim)

(*a*) Calculate the control signals from the base level transmitter, temperature transmitter, steam flow transmitter, bottoms flow transmitter, temperature controller, steam flow controller, and base level controller.
(*b*) What is the instantaneous effect of a +5°C step change in tray 4 temperature on the control signals and flow rates?

3.13. A reactor is cooled by a circulating jacket water system. The system employs a double cascade reactor temperature control to jacket temperature control to makeup cooling water flow control. Instrumentation details are as follows (electronic, 4–20 mA):

Reactor temperature transmitter range: 50–250°F
Circulating jacket water temperature transmitter range: 50–150°F
Makeup cooling water flow transmitter range: 0–250 gpm
(orifice plate + differential pressure transmitter)
Control valve: linear trim, constant 35-psi pressure drop

Normal operating conditions are:

Reactor temperature = 140°F
Circulating water temperature = 106°
Makeup water flow rate = 63 gpm
Control valve 25% open

(*a*) Specify the action and size of the makeup cooling water control valve.
(*b*) Calculate the milliampere control signals from all transmitters and controllers at normal operating conditions.
(*c*) Specify whether each controller is reverse or direct-acting.
(*d*) Calculate the instantaneous values of all control signals if reactor temperature increases suddenly by 10°F.

Proportional band settings of the reactor temperature controller, circulating jacket water temperature controller, and cooling water flow controller are 20, 67, and 200, respectively.

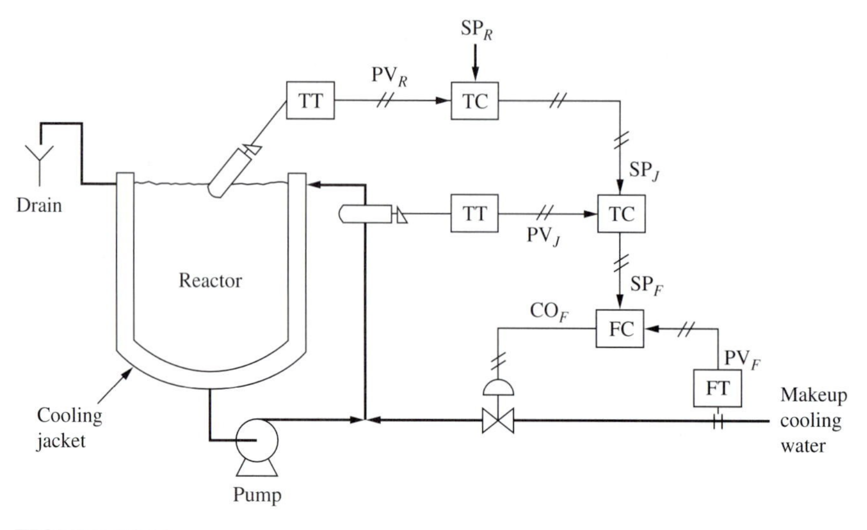

FIGURE P3.13

3.14. Three vertical cylindrical tanks (10 feet high, 10 feet in diameter) are used in a process. Two tanks are process tanks and are level-controlled by manipulating outflows using proportional-only level controllers (PB = 100). Level transmitter spans are 10 feet. Control valves are linear, 50 percent open at the normal liquid rate of 1000 gpm, and air-to-open, with constant pressure drop. These two process tanks are 50 percent full at the normal liquid rate of 1000 gpm.

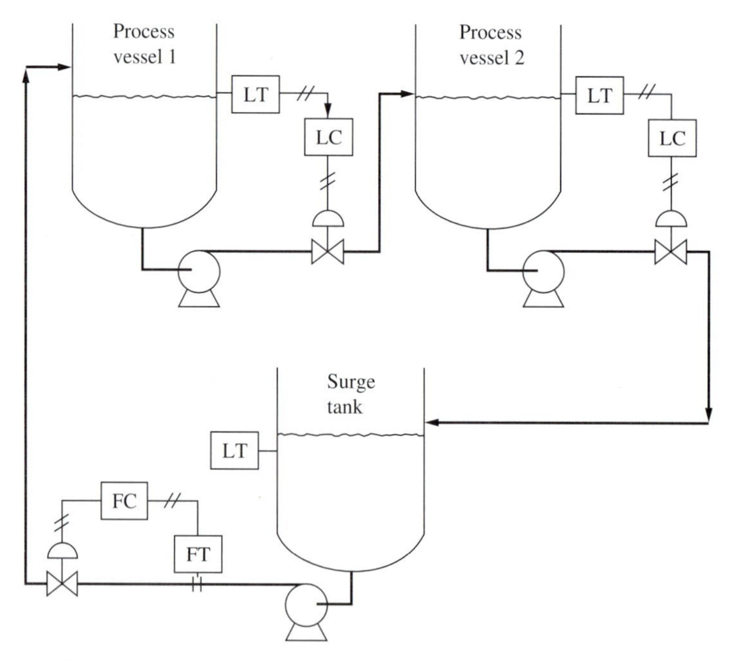

FIGURE P3.14

The third tank is a surge tank whose level is uncontrolled. Liquid is pumped from this tank to the first process vessel, on to the second tank in series, and then back to the surge tank. If the surge tank is half full when 1000 gpm of liquid are circulated, how full will the surge tank be, at the new steady state, when the circulating rate around the system is cut to 500 gpm?

3.15. Liquid (sp gr = 1) is pumped from a tank at atmospheric pressure through a heat exchanger and a control valve into a process vessel held at 100 psig pressure. The system is designed for a maximum flow rate of 400 gpm. At this maximum flow rate the pressure drop across the heat exchanger is 50 psi.

A centrifugal pump is used with a performance curve that can be approximated by the relationship

$$\Delta P_p = 198.33 - 1.458 \times 10^{-4} F^2$$

where ΔP_p = pump head in psi
F = flow rate in gpm

The control valve has linear trim.

(*a*) Calculate the fraction that the control valve is open when the throughput is reduced to 200 gpm by pinching down on the control valve.

(*b*) An orifice-plate differential-pressure transmitter is used for flow measurement. If the maximum full-scale flow reading is 400 gpm, what will the output signal from the electronic flow transmitter be when the flow rate is reduced to 150 gpm?

3.16. Design liquid level control systems for the base of a distillation column and for the vaporizer shown. Steam flow to the vaporizer is held constant and cannot be used to control level. Liquid feed to the vaporizer can come from the column and/or from the surge tank. Liquid from the column can go to the vaporizer and/or to the surge tank.

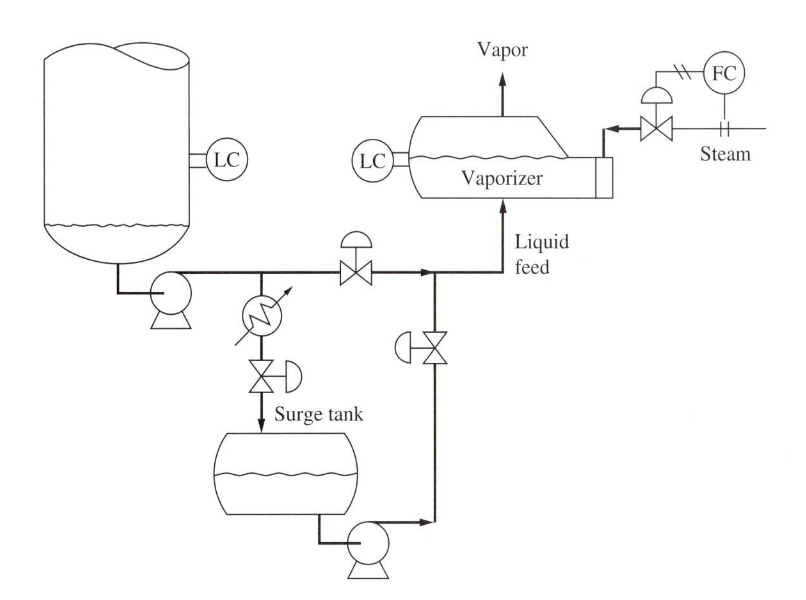

FIGURE P3.16

Since the liquid must be cooled if it is sent to the surge tank and then reheated in the vaporizer, there is an energy cost penalty associated with sending more material to the surge tank than is absolutely necessary. Your level control system should therefore hold both levels and also minimize the amount of material sent to the surge tank. (*Hint:* One way to accomplish this is to make sure that the valves in the lines to and from the surge tank cannot be open simultaneously.)

3.17. A chemical reactor is cooled by a circulating oil system as shown. Oil is circulated through a water-cooled heat exchanger and through control valve V_1. A portion of the oil stream can be bypassed around the heat exchanger through control valve V_2. The system is to be designed so that at design conditions:

- The oil flow rate through the heat exchanger is 50 gpm (sp gr = 1) with a 10-psi pressure drop across the heat exchanger and with the V_1 control valve 25 percent open.
- The oil flow rate through the bypass is 100 gpm with the V_2 control valve 50 percent open.

Both control valves have linear trim. The circulating pump has a flat pump curve. A maximum oil flow rate through the heat exchanger of 100 gpm is required.

(*a*) Specify the action of the two control valves and the two temperature controllers.

(*b*) Calculate the size (C_v) of the two control valves and the design pressure drops over the two valves.

(*c*) How much oil will circulate through the bypass valve if it is wide open and the valve in the heat exchanger loop is shut?

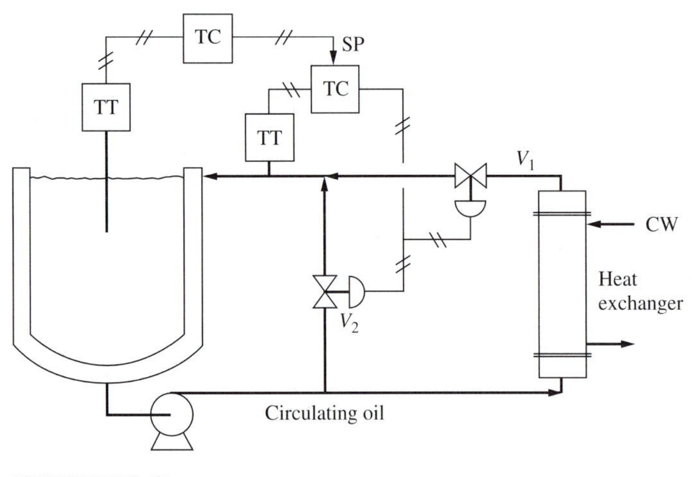

FIGURE P3.17

3.18. The formula for the flow of saturated steam through a control valve is

$$W = 2.1 C_v f_{(x)} \sqrt{(P_1 + P_2)(P_1 - P_2)}$$

where W = lb_m/hr steam
P_1 = upstream pressure, psia
P_2 = downstream pressure, psia

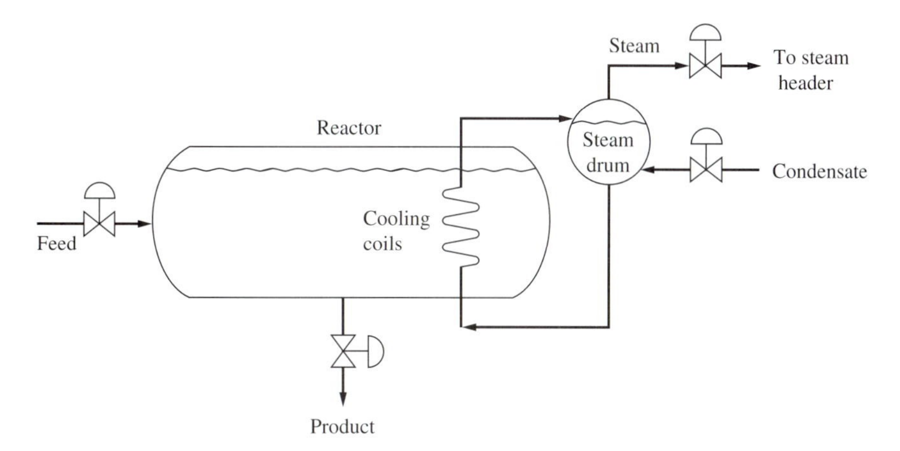

FIGURE P3.18

The temperature of the steam-cooled reactor shown is 285°F. The heat that must be transferred from the reactor into the steam generation system is 25×10^6 Btu/hr. The overall heat transfer coefficient for the cooling coils is 300 Btu/hr ft^2 °F. The steam discharges into a 25-psia steam header. The enthalpy difference between saturated steam and liquid condensate is 1000 Btu/lb$_m$. The vapor pressure of water can be approximated over this range of pressure by a straight line.

$$T(°\text{F}) = 195 + 1.8P \text{ (psia)}$$

Design two systems, one where the steam drum pressure is 40 psia at design and another where it is 30 psia.

(*a*) Calculate the area of the cooling coils for each case.

(*b*) Calculate the C_v value for the steam valve in each case, assuming that the valve is half open at design conditions: $f_{(x)} = 0.5$.

(*c*) What is the maximum heat removal capacity of the system for each case?

3.19. Cooling water is pumped through the jacket of a reactor. The pump and the control valve must be designed so that:

(*a*) The normal cooling water flow rate is 250 gpm.

(*b*) The maximum emergency rate is 500 gpm.

(*c*) The valve cannot be less than 10 percent open when the flow rate is 100 gpm.

Pressure drop through the jacket is 10 psi at design. The pump curve has a linear slope of −0.1 psi/gpm.

Calculate the C_v value of the control valve, the pump head at design rate, the size of the motor required to drive the pump, the fraction that the valve is open at design, and the pressure drop over the valve at design rate.

3.20. A C_2 splitter column uses vapor recompression. Because of the low temperature required to stay below the critical temperatures of ethylene and ethane, the auxiliary condenser must be cooled by a propane refrigeration system.

(*a*) Specify the action of all control valves.

(*b*) Sketch a control concept diagram that accomplishes the following objectives:

Level in the propane vaporizer is controlled by the liquid propane flow from the refrigeration surge drum.

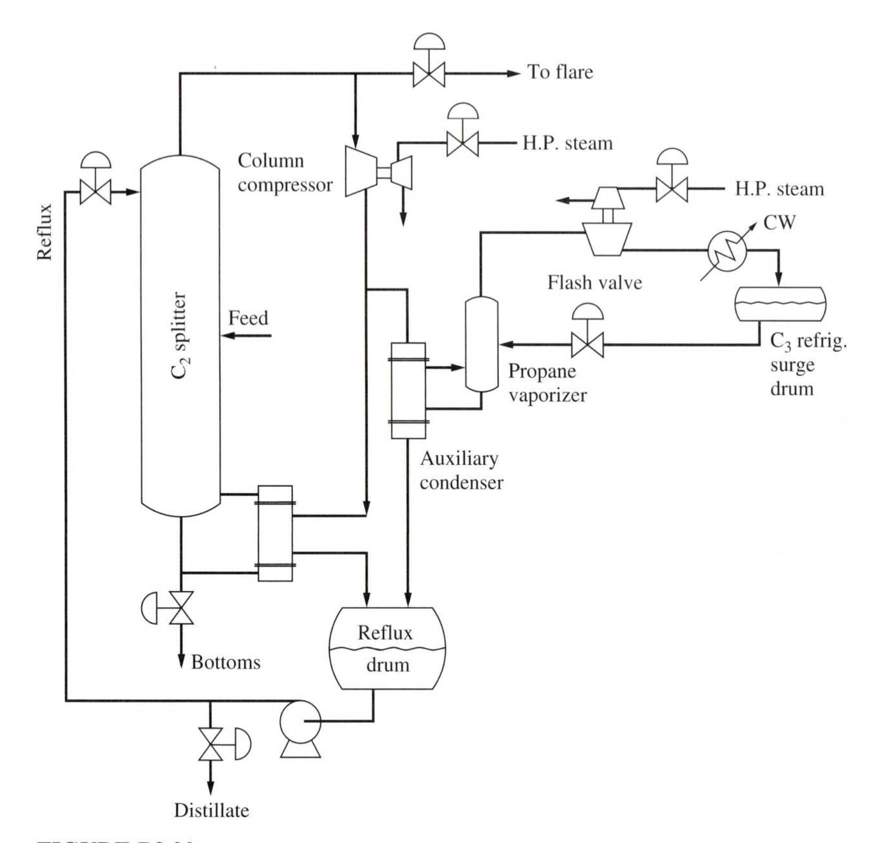

FIGURE P3.20

Column pressure is controlled by adjusting the speed of the column compressor through a steam flow control–speed control–pressure control cascade system.

Reflux is flow controlled. Reflux drum level sets distillate flow. Base level sets bottoms flow.

Column tray 10 temperature is controlled by adjusting the pressure in the propane vaporizer, which is controlled by refrigeration compressor speed.

High column pressure opens the valve to the flare.

(*c*) How effective do you think the column temperature control will be? Suggest an improved control system that still achieves minimum energy consumption in the two compressors.

3.21. Hot oil from the base of a distillation column is used to reboil two other distillation columns that operate at lower temperatures. The design flow rates through reboilers 1 and 2 are 100 gpm and 150 gpm, respectively. At these flow rates, the pressure drops through the reboilers are 20 psi and 30 psi. The hot oil pump has a flat pump curve.

Size the two control valves and the pump so that:

- Maximum flow rates through each reboiler can be at least twice design.
- At minimum turndown rates, where only half the design flow rates are required, the control valves are no less than 10 percent open.

What is the fraction of valve opening for each valve at design rates?

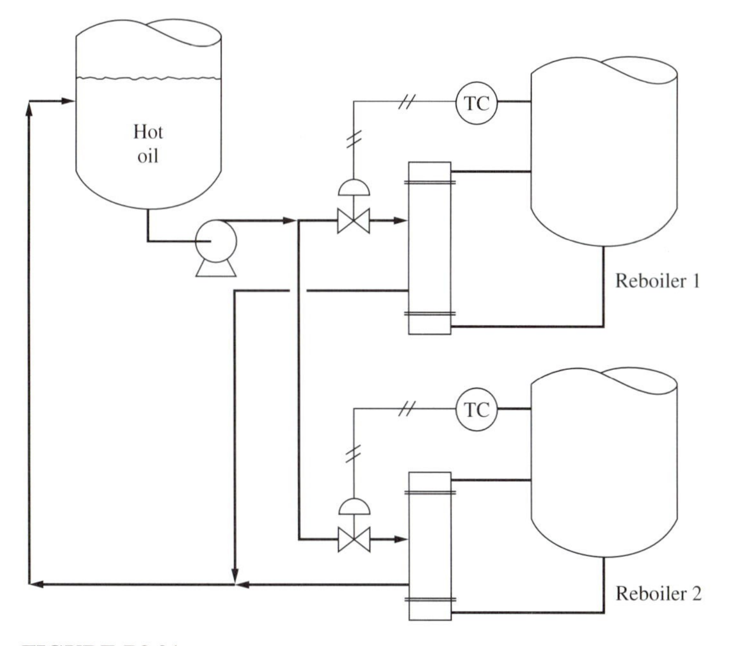

FIGURE P3.21

3.22. A reactor is cooled by circulating liquid through a heat exchanger that produces low-pressure (10 psig) steam. This steam is then split between a compressor and a turbine. The portion that goes through the turbine drives the compressor. The portion that goes through the compressor is used by 50-psig steam users. The turbine can also use 100-psig steam to provide power required beyond that available in the 10-psig steam. (See the figure on the next page.)

Sketch a control concept diagram that includes all valve actions and the following control strategies:

- Reactor temperature is controlled by changing the setpoint of the turbine speed controller.
- Turbine speed is controlled by two split-range valves, one on the 10-psig inlet to the turbine and the other on the 100-psig inlet. Your instrumentation system should be designed so that the valve on the 10-psig steam is wide open before any 100-psig steam is used.
- Liquid circulation from the reactor to the heat exchanger is flow controlled.
- Condensate level in the condensate drum is controlled by manipulating BFW (boiler feed water).
- Condensate makeup to the steam drum is ratioed to the 10-psig steam flow rate from the steam drum. This ratio is then reset by the steam drum level controller.
- Pressure in the 50-psig steam header is controlled by adding 100-psig steam.
- A high-pressure controller opens the vent valve on the 10-psig header when the pressure in the 10-psig header is too high.
- Compressor surge is prevented by using a low-flow controller that opens the valve in the spillback line from compressor discharge to compressor suction.

3.23. Water is pumped from an atmosphere tank, through a heat exchanger and a control valve, into a pressurized vessel. The operating pressure in the vessel can vary from

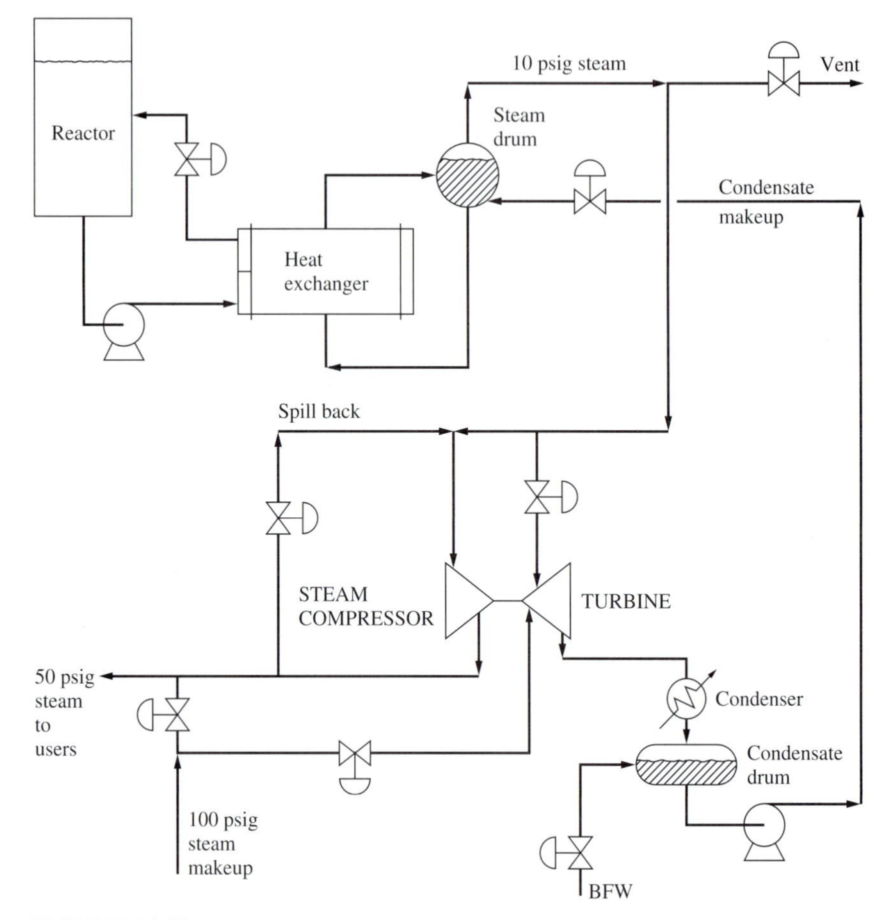

FIGURE P3.22

200 to 300 psig, but is 250 psig at design. Design flow rate is 100 gpm with a 20-psi pressure drop through the heat exchanger. Maximum flow rate is 250 gpm. Minimum flow rate is 25 gpm. A centrifugal pump is used that has a straight-line pump curve with a slope of −0.1 psi/gpm.

Design the control valve and pump so that the maximum and minimum flow rates can be handled with the valve never less than 10 percent open.

3.24. Reactant liquid is pumped into a batch reactor at a variable rate. The reactor pressure also varies during the batch cycle. Specify the control valve size and the centrifugal pump head required. Assume a flat pump curve.

The initial flow rate into the reactor is 20 gpm (sp gr = 1). It is decreased linearly with time down to 5 gpm at 5 hours into the batch cycle. The initial reactor pressure is 50 psig. It increases linearly with time up to 350 psig at 5 hours. The reactant liquid comes from a tank at atmospheric pressure.

3.25. Water is pumped from an atmospheric tank into a vessel at 50 psig through a heat exchanger. There is a bypass around the heat exchanger. The pump has a flat curve. The heat exchanger pressure drop is 30 psi with 200 gpm of flow through it.

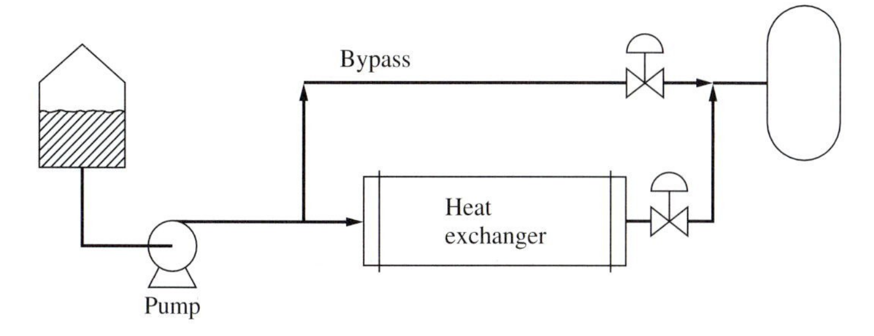

FIGURE P3.25

Size the pump and the two control valves so that:

- 200 gpm can be bypassed.
- Flow through the heat exchanger can be varied from 75 gpm to 300 gpm.

3.26. An engineer from Catastrophic Chemical Company has designed a system in which a positive-displacement pump is used to pump water from an atmospheric tank into a pressurized tank operating at 150 psig. A control valve is installed between the pump discharge and the pressurized tank.

With the pump running at a constant speed and stroke length, 350 gpm of water is pumped when the control valve is wide open and the pump discharge pressure is 200 psig.

If the control valve is pinched back to 50 percent open, what will be the flow rate of water and the pump discharge pressure?

3.27. Hot oil from a tank at 400°F is pumped through a heat exchanger to vaporize a liquid boiling at 200°F. A control valve is used to set the flow rate of oil through the loop. Assume the pump has a flat pump curve. The pressure drop over the control valve is 30 psi and the pressure drop over the heat exchanger is 35 psi under the normal design conditions given below:

Heat transferred in heat exchanger = 17×10^6 Btu/hr
Hot oil inlet temperature = 400°F
Hot oil exit temperature = 350°F
Fraction valve open = 0.8

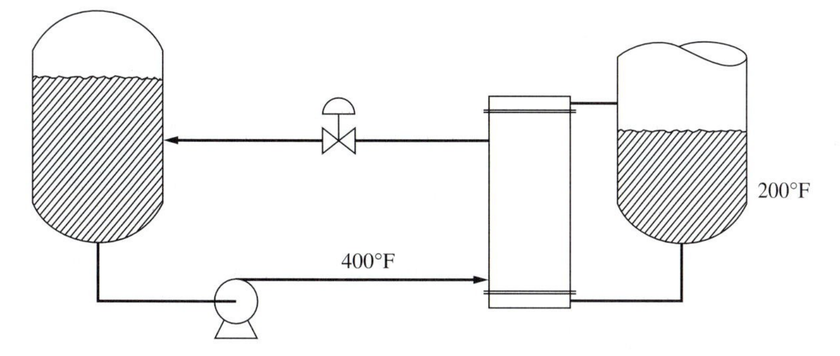

FIGURE P3.27

The hot oil gives off sensible heat only (heat capacity = 0.5 Btu/lb_m °F, density = 4.58 lb_m/gal). The heat transfer area in the exhanger is 652 ft^2. Assume the temperature on the tube side of the heat exchanger stays constant at 200°F and the inlet hot oil temperature stays constant at 400°F. A log mean temperature difference must be used.

Assuming the heat transfer coefficient does not change with the flow rate, what will the valve opening be when the heat transfer rate in the heat exchanger is half the normal design value?

3.28. A control valve–pump system proposed by Connell (*Chemical Engineering,* September 28, 1987, p. 123) consists of a centrifugal pump, several heat exchangers, a furnace, an orifice, and a control valve. Liquid is pumped through this circuit and up into a column that operates at 20 psig. Because the line running up the column is full of liquid, there is a hydraulic pressure differential between the base of the column and the point of entry into the column of 15 psi.

The pump suction pressure is constant at 10 psig. The design flow rate is 500 gpm. At this flow rate the pressure drop over the flow orifice is 2 psi, through the piping is 30 psi, over three heat exchangers is 32 psi, and over the furnace is 60 psi. Assume a flat pump curve and a specific gravity of 1.

Connell recommends that a control valve be used that takes a 76-psi pressure drop at design flow rate. The system should be able to increase flow to 120 percent of design.

(*a*) Calculate the pressure drop over the valve at the maximum flow rate.
(*b*) Calculate the pump discharge pressure and the control valve C_v.
(*c*) Calculate the fraction that the valve is open at design.
(*d*) If turndown is limited to a valve opening of 10 percent, what is the minimum flow rate?

3.29. A circulating-water cooling system is used to cool a chemical reactor. Treated water is pumped through a heat exchanger and then through the cooling coils inside the reactor. Some of the circulating water is bypassed around the heat exchanger. Cooling tower water is used on the shell side of the heat exchanger to cool the circulating water.

Two linear-trim control valves are used. Valve 1 in the bypass line is AO, and valve 2 in the heat exchanger line is AC. Both valves get their inputs from the CO signal from a temperature controller.

Design conditions are: CO is 75 percent of scale, flow through valve 1 is 300 gpm, flow through valve 2 is 100 gpm, pressure drop through the coil is 20 psi, and pressure drop through the heat exchanger is 5 psi. The centrifugal pump has a flat pump curve.

If the maximum flow rate through the heat exchanger is 300 gpm when the CO signal is 0 percent of scale, calculate the C_v value of both control valves and the required pump head. What is the flow rate through valve 1 when the CO signal is 100 percent of scale?

3.30. A gravity-flow condenser uses the hydraulic head of the liquid in the line from the condenser to overcome the pressure drop over the control valve and the difference between the pressure at the top of the distillation column P_1 and the pressure at the bottom of the condenser P_2. The pressure difference is due to the flow of vapor through the vapor line and condenser. When the flow rate of vapor from the top of the column is 141.6 lb_m/min, the pressure drop $P_1 - P_2$ is 2 psi. The pressure drop due to the liquid flowing through the liquid return line is negligible. Liquid density is 62.4 lb_m/ft^3.

(*a*) If we want the height of liquid in the return line to be 5 ft at design conditions (141.6 lb_m/min of liquid with the control valve half open), what is the required control valve C_v?

(*b*) If the vapor and liquid flow rates both increase to 208.2 lb_m/min when the control valve is wide open, what is the height of liquid in the liquid return line?

3.31. A hot vapor bypass pressure control system is used on a distillation column. Some of the vapor from the top of the column passes through a control valve and is added to the vapor space in the top of the reflux drum. The column operates at 7 atm and the overhead vapor is essentially pure isobutane. The vapor pressure of isobutane is given by the following equation:

$$\ln P = 9.91552 - 2586.8/(T + 273)$$

where P is in atmospheres and T is in degrees Celsius.

Since the flow through the valve is isenthalpic, the temperature in the hot vapor space in the reflux drum is the same as the temperature in the top of the column (assuming isobutane is an ideal gas at this pressure).

Most of the vapor from the top of the column is condensed and subcooled in a condenser. This subcooled liquid then flows into the base of the reflux drum. There is heat transfer between the hot vapor and the cooler vapor–liquid interface (at temperature T) and between the vapor–liquid interface and the cooler-subcooled reflux in the tank (at temperature T_R). The vapor film coefficient is 10 Btu/hr °C ft^2, and the liquid film coefficient is 30 Btu/hr °C ft^2. The heat transfer area on the surface of the liquid is 72 ft^2. The heat of vaporization of isobutane is 120 Btu/lb_m.

The control valve is sized to be 10 percent open during summer operation, when the temperature of the subcooled liquid in the tank is 45°C. Use the control valve sizing formula

$$F = C_v f_{(x)} \sqrt{P_{col} - P}$$

where F = lb/hr of vapor flow
P_{col} = pressure in the column = 7 atm
P = pressure in the reflux drum = saturation pressure of isobutane at the temperature T of the interface between the liquid and vapor phases.

(*a*) Calculate the C_v value of the control valve.
(*b*) Calculate the fraction that the valve will be open during winter operation, when the temperature of the subcooled reflux is 15°C. Column pressure is constant at 7 atm.

3.32. The steam supply to a sterilizer comes from an 84.7-psia header and is saturated vapor with an enthalpy of 1184.1 Btu/lb. It flows through a control valve into the sterilizer, where a temperature of 250°F is desired (saturation pressure of 29.82 psia and saturated liquid enthalpy of 218.48 Btu/lb). Condensate leaves through a steam trap. The heat required to maintain the sterilizer at its desired temperature is 200,000 Btu/hr. The control valve should be 25 percent open at these steady-state conditions. A pressure controller is used to control pressure in the sterilizer. The pressure transmitter has a range of 0–75 psig. All instrumentation is electronic with a signal range of 4 to 20 mA.

The equation for steam flow through a control valve when the upstream pressure is more than twice the downstream pressure is

$$F_S = \frac{3C_V}{2} P_S f_{(x)}$$

where F_S = steam flow rate, lb/hr
P_S = upstream pressure, psia

(*a*) Should the steam control valve be AO or AC?
(*b*) Calculate the C_v value of the control valve.
(*c*) Calculate the PV signal from the pressure transmitter and the CO signal from the pressure controller under steady-state conditions.
(*d*) If the proportional band of the controller is 75 and the pressure in the sterilizer suddenly drops by 5 psi, calculate the instantaneous value of the controller output and the new value of the steam flow rate.

3.33. Design a centrifugal pump and control valve system so that a maximum flow rate of 75 gpm and a minimum flow rate of 25 gpm are achievable with the control valve at 100 percent and 10 percent open, respectively. Liquid is pumped from a tank whose pressure can vary from 50 to 75 psia. The material is pumped through a heat exchanger (which takes 30-psi pressure drop at 50 gpm) and a control valve into a tank whose pressure can vary from 250 to 300 psia. Assume a flat pump curve.

CHAPTER 4

Advanced Control Systems

In the previous chapter we discussed the elements of a conventional single-input, single-output (SISO) feedback control loop. This configuration forms the backbone of almost all process control structures.

However, over the years a number of slightly more complex structures have been developed that can, in some cases, significantly improve the performance of a control system. These structures include ratio control, cascade control, and override control.

4.1 RATIO CONTROL

As the name implies, ratio control involves keeping constant the ratio of two or more flow rates. The flow rate of the "wild" or uncontrolled stream is measured, and the flow rate of the manipulated stream is changed to keep the two streams at a constant ratio with each other. Common examples include holding a constant reflux ratio on a distillation column, keeping stoichiometric amounts of two reactants being fed into a reactor, and purging off a fixed percentage of the feed stream to a unit. Ratio control is often part of a *feedforward* control structure, which we will discuss in Section 4.7.

Ratio control is achieved by two alternative schemes, shown in Fig. 4.1. In the scheme shown in Fig. 4.1*a*, the two flow rates are measured and their ratio is computed (by the divider). This computed ratio signal is fed into a conventional PI controller as the process variable (PV) signal. The setpoint of the ratio controller is the desired ratio. The output of the controller goes to the valve on the manipulated variable stream, which changes its flow rate in the correct direction to hold the ratio of the two flows constant. This computed ratio signal can also be used to trigger an alarm or an interlock.

In the scheme shown in Fig. 4.1*b*, the wild flow is measured and this flow signal is multiplied by a constant, which is the desired ratio. The output of the multiplier is the setpoint of a remote-set flow controller on the manipulated variable.

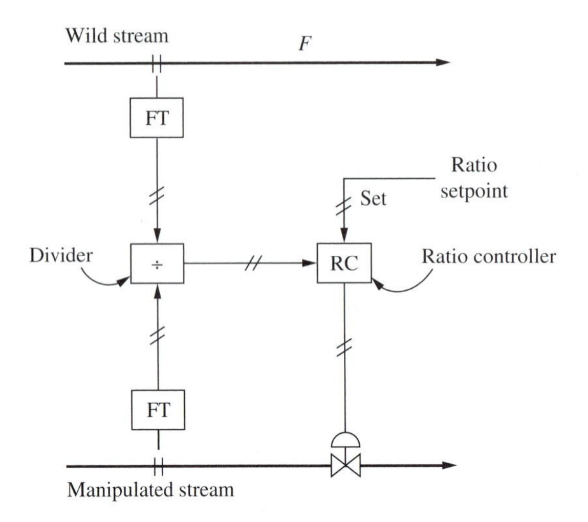

(*a*)

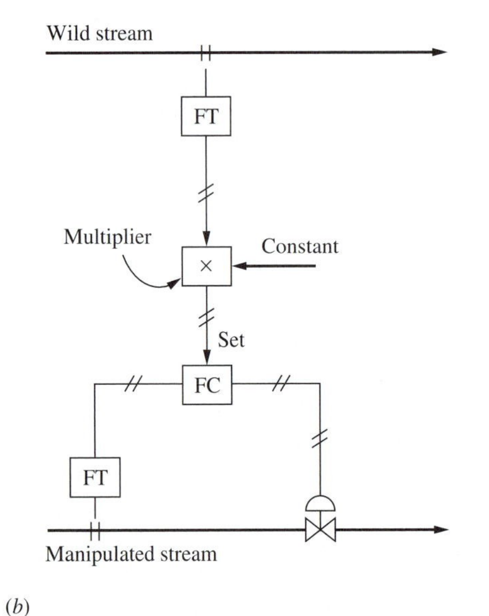

(*b*)

FIGURE 4.1
Ratio control. (*a*) Ratio compute. (*b*) Flow set.

If orifice plates are used as flow sensors, the signals from the differential-pressure transmitters are really the squares of the flow rates. Some instrument engineers prefer to put in square-root extractors and convert everything to linear flow signals.

4.2 CASCADE CONTROL

One of the most useful concepts in advanced control is cascade control. A cascade control structure has two feedback controllers, with the output of the primary (or master) controller changing the setpoint of the secondary (or slave) controller. The output of the secondary goes to the valve, as shown in Fig. 4.2.

There are two purposes for cascade control: (1) to eliminate the effects of some disturbances, and (2) to improve the dynamic performance of the control loop.

To illustrate the disturbance rejection effect, consider the distillation column reboiler shown in Fig. 4.2*a*. Suppose the steam supply pressure increases. The pressure drop over the control valve will be larger, so the steam flow rate will increase. With

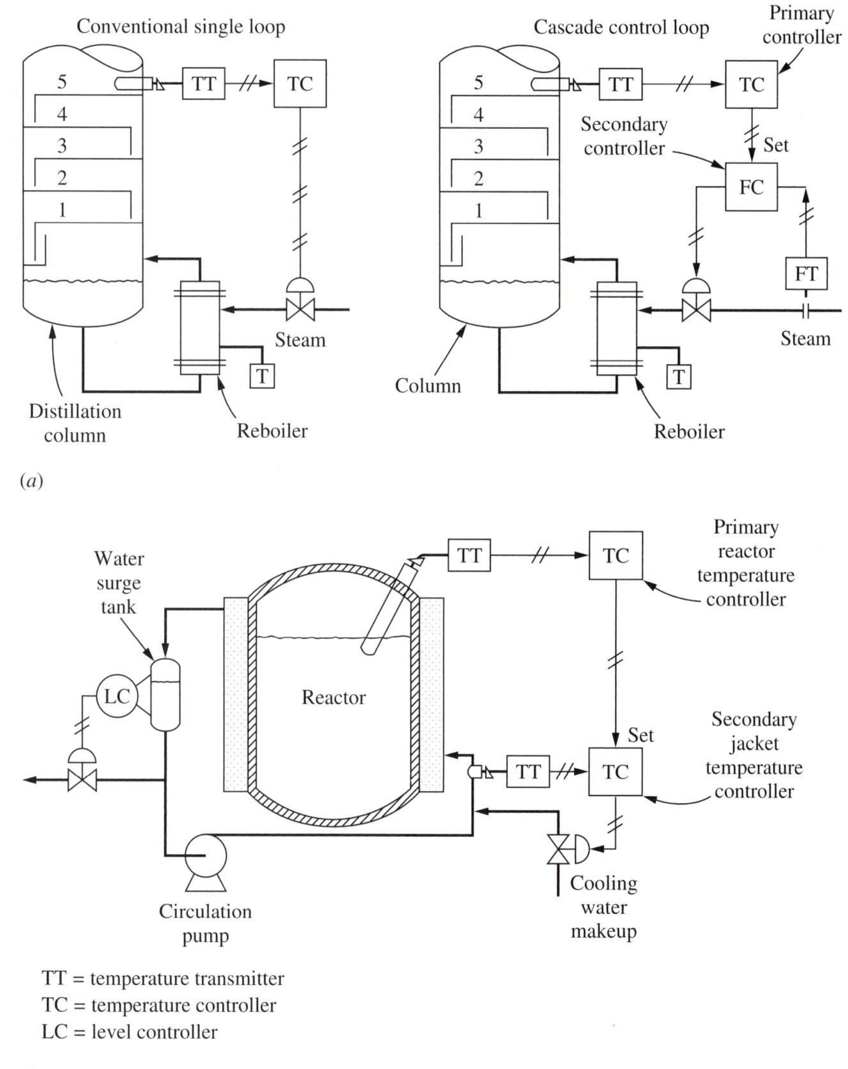

FIGURE 4.2
Conventional versus cascade control. (*a*) Distillation column–reboiler temperature control. (*b*) CSTR temperature control.

the single-loop temperature controller, no correction will be made until the higher steam flow rate increases the vapor boilup and the higher vapor rate begins to raise the temperature on tray 5. Thus, the whole system is disturbed by a supply steam pressure change.

With the cascade control system, the steam flow controller will immediately see the increase in steam flow and will pinch back on the steam valve to return the steam flow rate to its setpoint. Thus, the reboiler and the column are only slightly affected by the steam supply pressure disturbance.

Figure 4.2*b* shows another common system where cascade control is used. The reactor temperature controller is the primary controller; the jacket temperature controller is the secondary controller. The reactor temperature control is isolated by the cascade system from disturbances in cooling-water inlet temperature and supply pressure.

This system is also a good illustration of the improvement in dynamic performance that cascade control can provide in some systems. As we show quantitatively in Chapter 9, the closedloop time constant of the reactor temperature will be smaller when the cascade system is used than when the reactor temperature sets the cooling water makeup valve directly. Therefore, performance is improved by using cascade control.

We also talk in Chapter 9 about the two types of cascade control: series cascade and parallel cascade. The two examples just discussed are both series cascade systems because the manipulated variable affects the secondary controlled variable, and then the secondary variable affects the primary variable. In a parallel cascade system the manipulated variable affects *both* the primary and the secondary controlled variables directly. Thus, the two processes are basically different and result in different dynamic characteristics. We quantify these ideas later.

4.3 COMPUTED VARIABLE CONTROL

One of the most logical and earliest extensions of conventional control was the idea of controlling the variable that was of real interest by computing its value from other measurements.

For example, suppose we want to control the mass flow rate of a gas. Controlling the pressure drop over the orifice plate gives only an approximate mass flow rate because gas density varies with temperature and pressure in the line. By measuring temperature, pressure, and orifice plate pressure drop and feeding these signals into a mass-flow-rate computer, the mass flow rate can be controlled as sketched in Fig. 4.3*a*.

Another example is shown in Fig. 4.3*b*, where a hot oil stream is used to reboil a distillation column. Controlling the flow rate of the hot oil does not guarantee a fixed heat input because the inlet oil temperature can vary and the ΔT requirements in the reboiler can change. The heat input Q can be computed from the flow rate and the inlet and outlet temperatures, and this Q can then be controlled.

As a final example, consider the problem of controlling the temperature in a distillation column where significant pressure changes occur. We really want to measure

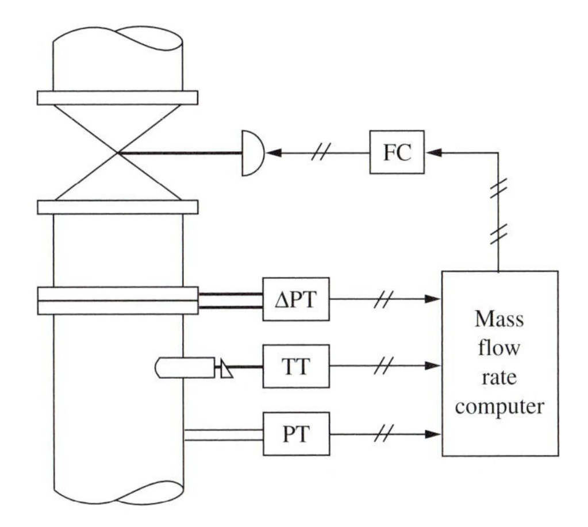

(*a*)

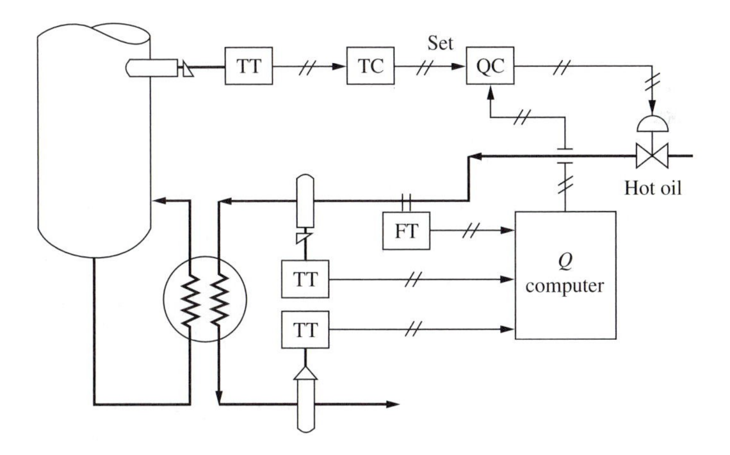

(*b*)

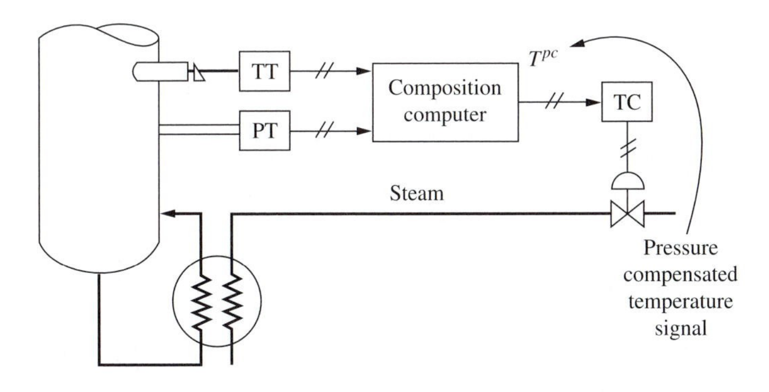

(*c*)

FIGURE 4.3
Computed variable control. (*a*) Mass flow rate. (*b*) Heat input. (*c*) Composition (pressure-compensated temperature).

and control composition, but temperature is used to infer composition because temperature measurements are much more reliable and inexpensive than composition measurements.

In a binary system, composition depends only on pressure and temperature:

$$x = f_{(T,P)} \tag{4.1}$$

Thus, changes in composition depend on changes in temperature and pressure.

$$\Delta x = \left(\frac{\partial x}{\partial P}\right)_T \Delta P + \left(\frac{\partial x}{\partial T}\right)_P \Delta T \tag{4.2}$$

where x = mole fraction of the more volatile component in the liquid.

The partial derivatives are usually assumed to be constants that are evaluated at the steady-state operating level from the vapor-liquid equilibrium data. Thus, pressure and temperature on a tray can be measured, as shown in Fig. 4.3*c*, and a composition signal or pressure-compensated temperature signal generated and controlled.

$$\Delta T^{PC} = K_1 \Delta P - K_2 \Delta T \tag{4.3}$$

where T^{PC} = pressure-compensated temperature signal
K_1 and K_2 = constants

Forty years ago these computed variables were calculated using pneumatic devices. Today they are much more easily done in the digital control computer. Much more complex types of computed variables can now be calculated. Several variables of a process can be measured, and all the other variables can be calculated from a rigorous model of the process. For example, the nearness to flooding in distillation columns can be calculated from heat input, feed flow rate, and temperature and pressure data. Another application is the calculation of product purities in a distillation column from measurements of several tray temperatures and flow rates by the use of mass and energy balances, physical property data, and vapor-liquid equilibrium information. Successful applications have been reported in the control of polymerization reactors.

The computer makes these "rigorous estimators" feasible. It opens up a number of new possibilities in the control field. The limitation in applying these more powerful methods is the scarcity of engineers who understand both control and chemical engineering processes well enough to apply them effectively. Hopefully, this book will help to remedy this shortage.

4.4 OVERRIDE CONTROL

There are situations where the control loop should monitor more than just one controlled variable. This is particularly true in highly automated plants, where the operator cannot be expected to make all the decisions that are required under abnormal conditions. This includes the startup and shutdown of the process.

Override control (or "selective control," as it is sometimes called) is a form of multivariable control in which a *manipulated* variable can be set at any time by one of a number of different *controlled* variables.

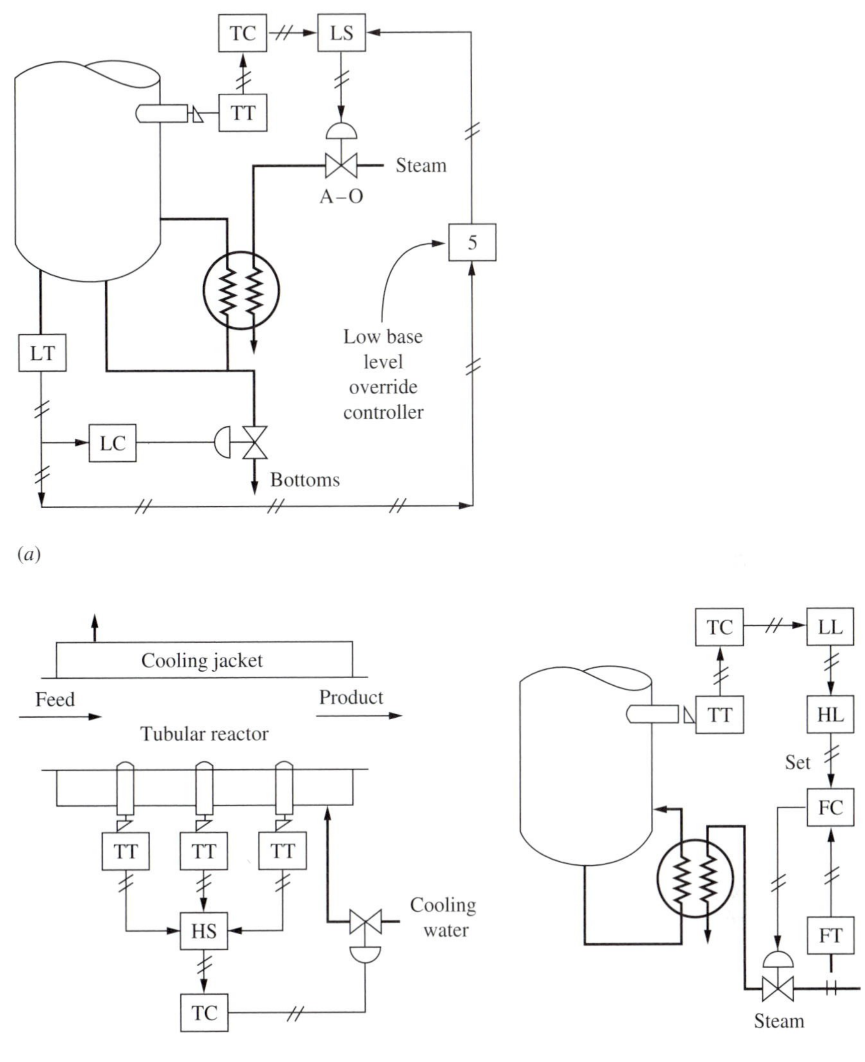

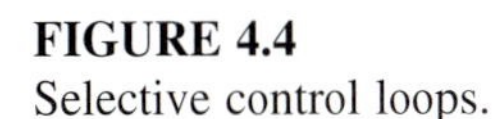

FIGURE 4.4
Selective control loops.

The idea is best explained with an example. Suppose the base level in a distillation column is normally held by bottoms product withdrawal as shown in Fig. 4.4*a*. A temperature in the stripping section is held by steam to the reboiler. Situations can arise where the base level continues to drop even with the bottoms flow at zero (vapor boilup is greater than the liquid rate from tray 1). If no corrective action is taken, the reboiler may boil dry (which could foul the tubes) and the bottoms pump could lose suction.

An operator who saw this problem developing would switch the temperature loop into "manual" and cut back on the steam flow. The control system in Fig. 4.4*a* will perform this "override" control automatically. The low selector (LS) sends to the steam valve the lower of the two signals. If the steam valve is air-to-open, the valve will be pinched back by either high temperature (through the reverse-acting temperature controller) or low base level (through the low-base-level override controller).

In level control applications, this override controller can be a simple fixed-gain relay that acts like a proportional controller. The gain of the controller shown in Fig. 4.4*a* is 5. It would be "zeroed" so that as the level transmitter dropped from 20 to 0 percent of full scale, the output of the relay would drop from 100 to 0 percent of scale. This means that under normal conditions when the level is above 20 percent, the output of the relay will be at 100 percent. This will be higher than the signal from the temperature controller, so the low selector will pass the temperature controller output signal to the valve. However, when the base level drops below 20 percent and continues to fall toward 0 percent, the signal from the relay will drop and at some point will become lower than the temperature controller output. At this point the temperature controller is overridden by the low-base-level override controller.

Other variables might also take over control of the steam valve. If the pressure in the column gets too high, we might want to pinch the steam valve. If the temperature in the base gets too high, we might want to do the same. So there could be a number of inputs to the low selector from various override controllers. The lowest signal will be the one that goes to the valve.

In temperature and pressure override applications the override controller usually must be a PI controller, not a P controller as used in the level override controller. This is because the typical change in the transmitter signal over which we want to take override action in these applications (high pressure, high temperature, etc.) is only a small part of the total transmitter span. A very high-gain P controller would have to be used to achieve the override control action, and the override control loop would probably be closedloop unstable at this high gain. Therefore, a PI controller must be used with a lower gain and a reasonably fast reset time to achieve the tightest control possible.

Figure 4.4*b* shows another type of *selective control* system. The signals from the three temperature transmitters located at various positions along a tubular reactor are fed into a high selector. The highest temperature is sent to the temperature controller, whose output manipulates cooling water. Thus, this system controls the peak temperature in the reactor, wherever it is located.

Another very common use of this type of system is in controlling two feed streams to a reactor where an excess of one of the reactants could move the composition in the reactor into a region where an explosion could occur. Therefore, it is vital that the flow rate of this reactant be less than some critical amount, relative to the other flow. Multiple, redundant flow measurements would be used, and the highest flow signal would be used for control. In addition, if the differences between the flow measurements exceeded some reasonable quantity, the whole system would be "interlocked down" until the cause of the discrepancy was found.

Thus, override and selective controls are widely used to handle safety problems and constraint problems. High and low limits on controller outputs, as illustrated in Fig. 4.4*c*, are also widely used to limit the amount of change permitted.

When a controller with integral action (PI or PID) sees an error signal for a long period of time, it integrates the error until it reaches a maximum (usually 100 percent of scale) or a minimum (usually 0 percent). This is called *reset windup.* A sustained error signal can occur for a number of reasons, but the use of override control is one major cause. If the main controller has integral action, it will wind up when the override controller has control of the valve. And if the override controller is a PI controller, it will wind up when the normal controller is setting the valve. So this reset windup problem must be recognized and solved.

This is accomplished in a number of different ways, depending on the controller hardware and software used. In pneumatic controllers, reset windup can be prevented by using *external reset feedback* (feeding back the signal of the control valve to the reset chamber of the controller instead of the controller output). This lets the controller integrate the error when its output is going to the valve, but breaks the integration loop when the override controller is setting the valve. Similar strategies are used in analog electronics. In computer control systems, the integration action is turned off when the controller does not have control of the valve.

4.5 NONLINEAR AND ADAPTIVE CONTROL

Since many of our chemical engineering processes are nonlinear, it would seem advantageous to use nonlinear controllers in some systems. The idea is to modify the controller action and/or settings in some way to compensate for the nonlinearity of the process.

For example, we could use a variable-gain controller in which the gain K_c varies with the magnitude of the error:

$$K_c = K_{c0}(1 + b|E|) \tag{4.4}$$

where K_{c0} = controller gain with zero error
$|E|$ = absolute magnitude of error
b = adjustable constant

This would permit us to use a low value of gain so that the system is stable near the setpoint over a broad range of operating levels with changing process gains. When the process is disturbed away from the setpoint, the gain will become larger. The system may even be closedloop unstable at some point. But the instability is in the direction of driving the loop rapidly back toward the stable setpoint region.

Another advantage of this kind of nonlinear controller is that the low gain at the setpoint reduces the effects of noise.

The parameter b can be different for positive and negative errors if the nonlinearity of the process is different for increasing or decreasing changes. For example, in distillation columns a change in a manipulated variable that moves product compositions in the direction of *higher* purity has less of an effect than a change in the direction toward *lower* purity. Thus, higher controller gains can be used as product purities rise, and lower gains can be used when purities fall.

Another type of nonlinear control can be achieved by using nonlinear transformations of the controlled variables. For example, in chemical reactor control the rate of reaction can be controlled instead of the temperature. The two are, of course, related through the exponential temperature relationship. In high-purity distillation columns, a logarithmic transformation of the type shown below can sometimes be useful to "linearize" the composition signal and produce improved control with a conventional linear controller still used.

$$(x_D)_{TR} = \log\left[\frac{1 - x_D}{1 - x_D^{set}}\right] \tag{4.5}$$

$$(x_B)_{TR} = \log\left[\frac{x_B}{x_B^{set}}\right] \tag{4.6}$$

where the subscript TR indicates transformed variables.

Adaptive control has been an active area of research for many years. The full-blown ideal adaptive controller continuously identifies (on-line) the parameters of the process as they change and retunes the controller appropriately. Unfortunately, this on-line adaptation is fairly complex and has some pitfalls that can lead to poor performance (instability or very sluggish control). Also, it takes considerable time for the on-line identification to be achieved, which means that the plant may have already changed to a different condition. These are some of the reasons on-line adaptive controllers are not widely used in the chemical industry.

However, the main reason for the lack of wide application of on-line adaptive control is the lack of economic incentive. On-line identification is rarely required because it is usually possible to predict with off-line tests how the controller must be retuned as conditions vary. The dynamics of the process are determined at different operating conditions, and appropriate controller settings are determined for all the different conditions. Then, when the process moves from one operating region to another, the controller settings are automatically changed. This is called "openloop-adaptive control" or "gain scheduling."

These openloop-adaptive controllers are really just another form of nonlinear control. They have been quite successfully used in many industrial processes, particularly in batch processes where operating conditions can vary widely and in processes where different grades of products are made in the same equipment.

The one notable case where on-line adaptive control *has* been widely used is in pH control. The wide variations in titration curves as changes in buffering occur make pH control ideal for on-line adaptive control methods. Several instrument vendors have developed commercial on-line adaptive controllers. Seborg, Edgar, and Shah (*AIChE Journal* 32:881, 1986) give a survey of adaptive control strategies in process control.

4.6 VALVE POSITION (OPTIMIZING) CONTROL

Shinskey [*Chem. Eng. Prog.* *72*(5):73, 1976; *Chem. Eng. Prog.* *74*(5):43, 1978] proposed the use of a type of control configuration that he called *valve position control.*

This strategy provides a very simple and effective method for achieving "optimizing control." The basic idea is illustrated by several important applications.

Since relative volatilities increase in most distillation systems as pressure decreases, the optimal operation would be to minimize the pressure at all times. One way to do this is to completely open the control valve on the cooling water. The pressure would then float up and down as cooling-water temperature changed.

However, if there is a sudden drop in cooling-water temperature (as can occur during a thundershower or "blue norther"), the pressure in the column can fall rapidly. This can cause flashing of the liquid on the trays, will upset the composition and level controls on the column, and could even cause the column to flood.

To prevent this rapid drop, Shinskey developed a "floating-pressure" control system, sketched in Fig. 4.5. A conventional PI pressure controller is used. The output of the pressure controller goes to the cooling-water valve, which is AC so that it will fail open. The pressure controller output is also sent to another controller, the "valve position controller" (VPC). This controller looks at the signal to the valve, compares it with the VPC setpoint signal, and sends out a signal that is the setpoint of the pressure controller. Since the valve is AC, the setpoint of the VPC is about 5 percent of scale to keep the cooling-water valve almost wide open.

The VPC scheme is a different type of cascade control system. The primary control is the position of the valve. The secondary control is the column pressure. The pressure controller is PI and is tuned fairly tightly so that it can prevent the sudden drops in pressure. Its setpoint is slowly changed by the VPC to drive the cooling-water valve nearly wide open. A slow-acting, integral-only controller should be used in the VPC.

Figure 4.6 shows another example of the application of VPC to optimize a process. We want to control the temperature of a reactor. The reactor is cooled by both cooling water flowing through a jacket surrounding the reactor and by condensing vapor that boils off the reactor in a heat exchanger cooled by a refrigerant. This form of cooling is called "autorefrigeration."

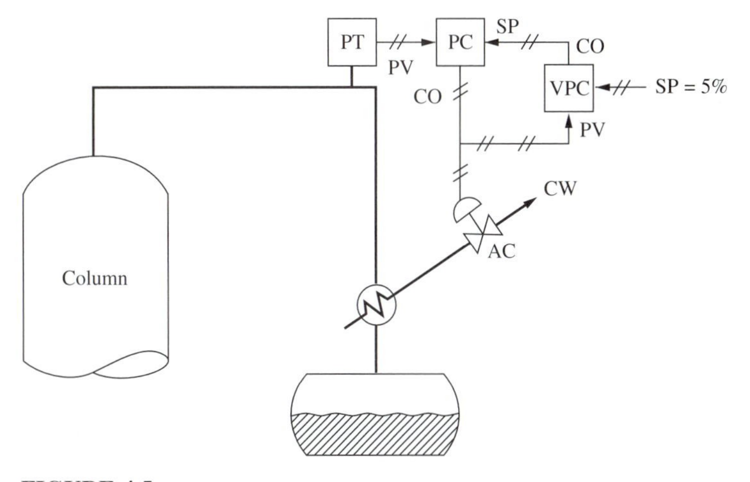

FIGURE 4.5
Floating pressure control (VPC).

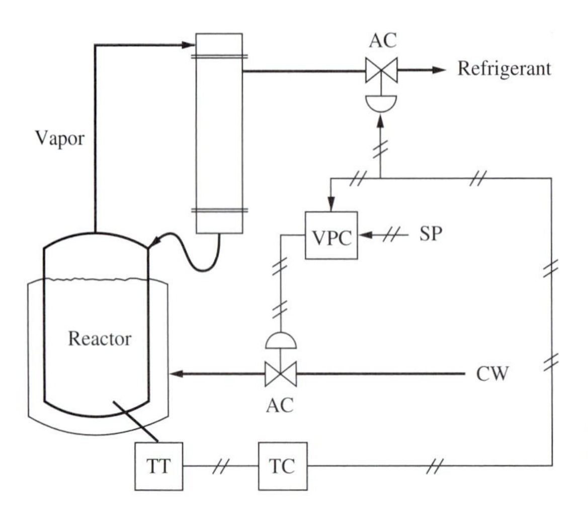

FIGURE 4.6
Use of VPC to minimize energy cost.

From an energy cost perspective, we would like to use cooling water and not refrigerant because water is much cheaper. However, the dynamic response of the temperature to a change in cooling water may be much slower than the response to a change in refrigerant flow. This is because the change in water flow must change the jacket temperature, which then changes the metal wall temperature, which then begins to change the reaction mass temperature. Changes in refrigerant flow quickly raise or lower the pressure in the condenser and change the amount of vaporization in the reactor, which is reflected in reactor temperature almost immediately.

So from a control point of view, we would like to use refrigerant to control temperature. Much tighter control could be achieved compared with the use of cooling water. The VPC approach handles this optimization problem very nicely. We simply control temperature with refrigerant, but send the signal that is going to the refrigerant valve (the temperature controller output) into a valve position controller, which will slowly move the cooling water valve to keep the refrigerant valve nearly closed. Since the refrigerant valve is AC, the setpoint signal to the VPC will be about 5 to 10 percent of full scale.

Note that in the floating-pressure application, there is only one manipulated variable (cooling-water flow) and one primary controlled variable (valve position). In the reactor temperature control application, there are two manipulated variables and two controlled variables (temperature and refrigerant valve position).

4.7 FEEDFORWARD CONTROL CONCEPTS

Up to this point we have used only feedback controllers. An error must be detected in a controlled variable before the feedback controller can take action to change the manipulated variable. So disturbances must upset the system before the feedback controller can do anything.

It seems reasonable that if we could detect a disturbance entering a process, we should begin to correct for it *before* it upsets the process. This is the basic idea of feedforward control. If we can measure the disturbance, we can send this signal through a feedforward control algorithm that makes appropriate changes in the manipulated variable to keep the controlled variable near its desired value.

We do not yet have all the tools to deal quantitatively with feedforward controller design. We will come back to this subject in Chapter 9, when our Russian lessons (Laplace transforms) have been learned.

However, we can describe the basic structure of several feedforward control systems. Figure 4.7 shows a blending system with one stream that acts as a disturbance; both its flow rate and its composition can change. In Fig. 4.7*a* the conventional feedback controller senses the controlled composition of the total blended stream and changes the flow rate of a manipulated flow. In Fig. 4.7*b* the manipulated flow is simply ratioed to the wild flow. This provides feedforward control for flow rate changes. Note that the disturbance must be measured to implement feedforward control.

In Figure 4.7*c* the ratio of the two flows is changed by the output of a composition controller. This system is a combination of feedforward and feedback control. Finally, in Fig. 4.7*d* a feedforward system is shown that measures both the flow rate and the composition of the disturbance stream and changes the flow rate of the manipulated variable appropriately. The feedback controller can also change the ratio. Note that *two* composition measurements are required, one measuring the disturbance and one measuring the controlled stream.

4.8 CONTROL SYSTEM DESIGN CONCEPTS

Having learned a little about hardware and about several strategies used in control, we are now ready to talk about some basic concepts for designing a control system. At this point the discussion will be completely qualitative. In later chapters we will quantify most of the statements and recommendations made in this section. Our purpose here is to provide a broad overview of how to go about finding an effective control structure *and* designing an easily controlled process.

A consideration of dynamics should be factored into the design of a plant at an early stage, preferably as early as the conceptual design stage. It is often easy and inexpensive in the early stages of a project to design a piece of process equipment so that it is easy to control. If the plant is designed with little or no consideration of dynamics, an elaborate control system may be required to make the most of a poor situation.

For example, it is important to have liquid holdups in surge vessels, reflux drums, column bases, etc. large enough to provide effective damping of disturbances (a much-used rule of thumb is 5 to 10 minutes). A sufficient excess of heat transfer area must be available in reboilers, condensers, cooling jackets, etc. to be able to handle the dynamic changes and upsets during operation. The same is true of flow rates of manipulated variables. Measurements and sensors should be located so that they can be used for effective control.

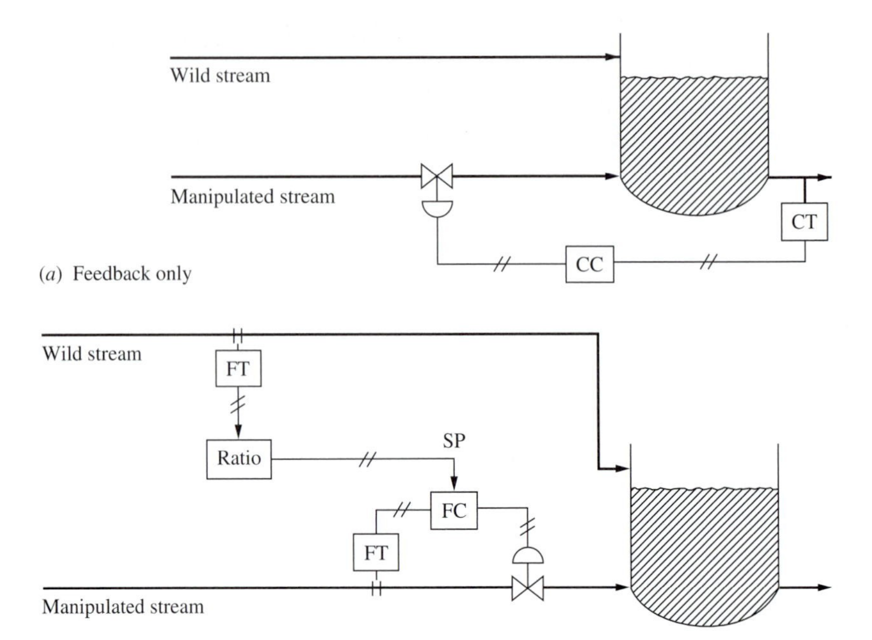

(*a*) Feedback only

(*b*) Feedforward for flow rate

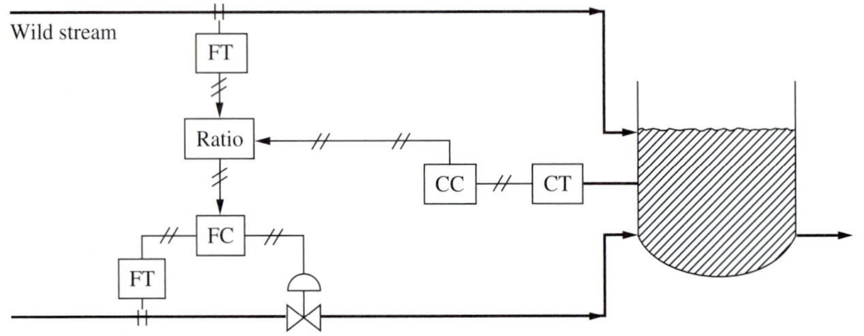

(*c*) Feedforward/feedback

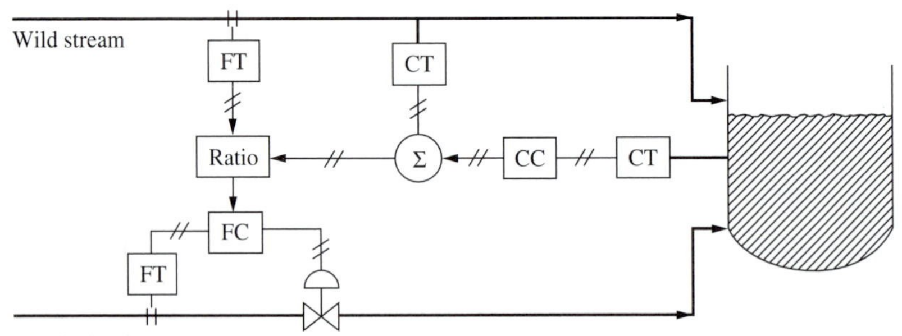

(*d*) Feedforward for both flow and composition

FIGURE 4.7
Feedforward control.

Some guidelines and recommendations are discussed below, together with a few examples of their application. The books by Buckley (*Techniques of Process Control,* 1964, Wiley, New York) and Shinskey (*Process-Control Systems,* 1967, McGraw-Hill, New York) are highly recommended for additional coverage of this important topic.

1. Keep the control system as simple as possible. Everyone involved in the process, from the operators up to the plant manager, should be able to understand the system, at least conceptually. Use as few pieces of control hardware as possible. Every additional gadget included in the system is one more item that can fail or drift. The instrument salesperson will never tell you this, of course.
2. Use feedforward control to compensate for large, frequent, and measurable disturbances.
3. Use override control to operate at or to avoid constraints.
4. Avoid large time lags and deadtimes in feedback loops. Control is improved by keeping the lags and deadtimes inside the loop as small as possible. This means that sensors should be located close to where the manipulated variable enters the process.

EXAMPLE 4.1. Consider the two blending systems shown in Fig. 4.8. The flow rate or composition of stream 1 is the disturbance. The flow rate of stream 2 is the manipulated variable. In Fig. 4.8*a* the sensor is located after the tank, and therefore the dynamic lag of the tank is included in the feedback control loop. In Fig. 4.8*b* the sensor is located at the inlet of the tank. The process lag is now very small since the tank is not inside the loop. The control performance in part *b*, in terms of speed of response and load rejection, would be better than the performance in part *a*. In addition, the tank now acts as a filter to average out any fluctuations in composition. ■

EXAMPLE 4.2. Composition control in distillation columns is frequently done by controlling a temperature somewhere in the column. The location of the best temperature control tray is a popular subject in the process control literature. The ideal location for controlling distillate composition x_D with reflux flow by using a tray temperature would be at the top of the column for a binary system (see Fig. 4.9*a*). This is desirable dynamically because it keeps the measurement lags as small as possible. It is also desirable from

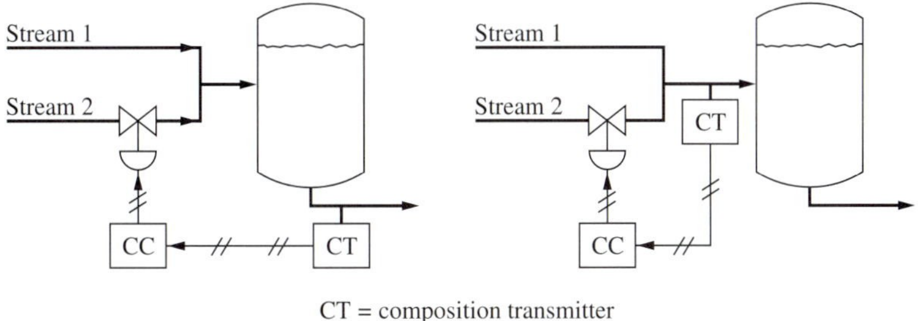

FIGURE 4.8
Blending systems. (*a*) With tank inside loop. (*b*) With tank outside loop.

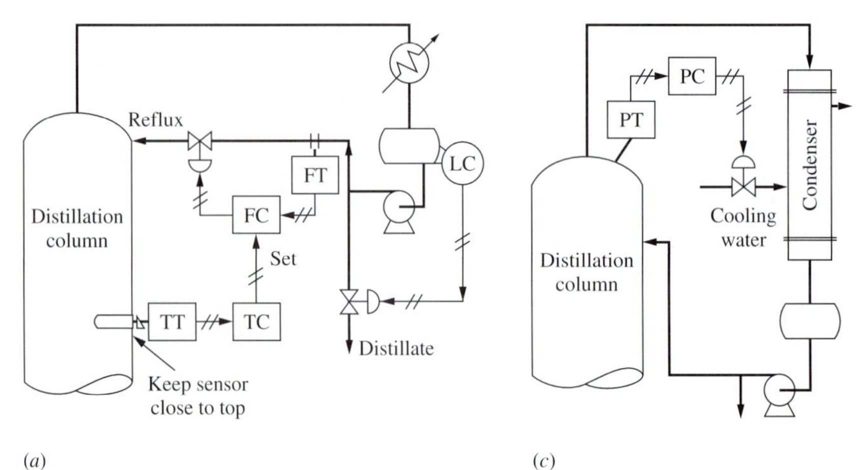

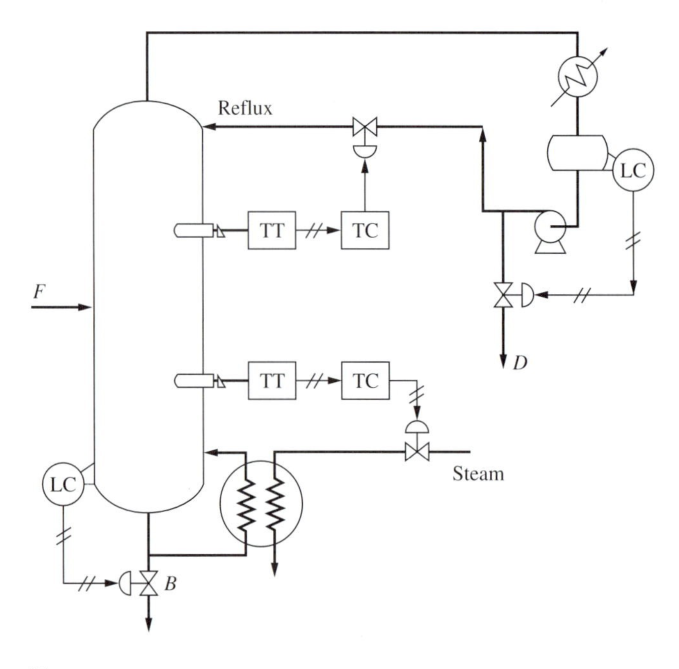

FIGURE 4.9
(*a*) Temperature control tray location. (*b*) Interaction. (*c*) Pressure control.

a steady-state standpoint because it keeps the distillate composition constant at steady state in a constant-pressure, binary system. Holding a temperature on a tray farther down in the column does not guarantee that x_D is constant, particularly when feed composition changes occur.

However, in many applications the temperature profile is quite flat (very little temperature change per tray) near the top of the column if the distillate product is of reasonable purity. The sensitivity of the temperature sensor may become limiting, but it is more probable that the limiting factor will be pressure changes swamping the effects of composition. In addition, if the system is not binary but has some lighter-than-light key components, these components will be at their highest concentration near the top of the column. In this case, the optimal temperature to hold constant is *not* at the top of the column, even from a steady-state standpoint.

For these reasons an intermediate tray is selected down the column where the temperature profile begins to break. Pressure compensation of the temperature signal should be used if column pressure or pressure drop varies significantly.

If bottoms composition is to be controlled by vapor boilup, the control tray should be located as close to the base of the column as possible in a binary system. In multicomponent systems where heavy components in the feed have their highest concentration in the base of the column, the optimal control tray is higher in the column.

5. Use proportional-only level controls where the absolute level is not important (surge tanks and the base of distillation columns) to smooth out disturbances.
6. Eliminate minor disturbances by using cascade control systems where possible.
7. Avoid control loop interaction if possible, but if not, make sure the controllers are tuned to make the entire system stable. Up to this point we have discussed tuning only single-input, single-output (SISO) control loops. Many chemical engineering systems are multivariable and inherently interacting, i.e., one control loop affects other control loops.

   The classic example of an interacting system is a distillation column in which two compositions or two temperatures are controlled. As shown in Fig. 4.9*b*, the upper temperature sets reflux and the lower temperature sets heat input. Interaction occurs because both manipulated variables affect both controlled variables.

   A common way to avoid interaction is to tune one loop very tight and the other loop loose. The performance of the slow loop is thus sacrificed. We discuss other approaches to this problem in Part Four.
8. Check the control system for potential dynamic problems during abnormal conditions or at operating conditions that are not the same as the design. The ability of the control system to work well over a range of conditions is called *flexibility.* Startup and shutdown situations should also be studied. Operation at low throughputs can also be a problem. Process gains and time constants can change drastically at low flow rate, and controller retuning may be required. Installation of dual control valves (one large and one small) may be required.

   Rangeability problems can also be caused by seasonal variations in cooling-water temperature. Consider the distillation column pressure control system shown in Fig. 4.9*c*. During the summer, cooling-water temperatures may be as high as 90°F and require a large flow rate and a big control valve. During the winter, the cooling-water temperatures may drop to 50°F, requiring much less water. The big valve may be almost on its seat, and poor pressure control may result. In addition, the water outlet temperature may get quite high under these low-flow conditions, presenting corrosion problems. In fact, if the process vapor temperature entering the condenser is above 212°F, the cooling water may even start to boil! Ambient effects can be even more severe in air-cooled condensers.
9. Avoid saturation of a manipulated variable. A good example of saturation is the level control of a reflux drum in a distillation column that has a very high reflux

Scheme A

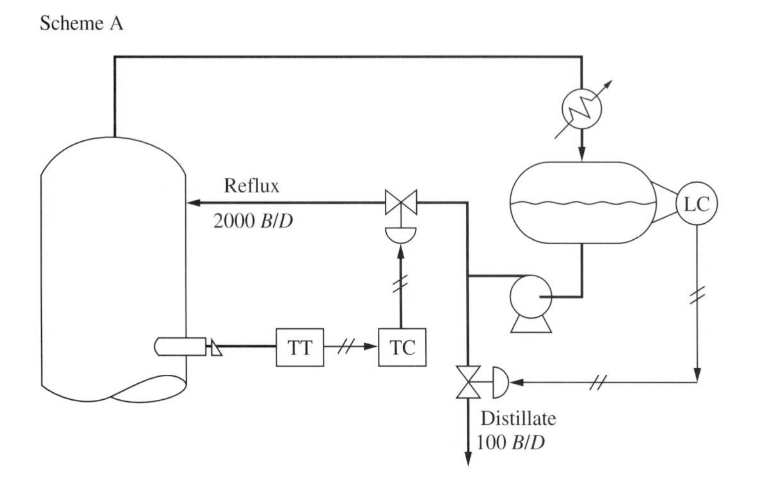

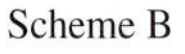
Scheme B

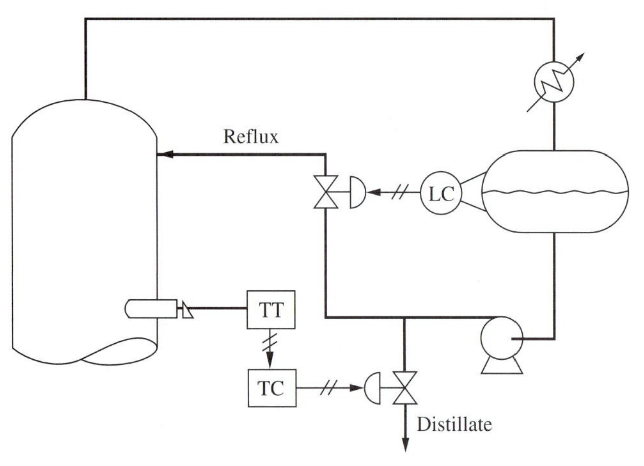

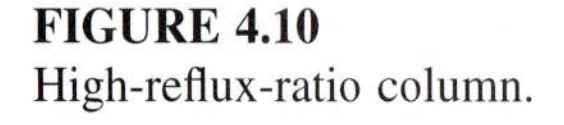
FIGURE 4.10
High-reflux-ratio column.

ratio. Suppose the reflux ratio (R/D) is 20, as shown in Fig. 4.10. Scheme A uses distillate flow rate D to control reflux drum level. If the vapor boilup dropped only 5 percent, the distillate flow would go to zero. Any bigger drop in vapor boilup would cause the drum to run dry (unless a low-level override controller were used to pinch back on the reflux valve). Scheme B is preferable for this high-reflux-ratio case.

10. Avoid "nesting" control loops. Control loops are nested if the operation of the external loop depends on the operation of the internal loop. Figure 4.11 illustrates a nested loop. A *vapor* sidestream is drawn off a column to hold the column base level, and a temperature higher up in the column is held by heat input to the reboiler. The base liquid level is affected only by the liquid stream entering and the vapor boiled off, and therefore is not directly influenced by the amount of vapor sidestream withdrawn. Thus, the base level cannot be held by the vapor

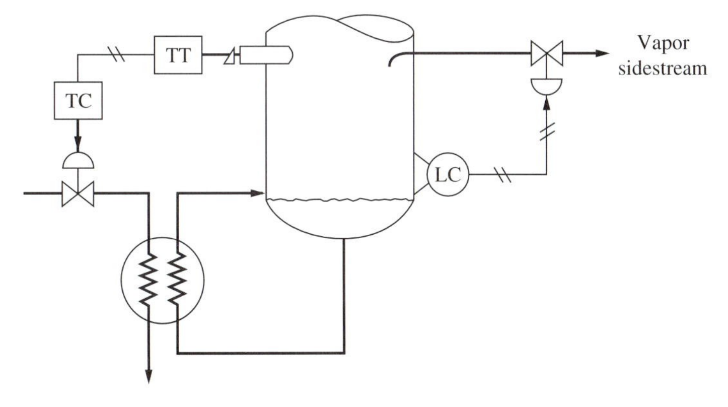

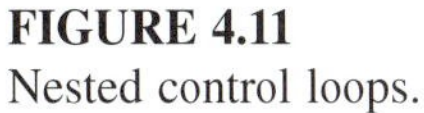

FIGURE 4.11
Nested control loops.

sidestream unless the temperature control loop is in operation. Then the change in the net vapor sent up the column will affect the temperature, and the vapor boilup will be changed by the temperature controller. This finally has an effect on the base level.

If the temperature controller is on "manual," the level loop cannot work. In this process it probably would be better to reverse the pairing of the loops: control temperature with vapor sidestream and control base level with heat input. Notice that if the sidestream were removed as a liquid, the control system would not be nested. Sometimes, of course, nested loops cannot be avoided. Notice that the recommended scheme B in Fig. 4.10 is just such a nested system. Distillate has no direct effect on tray temperature. It is only through the level loop and its changes in reflux that the temperature is affected.

4.9 CONCLUSION

In this chapter we have discussed in a qualitative way some control structures that go beyond the single-input, single-output structure. These ideas are widely applied in the chemical processing industries. They offer significant advantages in performance, safety, and flexibility. However, you should keep in mind that the use of more sensors and more components in a control loop has the effect of reducing reliability because if any one of the components fails, the whole loop will fail. One of the important features of any control system is its failure performance: will it degrade gracefully or catastrophically?

PROBLEMS

4.1. The suction pressure of an air compressor is controlled by manipulating an air stream from an off-site process. An override system is to be used in conjunction with the

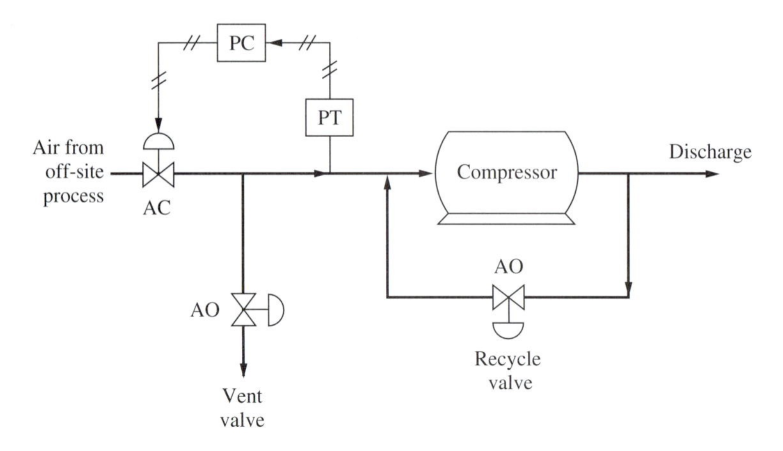

FIGURE P4.1

basic loop to prevent overpressuring or underpressuring the compressor suction during upsets. Valve actions are indicated on the sketch above.

The pressure transmitter span is 0 to 20 psig. The pressure controller setpoint is 10 psig. If the pressure gets above 15 psig, the vent valve is to start opening and is to be wide open at 20 psig. If the pressure drops below 5 psig, the recycle valve is to start opening and is to be wide open at 0 psig.

Specify the range and action of the override control elements required to achieve this control strategy.

4.2. Design an override control system that will prevent the liquid level in a reflux drum from dropping below 5 percent of the level transmitter span by pinching the reflux control valve. The system must also prevent the liquid level from rising above 90 percent of the level transmitter span by opening the reflux control valve. Normal level control is achieved by manipulating distillate flow over the middle 50 percent of the level transmitter span using a proportional level controller (proportional band setting is 50).

4.3. Design an override control system for the chilled-water loop considered in Problem 3.11. The flow rate of chilled water is not supposed to drop below 500 gpm. Your override control circuit should open the chilled-water control valve if chilled-water flow gets below 500 gpm, overriding the temperature controller.

4.4. Vapor feed to an adiabatic tubular reactor is heated to about 700°F in a furnace. The reaction is endothermic. The exit temperature of gas leaving the reactor is to be controlled at 600°F.

Draw an instrumentation and control diagram that accomplishes the following objectives:

- Feed is flow controlled.
- Fuel gas is flow controlled and ratioed to feed rate.
- The fuel to feed ratio is to be adjusted by a furnace exit temperature controller.
- The setpoint of the furnace exit temperature controller is to be adjusted by a reactor exit temperature controller.
- Furnace exit temperature is not to exceed 750°F.
- High furnace stack-gas temperature should pinch the fuel gas control valve.

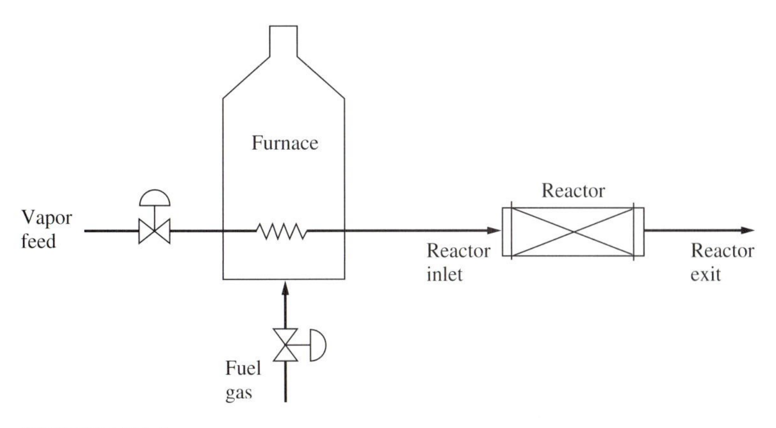

FIGURE P4.4

4.5. Sketch control system diagrams for the following systems.

(*a*) Temperature in a reactor is controlled by manipulating cooling water to a cooling coil. Pressure in the reactor is controlled by admitting a gas feed into the reactor. A high-temperature override pinches reactant feed gas.

(*b*) Reflux drum level is controlled by a reflux flow rate back to a distillation column. Distillate flow is manipulated to maintain a specified reflux ratio (reflux/distillate). This specified reflux ratio can be changed by a composition controller in the top of the column.

4.6. Sketch a control concept diagram for the distillation column shown. The objectives of your control system are:

- Reflux is flow controlled, ratioed to feed rate, and overridden by low reflux drum level.

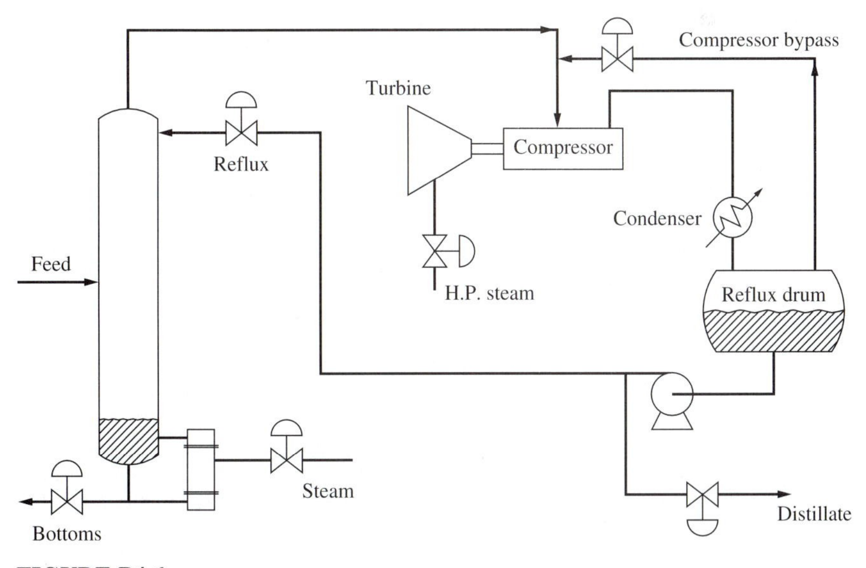

FIGURE P4.6

- Steam is flow controlled, with the flow controller setpoint coming from a temperature controller that controls a tray temperature in the stripping section of the column. Low base level or high column pressure pinches the steam valve.
- Base level is controlled by bottom product flow rate.
- Reflux drum level is controlled by distillate product flow rate.
- Column pressure is controlled by changing the setpoint of a speed controller on the compressor turbine. The speed controller output sets a flow controller on the high-pressure steam to the turbine.
- A minimum flow controller ("antisurge") sets the valve in the compressor bypass line to prevent the flow rate through the compressor from dropping below some minimum flow rate.

4.7. A distillation column operates with vapor recompression.
(*a*) Specify the action of all control valves.
(*b*) Sketch a control concept diagram with the following loops:

- Reflux is flow controlled and overridden by low reflux drum level.
- Reflux drum level is controlled by distillate flow.
- Steam to the turbine driving the compressor is flow controlled, reset by a speed controller, which is reset by a distillate composition controller.

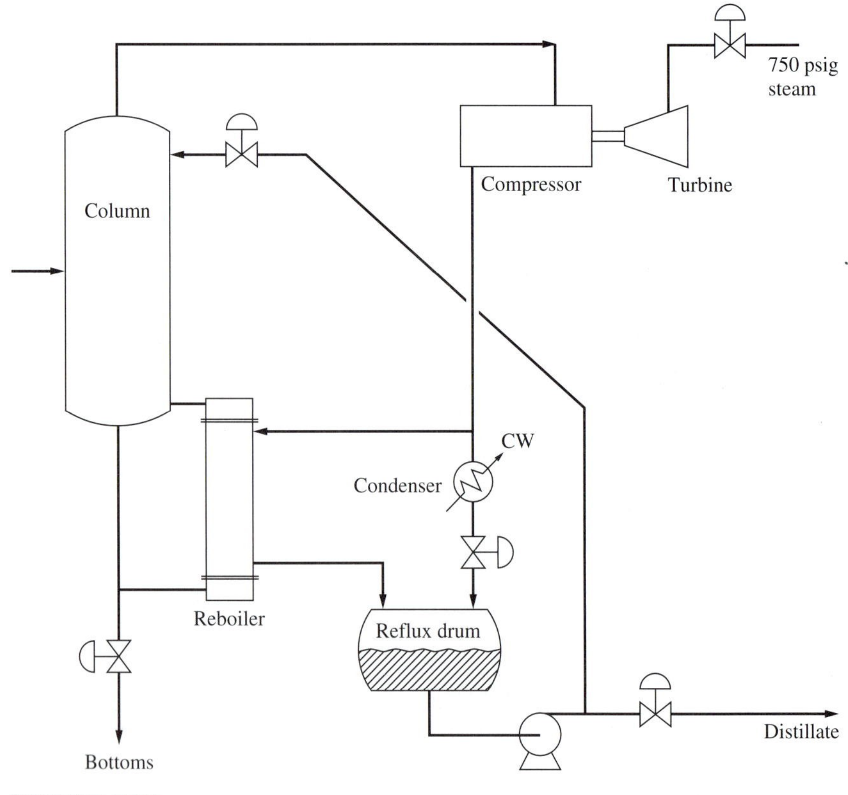

FIGURE P4.7

- Column pressure is controlled by the valve under the condenser, which floods or exposes tubes in the condenser to vary the heat transfer area.
- High reflux drum pressure overrides the steam valve on the compressor turbine.

(*c*) At design conditions, the total flow through the compressor is 120,000 lb_m/hr, the flow through the hot side of the reboiler is 80,000 lb_m/hr, the pressure drop through the condenser is 2 psi, the pressure drop through the hot side of the reboiler is 10 psi, and the control valve below the condenser is 25 percent open. Density of the liquid is 5 lb_m/gal.

(i) What is the C_v value of the control valve?

(ii) What is the flow rate through the hot side of the reboiler when the control valve is wide open? Assume that the pressure in the reflux drum is constant and that the distillate composition controller will adjust compressor speed to keep the total flow through the compressor constant.

4.8. Sketch a control concept diagram for a chemical reactor that is cooled by generating steam (see Problem 3.18).

- Steam drum pressure is controlled by the valve in the steam exit line.
- Condensate flow is ratioed to steam flow.
- Steam drum liquid level is controlled by adjusting the condensate to steam ratio.
- Feed is flow controlled.
- Reactor liquid level is controlled by product withdrawal.
- Reactor temperature is controlled by resetting the setpoint of the steam pressure controller.
- The override controls are as follows:

  High reactor temperature pinches the reactor feed valve.
  Low steam drum level pinches reactor feed.

4.9. A chemical plant has a four-header steam system. A boiler generates 900-psig steam, which is let down through turbines to a 150-psig header and to a 25-psig header. There are several consumers at each pressure level. There are also other producers of 25-psig steam and 10-psig steam. (See the figure on the next page.) Sketch a control system that:

- Controls pressure in the 900-psig header by fuel firing rate to the boiler.
- Controls pressure in the 150-psig header by valve A.
- Controls pressure in the 25-psig header by opening valve B if pressure is low and opening valve C if pressure is high.
- Controls pressure in the 10-psig header by opening valve C if pressure is low and opening valve D if pressure is high.

4.10. Sketch a control system for the two-column heat-integrated distillation system shown.

- Reflux drum levels are controlled by distillate flows on each column.
- Reflux flow is ratioed to distillate flow on each column.
- Column pressure drop is controlled on the first column by manipulating steam flow to the auxiliary reboiler.
- Temperature in the second column is controlled by steam flow to its reboiler.
- Base levels are controlled by bottoms flow rates on each column.
- High or low base level in the first column overrides both steam valves.
- High or low base level in the second column overrides the steam valve on its reboiler.

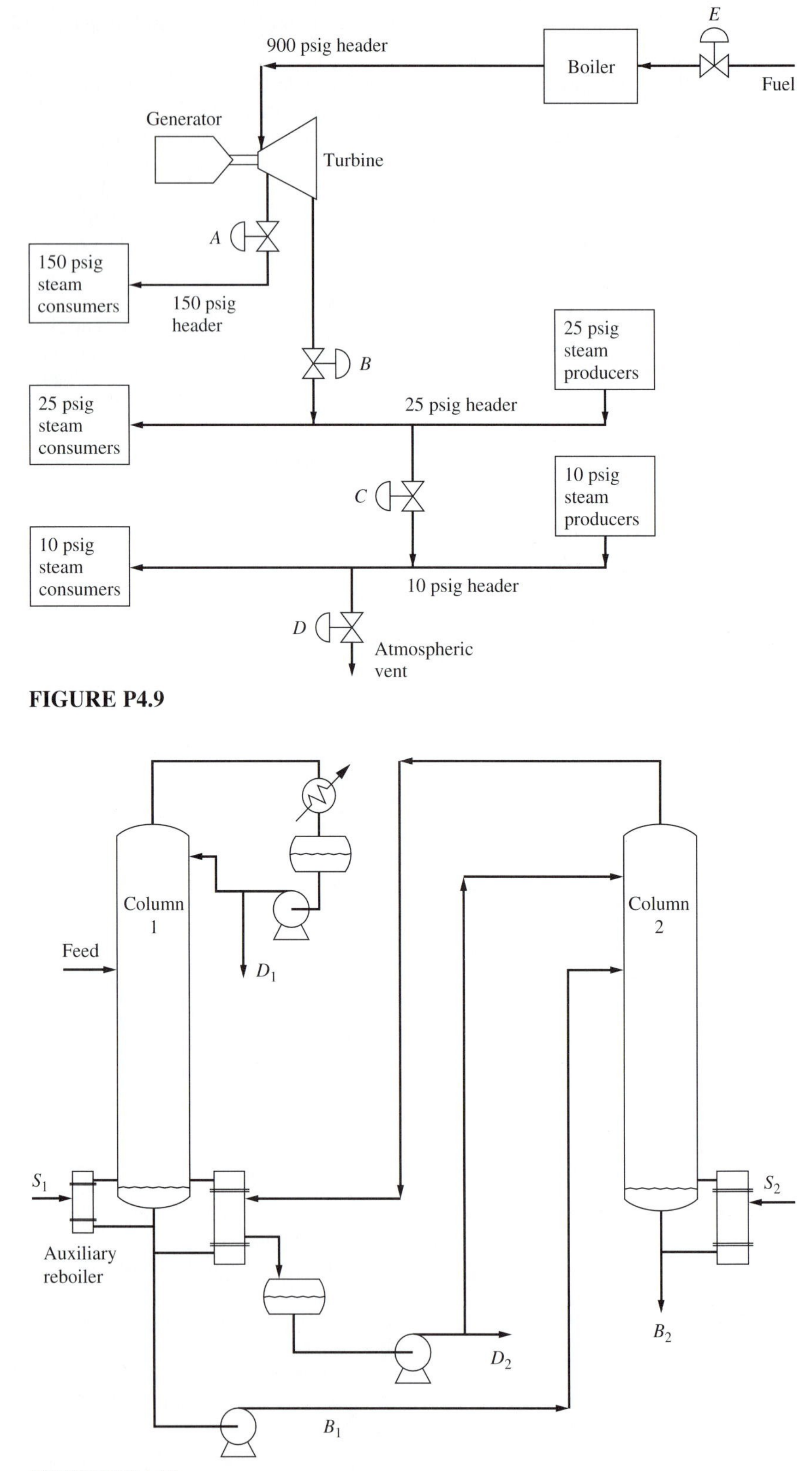

FIGURE P4.9

FIGURE P4.10

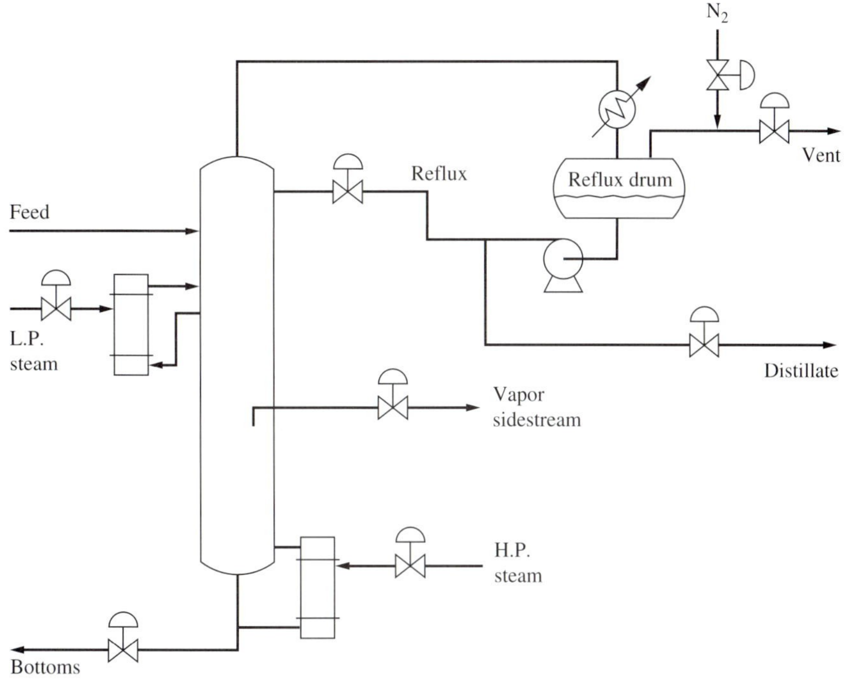

FIGURE P4.11

4.11. The distillation column sketched above has an intermediate reboiler and a vapor sidestream. Sketch a control concept diagram showing the following control objectives:

- Column top pressure is controlled by a vent/bleed system; i.e., inert gas is added if pressure is low, and gas from the reflux drum is vented if pressure is high.
- Reflux drum level is controlled by distillate flow.
- Reflux is ratioed to feed rate.
- Low-pressure steam flow rate to the intermediate reboiler is ratioed to feed rate.
- Vapor sidestream flow rate is set by a temperature controller holding tray 10 temperature.
- Base level is controlled by high-pressure steam flow rate to the base reboiler.
- Bottoms purge flow is flow controlled.
- Low base level overrides the low-pressure steam flow to the intermediate reboiler.
- High column pressure overrides both steam valves.

4.12. Hot oil is heated in a furnace in three parallel passes. The oil is used as a heat source in four parallel process heat exchangers. Draw a control concept diagram that achieves the following objectives:

- The temperature of each process stream leaving the four process heat exchangers is controlled by manipulating the hot oil flow rate through each exchanger.
- Furnace hot oil exit temperature is controlled by manipulating fuel flow rate.
- A valve position controller is used to reset the setpoint of the furnace exit temperature controller such that the most open of the four control valves on the hot oil streams flowing through the process heat exchangers is 80 percent open.

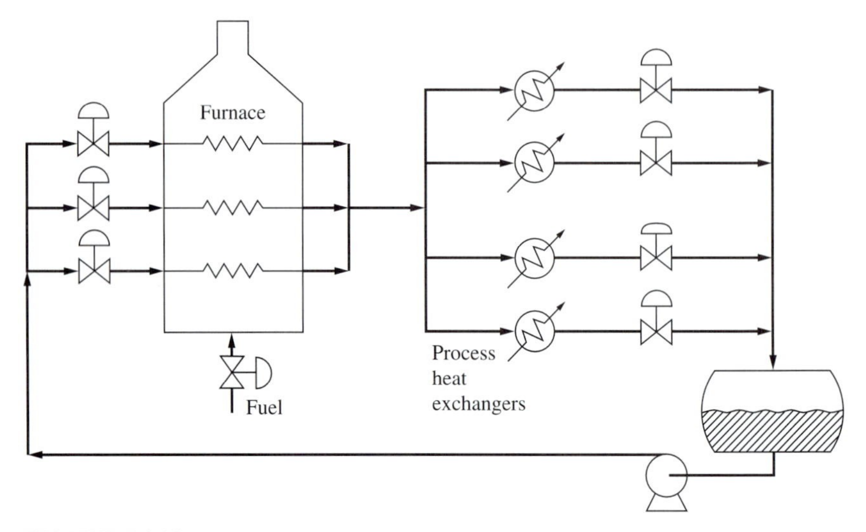

FIGURE P4.12

- The flow rates of hot oil through the three parallel passes in the furnace are maintained equal.

4.13. Terephthalic acid (TPA) is produced by air oxidation of paraxylene at 200 psig. The reaction is exothermic. Nitrogen plus excess oxygen leaves the top of the reactor.

$$\text{xylene} + O_2 \rightarrow \text{TPA} + \text{water}$$

Liquid products (TPA and water) are removed from the bottom of the reactor. Heat is removed by circulating liquid from the base of the reactor through a water-cooled heat exchanger. The air compressor is driven by a steam turbine.

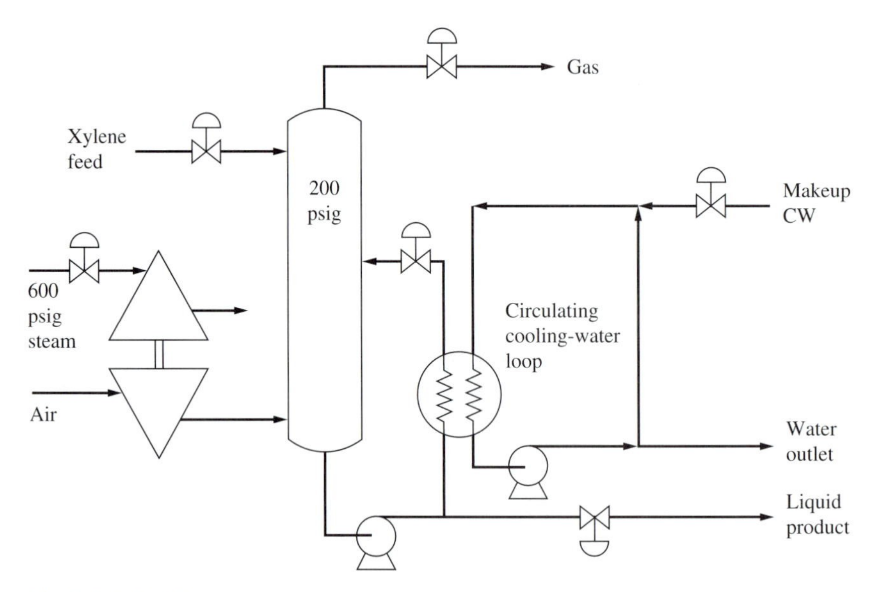

FIGURE P4.13

Specify the action of all control valves and sketch a control system that will achieve the following objectives:

- Xylene feed is flow controlled.
- Air feed is flow controlled by adjusting the speed of the turbine.
- Reactor temperature is controlled through a cascade system. Circulating water temperature is controlled by makeup cooling water. The setpoint of this temperature controller is set by the reactor temperature controller. The circulation rate of process liquid through the cooler is flow controlled.
- Reactor pressure is controlled by the gas leaving the reactor.
- Base liquid level is controlled by liquid product flow.
- Air flow rate is adjusted by a composition controller that holds 2 percent oxygen in the gas leaving the reactor.
- High reactor pressure overrides steam flow to the air compressor turbine.

4.14. Overhead vapor from a distillation column passes through a partial condenser. The uncondensed portion is fed into a vapor-phase reactor. The condensed portion is used for reflux in the distillation column.

The vapor fed to the reactor can also come from a vaporizer, which is fed from a surge tank. To conserve energy, it is desirable to feed the reactor with vapor directly from the column instead of from the vaporizer. The only time the vaporizer should be used is when there is not enough vapor produced by the column.

Sketch a control concept diagram showing the following characteristics:

- Total vapor flow rate to the reactor is flow controlled by valve V_1.
- Reflux drum pressure is controlled by valve V_2.
- Vaporizer pressure is controlled by valve V_3.
- Vaporizer liquid level is controlled by valve V_4.
- Column reflux is flow controlled by valve V_5.
- Reflux drum level is controlled by valve V_6.
- High reflux drum level opens valve V_7.
- High vaporizer pressure overrides valve V_2.
- High reflux drum pressure overrides valve V_6.

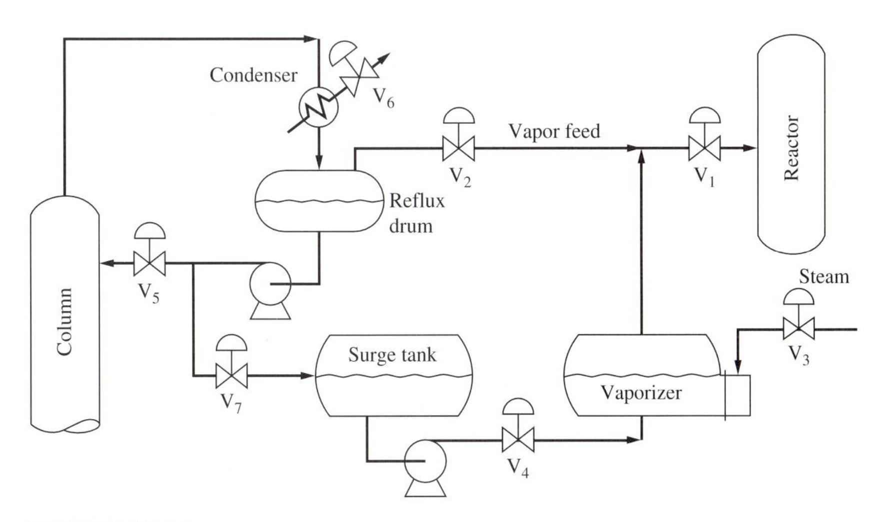

FIGURE P4.14

4.15. Two distillation columns are heat integrated, as shown in the sketch. The first column has an auxiliary condenser to take any excess vapor that the second column does not need. The second column has an auxiliary reboiler that provides additional heat if required.

Prepare a control concept diagram that includes the following control objectives:

- Base levels are controlled by bottoms flows.
- Reflux drum levels are controlled by distillate flows.
- Reflux flows are flow controlled.
- The pressure in the first column is controlled by vapor flow rate to the auxiliary condenser. A low-pressure override pinches the vapor valve to the second column reboiler.
- The pressure in the second column is controlled by manipulating cooling water to the condenser.
- A temperature in the stripping section of the first column is controlled by manipulating steam to the reboiler.
- A temperature in the stripping section of the second column is controlled by manipulating the vapor to the reboiler of the second column that comes from the first

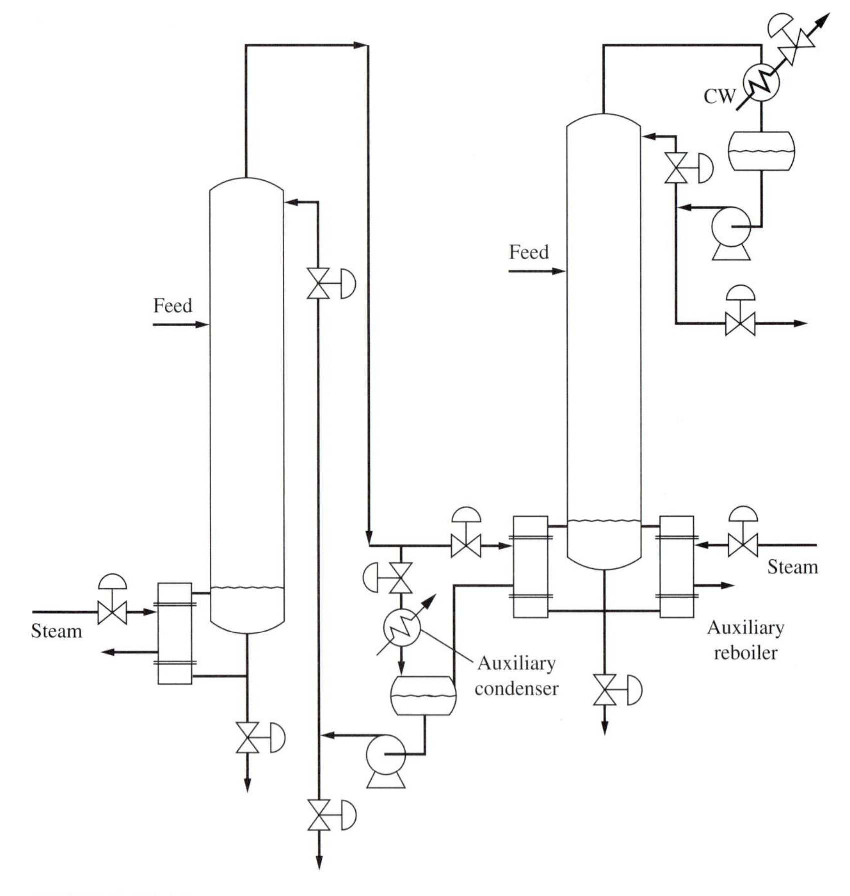

FIGURE P4.15

column and by manipulating the steam to the auxiliary reboiler. A split-range system is used so that steam to the auxiliary reboiler is used only when insufficient heat is available from the vapor from the first column.

- High column pressures in both columns pinch reboiler steam.

4.16. The sketch below shows a distillation column that is heat integrated with an evaporator. Draw a control concept diagram that accomplishes the following objectives:

- In the evaporator, temperature is controlled by steam, level by liquid product, and pressure by auxiliary cooling or vapor to the reboiler. Level in the condensate receiver is controlled by condensate.

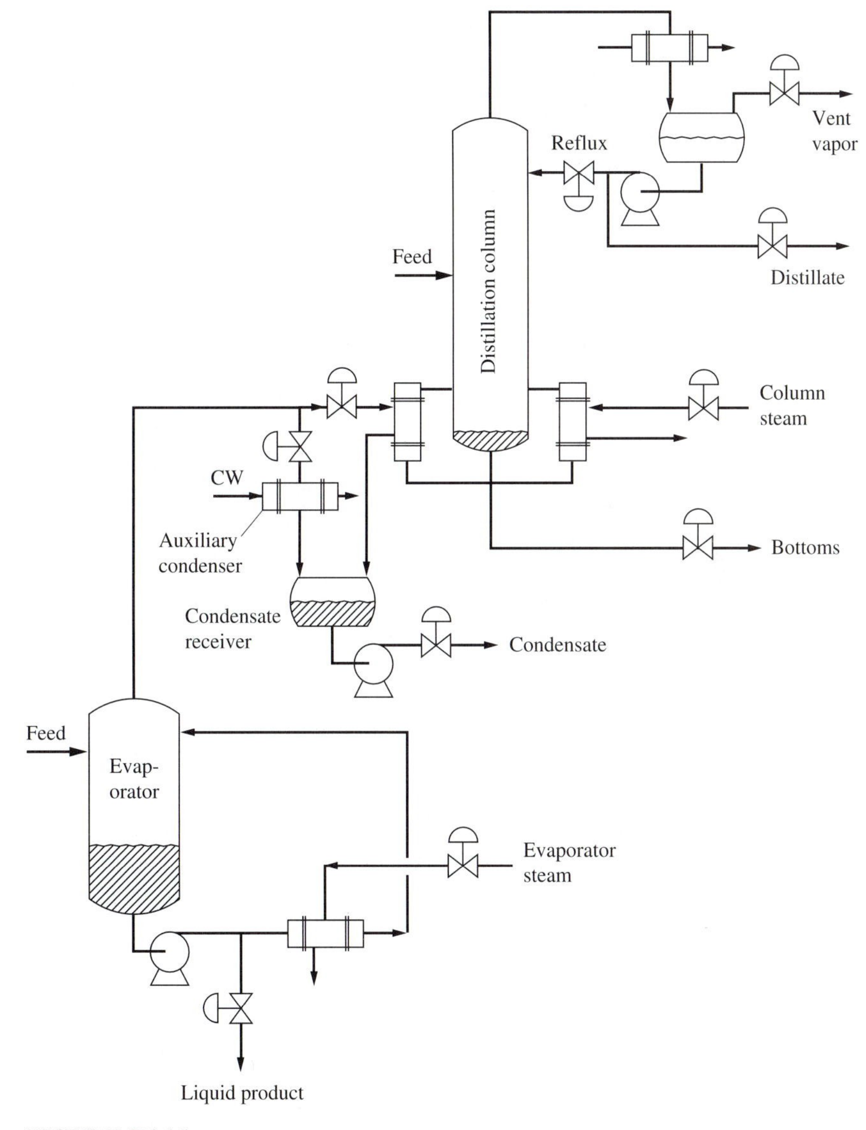

FIGURE P4.16

- In the column, reflux is flow controlled, reflux drum level is controlled by distillate, base level by bottoms, pressure by vent vapor, and temperature by steam to the auxiliary reboiler or vapor from the evaporator.
- A high-column-pressure override controller pinches steam to both the evaporator and the auxiliary reboiler.

4.17. Sketch a control scheme for the cryogenic stripper shown on the next page, which is used for removing small amounts of propane from natural gas.

- Cooling-water valve V_1 is manipulated to control the gas temperature leaving the cooler.
- Valve V_2 controls a temperature on tray 15 in the stripper.
- Valve V_3 controls the total flow rate of gas into the compressor.
- Valve V_4 controls the temperature of the propane bottoms product leaving the unit.
- Valve V_5 controls the column base level.
- Valve V_6 controls the liquid level in tank 1.
- Valve V_7 controls the speed of the expander turbine.
- A valve position controller is used to keep valve V_7 nearly wide open by adjusting the setpoint of the expander speed controller.
- A pressure controller opens valve V_8 if the pressure in tank 1 gets too high.
- Valve V_9 controls tank 2 level.
- Valve V_{10} controls tank 2 pressure.
- Valve V_{11} controls the pressure in the stripper.
- If the pressure in the stripper gets too high, an override controller pinches both valves V_2 and V_3.

4.18. A distillation column is used to separate two close-boiling components that have a relative volatility close to 1. The reflux ratio is quite high (15), and many trays are required (150). To control the compositions of both products, the flow rates of the product streams (distillate D and bottoms B) are manipulated. Gas chromatographs are used to measure the product compositions. Base level is controlled by steam flow rate to the reboiler, and reflux drum level is controlled by reflux flow rate.

(*a*) Design an override control system that will use bottoms flow rate to control base level if a high limit is reached on the steam flow.

(*b*) Design an override control system that will use very low composition transmitter output signals to detect if either of the chromatographs has failed or is out of service. The control system should switch the structure of the control loops as follows:

- If the distillate composition signal is not available, reflux should be held constant at the given flow rate and reflux drum level should be held by distillate flow rate.
- If the bottoms composition signal is not available, steam should be held constant and base level should be held by bottoms flow rate.

4.19. A furnace control system consists of the following loops:

- Temperature of the process stream leaving the furnace changes the setpoint of a flow controller on fuel.
- Air flow is ratioed to fuel flow, with the ratio changed by a stack-gas excess air controller.

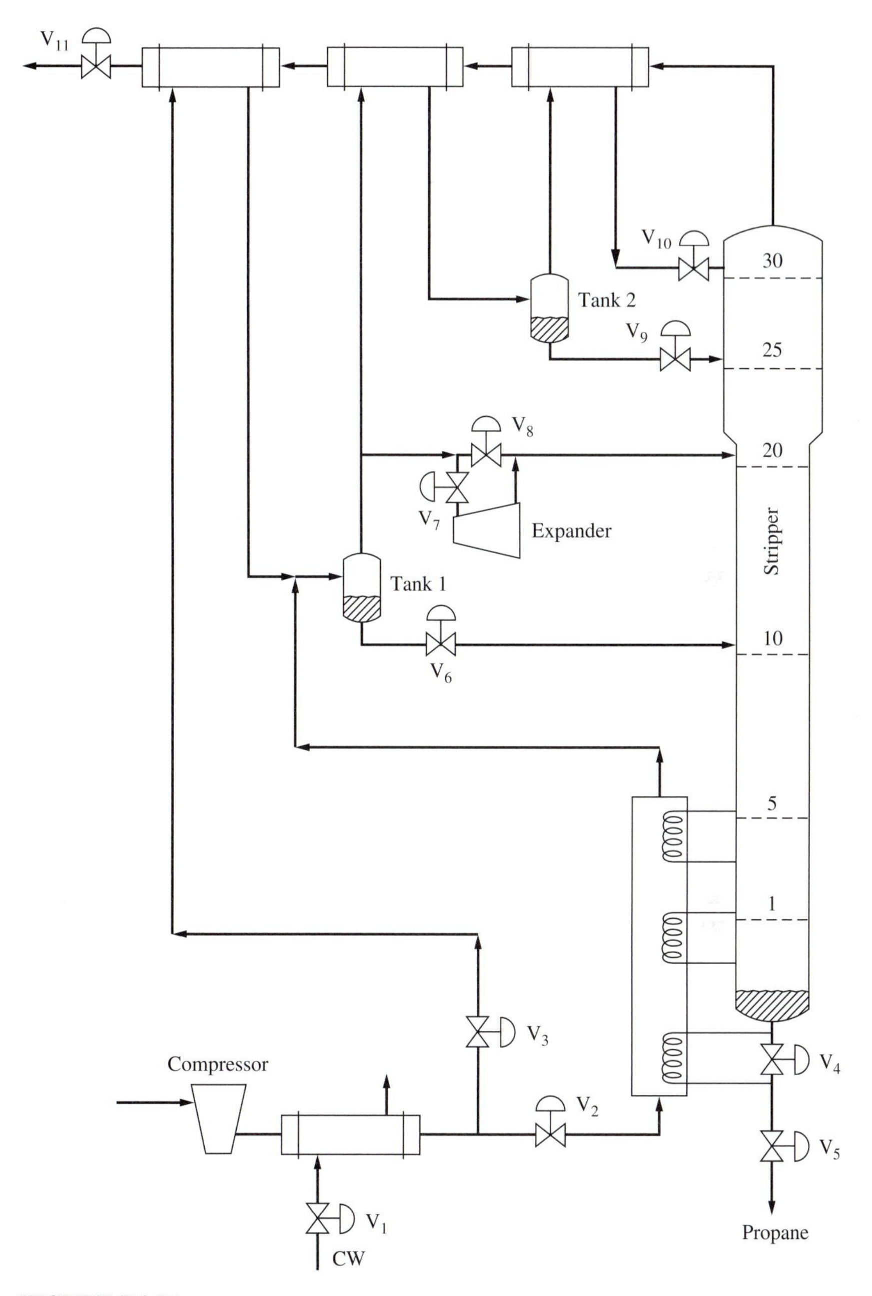

FIGURE P4.17

Show how some simple first-order lags and selectors can be used to produce a control system in which:

- The air flow *leads* the fuel flow when an increase in fuel flow occurs.
- The air flow *lags* the fuel flow when a decrease in fuel flow occurs.

(*Hint:* Sketch the responses for both positive and negative step changes in the input to a circuit that consists of a first-order lag with unity gain and a low-selector. The input signal goes in parallel to the lag and to the low-selector. The output of the lag goes to the other input of the low-selector.)

4.20. Figure P4.20 shows the Amoco Model 4 fluidized-bed catalytic cracking unit. Several hydrocarbon feeds (gas oil, slurry recycle, etc.) are fed into the reactor along with hot solid catalyst from the regenerator. The endothermic cracking reactions cool the catalyst and deposit coke on it. The catalyst from the reactor is circulated back to the regenerator, where air is added to burn off the coke. The heat from the combustion reaction heats the catalyst.

This model cat cracker does not use slide valves to control catalyst circulation. Instead, the lift air (valve V_1) is used to transport catalyst from the reactor into the regenerator. The flow rate of catalyst from the regenerator to the reactor is controlled by changing the pressure differential between the reactor and the regenerator using valve V_2. Pressure in the reactor is controlled by valve V_3 on the suction of the wet gas compressor.

Sketch a control concept diagram that shows the following features:

- Reactor temperature is controlled by changing the setpoint of the ΔP controller.
- The level of catalyst in the regenerator standpipe is controlled by manipulating valve V_1.
- Oxygen concentration in the flue gas from the regenerator changes the setpoint of a total air controller. This total air controller looks at the sum of the lift air, spill air, and combustion air and manipulates the suction valve V_4 on the combustion air compressor.

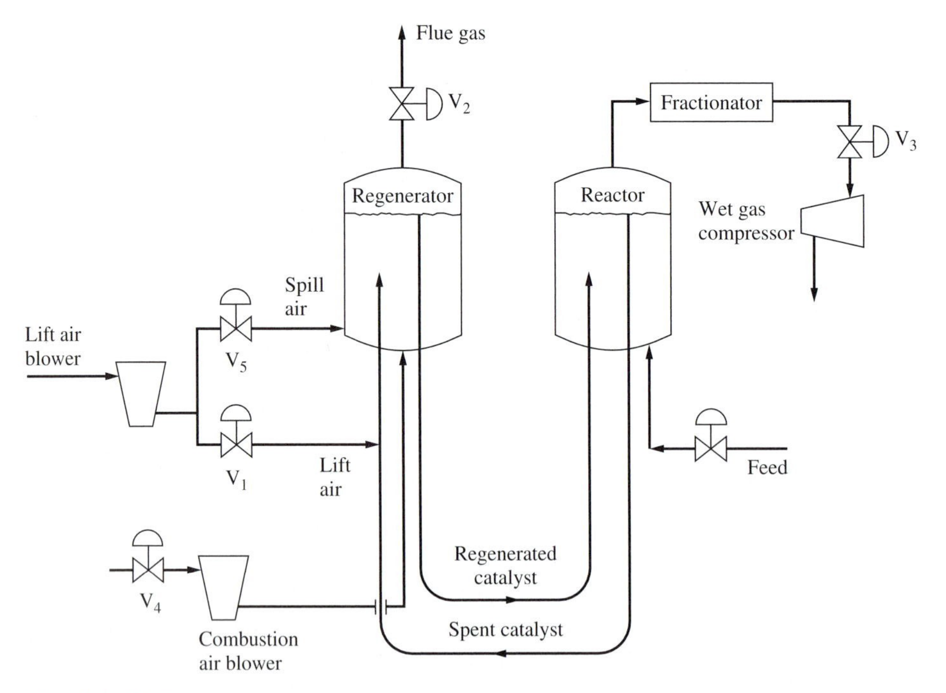

FIGURE P4.20

- If valve V_4 starts to approach the wide-open position, the valve on the spill air line (V_5) should be opened enough to keep V_4 about 90 percent open.
- If valve V_3 on the wet gas compressor is less than 90 percent open, the setpoint to the feed flow controller should be increased (i.e., bring in as much feed as the wet gas compressor can handle).

4.21. A process stream is fed into a flash tank that is heated by adding steam to an internal coil in the tank. The feed to the tank comes from an upstream process and is a load disturbance to the tank. Vapor from the tank flows into a pipeline that is at constant pressure. Liquid from the tank is pumped through a control valve into another vessel.

Sketch a control scheme that will accomplish the objective of taking 10 percent of the feed stream out of the top of the flash tank as vapor. Assume all the necessary measurements can be made.

4.22. International Paper operates a 200-ton/day lime kiln at one of its pulp and paper mills. A $CaCO_3/H_2O$ slurry (70 percent solids) is fed into one end of the kiln. It is dried and CO_2 is driven off to produce CaO. Heat is supplied by burning natural gas in the other end of the kiln. The hot gases flow countercurrent to the solids and leave through an adjustable damper.

Sketch a control system that accomplishes the following:

- Fuel gas controls the temperature of the solid CaO leaving the hot end of the kiln.
- Damper opening controls the temperature of the gas leaving the cool end of the kiln by changing the amount of combustion air drawn into the kiln.
- If the oxygen content of the gas leaving the kiln drops below 2 percent, an override controller should begin to open the damper.
- It is more important to keep exit gas temperature near its setpoint than exit solid temperature. Therefore, if the exit gas temperature cannot be held by the damper opening, it should be held by fuel gas flow rate.

4.23. The temperature in a chemical reactor can be controlled by manipulating cooling-water flow rate to the cooling coil. The flow rate of a reactant fed into the reactor also affects the temperature. The reaction is exothermic.

Sketch a control scheme that embodies the following concepts:

- Reactant is flow controlled.
- The flow rate of cooling water is ratioed to reactant flow.
- Reactor temperature is controlled by changing the water to reactant ratio.
- High reactor temperature *or* high reactor pressure overrides the reactant feed valve.

4.24. In a cell harvesting operation, a mixture of cells and liquid is fed into a tank. The material in the tank is continuously pumped by a positive displacement pump through the high-pressure side of a cross-flow filter. The material that does not flow through the filter membrane to the low-pressure side recycles back to the tank. There is a control valve in the return line to the tank downstream from the filter. The liquid that passes through the filter membrane leaves through another control valve. Sketch a control concept diagram that achieves the following objectives:

- Flow rate of recycled material back to the tank is controlled by manipulating pump speed.
- Liquid level in the tank is controlled by feed to the tank.

- Pressure on the high-pressure side of the filter is controlled by the control valve in the recycle line back to the tank.
- The pressure differential between the high- and low-pressure sides of the filter (the transmembrane pressure) is controlled by the control valve on the filtrate liquid leaving the filter.
- A high-pressure override controller on the high-pressure side of the filter reduces the setpoint to the pump speed controller.
- A low-flow override controller on the filtrate flow rate increases the setpoint to the differential-pressure controller.
- A low-level override controller on tank level reduces the setpoint of the pump speed controller and reduces the opening of the control valve on the filtrate stream.

4.25. There are two ways to control the idling speed of a gasoline engine: manipulate the throttle opening or manipulate the spark advance. Because of the electronic ignition, the dynamic response of idle speed is much faster to spark advance (0.01 sec deadtime) than to throttle speed (0.1 sec). Therefore, tighter idle speed control can be achieved by the use of spark advance. However, using spark advance degrades fuel efficiency, so throttle opening should be used under normal steady-state conditions. Develop a control strategy to handle this problem, and sketch a block diagram of your closedloop system.

CHAPTER 5

Interaction between Steady-State Design and Dynamic Controllability

In this chapter we discuss some important concepts that are central to the historical mission of chemical engineers: developing, designing, building, and operating chemical plants. We give some specific examples that illustrate the trade-offs between steady-state design and dynamic controllability, and some general guidelines for improving controllability are presented.

5.1 INTRODUCTION

The traditional approach to developing a process has been to perform the design and control analyses sequentially. The task of the process design engineer focused on determining (synthesizing) the flowsheet structure, parameter values, and steady-state operating conditions required to meet the production goals. The objective was to optimize economics (minimize annual cost, maximize annual profit, maximize net present value, etc.) considering only steady-state operation in evaluating the large number of alternative flowsheets and design conditions and parameters satisfying the operational requirements. Little attention was given to dynamic controllability during the early stages of design. After a final design had been developed, the detailed plans were "thrown over the wall" to the process control engineers.

The task of the process control engineer then centered on establishing a control strategy to ensure stable dynamic performance and to satisfy product quality requirements. These objectives must be met in the face of potential disturbances, equipment failures, production rate changes, and transitions from one product to another.

It has long been recognized that this traditional sequential approach is a poor way to do business. The design of a process determines its inherent controllability, which qualitatively means how well the process rejects disturbances and how easily it moves from one operating condition to another. Consideration of the dynamic controllability of a process should be an integral part of the design synthesis activity. Over the past two decades competitive pressures, safety issues, energy costs, and environmental concerns have increased the complexity and sensitivity of processes. Plants have become more highly integrated with respect to both material flows (complex configurations, recycle streams, etc.) and energy (heat integration). These high-performance processes are often difficult to control.

There are several reasons the traditional approach was taken. One involved the lack of computer-aided design tools for plantwide dynamic studies. Over the past several years, many advances in this area have occurred. We anticipate that easy-to-use software tools will soon be available to permit the design engineer and the control engineer to move seamlessly between steady-state flowsheet development and dynamic controllability analysis. Another reason the traditional approach has been used is the technology barriers separating design engineers (who know little about control) from the control engineers (who know little about design). We hope this book in some small way helps to break down these barriers.

During the last decade the concept of reducing product quality variability has become widespread. In the past, *good* control meant meeting or exceeding product specifications. Now customers often demand "on-aim" control, where variability in either direction from a specification means *poor* control. This makes the job of a process control engineer significantly more difficult and emphasizes the importance of developing an easily controllable process, i.e., one that almost flies itself and has low sensitivity to disturbances.

A growing amount of evidence points to the desirability of incorporating dynamic controllability considerations into all phases of the plant design. It may be better in the long run to build a process that has higher capital and energy costs if the plant provides more stable operation and achieves less variability in product quality. We present several examples in this chapter illustrating the inherent trade-off between steady-state economics and dynamic controllability.

5.2 QUALITATIVE EXAMPLES

5.2.1 Liquid Holdups

The most common and important trade-off involves specifying liquid holdup volumes in tanks, column bases, reflux drums, etc. From a steady-state viewpoint, these volumes should be kept as small as possible to minimize capital investment and reduce potential safety and pollution problems. For example, the more holdup needed in the base of a distillation column, the taller the column must be. In addition, if the material in the base of the column is heat sensitive, it is very desirable to keep holdup as small as possible to reduce the time that the material is at the high base temperature. Large holdups also increase safety and pollution risks in the handling of hazardous or toxic materials because equipment failures have the potential to release large amounts of material.

All these considerations suggest the use of small liquid holdups. However, from a dynamics and control viewpoint we want enough holdup to be able to handle sudden large changes in inlet or outlet flow rates. For example, in a distillation column we want to be able to ride through disturbances without losing liquid levels in the base or reflux drum. Suppose a large amount of light noncondensable material enters the column and reduces condenser heat transfer rates because of lower temperatures. The liquid level in the reflux drum will drop and, unless enough surge capacity exists, we may lose reflux flow and have to shut down the column.

Over the years some heuristics have been developed for estimating liquid holdups in most systems. Holdup times (based on *total* flow in and out of the surge volume) of about 5 to 10 minutes work well. If a distillation column has a fired reboiler, the base holdup should be made larger. If a downstream unit is particularly sensitive to rate or composition changes, then holdup volumes in upstream equipment should be increased. Such matters should be considered when the vessels are designed.

5.2.2 Gravity-Flow Condenser

Another trade-off is exemplified by a gravity-flow condenser design. Suppose we are cooling a reactor by removing vapor overhead and condensing it in a heat exchanger located above the reactor. The condensed material returns to the reactor as liquid through a return line. The pressure in the reactor must be higher than the pressure in the condenser for vapor to flow. The return liquid stream must flow back into the reactor against this positive pressure gradient. This is achieved by building up a leg of liquid in the liquid return line of sufficient height to overcome the pressure gradient and any frictional pressure drop through the line.

Under normal operating conditions, the liquid height may be several feet. But since pressure drop through the vapor line increases as the square of the vapor flow rate, the liquid height must increase by a factor of 4 if the vapor rate doubles. Thus, the control engineer pushes the design engineer to mount the condenser much higher in the air than the steady-state conditions require, and this increases the cost of equipment.

5.3 SIMPLE QUANTITATIVE EXAMPLE

To illustrate the design/control trade-off more quantitatively, let us consider a simple chemical engineering system: a series of continuous stirred-tank reactors (CSTRs) with jacket cooling. This type of reactor system is widely used in industry. The reactions and the reactors are quite simple, but they provide some important insights into evaluating the trade-offs between steady-state design and control.

In this section we consider the simplest possible reaction case: one irreversible reaction $A \rightarrow B$, equal-size reactors, and given (and equal) temperatures in all reactors. This serves to show that a process with a higher capital cost (one large CSTR) provides much better temperature control than a less expensive process (multiple CSTRs in series). The next section deals with the problem of selecting the "best" reactor temperature to illustrate how a more complex kinetic system leads to conflicts between steady-state economics and dynamic controllability that affect both capital cost and product yield.

The main objective of this section is to provide a quantitative example in which the "best" process is not the optimal steady-state economic design. In this example reactor system, temperature control is the measure of product quality. For this type of system, dynamic controllability is improved by increasing the heat transfer area in the reactor.

5.3.1 Steady-State Design

The first-order irreversible exothermic reaction A → B occurs in the liquid phase of one or more stirred-tank reactors in series. In a series configuration all reactors have the same volume and operate at the same specified temperature. A height-to-diameter ratio of 2 is assumed in the design. The reaction rate is

$$\mathcal{R}_n = V_n k_n z_n \tag{5.1}$$

where $\mathcal{R}_n$ = rate of consumption of reactant A (lb-mol/hr) in the nth reactor stage
V_n = holdup of nth reactor (lb-mol)
k_n = specific reaction rate (hr^{-1}) in the nth reactor stage
z_n = concentration of reactant A in the nth reactor (mole fraction A)

In the dynamic simulation, the following data are used. The steady-state operating temperature of all reactors is 140°F with a specific reaction rate of 0.5 hr^{-1} and an activation energy of 30,000 Btu/lb-mol. The fresh feed enters the first reactor at a flow rate F = 100 lb-mol/hr with temperature T_0 = 70°F and concentration z_0 = 1 (mole fraction A). Molecular weight is 50 lb/lb-mol, and liquid density is 50 lb/ft^3. The heat of reaction is −30,000 Btu/lb-mol unless otherwise noted.

The reactor is cooled by flowing cooling water at a rate F_J (ft^3/hr) through a jacket that surrounds the vertical walls of the reactor. Perfect mixing in the jacket is assumed with a 4-in jacket clearance. Inlet cooling water has a temperature of 70°F. The overall heat transfer coefficient U is assumed to be constant (300 Btu/hr °F ft^2).

Given the fresh feed flow rate and composition, the steady-state design procedure is:

1. Specify the conversion χ.
2. Calculate the concentration of the product stream leaving the last ($n = N$) reactor stage z_N.

$$z_N = z_0(1 - \chi) \tag{5.2}$$

3. Calculate the required reactor size for N = 1, 2, and 3.
   a. one CSTR ($N = 1$):

$$V_1 = \frac{F\chi}{k(1 - \chi)} \tag{5.3}$$

$$z_1 = z_0(1 - \chi) \tag{5.4}$$

b. Two CSTRs ($N = 2$):

$$V_1 = V_2 = \frac{F(1 - \sqrt{1 - \chi})}{k\sqrt{1 - \chi}} \tag{5.5}$$

$$z_2 = z_0(1 - \chi) \tag{5.6}$$

$$z_1 = \frac{Fz_0}{F + V_1 k} \tag{5.7}$$

c. Three CSTRs ($N = 3$):

$$V_1 = V_2 = V_3 = \frac{F[1 - (1 - \chi)^{1/3}]}{k(1 - \chi)^{1/3}} \tag{5.8}$$

$$z_3 = z_0(1 - \chi) \tag{5.9}$$

$$z_1 = \frac{Fz_0}{F + V_1 k} \tag{5.10}$$

$$z_2 = \frac{Fz_1}{F + V_2 k} \tag{5.11}$$

4. Calculate the diameter, length, and heat transfer area of each reactor, and determine its capital cost (J. M. Douglas, *Conceptual Design of Chemical Processes*, 1988, McGraw-Hill, New York).

$$D_n = (2V_n/\pi)^{1/3} \tag{5.12}$$

$$L_n = 2D_n \tag{5.13}$$

$$\text{Cost} = 1916.9(D_n)^{1.066}(L_n)^{0.802} \tag{5.14}$$

This is the cost of a tank. In some of our later cases we assume that the cost of a reactor is three times the cost of a simple tank. Clearly, reactor cost depends on the complexity of the reactor, its materials of construction, and the cost of catalyst.

$$A_{Hn} = 2\pi(D_n)^2 \tag{5.15}$$

5. Calculate the heat-removal rate Q_n, the jacket temperature T_{Jn}, and the cooling-water flow rate F_{Jn} for each reactor stage.

$$Q_1 = -(z_0 - z_1)F\lambda - c_p MF(T_1 - T_0) \tag{5.16}$$

$$Q_n = -(z_{n-1} - z_n)F\lambda \tag{5.17}$$

where
λ = heat of reaction (Btu/lb-mol)
c_p = heat capacity reaction liquid = 0.75 (Btu/lb °F)
T_1 = reactor temperature = 140 (°F)
T_0 = feed temperature = 70 (°F)

$$T_{Jn} = T_n - \frac{Q_n}{UA_{Hn}} \tag{5.18}$$

$$F_{Jn} = \frac{Q_n}{\rho_J c_J (T_{Jn} - T_{J0})} \tag{5.19}$$

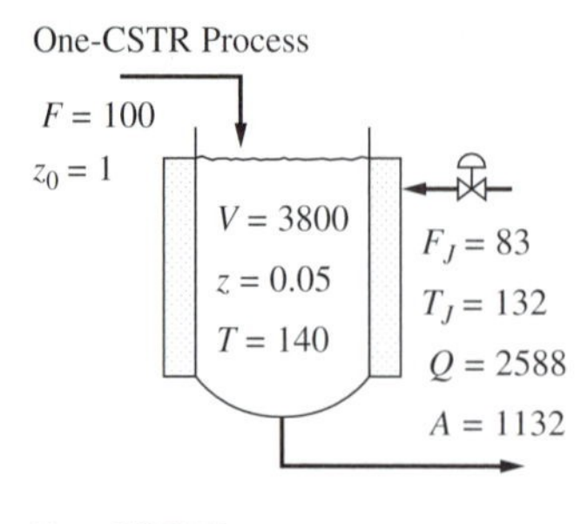

k = 0.5
Conversion = 0.95

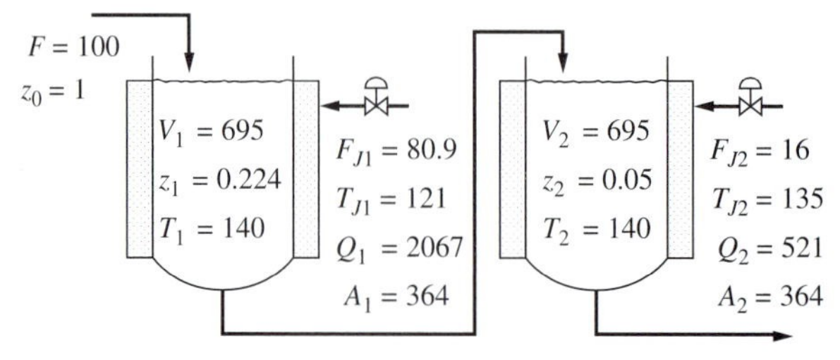

Units

Holdup: lb-mol
Composition: Mole fraction
Flow: lb-mol/hr
Coolant flow: gpm
Q: 1000 Btu/hr
Area: ft^2
Temperatures: °F

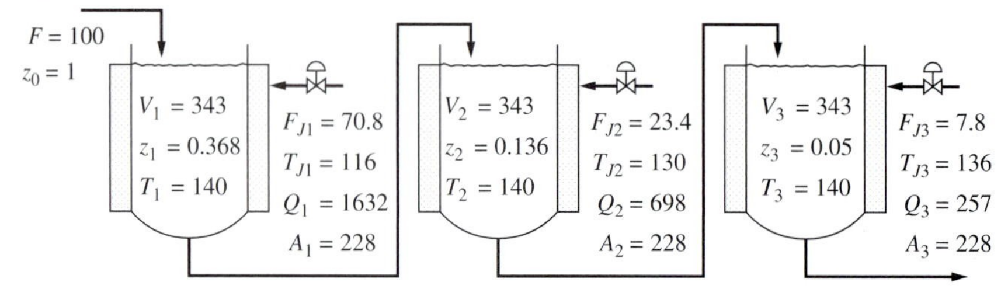

FIGURE 5.1
Alternative designs.

where F_J = flow rate of cooling water (ft^3/hr)
ρ_J = density of cooling water = 62.3 (lb/ft^3)
c_J = heat capacity of coolant = 1 (Btu/lb °F)
T_{J0} = inlet temperature of cooling water = 70 (°F)

Figure 5.1 gives flowsheet conditions for three alternative processes for the 95 percent conversion case. Table 5.1 gives more steady-state design results for a number of cases. Conversion is varied from 99.5 to 75 percent, and specific reaction rate is varied from 0.1 to 2.5 hr^{-1}. These results show some interesting trends that provide us with important insights about the controllability of these various design cases.

1. The heat removal rate is much higher in the first stage than in the later stages. This occurs because the concentration of reactant is the highest at this point. This makes control of the first stage the most difficult.
2. As conversion increases, the required reactor holdups increase. This increases the size of the reactors, giving larger heat transfer areas. Therefore, jacket temperatures are higher and cooling-water flow rates are lower. As we will explain,

the larger heat transfer area improves the controllability of the system. So reactor systems with high conversions are easier to control than reactor systems with low conversions.

3. As specific reaction rates increase, reactor sizes decrease. This decreases the heat transfer area and degrades the controllability of the process. So reactor systems with small specific reaction rates are easier to control than reactor systems with large specific reaction rates (at the same level of conversion).

TABLE 5.1
Steady-state designs of CSTR processes

k	0.10	0.10	0.10	0.10	0.50	0.50	0.50	0.50
χ	0.995	0.975	0.95	0.75	0.995	0.975	0.95	0.75
				1 CSTR				
V	199,000	39,000	19,000	3000	39,800	7800	3800	600
D	50.2	29.16	76.82	12.4	29.4	17.06	13.42	7.25
A_H	15,800	5344	3309	967	5417	1828	1131	228
z	0.005	0.025	0.05	0.25	0.005	0.025	0.05	0.25
Q	2722	2662	2588	1988	2722	2662	2588	1988
T_J	139.43	138.34	137.39	133.15	138.32	135.14	132.38	119.96
F_J	78.46	77.95	76.82	62.97	79.72	81.77	82.99	79.59
				2 CSTRs				
V_n	13,142	5325	3472	1000	2628	1065	694	400
D_n	20.3	15.02	13.02	8.60	11.87	8.78	7.62	5.03
A_H	2588	1417	1066	465	885	485	364	159
z_1	0.0707	0.1581	0.2236	0.50	0.0707	0.1581	0.2236	0.50
z_2	0.005	0.025	0.05	0.25	0.005	0.025	0.05	0.25
Q_1	2525	2263	2067	1238	2525	2263	2067	1238
Q_2	197	399	521	750	197	399	521	750
T_{J1}	136.75	134.68	133.54	131.12	130.49	124.43	121.1	114.05
T_{J2}	139.75	139.06	138.37	134.62	139.26	137.25	135.24	124.27
F_{J1}	75.7	70.01	65.08	40.51	83.53	83.18	80.92	56.21
F_{J2}	5.66	11.57	15.24	23.22	5.69	11.88	15.97	27.65
				3 CSTRs				
V_n	4847	2420	1714	587	969	484	343	117.5
D_n	14.55	11.55	10.29	7.2	8.51	6.75	6.02	4.21
A_H	1331	838	666	326	455	286	228	112
z_1	0.171	0.2924	0.3684	0.63	0.171	0.2924	0.3684	0.63
z_2	0.0293	0.0855	0.1357	0.3969	0.0293	0.0855	0.1357	0.3969
z_3	0.005	0.025	0.05	0.25	0.005	0.025	0.05	0.25
Q_1	2224	1860	1632	848	2224	1860	1632	848
Q_2	425	621	698	699	425	621	698	699
Q_3	72.7	182	257	441	72.7	182	257	441
T_{J1}	134.43	132.6	131.83	131.33	123.71	118.36	116.1	114.66
T_{J2}	138.93	137.53	136.5	132.85	136.89	132.78	129.78	119.09
T_{J3}	139.82	139.28	138.71	135.49	139.47	137.89	136.23	126.83
F_{J1}	69.1	59.5	52.8	27.6	82.9	77.0	70.8	38.0
F_{J2}	12.3	18.4	21.0	22.3	12.7	19.8	23.4	28.5
F_{J3}	2.1	5.2	7.5	13.5	2.1	5.4	7.8	15.5

TABLE 5.1 (CONTINUED)
Steady-state designs of CSTR processes

k	2.5	2.5	2.5	2.5
χ	0.995	0.975	0.95	0.75
			1 CSTR	
V	7960	1560	760	120
D	17.17	9.97	7.85	4.24
A_H	1853	625	387	113
z	0.005	0.025	0.05	0.25
Q	2722	2662	2588	1988
T_J	135.10	125.80	117.72	81.42
F_J	83.67	95.46	108.49	348.24
			2 CSTRs	
V_n	1051	426	139	*
D_n	6.94	5.14	4.45	
A_H	303	166	125	
z_1	0.0707	0.1581	0.2236	
z_2	0.005	0.025	0.05	
Q_1	2525	2263	2067	
Q_2	197	399	521	
T_{J1}	112.19	94.49	84.74	
T_{J2}	137.83	131.97	126.07	
F_{J1}	119.75	184.88	280.6	
F_{J2}	5.81	12.89	18.58	
			3 CSTRs	
V_n	194	96.8	68.6	
D_n	4.98	3.95	3.52	
A_H	156	98	77.9	
z_1	0.171	0.2924	0.3684	
z_2	0.0293	0.0855	0.1357	
z_3	0.005	0.025	0.05	
Q_1	2224	1860	1632	
Q_2	425	621	698	
Q_3	72.7	182	257	
T_{J1}	130.89	76.72	70.13	
T_{J2}	130.89	118.88	110.12	
T_{J3}	138.44	133.82	128.99	
F_{J1}	198.9	553.9	24,330	
F_{J2}	14.0	25.4	34.8	
F_{J3}	2.1	5.7	8.7	

*Design not feasible.
Note: F_J in gpm, Q in 10^3 Btu/hr.

Note that for very high specific reaction rates, a multistage reactor system is not feasible for the numerical case studied. This is because the reactor size gives a transfer area that is so small that the required jacket temperature is lower than the temperature of the available cooling water. Refrigeration could be used or heat transfer area could be increased, as discussed later.

The capital costs of some of the alternative processes are summarized as follows:

Case 1: $k = 0.5$ and 95 percent conversion
 Cost of one-CSTR process = \$427,300
 Cost of two-CSTR process = \$296,600
 Cost of three-CSTR process = \$286,700
Case 2: $k = 0.5$ and 99 percent conversion
 Cost of one-CSTR process = \$1,194,000
 Cost of two-CSTR process = \$536,700
 Cost of three-CSTR process = \$458,200

These numbers show that the optimal economic steady-state design is the multiple-stage reactor system. The higher the conversion, the larger the economic incentive to have multiple stages. At 95 percent conversion, the capital cost of a three-CSTR process is 67 percent that of a one-CSTR process. At 99 percent conversion, the cost is only 38 percent. Thus, if only steady-state economics is considered, the design of choice in these numerical cases is a process with two or three CSTRs in series.

5.3.2 Dynamic Controllability

Now let us examine the dynamic aspects of these alternative designs using rigorous nonlinear simulations to study their controllability. Three differential equations describe each stage: reactor component balance, reactor energy balance, and jacket energy balance (see Appendix A). Two 1-minute first-order lags (we define these in Chapter 8) in the temperature measurement and PI temperature controllers are used. Controller tuning uses the TLC method with some empirical adjustments to give reasonable closedloop damping coefficients with the tightest control of temperature possible.

Figure 5.2 gives the response of a one-CSTR process for step changes in feed rate from 100 to 150 lb-mol/hr and from 100 to 50 lb-mol/hr for the system with $k = 0.5$ and 95 percent conversion. When feed rate is increased, the temperature in the reactor initially decreases. This is due to the sensible heat effect of the colder feed (70°F versus 140°F). After about five minutes, the temperature starts to increase because the concentration of reactant has increased, which increases the rate of reaction. The maximum temperature deviation is only 0.06°F, but it takes over five hours to return to the setpoint because of the slow change in reactor concentration and the large reset time.

Figure 5.3 shows the response of the two-CSTR process for the same disturbances. Now the maximum temperature deviation in the first reactor is about 0.6°F. This is *10 times larger* than the deviation experienced in the one-CSTR process. Thus, we find that the control of the two-CSTR process is considerably worse than the control of the one-CSTR process. Things get even worse when the three-CSTR process is evaluated. The responses to changes in feed flow rate exhibit larger temperature deviations (1.2°F).

To test these alternative reactor systems with a fairly severe upset, the heat of reaction is increased at time zero. This disturbance could correspond to a sudden change in catalyst activity or to the initiation of a side reaction with a higher heat of

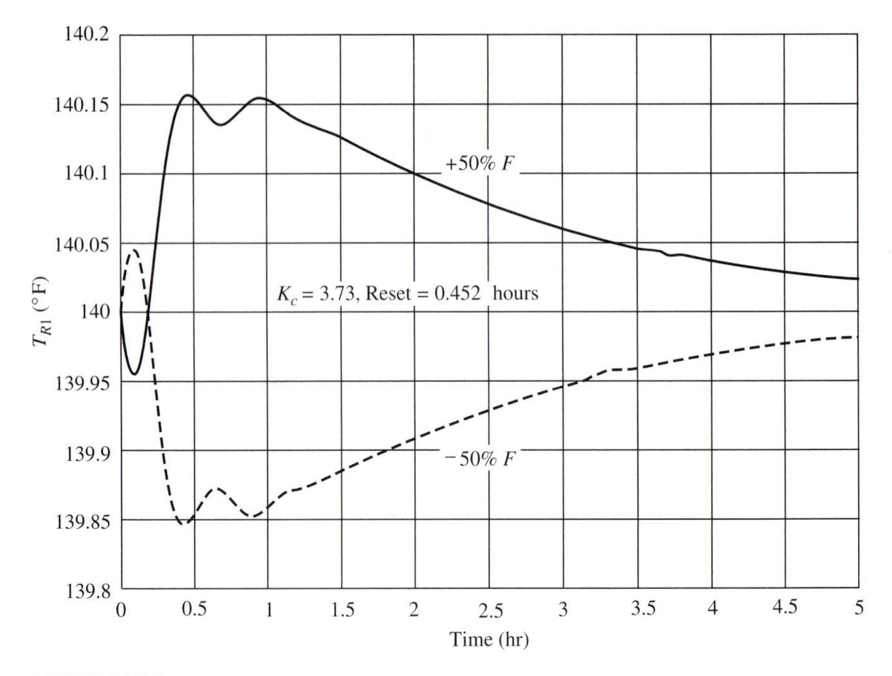

FIGURE 5.2
One CSTR, 95 percent conversion.

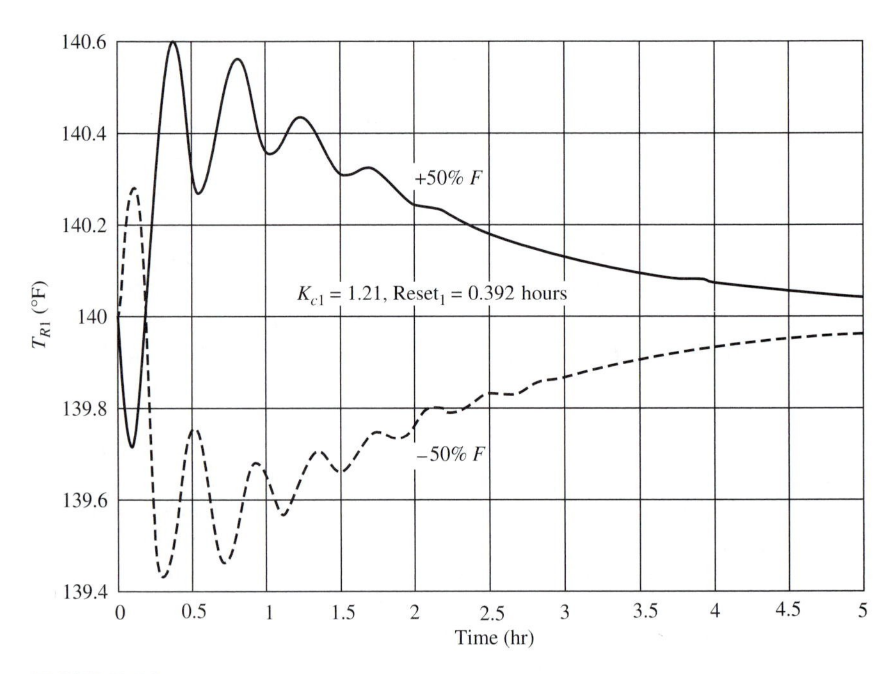

FIGURE 5.3
Two CSTRs.

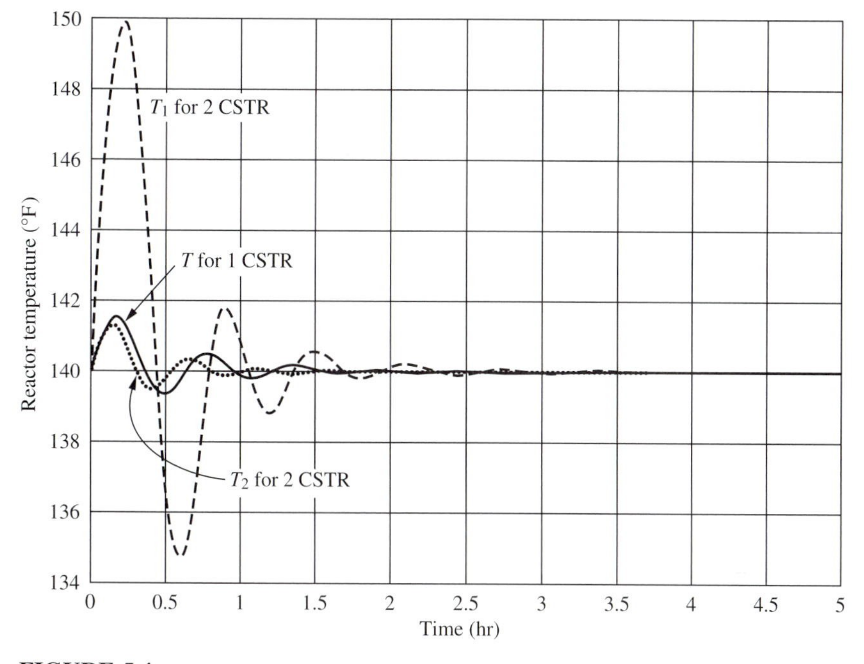

FIGURE 5.4
50 percent reaction heat increase.

reaction. In Fig. 5.4 the heat of reaction increases from 30,000 to 45,000 Btu/lb-mol. Now the better controllability and rangeability of the one-CSTR process is striking compared to the two-CSTR process. A single CSTR handles this disturbance with only a 1.5°F temperature deviation. In a two-reactor system, this disturbance causes a 10°F jump in the temperature in the first reactor. In many reaction systems, product quality may be sensitive to temperature control, so these large temperature deviations would yield poor-quality product.

These results clearly demonstrate that the most economical process from a steady-state point of view is not the best from a dynamic point of view.

The explanation of why the one-CSTR process gives better temperature control than the two-CSTR process involves heat transfer. The total amount of heat generated in the reaction system is the same whether there is one reactor or two reactors. But in the one-CSTR process the heat transfer area is quite large, so the jacket temperature is only a little bit lower than the reactor temperature. Since the inlet cooling-water temperature is much lower than the jacket temperature, a modest increase in the cooling-water flow rate drops the jacket temperature appreciably and gives a large increase in the heat transfer rate. We can easily double or triple the temperature difference between the reactor and the jacket and achieve large increases in heat transfer rates. This means that we can achieve good temperature control because the manipulated variable (cooling-water flow rate) has a big effect on the controlled variable (reactor temperature).

On the other hand, the two-CSTR process has two small reactors with small heat transfer areas. The heat transfer rate in the first reactor (where the reactant concentration is higher) is about 80 percent of the total heat generated in the process. The large amount of heat must be transferred through a smaller area, and this requires a larger temperature difference between the reactor and the jacket. The design temperature difference between the reactor and the jacket is a larger fraction of the maximum available temperature difference (reactor temperature minus inlet cooling-water temperature). Now when the cooling-water flow rate is increased, the jacket temperature can be decreased only slightly, and large changes in heat transfer rates are not possible. This translates into poorer temperature control.

5.3.3 Maximum Heat Removal Rate Criterion

The next logical question is, how do we reconcile these conflicts between steady-state economics and dynamic controllability? In Section 5.6 we present a general procedure that permits quantitative comparisons between alternative designs, taking into account both steady-state economics and dynamic controllability for any chemical process.

For the specific jacket-cooled CSTR process considered in this section and the next, a simple heuristic approach can be used to incorporate quantitatively the limitations of controllability into the steady-state design. The idea is to specify a design criterion that ensures good controllability. In the reactor temperature control problem we use the criterion of a specified ratio of the maximum heat removal rate to the heat removal rate at design conditions. This simple approach is easily understood by designers and operators, and it requires no dynamic simulation or control analysis. We illustrate its usefulness in the following section to determine the "best" reactor operating temperature.

A vital issue in the design of reactors is the ability of the cooling system to handle momentary or sustained heat removal rates that are larger than the nominal design heat removal rate Q. This maximum heat removal rate Q_{max} may have to be only slightly higher than normal if disturbances and uncertainties in kinetic and thermodynamic properties are small. However, this is seldom true, particularly in new processes. So the ratio of Q_{max} to Q may have to be quite large in some kinetic systems. Typical numbers range from 2 to 4 depending on the severity of reaction disturbances and the consequences of a reactor runaway. Specifying this ratio limits the reactor design and establishes the maximum feed rate that can be achieved in a single CSTR with only jacket cooling for a given inlet coolant temperature.

To calculate the maximum steady-state heat removal capacity, we assume that the cooling-water flow rate at design is 25 percent of its maximum. Thus, when the control valve is wide open, four times the design flow rate of coolant is available. We also assume that the reactor temperature is held at its specified value. The steady-state heat transfer rate under these conditions can be calculated from the following equations:

$$Q_{\text{max}} = UA_H(T_R - T_J^{\text{max}}) \tag{5.20}$$

$$Q_{\text{max}} = F_J^{\text{max}} \rho_J c_J (T_J^{\text{max}} - T_{J0}) \tag{5.21}$$

$$F_J^{\text{max}} = 4F_J^{\text{design}} \tag{5.22}$$

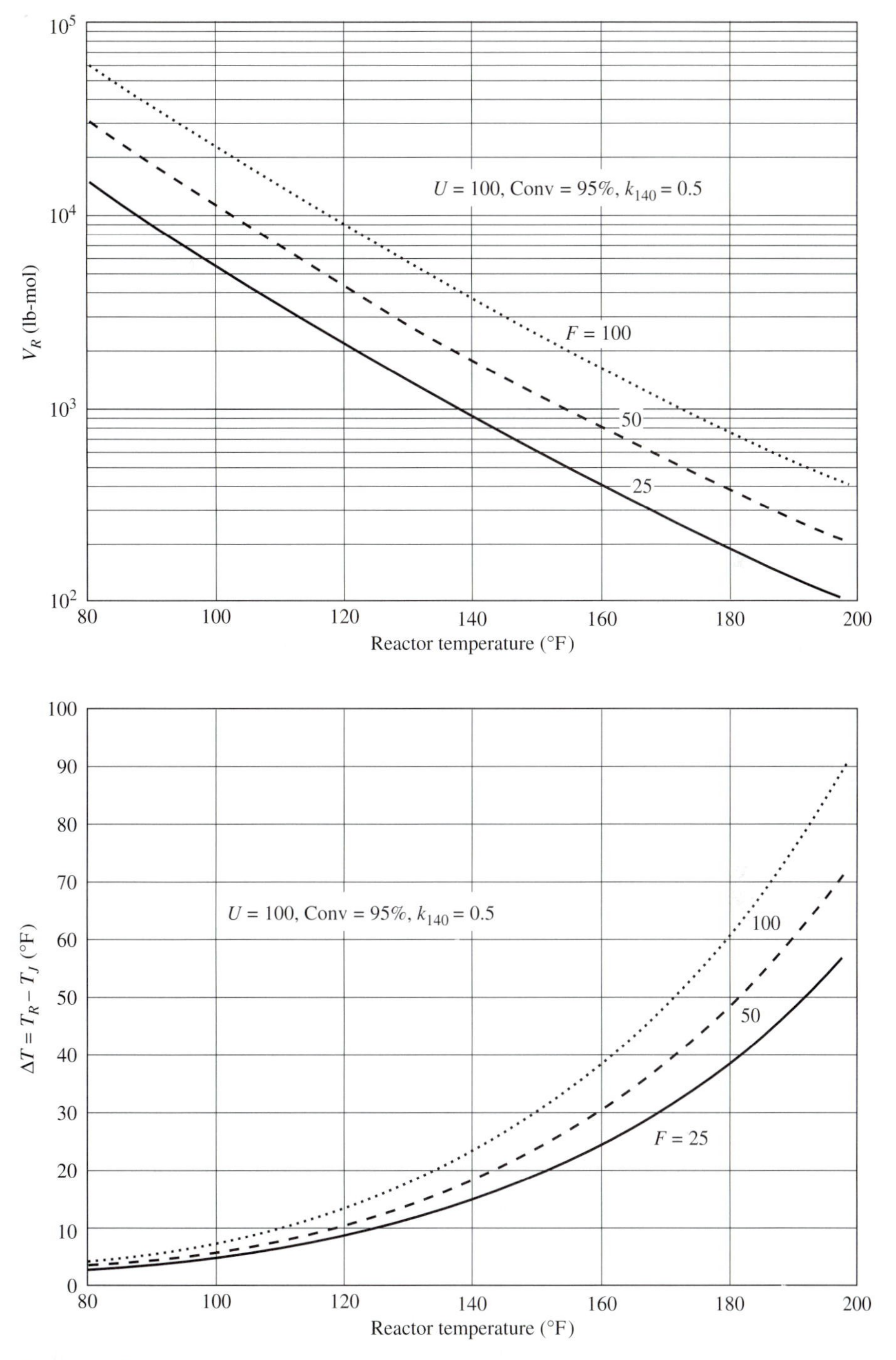

FIGURE 5.5
CSTR, feed flow.

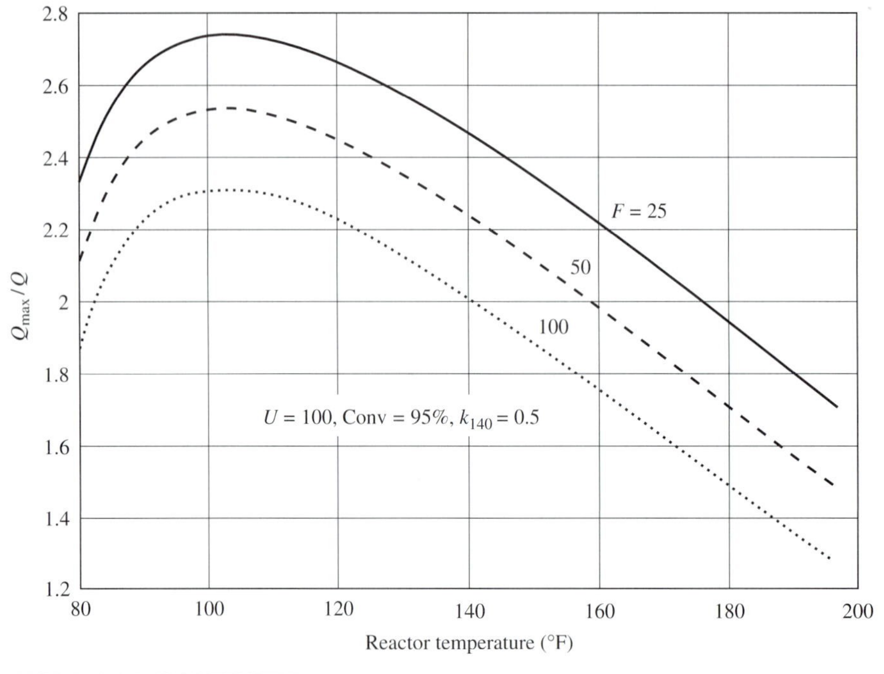

FIGURE 5.5 (CONTINUED)
CTSR, feed flow.

Combining gives

$$T_J^{max} = \frac{UA_H T_R + F_J^{max} c_J \rho_J T_{J0}}{F_J^{max} c_J \rho_J + UA_H} \tag{5.23}$$

Then Q_{max} can be found from Eq. (5.20).

Figure 5.5 shows how reactor holdup V_R, the ratio Q_{max}/Q, and the temperature difference between the reactor and the jacket, ΔT, vary with reactor temperature for three feed flow rates (95 percent conversion and a specific reaction rate at 140°F of 0.5 hr^{-1}). The higher the temperature and the lower the feed rate, the smaller the reactor holdup is. The higher the temperature and the higher the feed rate, the larger the temperature difference is. The Q_{max}/Q curves are not monotonic. At low reactor temperatures the reactor is large, the heat transfer area is large, and ΔT is small. But the jacket temperature is only slightly above the inlet cooling-water temperature (70°F). This means that the cooling-water flow rate at design conditions is quite large, and even if it is increased to four times the design flow rate, the jacket temperature cannot be lowered below the inlet cooling-water temperature. Therefore, as we move to higher temperatures, we see an increase in the Q_{max}/Q ratio because the jacket temperature at design conditions is increasing. However, as reactor temperature is increased still further, the decrease in reactor heat transfer area begins to rapidly increase ΔT at design conditions. When cooling water is increased by a factor of 4 in these high-temperature designs, the jacket temperature cannot be decreased enough to change ΔT (and Q) that much. At low temperatures, maximum heat removal is limited by inlet cooling water temperature. At high temperatures, it is limited by heat transfer area. The nonmonotonic curve means that for a given feed rate and a specified ratio Q_{max}/Q , there are two possible designs. The first would have a low reactor temperature and a large reactor volume. The second would have a high reactor temperature and a small reactor volume. Naturally, the latter is the design of choice since it offers the same controllability at a lower capital cost.

There is no guarantee that the specified Q_{max}/Q ratio can be achieved at the specified feed flow rate. Figure 5.5 shows that the highest ratio obtainable for a feed rate of 100 lb-mol/hr under the base-case conditions is 2.3. If we desired a ratio greater than this, the process would have to be modified to provide more heat transfer area than is given by just jacket cooling. This can be accomplished in several ways: use an external heat exchanger with circulation of reaction liquid, use evaporative cooling with a condenser mounted above the reactor, use internal cooling coils, or use two (or more) reactors *in parallel.*

5.4 IMPACT OF CONTROLLABILITY ON CAPITAL INVESTMENT AND YIELD

Let us consider the situation where reactor temperature is a design parameter. We explore the impact of controllability questions on the choice of the "best" temperature for two kinetic cases: a simple reaction case A $\rightarrow$ B, and a consecutive reaction case A $\rightarrow$ B $\rightarrow$ C. Only single-CSTR processes are discussed.

5.4.1 Single-Reaction Case

The optimal reactor temperature from the standpoint of steady-state economics is the highest possible temperature since this minimizes reactor holdup for a given conversion, which results in the smallest capital cost. The upper temperature limitation may be due to metallurgical constraints, product thermal degradation, safety, undesirable side reactions, or other factors.

We demonstrate in this section that the "best" temperature from the standpoint of controllability is *not* the highest possible. This results from the reduction in cooling-jacket heat transfer area that occurs as the size of the reactor is reduced. The temperature difference between the reactor and the jacket becomes bigger, resulting in a reactor that is more difficult to control.

The numerical values of parameters in this simulation are the same as those used previously. Several different values of conversion, feed flow rate, and heat transfer coefficient were studied. The specific reaction rate at a base temperature of 140°F (k_{140}) is assumed to be 0.5 hr^{-1}.

The steady-state design procedure is outlined for a given fresh feed flow rate F and fresh feed composition ($z_0 = 1$, pure reactant) and a specified conversion χ, overall heat transfer coefficient U, and k_{140}.

1. Calculate the concentration of the reactant in the reactor, z.

$$z = z_0(1 - \chi) \tag{5.24}$$

2. Select a reactor temperature, T_R.
3. Calculate the specific reaction rate at this temperature.

$$k = k_0 e^{-E/R(T_R+460)} \tag{5.25}$$

where $k_0 = k_{140} e^{E/600R}$ (5.26)

E = activation energy = 30,000 Btu/lb-mol

R = 1.99 Btu/lb-mol °R

4. Calculate the required reactor holdup.

$$V_R = \frac{F(z_0 - z)}{kz} \tag{5.27}$$

5. Calculate the diameter D_R and length L_R of the reactor, the heat transfer area A_H, and the installed capital cost.
6. Calculate the heat removal rate Q, the jacket temperature T_J, and the cooling water flow rate F_J.
7. Repeat steps 3 through 6 for a range of reactor temperatures.

Detailed steady-state design results for two levels of conversion are given in Table 5.2, and the significant results are illustrated in Fig. 5.5. These steady-state design calculations all indicate that the reactor should be operated at the highest

TABLE 5.2
Steady-state designs of CSTR processes (U = 100, k_{140} = 0.5)

Conversion = 95% (z = 0.05)						
T_R	100	120	140	160	180	200
V_R	22, 867	9038	3800	1690	790	387
D_R	24.41	17.91	13.42	10.24	7.95	6.27
A_H	3744	2016	1132	659.3	397.3	246.9
Q	2737	2662	2587	2512	2437	2362
T_J	92.69	106.80	117.14	121.89	118.65	104.30
F_J	241.4	144.8	109.8	96.9	100.3	137.8
ΔT	7.31	13.2	22.86	38.11	61.35	95.70
T_J^{max}	83.1	90.53	93.81	92.86	88.20	80.69
Q_{max}	6325	5942	5228	4427	3647	2945
Ratio Q_{max}/Q	2.311	2.232	2.021	1.762	1.497	1.247
Cost	1306	733	427	258	161	103
Conversion = 99% (z = 0.01)						
T_R	120	140	160	180	200	
V_R	47, 091	19, 800	8804	4118	2017	
D_R	31.05	23.27	17.76	13.78	10.87	
A_H	6059	3401	1981	1194	742	
Q	2782	2707	2632	2557	2482	
T_J	115.41	132.04	146.71	158.58	166.54	
F_J	122.6	87.3	68.7	57.8	51.5	
ΔT	4.59	7.96	13.29	21.42	33.46	
T_J^{max}	105.6	116.26	123.17	125.91	124.47	
Q_{max}	8726	8075	7298	6457	5603	
Ratio Q_{max}/Q	3.137	2.983	2.773	2.525	2.257	
Cost	2048	1194	721	449	288	

Note: Q in 10^3 Btu/hr; F_J in gpm; capital cost in \$1000; temperatures in °F.

possible temperature to minimize reactor size. However, as reactor temperature increases and reactor volume decreases, the heat transfer area A_H also decreases. The rate of heat transfer decreases only very slightly as reactor temperature increases because most of the heat to be removed comes from the heat of reaction, and sensible heat effects are quite small. This means that the heat transfer differential temperature ΔT between the reactor and the cooling jacket must increase.

As we have already demonstrated, this increase in ΔT indicates that the control of the reactor is more difficult. Anything that increases ΔT gives a less controllable reactor. Therefore, reactors with low conversions, high feed rates, and low overall heat transfer coefficients are difficult to control.

5.4.2 Consecutive Reactions Case

Here we examine reactor design and stability when the kinetic system consists of consecutive reactions A → B → C with component B as the desired product. When the activation energy is larger for the first reaction than for the second, the highest yield of component B is obtained by operating at the highest possible temperature. However, dynamic controllability limits this maximum temperature for a given reactor size. Thus, controllability constraints not only give higher capital costs, but they also result in lower yields of product B. Yields have a significant effect on process economics. First we look at the reactor by itself, and then we look at a reactor-column system with recycling of unreacted component A back to the reactor.

Two consecutive reactions occur in a single isothermal CSTR.

$$\mathrm{A} \xrightarrow{k_1} \mathrm{B} \xrightarrow{k_2} \mathrm{C}$$

$$k_i = k_{i0} e^{-E_i/RT_R} \tag{5.28}$$

where k_i = specific reaction rate of ith reaction (hr^{-1})
k_{i0} = preexponential factor
E_i = activation energy
T_R = reactor temperature (°R)

The reaction rates for the two reactions are each first-order in the concentrations of components A and B. The activation energy for the first reaction is 30,000 Btu/lb-mol and for the second is 15,000 Btu/lb-mol. The specific reaction rate of the first reaction is 0.5^{-1} and of the second reaction is 0.05 hr^{-1} at a temperature of 140°F.

$$\mathcal{R}_1 = V_R k_1 z_A \tag{5.29}$$

$$\mathcal{R}_2 = V_R k_2 z_B \tag{5.30}$$

where $\mathcal{R}_i$ = reaction rate of ith reaction (lb-mol/hr of A or B)
z_A = concentration of reactant A in the reactor (mole fraction A)
z_B = concentration of reactant B in the reactor (mole fraction B)

In all the numerical cases studied the following design parameters are assumed to be constant:

U = overall heat transfer coefficient = 100 Btu/hr ft^2 °F
T_{J0} = inlet cooling-water temperature = 70°F

ΔH_R = heat of reaction for each reaction = 20,000 Btu/lb-mol of component A or component B
T_0 = fresh feed temperature = 70°F
z_0 = fresh feed composition = 1 mole fraction component A

To gain some understanding of the steady-state design aspects of the reactor by itself, the following procedure is used:

1. Pick a reactor holdup V_R, a reactor temperature T_R, and a fresh feed rate F.
2. Calculate the concentrations z_A and z_B in the reactor from component balances.

$$z_A = \frac{F z_0}{F + V_R k_1} \tag{5.31}$$

$$z_B = \frac{V_R k_1 z_A}{F + V_R k_2} \tag{5.32}$$

3. Calculate the heat removal rate Q, the jacket temperature T_J, and the coolant flow rate F_J under these design conditions.

$$Q = -\lambda V_R(k_1 z_A + k_2 z_B) - c_p M F(T_R - T_0) \tag{5.33}$$

   The heat of reaction λ is assumed to be the same for the two reactions in Eq. (5.33).
4. Calculate the Q_{max}/Q ratio using Eqs. (5.20) and (5.23).

Results for several numerical cases are shown in Fig. 5.6. The fresh feed rate is set at 50 lb-mol/hr. The concentration of the desired component B is plotted for three reactor temperatures (150, 175, and 200°F) over a range of reactor holdups. It is clear that for the highest yields, the reactor should operate at the highest temperature and the optimal reactor holdup is small. This prevents too much of component B from reacting to form the undesired component C.

However, we can see that the small reactors have small Q_{max}/Q ratios and are more difficult to control. For example, if we could operate a 60-lb-mol reactor at 200°F, we could achieve 72 percent yield of component B. However, the Q_{max}/Q ratio of this reactor is only 1.15, indicating that temperature control will be poor and reactor runaways can easily occur. A 120-lb-mole reactor at 175°F has a Q_{max}/Q ratio of 1.3 and will show better dynamic controllability, but the yield is only 66 percent.

Suppose our design criterion is a Q_{max}/Q ratio of 1.5. Figure 5.6 shows that this requires the 200°F reactor to have a holdup of about 170 lb-mol, giving a yield of 62 percent (point C on the figure). A 175°F reactor would have a holdup of 220 lb-mol, giving a yield of about 63 percent (point B). A 150°F reactor would have a holdup of 270 lb-mol, giving a yield of 61 percent. Thus, there is an optimal reactor temperature and a maximum attainable yield that is limited by the controllability criterion.

So far we have looked at just the reactor by itself. In a real chemical plant the reactor is typically part of a reaction-separation system to increase the yield of the desired product. The concentration of component B in the reactor is kept low so that little component C is produced. But the higher concentration of component A in

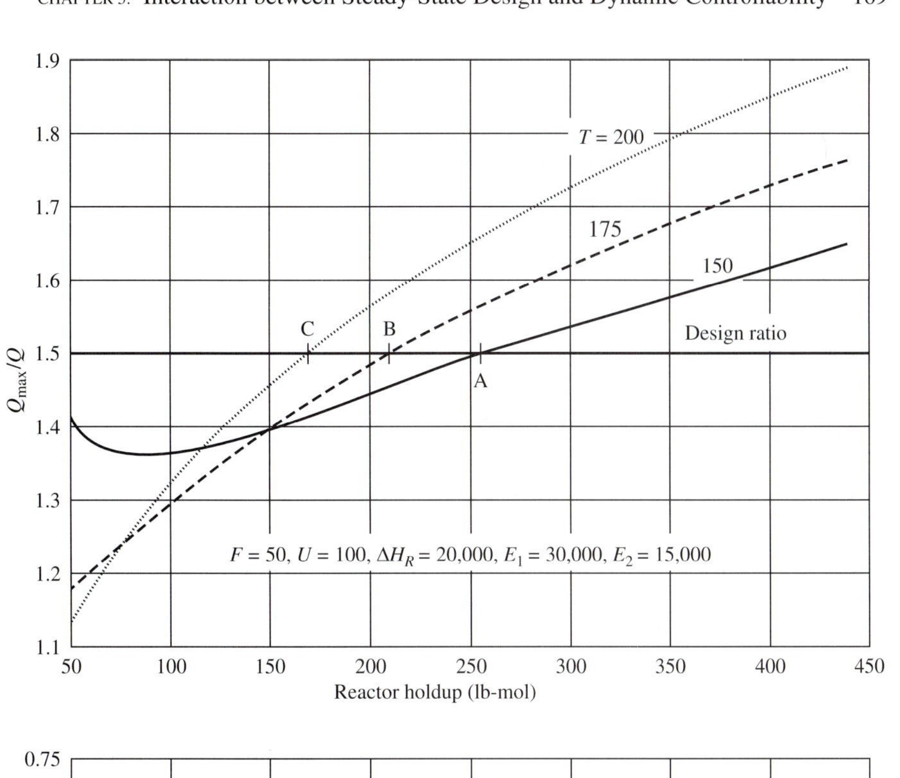

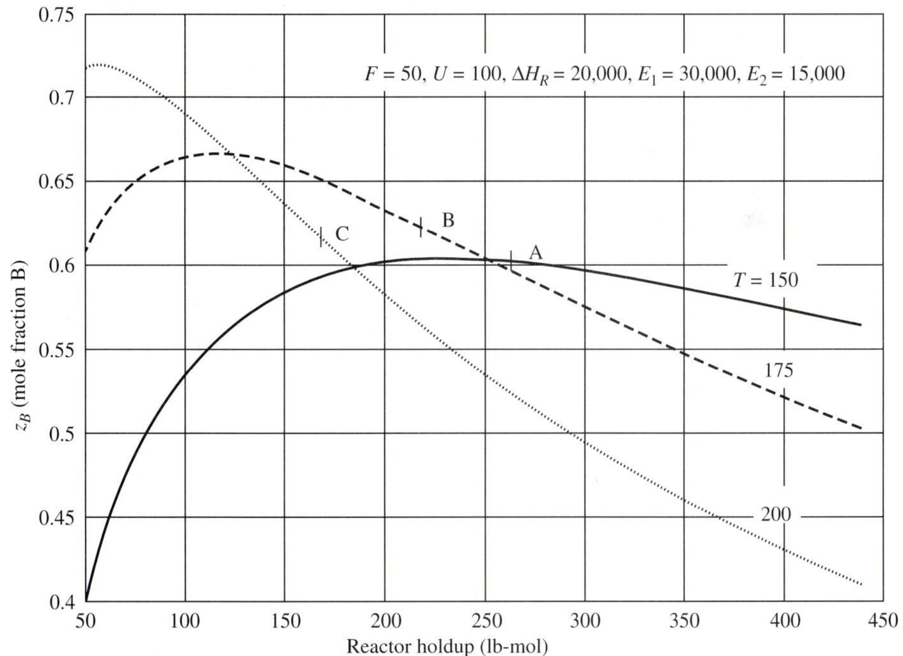

FIGURE 5.6
Consecutive reactions.

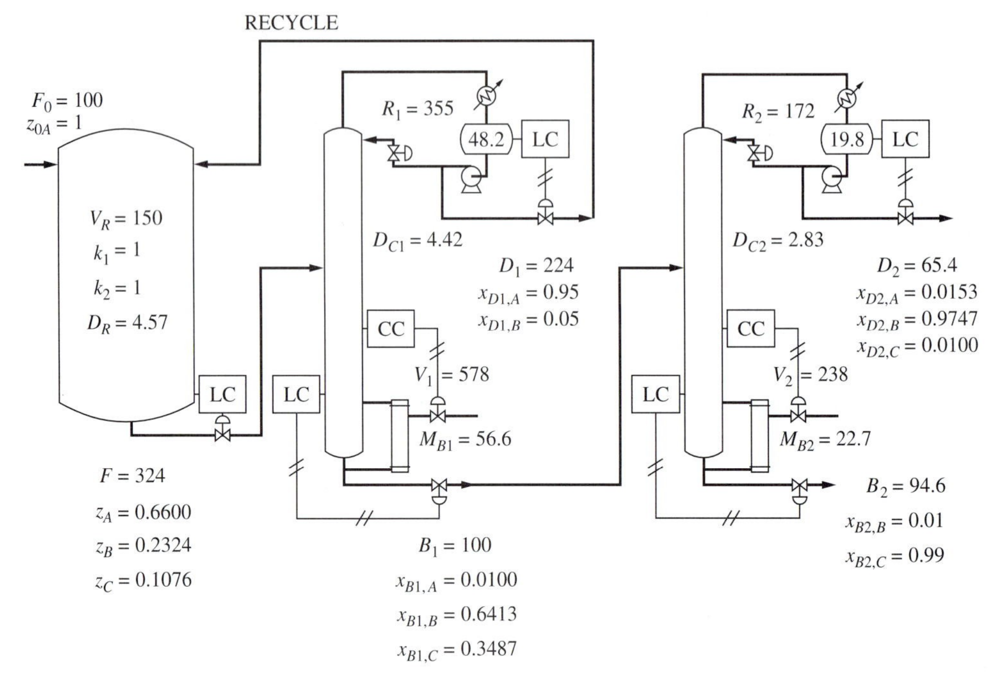

FIGURE 5.7
Ternary reactor, two-column process ($\alpha = 4/2/1$).

the reactor effluent requires a recycle of this component back to the reactor. Much higher yields can be obtained from this type of recycle system. However, the reactor still encounters controllability limitations.

Figure 5.7 is a sketch of the plant under consideration. Fresh feed enters the reactor at a flow rate F_0 and composition $z_{0,A} = 1$ (pure component A in the fresh feed). We assume that the relative volatilities of components A, B, and C, $\alpha_A/\alpha_B/\alpha_C$, are 4/2/1, respectively, so unreacted component A comes overhead in the first distillation column and is recycled back to the reactor at a rate D_1 and composition x_{D1_j}. Reactor effluent F is fed into the first distillation column. The flow rates of reflux and vapor boilup in this column are R_1 and V_1. Bottoms B_1 from the first column is fed into the second column, in which components B and C are separated into product streams with about 1 percent impurity levels.

Steady-state component balances around the whole system and around each of the units are used to solve for the conditions throughout the plant for a given recycle flow rate D_1. The reactor holdup V_R and the reactor temperature T_R necessary to achieve a specified $Q_{\max}/Q$ ratio are calculated as part of the design procedure. The other fixed design parameters are the kinetic constants (preexponential factors, activation energies, and heats of reaction for both reactions), the fresh feed flow rate and composition, the overall heat transfer coefficient in the reactor, the inlet coolant temperature, and thc following product purity specifications:

$$x_{D1,C} = 0 \qquad x_{D1,B} = 0.05 \qquad x_{B1,A} = 0.01$$
$$x_{D2,C} = 0.01 \qquad x_{B2,B} = 0.01 \qquad x_{B2,A} = 0$$

The design procedure is as follows:

1. Set D_1. There is an optimal value of this recycle flow rate that maximizes the yield of B.
2. Calculate $F = F_0 + D_1$ and $B_1 = F_0$.
3. Calculate z_A.

$$z_A = \frac{B_1 x_{B1,A} + D_1 x_{D1,A}}{F} \tag{5.34}$$

4. Calculate the product of V_R and k_1.

$$[V_R k_1] = F_0(z_{0,A} - x_{B1,A})/z_A \tag{5.35}$$

5. Guess a value of reactor temperature T_R.
   a. Calculate k_1 and k_2.
   b. Calculate V_R from the $[V_R k_1]$ product calculated in Eq. (5.35).
   c. Calculate the diameter D_R and length L_R of the reactor.
   d. Calculate the circumferential heat transfer area of the jacket.
   e. Calculate z_B.

$$z_B = \frac{V_R k_1 z_A - D_1 x_{D1,B}}{F + V_R k_2} \tag{5.36}$$

   f. Calculate Q, T_J, F_J, and $Q_{\max}$.
   g. If the $Q_{\max}/Q$ ratio is not equal to the desired value, reguess reactor temperature.

6. Calculate remaining flow rates and compositions in the columns from component balances.

$$\begin{aligned} x_{B1,B} &= (F z_B - D_1 x_{D1,B})/B_1 \\ x_{B1,C} &= 1 - x_{B1,A} - x_{B1,B} \end{aligned} \tag{5.37}$$

$$B_2 = \frac{B_1(1 - x_{D2,C} - x_{B1,A} - x_{B1,B})}{1 - x_{D2,C} - x_{B2,B}} \tag{5.38}$$

$$D_2 = B_1 - B_2 \tag{5.39}$$

$$x_{D2,A} = B_1 x_{B1,A}/D_2 \tag{5.40}$$

7. Size the reactor and the distillation columns (using 1.5 times the minimum number of trays and 1.2 times the minimum reflux ratio for each column).
8. Calculate the capital cost, the energy cost, and the total annual cost.

Table 5.3 gives detailed results of these calculations for several feed rates, and Fig. 5.8 shows some of the important results.

TABLE 5.3
Reactor-column designs

F_0	40	50	60
	Reactor		
D_1^{opt}	20	17	15
T_R	126.76	130.63	134.90
V_R	431.9	593.6	744.1
z_A	0.3233	0.2485	0.1980
z_B	0.5325	0.5514	0.5548
T_J	95.22	96.94	98.85
F_J	66.43	82.12	95.46
A_H	265.6	328.3	381.7
D_R	6.50	7.23	7.79
Q	837.4	1106	1376
Q_{max}	1256	1659	2064
	Column 1		
N_T/N_F	15/7	15/7	15/6
B_1	40	50	60
D_1	20	17	15
R_1	59.0	63.6	68.5
V_1	79.0	80.6	83.5
D_{C1}	1.63	1.65	1.68
	Column 2		
N_T/N_F	19/12	19/12	19/10
B_2	8.43	13.17	18.30
D_2	31.57	36.83	41.70
R_2	44.0	56.0	67.85
V_2	75.6	92.8	109.6
D_{C2}	1.60	1.77	1.92
Energy	1.93	2.17	2.41
	Costs		
Reactor	110.3	134.5	154.8
Column 1	57.3	57.9	59.0
Column 2	67.6	75.5	82.4
HtEx.	141.3	152.5	163.5
Capital	379.3	423.2	462.8
Total annual cost	211.1	236.0	260.0
Yield	78.93	73.66	69.85

Notes: Total annual cost = \$1000/yr, capital cost = \$1000, diameter = feet, energy = 10^6 Btu/hr, composition = mole fraction, flow rate = lb-mol/hr, holdup = lb-mol

1. There is an optimal recycle flow rate for a given feed rate that maximizes yield.
2. The maximum attainable yield in the reactor-column recycle process is higher than for just a reactor for the same controllability. The reactor alone with a feed rate of 50 lb-mol/hr gives a maximum yield of about 63 percent for a Q_{max}/Q ratio of 1.5. For the same ratio, the reactor-column system with the same fresh

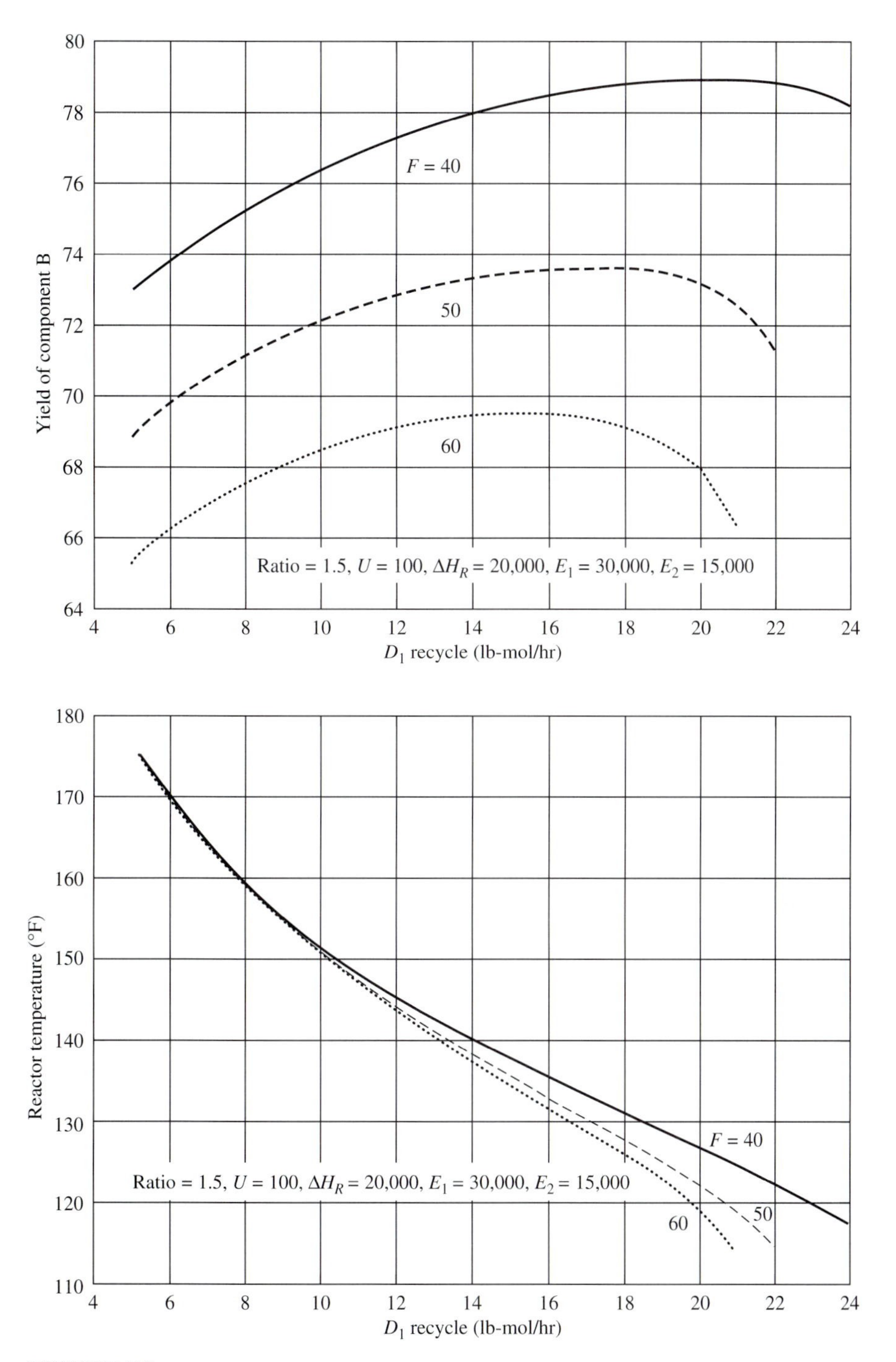

FIGURE 5.8
CSTR-column design.

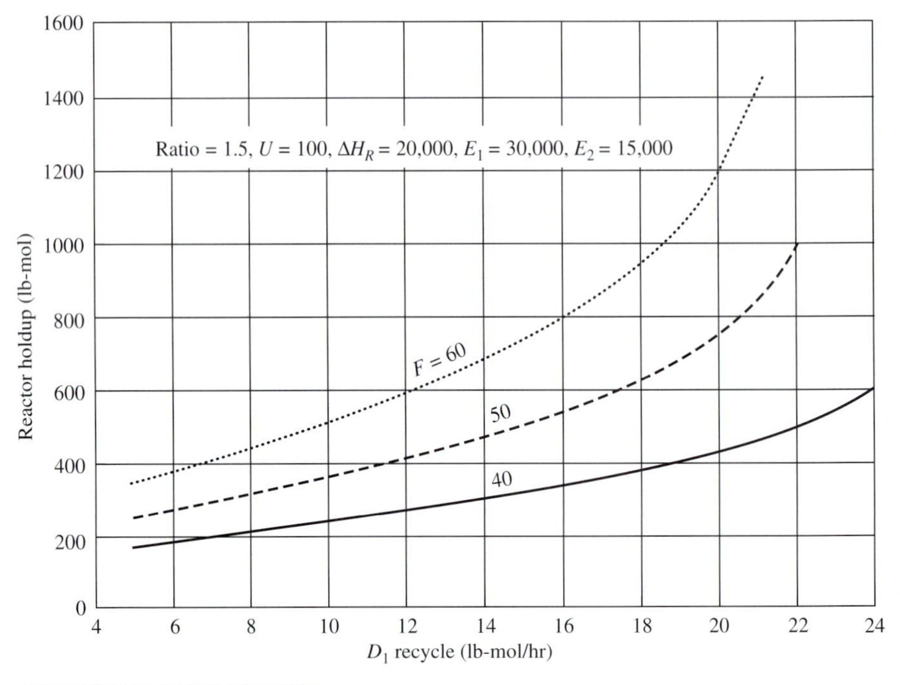

FIGURE 5.8 (CONTINUED)
CSTR-column design.

feed rate gives a yield of 73.5 percent. Of course, the energy and capital costs are higher.

3. The maximum yield depends strongly on the feed flow rate.

The last result suggests that we may want to modify the process to achieve better yields but, at the same time, maintain controllability. This can be done by increasing the heat transfer area in the reactor.

5.5 GENERAL TRADE-OFF BETWEEN CONTROLLABILITY AND THERMODYNAMIC REVERSIBILITY

The field of engineering contains many examples of trade-offs. You have seen some of them in previous courses. In distillation there is the classical trade-off between the number of trays (height) and the reflux ratio (energy and diameter). In heat transfer there is the trade-off between heat exchanger size (area) and pressure drop (pump or compressor work); more pressure drop gives higher heat transfer coefficients and smaller areas but increases energy cost. We have mentioned several trade-offs in this book: control valve pressure drop versus pump head, robustness versus performance, etc.

The results from the design-control interaction examples discussed in previous sections hint at the existence of another important trade-off: dynamic controllability versus thermodynamic reversibility. As we make a process more and more efficient (reversible in a thermodynamic sense), we are reducing driving forces, i.e., pressure drops, temperature differences, etc. These smaller driving forces mean that we have weaker handles to manipulate, so that it becomes more difficult to hold the process at the desired operating point when disturbances occur or to drive the process to a new operating point.

Control valve pressure drop design illustrated this clearly. Low-pressure-drop designs are more efficient because they require less pump energy. But low-pressure-drop designs have limited ability to change the flow rates of manipulated variables.

The jacketed reactor process also illustrates the principle. The big reactor has a lot of heat transfer area, so only a fraction of the available temperature difference between the inlet cooling water and the reactor is used. A thermodynamically reversible process has no temperature difference between the source (the reactor) and the sink (the inlet cooling water). So the big reactor is thermodynamically inefficient, but it gives better control.

We could cite many other examples of this controllability/reversibility trade-off, but the simple ones mentioned above should convey the point: the more efficient the process, the more difficult it is to control. This general concept helps to explain in a very general way why the steady-state process engineer and the dynamic control engineer are almost always on opposite sides in process synthesis discussions.

5.6 QUANTITATIVE ECONOMIC ASSESSMENT OF STEADY-STATE DESIGN AND DYNAMIC CONTROLLABILITY

One of the most important problems in process design and process control is how to incorporate dynamic controllability quantitatively into conventional steady-state design. Normally, steady-state economics considers capital and energy costs to calculate a total annual cost, a net present value, etc. If the value of products and the costs of raw materials are included, the annual profit can be calculated. The process that minimizes total annual cost or maximizes annual profit is the "best" design.

However, as we have demonstrated in our previous examples, this design is usually not the one that provides the best control, i.e., the least variability of product quality. What we need is a way to incorporate quantitatively (in terms of dollars/year) this variability into the economic calculations. We discuss in this section a method called the capacity-based approach that accomplishes this objective. It should be emphasized that the method provides an analysis tool, not a synthesis tool. It can provide a quantitative assessment of a proposed flowsheet or set of parameter values or even a proposed control structure. But it does not generate the "best" flowsheet or parameter values; it only evaluates proposed systems.

5.6.1 Alternative Approaches

A. Constraint-based methods

The basic idea behind constraint-based approaches is to take the optimal steady-state design and determine how far away from this optimal point the plant must operate in order not to violate constraints during dynamic upsets. The steady-state economics are then calculated for this new operating point. Alternative designs are compared on the basis of their economics at their dynamically limited operating point. This method yields realistic comparisons, but it is computationally intensive and is not a simple, fast tool that can be used for screening a large number of alternative conceptual designs.

B. Weighting-factor methods

With weighting-factor methods, the basic idea is to form a multiobjective optimization problem in which some factor related to dynamic controllability is added to the traditional steady-state economic factors. These two factors are suitably weighted, and the sum of the two is minimized (or maximized). The dynamic controllability factor can be some measure of the "goodness" of control (integral of the squared error), the cost of the control effort, or the value of some controllability measure (such as the plant condition number, to be discussed in Chapter 9). One real problem with these approaches is the difficulty of determining suitable weighting factors. It is not clear how to do this in a general, easily applied way.

5.6.2 Basic Concepts of the Capacity-Based Method

The basic idea of the capacity-based approach is illustrated in Fig. 5.9 for three plant designs. The dynamic responses of these three hypothetical processes to the same set of disturbances can be quite different. The variable plotted indicates the quality of the product stream leaving the plant. Better control of product quality is achieved in plant design 3 than in the other designs.

Suppose the dashed lines in Fig. 5.9 indicate the upper and lower limits for "on-aim" control of product quality. Plant 3 is always within specification, and therefore all of its production can be sold as top-quality product. Plant 1 has extended periods when its product quality is outside the specification range. During these periods the production would have to be diverted from the finished-product tank and sent to another tank for reworking or disposal. This means that the capacity of plant 1 is reduced by the fraction of the time its products are outside the specification range. This has a direct effect on economics. Thus, the three plant designs can be directly and quantitatively compared using the appropriate capacity factors for each plant.

The annual profit for each plant is calculated by taking the value of the on-specification products and subtracting the cost of reprocessing off-specification material, the cost of raw materials, the cost of energy, and the cost of capital. Plant 3 may have higher capital cost and higher energy cost than plant 1, but since its product is on-specification all the time, its annual profit may be higher than that of plant 1.

Two approaches can be used to calculate the capacity factors (the fraction of time that the plant is producing on-specification product). The more time-consuming approach is to use dynamic simulations of the plant and impose a series of disturbances. The other approach is more efficient and more suitable for screening a large

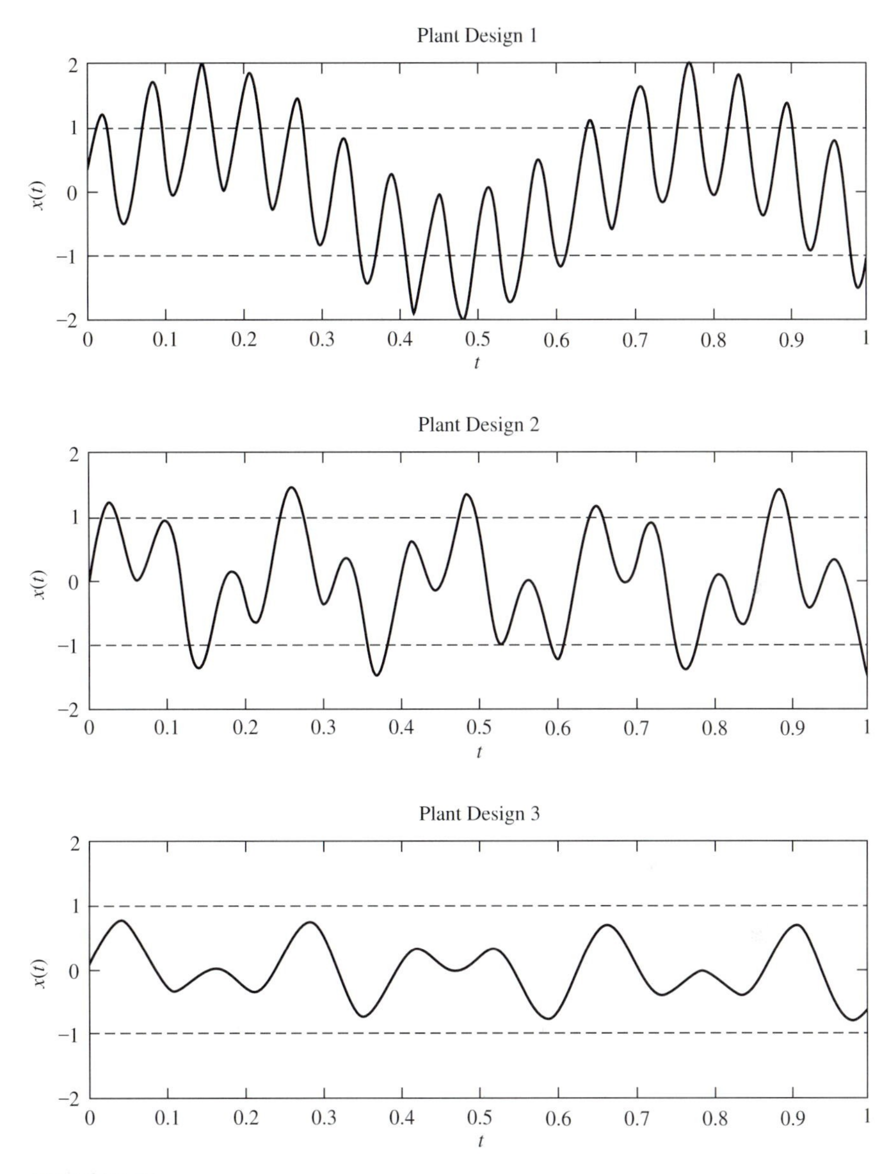

FIGURE 5.9

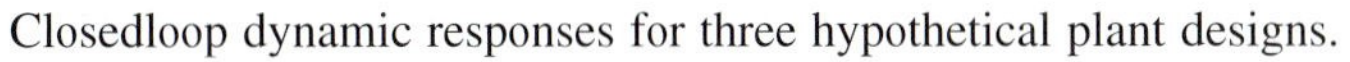
Closedloop dynamic responses for three hypothetical plant designs.

number of alternative designs. It uses frequency-domain methods, which we discuss in Chapter 10.

We illustrate the method in the next section, considering a simple reactor-column process with recycle. In this example the flowsheet is fixed, and we wish to determine the "best" values of two design optimization parameters.

5.6.3 Reactor-Column-Recycle Example

A first-order, irreversible liquid-phase reaction $A \xrightarrow{k} B$ occurs in a single CSTR with constant holdup V_R. The reactor operates at 140°F with a specific reaction rate k of 0.34086 hr^{-1}. The activation energy E is 30,000 Btu/lb-mol.

Figure 5.10 gives the flowsheet of the process and defines the nomenclature. Fresh feed to the reactor has a flow rate of $F_0 = 239.5$ lb-mol/hr and a composition $z_0 = 0.9$ mole fraction component A (and 0.1 mole fraction component B). A recycle stream D from the stripping column is also fed into the reactor. The reactor is cooled by the addition of cooling water into the jacket surrounding the vertical reactor walls.

The reactor effluent is a binary mixture of components A and B. Its flow rate is F lb-mol/hr and its composition is z mole fraction component A. It is fed as saturated liquid onto the top tray of a stripping column. The volatility of component A to component B is $\alpha = 2$, so the bottoms from the stripper is a product stream of mostly component B, and the overhead from the stripper is condensed and recycled back to the reactor. Product quality is measured by the variability of x_B, the mole fraction of component A impurity in the bottom. The nominal steady-state value of x_B is 0.0105 mole fraction component A.

We assume constant density, equimolal overflow, theoretical trays, total condenser, partial reboiler, and five-minute holdups in the column base and the overhead receiver. Tray holdups and the liquid hydraulic constants are calculated from the Francis weir formula using a one-inch weir height.

Given the fresh feed flow rate F_0, the fresh feed composition z_0, the specific reaction rate k, and the desired product purity x_B, this process has 2 design degrees of freedom; i.e., setting two parameters completely specifies the system. Therefore, there are two design parameters that can be varied to find the "best" plant design. Let us select reactor holdup V_R and number of trays in the stripper N_T as the design

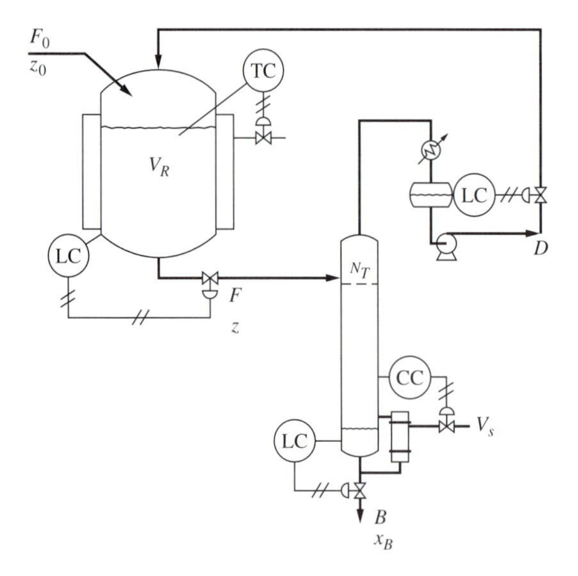

FIGURE 5.10
Reactor/stripper process.

parameters. The following steady-state design procedure can be used to calculate the values of all other variables given V_R and N_T.

1. Calculate the reactor composition from an overall component balance for A.

$$z = \frac{F_0(z_0 - x_B)}{kV_R} \tag{5.41}$$

2. Use a steady-state tray-to-tray rating program for the stripper to calculate the vapor boilup V_S. First guess a value for V_S, and then calculate F from Eq. (5.42) (since $B = F_0$).

$$F = V_S + F_0 \tag{5.42}$$

Then calculate tray by tray from x_B up N_T trays (using component balances and constant relative volatility vapor-liquid equilibrium relationships) to obtain the vapor composition on the top tray y_{NT}. Compare this value with that obtained from a component balance around the reactor, Eq. (5.43).

$$y_{NT} = \frac{Fz + V_R kz - F_0 z_0}{V_S} \tag{5.43}$$

If the two values of y_{NT} are not the same, guess a new value of V_S.

3. Calculate the size of the reactor (from V_R) and the size of the column (from V_S and N_T). Then calculate their capital costs.
4. Calculate the size and the capital costs of the reboiler and condenser (from V_S). Calculate the annual cost of energy (also from V_S).
5. Calculate the total annual cost (TAC) for each V_R-N_T pair of design parameters.

$$\begin{aligned} \text{TAC} &= \text{annual energy cost} \\ &\quad + \text{total capital cost (column + reactor + heat exchangers)/3} \end{aligned}$$

A three-year payback on capital costs is assumed.

By calculating TACs for a range of values of V_R and N_T, the minimum steady-state optimal plant turns out to have a reactor holdup of 3000 lb-mol and a stripper with 19 trays. With no consideration of dynamic controllability, this is the "best" plant.

Now let us apply the capacity-based approach. Positive and negative 10 percent disturbances are made in the fresh feed flow rate F_0 and in the fresh feed composition z_0. Dynamic simulations (confirmed by frequency-domain analysis, to be discussed in Chapter 10) show that the variability in product quality x_B is decreased by increasing reactor volume or by decreasing the number of trays in the stripper.

For a very large specification range on x_B (3 mol%), all the process designs produce on-specification products 100 percent of the time, so the maximum profit plant naturally corresponds to the minimum-TAC plant. However, as the specification range is reduced, the most profitable plant is not the minimum-TAC plant. The less controllable plants produce more off-specification products because they have more variability in x_B, and this reduces their annual profit.

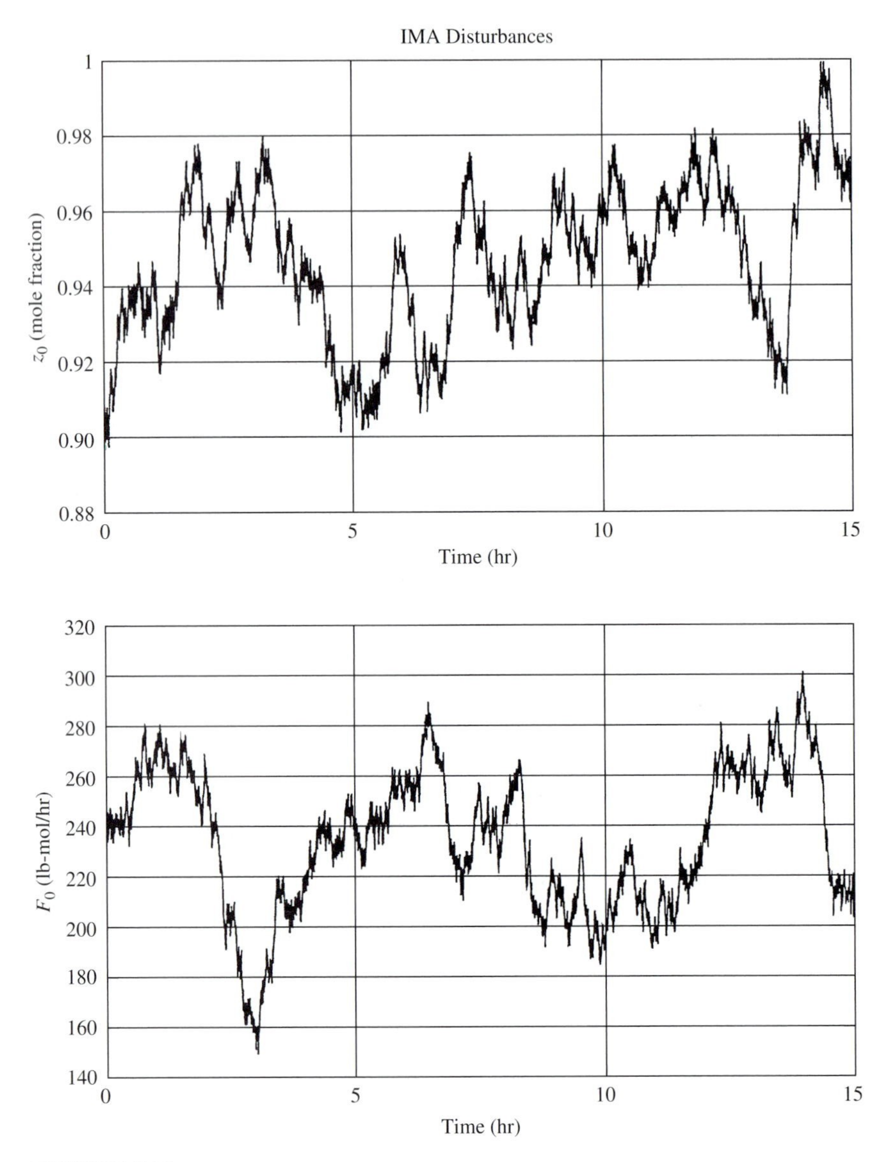

FIGURE 5.11
Disturbances in z_0 and F_0.

For example, with a specification range of 0.72 mol% (±0.36 mol%), the most profitable design has a reactor holdup V_R = 5000 lb-mol and a 12-tray stripper. The total annual cost of this plant (energy plus capital) is \$725,800/yr, which is higher than the \$693,000/yr TAC of the V_R = 3000 and N_T = 19 plant. However, the annual profit for the 5000/12 design is \$1,524,000/yr, which is larger than the annual profit of the 3000/19 design (\$737,000/yr). This is caused by the differences in the capacity factors. The 5000/19 design produces product that is inside the specification

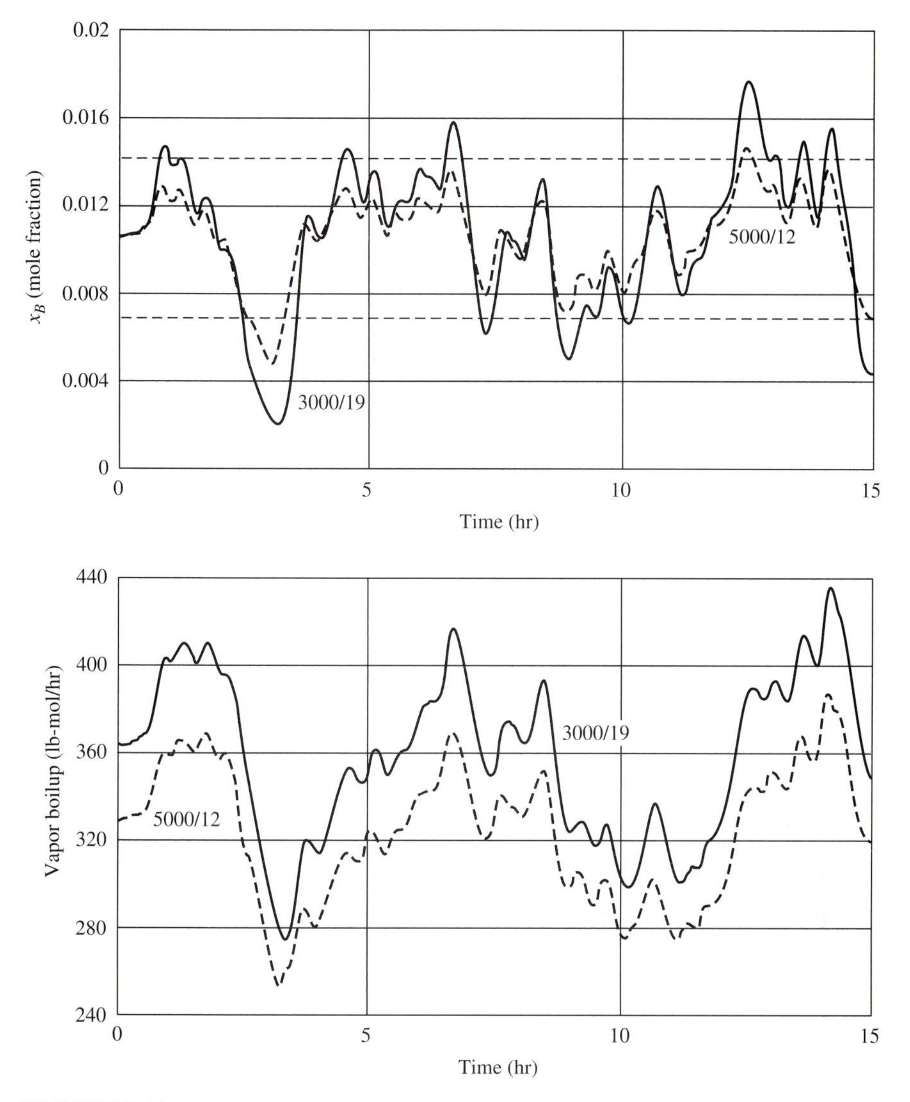

FIGURE 5.12
Responses of two designs.

range 71.2 percent of the time; i.e., its capacity factor is 0.712. The 3000/14 design produces product that is inside the specification range 92.9 percent of the time. The sequences of disturbances in feed flow rate and feed composition are shown in Fig. 5.11. The responses of product purity (x_B) for the two designs are shown in Fig. 5.12, along with the changes in the vapor boilup in the column.

The method just discussed permits a quantitative comparison of alternative designs that incorporates both steady-state economics and dynamic controllability in a logical and natural way. This approach handles the very important question of product quality variability in an explicit way.

However, a control structure must be chosen, controllers must be tuned, and a series of disturbances must be specified. The closedloop system is then simulated, and the capacity factors are calculated for each design. Using dynamic simulation can require a lot of computer time. In Chapter 10 we describe how the procedure can be made much easier and quicker using a frequency-domain approach.

5.7 CONCLUSION

This chapter has discussed some very important concepts that are basic to the practice of chemical engineering. Plant designs should not be developed only on the basis of steady-state operation. If we had no disturbances coming into the process, such an approach would be fine. But all chemical processes have disturbances, upsets, or changes in operating conditions. Therefore, it is vital to consider the dynamic effects of these disturbances. By modifying some of the design parameters, we may be able to develop a process that has only slightly higher capital and energy cost but is less sensitive to disturbances and therefore produces products with less variability.

CHAPTER 6

Plantwide Control

In this chapter we study the important question of how to develop a control system for an entire plant consisting of many interconnected unit operations.

6.1 SERIES CASCADES OF UNITS

Effective control schemes have been developed for many of the traditional chemical unit operations over the last three or four decades. If the structure of the plant is a sequence of units in series, this knowledge can be directly applied to the plantwide control problem. Each downstream unit simply sees disturbances coming from its upstream neighbor.

The design procedure for series cascades of units was proposed three decades ago (P. S. Buckley, *Techniques of Process Control,* 1964, Wiley, New York) and has been widely used in industry for many years. The first step is to lay out a logical and consistent "material balance" control structure that handles the inventory controls (liquid levels and gas pressures). The "hydraulic" structure provides gradual, smooth flow rate changes from unit to unit. This filters flow rate disturbances so that they are attenuated and not amplified as they work their way down through the cascade of units. Slow-acting, proportional level controllers provide the most simple and the most effective way to achieve this flow smoothing.

Then "product quality" loops are closed on each of the individual units. These loops typically use fast proportional-integral controllers to hold product streams as close to specification values as possible. Since these loops are typically quite a bit faster than the slow inventory loops, interaction between the two is often not a problem. Also, since the manipulated variables used to hold product qualities are quite often streams internal to each individual unit, changes in these manipulated variables may have little effect on downstream processes. The manipulated variables frequently are utility streams (cooling water, steam, refrigerant, etc.), which are provided by the plant utility system. Thus, the boiler house may be disturbed, but the other process units in the plant do not see disturbances coming from other process units. This is, of course, true only when the plant utility systems themselves have effective control systems that can respond quickly to the many disturbances coming in from units all over the plant.

The preceding discussion applies to cascades of units in series. If recycle streams occur in the plant, which frequently happens, the procedure for designing an effective "plantwide" control system becomes much less clear, and the literature provides much less guidance. Since processes with recycle streams are quite common, the heart of the plantwide control problem centers on how to handle recycles. The typical approach in the past for plants with recycle streams has been to install large surge tanks. This isolates sequences of units and permits the use of conventional cascade process design procedures. However, this practice can be very expensive in terms of tankage capital costs and working capital investment. In addition, and increasingly more important, the large inventories of chemicals, if dangerous or environmentally unfriendly, can greatly increase safety and environmental hazards.

The purpose of this chapter is to present some evolving ideas about plantwide control by looking at both the dynamic and the steady-state effects of recycles.

6.2 EFFECT OF RECYCLE ON TIME CONSTANTS

One of the most important effects of recycle is to slow down the response of the process, i.e., increase the process time constant. Consider the simple two-unit system shown in Fig. 6.1. The input to the process u is added to the output x from the unit in the recycle loop, giving z ($z = u + x$). The variable z is fed into the unit in the forward path, and the output of this unit is y. Thus, if there is no recycle, u simply affects y through the forward unit. However, the presence of the recycle means that there is a feedback loop from y back through the recycle unit, which again affects y.

The unit in the forward path has a steady-state gain K_F and a time constant τ_F. The unit in the recycle path has a gain K_R and time constant τ_R. The load disturbance

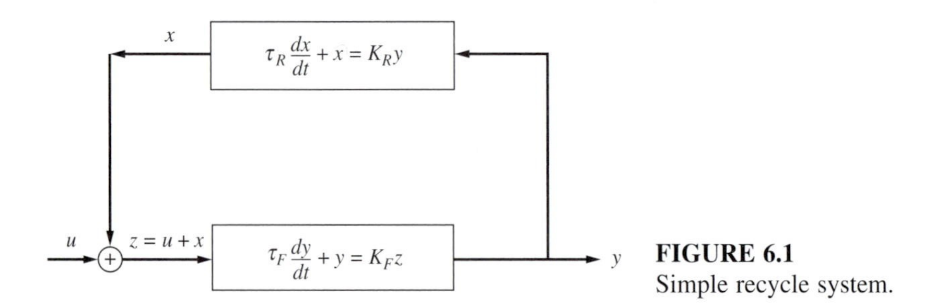

FIGURE 6.1
Simple recycle system.

into the plant is u, and the output of the plant is y. Suppose the dynamics of the two units can be described by simple first-order ODEs.

$$\tau_F \frac{dy}{dt} + y = K_F(u + x) \tag{6.1}$$

$$\tau_R \frac{dx}{dt} + x = K_R y \tag{6.2}$$

Differentiating Eq. (6.1) with respect to time and combining with Eq. (6.2) give

$$\left(\frac{\tau_F \tau_R}{1 - K_F K_R}\right)\frac{d^2 y}{dt^2} + \left(\frac{\tau_F + \tau_R}{1 - K_F K_R}\right)\frac{dy}{dt} + y = \left(\frac{K_F}{1 - K_F K_R}\right)\left(\tau_R \frac{du}{dt} + u\right) \tag{6.3}$$

Remember from Chapter 2 that the characteristic equation of this system is

$$\tau^2 s^2 + 2\tau\zeta s + 1 = 0 \tag{6.4}$$

where the overall time constant of the process τ is given by

$$\tau = \sqrt{\frac{\tau_F \tau_R}{1 - K_F K_R}} \tag{6.5}$$

Equation (6.5) clearly shows that the time constant of the overall process depends very strongly on the product of the gains around the recycle loop, $K_F K_R$. When the effect of the recycle is small (K_R is small), the time constant of the process is near the geometric average of τ_F and τ_R. However, as the product of the gains around the loop $K_F K_R$ gets closer and closer to unity, the time constant of the overall process becomes larger and larger. This simple process illustrates mathematically why time constants in recycle systems are typically much larger than the time constants of the individual units. The dynamics slow down as the recycle loop gain increases.

It should be noted that this system has positive feedback, so if the loop gain is greater than unity, the process is unstable.

6.3 SNOWBALL EFFECTS IN RECYCLE SYSTEMS

An important phenomenon has been observed in the operation of many chemical plants with recycle streams. The same phenomenon has been observed and quantified in numerical simulation studies of industrial processes with recycles. A small change in a load variable causes a very large change in the flow rates around the recycle loop. We call this the "snowball" effect.

It is important to note that snowballing is a steady-state phenomenon and has nothing to do with dynamics. It does, however, depend on the structure of the control system, as we illustrate in a mathematical analysis of the problem. Large changes in recycle flows mean large load changes for the distillation separation section. These are very undesirable because a column can tolerate only a limited turndown ratio, which is the ratio of the maximum vapor boilup (usually limited by column flooding) to the minimum vapor boilup (usually limited by poor liquid distribution or weeping).

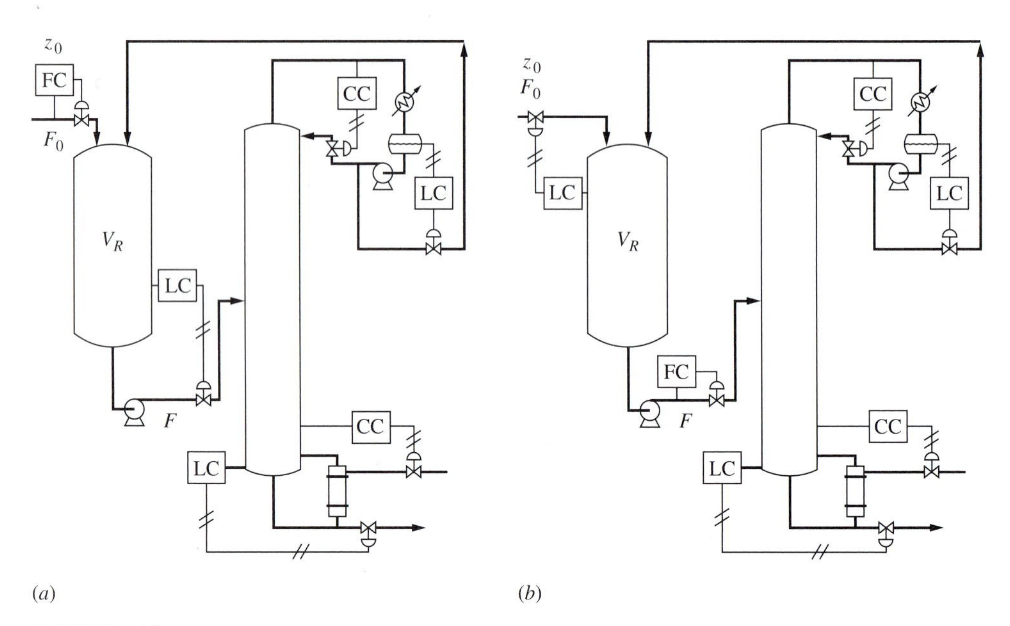

FIGURE 6.2
Reactor-column process with recycle. (*a*) Constant V_R control structure. (*b*) Constant F control structure.

To illustrate snowballing quantitatively, let us consider the simple process with one reactor and one column sketched in Fig. 6.2. The reaction is the simple irreversible A → B. The reactor effluent is a mixture of A and B. Its flow rate is F and its composition is z (mole fraction component A). Component A is more volatile than component B, so in the distillation column the bottoms is mostly component B and the distillate is mostly unreacted component A. First-order kinetics and isothermal operation are assumed in the reactor.

$$\mathcal{R} = V_R k z \tag{6.6}$$

where $\mathcal{R}$ = rate of consumption of reactant A (mol/hr)
V_R = reactor holdup (mol)
k = specific reaction rate (hr^{-1})
z = concentration of reactant A in the reactor (mole fraction A)

Two different control structures are explored. The conventional control is called the *constant* V_R structure.

- Control reactor holdup V_R by manipulating reactor effluent flow rate F.
- Flow-control fresh feed flow rate F_0.
- Control the impurity of component *A* in the base of the column x_B by manipulating heat input.
- Control reflux drum level in the column by manipulating distillate flow rate D.
- Control the impurity of component B in the distillate $(1 - x_D)$ from the column by manipulating reflux.

Note that this control structure has both of the flow rates in the recycle loop (reactor effluent F and distillate from the column D) set by level controllers.

The second control structure is called the *constant F* scheme. It switches the first two loops in the conventional structure.

- Flow-control reactor effluent.
- Control reactor level by manipulating fresh feed flow rate.

A. Conventional structure (constant reactor holdup)

The variables that are constant are V_R, k, x_D, and x_B. The variables that change when disturbances occur are F, z, and the recycle flow rate D. The steady-state equations that describe the system are as follows.

Process overall:

$$F_0 = B \tag{6.7}$$

$$F_0 z_0 = B x_B + V_R k z \tag{6.8}$$

Reactor:

$$F_0 + D = F \tag{6.9}$$

$$F_0 z_0 + D x_D = F z + V_R k z \tag{6.10}$$

Equations (6.7) and (6.8) can be combined to yield Eq. (6.11), which shows how reactor composition z must change as fresh feed flow rate F_0 and fresh feed composition z_0 change when the conventional control structure is employed (i.e., reactor volume is constant).

$$z = \frac{F_0(z_0 - x_B)}{kV_R} \tag{6.11}$$

Equations (6.9) and (6.10) can be combined to give recycle flow rate D.

$$D = \frac{z(F_0 + kV_R) - F_0 z_0}{x_D - z} \tag{6.12}$$

Substituting Eq. (6.11) into Eq. (6.12) gives an analytical expression showing how the recycle flow rate D changes with disturbances in fresh feed flow rate F_0 and fresh feed composition z_0.

$$D = \frac{F_0 - \beta x_B}{\beta x_D / F_0 - 1} \tag{6.13}$$

where $\beta \equiv \dfrac{kV_R}{z_0 - x_B}$ (6.14)

It is useful to look at the limiting (high-purity) case in which $x_D \simeq 1$ and $x_B \simeq 0$. Under these conditions, Eq. (6.13) becomes

$$D = \frac{z_0(F_0)^2}{kV_R - z_0 F_0} \tag{6.15}$$

This equation clearly shows the strong dependence of the recycle flow rate on the fresh feed flow rate: increasing F_0 increases the numerator (as the square) and decreases the denominator. Both effects tend to increase D sharply as F_0 increases. Results from a numerical example are given later.

B. Reactor effluent fixed structure (variable reactor holdup)

The variables that are constant with the alternative structure are F, k, x_D, and x_B. The variables that change when disturbances occur are V_R, z, and the recycle flow rate D. Equations (6.7) through (6.10) still describe the system, but now F is constant while V_R varies. Combining these equations gives

$$D = F - F_0 \tag{6.16}$$

$$kV_R = \frac{FF_0(z_0 - x_B)}{Fx_D - F_0(x_D - x_B)} \tag{6.17}$$

Equation (6.16) shows that the recycle flow rate D changes in direct proportion to the change in fresh feed flow rate and does not change at all when fresh feed composition changes. Equation (6.17) shows that reactor holdup V_R changes as fresh feed flow rate and fresh feed composition change.

It is useful to look at the limiting, high-purity case in which $x_D \simeq 1$ and $x_B \simeq 0$. Under these conditions, Eq. (6.17) becomes

$$kV_R = \frac{FF_0z_0}{F - F_0} \tag{6.18}$$

This equation shows that reactor holdup changes in direct proportion to fresh feed composition and is less dependent on fresh feed flow rate since the F_0 term in the numerator is now only to the first power. Keep in mind that F_0 is not really a disturbance with this structure since fresh feed is used to control reactor holdup. However, the changes required in the setpoint of the level controller to accomplish a desired change in fresh feed flow rate can be calculated from Eq. (6.18). Note that the fresh feed flow rate changes as fresh feed composition changes for a constant reactor holdup.

Figure 6.3 gives numerical results for a system with the values of design parameters given in Table 6.1. The large changes in recycle flow rates when the conventional (constant V_R) control structure is used are clearly shown.

The fundamental reason for the occurrence of snowballing in recycle systems is the large changes in reactor composition that some control structures produce when disturbances occur. The final steady-state values of the reactor composition must satisfy the steady-state component balances. These composition changes represent load disturbances to the separation section, and separation units usually cannot handle excessively large throughput changes.

A very useful heuristic rule has been developed as a result of our studies of recycle systems:

A stream somewhere in a liquid recycle loop should be flow controlled.

Note that the constant-reactor-effluent structure used in the simple process just discussed follows this rule and does indeed prevent snowballing. The control structures discussed in examples presented later in this chapter follow this rule.

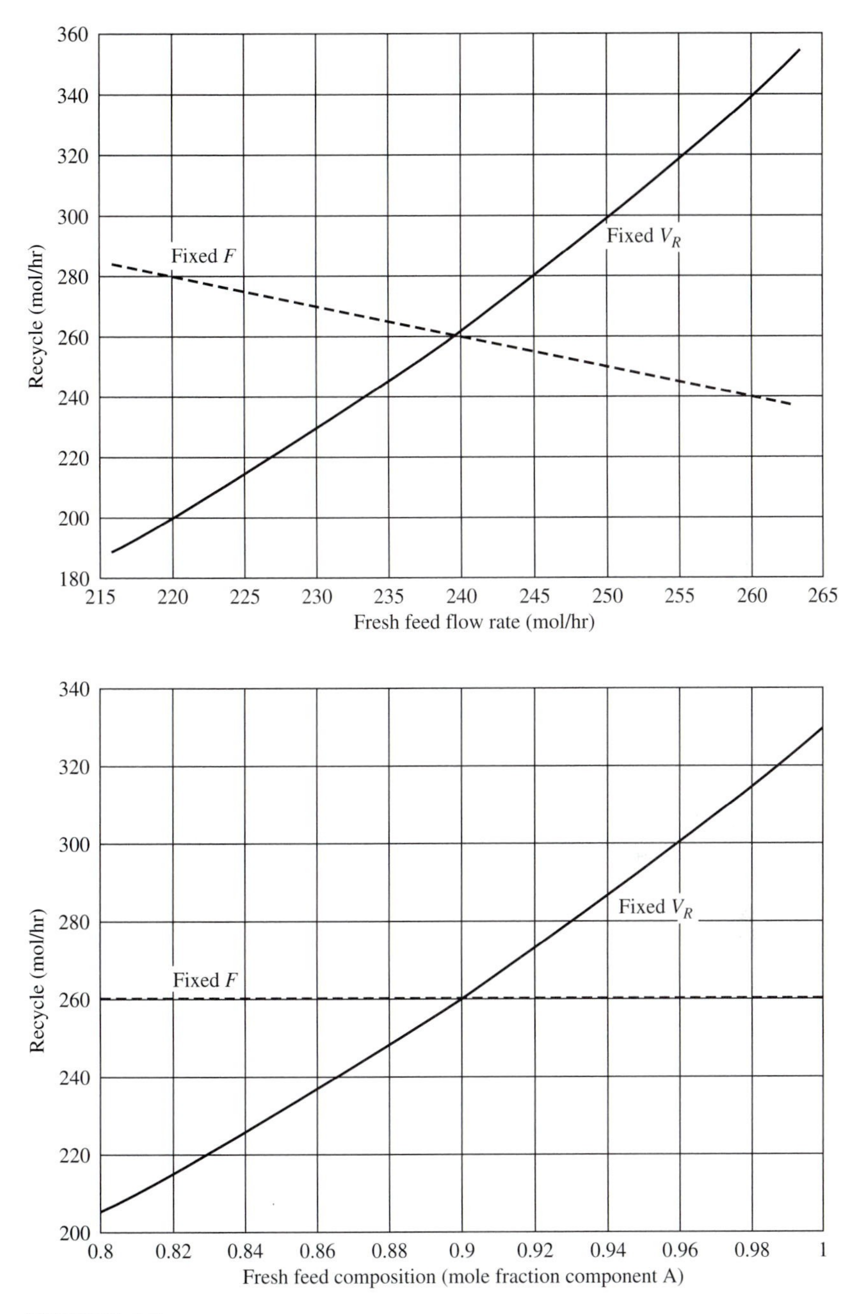

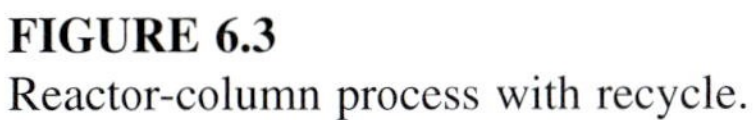

FIGURE 6.3
Reactor-column process with recycle.

TABLE 6.1
Parameter values for reactor-column process

At normal design conditions:
Fresh feed composition = z_0 = 0.9 mole fraction component A
Fresh feed flow rate = F_0 = 239.5 mol/hr
Reactor holdup = V_R = 1250 mol
Reactor effluent flow rate = F = 500 mol/hr
Recycle flow rate = distillate flow rate = D = 260.5 mol/hr
Parameter values:
Specific reaction rate = k = 0.34086 hr^{-1}
Bottoms composition = x_B = 0.0105 mole fraction component A
Distillate composition = x_D = 0.95 mole fraction component A

6.4
USE OF STEADY-STATE SENSITIVITY ANALYSIS TO SCREEN PLANTWIDE CONTROL STRUCTURES

A chemical plant typically has a large number of units with multiple recycle streams. Many different control strategies are possible, and it would be impractical to perform a detailed dynamic study for each alternative. We would like to have an analysis procedure to screen out poor control structures. The steady-state snowball analysis of the simple process in the previous section logically suggests that a similar steady-state analysis may be useful for screening out poor control structures in more realistically complex processes. If a steady-state analysis can reveal structures that require large changes in manipulated variables when load disturbances occur or when a change in throughput is made, these structures can be eliminated from further study.

The idea is to specify a control structure (fix the variables that are held constant in the control scheme) and specify a disturbance. Then solve the nonlinear algebraic equations to determine the values of all variables at the new steady-state condition. The process considered in the previous section is so simple that an analytical solution can be found for the dependence of the recycle flow rate on load disturbances. For realistically complex processes, analytical solution is out of the question and numerical methods must be used. Modern software tools (such as SPEEDUP, HYSYS, or GAMS) make these calculations relatively easy to perform.

To illustrate the procedure, we consider a fairly complex process sketched in Fig. 6.4, which shows the process flowsheet and the nomenclature used. In the continuous stirred-tank reactor, a multicomponent, reversible, second-order reaction occurs in the liquid phase: A + B $\rightleftarrows$ C + D. The component volatilities are such that reactant A is the most volatile, product C is the next most volatile, reactant B has intermediate volatility, and product D is the heaviest component: $\alpha_A > \alpha_C > \alpha_B > \alpha_D$. The process flowsheet consists of a reactor that is coupled with a stripping column to keep reactant A in the system, and two distillation columns to achieve the removal of products C and D and the recovery and recycle of reactant B.

The two recycle streams are D_1 from the first column (mostly component A) and D_3 from the third column (mostly component B). The two product streams are the distillate from the second column, D_2, and the bottoms from the third column, B_3.

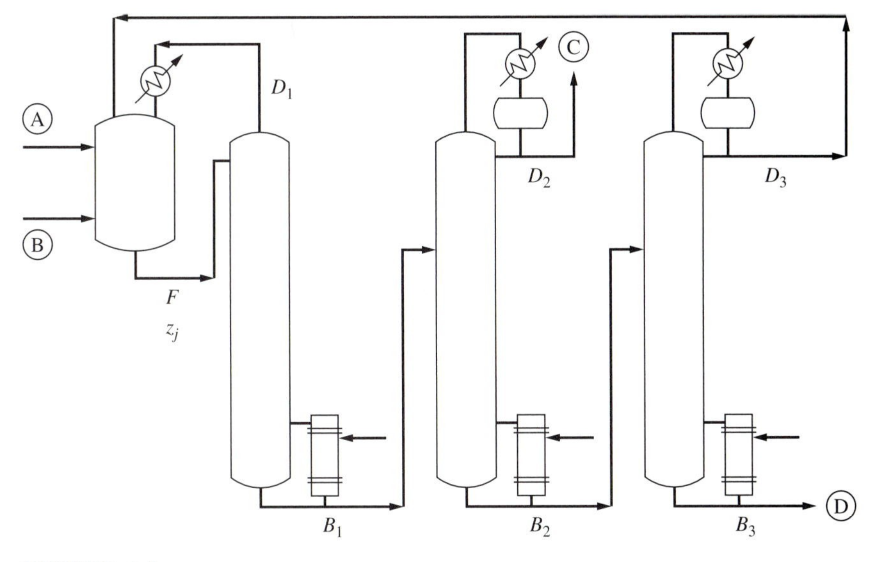

FIGURE 6.4
Reactor–three-column–two-recycle–four-component process.

The former is mostly component C with impurities of component A ($x_{D2,A}$) and of component B ($x_{D2,B}$). The latter is mostly component D with impurity of component B ($x_{B3,B}$).

6.4.1 Control Structures Screened

This process has 15 control valves, so there is an enormous number (16 factorial) of possible simple SISO control structures. The nine inventories (six levels and the three pressures) *must* be controlled. The three impurities in the two product streams must also be controlled. The production rate must also be set. This leaves $15 - 9 - 3 - 1 = 2$ control valves that can be set to accomplish other objectives (typically economic objectives, such as minimizing energy costs).

For purposes of illustration, let us consider the three alternative control structures shown in Fig. 6.5. The following loops are used in all three structures:

- Reactor effluent is flow controlled.
- Column base levels are held by bottoms flows.
- Component A impurity in D_2 is held by controlling $x_{B1,A}$ by manipulating heat input V_1.
- Component B impurity in D_2 is held by manipulating heat input V_2.
- Component B impurity in B_3 is held by manipulating heat input V_3.
- Pressures are controlled by coolant flow rates in condensers.
- The reflux in the second column, R_2, is flow controlled.
- Reflux drum level in the second column is controlled by distillate D_2.

These decisions naturally eliminate certain alternative control strategies.

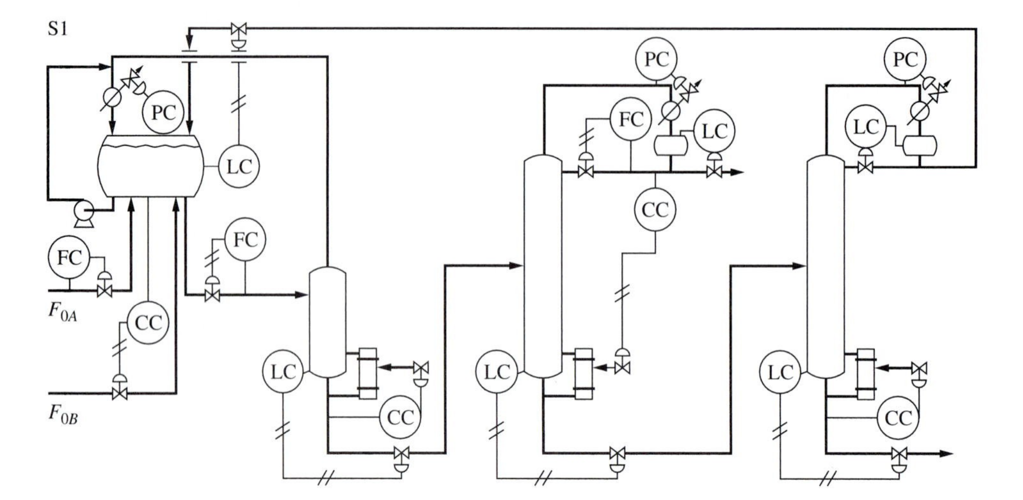

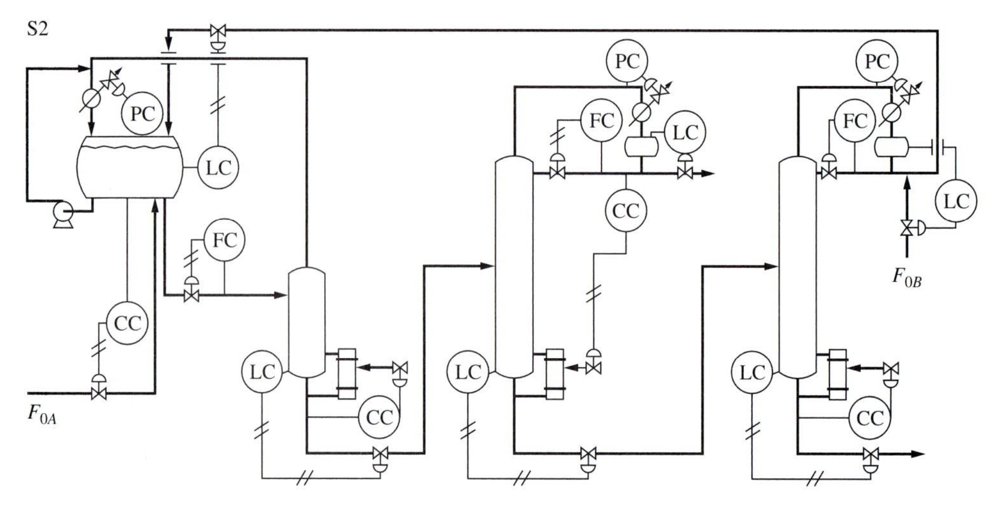

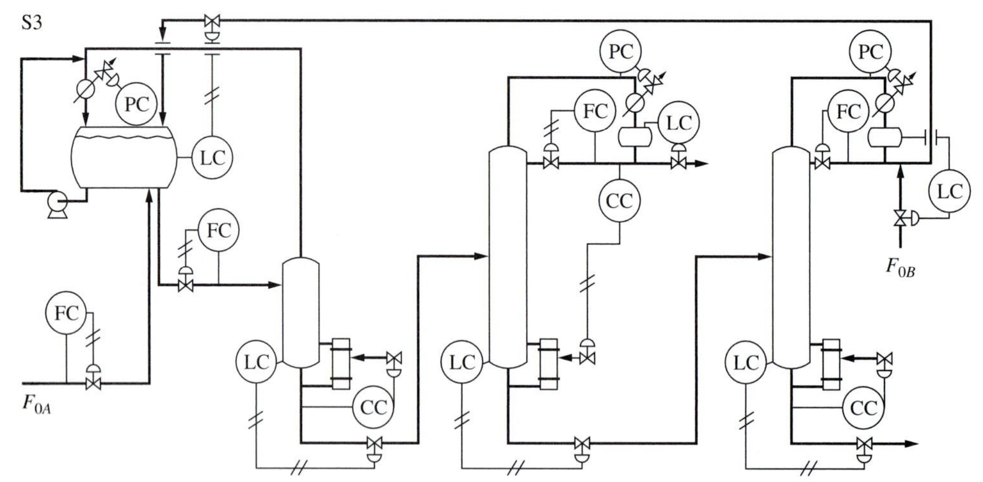

FIGURE 6.5
Control structures.

A. Structure S1

Fresh feed of component A (F_{0A}) is flow controlled. Fresh feed of component B (F_{0B}) is manipulated to control the composition of component B in the reactor (z_B). Reactor level is held by D_3 recycle. Level in the reflux drum of the third column is controlled by reflux R_3.

When a very small change is made in the composition of the F_{0B} stream ($z_{0B,B}$ changed from 1 to 0.999 and $z_{0B,A}$ changed from 0 to 0.001), the process can barely handle it. The steady-state value of R_3 in the third column changes 15 percent for this very small disturbance. Thus, the steady-state analysis predicts that this structure will not work. Dynamic simulations confirmed this; very small disturbances drive the control valves on R_3 and V_3 wide open, and product quality cannot be maintained.

B. Structure S2

Fresh feed of component A (F_{0A}) is manipulated to control the composition of component A in the reactor (z_A).The level in the reflux drum in the third column is controlled by manipulating the fresh feed F_{0B}, which is added to the distillate from the third column, D_3. The total of D_3 and F_{0B} is manipulated to control reactor level. The reflux R_3 is flow controlled.

When a large change is made in $z_{0B,B}$ (to 0.90), the new steady-state values of the manipulated variables were only slightly different from the base-case values. The makeup flow rates of fresh feed change: F_{0A} increases 10 percent and F_{0B} decreases 10 percent. Production rates of D_2 and B_3 stay the same, as do other flow rates and compositions throughout the process. Thus, the steady-state sensitivity analysis suggests that this structure should handle disturbances easily. Dynamic simulations confirm that this control structure works quite well.

C. Structure S3

The loops are the same as in S2 except that the fresh feed F_{0A} is flow controlled. There is no control of any reactor composition.

A small change in the composition of the fresh feed of component A from $z_{0A,A} = 1$ to $z_{0A,A} = 0.99$ and $z_{0A,B} = 0.01$ produces 15 to 20 percent changes in the recycle flow rates V_1 and D_3. Therefore, the steady-state sensitivity analysis predicts that this control structure will not be able to handle large disturbances.

It is interesting to note that this control structure exhibits multiple steady-state solutions. There are two sets of recycle flow rates, reactor temperatures, and reactor compositions that give the same production rates for the same feed rates. Structures that give multiple steady states should be avoided because the operation of the plant may be quite erratic.

These results suggest that we need to have direct (or indirect) measurement of compositions in the reaction section. This is discussed more fully in the next section, where a generic rule is proposed:

A reactant cannot be flow controlled into a reactor in which a second-order reaction takes place unless the concentration of this reactant in the reactor is much smaller than that of the other reactant.

6.5 SECOND-ORDER REACTION EXAMPLE

As our last plantwide control example, let us consider a process in which a second-order reaction A + B → C occurs. There are two fresh feed makeup streams. The process flowsheet consists of a single isothermal, perfectly mixed reactor followed by a separation section. One distillation column is used if there is only one recycle stream. Two are used if two recycle streams exist.

Two cases are considered: (1) instantaneous and complete one-pass conversion of one of the two components in the reactor so that there is an excess of only one component that must be recycled, and (2) incomplete conversion per pass so that there are two recycle streams.

6.5.1 Complete One-Pass Conversion

Figure 6.6 shows the process for complete one-pass conversion. Pure component B is fed into the reactor on flow control. The concentration of B in the reactor, z_B, is zero (or very small) because we assume complete one-pass conversion of B. A large recycle stream of component A is fed into the reactor. The reactor effluent is a mixture of unreacted component A and product C. This binary mixture is separated in a distillation column. We assume that the relative volatility of A is greater than that of C, so the distillate product is recycled back to the reactor.

This simple system is encountered in a number of commercial processes. It occurs when reaction rates are so fast that B reacts quickly with A, but a large excess

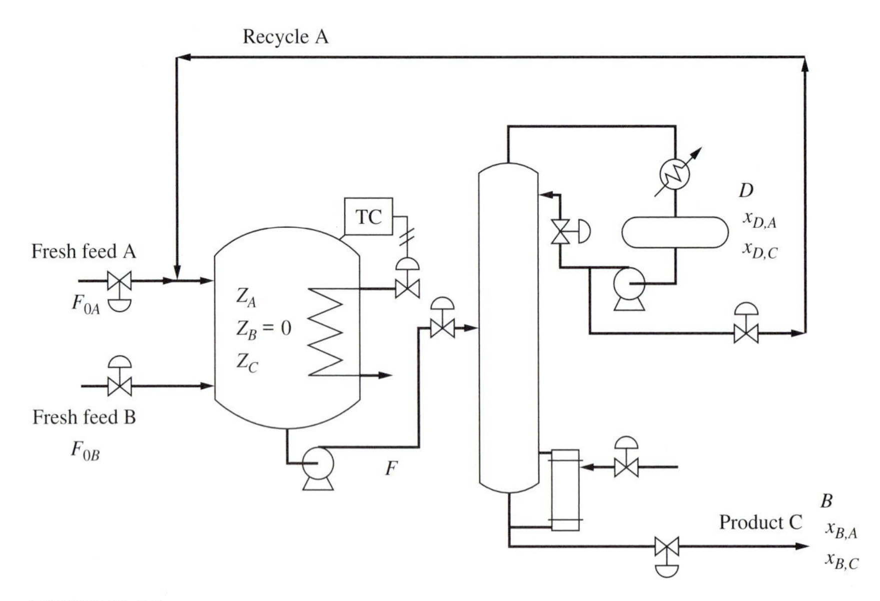

FIGURE 6.6
Process with complete one-pass conversion of component B.

of A is needed. There are several reasons that a large recycle of one component may be needed. An important one is to prevent the occurrence of undesirable side reactions. The alkylation process in petroleum refining is a common example. Another reason for recycle may be the need to limit the temperature rise through an adiabatic reactor by providing a thermal sink for the heat of reaction. Avoiding explosivity composition regions, particularly for oxidation reactions, often requires an excess of one reactant.

A. Steady-state design

The purity of the product stream B leaving the bottom of the distillation column, $x_{B,A}$, is set at 0.01 mole fraction component A, with the concentration of component C being 0.99. There is no component B in the feed to the column because complete one-pass conversion is assumed. The column is designed for a distillate purity $x_{D,A}$ of 0.95 mole fraction component A. In the base case the fresh feed flow rate is 100 lb-mol/hr (F_{0B}), and it is pure component B. Since component B is completely converted, the amount of component C in the product stream must also be 100 lb-mol/hr at steady state. Therefore, the total flow rate of the product stream and the flow rate of the makeup fresh feed of component A, F_{0A}, can both be calculated.

$$B = F_{0B}/(1 - x_{B,A}) \tag{6.19}$$

$$F_{0A} = F_{0B} + Bx_{B,A} \tag{6.20}$$

Under the assumption of complete one-pass conversion of component B, in theory both the recycle flow rate D and the holdup of the reactor V_R can be set at any arbitrary values. Once these are selected, the system can be designed. The feed flow rate F and composition z_A to the column can be calculated once D has been specified.

$$F = B + D \tag{6.21}$$

$$Fz_A = Dx_{D,A} + Bx_{B,A} \tag{6.22}$$

The separation in the distillation column is binary between A and C, so the design of the column is straightforward. Typically, the reflux ratio is set at 1.2 times the minimum, and tray-to-tray calculations give the total number of trays N_T and the optimal feed tray N_F.

These steady-state design calculations show that as recycle flow rate D is increased, the concentration of A in the reactor, z_A, increases. This causes the reflux flow rate to increase initially, and then decrease. At very high recycle flow rates, the reflux rate goes to zero, indicating that the distillation column becomes just a stripping column. As recycle flow rate increases, energy consumption, capital investment, and total cost all increase. Thus, the recycle flow rate should be kept as low as possible, subject to the constraints on the minimum recycle flow, e.g., preventing undesirable side reactions from occurring or limiting adiabatic temperature rise through the reactor.

B. Dynamics and control

Two alternative control schemes are evaluated. In scheme A both of the fresh feed streams, F_{0A} and F_{0B}, are flow controlled (or one is ratioed to the other). This

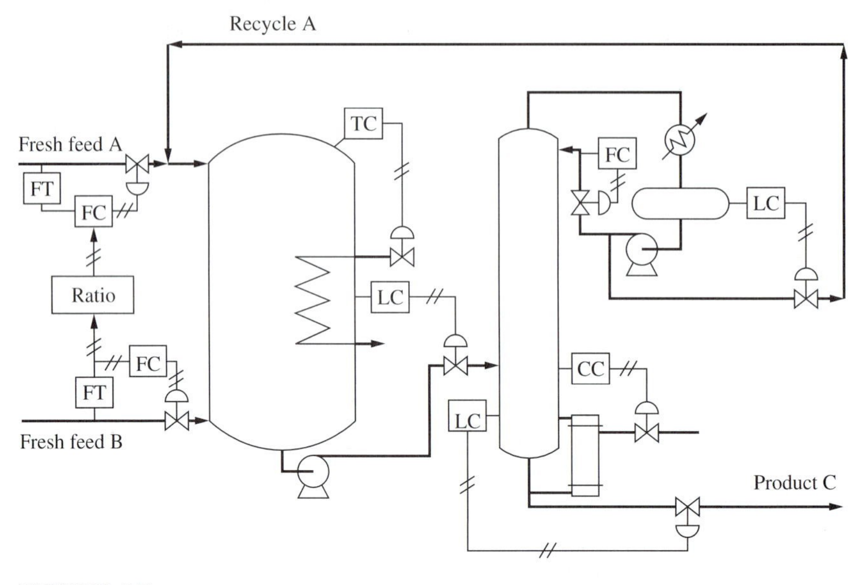

FIGURE 6.7
Scheme A: Makeup feeds flow controlled.

is a control strategy seen quite commonly in plants, but it has major weaknesses, as will be demonstrated. We think it is important to point out these problems clearly and to illustrate them quantitatively by means of a numerical example.

As sketched in Fig. 6.7, control structure scheme A has reactor level controlled by column feed. Column base level is held by bottoms. Reflux drum level is held by distillate recycle back to the reactor. Reflux flow rate is flow controlled. Distillate composition is not controlled since the recycle is an internal stream within the process. Bottoms product purity is controlled by manipulating heat input. Note that this scheme violates the rule for liquid recycles since the streams in the recycle loop (F and D) are both on level control.

In control scheme B, sketched in Fig. 6.8, the total recycle flow rate to the reactor (distillate plus makeup A) is flow controlled. The makeup of reactant A is used to hold the level in the reflux drum. This level indicates the inventory of component A in the system.

Dynamic simulations of the process were made using these two control structures. Figure 6.9 shows what happens using scheme A when the flow rate of makeup A is incorrect by only 1 percent, which is smaller than any real flow measurement device can achieve in most plants. The recycle flow rate D and the composition of component A in the reactor ramp up with time, and it takes more and more vapor boilup in the column to keep component A from dropping out the bottom. These trends would continue until the column floods; some other constraint in heat input, heat removal, or pumping capacity is encountered; or the operator intervenes manually. Figure 6.10 shows that when a pulse is made in F_{0A}, the system goes through a transient and lines out with a new recycle flow rate and a new reactor composition.

These results quantitatively illustrate the basic flaw in control structure A: both reactants cannot be flow controlled because flow measurement inaccuracy makes

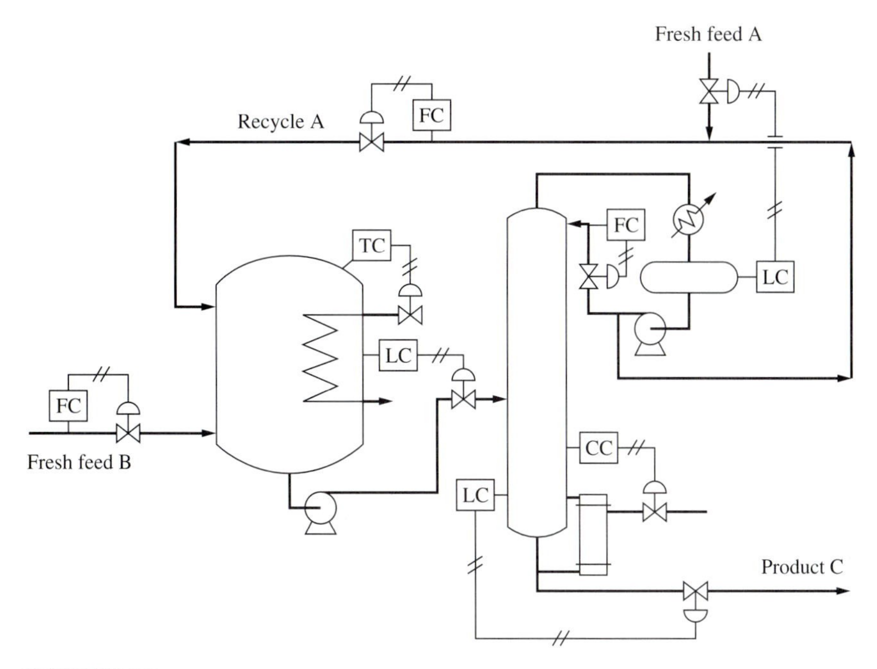

FIGURE 6.8
Scheme B: Component A feed on level control.

it impossible to achieve perfect stoichiometric amounts of the two reactants in an openloop system. Thus, this is *not* a workable scheme. Somehow the amount of component A in the system must be determined, and the makeup of component A must be adjusted to maintain its inventory at a reasonable level. The problem cannot be solved by the use of other types of controllers.

Control structure B provides good control of the system. Figure 6.11 shows what happens using scheme B when the total recycle flow rate is reduced from 500 to 400 lb-mol/hr. The system goes through a transient and ends up at the same fresh feed flow rate for reactant A. The reflux drum level controller adjusts the flow rate of F_{0A} to maintain the correct inventory of component A in the system. Note that the concentration of component A in the reactor, z_A, decreases when the recycle flow rate is decreased. This has no effect on the reaction rate because we have assumed the instantaneous reaction of component B. Figure 6.12 shows the response for a step increase in F_{0B}. The control system automatically increases the makeup of component A to satisfy the stoichiometry of the reaction.

6.5.2 Incomplete Conversion Case

We now look at the more common situation in which both reactants are present in the reactor since one-pass conversion of neither reactant is 100 percent, and therefore both reactants must be recycled. The concentrations of the two reactants in the reactor are z_A and z_B.

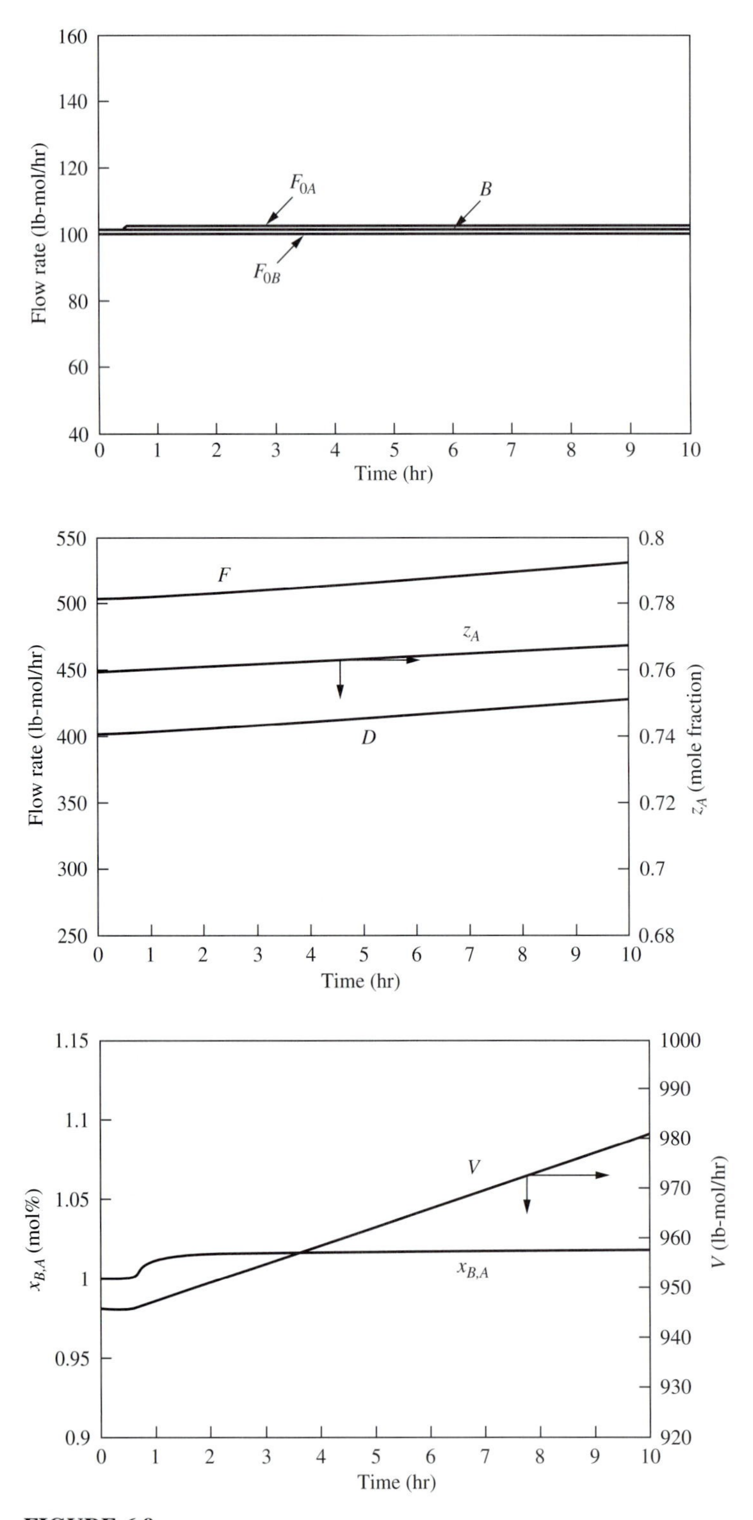

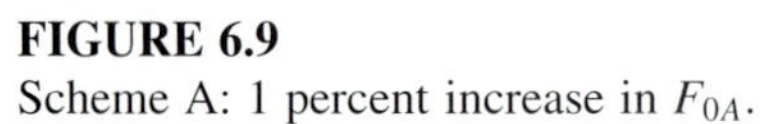

FIGURE 6.9
Scheme A: 1 percent increase in F_{0A}.

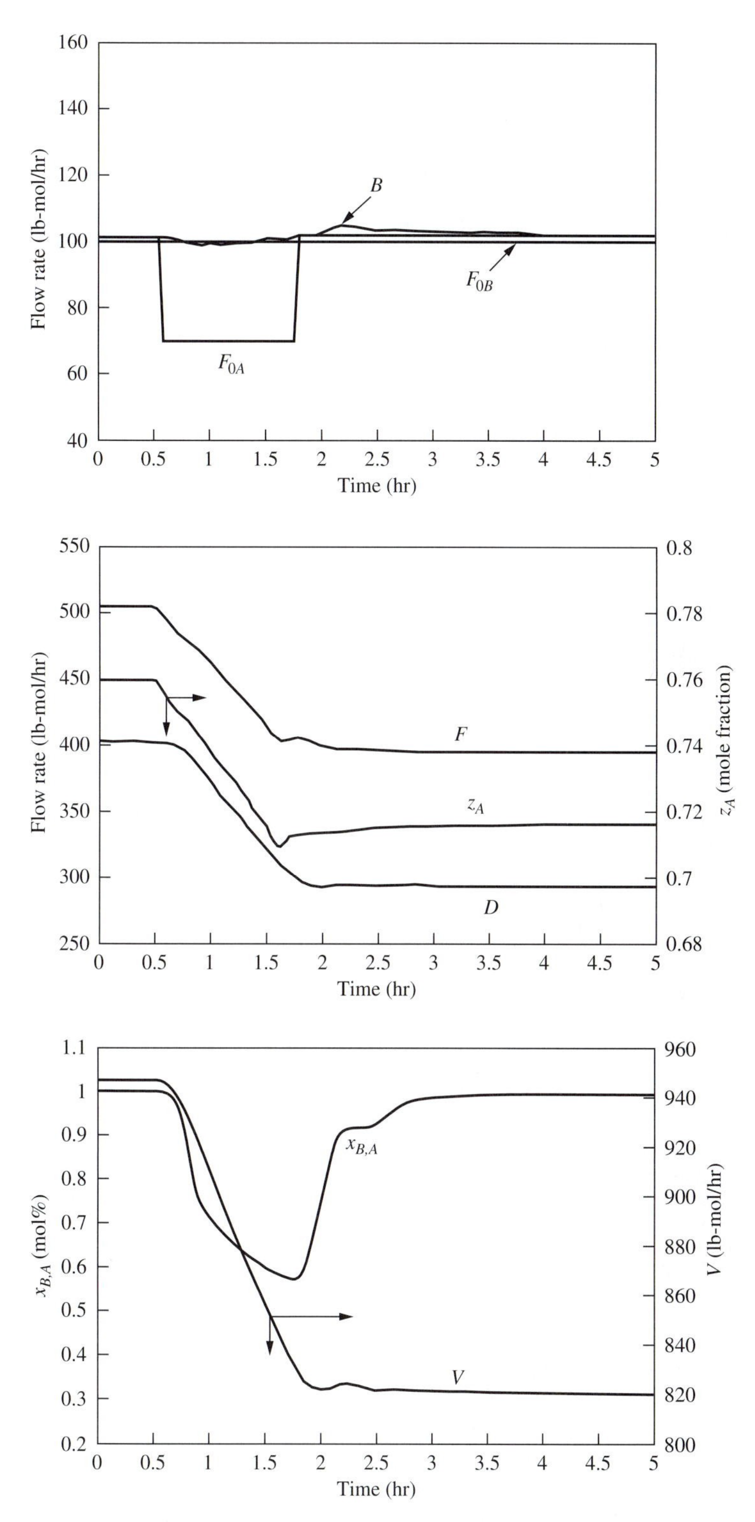

FIGURE 6.10
Scheme A: Pulse in F_{0A}.

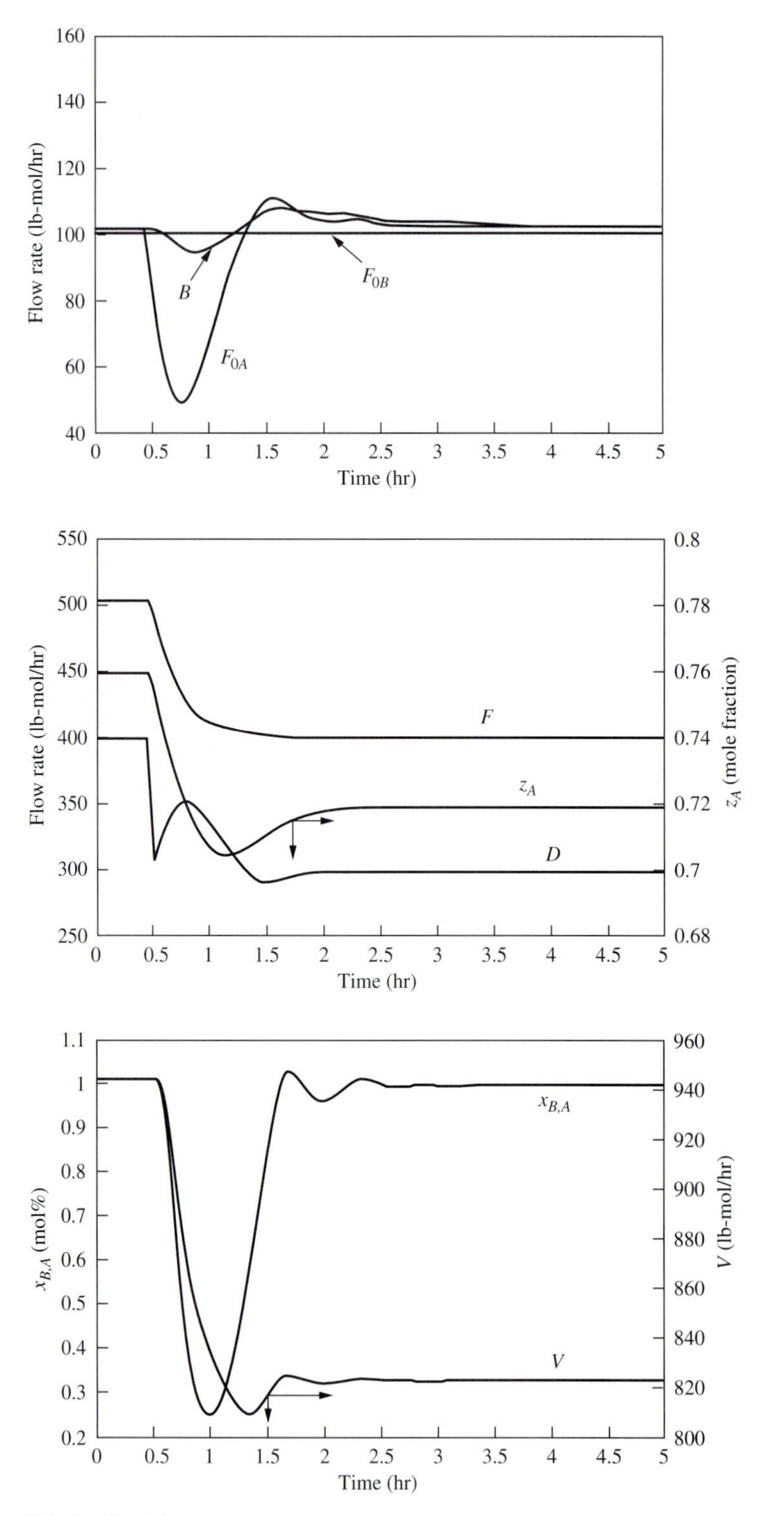

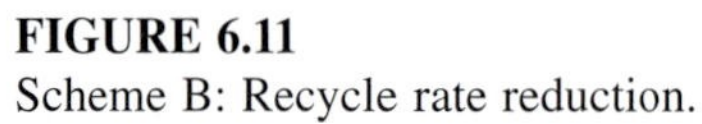

FIGURE 6.11
Scheme B: Recycle rate reduction.

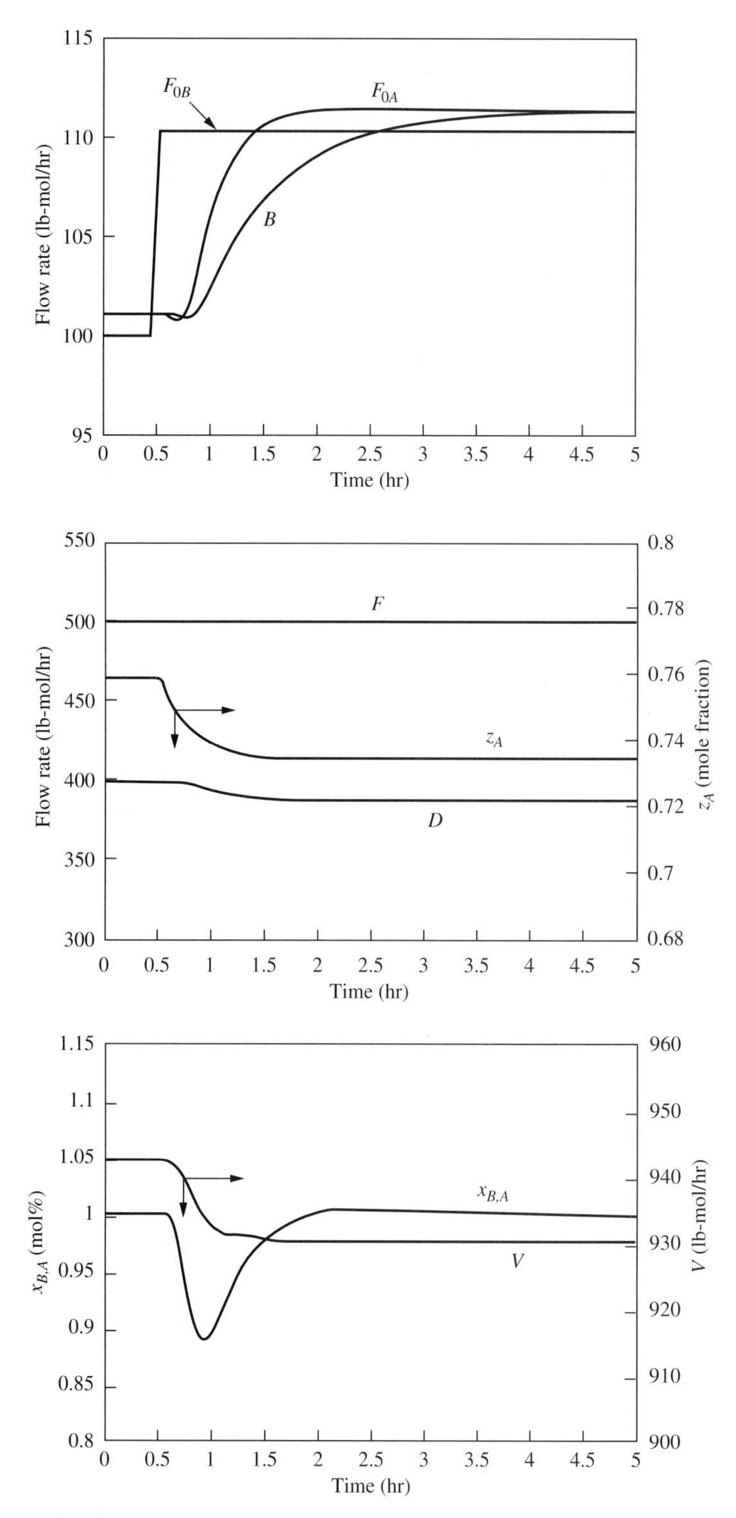

FIGURE 6.12
Scheme B: Increase in F_{0B}.

The volatilities of the A, B, and C components dictate what the recycle streams will be. If components A and B are both lighter or heavier than component C, a single column can be used, recycling a mixture of components A and B from either the top or the bottom of the column back to the reactor and producing product at the other end. If the volatility of component C is intermediate between components A and B, two columns and two recycle streams are required.

As sketched in Fig. 6.13, the process studied has volatilities that are $\alpha_A = 4$, $\alpha_B = 1$, and $\alpha_C = 2$. Component B, the heaviest, is recycled from the bottom of the first column back to the reactor. Component A, the lightest, is recycled from the top of the second column back to the reactor. The alternative flowsheet (recycling A from the top of the first column and recycling B from the bottom of the second column) would give similar results. We use the first flowsheet because in many processes we want to keep column base temperatures as low as possible, and this is accomplished by removing the heaviest component first.

Second-order isothermal kinetics are assumed in the reaction A + B → C:

$$\mathcal{R} = V_R k z_A z_B \tag{6.23}$$

where $\mathcal{R}$ = reaction rate (lb-mol/hr)
V_R = holdup in reactor (lb-mol)
k = specific reaction rate (hr^{-1})

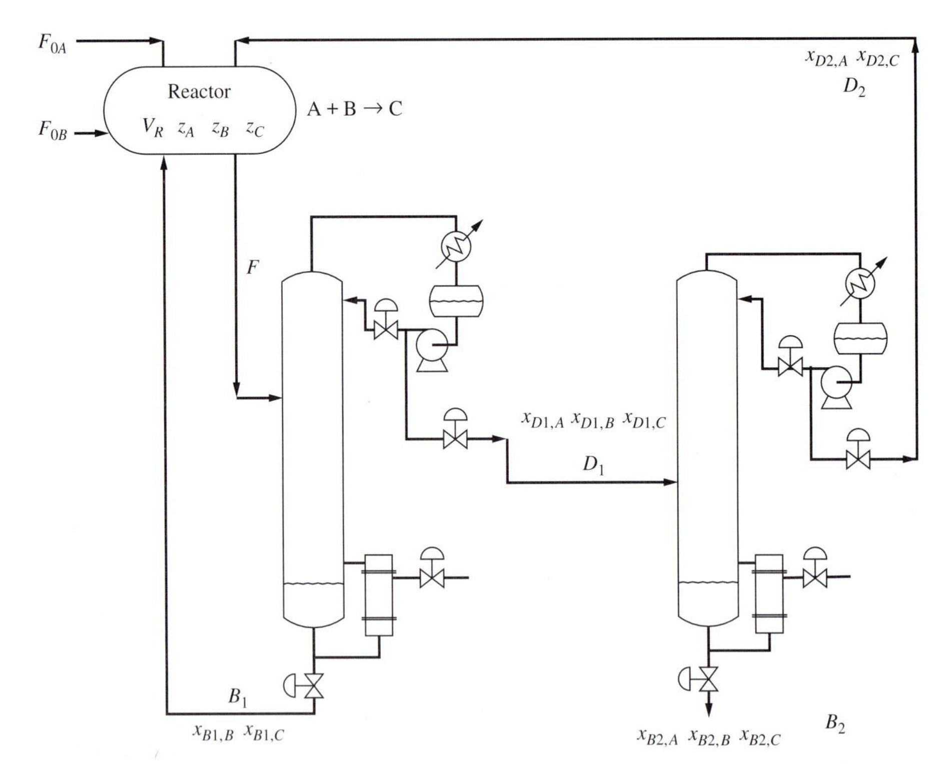

FIGURE 6.13
Process with two recycles and second-order reaction.

z_A = concentration of component A in the reactor (mole fraction A)
z_B = concentration of component B in the reactor (mole fraction B)

Note that "moles" are not conserved in this system.

A. Steady-state design

The process is described by three component balances for each unit: reactor, column 1, and column 2. There are a number of ways to solve the nine nonlinear algebraic equations, but the following procedure is straightforward and involves no iteration.

1. The flow rate of the product stream B_2 leaving the bottom of the second column is fixed at 100 lb-mol/hr.
2. The composition of this stream is specified to be $x_{B2,A} = 0.01$, $x_{B2,B} = 0.01$, and $x_{B2,C} = 0.98$.
3. The flow rate of the light recycle stream D_2 from the top of the second column is fixed (to be varied later to determine the optimal flow rate). The composition of this recycle stream is specified to be $x_{D2,A} = 0.99$, $x_{D2,B} = 0$, and $x_{D2,C} = 0.01$.
4. The flow rate of the heavy recycle stream B_1 from the bottom of the first column is fixed (to be varied later to determine the optimal flow rate). The composition of this recycle stream is specified to be $x_{B1,A} = 0$, $x_{B1,B} = 0.99$, and $x_{B1,C} = 0.01$.
5. Calculate the feed to the second column:

$$D_1 = B_2 + D_2 \tag{6.24}$$

$$D_1 x_{D1,A} = B_2 x_{B2,A} + D_2 x_{D2,A} \tag{6.25}$$

$$D_1 x_{D1,B} = B_2 x_{B2,B} + D_2 x_{D2,B} \tag{6.26}$$

6. Calculate the feed to the first column:

$$F = B_1 + D_1 \tag{6.27}$$

$$F z_A = B_1 x_{B1,A} + D_1 x_{D1,A} \tag{6.28}$$

$$F z_B = B_1 x_{B1,B} + D_1 x_{D1,B} \tag{6.29}$$

7. The rate of production of product C is equal to $B_2 x_{B2,C}$, and this must be equal to the rate of generation of component C in the reactor (assuming the two fresh feed streams contain no component C). Therefore, the reactor volume can be calculated:

$$V_R = \frac{B_2 x_{B2,C}}{k z_A z_B} \tag{6.30}$$

   Note that there is a unique reactor size for each selected pair of light and heavy recycle flow rates.
8. Calculate the fresh feed makeup flows of both component A (F_{0A}) and component B (F_{0B}).

$$F_{0A} = F z_A + V_R k z_A z_B - D_2 x_{D2,A} \tag{6.31}$$

$$F_{0B} = F z_B + V_R k z_A z_B - B_1 x_{B1,B} \tag{6.32}$$

9. Calculate the minimum number of trays for each column from the Fenske equation and the minimum reflux ratio from the Underwood equations (see E. J. Henley and J. D. Seader, *Equilibrium-Stage Separation Operations in Chemical Engineering,* 1981, Wiley, New York). Set the actual number of trays equal to 1.5 times the minimum, and set the actual reflux ratio equal to 1.2 times the minimum.
10. Calculate capital and energy costs for each set of recycle flows chosen (D_2 and B_1).

Figure 6.14 shows how several design parameters vary as the heavy and light recycle flow rates are varied. Low flow rates of either light or heavy recycle result in very large reactor size and high capital cost but low energy cost. As heavy recycle B_1 is increased, reactor holdup V_R decreases, the concentration of component B in the reactor increases, energy cost increases slightly, and capital cost decreases slightly. There is a minimum in the total cost curve at some heavy recycle flow rate because of the trade-off between increasing energy cost and decreasing capital cost (due to the decrease in reactor size). As light recycle D_2 increases, reactor holdup decreases, the concentration of component A in the reactor increases, energy cost increases rapidly, and capital cost decreases.

There is an optimal pair of light and heavy recycle flow rates. As can be seen in Fig. 6.15, the minimum total cost ($798,900/yr) occurs with a light recycle flow rate $D_2 = 40$ lb-mol/hr and a heavy recycle flow rate $B_1 = 60$ lb-mol/hr.

A specific reaction rate $k = 1$ hr^{-1} was used in the preceding calculations. If other values of k are used, the optimal light and heavy recycle flow rates and the minimum total cost change. This occurs primarily because the size of the reactor depends directly on the value of k. The smaller the value of k, the larger the optimal recycle flow rates and the higher the total cost.

B. Dynamics and control

A number of alternative control structures can be proposed for this process. No control structure that violates the recycle rule has been found to work. Unless one flow somewhere in the recycle loop is fixed, the recycle flow grows to very high rates when a disturbance occurs or when additional throughput is desired. Four typical control structures are sketched in Fig. 6.16.

In all the schemes, the following control loops are used:

1. The reflux flow rates on both columns are fixed; dual composition control is not used to keep the column control systems as simple as possible. In some situations better control may be achieved by using dual composition control in the columns to prevent the impurity levels in the recycle streams from becoming so large that they affect conditions in the reactor.
2. Vapor boilup in the first column controls the impurity of B in the final product stream (the bottoms from the second column) through a composition/composition cascade strategy.
3. Impurity of component A in the final product is controlled by vapor boilup in the second column.
4. Base levels in both columns are controlled by bottoms flow rates.

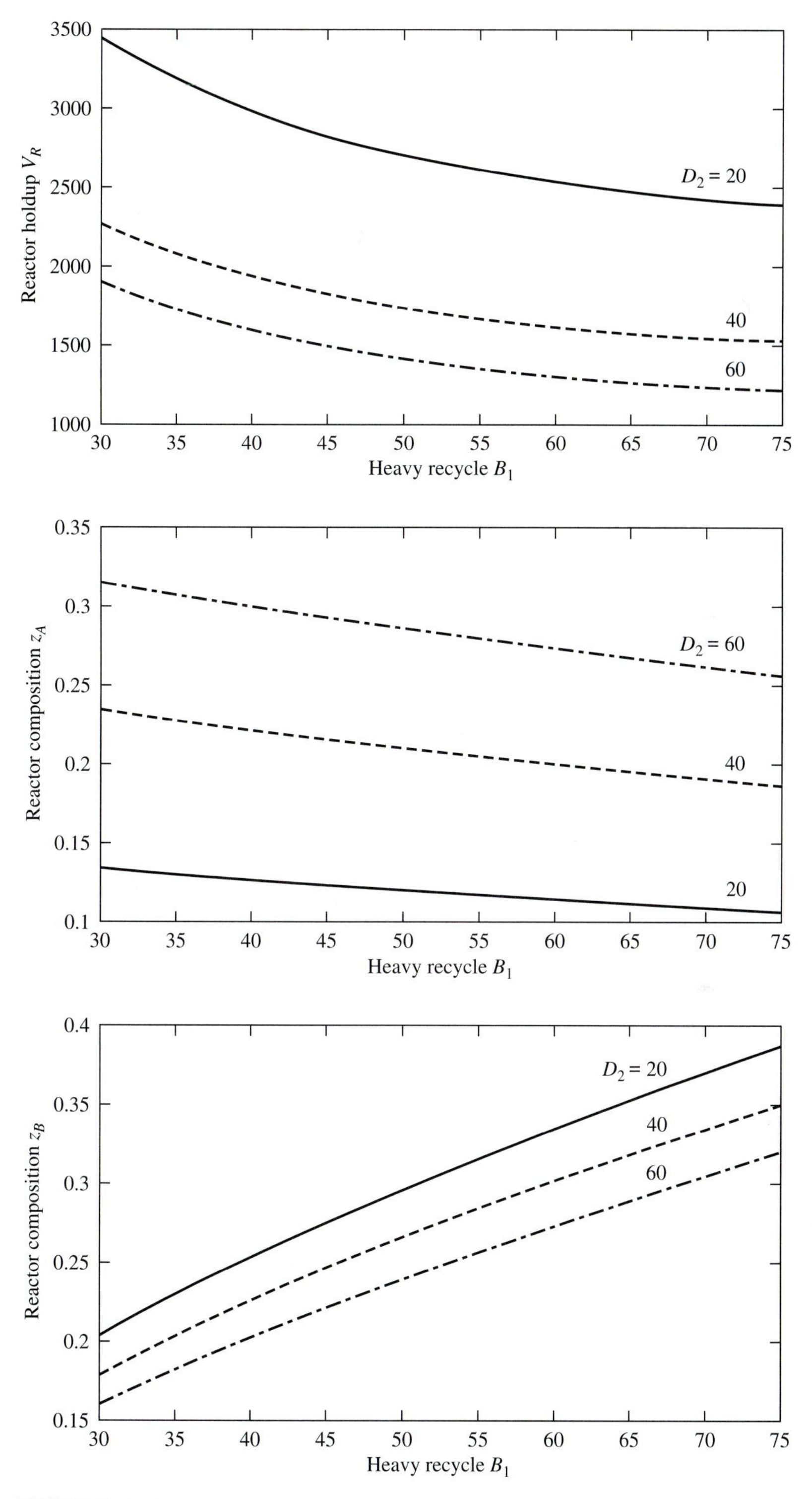

FIGURE 6.14
Effect of two recycles.

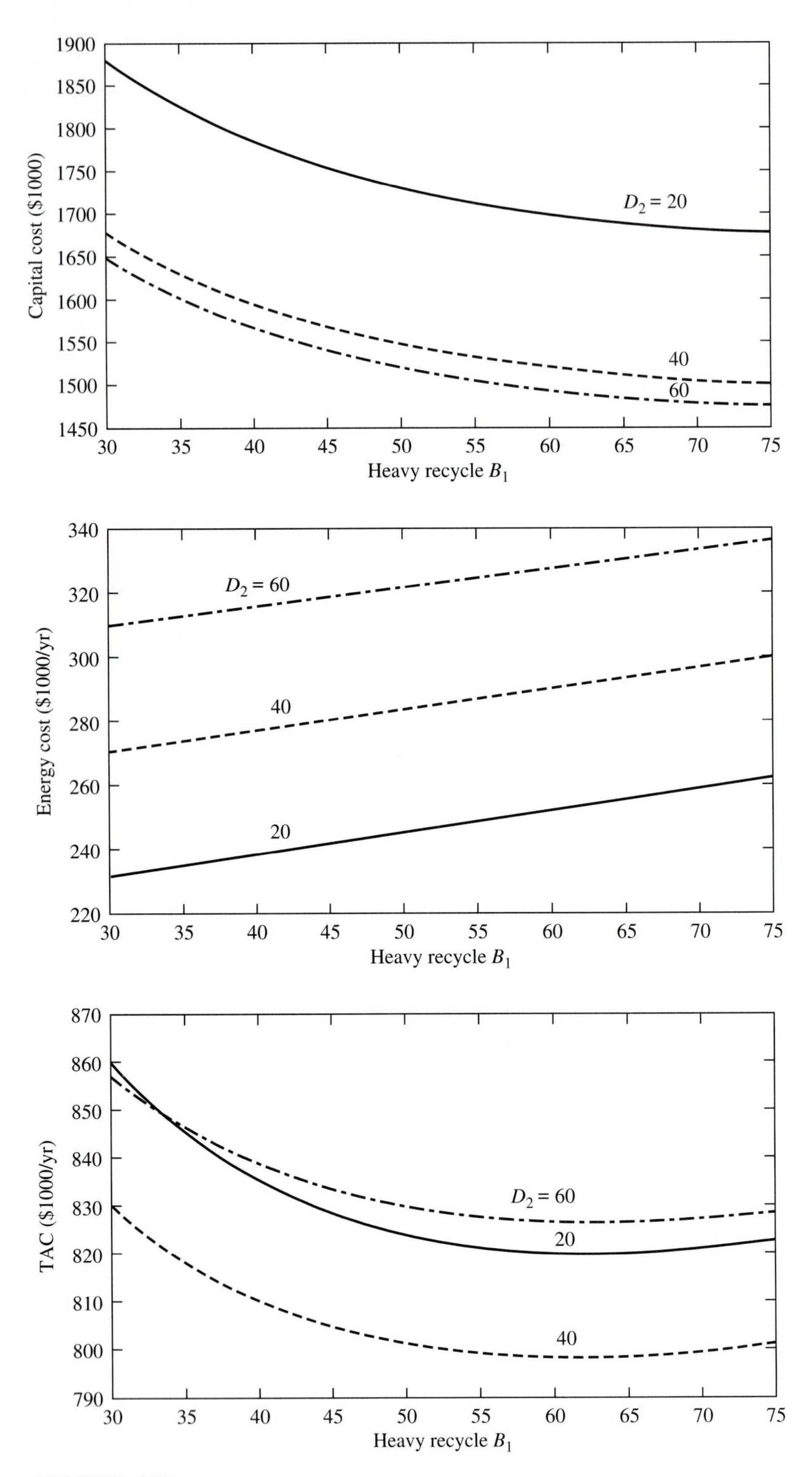

FIGURE 6.15
Effect of two recycles.

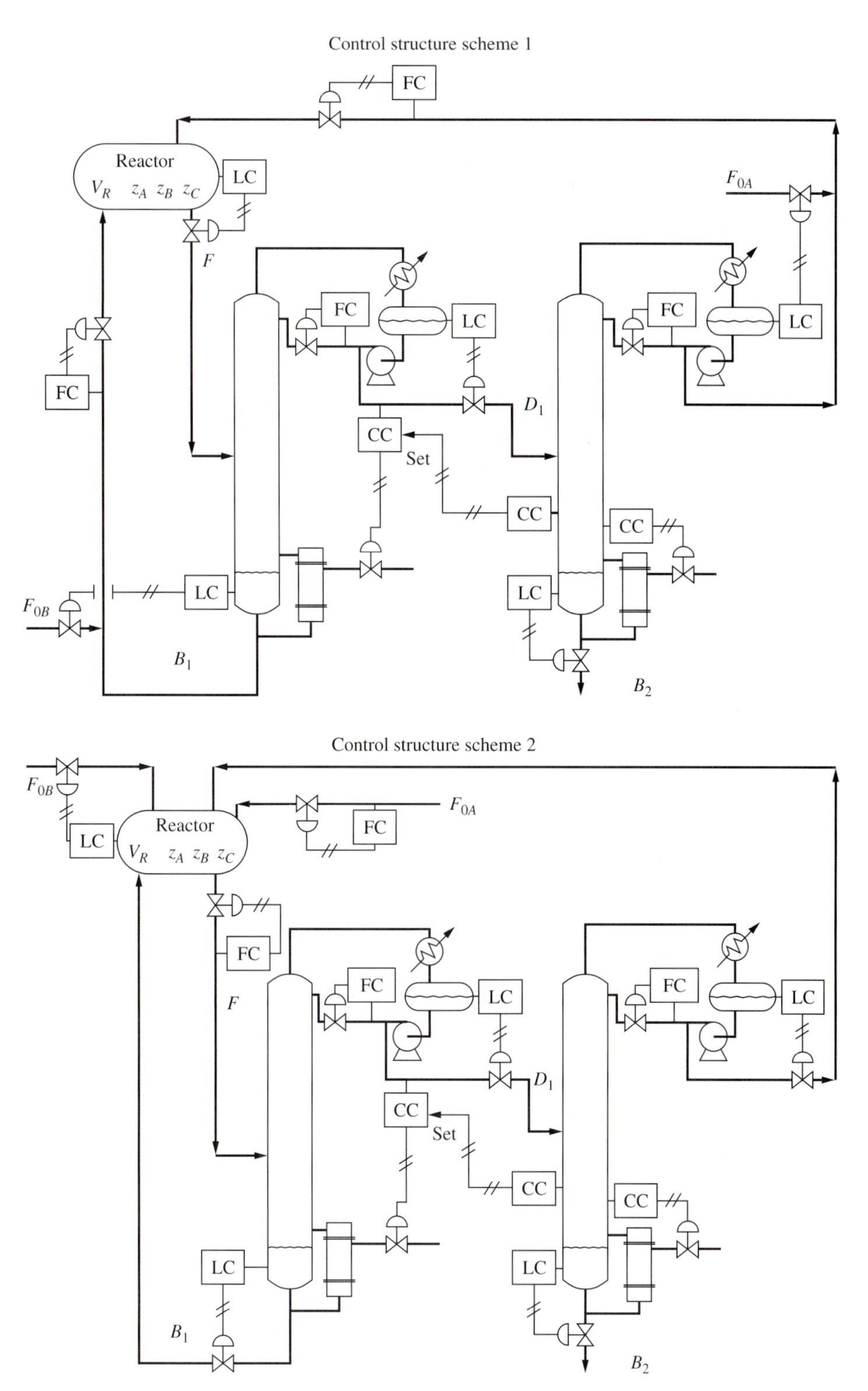

FIGURE 6.16
Alternative control structures.

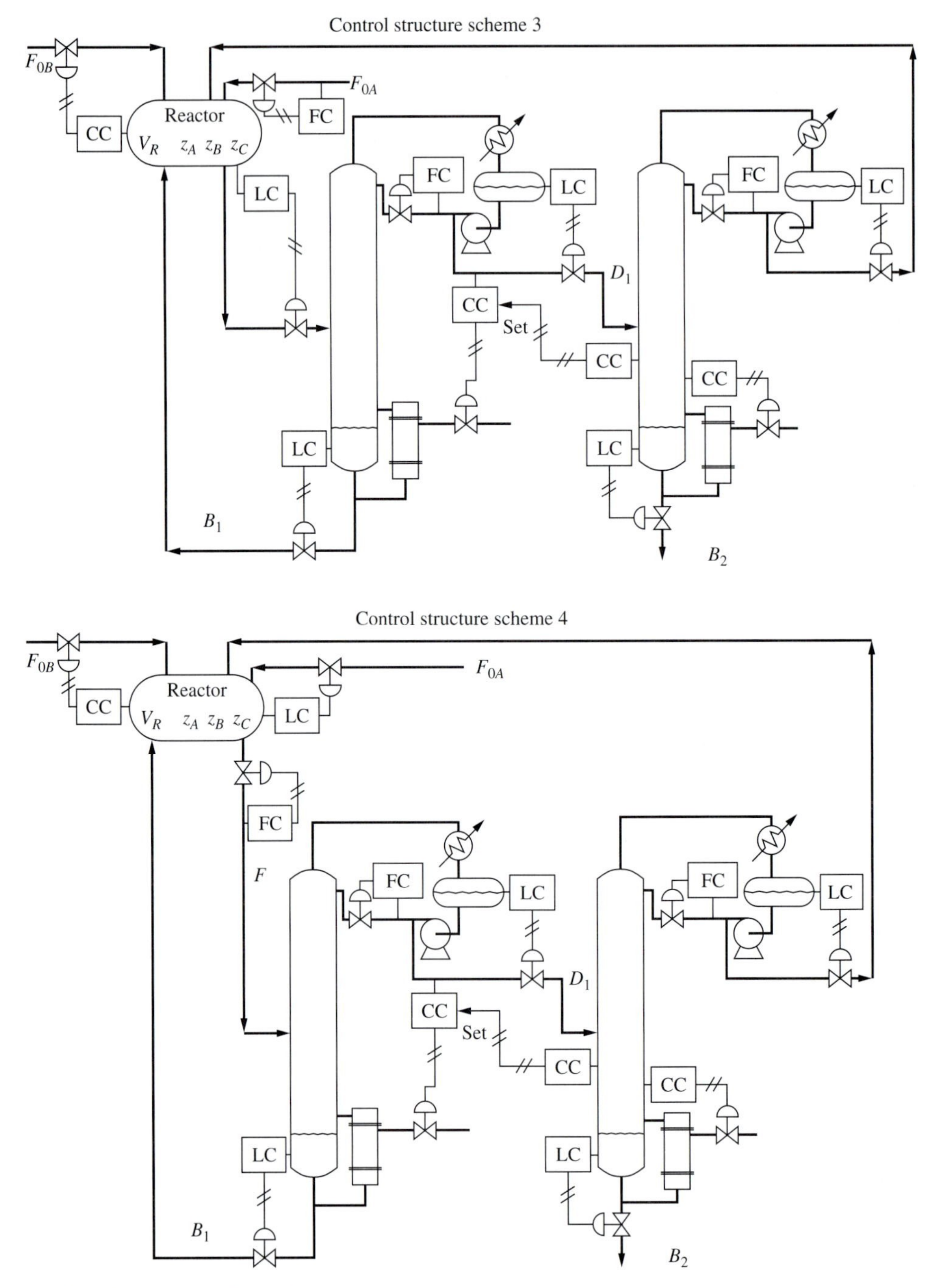

FIGURE 6.16 (CONTINUED)
Alternative control structures.

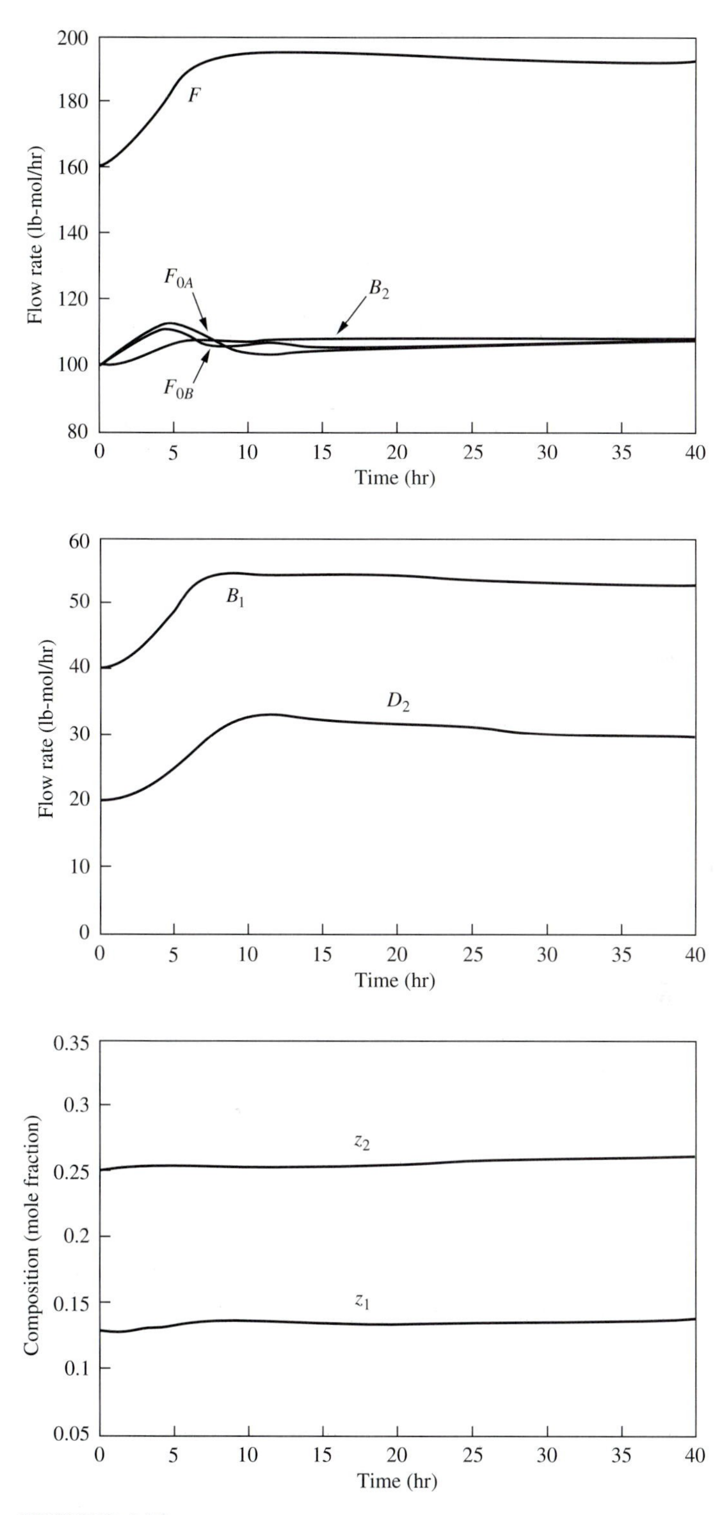

FIGURE 6.17
Scheme 1: Increase in reactor feeds.

In control structure scheme 1, both total recycle streams ($F_{0A} + D_2$ and $F_{0B} + B_1$) are flow controlled. Reactor level is controlled by reactor effluent. Fresh feed makeup of component A controls the level in the reflux drum in the second column. Fresh feed makeup of component B controls the level in the base of the first column. This scheme works well. Throughput can be changed by changing either (or better yet, both) of the recycle flow rates or the reactor temperature. Figure 6.17 shows how this control structure handles an increase in reactor feeds.

In control structure scheme 2 the fresh feed makeup F_{0A} is fixed. Fresh feed makeup F_{0B} of component B is used to control reactor level. Reactor effluent is fixed. This scheme is similar to scheme B in the previous section in that the limiting component is flow controlled into the reactor. The difference here is that the limiting component is not completely converted per pass through the reactor. This scheme does not work. Figure 6.18 shows what happens when a 5 percent increase is made in F_{0A}. The process shuts down in about 150 hours. Component A builds up in the reactor, and component B becomes depleted. The reactor level controller shuts down on the makeup F_{0B} as the reactor level increases because component A is building up.

In control structure scheme 3 fresh feed makeup of component A is fixed. Fresh feed makeup of component B is used to control the composition of component B

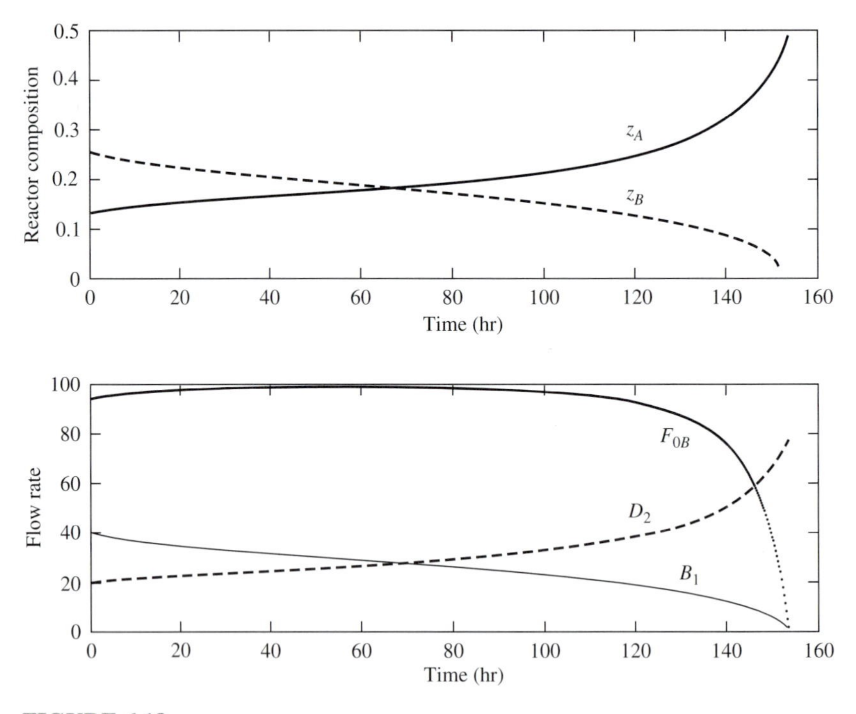

FIGURE 6.18
Base-case V_R, disturbance of $1.05 \times F_{0A}$.

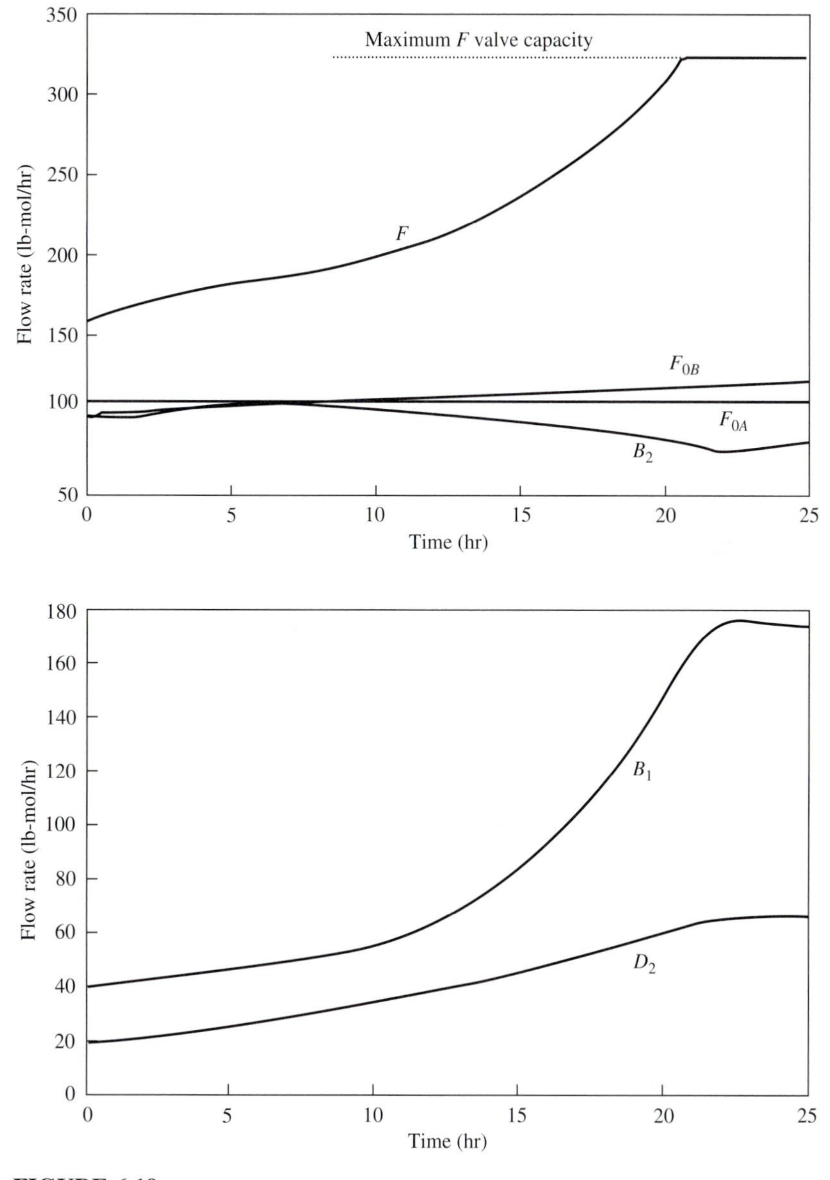

FIGURE 6.19
Scheme 3: Increase in F_{0A}.

in the reactor. Reactor level is controlled by reactor effluent. This scheme does not work because of the "snowball" effect. As shown in Fig. 6.19, the flow rates of reactor effluent, light recycle, and heavy recycle change drastically when the feed rate F_{0A} is changed. This control structure has liquid level controllers in both recycle feedback loops, so snowballing can easily occur.

In control structure scheme 4 fresh feed makeup F_{0A} controls reactor level. Fresh feed makeup F_{0B} controls the reactor composition z_B. Reactor effluent is flow controlled. This scheme works well. However, it requires an additional composition analyzer. This may be a serious drawback in many processes because of the high capital and operating costs of some analyzers and because of their poor reliability.

6.5.3 Interaction between Design and Control

The second-order, two-column, two-recycle process studied above provides another good example of the interaction between design and control. In this section we analyze control structure scheme 2 in more detail to understand why it does not work and how it can be made to work by modifying our steady-state design parameters.

The two control structures that were found to work both lack the convenient feature of being able to set the production rate directly. Structure 4 has the additional major problem of requiring a composition measurement. Composition analyzers are often both expensive and unreliable. It would be nice to be able to find a control structure that does not have these drawbacks.

Control structure 2 overcomes these problems. No reactor composition measurement is used, and throughput is directly fixed by flow-controlling the fresh feed F_{0A}. This control scheme is intuitively appealing and is often proposed in developing control strategies for this type of process. Unfortunately, as we demonstrated in the previous section, it does not work.

Figure 6.20 shows that it will handle a very small (2 percent) increase in F_{0A}. But we already demonstrated in Fig. 6.18 that it cannot handle a 5 percent increase; the process shuts down in about 150 hours. Note the very slow drift in the reactor compositions z_A and z_B. The initial concentrations of components A and B are 13.00 and 25.38 mol%, respectively. At about 70 hours these concentration trajectories cross, and 70 hours later a shutdown occurs. We will show the important insights that can be gained from examining these reactor composition effects.

To understand why scheme 2 does not work we can look at the steady-state effects of various parameters under certain design conditions. This will provide valuable insights into the physically realizable regions of operation. The following assumptions are made for values of recycle compositions (x_{B1} and x_{D2}) and losses of reactants in the product stream B_2.

1. There is no component A leaving the bottoms of the first column ($x_{B1,A} = 0$). Therefore, $x_{B1,B} = 1 - x_{B1,C}$.
2. There is no component B leaving the top of the second column ($x_{D2,B} = 0$). Therefore, $x_{D2,A} = 1 - x_{D2,C}$.
3. The composition of component C in B_1 is $x_{B1,C} = 0.01$.
4. The composition of component C in D_2 is $x_{D2,C} = 0.01$.
5. The losses of reactants A and B in the product stream B_2 are $A_{loss} = 1$ lb-mol/hr and $B_{loss} = 1$ lb-mol/hr based on a production rate of component C of 98 lb-mol/hr ($x_{B2,A} = x_{B2,B} = 0.01$ mole fraction).

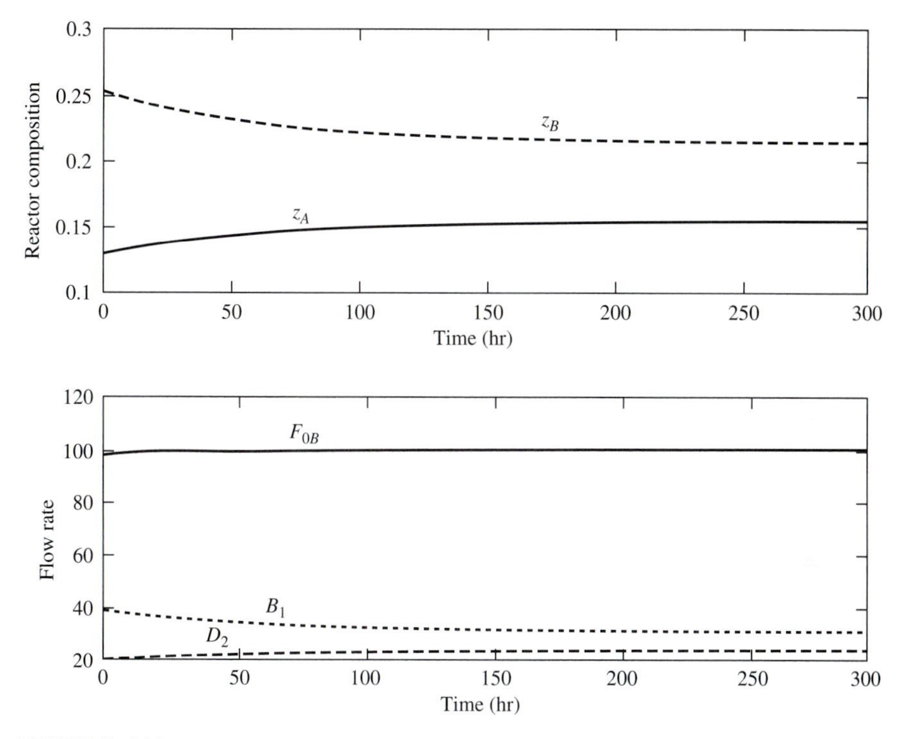

FIGURE 6.20
Base-case V_R, disturbance of $1.02 \times F_{0A}$.

A. Steady-state analysis at fixed production rate

We first look at alternative designs for a fixed production rate of component C. Our calculation procedure is:

1. Fix the value of reactor holdup (V_R = 2971 lb-mol), production rate of component C ($\mathcal{R}_C$ = 98 lb-mol/hr), specific reaction rate (k = 1 hr^{-1}), A_{loss} = 1 lb-mol/hr, and B_{loss} = 1 lb-mol/hr.
2. Specify a value of reactor composition z_B (to be varied later).
3. Calculate z_A from Eq. (6.33).

$$z_A = \mathcal{R}_C / V_R k z_B \tag{6.33}$$

4. Calculate values for the fresh feed makeups and product streams.

$$F_{0A} = \mathcal{R}_C + A_{\text{loss}} \tag{6.34}$$

$$F_{0B} = \mathcal{R}_C + B_{\text{loss}} \tag{6.35}$$

$$B_2 = F_{0A} + F_{0B} - \mathcal{R}_C \tag{6.36}$$

5. From the two steady-state component balances and steady-state total molar balance (remember that the reaction is nonequimolar) around the reactor, we can solve for the three unknowns: the reactor effluent flow rate F, the heavy recycle B_1, and the light recycle D_2.

Component A balance:

$$F_{0A} + D_2 x_{D2,A} = F z_A + \mathcal{R}_C \tag{6.37}$$

Component B balance:

$$F_{0B} + B_1 x_{B1,B} = F z_B + \mathcal{R}_C \tag{6.38}$$

Total molar balance:

$$F_{0A} + F_{0B} + B_1 + D_2 = F + \mathcal{R}_C \tag{6.39}$$

Combining Eqs. (6.37), (6.38), and (6.39) gives

$$F = \frac{F_{0A} + F_{0B} + \dfrac{\mathcal{R}_C - F_{0B}}{x_{B1,B}} + \dfrac{\mathcal{R}_C - F_{0A}}{x_{D2,A}} - \mathcal{R}_C}{1 - z_A/x_{D2,A} - z_B/x_{B1,B}} \tag{6.40}$$

$$D_2 = \frac{F z_A - F_{0A} + \mathcal{R}_C}{x_{D2,A}} \tag{6.41}$$

$$B_1 = \frac{F z_B - F_{0B} + \mathcal{R}_C}{x_{B1,B}} \tag{6.42}$$

All variables on the right side of Eq. (6.40) are known, so F can be calculated. Then Eqs. (6.41) and (6.42) give the two recycle flow rates. Note that the design is physically realizable if the flow rates are all positive and the compositions z_A and z_C ($z_C = 1 - z_A - z_B$) are between 0 and 1 for the specified value of z_B. Not all regions of the z_B-z_A plane give operable plants.

Figure 6.21 gives results for the case where $V_R = 2971$ lb-mol and $\mathcal{R}_C = 98$ lb-mol/hr. Figure 6.21*a* shows the hyperbolic relationship between the selected value of z_B and the required value of z_A. The process can be operated at any point on this curve, giving the same production rate for the given reactor volume, but the reactor effluent and recycle flow rates will be different, as shown in Figure 6.21*b*. For small values of z_B, there is more component A in the reactor, so the D_2 recycle is large. For large values of z_B, there is more component B in the reactor, so the B_1 recycle is large. The reactor effluent flow rate F becomes large for either large or small values of z_B. There is a value of z_B that gives a minimum reactor effluent flow rate, and this occurs in the region where z_A and z_B are similar in value. As Eq. (6.33) shows, the reaction rate depends on the product of the two concentrations, so designs with similar z_A and z_B concentrations would be expected to be the optimum. As we will show, these designs will not provide the best dynamic controllability when control structure 2 is used, so there is a trade-off between design and control.

Figure 6.21*b* also demonstrates that there are two different values of z_B that give the same production rate in the same reactor for the same value of reactor effluent flow rate F. These two alternative operating points have different light and heavy recycle flow rates, D_2 and B_1. This type of multiplicity was pointed out in the previous section.

B. Steady-state analysis for variable production rate

Now we vary the production rate ($\mathcal{R}_C$) with reactor volume held constant and look at the physically realizable ranges of parameters. Figure 6.22*a* shows how z_A must increase as the production rate is increased for a given value of z_B. Figure 6.22*b*

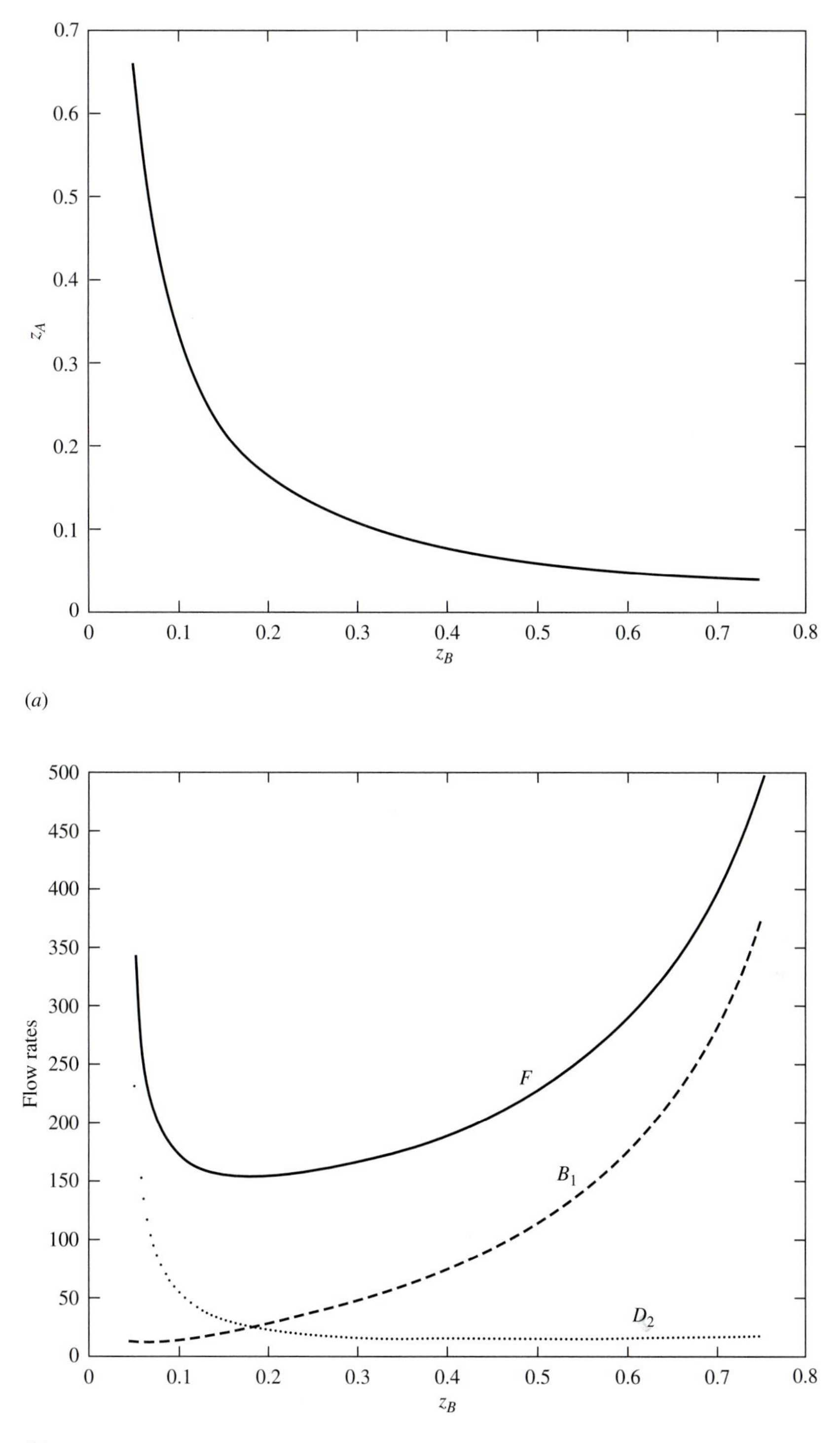

FIGURE 6.21
Production rate = 98, V_R = 2971.

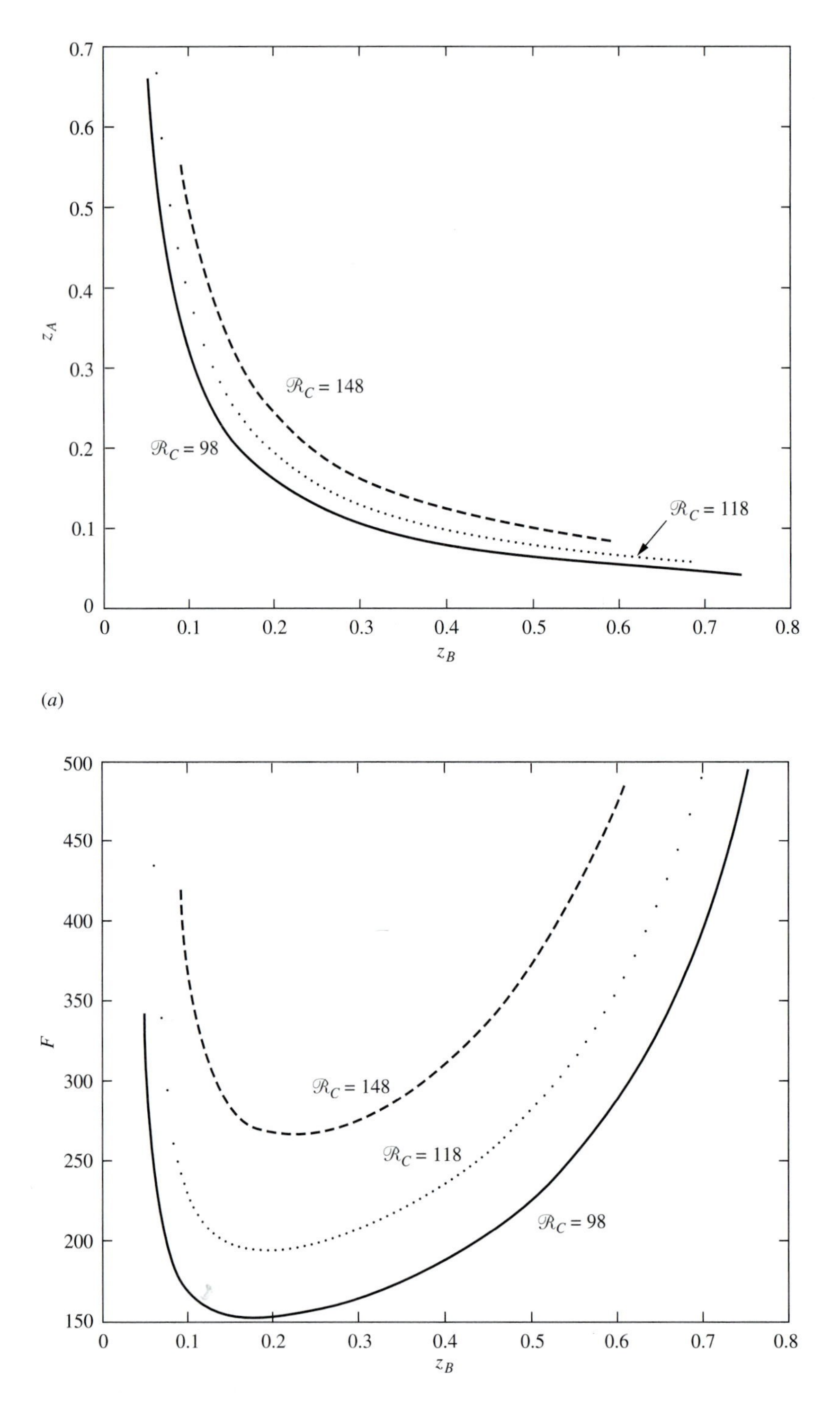

FIGURE 6.22
Base-case V_R.

gives the required changes in reactor effluent flow rate F. Remember that control structure 2 fixes the flow rate of the reactor effluent. If F is fixed at the design point of 160 lb-mol/hr, Fig. 6.22*b* clearly shows that there is no way that a production rate of 120 lb-mol/hr can be attained. To get to this production rate, F must be at least 210 lb-mol/hr. This steady-state analysis provides a simple explanation of why control structure scheme 2 cannot handle throughput changes and confirms the dynamic simulation results.

Of course, this suggests that it may be possible to modify the process operating conditions (recycle flow rates) or the process design parameters (reactor holdup) and move away from the steady-state economic optimum point to be able to use control structure 2 with its advantages of allowing production rate to be directly set and not requiring a composition measurement.

Figure 6.23 gives results for a design with a larger reactor (twice the base case). Now higher production rates can be achieved for the same reactor effluent flow rate. Dynamic simulations of the system with a larger reactor demonstrate the improved rangeability of the process. Figure 6.24 shows that now a 10 percent increase in F_{0A} can be handled with scheme 2. But the system cannot handle a 20 percent increase. Increasing the reactor holdup for the same reactor effluent flow rate changes the reactor compositions in the direction of increasing z_B (25 to 33 percent) and decreasing z_A (13 to 5 percent). The larger difference between the two compositions makes the reactor more stable. We explain this mathematically in the next section.

This case study is another illustration of the important linkage between steady-state design and dynamic controllability. The lesson to be taken away from the preceding steady-state analysis is that there are a number of possible combinations of reactor compositions z_A and z_B for a given production rate. Each set of compositions puts different loads on the separation section. The possible composition combinations vary with reactor size and temperature.

6.5.4 Stability Analysis

The dynamic simulation study of the system with control structure 2 indicates that a reactor shutdown occurs when the disturbance in F_{0A} drives the reactor compositions into a region where z_A becomes greater than z_B. To understand the fundamental reason for this observed phenomenon, we develop a linearized model of a simplified process. The separation section is assumed to be at steady state, and the only dynamics are in the reactor compositions. Perfect reactor level control is assumed. The disturbance is the fresh feed flow rate F_{0A}. Reactor effluent flow rate F is fixed. State variables are z_A and z_B. Algebraic dependent variables at any point in time are the flow rates B_1, D_2, and F_{0B}. To simplify the analysis we assume that the losses of components A and B (A_{loss} and B_{loss}) in the product B_2 stream are constant.

Note that we are looking at the *closedloop* system with control structure scheme 2 in place. Therefore, we are exploring the closedloop stability of the system. The closedloop process is described by two nonlinear ordinary differential equations and three algebraic equations.

$$V_R \frac{dz_A}{dt} = F_{0A} + D_2 x_{D2,A} - F z_A - V_R k z_A z_B \tag{6.43}$$

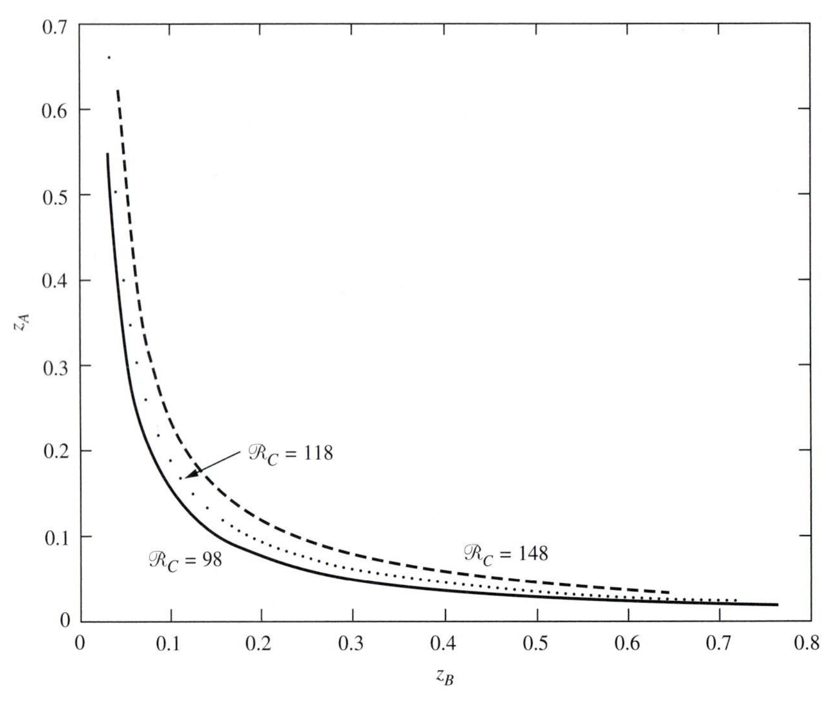

(*a*)

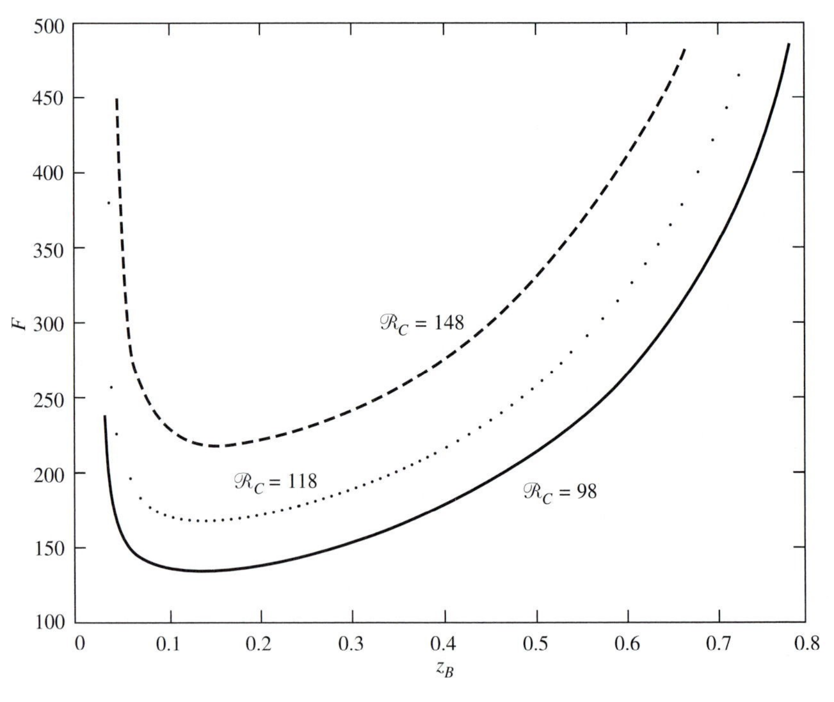

(*b*)

FIGURE 6.23
Twice the base-case V_R.

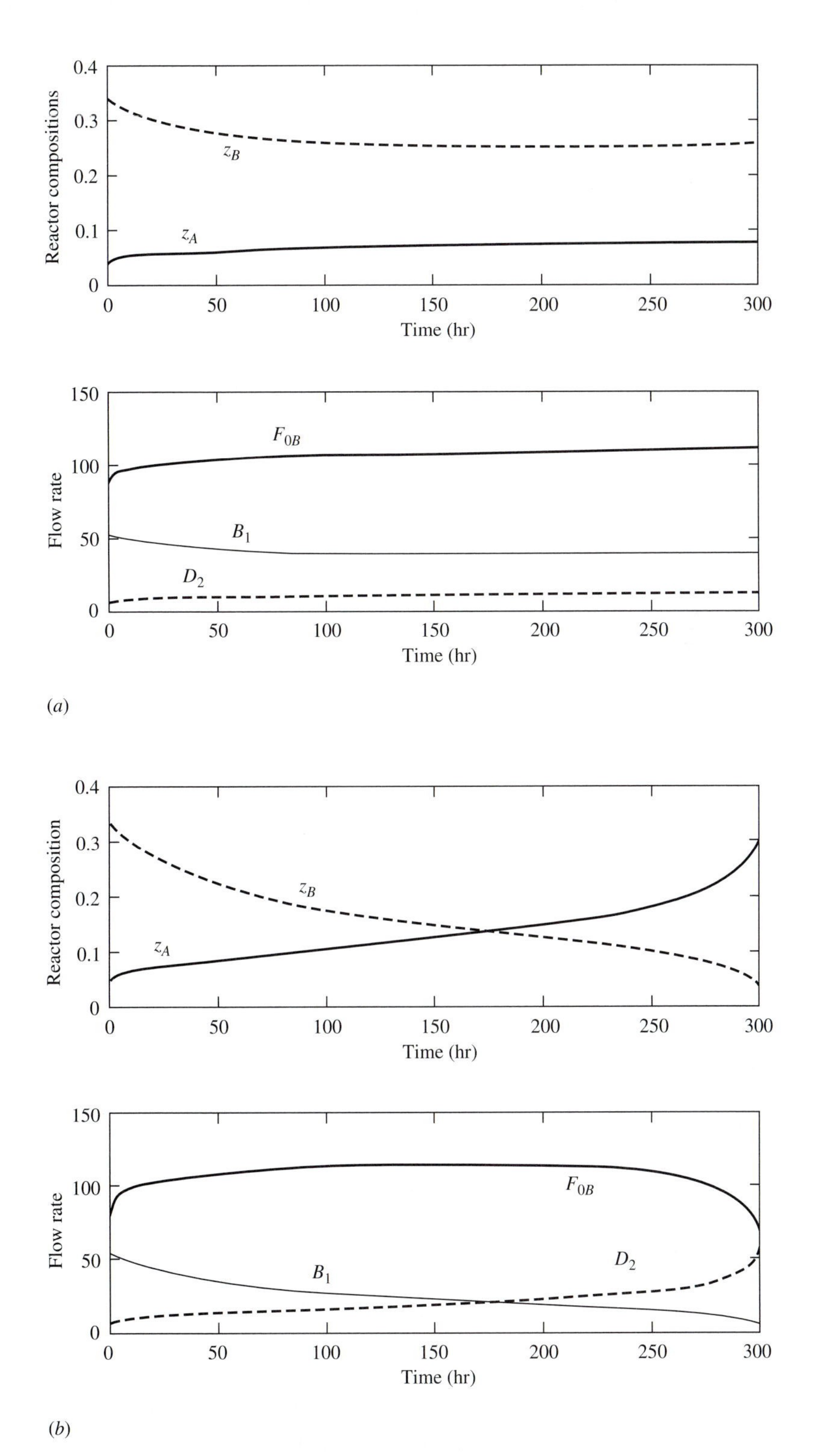

FIGURE 6.24
(*a*) Twice the base-case V_R, disturbance of $1.1 \times F_{0A}$. (*b*) Twice the base-case V_R, disturbance of $1.2 \times F_{0A}$.

$$V_R \frac{dz_B}{dt} = F_{0B} + B_1 x_{B1,B} - F z_B - V_R k z_A z_B \tag{6.44}$$

$$F_{0A} + F_{0B} + D_2 + B_1 = F + V_R k z_A z_B \tag{6.45}$$

$$D_2 = \frac{F z_A - A_{\text{loss}}}{x_{D2,A}} \tag{6.46}$$

$$B_1 = \frac{F z_B - B_{\text{loss}}}{x_{B1,B}} \tag{6.47}$$

Substituting Eqs. (6.45), (6.46), and (6.47) into Eqs. (6.43) and (6.44), linearizing around the steady-state values $\bar{z}_A$ and $\bar{z}_B$, and combining the two first-order ODEs into one second-order ODE give the characteristic equation of the closedloop system.

$$s^2 + s\left(k\bar{z}_B + \frac{F}{V_R x_{B1,B}}\right) + \frac{kF}{V_R}\left(\frac{\bar{z}_B}{x_{B1,B}} - \frac{\bar{z}_A}{x_{D2,A}}\right) \tag{6.48}$$

The linear analysis predicts that the process will become unstable whenever

$$\frac{\bar{z}_B}{x_{B1,B}} < \frac{\bar{z}_A}{x_{D2,A}} \tag{6.49}$$

If the two recycle purity levels are about the same ($x_{B1,B} = x_{D2,A}$), which is assumed in the process studied, the linear analysis predicts that instability will occur when the reactor compositions cross. Remember that the initial steady-state values are $z_A = 0.1300$ and $z_B = 0.2538$; i.e., the concentration of component A is lower than that of component B. Thus, any disturbance that pushes the compositions in the direction of increasing z_A and decreasing z_B tends to make the system unstable.

The linear analysis predicts the observed behavior of the system.

6.6 PLANTWIDE CONTROL DESIGN PROCEDURE

So far in this chapter we have considered some specific features and problems concerning plantwide control. In this section we present a design procedure that can be used to generate an effective plantwide control structure.

The number of variables that can be controlled in any plant equals the number of control valves. Most of these "control degrees of freedom" must be used to set production rate, control product quality, account for safety and environmental constraints, control liquid levels, and control gas pressures. Any remaining degrees of freedom can be used to achieve economic or dynamic objectives.

The method consists of six basic steps. The steps may require some iteration through the procedure.

1. Count the number of control valves (make sure all are legitimate, i.e., only one valve in a liquid-filled line). This is the number of control degrees of freedom.
2. Determine what valve will be used to set production rate. Often this decision is established by the design basis of the plant. Production rate is fixed by a feed stream to the plant if an upstream process establishes the flow rate of material

sent to the plant. Production rate is fixed by the flow rate of a product stream if a downstream process can demand an arbitrary flow rate. If neither of these design requirements is specified, we are free to select the valve that provides smooth and stable production rate transitions and rejects disturbances. This may be the flow rate of the feed stream to a separation section, the flow rate of a recycle stream, the flow rate of a catalyst to a reactor, or a reactor heat removal rate. Controller setpoints of reactor temperature, level, or pressure can also be used to change production rate. Dynamic simulations of alternatives are required to select the best structure.

3. Select the "best" manipulated variables to control each of the product quality variables and the safety and environmental variables. These manipulated variables are selected to give the tightest possible control of these important variables. The dynamic relationships between controlled and manipulated variables should feature small time constants and deadtimes, sufficiently large gains, and wide rangeability of manipulated variables. These decisions must consider some other factors, such as the magnitudes of flow rates. For example, in a high–reflux ratio column, the distillate flow rate should be used to control product quality because the reflux flow rate must be used to control reflux drum liquid level. Another common example is the control of temperature and base level in a distillation column with a very small bottoms flow rate. In most columns, temperature should be controlled by reboiler heat input, and base level controlled by bottoms flow rate. When the bottoms flow rate is less than about 20 percent of the rate of liquid entering the column base, these loops should be reversed (even though this results in "nested" loops).
4. Determine the valves to use for inventory control: all liquid levels (except for surge volumes in liquid recycle systems) and gas pressures must be controlled. Select the largest stream to control levels whenever possible. Use proportional-only level control in nonreactive level loops for cascaded units. Fresh feed makeup streams are often used to hold levels or pressures when these variables reflect the inventory of specific components in the process. There should be a flow controller somewhere in all liquid recycle loops.
5. Make sure that the overall component balances for all chemical components can be satisfied. Light, heavy, and intermediate inert components must have a way to exit the system. Reactant components must be consumed in the reaction section or leave the system as impurities in product streams. Therefore, either reaction rates (temperature, pressure, catalyst addition rate, etc.) must be changed or the flow rates of the fresh feed makeup streams must be manipulated somehow. Makeups can be used to control compositions in the reactor or in recycle streams, or to control inventories that reflect the amount of the specific components contained in the process. For example, bring in a gaseous fresh feed to hold the pressure somewhere in the system, or bring in a liquid fresh feed to hold the level in a reflux drum or column base where the component is in fairly high concentration (typically in a recycle stream).
6. Use the remaining control valves for either steady-state optimization (minimize energy, maximize yield, etc.) or to improve dynamic controllability. A common example is controlling purities of recycle streams. Even though these streams are not products and do not have any quality specifications, the steady-state and

dynamic ability of the process to handle load and production rate changes is sometimes improved by controlling recycle purities (perhaps not at precise setpoints, but sufficiently to prevent excessive build-up of impurities in the recycles).

The examples presented in this chapter illustrate these various steps. Once the control structure has been selected, dynamic simulations of the entire process can be used to evaluate controller performance. Commercial software is being developed that will facilitate plantwide dynamic simulation studies. To tune controllers, each individual unit operation can be isolated and controllers tuned using the relay-feedback test (discussed in Chapter 16).

6.7 CONCLUSION

The development of control structures for processes with multiple interconnected units has been discussed in this chapter. Two important heuristic rules have been presented:

1. A stream somewhere in a liquid recycle loop should be flow controlled.
2. A reactant cannot be flow controlled into a reactor in which a second-order reaction takes place unless its concentration in the reactor is much smaller than that of the other reactant.

The use of steady-state sensitivity analysis for screening out poor control structures was illustrated.

The area of plantwide control is the active focus of research in process control. Improved methodologies and approaches to the problem should be developed in the future. The incentives are great, as are the challenges.

PROBLEMS

6.1. A second-order reaction occurs in a continuous stirred-tank reactor (CSTR).

$$A + B \xrightarrow{k} C$$

Pure reactant A is fed into the reactor at a flow rate of F_{0A} kg-mol/hr. Pure reactant B is fed into the reactor at a flow rate of F_{0B} kg-mol/hr. The mole fractions of components A, B, and C inside the reactor are z_A, z_B, and z_C, respectively. The rate of production of product C is given by the kinetic expression

$$\mathcal{R}_C = V_R k z_A z_B$$

where $\mathcal{R}_C$ = kg-mol/hr of component C produced
V_R = holdup in the reactor = 337.1 kg-mol
k = specific reaction rate constant = 2 hr^{-1}

The reactor effluent is fed into a distillation column. The volatilities of components A, B, and C are α_A, α_B, and α_C, respectively. Since $\alpha_A > \alpha_B > \alpha_C$, the distillate stream from the column contains all the component A in the column feed and most of component B, with a small amount of component C: $x_{D,A} = 0.60$ mole fraction A, $x_{D,B} = 0.35$

mole fraction B, and $x_{D,C} = 0.05$ mole fraction C. The distillate is recycled back to the reactor.

The composition of the bottoms stream from the column is 0.01 mole fraction component B and 0.99 mole fraction component C. The flow rate of the bottoms product from the column is 100 kg-mol/hr.

(*a*) Calculate how much component C is produced in the reactor.
(*b*) Calculate the flow rates of the two fresh feed makeup streams F_{0A} and F_{0B}.
(*c*) Calculate the flow rates of the reactor effluent and the recycle streams.
(*d*) Calculate the z_A and z_B compositions in the reactor.
(*e*) Sketch a plantwide control concept diagram for this process.

6.2. We want to develop a process flowsheet and plantwide control structure for a plant in which the reaction A + B $\rightarrow$ C + D occurs in a CSTR. Fresh feed makeup streams of pure component A (F_{0A}) and pure component B (F_{0B}) are used. The per-pass conversion is not 100 percent, so the reactor effluent contains all four components.

Distillation is used to achieve the required separation of components into fairly pure product streams and the necessary recycle of the reactants. The volatilities of the chemical components are $\alpha_A > \alpha_C > \alpha_B > \alpha_D$.

(*a*) Sketch a reasonable flowsheet for this process showing major pieces of equipment and process flows.
(*b*) Propose a control structure for your process that incorporates the basic principles of plantwide control.

6.3. The Eastman plantwide control process contains seven components, a liquid-phase reactor (with only a vapor stream leaving the reactor), a stripper, and a gas recycle stream. Figure P6.3 gives a sketch of the process. There are four reactions occurring:

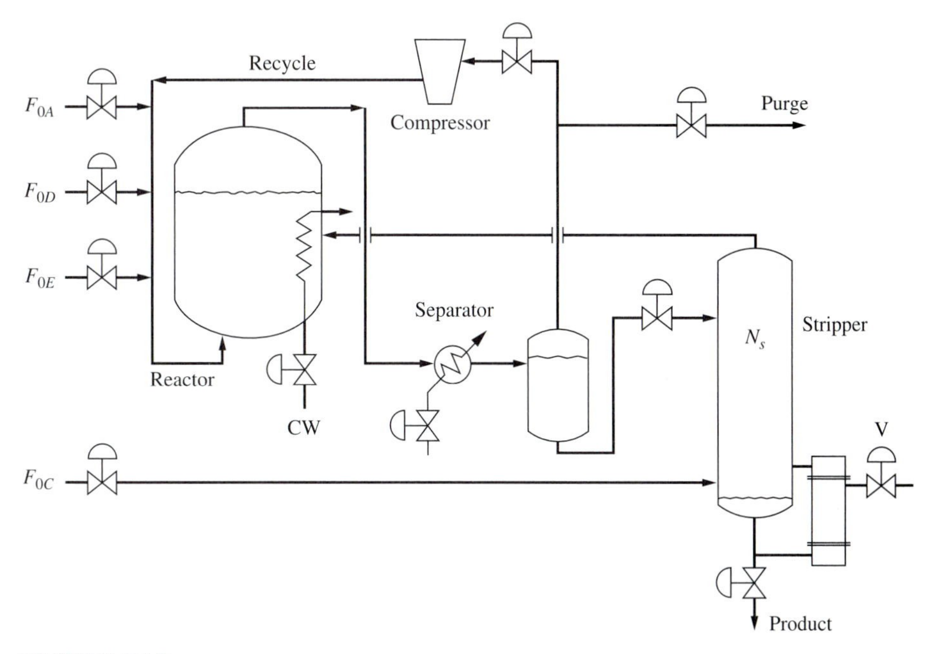

FIGURE P6.3

$A + C + D \rightarrow G$, $A + C + E \rightarrow H$, $A + E \rightarrow F$, $3D \rightarrow 2F$. Four gas fresh feed streams enter the process. One of these is fed into the bottom of the stripper. The vapor stream leaving the reactor is cooled before entering a separator. Liquid from the separator is fed into a stripper, which produces a bottoms product stream. Vapor from the separator is split between a gas purge and a recycle stream back to the reactor. The vapor from the stripper is also fed into the reactor. The reactor is cooled by manipulating cooling-water flow rate to cooling coils.

Develop a plantwide control system for this process.

6.4. The vinyl acetate process shown contains seven components. The chemistry consists of two reactions, which produce vinyl acetate ($C_4H_6O_2$) from ethylene, oxygen, and acetic acid ($C_2H_4O_2$) and form by-products of water and carbon dioxide.

$$C_2H_4 + C_2H_4O_2 + \tfrac{1}{2}O_2 \rightarrow C_4H_6O_2 + H_2O$$

$$C_2H_4 + 3O_2 \rightarrow 2CO_2 + 2H_2O$$

Fresh acetic acid and a liquid acetic acid recycle stream are fed into a vaporizer along with a gas recycle stream and fresh ethylene feed. Oxygen is added after the vaporizer, and reactions occur in a gas-phase reactor. Reactor effluent is cooled and fed into a separator. Vapor from the separator is compressed and fed into an absorber to recover vinyl acetate. Recycle acetic acid is used as lean oil in the absorber. Exit gas from the absorber is sent to a CO_2 removal unit, which produces a CO_2 purge stream and gas recycle. Another purge stream is used to remove the small amount of ethane that is in the fresh ethylene makeup feed.

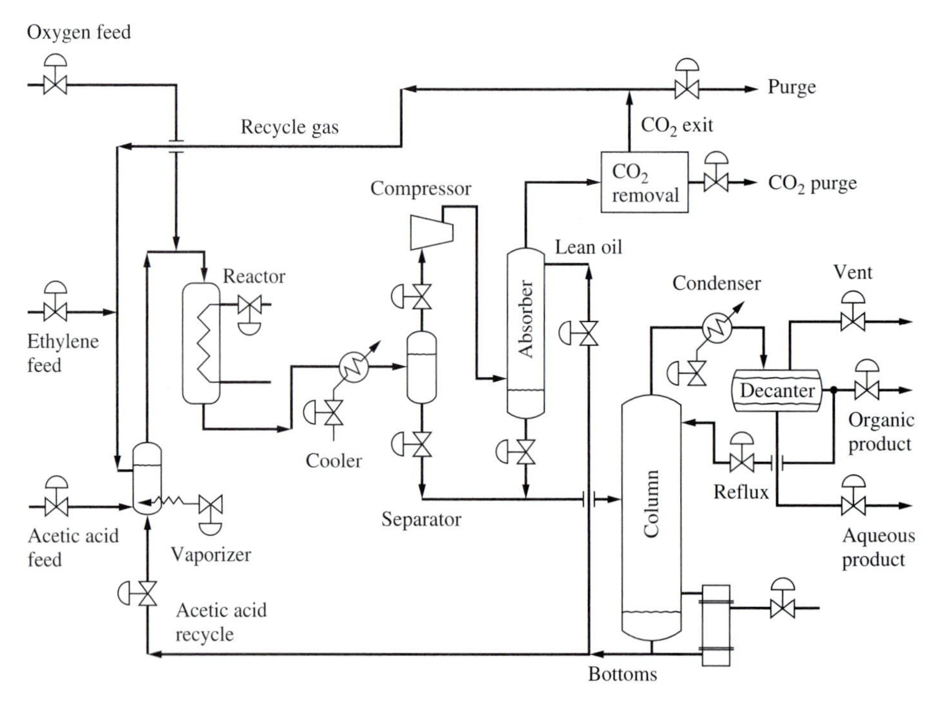

FIGURE P6.4

The liquid streams from the separator and the bottom of the absorber are combined and fed into a distillation column. The bottoms from the column is split into two streams: absorber lean oil and recycle acetic acid. The overhead vapor condenses into two liquid phases because of the nonideality of the phase equilibrium. The aqueous phase from the decanter is removed as product and sent to further processing, which we do not consider here. Some of the organic phase (mostly vinyl acetate) is refluxed back to the column, and some is removed for further processing.

Develop a plantwide control system for this process.

6.5. The following consecutive chemical reactions occur in a CSTR.

$$A + B \rightarrow M + C$$
$$M + B \rightarrow D + C$$
$$D + B \rightarrow T + C$$

Fresh feed streams of A and B are fed into a reactor. The reactor effluent, a mixture of unreacted A and B with products M, D, T, and C, is fed into a sequence of distillation columns. The light-out-first (direct) sequence is used, i.e., the lightest component is removed overhead in the first column, the next lightest component is removed overhead in the second column, and so on. The relative volatilities are $A > M > D > T > B > C$.

Unreacted A and B are recycled back to the reactor. Product streams of components M, D, T, and C are removed from the distillation train.

(*a*) Sketch a reasonable flowsheet for this process showing major pieces of equipment and process flows.

(*b*) Propose a plantwide control system for this process.

PART TWO

Laplace-Domain Dynamics and Control

It is finally time for our Russian lessons! We have explored dynamics and control in the "English" time domain, using differential equations and finding exponential solutions. We saw that the important parameters are the time constant, the steady-state gain, and the damping coefficient of the system. If the process has no controllers, the system is openloop and we look at openloop time constants and openloop damping coefficients. If controllers are included with the process, the analysis considers the closedloop system. We tune the controller to achieve certain desired closedloop time constants and closedloop damping coefficients. It is important that you keep track of the kind of system you are looking at, openloop or closedloop. We do not want to compare apples and oranges in our studies.

In the next three chapters we develop methods of analysis of dynamic systems, both openloop and closedloop, using Laplace transformation. This form of system representation is much more compact and convenient than the time-domain representation. The Laplace-domain description of a process is a "transfer function." This is a relationship between the input to a system and the output of the system. Transfer functions contain all the steady-state and dynamic information about a process in a very compact form.

You will find it very useful to learn a little "Russian." Don't get too concerned about having to learn an extensive Russian vocabulary. As you will see, there are only nine "words" that you have to learn in Russian: nine transfer functions can describe almost all chemical engineering processes.

Laplace transformation can be applied only to *linear* ordinary differential equations. So for most of the rest of this book, we will be dealing with linear systems.

CHAPTER 7

Laplace-Domain Dynamics

The use of Laplace transformations yields some very useful simplifications in notation and computation. Laplace-transforming the linear ordinary differential equations describing our processes in terms of the independent variable t converts them into algebraic equations in the Laplace transform variable s. This provides a very convenient representation of system dynamics.

Most of you have probably been exposed to Laplace transforms in a mathematics course. We lead off this chapter with a brief review of some of the most important relationships. Then we derive the Laplace transformations of commonly encountered functions. Next we develop the idea of transfer functions by observing what happens to the differential equations describing a process when they undergo Laplace transformation. Finally, we apply these techniques to some chemical engineering systems.

7.1 LAPLACE TRANSFORMATION FUNDAMENTALS

7.1.1 Definition

The Laplace transformation of a function of time $f_{(t)}$ consists of "operating on" the function by multiplying it by e^{-st} and integrating with respect to time t from 0 to infinity. The operation of Laplace transforming is indicated by the notation

$$\mathscr{L}[f_{(t)}] \equiv \int_0^\infty f_{(t)} e^{-st}\, dt \tag{7.1}$$

where $\mathscr{L}$ = Laplace transform operator
s = Laplace transform variable

In integrating between the definite limits of 0 and infinity, we "integrate out" the time variable t and are left with a new quantity that is a function of s. We use the notation

$$\mathscr{L}[f_{(t)}] \equiv F_{(s)} \tag{7.2}$$

The variable s is a complex number.

Thus, Laplace transformation converts functions from the time domain (where t is the independent variable) to the Laplace domain (where s is the independent variable). The advantages of using this transformation will become clear later in this chapter.

7.1.2 Linearity Property

One of the most important properties of the Laplace transformation is that it is linear.

$$\mathscr{L}[f_{1(t)} + f_{2(t)}] = \mathscr{L}[f_{1(t)}] + \mathscr{L}[f_{2(t)}] \tag{7.3}$$

This property is easily proved:

$$\begin{aligned} \mathscr{L}[f_{1(t)} + f_{2(t)}] &= \int_0^\infty [f_{1(t)} + f_{2(t)}]e^{-st}\,dt \\ &= \int_0^\infty f_{1(t)}e^{-st}\,dt + \int_0^\infty f_{2(t)}e^{-st}\,dt \\ &= \mathscr{L}[f_{1(t)}] + \mathscr{L}[f_{2(t)}] = F_{1(s)} + F_{2(s)} \end{aligned} \tag{7.4}$$

7.2 LAPLACE TRANSFORMATION OF IMPORTANT FUNCTIONS

Let us now apply the definition of the Laplace transformation to some important time functions: steps, ramps, exponential, sines, etc.

7.2.1 Step

Consider the function

$$f_{(t)} = Ku_{n(t)} \tag{7.5}$$

where K is a constant and $u_{n(t)}$ is the unit step function defined in Section 2.1 as

$$\begin{aligned} u_{n(t)} &= 1 \quad t > 0 \\ &= 0 \quad t \leq 0 \end{aligned} \tag{7.6}$$

Note that the step function is just a constant (for time greater than zero).

Laplace-transforming this function gives

$$\mathscr{L}[Ku_{n(t)}] \equiv \int_0^\infty [Ku_{n(t)}]e^{-st}\,dt = K\int_0^\infty e^{-st}\,dt$$

since $u_{n(t)}$ is just equal to unity over the range of integration.

$$\mathscr{L}[K u_{n(t)}] = \left[-\frac{K}{s} e^{-st} \right]_{t=0}^{t=\infty} = -\frac{K}{s}[0 - 1] = \frac{K}{s} \tag{7.7}$$

Therefore, the Laplace transformation of a step function (or a constant) of magnitude K is simply $K(1/s)$.

7.2.2 Ramp

The ramp function is one that changes continuously with time at a constant rate K.

$$f_{(t)} = Kt \tag{7.8}$$

Then the Laplace transformation is

$$\mathscr{L}[Kt] \equiv \int_0^\infty [Kt] e^{-st}\, dt$$

Integrating by parts, we let

$$u = t \quad \text{and} \quad dv = e^{-st}\, dt$$

Then

$$du = dt \quad \text{and} \quad v = -\frac{1}{s} e^{-st}$$

Since

$$\int_0^\infty u\, dv = [uv]_0^\infty - \int_0^\infty v\, du \tag{7.9}$$

$$K \int_0^\infty t e^{-st}\, dt = \left[-\frac{Kt}{s} e^{-st} \right]_{t=0}^{t=\infty} + \int_0^\infty \frac{K}{s} e^{-st}\, dt$$

$$= [0 - 0] - \left[\frac{K}{s^2} e^{-st} \right]_{t=0}^{t=\infty} = K\left(\frac{1}{s^2}\right)$$

Therefore, the Laplace transformation of a ramp function is

$$\mathscr{L}[Kt] = K\left(\frac{1}{s^2}\right) \tag{7.10}$$

7.2.3 Sine

For the function

$$f_{(t)} = \sin(\omega t) \tag{7.11}$$

where ω = frequency (radians/time),

$$\mathcal{L}[\sin(\omega t)] \equiv \int_0^\infty [\sin(\omega t)]e^{-st}\,dt$$

Using

$$\sin(\omega t) = \frac{e^{i\omega t} - e^{-i\omega t}}{2i} \tag{7.12}$$

$$\mathcal{L}[\sin(\omega t)] \equiv \int_0^\infty \left[\frac{e^{i\omega t} - e^{-i\omega t}}{2i}\right] e^{-st}\,dt$$

$$= \int_0^\infty \left[\frac{e^{-(s-i\omega)t} - e^{-(s+i\omega)t}}{2i}\right] dt$$

$$= \frac{1}{2i}\left[\frac{-e^{-(s-i\omega)t}}{s - i\omega} + \frac{e^{-(s+i\omega)t}}{s + i\omega}\right]_{t=0}^{t=\infty} = \frac{1}{2i}\left[\frac{1}{s - i\omega} - \frac{1}{s + i\omega}\right]$$

Therefore,

$$\mathcal{L}[\sin(\omega t)] = \frac{\omega}{s^2 + \omega^2} \tag{7.13}$$

7.2.4 Exponential

Since we found in Chapter 2 that the responses of linear systems are a series of exponential terms, the Laplace transformation of the exponential function is the most important of any of the functions.

$$f_{(t)} = e^{-at} \tag{7.14}$$

$$\mathcal{L}[e^{-at}] \equiv \int_0^\infty [e^{-at}]e^{-st}\,dt = \int_0^\infty [e^{-(s+a)t}]\,dt$$

$$= \left[\frac{-1}{s+a} e^{-(s+a)t}\right]_{t=0}^{t=\infty} = \frac{1}{s+a}$$

Therefore

$$\boxed{\mathcal{L}[e^{-at}] = \frac{1}{s+a}} \tag{7.15}$$

We will use this function repeatedly throughout the rest of this book.

7.2.5 Exponential Multiplied by Time

In Chapter 2 repeated roots of the characteristic equation yielded time functions that contained an exponential multiplied by time.

$$f_{(t)} = te^{-at} \tag{7.16}$$

$$\mathcal{L}[te^{-at}] \equiv \int_0^\infty [te^{-at}]e^{-st}\,dt = \int_0^\infty [te^{-(s+a)t}]\,dt$$

Integrating by parts,

$$u = t \qquad dv = e^{-(s+a)t}\,dt$$

$$du = dt \qquad v = -\frac{1}{s+a}e^{-(s+a)t}$$

$$\int_0^\infty [te^{-(s+a)t}]\,dt = \left[\frac{-te^{-(s+a)t}}{s+a}\right]_{t=0}^{t=\infty} + \int_0^\infty \frac{e^{-(s+a)t}}{s+a}\,dt$$

$$= [0-0] - \left[\frac{1}{(s+a)^2}e^{-(s+a)t}\right]_{t=0}^{t=\infty}$$

Therefore,

$$\mathcal{L}[te^{-at}] = \frac{1}{(s+a)^2} \tag{7.17}$$

Equation (7.17) can be generalized for a repeated root of nth order to give

$$\mathcal{L}[t^n e^{-at}] = \frac{n!}{(s+a)^{n+1}} \tag{7.18}$$

7.2.6 Impulse (Dirac Delta Function $\delta_{(t)}$)

The impulse function is an infinitely high spike that has zero width and an area of 1 (see Fig. 7.1*a*). It cannot occur in any real system, but it is a useful mathematical function that is used in several spots in this book.

One way to define $\delta_{(t)}$ is to call it the derivative of the unit step function, as sketched in Fig. 7.1*b*.

$$\delta_{(t)} = \frac{du_{n(t)}}{dt} \tag{7.19}$$

Now the unit step function can be expressed as the limit of the first-order exponential step response as the time constant goes to zero.

$$u_{n(t)} = \lim_{\tau \to 0}\left(1 - e^{-t/\tau}\right) \tag{7.20}$$

Now

$$\mathcal{L}[\delta_{(t)}] = \mathcal{L}\left[\frac{d}{dt}\left\{\lim_{\tau \to 0}(1 - e^{-t/\tau})\right\}\right] = \lim_{\tau \to 0}\mathcal{L}\left[\frac{1}{\tau}e^{-t/\tau}\right]$$

$$= \lim_{\tau \to 0}\left(\frac{1/\tau}{s + 1/\tau}\right) = \lim_{\tau \to 0}\left(\frac{1}{\tau s + 1}\right) = 1$$

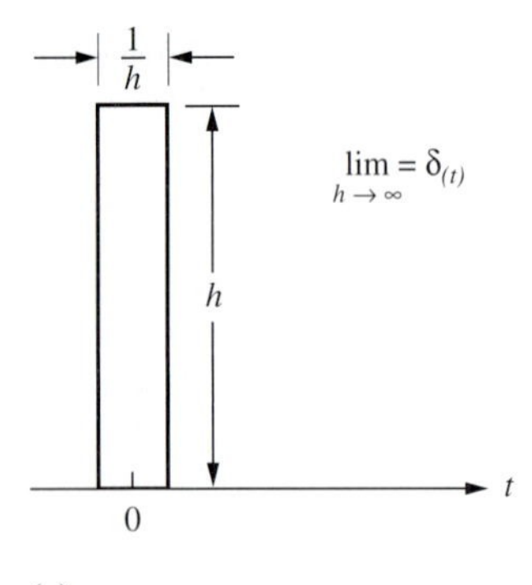

(a)

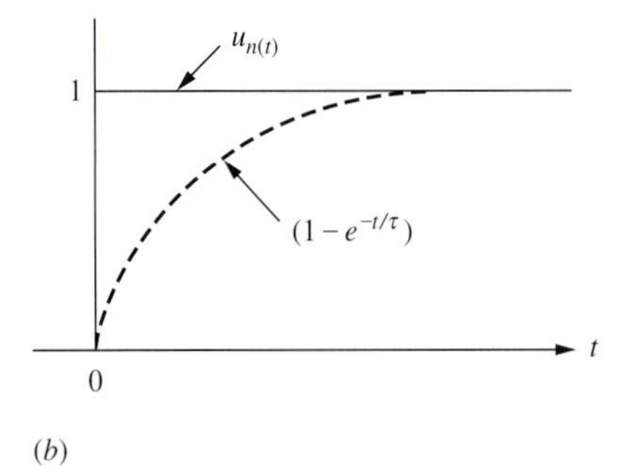

(b)

FIGURE 7.1
$\delta_{(t)}$ function.

Therefore,

$$\mathscr{L}[\delta_{(t)}] = 1 \tag{7.21}$$

7.3 INVERSION OF LAPLACE TRANSFORMS

After transforming equations into the Laplace domain and solving for output variables as functions of s, we sometimes want to transform back into the time domain. This operation is called *inversion* or *inverse Laplace transformation.* We are translating from Russian into English. We will use the notation

$$\mathscr{L}^{-1}[F_{(s)}] = f_{(t)} \tag{7.22}$$

There are several ways to invert functions of s into functions of t. Since s is a complex number, a contour integration in the complex s plane can be used.

$$f_{(t)} = \frac{1}{2\pi i}\int_{\alpha-i\omega}^{\alpha+i\omega} e^{st}F_{(s)}\,ds \tag{7.23}$$

Another method is simply to look up the function in mathematics tables.

The most common inversion method is called *partial fractions expansion.* The function to be inverted, $F_{(s)}$, is merely rearranged into a sum of simple functions:

$$F_{(s)} = F_{1(s)} + F_{2(s)} + \cdots + F_{N(s)} \tag{7.24}$$

Then each term is inverted (usually by inspection, because they are simple). The total time-dependent function is the sum of the individual time-dependent functions [see Eq. (7.3)].

$$\begin{aligned} f_{(t)} &= \mathscr{L}^{-1}[F_{1(s)}] + \mathscr{L}^{-1}[F_{2(s)}] + \cdots + \mathscr{L}^{-1}[F_{N(s)}] \\ &= f_{1(t)} + f_{2(t)} + \cdots + f_{N(t)} \end{aligned} \tag{7.25}$$

As we will shortly find out, the $F_{(s)}$'s normally appear as ratios of polynomials in s.

$$F_{(s)} = \frac{Z_{(s)}}{P_{(s)}} \tag{7.26}$$

where $Z_{(s)}$ = Mth-order polynomial in s
$P_{(s)}$ = Nth-order polynomial in s

Factoring the denominator into its roots (or *zeros*) gives

$$F_{(s)} = \frac{Z}{(s - p_1)(s - p_2)(s - p_3)\cdots(s - p_N)} \tag{7.27}$$

where the p_i are the roots of the polynomial $P_{(s)}$, which may be distinct or repeated.

If all the p_i are different (i.e., distinct roots), we can express $F_{(s)}$ as a sum of N terms:

$$F_{(s)} = \frac{A}{s - p_1} + \frac{B}{s - p_2} + \frac{C}{s - p_3} + \cdots + \frac{W}{s - p_N} \tag{7.28}$$

The numerators of each of the terms in Eq. (7.28) can be evaluated as shown below and then each term inverted.

$$\begin{aligned} A &= \lim_{s \to p_1} \left[(s - p_1)F_{(s)}\right] \\ B &= \lim_{s \to p_2} \left[(s - p_2)F_{(s)}\right] \\ &\vdots \\ W &= \lim_{s \to p_N} \left[(s - p_N)F_{(s)}\right] \end{aligned} \tag{7.29}$$

EXAMPLE 7.1. Given the following $F_{(s)}$, find its inverse $f_{(t)}$ by partial fractions expansion.

$$F_{(s)} = \frac{K_p \overline{C}_{A0}}{s(\tau_o s + 1)} \tag{7.30}$$

$$F_{(s)} = \frac{K_p \overline{C}_{A0}/\tau_o}{s(s + 1/\tau_o)} = \frac{A}{s} + \frac{B}{s + 1/\tau_o}$$

The roots of the denominator are 0 and $-1/\tau_o$.

$$A = \lim_{s \to 0}\left[sF_{(s)}\right] = \lim_{s \to 0}\left[s\frac{K_p \overline{C}_{A0}}{s(\tau_o s + 1)}\right] = K_p \overline{C}_{A0}$$

$$B = \lim_{s \to -1/\tau_o}\left[(s + 1/\tau_o)F_{(s)}\right] = \lim_{s \to -1/\tau_o}\left[\frac{K_p \overline{C}_{A0}/\tau_o}{s}\right] = -K_p \overline{C}_{A0}$$

Therefore,

$$F_{(s)} = K_p \overline{C}_{A0} \left[\frac{1}{s} - \frac{1}{s + 1/\tau_o} \right] \tag{7.31}$$

The two simple functions in Eq. (7.31) can be inverted by using Eqs. (7.7) and (7.15).

$$f_{(t)} = K_p \overline{C}_{A0} [1 - e^{-t/\tau_o}] \tag{7.32}$$

■

If there are some repeated roots in the denominator of Eq. (7.27), we must expand $F_{(s)}$ as a sum of N terms:

$$F_{(s)} = \frac{Z}{(s - p_1)^2 (s - p_3)(s - p_4) \cdots (s - p_N)} \tag{7.33}$$

$$F_{(s)} = \frac{A}{(s - p_1)^2} + \frac{B}{s - p_1} + \frac{C}{s - p_3} + \cdots + \frac{W}{s - p_N} \tag{7.34}$$

This is for a repeated root of order 2. If the root is repeated three times (of order 3), the expansion would be

$$F_{(s)} = \frac{Z}{(s - p_1)^3 (s - p_4)(s - p_5) \cdots (s - p_N)} \tag{7.35}$$

$$F_{(s)} = \frac{A}{(s - p_1)^3} + \frac{B}{(s - p_1)^2} + \frac{C}{s - p_1} + \cdots + \frac{W}{s - p_N} \tag{7.36}$$

The numerators of the terms in Eq. (7.34) are found from the following relationships. These are easily proved by merely carrying out the indicated operations on Eq. (7.34).

$$\begin{aligned} A &= \lim_{s \to p_1} [(s - p_1)^2 F_{(s)}] \\ B &= \lim_{s \to p_1} \left\{ \frac{d}{ds} [(s - p_1)^2 F_{(s)}] \right\} \\ C &= \lim_{s \to p_3} [(s - p_3) F_{(s)}] \end{aligned} \tag{7.37}$$

To find the C numerator in Eq. (7.36) a second derivative with respect to s would have to be taken. Generalizing to the mth term A_m of an Nth-order root at p_1,

$$A_m = \lim_{s \to p_1} \left\{ \frac{d^{m-1}}{ds^{m-1}} \left[(s - p_1)^N F_{(s)} \right] \right\} \frac{1}{(m - 1)!} \tag{7.38}$$

EXAMPLE 7.2. Given the following $F_{(s)}$, find its inverse.

$$F_{(s)} = \frac{K_p}{s(\tau_o s + 1)^2} = \frac{K_p/\tau_o^2}{s(s + 1/\tau_o)^2} = \frac{A}{s} + \frac{B}{(s + 1/\tau_o)^2} + \frac{C}{s + 1/\tau_o}$$

$$A = \lim_{s \to 0} \frac{K_p}{(\tau_o s + 1)^2} = K_p$$

$$B = \lim_{s \to -1/\tau_o} \frac{K_p/\tau_o^2}{s} = -\frac{K_p}{\tau_o}$$

$$C = \lim_{s \to -1/\tau_o} \left[\frac{d}{ds} \left(\frac{K_p/\tau_o^2}{s} \right) \right] = \lim_{s \to -1/\tau_o} \frac{-K_p/\tau_o^2}{s^2} = -K_p$$

Therefore,

$$F_{(s)} = K_p \left[\frac{1}{s} - \frac{1/\tau_o}{(s + 1/\tau_o)^2} - \frac{1}{s + 1/\tau_o} \right] \tag{7.39}$$

Inverting term by term yields

$$f_{(t)} = K_p \left(1 - \frac{t}{\tau_o} e^{-t/\tau_o} - e^{-t/\tau_o} \right) \tag{7.40}$$

■

7.4 TRANSFER FUNCTIONS

Our primary use of Laplace transformations in process control involves representing the dynamics of the process in terms of "transfer functions." These are output-input relationships and are obtained by Laplace-transforming algebraic and differential equations. In the following discussion, the output variable of the process is $y_{(t)}$. The input variable or the forcing function is $u_{(t)}$.

7.4.1 Multiplication by a Constant

Consider the algebraic equation

$$y_{(t)} = K u_{(t)} \tag{7.41}$$

Laplace-transforming both sides of the equation gives

$$\int_0^\infty y_{(t)} e^{-st}\, dt = K \int_0^\infty u_{(t)} e^{-st}\, dt$$
$$Y_{(s)} = K U_{(s)} \tag{7.42}$$

where $Y_{(s)}$ and $U_{(s)}$ are the Laplace transforms of $y_{(t)}$ and $u_{(t)}$. Note that $u_{(t)}$ is an arbitrary function of time. We have not specified at this point the exact form of the input. Comparing Eqs. (7.41) and (7.42) shows that the input and output variables are related in the Laplace domain in exactly the same way as they are related in the time domain. Thus, the English and Russian words describing this situation are the same.

Equation (7.42) can be put into transfer function form by finding the output/input ratio:

$$\frac{Y_{(s)}}{U_{(s)}} = K \tag{7.43}$$

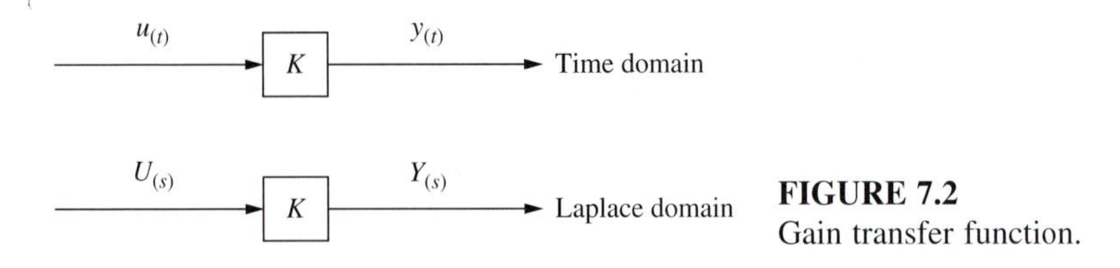

FIGURE 7.2
Gain transfer function.

For any input $U_{(s)}$ the output $Y_{(s)}$ is found by simply multiplying $U_{(s)}$ by the constant K. Thus, the transfer function relating $Y_{(s)}$ and $U_{(s)}$ is a constant or a "gain." We can represent this in block-diagram form as shown in Fig. 7.2.

7.4.2 Differentiation with Respect to Time

Consider what happens when the time derivative of a function $y_{(t)}$ is Laplace transformed.

$$\mathscr{L}\left[\frac{dy}{dt}\right] = \int_0^\infty \left(\frac{dy}{dt}\right) e^{-st}\, dt \tag{7.44}$$

Integrating by parts gives

$$u = e^{-st} \qquad dv = \frac{dy}{dt}\,dt$$

$$du = -se^{-st}\,dt \qquad v = y$$

Therefore,

$$\int_0^\infty \left(\frac{dy}{dt}\right) e^{-st}\, dt = \left[ye^{-st}\right]_{t=0}^{t=\infty} + \int_0^\infty sye^{-st}\, dt$$

$$= 0 - y_{(t=0)} + s\int_0^\infty y_{(t)} e^{-st}\, dt$$

The integral is, by definition, just the Laplace transformation of $y_{(t)}$, which we call $Y_{(s)}$.

$$\mathscr{L}\left[\frac{dy}{dt}\right] = sY_{(s)} - y_{(t=0)} \tag{7.45}$$

The result is the most useful of all the Laplace transformations. It says that the operation of differentiation in the time domain is replaced by multiplication by s in the Laplace domain, minus an initial condition. This is where perturbation variables become so useful. If the initial condition is the steady-state operating level, all the initial conditions like $y_{(t=0)}$ are equal to zero. Then simple multiplication by s is equivalent to differentiation. An ideal derivative unit or a perfect differentiator can be represented in block diagram form as shown in Fig. 7.3.

The same procedure applied to a second-order derivative gives

$$\mathscr{L}\left[\frac{d^2y}{dt^2}\right] = s^2Y_{(s)} - sy_{(t=0)} - \left(\frac{dy}{dt}\right)_{t=0} \tag{7.46}$$

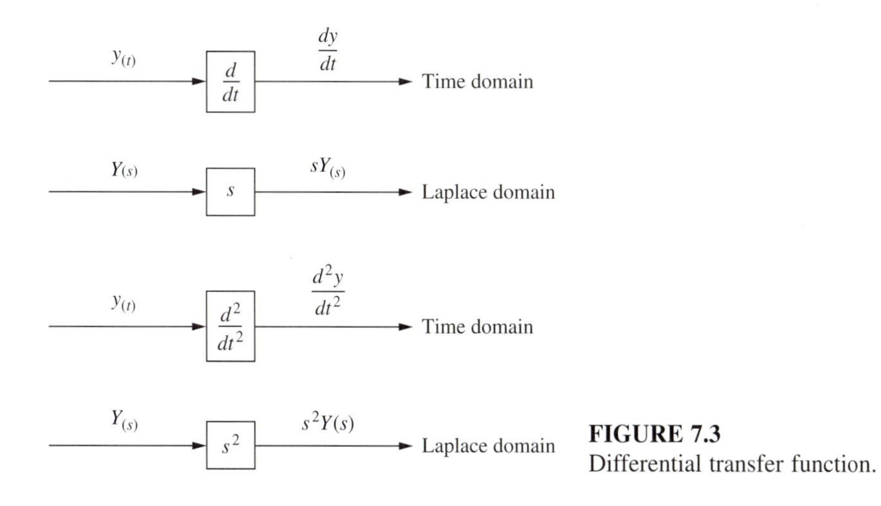

FIGURE 7.3
Differential transfer function.

Thus, differentiation twice is equivalent to multiplying twice by s, if all initial conditions are zero. The block diagram is shown in Fig. 7.3.

The preceding can be generalized to an Nth-order derivative with respect to time. In going from the time domain into the Laplace domain, $d^N x/dt^N$ is replaced by s^N. Therefore, an Nth-order differential equation becomes an Nth-order algebraic equation.

$$a_N \frac{d^N y}{dt^N} + a_{N-1} \frac{d^{N-1} y}{dt^{N-1}} + \cdots + a_1 \frac{dy}{dt} + a_0 y = u_{(t)} \tag{7.47}$$

$$a_N s^N Y_{(s)} + a_{N-1} s^{N-1} Y_{(s)} + \cdots + a_1 s Y_{(s)} + a_0 Y_{(s)} = U_{(s)} \tag{7.48}$$

$$(a_N s^N + a_{N-1} s^{N-1} + \cdots + a_1 s + a_0) Y_{(s)} = U_{(s)} \tag{7.49}$$

Notice that the polynomial in Eq. (7.49) looks exactly like the characteristic equation discussed in Chapter 2. We return to this not-accidental similarity in the next section.

7.4.3 Integration

Laplace-transforming the integral of a function $y_{(t)}$ gives

$$\mathcal{L}\left[\int y_{(t)}\,dt\right] = \int_0^\infty \left(\int y_{(t)}\,dt\right) e^{-st}\,dt$$

Integrating by parts,

$$u = \int y\,dt \qquad dv = e^{-st}\,dt$$

$$du = y\,dt \qquad v = -\frac{1}{s}e^{-st}$$

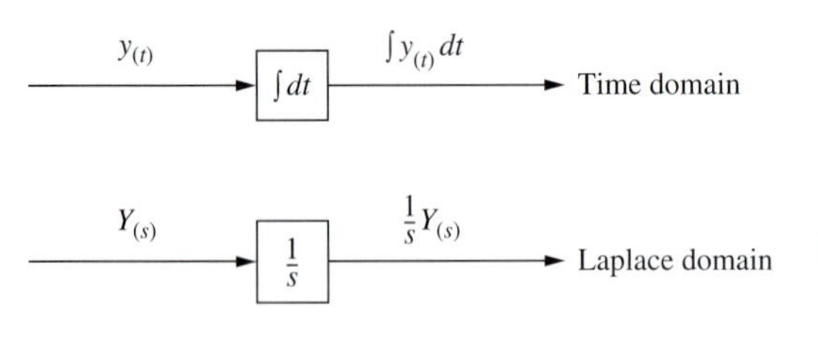

FIGURE 7.4
Integration transfer function.

$$\int_0^\infty \left(\int y_{(t)}\,dt\right) e^{-st}\,dt = \left[-\frac{1}{s}e^{-st}\int y\,dt\right]_{t=0}^{t=\infty} + \frac{1}{s}\int_0^\infty y_{(t)}e^{-st}\,dt$$

Therefore,

$$\mathcal{L}\left[\int y_{(t)}\,dt\right] = \frac{1}{s}Y_{(s)} + \frac{1}{s}\left(\int y\,dt\right)_{(t=0)} \tag{7.50}$$

The operation of integration is equivalent to division by s in the Laplace domain, using zero initial conditions. Thus integration is the inverse of differentiation. Figure 7.4 gives a block diagram representation.

The $1/s$ is an operator or a transfer function showing what operation is performed on the input signal. This is a completely different idea from the simple Laplace transformation of a function. Remember, the Laplace transform of the unit step function is also equal to $1/s$. But this is the Laplace transformation of a function. The $1/s$ operator discussed above is a transfer function, not a function.

7.4.4 Deadtime

Delay time, transportation lag, or deadtime is frequently encountered in chemical engineering systems. Suppose a process stream is flowing through a pipe in essentially plug flow and that it takes D minutes for an individual element of fluid to flow from the entrance to the exit of the pipe. Then the pipe represents a deadtime element.

If a certain dynamic variable $f_{(t)}$, such as temperature or composition, enters the front end of the pipe, it will emerge from the other end D minutes later with exactly the same shape, as shown in Fig. 7.5.

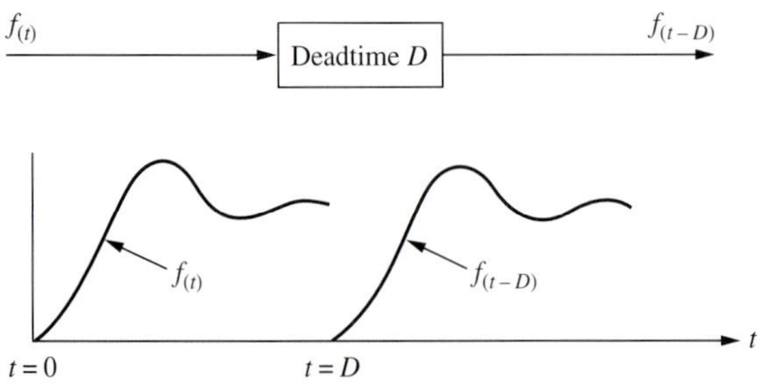

FIGURE 7.5
Effect of a dead-time element.

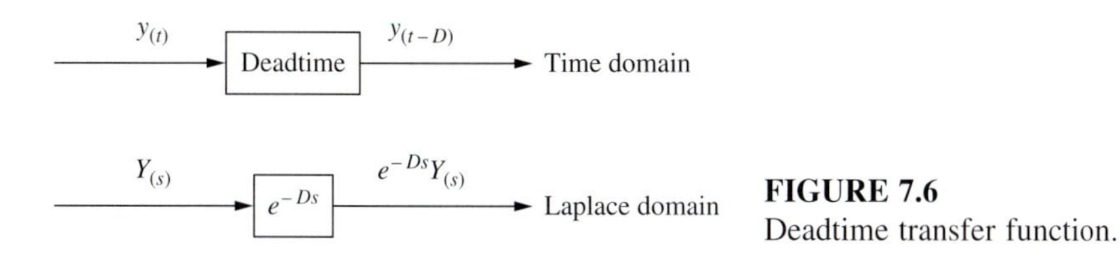

FIGURE 7.6
Deadtime transfer function.

Let us see what happens when we Laplace-transform a function $f_{(t-D)}$ that has been delayed by a deadtime. Laplace transformation is defined in Eq. (7.51).

$$\mathcal{L}[f_{(t)}] \equiv \int_0^\infty f_{(t)} e^{-st}\, dt = F_{(s)} \tag{7.51}$$

The variable t in this equation is just a "dummy variable" of integration. It is integrated out, leaving a function of only s. Thus, we can write Eq. (7.51) in a completely equivalent mathematical form:

$$F_{(s)} = \int_0^\infty f_{(y)} e^{-sy}\, dy \tag{7.52}$$

where y is now the dummy variable of integration. Now let $y = t - D$.

$$F_{(s)} = \int_0^\infty f_{(t-D)} e^{-s(t-D)}\, d(t-D) = e^{Ds} \int_0^\infty f_{(t-D)} e^{-st}\, dt$$

$$F_{(s)} = e^{Ds} \mathcal{L}[f_{(t-D)}] \tag{7.53}$$

Therefore,

$$\mathcal{L}[f_{(t-D)}] = e^{-Ds} F_{(s)} \tag{7.54}$$

Thus, time delay or deadtime in the time domain is equivalent to multiplication by e^{-Ds} in the Laplace domain.

If the input into the deadtime element is $u_{(t)}$ and the output of the deadtime element is $y_{(t)}$, then u and y are related by

$$y_{(t)} = u_{(t-D)}$$

And in the Laplace domain,

$$Y_{(s)} = e^{-Ds} U_{(s)} \tag{7.55}$$

Thus, the transfer function between output and input variables for a pure deadtime process is e^{-Ds}, as sketched in Fig. 7.6.

7.5 EXAMPLES

Now we are ready to apply all these Laplace transformation techniques to some typical chemical engineering processes.

EXAMPLE 7.3. Consider the isothermal CSTR of Example 2.6. The equation describing the system in terms of perturbation variables is

$$\frac{dC_A}{dt} + \left(\frac{1}{\tau} + k\right)C_{A(t)} = \frac{1}{\tau}C_{A0(t)} \tag{7.56}$$

where k and τ are constants. The initial condition is $C_{A(0)} = 0$. We do not specify what $C_{A0(t)}$ is for the moment, but just leave it as an arbitrary function of time. Laplace-transforming each term in Eq. (7.56) gives

$$sC_{A(s)} - C_{A(t=0)} + \left(\frac{1}{\tau} + k\right)C_{A(s)} = \frac{1}{\tau}C_{A0(s)} \tag{7.57}$$

The second term drops out because of the initial condition. Grouping like terms in $C_{A(s)}$ gives

$$\left(s + \frac{1}{\tau} + k\right)C_{A(s)} = \frac{1}{\tau}C_{A0(s)}$$

Thus, the ratio of the output to the input (the "transfer function" $G_{(s)}$) is

$$G_{(s)} \equiv \frac{C_{A(s)}}{C_{A0(s)}} = \frac{1/\tau}{s + k + 1/\tau} \tag{7.58}$$

The denominator of the transfer function is exactly the same as the polynomial in s that was called the characteristic equation in Chapter 2. The roots of the denominator of the transfer function are called the *poles* of the transfer function. These are the values of s at which $G_{(s)}$ goes to infinity.

The roots of the characteristic equation are equal to the poles of the transfer function.

This relationship between the poles of the transfer function and the roots of the characteristic equation is extremely important and useful.

The transfer function given in Eq. (7.58) has one pole with a value of $-(k + 1/\tau)$. Rearranging Eq. (7.58) into the standard form of Eq. (2.51) gives

$$G_{(s)} = \frac{\left(\dfrac{1}{1 + k\tau}\right)}{\left(\dfrac{1}{k + 1/\tau}\right)s + 1} = \frac{K_p}{\tau_o s + 1} \tag{7.59}$$

where K_p is the process steady-state gain and τ_o is the process time constant. The pole of the transfer function is the reciprocal of the time constant.

This particular type of transfer function is called a *first-order lag.* It tells us how the input C_{A0} affects the output C_A, both dynamically and at steady state. The form of the transfer function (polynomial of degree 1 in the denominator, i.e., one pole) and the numerical values of the parameters (steady-state gain and time constant) give a complete picture of the system in a very compact and usable form. The transfer function is a property of the system only and is applicable for any input. We can determine the dynamics and the steady-state characteristics of the system without having to pick any specific forcing function.

If the same input as used in Example 2.6 is imposed on the system, we should be able to use Laplace transforms to find the response of C_A to a step change of magnitude $\overline{C}_{A0}$.

$$C_{A0(t)} = \overline{C}_{A0} u_{n(t)} \tag{7.60}$$

We take the Laplace transform of $C_{A0(t)}$, substitute into the system transfer function, solve for $C_{A(s)}$, and invert back into the time domain to find $C_{A(t)}$.

$$\mathscr{L}[C_{A0(t)}] = C_{A0(s)} = \overline{C}_{A0}\frac{1}{s} \tag{7.61}$$

$$C_{A(s)} = G_{(s)}C_{A0(s)} = \left(\frac{K_p}{\tau_o s + 1}\right)\left(\overline{C}_{A0}\frac{1}{s}\right) = \frac{K_p\overline{C}_{A0}}{s(\tau_o s + 1)} \tag{7.62}$$

Using partial fractions expansion to invert (see Example 7.1) gives

$$C_{A(t)} = K_p\overline{C}_{A0}\left(1 - e^{-t/\tau_o}\right)$$

This is exactly the solution obtained in Example 2.6 [Eq. (2.53)]. ■

EXAMPLE 7.4. The ODE of Example 2.8 with an arbitrary forcing function $u_{(t)}$ is

$$\frac{d^2y}{dt^2} + 5\frac{dy}{dt} + 6y = u_{(t)} \tag{7.63}$$

with the initial conditions

$$y_{(0)} = \left(\frac{dy}{dt}\right)_{(0)} = 0 \tag{7.64}$$

Laplace transforming gives

$$s^2Y_{(s)} + 5sY_{(s)} + 6Y_{(s)} = U_{(s)}$$
$$Y_{(s)}(s^3 + 5s + 6) = U_{(s)}$$

The process transfer function $G_{(s)}$ is

$$\frac{Y_{(s)}}{U_{(s)}} = G_{(s)} = \frac{1}{s^2 + 5s + 6} = \frac{1}{(s+2)(s+3)} \tag{7.65}$$

Notice that the denominator of the transfer function is again the same polynomial in s as appeared in the characteristic equation of the system [Eq. (2.73)]. The poles of the transfer function are located at $s = -2$ and $s = -3$. So the poles of the transfer function are the roots of the characteristic equation.

If $u_{(t)}$ is a ramp input as in Example 2.13,

$$\mathscr{L}[u_{(t)}] = \mathscr{L}[t] = \frac{1}{s^2} \tag{7.66}$$

$$Y_{(s)} = G_{(s)}U_{(s)} = \left(\frac{1}{s^2 + 5s + 6}\right)\left(\frac{1}{s^2}\right) = \frac{1}{s^2(s+2)(s+3)} \tag{7.67}$$

Partial fractions expansion gives

$$Y_{(s)} = \frac{A}{s^2} + \frac{B}{s} + \frac{C}{s+2} + \frac{D}{s+3}$$

$$A = \lim_{s\to 0}\left(s^2Y_{(s)}\right) = \lim_{s\to 0}\left[\frac{1}{(s+2)(s+3)}\right] = \frac{1}{6}$$

$$B = \lim_{s\to 0}\left[\frac{d}{ds}\left(s^2Y_{(s)}\right)\right] = \lim_{s\to 0}\left[\frac{d}{ds}\left(\frac{1}{s^2 + 5s + 6}\right)\right]$$
$$= \lim_{s\to 0}\left[\frac{-(2s+5)}{(s^2 + 5s + 6)^2}\right] = -\frac{5}{36}$$

$$C = \lim_{s \to -2} \left[(s+2)Y_{(s)}\right] = \lim_{s \to -2} \left[\frac{1}{s^2(s+3)}\right] = \frac{1}{4}$$

$$D = \lim_{s \to -3} \left[(s+3)Y_{(s)}\right] = \lim_{s \to -3} \left[\frac{1}{s^2(s+2)}\right] = -\frac{1}{9}$$

Therefore,

$$Y_{(s)} = \frac{\frac{1}{6}}{s^2} - \frac{\frac{5}{36}}{s} + \frac{\frac{1}{4}}{s+2} - \frac{\frac{1}{9}}{s+3} \tag{7.68}$$

Inverting into the time domain gives the same solution as Eq. (2.109).

$$y_{(t)} = \tfrac{1}{6}t - \tfrac{5}{36} + \tfrac{1}{4}e^{-2t} - \tfrac{1}{9}e^{-3t} \tag{7.69}$$

■

EXAMPLE 7.5. An isothermal three-CSTR system is described by the three linear ODEs

$$\begin{aligned}
\frac{dC_{A1}}{dt} + \left(k_1 + \frac{1}{\tau_1}\right)C_{A1} &= \frac{1}{\tau_1}C_{A0} \\
\frac{dC_{A2}}{dt} + \left(k_2 + \frac{1}{\tau_2}\right)C_{A2} &= \frac{1}{\tau_2}C_{A1} \\
\frac{dC_{A3}}{dt} + \left(k_3 + \frac{1}{\tau_3}\right)C_{A3} &= \frac{1}{\tau_3}C_{A2}
\end{aligned} \tag{7.70}$$

The variables can be either total or perturbation variables since the equations are linear (all k's and τ's are constant). Let us use perturbation variables, and therefore the initial conditions for all variables are zero.

$$C_{A1(0)} = C_{A2(0)} = C_{A3(0)} = 0 \tag{7.71}$$

Laplace transforming gives

$$\begin{aligned}
\left(s + k_1 + \frac{1}{\tau_1}\right)C_{A1(s)} &= \frac{1}{\tau_1}C_{A0(s)} \\
\left(s + k_2 + \frac{1}{\tau_2}\right)C_{A2(s)} &= \frac{1}{\tau_2}C_{A1(s)} \\
\left(s + k_3 + \frac{1}{\tau_3}\right)C_{A3(s)} &= \frac{1}{\tau_3}C_{A2(s)}
\end{aligned} \tag{7.72}$$

These can be rearranged to put them in terms of transfer functions for each tank.

$$\begin{aligned}
G_{1(s)} &\equiv \frac{C_{A1(s)}}{C_{A0(s)}} = \frac{1/\tau_1}{s + k_1 + 1/\tau_1} \\
G_{2(s)} &\equiv \frac{C_{A2(s)}}{C_{A1(s)}} = \frac{1/\tau_2}{s + k_2 + 1/\tau_2} \\
G_{3(s)} &\equiv \frac{C_{A3(s)}}{C_{A2(s)}} = \frac{1/\tau_3}{s + k_3 + 1/\tau_3}
\end{aligned} \tag{7.73}$$

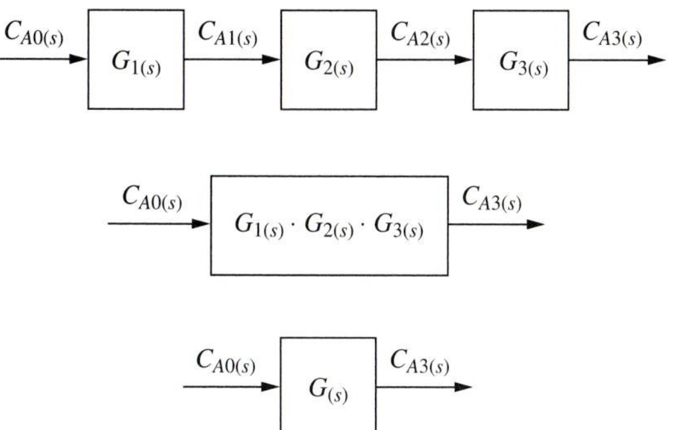

FIGURE 7.7
Transfer functions in series.

If we are interested in the total system and want only the effect of the input C_{A0} on the output C_{A3}, the three equations can be combined to eliminate C_{A1} and C_{A2}.

$$C_{A3(s)} = G_3 C_{A2(s)} = G_3 \left(G_2 C_{A1(s)}\right) = G_3 G_2 \left(G_1 C_{A0(s)}\right) \tag{7.74}$$

The overall transfer function $G_{(s)}$ is

$$G_{(s)} \equiv \frac{C_{A3(s)}}{C_{A0(s)}} = G_{1(s)} G_{2(s)} G_{3(s)} \tag{7.75}$$

This demonstrates one very important and useful property of transfer functions. The total effect of a number of transfer functions connected in series is just the product of all the individual transfer functions. Figure 7.7 shows this in block diagram form. The overall transfer function is a third-order lag with three poles.

$$G_{(s)} = \frac{1/\tau_1 \tau_2 \tau_3}{(s + k_1 + 1/\tau_1)(s + k_2 + 1/\tau_2)(s + k_3 + 1/\tau_3)} \tag{7.76}$$

Further rearrangement puts the above expression in the standard form with time constants τ_{oi} and a steady-state gain K_p.

$$G = \frac{\dfrac{1}{(1 + k_1\tau_1)} \dfrac{1}{(1 + k_2\tau_2)} \dfrac{1}{(1 + k_3\tau_3)}}{\left(\dfrac{\tau_1}{1 + k_1\tau_1} s + 1\right)\left(\dfrac{\tau_2}{1 + k_2\tau_2} s + 1\right)\left(\dfrac{\tau_3}{1 + k_3\tau_3} s + 1\right)}$$
$$G_{(s)} = \frac{K_p}{(\tau_{o1}s + 1)(\tau_{o2}s + 1)(\tau_{o3}s + 1)} \tag{7.77}$$

Let us assume a unit step change in the feed concentration C_{A0} and solve for the response of C_{A3}. We will take the case where all the τ_{oi}'s are the same, giving a repeated root of order 3 (a third-order pole at $s = -1/\tau_o$).

$$C_{A0(t)} = u_{n(t)} \quad \Rightarrow \quad C_{A0(s)} = \frac{1}{s}$$
$$C_{A3(s)} = G_{(s)} C_{A0(s)} = \frac{K_p}{(\tau_o s + 1)^3} \frac{1}{s} = \frac{K_p/\tau_o^3}{s(s + 1/\tau_o)^3} \tag{7.78}$$

Applying partial fractions expansion,

$$C_{A3(s)} = \frac{A}{s} + \frac{B}{(s+1/\tau_o)^3} + \frac{C}{(s+1/\tau_o)^2} + \frac{D}{s+1/\tau_o} \tag{7.79}$$

$$A = \lim_{s\to 0}\left(\frac{K_p/\tau_o^3}{(s+1/\tau_o)^3}\right) = K_p$$

$$B = \lim_{s\to -1/\tau_o}\left(\frac{K_p/\tau_o^3}{s}\right) = -\frac{K_p}{\tau_o^2}$$

$$C = \lim_{s\to -1/\tau_o}\left[\frac{d}{ds}\left(\frac{K_p/\tau_o^3}{s}\right)\right] = \lim_{s\to -1/\tau_o}\left[-\frac{K_p/\tau_o^3}{s^2}\right] = -\frac{K_p}{\tau_o}$$

$$D = \lim_{s\to -1/\tau_o}\left[\frac{1}{2!}\frac{d^2}{ds^2}\left(\frac{K_p/\tau_o^3}{s}\right)\right] = \lim_{s\to -1/\tau_o}\left[\frac{1}{2}\frac{2K_p/\tau_o^3}{s^3}\right] = -K_p$$

Inverting Eq. (7.79) with the use of Eq. (7.18) yields

$$C_{A3(t)} = K_p\left[1 - \frac{1}{2}\left(\frac{t}{\tau_o}\right)^2 e^{-t/\tau_o} - \frac{t}{\tau_o}e^{-t/\tau_o} - e^{-t/\tau_o}\right] \tag{7.80}$$

■

EXAMPLE 7.6. A nonisothermal CSTR can be linearized (see Problem 2.8) to give two linear ODEs in terms of perturbation variables.

$$\begin{aligned}\frac{dC_A}{dt} &= a_{11}C_A + a_{12}T + a_{13}C_{A0} + a_{15}F \\ \frac{dT}{dt} &= a_{21}C_A + a_{22}T + a_{24}T_0 + a_{25}F + a_{26}T_J\end{aligned} \tag{7.81}$$

where

$$\begin{aligned} a_{11} &= -\frac{\bar{F}}{\bar{V}} - \bar{k} & a_{12} &= \frac{-\bar{C}_A E\bar{k}}{R\bar{T}^2} & a_{13} &= \frac{\bar{F}}{V} \\ a_{15} &= \frac{\bar{C}_{A0} - \bar{C}_A}{V} & a_{21} &= \frac{-\lambda\bar{k}}{\rho C_p} & a_{22} &= \frac{-\lambda\bar{k}E\bar{C}_A}{\rho C_p R\bar{T}^2} - \frac{\bar{F}}{V} - \frac{UA}{V\rho C_p} \\ a_{24} &= \frac{\bar{F}}{V} & a_{25} &= \frac{\bar{T}_0 - \bar{T}}{V} & a_{26} &= \frac{UA}{V\rho C_p}\end{aligned} \tag{7.82}$$

The variables C_{A0}, T_0, F, and T_J are all considered inputs. The output variables are C_A and T. Therefore, eight different transfer functions are required to describe the system completely. This multivariable aspect is the usual situation in chemical engineering systems.

$$\begin{aligned} C_{A(s)} &= G_{11(s)}C_{A0(s)} + G_{12(s)}T_{0(s)} + G_{13(s)}F_{(s)} + G_{14(s)}T_{J(s)} \\ T_{(s)} &= G_{21(s)}C_{A0(s)} + G_{22(s)}T_{0(s)} + G_{23(s)}F_{(s)} + G_{24(s)}T_{J(s)}\end{aligned} \tag{7.83}$$

The G_{ij} are, in general, functions of s and are the transfer functions relating inputs and outputs. Since the system is linear, the output is the sum of the effects of each individual input. This is called the principle of *superposition.*

To find these transfer functions, Eqs. (7.81) are Laplace transformed and solved simultaneously.

$$\begin{aligned} sC_A &= a_{11}C_A + a_{12}T + a_{13}C_{A0} + a_{15}F \\ sT &= a_{21}C_A + a_{22}T + a_{24}T_0 + a_{25}F + a_{26}T_J \\ (s - a_{11})C_A &= a_{12}T + a_{13}C_{A0} + a_{15}F \\ (s - a_{22})T &= a_{21}C_A + a_{24}T_0 + a_{25}F + a_{26}T_J \end{aligned}$$

Combining,

$$(s - a_{11})C_A = a_{12}\left(\frac{a_{21}C_A + a_{24}T_0 + a_{25}F + a_{26}T_J}{s - a_{22}}\right) + a_{13}C_{A0} + a_{15}F$$

$$\left(s - a_{11} - \frac{a_{12}a_{21}}{s - a_{22}}\right)C_A = \left(\frac{a_{12}a_{24}}{s - a_{22}}\right)T_0 + \left(\frac{a_{12}a_{25}}{s - a_{22}} + a_{15}\right)F + \left(\frac{a_{12}a_{26}}{s - a_{22}}\right)T_J + a_{13}C_{A0}$$

Finally,

$$\begin{aligned} C_{A(s)} = &\left[\frac{a_{13}(s - a_{22})}{s^2 - (a_{11} + a_{22})s + a_{11}a_{22} - a_{12}a_{21}}\right]C_{A0(s)} \\ &+ \left[\frac{a_{12}a_{24}}{s^2 - (a_{11} + a_{22})s + a_{11}a_{22} - a_{12}a_{21}}\right]T_{0(s)} \\ &+ \left[\frac{a_{12}a_{25} + a_{15}(s - a_{22})}{s^2 - (a_{11} + a_{22})s + a_{11}a_{22} - a_{12}a_{21}}\right]F_{(s)} \\ &+ \left[\frac{a_{12}a_{26}}{s^2 - (a_{11} + a_{22})s + a_{11}a_{22} - a_{12}a_{21}}\right]T_{J(s)} \end{aligned} \quad (7.84)$$

$$\begin{aligned} T_{(s)} = &\left[\frac{a_{13}a_{21}}{s^2 - (a_{11} + a_{22})s + a_{11}a_{22} - a_{12}a_{21}}\right]C_{A0(s)} \\ &+ \left[\frac{a_{24}(s - a_{11})}{s^2 - (a_{11} + a_{22})s + a_{11}a_{22} - a_{12}a_{21}}\right]T_{0(s)} \\ &+ \left[\frac{a_{15}a_{21} + a_{25}(s - a_{11})}{s^2 - (a_{11} + a_{22})s + a_{11}a_{22} - a_{12}a_{21}}\right]F_{(s)} \\ &+ \left[\frac{a_{26}(s - a_{11})}{s^2 - (a_{11} + a_{22})s + a_{11}a_{22} - a_{12}a_{21}}\right]T_{J(s)} \end{aligned} \quad (7.85)$$

The system is shown in block diagram form in Fig. 7.8.

Notice that the G's are ratios of polynomials in s. The $s - a_{11}$ and $s - a_{22}$ terms in the numerators are called *first-order leads.* Notice also that the denominators of all the G's are exactly the same. ■

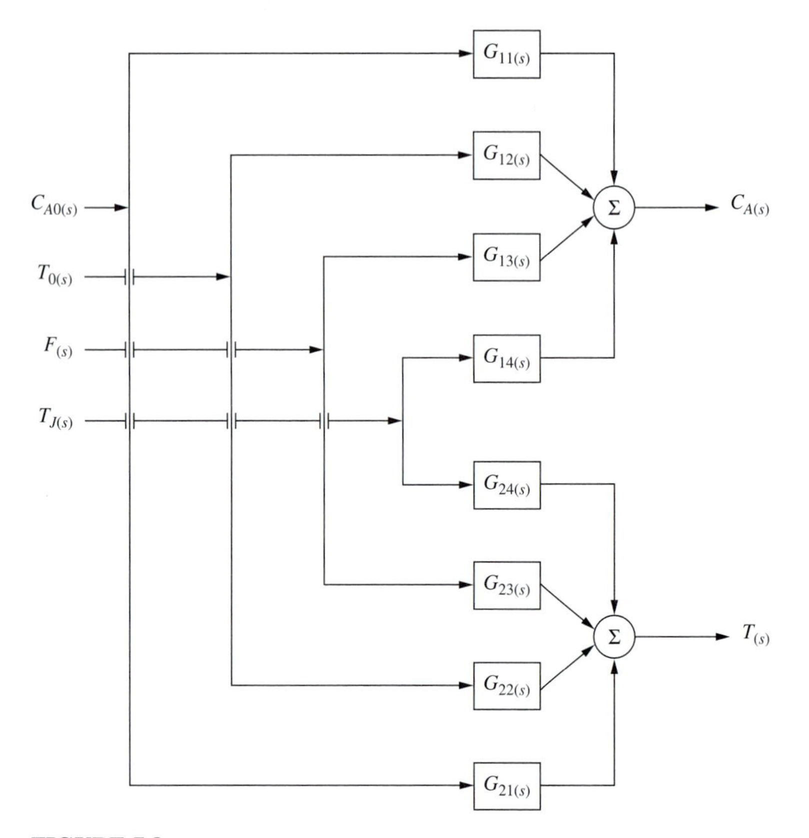

FIGURE 7.8
Block diagram of a multivariable linearized nonisothermal CSTR system.

EXAMPLE 7.7. A two-heated-tank process is described by two linear ODEs:

$$\rho C_p V_1 \frac{dT_1}{dt} = \rho C_p F(T_0 - T_1) + Q_1 \tag{7.86}$$

$$\rho C_p V_2 \frac{dT_2}{dt} = \rho C_p F(T_1 - T_2) \tag{7.87}$$

The numerical values of variables are:

$$F = 90 \text{ ft}^3/\text{min} \qquad \rho = 40 \text{ lb}_\text{m}/\text{ft}^3 \qquad C_p = 0.6 \text{ Btu/lb}_\text{m}\ ^\circ\text{F}$$
$$V_1 = 450 \text{ ft}^3 \qquad V_2 = 90 \text{ ft}^3$$

Plugging these into Eqs. (7.86) and (7.87) gives

$$(40)(0.6)(450)\frac{dT_1}{dt} = (40)(90)(0.6)(T_0 - T_1) + Q_1 \tag{7.88}$$

$$(40)(0.6)(90)\frac{dT_2}{dt} = (40)(90)(0.6)(T_1 - T_2) \tag{7.89}$$

$$5\frac{dT_1}{dt} + T_1 = T_0 + \frac{Q_1}{2160} \tag{7.90}$$

$$\frac{dT_2}{dt} + T_2 = T_1 \tag{7.91}$$

Laplace transforming gives

$$(5s + 1)T_{1(s)} = T_{0(s)} + \tfrac{1}{2160}Q_1$$

$$(s + 1)T_{2(s)} = T_{1(s)}$$

Rearranging and combining to eliminate T_1 give the output variable T_2 as a function of the two input variables, T_0 and Q_1.

$$T_{2(s)} = \left[\frac{1}{(s+1)(5s+1)}\right]T_{0(s)} + \left[\frac{1/2160}{(s+1)(5s+1)}\right]Q_{1(s)} \tag{7.92}$$

The two terms in the brackets represent the transfer functions of this openloop process. In the next chapter we look at this system again and use a temperature controller to control T_2 by manipulating Q_1. The transfer function relating the controlled variable T_2 to the *manipulated* variable Q_1 is defined as $G_{M(s)}$. The transfer function relating the controlled variable T_2 to the *load* disturbance T_0 is defined as $G_{L(s)}$.

$$T_{2(S)} = G_{L(s)}T_{0(s)} + G_{M(s)}Q_{1(s)} \tag{7.93}$$

Both of these transfer functions are second-order lags with time constants of 1 and 5 minutes. ■

7.6 PROPERTIES OF TRANSFER FUNCTIONS

An *N*th-order system is described by the linear ODE

$$a_N\frac{d^N y}{dt^N} + a_{N-1}\frac{d^{N-1}y}{dt^{N-1}} + \cdots + a_1\frac{dy}{dt} + a_0 y = b_M\frac{d^M u}{dt^M} + b_{M-1}\frac{d^{M-1}u}{dt^{M-1}} + \cdots + b_1\frac{du}{dt} + b_0 u \tag{7.94}$$

where a_i and b_i = constant coefficients
y = output
u = input or forcing function

7.6.1 Physical Realizability

For Eq. (7.94) to describe a real physical system, the order of the right-hand side, *M*, cannot be greater than the order of the left-hand side, *N*. This criterion for physical realizability is

$$N \geq M \tag{7.95}$$

This requirement can be proved intuitively from the following reasoning. Take a case where $N = 0$ and $M = 1$.

$$a_0 y = b_1 \frac{du}{dt} + b_0 u \tag{7.96}$$

This equation says that we have a process whose output y depends on the value of the input and the value of the derivative of the input. Therefore, the process must be able to differentiate, perfectly, the input signal. But it is impossible for any real system to differentiate perfectly. This would require that a step change in the input produce an infinite spike in the output, which is physically impossible.

This example can be generalized to any case where $M \geq N$ to show that differentiation would be required. Therefore, N must always be greater than or equal to M. Laplace-transforming Eq. (7.96) gives

$$\frac{Y_{(s)}}{U_{(s)}} = \frac{b_1}{a_0} s + \frac{b_0}{a_0}$$

This is a first-order lead. It is physically unrealizable; i.e., a real device cannot be built that has exactly this transfer function.

Consider the case where $M = N = 1$.

$$a_1 \frac{dy}{dt} + a_0 y = b_1 \frac{du}{dt} + b_0 u \tag{7.97}$$

It appears that a derivative of the input is again required. But Eq. (7.97) can be rearranged, grouping the derivative terms:

$$\frac{d}{dt}(a_1 y - b_1 u) = \frac{dz}{dt} = b_0 u - a_0 y \tag{7.98}$$

The right-hand side of this equation contains functions of time but no derivatives. This ODE can be integrated by evaluating the right-hand side (the derivative) at each point in time and integrating to get z at the new point in time. Then the new value of y is calculated from the known value of u: $y = (z + b_1 u)/a_1$. Differentiation is not required, and this transfer function is physically realizable. Remember, nature always integrates, it never differentiates!

Laplace-transforming Eq. (7.97) gives

$$\frac{Y_{(s)}}{U_{(s)}} = \frac{b_1 s + b_0}{a_1 s + a_0}$$

This is called a *lead-lag* element and contains a first-order lag and a first-order lead. See Table 7.1 for some commonly used transfer function elements.

7.6.2 Poles and Zeros

Returning now to Eq. (7.94), let us Laplace transform and solve for the ratio of output $Y_{(s)}$ to input $U_{(s)}$, the system transfer function $G_{(s)}$.

TABLE 7.1
Common transfer functions

Terminology	$G_{(s)}$
Gain	K
Derivative	s
Integrator	$\frac{1}{s}$
First-order lag	$\frac{1}{\tau s + 1}$
First-order lead	$\tau s + 1$
Second-order lag	
Underdamped, $\zeta < 1$	$\frac{1}{\tau^2 s^2 + 2\tau\zeta s + 1}$
Critically damped, $\zeta = 1$	$\frac{1}{(\tau s + 1)^2}$
Overdamped, $\zeta > 1$	$\frac{1}{(\tau_1 s + 1)(\tau_2 s + 1)}$
Deadtime	e^{-Ds}
Lead-lag	$\frac{\tau_z s + 1}{\tau_p s + 1}$

$$G_{(s)} = \frac{Y_{(s)}}{U_{(s)}} = \frac{b_M s^M + b_{M-1} s^{M-1} + \cdots + b_1 s + b_0}{a_N s^N + a_{N-1} s^{N-1} + \cdots + a_1 s + a_0} \tag{7.99}$$

The denominator is a polynomial in s that is the same as the characteristic equation of the system. Remember, the characteristic equation is obtained from the homogeneous ODE, that is, considering the right-hand side of Eq. (7.94) equal to zero.

The roots of the denominator are called the *poles* of the transfer function. The roots of the numerator are called the *zeros* of the transfer function (these values of s make the transfer function equal zero). Factoring both numerator and denominator yields

$$G_{(s)} = \left(\frac{b_M}{a_N}\right) \frac{(s - z_1)(s - z_2)\cdots(s - z_M)}{(s - p_1)(s - p_2)\cdots(s - p_N)} \tag{7.100}$$

where z_i = zeros of the transfer function
p_i = poles of the transfer function

As noted in Chapter 2, the roots of the characteristic equation, which are the poles of the transfer function, must be real or must occur as complex conjugate pairs. In addition, the real parts of all the poles must be negative for the system to be stable.

A system is stable if all its poles lie in the left half of the s plane.

The locations of the zeros of the transfer function have *no* effect on the stability of the system! They certainly affect the dynamic response, but they do not affect stability.

7.6.3 Steady-State Gains

One final point should be made about transfer functions. The steady-state gain K_p for all the transfer functions derived in the examples was obtained by expressing the transfer function in terms of time constants instead of in terms of poles and zeros. For the general system of Eq. (7.94) this would be

$$G_{(s)} = K_p \frac{(\tau_{z1}s + 1)(\tau_{z2}s + 1)\cdots(\tau_{zM}s + 1)}{(\tau_{p1}s + 1)(\tau_{p2}s + 1)\cdots(\tau_{pN}s + 1)} \tag{7.101}$$

The steady-state gain is the ratio of output steady-state perturbation to the input perturbation.

$$K_p = \left(\frac{y^p}{u^p}\right)_{(t\to\infty)} = \frac{\overline{y}^p}{\overline{u}^p} \tag{7.102}$$

In terms of total variables,

$$K_p = \left(\frac{y - \overline{y}}{u - \overline{u}}\right)_{(t\to\infty)} = \frac{\Delta\overline{y}}{\Delta\overline{u}}$$

Thus, for a step change in the input variable of $\Delta\overline{u}$, the steady-state gain is found simply by dividing the steady-state change in the output variable $\Delta\overline{y}$ by $\Delta\overline{u}$, as sketched in Fig. 7.9.

Instead of rearranging the transfer function to put it into the time-constant form, it is sometimes more convenient to find the steady-state gain by an alternative method that does not require factoring of polynomials. This consists of merely letting $s = 0$ in the transfer function.

$$K_p = \lim_{s\to 0} G_{(s)} \tag{7.103}$$

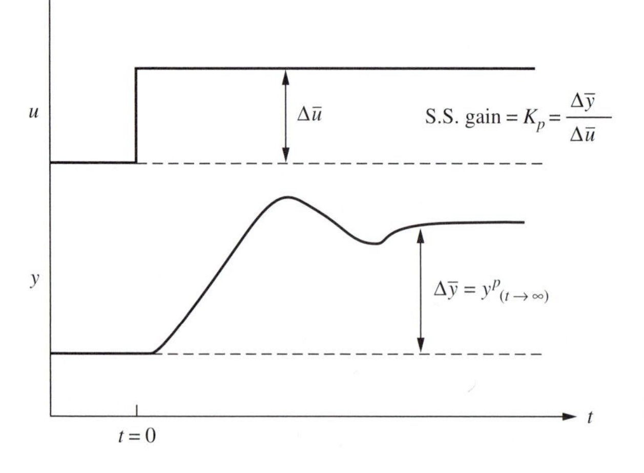

FIGURE 7.9
Steady-state gain.

By definition, steady state corresponds to the condition that all time derivatives are equal to zero. Since the variable s replaces d/dt in the Laplace domain, letting s go to zero is equivalent to the steady-state gain.

This can be proved more rigorously by using the *final-value theorem* of Laplace transforms:

$$\lim_{t \to \infty}[f_{(t)}] = \lim_{s \to 0}[sF_{(s)}] \tag{7.104}$$

Consider the arbitrary transfer function

$$G_{(s)} = \frac{Y_{(s)}}{U_{(s)}}$$

If a unit step disturbance is used,

$$U_{(s)} = \frac{1}{s}$$

This means that the output is

$$Y_{(s)} = G_{(s)}\frac{1}{s}$$

The final steady-state value of the output will be equal to the steady-state gain since the magnitude of the input was 1.

$$K_p = \lim_{t \to \infty}[y_{(t)}] = \lim_{s \to 0}[sY_{(s)}] = \lim_{s \to 0}\left[sG_{(s)}\frac{1}{s}\right] = \lim_{s \to 0}[G_{(s)}]$$

For example, the steady-state gain for the transfer function given in Eq. (7.99) is

$$K_p = \lim_{s \to 0}\left[\frac{b_M s^M + b_{M-1}s^{M-1} + \cdots + b_1 s + b_0}{a_N s^N + a_{N-1}s^{N-1} + \cdots + a_1 s + a_0}\right] = \frac{b_0}{a_0} \tag{7.105}$$

It is obvious that this must be the right value of gain since at steady state Eq. (7.94) reduces to

$$a_0\overline{y} = b_0\overline{u} \tag{7.106}$$

For the two-heated-tank process of Example 7.7, the two transfer functions were given in Eq. (7.92). The steady-state gain between the inlet temperature T_0 and the output T_2 is found to be 1°F/°F when s is set equal to zero. This says that a 1° change in the inlet temperature raises the outlet temperature by 1°, which seems reasonable. The steady-state gain between T_2 and the heat input Q_1 is 1/2160 °F/Btu/min. You should be careful about the units of gains. Sometimes they have engineering units, as in this example. At other times dimensionless gains are used. We discuss this in more detail in Chapter 8.

7.7 TRANSFER FUNCTIONS FOR FEEDBACK CONTROLLERS

As discussed in Chapter 3, the three common commercial feedback controllers are proportional (P), proportional-integral (PI), and proportional-integral-derivative (PID). The transfer functions for these devices are developed here.

The equation describing a proportional controller in the time domain is

$$CO_{(t)} = \text{Bias} \pm K_c(SP_{(t)} - PV_{(t)}) \tag{7.107}$$

where CO = controller output signal sent to the control valve
Bias = constant
SP = setpoint
PV = process variable signal from the transmitter

Equation (7.107) is written in terms of total variables. If we are dealing with perturbation variables, we simply drop the Bias term. Laplace transforming gives

$$CO_{(s)} = \pm K_c(SP_{(s)} - PV_{(s)}) = \pm K_c E_{(s)} \tag{7.108}$$

where E = error signal = SP − PV. Rearranging to get the output over the input gives the transfer function $G_{C(s)}$ for the controller.

$$\frac{CO_{(s)}}{E_{(s)}} = \pm K_c \equiv G_{C(s)} \tag{7.109}$$

So the transfer function for a proportional controller is just a gain.

The equation describing a proportional-integral controller in the time domain is

$$CO_{(t)} = \text{Bias} \pm K_c\left[E_{(t)} + \frac{1}{\tau_I}\int E_{(t)}\,dt\right] \tag{7.110}$$

where τ_I = reset time, in units of time. Equation (7.110) is in terms of total variables. Converting to perturbation variables and Laplace transforming give

$$CO_{(s)} = \pm K_c\left[E_{(s)} + \frac{1}{\tau_I s}E_{(s)}\right]$$

$$\frac{CO_{(s)}}{E_{(s)}} = G_{C(s)} = \pm K_c\left[1 + \frac{1}{\tau_I s}\right] = \pm K_c\left(\frac{\tau_I s + 1}{\tau_I s}\right) \tag{7.111}$$

Thus, the transfer function for a PI controller contains a first-order lead and an integrator. It is a function of s, having numerator and denominator polynomials of order 1.

The transfer function of a "real" PID controller, as opposed to an "ideal" one, is the PI transfer function with a lead-lag element placed in series.

$$\frac{CO_{(s)}}{E_{(s)}} = G_{C(s)} = \pm K_c\left(\frac{\tau_I s + 1}{\tau_I s}\right)\left(\frac{\tau_D s + 1}{\alpha\tau_D s + 1}\right) \tag{7.112}$$

where τ_D = derivative time constant, in units of time
α = constant = 0.1 to 0.05 for many commercial controllers

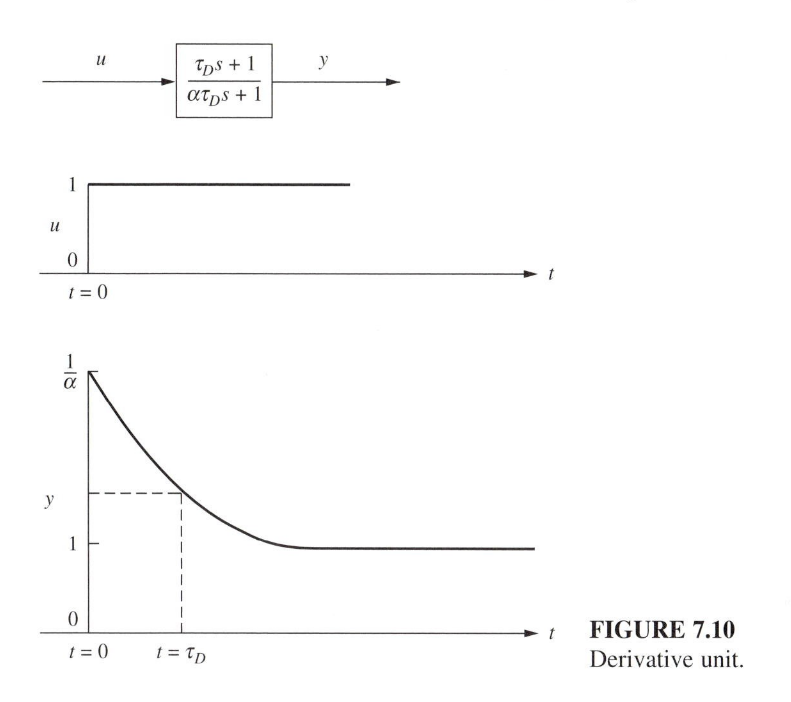

FIGURE 7.10
Derivative unit.

The lead-lag unit is called a "derivative unit," and its step response is sketched in Fig. 7.10. For a unit step change in the input, the output jumps to $1/\alpha$ and then decays at a rate that depends on τ_D. So the derivative unit approximates an ideal derivative. It is physically realizable since the order of its numerator polynomial is the same as the order of its denominator polynomial.

7.8 CONCLUSION

In this chapter we have developed the mathematical tools (Laplace transforms) that facilitate the analysis of dynamic systems. The usefulness of these tools will become apparent in the next chapter.

PROBLEMS

7.1. Prove that the Laplace transformations of the following functions are as shown.

(*a*) $\mathcal{L}\left[\frac{d^2 f}{dt^2}\right] = s^2 F_{(s)} - s f_{(0)} - \left(\frac{df}{dt}\right)_{(t=0)}$

(*b*) $\mathcal{L}[\cos(\omega t)] = \frac{s}{s^2 + \omega^2}$

(*c*) $\mathcal{L}[e^{-at}\sin(\omega t)] = \frac{\omega}{(s+a)^2 + \omega^2}$

7.2. Find the Laplace transformation of a rectangular pulse of height H_p and duration T_p.

7.3. An isothermal perfectly mixed batch reactor has consecutive first-order reactions

$$\mathrm{A} \xrightarrow{k_1} \mathrm{B} \xrightarrow{k_2} \mathrm{C}$$

The initial material charged to the vessel contains only A at a concentration C_{A0}. Use Laplace transform techniques to solve for the changes in C_A and C_B with time during the batch cycle for:

(*a*) $k_1 > k_2$
(*b*) $k_1 = k_2$

7.4. Two isothermal CSTRs are connected by a long pipe that acts as a pure deadtime of D minutes at the steady-state flow rates. Assume constant throughputs and holdups and a first-order irreversible reaction $\mathrm{A} \xrightarrow{k} \mathrm{B}$ in each tank. Derive the transfer function relating the feed concentration to the first tank, C_{A0}, and the concentration of A in the stream leaving the second tank, C_{A2}. Use inversion to find $C_{A2(t)}$ for a unit step disturbance in C_{A0}.

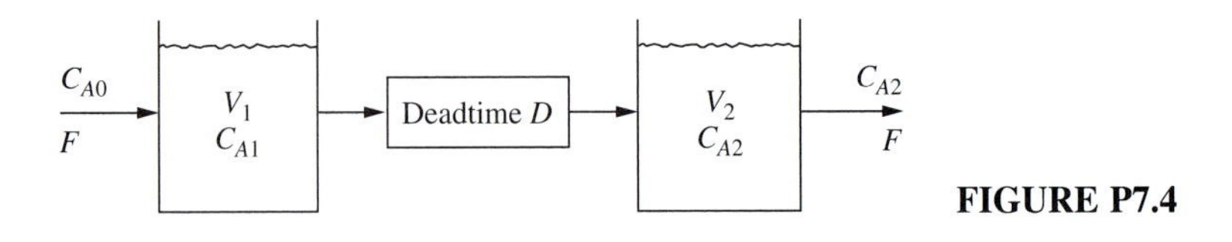

FIGURE P7.4

7.5. A general second-order system is described by the ODE

$$\tau_o^2 \frac{d^2x}{dt^2} + 2\tau_o\zeta\frac{dx}{dt} + x = K_p m_{(t)}$$

If $\zeta > 1$, show that the system transfer function has two first-order lags with time constants τ_{o1} and τ_{o2}. Express these time constants in terms of τ_o and ζ.

7.6. Use Laplace transform techniques to solve Example 2.7, where a ramp disturbance drives a first-order system.

7.7. The imperfect mixing in a chemical reactor can be modeled by splitting the total volume into two perfectly mixed sections with circulation between them. Feed enters and leaves one section. The other section acts like a "side-capacity" element.

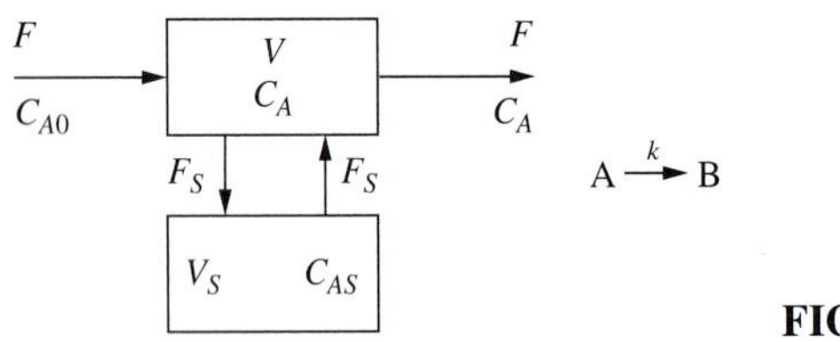

FIGURE P7.7

Assume holdups and flow rates are constant. The reaction is an irreversible, first-order consumption of reactant A. The system is isothermal. Solve for the transfer function relating C_{A0} and C_A. What are the zeros and poles of the transfer function? What is the steady-state gain?

7.8. One way to determine the rate of change of a process variable is to measure the differential pressure $\Delta P = P_{out} - P_{in}$ over a device called a derivative unit that has the transfer function

$$\frac{P_{out(s)}}{P_{in(s)}} = \frac{\tau s + 1}{(\tau/6)s + 1}$$

(*a*) Derive the transfer function between ΔP and P_{in}.
(*b*) Show that the ΔP signal will be proportional to the rate of rise of P_{in}, after an initial transient period, when P_{in} is a ramp function.

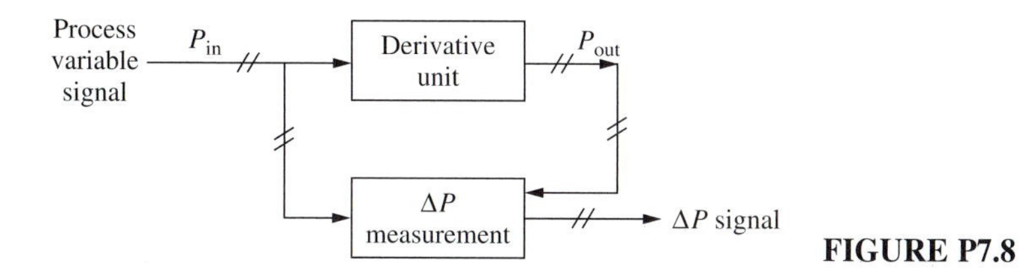

FIGURE P7.8

7.9. A convenient way to measure the density of a liquid is to pump it slowly through a vertical pipe and measure the differential pressure between the top and the bottom of the pipe. This differential head is directly related to the density of the liquid in the pipe if frictional pressure losses are negligible.

Suppose the density can change with time. What is the transfer function relating a perturbation in density to the differential-pressure measurement? Assume the fluid moves up the vertical column in plug flow at constant velocity.

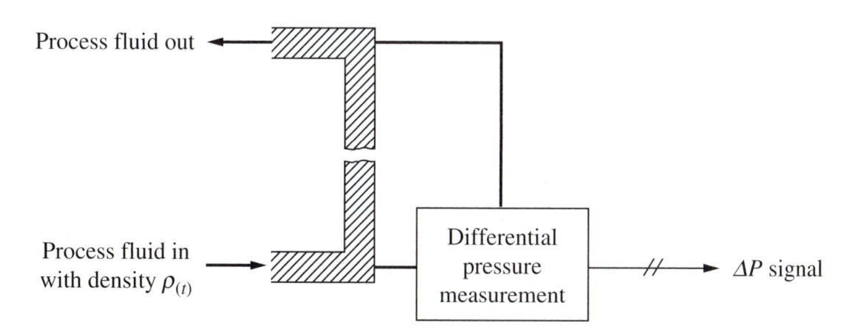

FIGURE P7.9

7.10. A thick-walled kettle of mass M_M, temperature T_M, and specific heat C_M is filled with a perfectly mixed process liquid of mass M, temperature T, and specific heat C. A heating fluid at temperature T_J is circulated in a jacket around the kettle wall. The heat transfer coefficient between the process fluid and the metal wall is U and between the metal outside wall and the heating fluid is U_M. Inside and outside heat transfer areas A are approximately the same. Neglecting any radial temperature gradients through the metal wall, show that the transfer function between T and T_J is two first-order lags.

$$G_{(s)} = \frac{K_p}{(\tau_{o1}s + 1)(\tau_{o2}s + 1)}$$

The value of the steady-state gain K_p is unity. Is this reasonable?

7.11. An ideal three-mode PID (proportional-integral-derivative) feedback controller is described by the equation

$$\text{CO}_{(t)} = \text{Bias} + K_c\left[E_{(t)} + \frac{1}{\tau_I}\int E_{(t)}\,dt + \tau_D\frac{dE}{dt}\right]$$

Derive the transfer function between $\text{CO}_{(s)}$ and $E_{(s)}$. Is this transfer function physically realizable?

7.12. Show that the linearized nonisothermal CSTR of Example 7.6 can be stable only if

$$\frac{UA}{V\rho C_p} > \frac{-\lambda\bar{k}\,\overline{C}_A E}{\rho C_p R\overline{T}^2} - 2\frac{\overline{F}}{V} - \bar{k}$$

7.13. A deadtime element is basically a distributed system. One approximate way to get the dynamics of distributed systems is to lump them into a number of perfectly mixed sections. Prove that a series of N mixed tanks is equivalent to a pure deadtime as N goes to infinity. (*Hint:* Keep the total volume of the system constant as more and more lumps are used.)

7.14. A feedback controller is added to the three-CSTR system of Example 7.5. Now C_{A0} is changed by the feedback controller to keep C_{A3} at its setpoint, which is the steady-state value of C_{A3}. The error signal is therefore just $-C_{A3}$ (the perturbation in C_{A3}). Find the transfer function of this closedloop system between the disturbance C_{AD} and C_{A3}. List the values of poles, zeros, and steady-state gain when the feedback controller is:

(*a*) Proportional: $C_{A0} = C_{AD} + K_c(-C_{A3})$

(*b*) Proportional-integral: $C_{A0} = C_{AD} + K_c\left[-C_{A3} + \frac{1}{\tau_I}\int(-C_{A3})\,dt\right]$

Note that these equations are in terms of perturbation variables.

7.15. The partial condenser sketched on the following page is described by two ODEs:

$$\left(\frac{\text{Vol}}{RT}\right)\frac{dP}{dt} = F - V - \frac{Q_c}{\Delta H}$$

$$\frac{dM_R}{dt} = \frac{Q_c}{\Delta H} - L$$

where
P = pressure
Vol = volume of condenser
M_R = liquid holdup
F = vapor feed rate
V = vapor product
L = liquid product

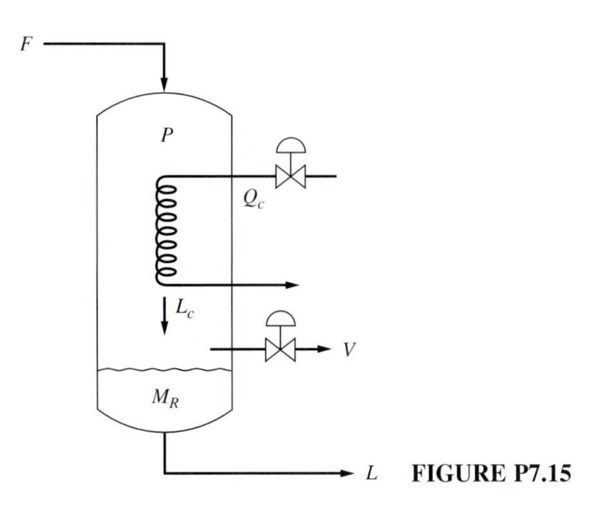

FIGURE P7.15

(*a*) Draw a block diagram showing the transfer functions describing the openloop system.

(*b*) Draw a block diagram of the closedloop system if a proportional controller is used to manipulate Q_c to hold M_R and a PI controller is used to manipulate V to hold P.

7.16. Show that a proportional-only level controller on a tank will give zero steady-state error for a step change in level setpoint.

7.17. Use Laplace transforms to prove mathematically that a P controller produces steady-state offset and that a PI controller does not. The disturbance is a step change in the load variable. The process openloop transfer functions, G_M and G_L, are both first-order lags with different gains but identical time constants.

7.18. Two 100-barrel tanks are available to use as surge volume to filter liquid flow rate disturbances in a petroleum refinery. Average throughput is 14,400 barrels per day. Should these tanks be piped up for parallel operation or for series operation? Assume proportional-only level controllers.

7.19. A perfectly mixed batch reactor, containing 7500 lb_m of liquid with a heat capacity of 1 Btu/lb_m °F, is surrounded by a cooling jacket that is filled with 2480 lb_m of perfectly mixed cooling water.

At the beginning of the batch cycle, both the reactor liquid and the jacket water are at 203°F. At this point in time, catalyst is added to the reactor and a reaction occurs that generates heat at a constant rate of 15,300 Btu/min. At this same moment, makeup cooling water at 68°F is fed into the jacket at a constant 832-lb_m/min flow rate.

The heat transfer area between the reactor and the jacket is 140 ft^2. The overall heat transfer coefficient is 70 Btu/hr °F ft^2. Mass of the metal walls can be neglected. Heat losses are negligible.

(*a*) Develop a mathematical model of the process.

(*b*) Use Laplace transforms to solve for the dynamic change in reactor temperature $T_{(t)}$.

(*c*) What is the peak reactor temperature and when does it occur?
(*d*) What is the final steady-state reactor temperature?

7.20. The flow of air into the regenerator on a catalytic cracking unit is controlled by two control valves. One is a large, slow-moving valve that is located on the suction of the air blower. The other is a small, fast-acting valve that vents to the atmosphere.

The fail-safe condition is to not feed air into the regenerator. Therefore, the suction valve is air-to-open and the vent valve is air-to-close. What action should the flow controller have, direct or reverse?

The device with the following transfer function $G_{(s)}$ is installed in the control line to the vent valve.

$$G_{(s)} = \frac{\tau s}{\tau s + 1} = \frac{P_{\text{valve}(s)}}{\text{CO}_{(s)}}$$

The purpose of this device is to cause the vent valve to respond quickly to changes in CO but to minimize the amount of air vented (since this wastes power) under steady-state conditions. What will be the dynamic response of the perturbation in P_{valve} for a step change of 10 percent of full scale in CO? What is the new steady-state value of P_{valve}.

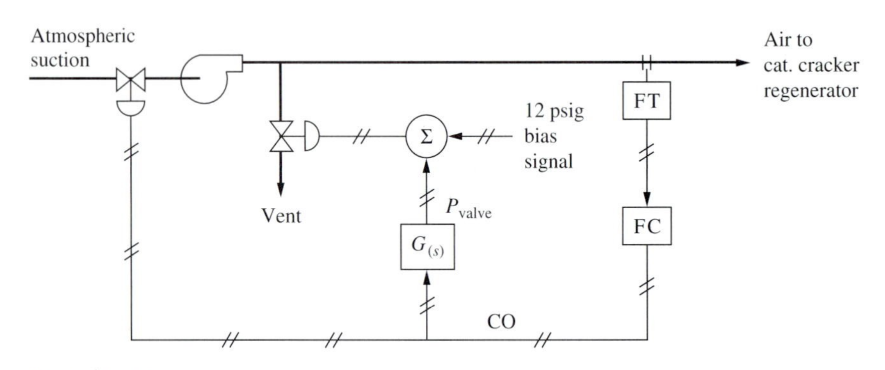

FIGURE P7.20

7.21. An openloop process has the transfer function

$$G_M = \frac{s}{\tau s + 1}$$

Calculate the openloop response of this process to a unit step change in its input. What is the steady-state gain of this process?

7.22. A chemical reactor is cooled by both jacket cooling and autorefrigeration (boiling liquid in the reactor). Sketch a block diagram, using appropriate process and control system transfer functions, describing the system. Assume these transfer functions are known, either from fundamental mathematical models or from experimental dynamic testing.

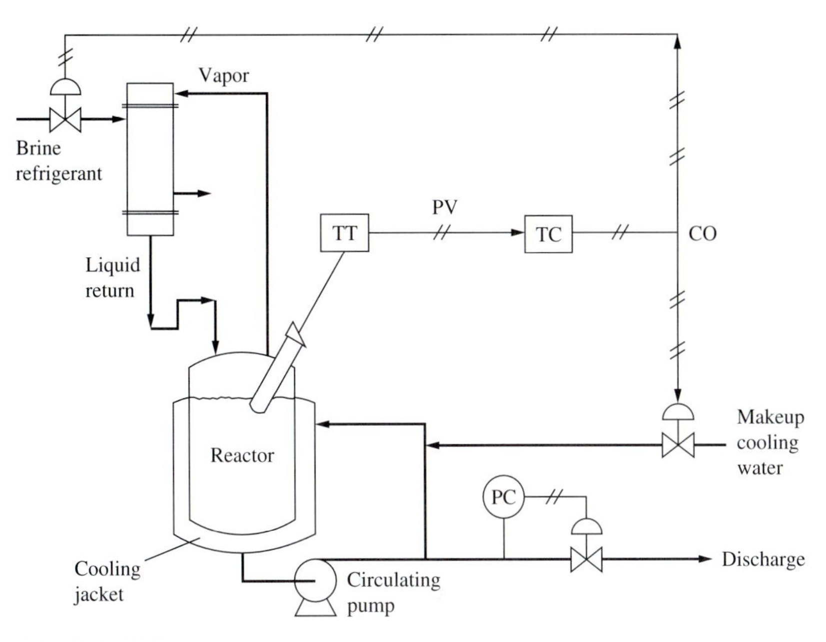

FIGURE P7.22

7.23. Solve the following problem, which is part of a problem given in Levenspiel's *Chemical Reaction Engineering* (1962, John Wiley, New York), using Laplace transform techniques. Find analytical expressions for the number of Nelson's ships $N_{(t)}$ and the number of Villeneuve's ships $V_{(t)}$ as functions of time.

> The great naval battle, to be known to history as the battle of Trafalgar (1805), was soon to be joined. Admiral Villeneuve proudly surveyed his powerful fleet of 33 ships stately sailing in single file in the light breeze. The British fleet under Lord Nelson was now in sight, 27 ships strong. Estimating that it would still be two hours before the battle, Villeneuve popped open another bottle of burgundy and point by point reviewed his carefully thought out battle strategy. As was the custom of naval battles at that time, the two fleets would sail in single file parallel to each other and in the same direction, firing their cannons madly. Now, by long experience in battles of this kind, it was a well-known fact that the rate of destruction of a fleet was proportional to the fire power of the opposing fleet. Considering his ships to be on a par, one for one, with the British, Villeneuve was confident of victory. Looking at his sundial, Villeneuve sighed and cursed the light wind; he'd never get it over with in time for his favorite television western. "Oh well," he sighed, "c'est la vie." He could see the headlines next morning. "British Fleet annihilated, Villeneuve's losses are . . . ." Villeneuve stopped short. How many ships would he lose? Villeneuve called over his chief bottle cork popper, Monsieur Dubois, and asked this question. What answer did he get?

7.24. While Admiral Villeneuve was doing his calculations about the outcome of the battle of Trafalgar, Admiral Nelson was also doing some thinking. His fleet was outnumbered 33 to 27, so it didn't take a rocket scientist to predict the outcome of the battle if the normal battle plan was followed (the opposing fleets sailing parallel to each other). So Admiral Nelson turned for help to his trusty young assistant Lt. Steadman, who fortunately was an innovative Lehigh graduate in chemical engineering (Class of 1796). Steadman opened up her textbook on Laplace transforms and did some back-of-the-envelope calculations to evaluate alternative battle strategies.

After several minutes of brainstorming and calculations (she had her PC on board, so she could use MATLAB to aid in the numerical calculations), Lt. Steadman devised the following plan: The British fleet would split the French fleet, taking on 17 ships first and then attacking the other 16 French ships with the remaining British ships. Admiral Nelson approved the plan, and the battle was begun.

Solve quantitatively for the dynamic changes in the number of British and French ships as functions of time during the battle. Assume that the rate of destruction of a fleet is proportional to the firepower of the opposing fleet and that the ships in both fleets are on a par with each other in firepower.

7.25. Consider a feed preheater-reactor process in which gas is fed at temperature T_0 into a heat exchanger. It picks up heat from the hot gas leaving the reactor and exits the heat exchanger at temperature T_1. The gas then enters the adiabatic tubular reactor, where an exothermic chemical reaction occurs. The heat of reaction heats the gas stream, and the gas leaves the reactor at temperature T_2.

The transfer functions describing the two units are

$$T_1 = G_{1(s)} T_0 + G_{2(s)} T_2$$

$$T_2 = G_{3(s)} T_1$$

Derive the openloop transfer function between T_1 and T_0.

7.26. A second-order reaction $A + B \xrightarrow{k} C$ occurs in an isothermal CSTR. The reaction rate is proportional to the concentrations of each of the reactants, z_A and z_B (mole fractions of components A and B).

$$\mathcal{R} = k V_R z_A z_B \quad \text{(moles of component C produced per hour)}$$

The reactor holdup is V_R (mol) and the specific reaction rate is k (hr^{-1}).

Two fresh feed streams at flow rates F_{0A} and F_{0B} (mol/hr) and compositions $z_{0A,j}$ and $z_{0B,j}$ (mole fraction component j, $j = A, B, C$) are fed into the reactor along with a recycle stream. The reactor effluent has composition z_j and flow rate F (mol/hr). It is fed into a flash drum in which a vapor stream is removed and recycled back to the reactor at a flow rate R (mol/hr) and composition $y_{R,j}$.

The liquid from the drum is the product stream with flow rate P (mol/hr) and composition $x_{P,j}$. The liquid and vapor streams are in phase equilibrium: $y_{R,A} = K_A x_{P,A}$ and $y_{R,B} = K_B x_{P,B}$, where K_A and K_B are constants. The vapor holdup in the flash drum is negligible. The liquid holdup is M_D (mol).

The control system consists of the following loops: fresh feed F_{0A} is manipulated to control reactor composition z_A, fresh feed F_{0B} is manipulated to control reactor holdup V_R, reactor effluent F is flow controlled, product flow rate P is manipulated to hold drum holdup M_D constant, and recycle vapor R is manipulated to control product composition $x_{P,B}$. Assume V_R and M_D are held perfectly constant. The molecular weight of component A is 30; the molecular weight of component B is 70.

(*a*) Write the dynamic equations describing the openloop process.

(*b*) Using the following numerical values for the parameters in the system, calculate the values of F_{0A}, F_{0B}, F, V_R, z_A, and z_B at steady state. The amount of component

C produced is 100 mol/hr, and it all leaves in the product stream P. Fresh feed F_{0A} is pure component A, and fresh feed F_{0B} is pure component B.

$$R = 300 \text{ mol/min} \qquad k = 10 \text{ hr}^{-1}$$
$$x_{P,A} = 0.01 \text{ mole fraction component A} \qquad K_A = 40$$
$$x_{P,B} = 0.01 \text{ mole fraction component B} \qquad K_B = 30$$

7.27. A first-order reaction A $\xrightarrow{k}$ B occurs in an isothermal CSTR. Fresh feed at a flow rate F_0 (mol/min) and composition z_0 (mole fraction A) is fed into the reactor along with a recycle stream. The reactor holdup is V_R (mol). The reactor effluent has composition z (mole fraction A) and flow rate F (mol/min). It is fed into a flash drum in which a vapor stream is removed and recycled back to the reactor at a flow rate R (mol/min) and composition y_R (mole fraction A).

The liquid from the drum is the product stream with flow rate P (mol/min) and composition x_P (mole fraction A). The liquid and vapor streams are in phase equilibrium: $y_R = Kx_P$, where K is a constant. The vapor holdup in the flash drum is negligible. The liquid holdup is M_D.

The control system consists of the following loops: fresh feed F_0 is flow controlled, reactor effluent F is manipulated to hold reactor holdup V_R constant, product flow rate P is manipulated to hold drum holdup M_D constant, and recycle vapor R is manipulated to control product composition x_P. Assume V_R and M_D are held perfectly constant.

(*a*) Write the dynamic equations describing the openloop process system.

(*b*) Linearize the ODEs describing the system, assuming V_R, M_D, F_0, k, K, z_0, and P are constant. Then Laplace transform and show that the openloop transfer function relating controlled variable x_P to manipulated variable R has the form

$$G_{M(s)} = \frac{K_p(\tau_1 s + 1)}{(\tau_2 s + 1)(\tau_3 s + 1)}$$

(*c*) Using the following numerical values for the parameters in the system, calculate the values of R and z at steady state.

$$F_0 = 100 \text{ mol/min} \qquad z_0 = 1 \text{ mole fraction component A} \qquad k = 1 \text{ min}^{-1}$$
$$V_R = 500 \text{ moles} \qquad x_P = 0.05 \text{ mole fraction component A} \qquad K = 10$$

7.28. The 1940 Battle of the North Atlantic is about to begin. The German submarine fleet, under the command of sinister Admiral von Dietrich, consists of 200 U-boats at the beginning of the battle. The British destroyer fleet, under the command of heroic Admiral Steadman (a direct descendant of the intelligence officer responsible for the British victory at the Battle of Trafalgar), consists of 150 ships at the beginning of the battle. The rate of destruction of submarines by destroyers is equal to the rate of destruction of destroyers by submarines: 0.25 ships/week/ship.

Germany is launching two new submarines per week and adding them to its fleet. President Roosevelt is trying to decide how many new destroyers per week must be sent to the British fleet under the Lend-Lease Program in order to win the battle. Admiral Steadman claims she needs 15 ships added to her fleet per week to defeat the U-boat fleet. The Secretary of the Navy, William Gustus, claims she only needs 5 ships per week. Who is correct?

7.29. Captain James Kirk is in command of a fleet of 16 starships. A Klingon fleet of 20 ships has been spotted approaching. The legendary Lt. Spock has recently retired, so Captain Kirk turns to his new intelligence officer, Lt. Steadman (Lehigh Class of 2196 in chemical engineering), for a prediction of the outcome of the upcoming battle. Steadman has been working with the new engineering officers in the fleet, Lt. Moquin and Lt. Walsh, who have replaced the retired Lt. Scott. These innovative officers have been able to increase the firepower of half of the vessels in Kirk's fleet by a factor of 2 over the firepower of the Klingon vessels, which all have the same firepower. The firepower of the rest of Kirk's fleet is on a par with that of the Klingons. But these officers have also been able to improve the defensive shields on this second half of the fleet. The more effective shields reduce by 50 percent the destruction rate of these vessels by the Klingon firepower.

Thus, there are two classes of starships: eight vessels are Class E_1 with increased firepower, and eight vessels are Class E_2 with improved defensive shields. Assume that half of the Klingon fleet is firing at each class at any point in time.

Calculate who wins the battle and how many vessels of each type survive.

CHAPTER 8

Laplace-Domain Analysis of Conventional Feedback Control Systems

Now that we have learned a little Russian, we are ready to see how useful it is in analyzing the dynamics and stability of systems. Laplace-domain methods provide a lot of insight into what is happening to the damping coefficients and time constants as we change the settings on the controller. The *root-locus* plots that we use are similar in value to the graphical McCabe-Thiele diagram in binary distillation: they provide a nice picture in which the effects of parameters can be easily seen.

In this chapter we demonstrate the significant computational and notational advantages of Laplace transforms. The techniques involve finding the transfer function of the openloop process, specifying the desired performance of the closedloop system (process plus controller), and finding the feedback controller transfer function that is required to do the job.

8.1 OPENLOOP AND CLOSEDLOOP SYSTEMS

8.1.1 Openloop Characteristic Equation

Consider the general openloop system sketched in Fig. 8.1*a*. The load variable $L_{(s)}$ enters through the openloop process transfer function $G_{L(s)}$. The manipulated variable $M_{(s)}$ enters through the openloop process transfer function $G_{M(s)}$. The controlled variable $Y_{(s)}$ is the sum of the effects of the manipulated variable and the load variable. Remember, we are working with linear systems in the Laplace domain, so superposition applies.

Figure 8.1*b* shows a specific example: the two-heated-tank process discussed in Example 7.7. The load variable is the inlet temperature T_0. The manipulated variable is the heat input to the first tank Q_1. The two transfer functions $G_{L(s)}$ and $G_{M(s)}$ were derived in Chapter 7.

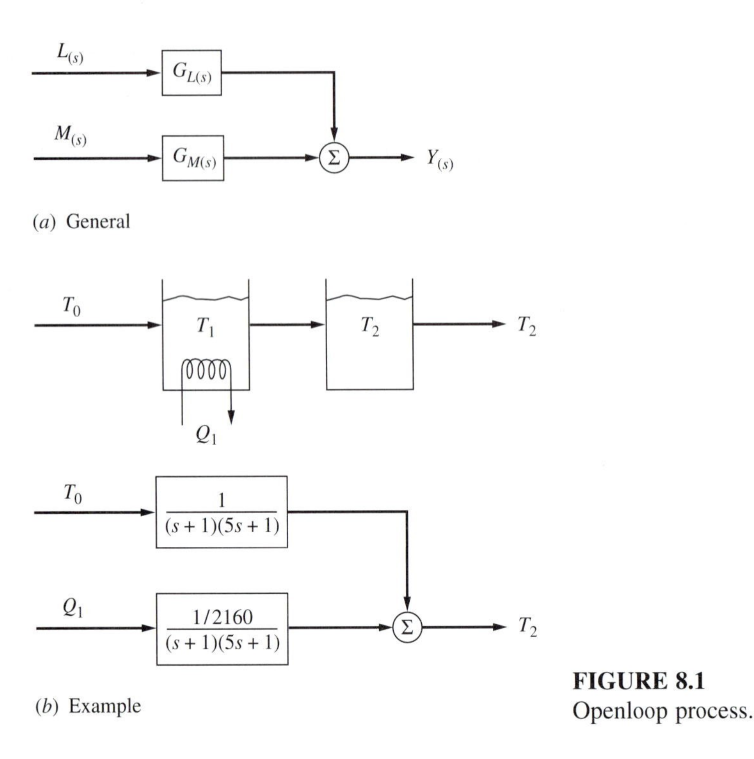

FIGURE 8.1
Openloop process.

The dynamics of this openloop system depend on the roots of the openloop characteristic equation, i.e., on the roots of the polynomials in the denominators of the openloop transfer functions. These are the poles of the openloop transfer functions. If all the roots lie in the left half of the s plane, the system is openloop stable. For the two-heated-tank example shown in Fig. 8.1*b*, the poles of the openloop transfer function are $s = -1$ and $s = -\frac{1}{5}$, so the system is openloop stable.

Note that the $G_{L(s)}$ transfer function for the two-heated tank process has a steady-state gain with units of °F/°F. The $G_{M(s)}$ transfer function has a steady-state gain with units of °F/Btu/min.

8.1.2 Closedloop Characteristic Equation and Closedloop Transfer Functions

Now let us put a feedback controller on the process, as shown in Fig. 8.2*a*. The controlled variable is converted to a process variable signal PV by the sensor/transmitter element $G_{T(s)}$. The feedback controller compares the PV signal to the desired setpoint signal SP, feeds the error signal E through a feedback controller transfer function $G_{C(s)}$, and sends out a controller output signal CO. The controller output signal changes the position of a control valve, which changes the flow rate of the manipulated variable M.

Figure 8.2*b* gives a sketch of the feedback control system and a block diagram for the two-heated-tank process with a controller. Let us use an analog electronic

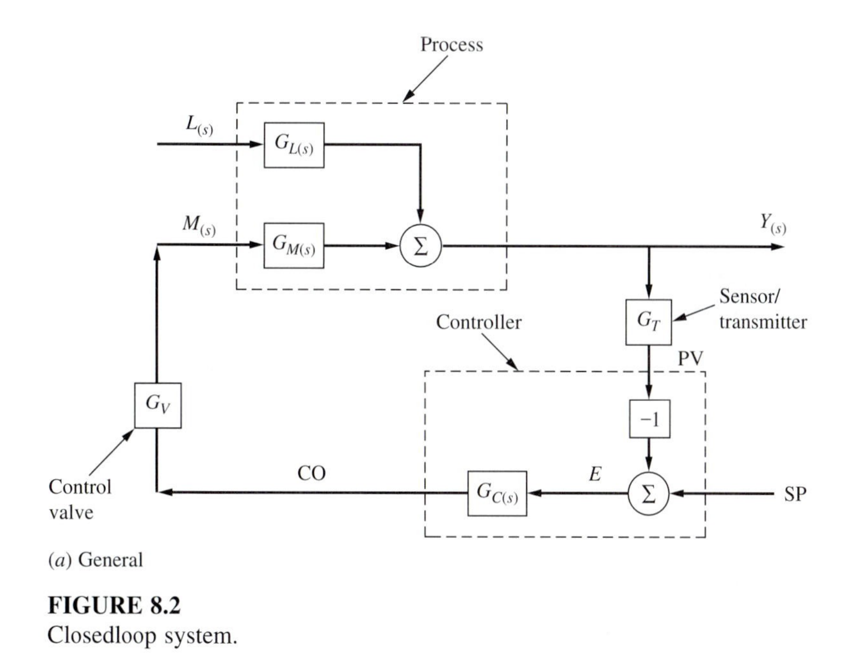

(*a*) General

FIGURE 8.2
Closedloop system.

system with 4 to 20 mA control signals. The temperature sensor has a span of 100°F, so the G_T transfer function (neglecting any dynamics in the temperature measurement) is

$$G_{T(s)} \equiv \frac{\text{PV}_{(s)}}{T_{2(s)}} = \frac{16 \text{ mA}}{100°\text{F}} \tag{8.1}$$

The controller output signal CO goes to an I/P transducer that converts 4 to 20 mA to a 3- to 15-psig air pressure signal to drive the control valve through which steam is added to the heating coil. Let us assume that the valve has linear installed characteristics (see Chapter 3) and can pass enough steam to add 500,000 Btu/min to the liquid in the tank when the valve is wide open. Therefore, the transfer function between Q_1 and CO (lumping together the transfer function for the I/P transducer and the control valve) is

$$G_{V(s)} \equiv \frac{Q_{1(s)}}{\text{CO}_{(s)}} = \frac{500{,}000 \text{ Btu/min}}{16 \text{ mA}} \tag{8.2}$$

Looking at the block diagram in Fig. 8.2*a*, we can see that the output $Y_{(s)}$ is given by

$$Y = G_{L(s)}L + G_{M(s)}M \tag{8.3}$$

But in this closedloop system, $M_{(s)}$ is related to $Y_{(s)}$:

$$\begin{aligned} M &= G_{V(s)}\text{CO} = G_{V(s)}G_{C(s)}E = G_{V(s)}G_{C(s)}(\text{SP} - \text{PV}) \\ M &= G_{V(s)}G_{C(s)}(\text{SP} - G_{T(s)}Y) \end{aligned} \tag{8.4}$$

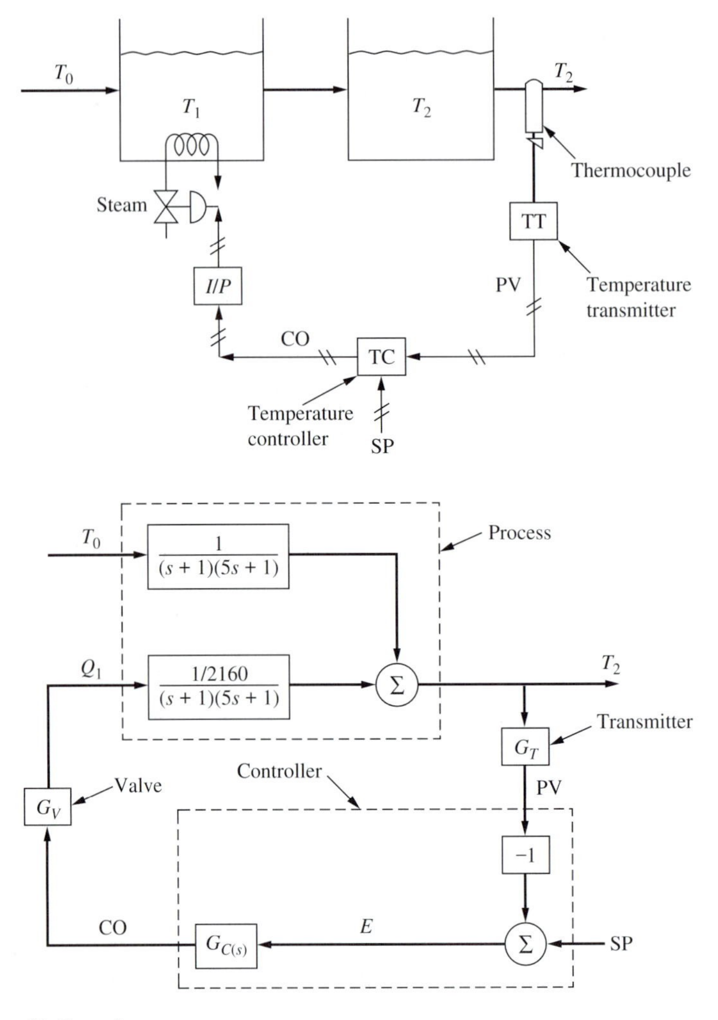

(*b*) Example

FIGURE 8.2 (CONTINUED)
Closedloop system.

Combining Eqs. (8.3) and (8.4) gives

$$Y = G_{L(s)}L + G_{M(s)}G_{V(s)}G_{C(s)}(\mathrm{SP} - G_{T(s)}Y)$$

$$[1 + G_{M(s)}G_{V(s)}G_{C(s)}G_{T(s)}]Y = G_{L(s)}L + G_{M(s)}G_{V(s)}G_{C(s)}\mathrm{SP}$$

$$Y_{(s)} = \left[\frac{G_{L(s)}}{1 + G_{M(s)}G_{V(s)}G_{C(s)}G_{T(s)}}\right]L_{(s)} + \left[\frac{G_{M(s)}G_{V(s)}G_{C(s)}}{1 + G_{M(s)}G_{V(s)}G_{C(s)}G_{T(s)}}\right]\mathrm{SP}_{(s)} \quad (8.5)$$

Equation (8.5) gives the transfer functions describing the *closedloop* system, so these are closedloop transfer functions. The two inputs are the load $L_{(s)}$ and the setpoint $\mathrm{SP}_{(s)}$. The controlled variable is $Y_{(s)}$. Note that the denominators of these closedloop transfer functions are identical.

EXAMPLE 8.1. The closedloop transfer functions for the two-heated-tank process can be calculated from the openloop process transfer functions and the feedback controller transfer function. We choose a proportional controller, so $G_{C(s)} = K_c$. Note that the dimensions of the gain of the controller are mA/mA, i.e., the gain is dimensionless. The controller looks at a milliampere signal (PV) and puts out a milliampere signal (CO).

$$G_{L(s)} = \frac{1°\text{F}/°\text{F}}{(s+1)(5s+1)}$$

$$G_{M(s)} = \frac{1/2160\ °\text{F/Btu/min}}{(s+1)(5s+1)}$$

$$G_{V(s)} = \frac{500{,}000\ \text{Btu/min}}{16\ \text{mA}}$$

$$G_{T(s)} = \frac{16\ \text{mA}}{100°\text{F}}$$

The closedloop transfer function for load changes is

$$\frac{T_2}{T_0} = \frac{G_{L(s)}}{1 + G_{M(s)}G_{V(s)}G_{C(s)}G_{T(s)}}$$

$$= \frac{\dfrac{1°\text{F}/°\text{F}}{(s+1)(5s+1)}}{1 + \left[\dfrac{1/2160\ °\text{F/Btu/min}}{(s+1)(5s+1)}\right]\left(\dfrac{500{,}000\ \text{Btu/min}}{16\ \text{mA}}\right)(K_c)\left(\dfrac{16\ \text{mA}}{100°\text{F}}\right)} \tag{8.6}$$

$$= \frac{1°\text{F}/°\text{F}}{(s+1)(5s+1) + 500K_c/216} = \frac{1°\text{F}/°\text{F}}{5s^2 + 6s + 1 + 500K_c/216}$$

The closedloop transfer function for setpoint changes is

$$\frac{T_2}{\text{SP}} = \frac{G_{M(s)}G_{V(s)}G_{C(s)}}{1 + G_{M(s)}G_{V(s)}G_{C(s)}G_{T(s)}}$$

$$= \frac{\left[\dfrac{1/2160\ °\text{F/Btu/min}}{(s+1)(5s+1)}\right]\left(\dfrac{500{,}000\ \text{Btu/min}}{16\ \text{mA}}\right)(K_c)}{1 + \left[\dfrac{1/2160\ °\text{F/Btu/min}}{(s+1)(5s+1)}\right]\left(\dfrac{500{,}000\ \text{Btu/min}}{16\ \text{mA}}\right)(K_c)\left(\dfrac{16\ \text{mA}}{100°\text{F}}\right)} \tag{8.7}$$

$$= \frac{50{,}000K_c/216/16\ °\text{F/mA}}{5s^2 + 6s + 1 + 500K_c/216}$$

If we look at the closedloop transfer function between PV and SP, we must multiply the above by G_T.

$$\frac{\text{PV}}{\text{SP}} = \frac{500K_c/216\ \text{mA/mA}}{5s^2 + 6s + 1 + 500K_c/216} \tag{8.8}$$

Notice that the denominators of all these closedloop transfer functions are identical. Notice also that the steady-state gain of the closedloop servo transfer function PV/SP is not unity; i.e., there is a steady-state offset. This is because of the proportional controller. We can calculate the PV/SP ratio at steady state by letting s go to zero in Eq. (8.8).

$$\lim_{t \to \infty} \left(\frac{\text{PV}}{\text{SP}} \right) = \frac{500K_c/216}{1 + 500K_c/216} = \frac{1}{\dfrac{216}{500K_c} + 1} \tag{8.9}$$

Equation (8.9) shows that the bigger the controller gain, the smaller the offset. ■

Since the characteristic equation of any system (openloop or closedloop) is the denominator of the transfer function describing it, the closedloop characteristic equation for this system is

$$\boxed{1 + G_{M(s)}G_{V(s)}G_{C(s)}G_{T(s)} = 0} \tag{8.10}$$

This equation shows that closedloop dynamics depend upon the process openloop transfer functions (G_M, G_V, and G_T) and on the feedback controller transfer function (G_C). Equation (8.10) applies for simple single-input, single-output systems. We derive closedloop characteristic equations for other systems in later chapters.

The first closedloop transfer function in Eq. (8.5) relates the controlled variable to the load variable. It is called the closedloop *regulator* transfer function. The second closedloop transfer function in Eq. (8.5) relates the controlled variable to the setpoint. It is called the closedloop *servo* transfer function.

Normally we design the feedback controller $G_{C(s)}$ to give some desired closed-loop performance. For example, we might specify the damping coefficient that we want the closedloop system to have. However, it is useful to consider the ideal situation. If we could design a controller without any regard for physical realizability, what would the ideal closedloop regulator and servo transfer functions be? Clearly, we would like a load disturbance to have *no* effect on the controlled variable. So the ideal closedloop regulator transfer function is zero. For setpoint changes, we would like the controlled variable to track the setpoint perfectly at all times. So the ideal servo transfer function is unity.

Equation (8.5) shows that both of these could be achieved if we could simply make $G_{C(s)}$ infinitely large. This would make the first term zero and the second term unity. However, as we saw in Chapter 1 in our simulation example, stability limitations prevent us from achieving this ideal situation.

Instead of considering the process, transmitter, and valve transfer functions separately, it is often convenient to combine them into just one transfer function. Then the closedloop block diagram, shown in Fig. 8.3, becomes simple and is described

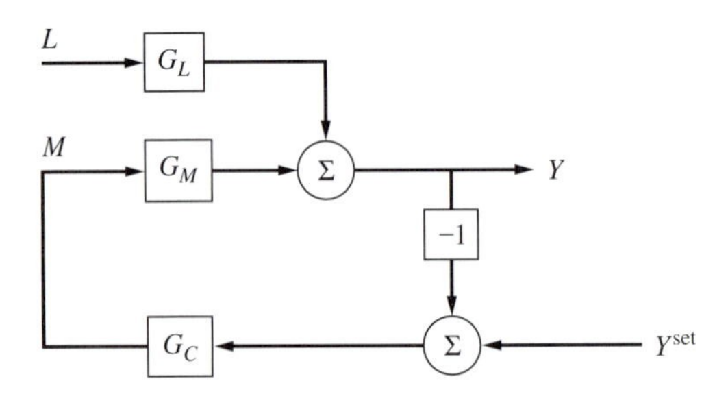

FIGURE 8.3
Simplified feedback loop.

by

$$Y_{(s)} = \left[\frac{G_{L(s)}}{1 + G_{M(s)}G_{C(s)}}\right] L_{(s)} + \left[\frac{G_{M(s)}G_{C(s)}}{1 + G_{M(s)}G_{C(s)}}\right] Y^{\text{set}}_{(s)} \tag{8.11}$$

This is the equation that we use in most cases because it is more convenient. Keep in mind that the $G_{M(s)}$ transfer function in Eq. (8.11) is a combination of the process, transmitter, and valve transfer functions. The closedloop characteristic equation is

$$\boxed{1 + G_{M(s)}G_{C(s)} = 0} \tag{8.12}$$

8.2 STABILITY

The most important dynamic aspect of any system is its stability. We learned in Chapter 2 that stability is dictated by the location of the roots of the characteristic equation of the system. In Chapter 7 we learned that the roots of the denominator of the system transfer function (poles) are exactly the same as the roots of the characteristic equation. Thus, for the system to be stable, the poles of the transfer function must lie in the left half of the s plane (LHP).

This stability requirement applies to any system, openloop or closedloop. The stability of an openloop process depends on the location of the poles of its *openloop* transfer function. The stability of a closedloop process depends on the location of the poles of its *closedloop* transfer function. These closedloop poles will naturally be different from the openloop poles because of the introduction of the feedback loop with the controller. Thus, the criteria for openloop and closedloop stability are different. Most systems are openloop stable but can be either closedloop stable or unstable, depending on the values of the controller parameters. We will show that any real process can be made closedloop unstable by making the gain of the feedback controller large enough. There are some processes that are openloop unstable. We will show that these systems can usually be made closedloop stable by the correct choice of the type of controller and its settings.

The most useful method for testing stability in the Laplace domain is direct substitution. This method is a simple way to find the values of parameters in the characteristic equation that put the system just at the limit of stability.

We know the system is stable if all the roots of the characteristic equation are in the LHP and unstable if any of the roots are in the RHP. Therefore, the imaginary axis represents the stability boundary. On the imaginary axis s is equal to some pure imaginary number: $s = i\omega$.

The technique consists of substituting $i\omega$ for s in the characteristic equation and solving for the values of ω and other parameters (e.g., controller gain) that satisfy the resulting equations. The method can be best understood by looking at the following example.

EXAMPLE 8.2. A three-CSTR system has an openloop transfer function $G_{M(s)}$ relating the controlled variable $C_{A3(s)}$ to the manipulated variable $C_{A0(s)}$.

$$G_{M(s)} = \frac{\frac{1}{8}}{(s+1)^3} = \left(\frac{C_{A3}}{C_{A0}}\right)_{(s)} \tag{8.13}$$

We want to look at the stability of the closedloop system with a proportional controller: $G_{C(s)} = K_c$. First, however, let us check the openloop stability of this system. The openloop characteristic equation is the denominator of the openloop transfer function set equal to zero.

$$(s+1)^3 = s^3 + 3s^2 + 3s + 1 = 0 \tag{8.14}$$

The roots of the openloop transfer function are $s_1 = s_2 = s_3 = -1$. These all lie in the LHP, so the process is openloop stable.

Now let us check for closedloop stability. The system is sketched in Fig. 8.4. The closedloop characteristic equation is

$$\begin{aligned} 1 + G_{M(s)}G_{C(s)} &= 0 = 1 + K_c\frac{\frac{1}{8}}{(s+1)^3} \\ s^3 + 3s^2 + 3s + 1 + \frac{K_c}{8} &= 0 \end{aligned} \tag{8.15}$$

Substituting $s = i\omega$ gives

$$\begin{aligned} -i\omega^3 - 3\omega^2 + 3i\omega + 1 + \frac{K_c}{8} &= 0 \\ \left(1 + \frac{K_c}{8} - 3\omega^2\right) + i(3\omega - \omega^3) &= 0 + i0 \end{aligned} \tag{8.16}$$

Equating the real and imaginary parts of the left- and right-hand sides of the equation gives two equations:

$$1 + \frac{K_c}{8} - 3\omega^2 = 0 \quad \text{and} \quad 3\omega - \omega^3 = 0$$

Therefore,

$$\omega^2 = 3 \quad \Rightarrow \quad \omega = \pm\sqrt{3} \tag{8.17}$$

$$\frac{K_c}{8} = 3\omega^2 - 1 = 3(3) - 1 \quad \Rightarrow \quad K_c = 64 \tag{8.18}$$

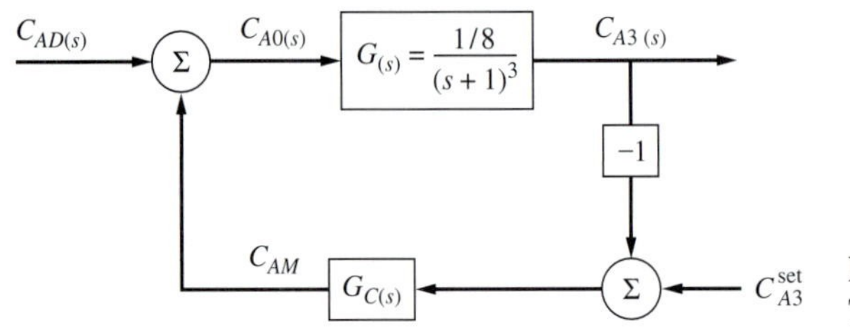

FIGURE 8.4
Three-CSTR system.

The value of the gain at the limit of stability is 64. It is called the *ultimate gain* K_u. The ω at this limit is the value of the imaginary part of s when the roots lie right on the imaginary axis. Since the real part of s is zero, the system will show a sustained oscillation with this frequency ω_u, called the *ultimate frequency,* in radians per time. The period of the oscillation is exactly the same as the ultimate period P_u that we defined in Chapter 2 in the Ziegler-Nichols tuning method.

$$P_u = \frac{2\pi}{\omega_u} \tag{8.19}$$

■

EXAMPLE 8.3. Suppose the closedloop characteristic equation for a system is

$$\tfrac{5}{2}s^3 + 8s^2 + \tfrac{13}{2}s + 1 + K_c = 0 \tag{8.20}$$

To find the ultimate gain K_u and ultimate frequency ω_u, we substitute $i\omega$ for s.

$$\begin{aligned} -\tfrac{5}{2}i\omega^3 - 8\omega^2 + \tfrac{13}{2}i\omega + 1 + K_c &= 0 \\ (1 + K_c - 8\omega^2) + i(-\tfrac{5}{2}\omega^3 + \tfrac{13}{2}\omega) &= 0 + i0 \end{aligned} \tag{8.21}$$

Solving the resulting two equations in two unknowns gives

$$\omega_u = \sqrt{\tfrac{13}{5}} \quad \text{and} \quad K_u = 19.8 \tag{8.22}$$

■

In Chapter 10 we show that we can convert from the Laplace domain (Russian) to the frequency domain (Chinese) by merely substituting $i\omega$ for s in the transfer function of the process. This is similar to the direct substitution method, but keep in mind that these two operations are different. In one we use the transfer function; in the other we use the characteristic equation.

8.3 PERFORMANCE SPECIFICATIONS

To design feedback controllers we must have some way to evaluate their effect on the performance of the closedloop system, both dynamically and at steady state.

8.3.1 Steady-State Performance

The usual steady-state performance specification is zero steady-state error. We will show that this steady-state performance depends on both the system (process and controller) and the type of disturbance. This is different from the question of stability of the system, which, as we have previously shown, is a function only of the system (roots of the characteristic equation) and does not depend on the input.

The error signal in the Laplace domain, $E_{(s)}$, is defined as the difference between the setpoint $Y^{set}_{(s)}$ and the process output $Y_{(s)}$.

$$E_{(s)} = Y^{set}_{(s)} - Y_{(s)} \tag{8.23}$$

Assuming that there is a change in the setpoint $Y_{(s)}^{set}$ but no change in the load disturbance ($L_{(s)} = 0$) and substituting for $Y_{(s)}$ from Eq. (8.11) give

$$E_{(s)} = Y_{(s)}^{set} - \left[\frac{G_{M(s)}G_{C(s)}}{1 + G_{M(s)}G_{C(s)}}\right] Y_{(s)}^{set}$$
$$\frac{E_{(s)}}{Y_{(s)}^{set}} = \frac{1}{1 + G_{M(s)}G_{C(s)}} \tag{8.24}$$

To find the steady-state value of the error, we will use the final-value theorem from Chapter 7.

$$\overline{E} \equiv \lim_{t \to \infty} E_{(t)} = \lim_{s \to 0} [sE_{(s)}] \tag{8.25}$$

Now let us look at two types of setpoint inputs: a step and a ramp.

A. Unit step input

$$Y_{(s)}^{set} = \frac{1}{s}$$

$$\overline{E} = \lim_{s \to 0} [sE_{(s)}] = \lim_{s \to 0} \left[s \frac{1}{1 + G_{M(s)}G_{C(s)}} \frac{1}{s} \right] = \lim_{s \to 0} \left[\frac{1}{1 + G_{M(s)}G_{C(s)}} \right]$$

If the steady-state error is to go to zero, the term $1/(1 + G_{M(s)}G_{C(s)})$ must go to zero as s goes to zero. This means that the term $G_{C(s)}G_{M(s)}$ must go to infinity as s goes to zero. Thus, $G_{C(s)}G_{M(s)}$ must contain a $1/s$ term, which is an integrator. If the process $G_{M(s)}$ does not contain integration, we must put it into the controller $G_{C(s)}$. So we add reset or integral action to eliminate steady-state error for step input changes in setpoint.

If we use a proportional controller, the steady-state error is

$$\overline{E} = \lim_{s \to 0} \left[\frac{1}{1 + G_{M(s)}G_{C(s)}} \right] = \frac{1}{1 + K_c \dfrac{z_1 z_2 \cdots z_M}{p_1 p_2 \cdots p_N}} \tag{8.26}$$

where z_i = zeros of $G_{M(s)}$
p_i = poles of $G_{M(s)}$

Thus the steady-state error is reduced by increasing K_c, the controller gain.

B. Ramp input

$$Y_{(s)}^{set} = \frac{1}{s^2}$$

$$\overline{E} = \lim_{s \to 0} \left[s \frac{1}{1 + G_{M(s)}G_{C(s)}} \frac{1}{s^2} \right] = \lim_{s \to 0} \left[\frac{1}{s(1 + G_{M(s)}G_{C(s)})} \right]$$

If the steady-state error is to go to zero, the term $1/s(1 + G_{M(s)}G_{C(s)})$ must go to zero as s goes to zero. This requires that $G_{C(s)}G_{M(s)}$ must contain a $1/s^2$ term. Double integration is needed to drive the steady-state error to zero for a ramp input (to make the output track the changing setpoint).

8.3.2 Dynamic Specifications

The dynamic performance of a system can be deduced by merely observing the location of the roots of the system characteristic equation in the s plane. The time-domain specifications of time constants and damping coefficients for a closedloop system can be used directly in the Laplace domain.

1. If all the roots lie in the LHP, the system is stable.
2. If all the roots lie on the negative real axis, we know the system is overdamped or critically damped (all real roots).
3. The farther out on the negative axis the roots lie, the faster the dynamics of the system will be (the smaller the time constants).
4. The roots that lie close to the imaginary axis will dominate the dynamic response since the ones farther out will die out quickly.
5. The farther any complex conjugate roots are from the real axis, the more underdamped the system will be.

There is a quantitative relationship between the location of roots in the s plane and the damping coefficient. Assume we have a second-order system or, if it is of higher order, assume it is dominated by the second-order roots closest to the imaginary axis. As shown in Fig. 8.5, the two roots are s_1 and s_2 and they are, of course, complex conjugates. From Eq. (2.68) the two roots are

$$s_1 = -\frac{\zeta}{\tau} + i\frac{\sqrt{1-\zeta^2}}{\tau}$$

$$s_2 = -\frac{\zeta}{\tau} - i\frac{\sqrt{1-\zeta^2}}{\tau}$$

τ and ζ are the time constant and damping coefficient of the system. If the system is openloop, these are the openloop time constant and openloop damping coefficient.

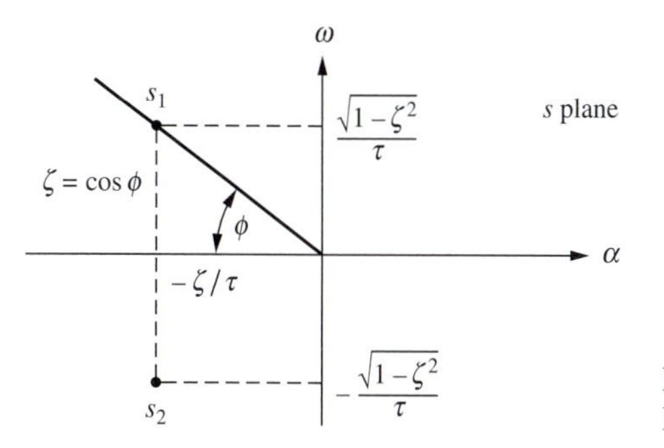

FIGURE 8.5
Dominant second-order root in the s plane.

If the system is closedloop, these are the closedloop time constant and closedloop damping coefficient.

The hypotenuse of the triangle shown in Fig. 8.5 is the distance from the origin out to the root s_1.

$$\sqrt{\left(\frac{\sqrt{1-\zeta^2}}{\tau}\right)^2 + \left(\frac{\zeta}{\tau}\right)^2} = \frac{1}{\tau} \tag{8.27}$$

The angle ϕ can be defined from the hypotenuse and the adjacent side of the triangle.

$$\cos\phi = \frac{\zeta/\tau}{1/\tau} = \zeta \tag{8.28}$$

Thus the location of a complex root can be converted directly to a damping coefficient and a time constant. The damping coefficient is equal to the cosine of the angle between the negative real axis and a radial line from the origin to the root. The time constant is equal to the reciprocal of the radial distance from the origin to the root.

Notice that lines of constant damping coefficient are radial lines in the s plane. Lines of constant time constant are circles.

8.4 ROOT LOCUS ANALYSIS

8.4.1 Definition

A *root locus* plot is a figure that shows how the roots of the closedloop characteristic equation vary as the gain of the feedback controller changes from zero to infinity. The abscissa is the real part of the closedloop root; the ordinate is the imaginary part. Since we are plotting closedloop roots, the time constants and damping coefficients that we pick off these root locus plots are all *closedloop time constants* and *closedloop damping coefficients.*

The following examples show the types of curves obtained and illustrate some important general principles.

EXAMPLE 8.4. Let us start with the simplest of all processes, a first-order lag. We choose a proportional controller. The system and controller transfer functions are

$$G_{M(s)}G_{C(s)} = \left(\frac{K_p}{\tau_o s + 1}\right)K_c \tag{8.29}$$

where K_p = steady-state gain of the openloop process
τ_o = time constant of the openloop process
K_c = controller gain

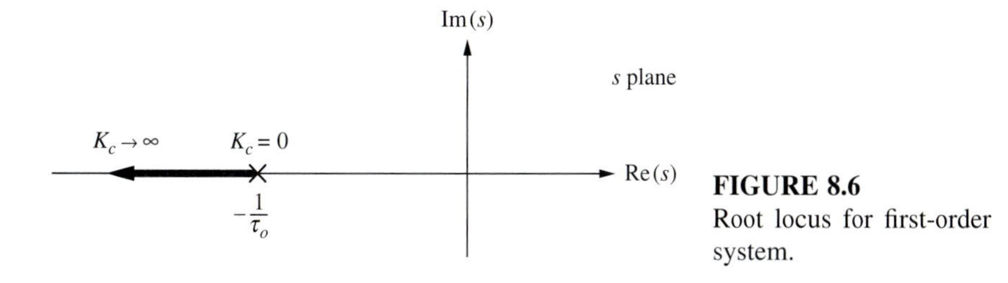

FIGURE 8.6
Root locus for first-order system.

The closedloop characteristic equation is

$$\begin{aligned} 1 + G_{M(s)}G_{C(s)} &= 0 \\ 1 + \frac{K_p K_c}{\tau_o s + 1} &= 0 \\ \tau_o s + 1 + K_p K_c &= 0 \end{aligned} \tag{8.30}$$

Solving for the closedloop root gives

$$s = -\frac{1 + K_p K_c}{\tau_o} \tag{8.31}$$

There is one root and there is only one curve in the s plane. Figure 8.6 gives the root locus plot. The curve starts at $s = -1/\tau_o$ when $K_c = 0$. The closedloop root moves out along the negative real axis as K_c is increased.

For a first-order system, the closedloop root is always real, so the system can never be underdamped or oscillatory. The closedloop damping coefficient of this system is always greater than 1. The larger the value of controller gain, the smaller is the closedloop time constant because the root moves farther away from the origin (remember, the time constant is the reciprocal of the distance from the root to the origin). If we wanted a closedloop time constant of $\frac{1}{10}\tau_o$ (i.e., the closedloop system is 10 times faster than the openloop system), we would set K_c equal to $9/K_p$. Equation (8.31) shows that at this value of gain the closedloop root is equal to $-10/\tau_o$.

This first-order system can never be closedloop unstable because the root always lies in the LHP. No real system is only first order. There are always small lags in the process, in the control valve, or in the instrumentation that make all real systems of order higher than first. ■

EXAMPLE 8.5. Now let's move up to a second-order system with a proportional controller.

$$G_{M(s)} = \frac{1}{(s+1)(5s+1)} \tag{8.32}$$

The closedloop characteristic equation is

$$1 + G_{M(s)}G_{C(s)} = 0 = 1 + \frac{1}{(s+1)(5s+1)}K_c$$

$$5s^2 + 6s + 1 + K_c = 0$$

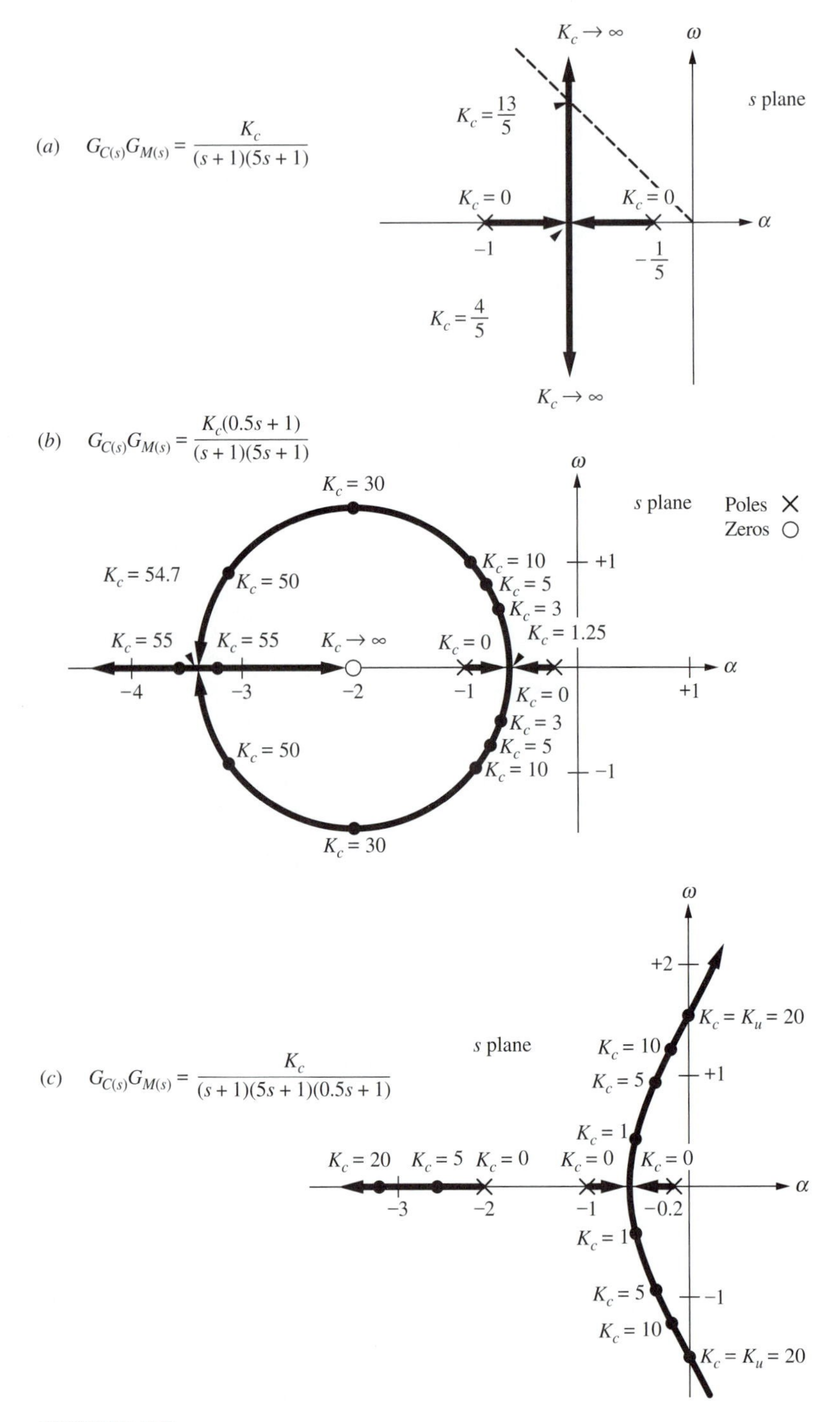

FIGURE 8.7
Root locus curves.

The quadratic formula gives the two closedloop roots:

$$s = \frac{-6 \pm \sqrt{(6)^2 - (4)(5)(1 + K_c)}}{(2)(5)}$$
$$s = -\tfrac{3}{5} \pm \tfrac{1}{5}\sqrt{4 - 5K_c} \tag{8.33}$$

The locations of these roots for various values of K_c are shown in Fig. 8.7*a*.

When K_c is zero, the closedloop roots are at $s = -\frac{1}{5}$ and $s = -1$. Notice that these values of s are the poles of the openloop transfer function. The root locus plot always starts at the poles of the openloop transfer function.

For K_c between zero and $\frac{4}{5}$, the two roots are real and lie on the negative real axis. The closedloop system is critically damped (the closedloop damping coefficient is 1) at $K_c = \frac{4}{5}$ since the roots are equal. For values of gain greater than $\frac{4}{5}$, the roots will be complex.

$$s = -\tfrac{3}{5} \pm i\tfrac{1}{5}\sqrt{5K_c - 4} \tag{8.34}$$

As the gain goes to infinity, the real parts of both roots are constant at $-\frac{3}{5}$ and the imaginary parts go to plus and minus infinity. Thus, the system becomes increasingly underdamped. The closedloop damping coefficient goes to zero as the gain becomes infinite.

However, this second-order system never becomes closedloop unstable since the roots are always in the LHP.

Suppose we wanted to design this system for a closedloop damping coefficient of 0.707. Equation (8.28) tells us that

$$\phi = \arccos 0.707 = 45^\circ$$

Therefore, we must find the value of gain on the root locus plot where it intersects a 45° line from the origin. At the point of intersection the real and imaginary parts of the roots must be equal. This occurs when $K_c = \frac{13}{5}$. The closedloop time constant τ_c of the system at this value of gain can be calculated from the reciprocal of the radial distance from the origin.

$$\tau_c = \frac{1}{\sqrt{(\frac{3}{5})^2 + (\frac{3}{5})^2}} = \frac{5}{3\sqrt{2}} \tag{8.35}$$

■

EXAMPLE 8.6. Let us change the system transfer function from the preceding example by adding a lead or a zero.

$$G_{M(s)}G_{C(s)} = \frac{K_c(\frac{1}{2}s + 1)}{(s + 1)(5s + 1)}$$

The closedloop characteristic equation becomes

$$1 + G_{M(s)}G_{C(s)} = 1 + \frac{K_c(\frac{1}{2}s + 1)}{(s + 1)(5s + 1)}$$
$$5s^2 + \left(6 + \frac{K_c}{2}\right)s + K_c + 1 = 0 \tag{8.36}$$

The roots are

$$s = -\left(\frac{3}{5} + \frac{K_c}{20}\right) \pm \frac{1}{10}\sqrt{\frac{K_c^2}{4} - 14K_c + 16} \tag{8.37}$$

For low values of K_c the term inside the square root will be positive, since the $+16$ will dominate; the two closedloop roots are real and distinct. For very large values of gain, the K_c^2 term will dominate and the roots will again be real. For intermediate values of K_c the term inside the square root will be negative and the roots will be complex.

The range of K_c values that give complex roots can be found from the roots of

$$\frac{K_c^2}{4} - 14K_c + 16 = 0 \tag{8.38}$$

$$K_{c1} = 28 - 12\sqrt{5}$$
$$K_{c2} = 28 + 12\sqrt{5} \tag{8.39}$$

where K_{c1} = smaller value of K_c where the square-root term is zero
K_{c2} = larger value of K_c where the square-root term is zero

The root locus plot is shown in Fig. 8.7*b*.

Note that the effect of adding a zero or a lead is to pull the root locus toward a more stable region of the s plane. The root locus starts at the poles of the openloop transfer function. As the gain goes to infinity, the two paths of the root locus go to minus infinity and to the zero of the transfer function at $s = -2$. We will find that this is true in general: the root locus plot ends at the zeros of the openloop transfer function.

The system is closedloop stable for all values of gain. The fastest-responding system would be obtained with $K_c = K_{c2}$, where the two roots are equal and real. ■

EXAMPLE 8.7. Now let us add a pole or a lag, instead of a zero, to the system of Example 8.5. The system is now third order.

$$G_{M(s)}G_{C(s)} = \frac{K_c}{(s+1)(5s+1)(\frac{1}{2}s+1)} \tag{8.40}$$

The closedloop characteristic equation becomes

$$1 + G_{M(s)}G_{C(s)} = 1 + \frac{K_c}{(s+1)(5s+1)(\frac{1}{2}s+1)}$$
$$\tfrac{5}{2}s^3 + 8s^2 + \tfrac{13}{2}s + 1 + K_c = 0 \tag{8.41}$$

We discuss how to solve for the roots of this cubic equation in the next section. The root locus curves are sketched in Fig. 8.7*c*. There are three curves because there are three roots. The root locus plot starts at the three openloop poles of the transfer function: -1, -2, and $-\frac{1}{5}$.

The effect of adding a lag or a pole is to pull the root locus plot toward the unstable region. The two curves that start at $s = -\frac{1}{5}$ and $s = -1$ become complex conjugates and curve off into the RHP. Therefore, this third-order system is closedloop unstable if K_c is greater than $K_u = 20$. This was the same result that we obtained in Example 8.3. ■

The preceding examples have illustrated a very important point: The higher the order of the system, the worse the dynamic response of the closedloop system. The first-order system is never underdamped and cannot be made closedloop unstable for any value of gain. The second-order system becomes underdamped as gain is increased but never goes unstable. Third-order (and higher) systems can be made closedloop unstable. If you remember, these are exactly the results we found in our simulation experiments in Chapter 1. Now we have shown mathematically why these various processes behave the way they do.

One of the basic limitations of root locus techniques is that deadtime cannot be handled conveniently. The *first-order Pade* approximation of deadtime is frequently used, but it is often not very accurate.

$$e^{-Ds} \simeq \frac{1 - (\frac{1}{2}D)s}{1 + (\frac{1}{2}D)s} \tag{8.42}$$

8.4.2 Construction of Root Locus Curves

Root locus plots are easy to generate for first- and second-order systems since the roots can be found analytically as explicit functions of controller gain. For higher-order systems things become more difficult. Both numerical and graphical methods are available. Root-solving subroutines can be easily used on any computer to do the job. The easiest way is to utilize some user-friendly software tools. We illustrate the use of MATLAB for making root locus plots.

A. Rules for root locus plots

There are several rules that enable engineers to quickly check the plots generated by the computer.

1. The root loci start ($K_c = 0$) at the poles of the system openloop transfer function $G_{M(s)}G_{C(s)}$.
2. The root loci end ($K_c = \infty$) at the zeros of $G_{M(s)}G_{C(s)}$.
3. The number of loci is equal to the order of the system, i.e., the number of poles of $G_{M(s)}G_{C(s)}$.
4. The complex parts of the curves always appear as complex conjugates.
5. The angle of the asymptotes of the loci (as $s \rightarrow \infty$) is equal to

$$\pm\frac{180^\circ}{N - M}$$

where N = number of poles of $G_{M(s)}G_{C(s)}$
M = number of zeros of $G_{M(s)}G_{C(s)}$

Rules 1 to 4 are fairly self-evident. Rule 5 comes from the fact that at a point on the root locus plot the complex number s must satisfy the equation

$$1 + G_{M(s)}G_{C(s)} = 0$$

$$G_{M(s)}G_{C(s)} = -1 + i0 \tag{8.43}$$

Therefore, the argument of $G_{M(s)}G_{C(s)}$ on a root locus must always be

$$\arg G_{M(s)}G_{C(s)} = \arctan \tfrac{0}{-1} = \pm\pi \tag{8.44}$$

Now $G_{M(s)}G_{C(s)}$ is a ratio of polynomials, Mth order in the numerator and Nth order in the denominator.

$$G_{M(s)}G_{C(s)} = \frac{b_M s^M + b_{M-1}s^{M-1} + \cdots + b_1 s + b_0}{a_N s^N + a_{N-1}s^{N-1} + \cdots + a_1 s + a_0}$$

On the asymptotes, s gets very big, so only the s^N and s^M terms remain significant.

$$\lim_{s\to\infty}[G_{M(s)}G_{C(s)}] = \frac{b_M s^M}{a_N s^N} = \frac{b_M/a_N}{s^{N-M}} \tag{8.45}$$

Putting s into polar form ($s = re^{i\theta}$) gives

$$\lim_{s\to\infty}[G_{M(s)}G_{C(s)}] = \frac{b_M/a_N}{r^{N-M}e^{i\theta(N-M)}}$$

The angle or argument of $G_{M(s)}G_{C(s)}$ is

$$\lim_{s\to\infty}[\arg G_{M(s)}G_{C(s)}] = -(N-M)\theta$$

Equation (8.44) must still be satisfied on the asymptote, and therefore, Q.E.D.,

$$(N-M)\theta = \pm\pi$$

Applying Rule 5 to a first-order process ($N = 1$ and $M = 0$) gives asymptotes that go off at 180° (see Example 8.4). Applying it to a second-order process ($N = 2$ and $M = 0$) gives asymptotes that go off at 90° (see Example 8.5). Example 8.6 has a second-order denominator ($N = 2$), but it also has a first-order numerator ($M = 1$). So this system has a "net order" ($N - M$) of 1, and the asymptotes go off at 180°. Example 8.7 shows that the asymptotes go off at 60° since the order of the system is third.*

B. Use of MATLAB software

The commercial software MATLAB makes it easy to generate root locus plots. The Control Toolbox contains programs that aid in this analysis. We illustrate in the following example the use of some simple MATLAB programs to generate root locus plots. Similar programs will be used in Chapter 11 to compute frequency response results.

EXAMPLE 8.8. Our three-heated-tank process considered in Chapter 1 is an interesting one to explore via root locus since it is third order. We use different types of feedback controllers and different settings and see how the root loci change.

Process. The process is described in Section 1.2.2. The temperature in the third tank T_3 is controlled by a feedback controller by manipulating the heat input Q_1 to the first tank.

*Rule 5 must be modified slightly for higher-order systems. See de Moivre's theorem.

The disturbance is a drop in inlet feed temperature T_0 from 90°F to 70°F at time $= 0$ hours. Proportional and proportional-integral controllers are used. The ODEs describing the system are

$$V_1 c_p \rho \frac{dT_1}{dt} = F c_p \rho (T_0 - T_1) + Q_1 \tag{8.46}$$

$$V_2 c_p \rho \frac{dT_2}{dt} = F c_p \rho (T_1 - T_2) \tag{8.47}$$

$$V_3 c_p \rho \frac{dT_3}{dt} = F c_p \rho (T_2 - T_3) \tag{8.48}$$

The numerical values of parameters are $V_n = 100$ ft^3, $c_p = 0.75$ Btu/lb °F, $\rho =$ 50 lb/ft^3, and $F = 1000$ ft^3/hr. The span of the temperature transmitter is 200°F. The control valve has linear installed characteristics and when wide open passes steam at a rate corresponding to a heat transfer rate of 10×10^6 Btu/hr.

Openloop process transfer function. These three ODEs are linear, so we do not have to linearize. Converting to perturbation variables, Laplace transforming, and solving for the transfer function between the controlled variable T_3 and the manipulated variable Q_1 give

$$G_{M(s)} = \frac{2.667 \times 10^{-5} \text{ °F/Btu/hr}}{(0.1s + 1)^3} \tag{8.49}$$

The transfer functions for the control valve and the temperature transmitter are

$$G_{V(s)} = \frac{10 \times 10^6 \text{ Btu/hr}}{16 \text{ mA}} \tag{8.50}$$

$$G_{T(s)} = \frac{16 \text{ mA}}{200\text{°F}} \tag{8.51}$$

The closedloop characteristic equation is

$$1 + G_{M(s)} G_{V(s)} G_{T(s)} G_{C(s)} = 0 \tag{8.52}$$

$$1 + \frac{1.333 \text{ mA/mA}}{(0.1s + 1)^3} G_{C(s)} = 0 \tag{8.53}$$

When a proportional controller is used,

$$G_{C(s)} = K_c$$

When a proportional-integral controller is used,

$$G_{C(s)} = K_c \frac{\tau_I s + 1}{\tau_I s} = K_c \frac{s + 1/\tau_I}{s}$$

This adds a lead (a zero at $s = -1/\tau_I$) and an integrator (a pole at $s = 0$) to the system.

Ultimate gain and frequency. Using a proportional controller, the closedloop characteristic equation is a third-order polynomial.

$$1 + \frac{1.333 K_c}{(0.1s + 1)^3} = 0 \tag{8.54}$$

TABLE 8.1

```
% Program "tempmat.m" uses Matlab to make root locus plots
%     for PI controllers with Ziegler-Nichols and Tyreus-Luyben settings
%
% Form transfer function of openloop process
num=1.333;
den=conv([0.1 1],[0.1 1]);
den=conv(den, [0.1 1]);
% Given ultimate gain and frequency
kult=6.;
wult=17.32;
%     (Ultimate period is in "hours"; ultimate frequency is in radians/hour)
pult=2*3.1416/wult;
% Calculate ZN and TL controller settings (k's and ti's)
kzn=kult/2.2;
tizn=pult/1.2;
ktl=kult/3.2;
titl=pult*2.2;
% Form transfer functions for total openloop (process times controller)
% Ziegler-Nichols
nzn=conv(num,[tizn 1]);
dzn=conv(den,[tizn 0]);
% Tyreus-Luyben
ntl=conv(num,[titl 1]);
dtl=conv(den,[titl 0]);
% Define range of controller gains (500 values on log scale from 0.001 to 100)
kc=logspace(-3,2,500);
%********************************
% P control
% Solve for roots of closedloop characteristic equation at all values of kc
% "sp" are the real and imaginary parts
[sp,kc]=rlocus(num,den,kc);
plot(sp,'o')
axis('square')
axis([-20 2 -1 21]);
title('3 Heated Tanks; P control')
xlabel ('Real(s)')
ylabel ('Imaginary(s)')
% Draw lines for a 0.3 damping coefficient
%   and time constants for 0.08, 0.1 and 0.12 hours
wn=[1/0.08 1/0.1 1/0.12];
sgrid(0.3,wn)
grid
text(-15,19,['Ku=',num2str(kult)])
text(-15,18,['wu=',num2str(wult)])
text(-19,16, 'Damping Coefficient Line at 0.3')
text(-19,15, 'Time Constant Lines at 0.08, 0.1 and 0.12 hours')
pause
% print -dps pfig108.ps
```

TABLE 8.1 (CONTINUED)

```
% PI control: ZN reset time
[szn,kc]=rlocus(nzn,dzn,kc);
plot(szn, 'o')
axis('square')
axis([-20 2 -1 21]);
title('3 Heated Tanks; PI control with ZN ')
xlabel('Real(s)')
ylabel('Imaginary(s)')
grid
sgrid(0.3,wn)
text(-15,19,['Kc=' ,num2str(kzn)])
text(-15,18,['Reset=' ,num2str(tizn)])
text(-19,16, 'Damping Coefficient Line at 0.3')
text(-19,15, 'Time Constant Lines at 0.08, 0.1 and 0.12 hours')
pause
% print -dps -append pfig108
% PI control: TL reset time
[stl,kc]=rlocus(ntl,dtl,kc);
plot(stl,'o')
axis('square')
axis([-20 2 -1 21]);
grid
sgrid(0.3,wn)
text(-15,19,['Kc=' ,num2str(ktl)])
text(-15,18,['Reset=' ,num2str(titl)])
text(-19,16, 'Damping Coefficient Line at 0.3')
text(-19,15, 'Time Constant Lines at 0.08, 0.1 and 0.12 hours')
title('3 Heated Tanks; PI control with TL ')
xlabel('Real(s)')
ylabel('Imaginary(s)')
pause
print -dps -append pfig108
```

Using the direct substitution method, we let $s = i\omega$ in Eq. (8.54) and solve the resulting two equations for the unknowns, K_u and ω_u.

$$K_u = 6 \qquad \omega_u = 17.3 \text{ rad/hr}$$

MATLAB program. Table 8.1 gives a MATLAB program for generating root locus plots for three controllers: a proportional controller, a PI controller using the Ziegler-Nichols value for reset, and a PI controller using the Tyreus-Luyben value for reset.

The key MATLAB functions used in the program are briefly described here. More details can be found by using the *help* feature.

1. *num* and *den* are defined as the numerator and denominator polynomials of the transfer function. They are given as row vectors of the numerical coefficients multiplying the powers of s in decreasing order. For example, consider a denominator polynomial

$$(3s + 1)(s + 1) = 3s^2 + 4s + 1 \tag{8.55}$$

We can define *den* = [3 4 1].

2. *conv* is used to multiply two polynomials together (from *convolution*). We could calculate the denominator polynomial given in Eq. (8.55) in the following way:

$$\text{den} = \text{conv}([3 \quad 1], [1 \quad 1]) \tag{8.56}$$

3. *axis* commands define the type of axis we want. For root locus plots we usually like the abscissa and ordinate scales to be the same, so we use the *axis('square')* command.
4. *sgrid* draws lines of constant damping coefficient (radial lines from the origin) and lines of constant time constant (circles around the origin).

Figure 8.8 gives the root locus plot for a proportional controller. The three curves start at -10 on the real axis (only one complex root is shown). Two of the loci go off at 60° angles and cross the imaginary axis at 17.3 (the ultimate frequency) when the gain is 6 (the ultimate gain). For a closedloop damping coefficient of 0.3 (the radial line), the closed-loop time constant is about 0.085 hours.

Figure 8.9 gives the root locus plot for a PI controller with a reset time $\tau_I = 0.302$ hours (the Ziegler-Nichols value). Now there are four loci. One starts at the origin

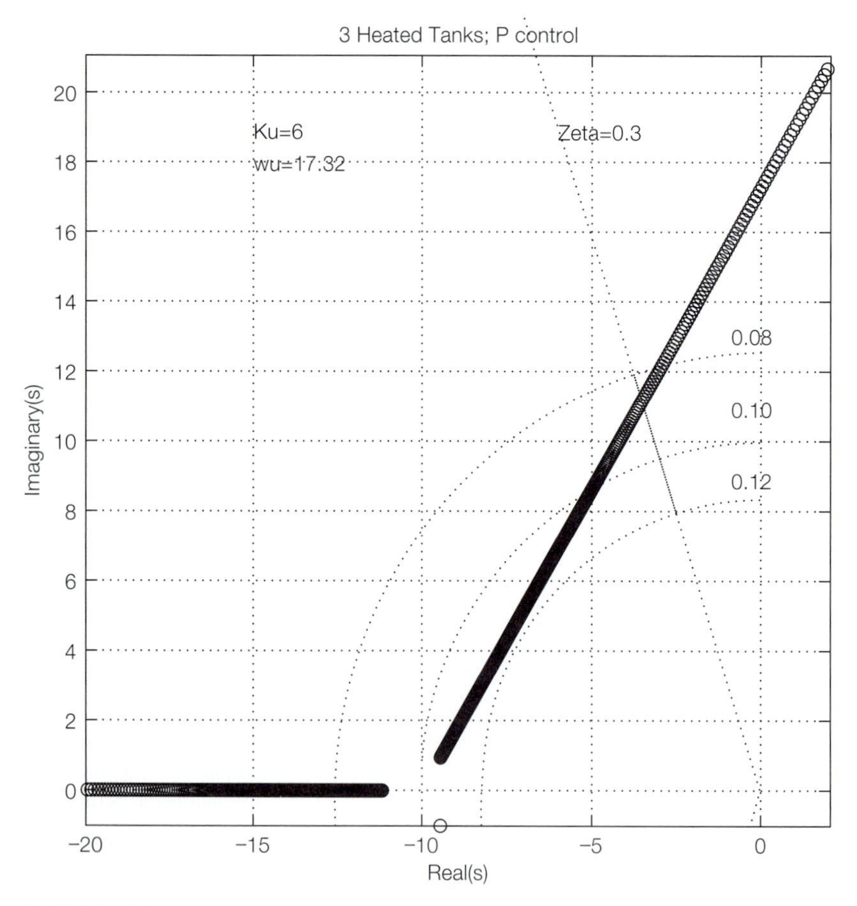

FIGURE 8.8

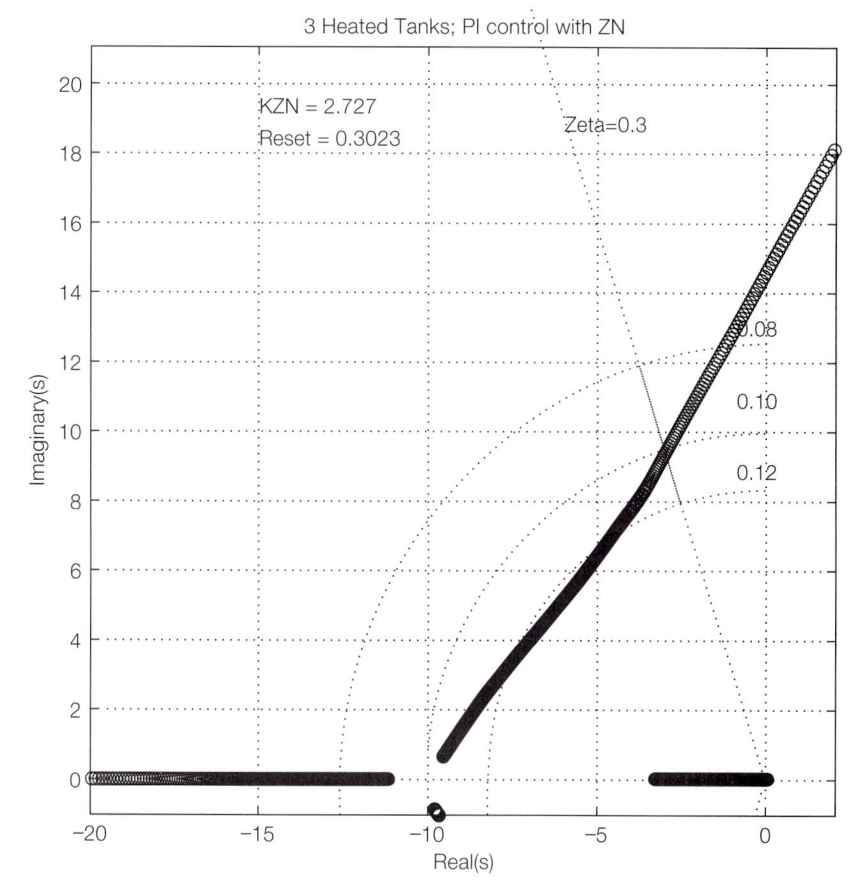

FIGURE 8.9

(because of the integrator giving a pole at $s = 0$) and goes to the zero at $s = -1/\tau_I$. For a closedloop damping coefficient of 0.3, the closedloop time constant is about 0.1 hours, which is slower than the P controller for the same damping coefficient.

Figure 8.10 gives the root locus plot for a PI controller with reset time $\tau_I = 0.798$ hours (the Tyreus-Luyben value). The zero moves closer to the origin.

The time-domain transient responses for this process with the three controllers are given in Fig. 3.16. ■

8.5 CONCLUSION

Plotting the roots of the closedloop characteristic equation as functions of the controller gain is an illuminating way to understand and visualize what happens to the dynamics of a closedloop process as we change the controller tuning constants. These

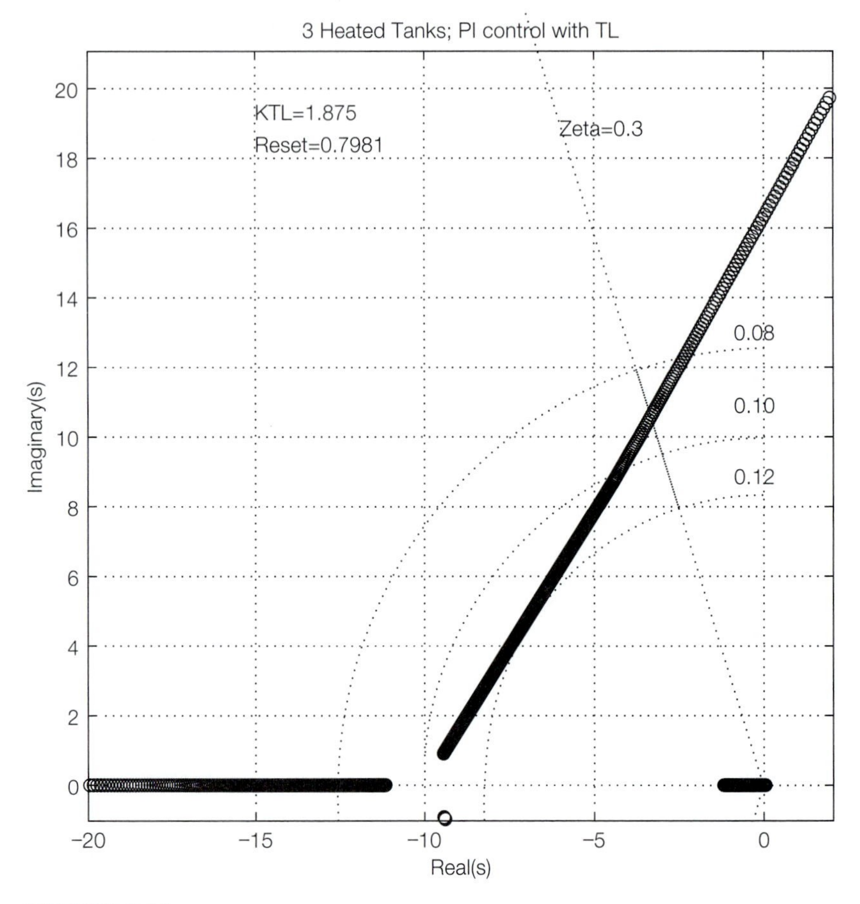

FIGURE 8.10

root locus plots provide valuable insights into how different types of processes, different types of controllers, and different values of controller tuning parameters affect closedloop performance. They provide us with a quantitative method for controller tuning.

PROBLEMS

8.1. Find the ultimate gain and period of the fourth-order system given below. The controller is proportional and the system openloop transfer function is

$$G_{M(s)} = \frac{(0.047)(112)(2)(0.12)}{(0.083s + 1)(0.017s + 1)(0.432s + 1)(0.024s + 1)}$$

8.2. Find the ultimate gain of the closedloop three-CSTR system with a PI controller:
(a) For $\tau_I = 3.03$.
(b) For $\tau_I = 4.5$.

8.3. Find the ultimate gain and period of a closedloop system with a proportional controller and openloop transfer function

$$G_{M(s)} = \frac{1}{(s+1)(5s+1)(\frac{1}{2}s+1)}$$

8.4. Find the value of feedback controller gain that gives a closedloop damping coefficient of 0.8 for the system with a proportional controller and openloop transfer function

$$G_{M(s)} = \frac{s+4}{s(s+2)}$$

8.5. The liquid level $h_{(t)}$ in a tank is held by a PI controller that changes the flow rate $F_{(t)}$ out of the tank. The flow rate into the tank $F_{0(t)}$ and the level setpoint $h^{set}_{(t)}$ are disturbances. The vertical cylindrical tank is 10 ft^2 in cross-sectional area. The transfer function of the feedback controller plus the control valve is

$$G_{C(s)} = \frac{F_{(s)}}{E_{(s)}} = -K_c\left(1 + \frac{1}{\tau_I s}\right) \quad \text{with units of} \quad \frac{ft^3/min}{ft}$$

(*a*) Write the equations describing the openloop system.
(*b*) Write the equations describing the closedloop system.
(*c*) Derive the openloop transfer functions of the system:

$$G_{M(s)} = \frac{H_{(s)}}{F_{(s)}} \quad \text{and} \quad G_{L(s)} = \frac{H_{(s)}}{F_{0(s)}}$$

(*d*) Derive the two closedloop transfer functions of the system:

$$\frac{H_{(s)}}{H^{set}_{(s)}} \quad \text{and} \quad \frac{H_{(s)}}{F_{0(s)}}$$

(*e*) Make a root locus plot of the closedloop system with a value of integral time $\tau_I = 10$ minutes.
(*f*) What value of gain K_c gives a closedloop system with a damping coefficient of 0.707? What is the closedloop time constant at this gain?
(*g*) What gain gives critical damping? What is the time constant with this gain?

8.6. Make a root locus plot of a system with openloop transfer function

$$G_{C(s)}G_{M(s)} = \frac{K_c}{(s+1)(5s+1)(\frac{1}{2}s+1)} \frac{\tau_D s + 1}{(\frac{1}{20}\tau_D s + 1)}$$

(*a*) For $\tau_D = 2.5$.
(*b*) For $\tau_D = 5$.
(*c*) For $\tau_D = 7.5$.

8.7. Find the ultimate gain and period of the closedloop three-CSTR system with a PID controller tuned at $\tau_I = \tau_D = 1$. Make a root locus plot of the system.

8.8. Make a root locus plot of a system with openloop transfer function

$$G_{C(s)}G_{M(s)} = \frac{K_c e^{-2s}}{s+1}$$

Use the first-order Pade approximation of a deadtime. Find the ultimate gain.

8.9. Make a root locus plot for a process with the openloop transfer function

$$G_{M(s)} = \frac{3(-3s+1)}{(s+1)(5s+1)}$$

when a proportional feedback controller is used.

8.10. A two-tank system with recycle is sketched below. Liquid levels are held by proportional controllers: $F_1 = K_1 h_1$ and $F_2 = K_2 h_2$. Flow into the system F_0 and recycle flow F_R can be varied by the operator.

(*a*) Derive the four closedloop transfer functions relating the two levels and the two load disturbances:

$$\left(\frac{H_1}{F_0}\right)_{(s)} \quad \left(\frac{H_1}{F_R}\right)_{(s)} \quad \left(\frac{H_2}{F_0}\right)_{(s)} \quad \left(\frac{H_2}{F_R}\right)_{(s)}$$

(*b*) Does the steady-state level in the second tank vary with the recycle flow rate F_R? Use the final-value theorem of Laplace transforms.

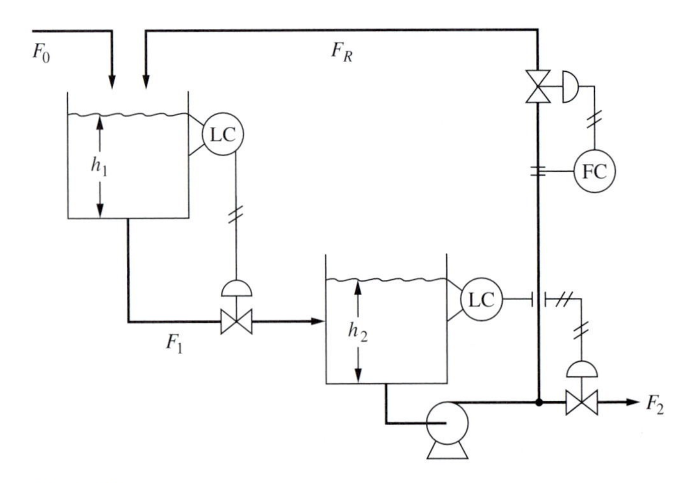

FIGURE P8.10

8.11. The system of Problem 8.3 is modified by using the cascade control system sketched below.

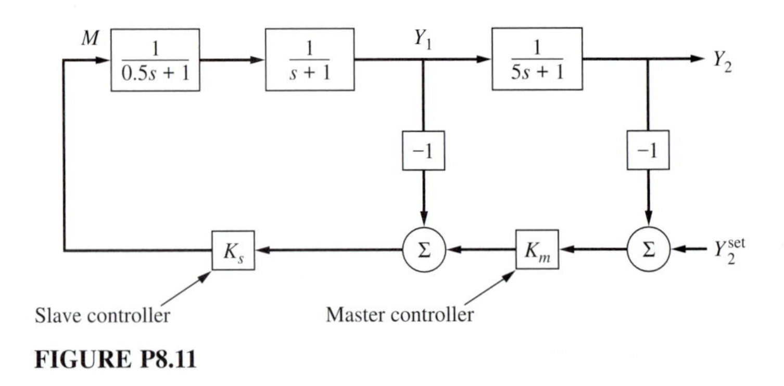

FIGURE P8.11

(*a*) Find the value of gain K_s in the proportional controller that gives a 0.707 damping coefficient for the closedloop slave loop.

(*b*) Using this value of K_s in the slave loop, find the maximum closedloop-stable value of the master controller gain K_M. Compare this with the ultimate gain found without cascade control in Problem 8.3. Also compare ultimate periods.

8.12. Repeat Problem 8.5 using a proportional feedback controller [parts (*b*) and (*d*)]. Will there be a steady-state error in the closedloop system for (*a*) a step change in setpoint h^{set} or (*b*) a step change in feed rate F_0?

8.13. We would like to compare the closedloop dynamic performance of two types of reboilers.

(*a*) In the first type, there is a control valve on the steam line to the reboiler and a steam trap on the condensate line leaving the reboiler. The flow transmitter acts like a first-order lag with a 6-second time constant. The control valve also acts like a 6-second lag. These are the only dynamic elements in the steam flow control loop. If a PI flow control is used with $\tau_I = 0.1$ minutes, calculate the closedloop time constant of the steam flow loop when a closedloop damping coefficient of 0.3 is used.

(*b*) In the second type of reboiler, the control valve is on the condensate line leaving the reboiler. There is no valve in the steam line, so the tubes in the reboiler see full steam-header pressure.

Changes in steam flow are achieved by increasing or decreasing the area used for condensing steam in the reboiler. This variable-area flooded reboiler is used in some processes because it permits the use of lower-pressure steam. However, as you will show in your calculations (we hope), the dynamic performance of this configuration is distinctly poorer than direct manipulation of steam flow.

The steam flow meter still acts like a first-order lag with a 6-second time constant, but the smaller control valve on the liquid condensate can be assumed to be instantaneous.

The condensing temperature of the steam is 300°F. The process into which the heat is transferred is at a constant temperature of 200°F. The overall heat transfer coefficient is 300 Btu/hr °F ft^2. The reboiler has 509 tubes that are 10 feet long with 1 inch inside

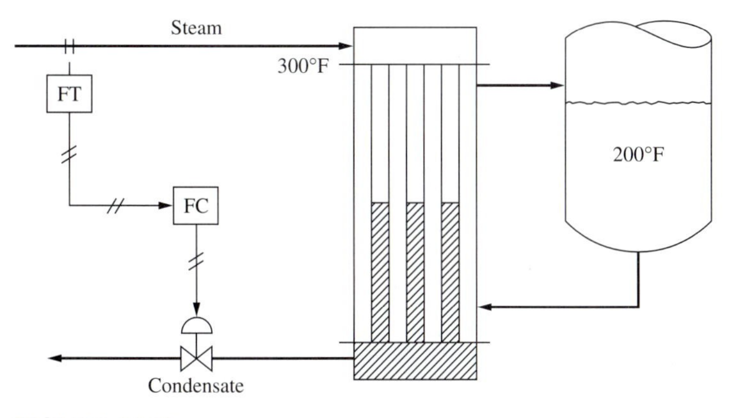

FIGURE P8.13

diameter. The steam and condensate are inside the tubes. The density of the condensate is 62.4 lb_m/ft^3, and the latent heat of condensation of the steam is 900 Btu/lb_m. Neglect any sensible heat transfer.

Derive a dynamic mathematical model of the flooded-condenser system. Calculate the transfer function relating steam flow rate to condensate flow rate. Using a PI controller with $\tau_I = 0.1$ minute, calculate the closedloop time constant of the steam flow control loop when a closedloop damping coefficient of 0.3 is used. Compare this with the result found in (*a*).

8.14. A chemical reactor is cooled by both jacket cooling water and condenser cooling water. A mathematical model of the system has yielded the following openloop transfer functions (time is in minutes):

$$\frac{T}{F_c} = \frac{-1}{s+1} \quad (°\text{F/gpm})$$

$$\frac{T}{F_J} = \frac{-5}{10s+1} \quad (°\text{F/gpm})$$

The range of the temperature transmitter is 100–200°F. Control valves have linear trim and constant pressure drop, and are half open under normal conditions. Normal condenser flow is 30 gpm. Normal jacket flow is 20 gpm. A temperature measurement lag of 12 seconds is introduced into the system by the thermowell.

If a proportional feedback temperature controller is used, calculate the controller gain K_c that yields a closedloop damping coefficient of 0.707, and calculate the closedloop time constant of the system when:

(*a*) Jacket water only is used.

(*b*) Condenser water only is used.

Derive the closedloop characteristic equation for the system when both jacket and condenser water are used.

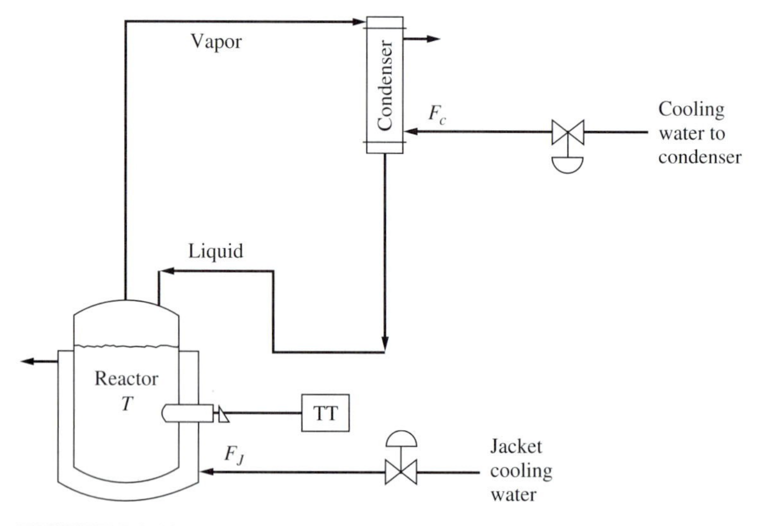

FIGURE P8.14

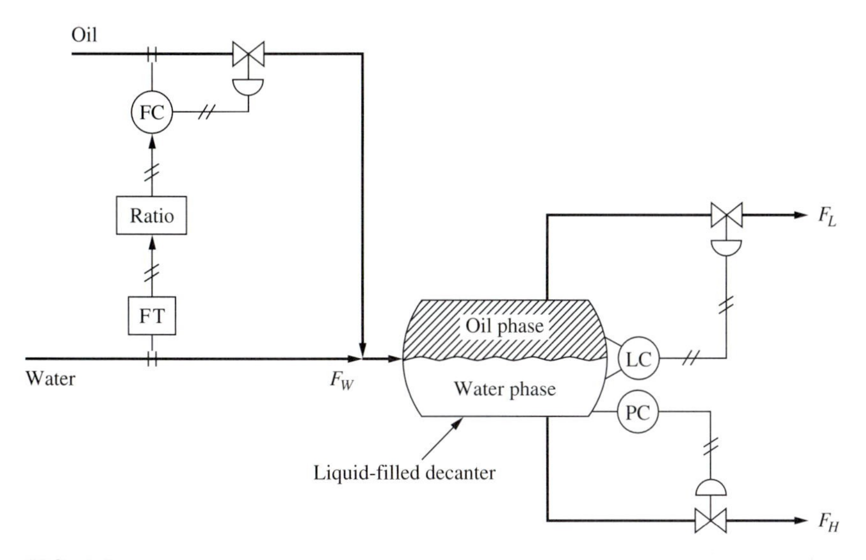

FIGURE P8.15

8.15. Oil and water are mixed together and then decanted. Oil flow rate is ratioed to water flow rate F_W. Interface is controlled by oil flow F_L from the decanter with a proportional level controller. Water flow (F_H) from the decanter, which is liquid full, is on pressure control (PI). Steady-state flow rates are

Oil	177.8 gpm
Water	448.2 gpm

Water and oil holdups in the decanter are each 1130 gallons at steady state.

(*a*) Derive the closedloop transfer function between F_H and F_W in terms of level controller gain K_c.

(b) Determine the transient response of F_H for a step change in F_W of 1 gpm when level controller gain is 6.36 gpm/gal.

8.16. The openloop transfer function relating steam flow rate to temperature in a feed preheater has been found to consist of a steady-state gain K_p and a first-order lag with time constant τ_o. The lag associated with temperature measurement is τ_m. A proportional-only temperature controller is used.

(*a*) Derive an expression for the roots of the closedloop characteristic equation in terms of the parameters τ_o, K_p, τ_m, and K_c.

(*b*) Solve for the value of controller gain that will give a critically damped closedloop system when $K_p = 1$, $\tau_o = 10$, and

(*i*) $\tau_m = 1$

(*ii*) $\tau_m = 5$

8.17. The liquid flow rate from a vertical cylindrical tank, 10 feet in diameter, is flow controlled. The liquid flow into the tank is manipulated to control liquid level in the tank. The control valve on the inflow stream has linear installed characteristics and can pass 1000 gpm when wide open. The level transmitter has a span of 6 feet of liquid. A proportional controller is used with a gain of 2. Liquid density is constant.

(*a*) Should the control valve be AO or AC?
(*b*) Should the controller be reverse or direct acting?
(*c*) What is the dimensionless openloop system transfer function relating liquid height and inflow rate?
(*d*) Solve for the time response of the inflow rate to a step change in the outflow rate from 500 to 750 gpm with the tank initially half full.

8.18. A process has a positive pole located at $(+1, 0)$ in the s plane (with time in minutes). The process steady-state gain is 2. An additional lag of 20 seconds exists in the control loop. Sketch root locus plots and calculate controller gains that give a closedloop damping coefficient of 0.707 when:
(*a*) A proportional feedback controller is used.
(*b*) A proportional-derivative feedback controller is used with the derivative time set equal to the lag in the control loop. The ratio of numerator to denominator time constants in the derivative unit is 6.

8.19. A process has an openloop transfer function that is a first-order lag with a time constant τ_o and a steady-state gain K_p. If a PI feedback controller is used with a reset time τ_I, sketch root locus plots for the following cases:
(*a*) $\tau_o < \tau_I$
(*b*) $\tau_o = \tau_I$
(*c*) $\tau_o > \tau_I$
What value of the τ_I/τ_o ratio gives a closedloop system that has a damping coefficient of $\frac{1}{2}\sqrt{2}$ for only one value of controller gain?

8.20. The openloop process transfer functions relating the manipulated and load variables (M and L) to the controlled variable (Y) are first-order lags with identical time constants (τ_o) but with different gains (K_M and K_L). Derive equations for the closedloop steady-state error and the closedloop time constant for step disturbances in load if a proportional feedback controller is used.

8.21. The liquid level in a tank is controlled by manipulating the flow out of the tank, using a PI controller. The outflow rate is a function of only the valve position. The valve has linear installed characteristics and passes 20 ft^3/min when wide open.

The tank is vertical and cylindrical with a cross-sectional area of 25 ft^2 and a 2-ft level transmitter span.
(*a*) Derive the relationship between the feedback controller gain K_c and the reset time τ_I that gives a critically damped closedloop system.
(*b*) For a critically damped system with $\tau_I = 5$ minutes, calculate the closedloop time constant.

8.22. A process has an openloop transfer function G_M relating controlled and manipulated variables that is a first-order lag τ_o and steady-state gain K_p. There is an additional first-order lag τ_m in the measurement of the controlled variable. A proportional-only feedback controller is used.

Derive an expression relating the controller gain K_c to the parameters τ_o, τ_m, and K_p such that the closedloop system damping coefficient is 0.707. What happens to K_c as τ_m gets very small or very large? What is the value of τ_m that provides the smallest value of K_c?

8.23. A fixed-gain relay is to be used as a low base-level override controller on a distillation column. The column is 7 ft in diameter and has a base-level transmitter span of 3 ft. The density of the liquid in the base is 50 lb_m/ft^3. Its heat of vaporization is 200 Btu/lb_m.

The reboiler steam valve has linear installed characteristics and passes 30,000 lb_m/hr when wide open. Steam latent heat is 1000 Btu/lb_m.

There is a first-order dynamic lag of τ minutes between a change in the signal to the steam valve and vapor boilup. The low base-level override controller pinches the reboiler steam valve over the lower 25 percent of the level transmitter span.

Solve for the value of gain that should be used in the relay as a function of τ to give a closedloop damping coefficient of 0.5 for the override level loop.

8.24. A process has openloop transfer functions

$$G_M = \frac{K_M}{(\tau_o s + 1)^2} \qquad G_L = \frac{K_L}{(\tau_o s + 1)^2}$$

If a PI controller is used with τ_I set equal to τ_o, calculate:

(*a*) The value of controller gain that gives a closedloop damping coefficient of 0.707.

(*b*) The closedloop time constant, using this value of gain.

(*c*) The closedloop transfer function between the load variable and the output variable.

(*d*) The steady-state error for a step change in the load variable.

8.25. Two tanks are connected by a pipe through which liquid can flow in either direction, depending on the difference in liquid levels.

$$F_c = K_B(h_1 - h_2)$$

where K_B is a constant with units of ft^2/min. Disturbances are the flow rates F_{01}, F_{02}, and F_2. The manipulated variable is the flow rate F_1. Cross-sectional areas of the vertical cylindrical tanks are A_1 and A_2.

(*a*) Derive the openloop transfer function between the height of liquid in tank 1 (h_1) and F_1.

(*b*) What are the poles and zeros of the openloop transfer function? What is the openloop characteristic equation?

(*c*) If a proportional-only level controller is used, derive the closedloop characteristic equation and sketch a root locus plot for the case where $A_1 = A_2$.

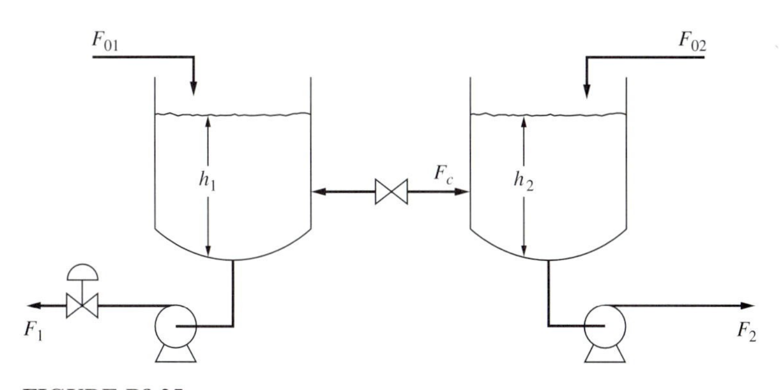

FIGURE P8.25

8.26. A heat exchanger has the openloop transfer function relating the controlled variable temperature T to the manipulated variable steam flow rate S

$$G_{M(s)} = \frac{5\ (°\text{F}/1000\ \text{lb}_\text{m}/\text{hr})}{(5s + 1)(s + 1)^2}$$

The span of the temperature transmitter is 50°F. The steam control valve has linear installed characteristics and passes 30,000 lb_m/hr of steam when wide open. A proportional-only temperature controller is used.

(*a*) What are the poles and zeros of the openloop system? Is it openloop stable?

(*b*) What is the minimum value of proportional band that gives a stable closedloop system?

(*c*) Using a controller gain of 2, what is the closedloop servo transfer function relating the temperature transmitter output PV to the setpoint SP?

(*d*) The openloop transfer function relating the controlled variable T to the load variable feed flow rate F is

$$G_{L(s)} = \frac{2(2s + 1)\ (°\text{F}/100\ \text{gpm})}{(5s + 1)(s + 1)^2}$$

Using a controller gain of 2, what is the closedloop regulatory transfer function relating temperature T to feed flow F?

8.27. A process has the following openloop transfer function relating the controlled variable Y and the manipulated variable M.

$$\frac{Y_{(s)}}{M_{(s)}} = \frac{1}{(s + 1)^2}$$

(*a*) What are the poles, zeros, and steady-state gain of this transfer function? Is the process openloop stable or unstable?

(*b*) If a proportional controller is used, what is the closedloop characteristic equation?

(*c*) Derive a relationship between the roots of the closedloop characteristic equation and the controller gain K_c.

(*d*) Derive equations that show how the closedloop time constant τ_{CL} and the closedloop damping coefficient ζ_{CL} vary with controller gain.

8.28. For the process considered in Problem 8.27, the openloop transfer function relating the controlled variable and the load variable L is

$$\frac{Y_{(s)}}{L_{(s)}} = \frac{0.5}{(s + 1)^2}$$

(*a*) If a proportional controller is used with a gain of 10.11, derive the closedloop transfer function relating the controlled variable Y and the load variable L.

(*b*) What is the steady-state gain of this closedloop transfer function?

8.29. A two-stage evaporator is used to concentrate a brine solution of NaCl in water. Assume that the NaCl is completely nonvolatile. Steam is fed into the reboiler in stage 1 at a rate V_0 (kg/hr). If we neglect sensible heat effects, this amount of steam will produce about the same amount of vapor in stage 1 (V_1, kg/hr), and condensing V_1 in a reboiler in stage 2 will produce about the same amount of vapor in stage 2 (V_2, kg/hr). Thus, for a simplified model we will assume that $V_0 = V_1 = V_2 = V$. Since NaCl is nonvolatile, these vapors are pure water.

Feed is introduced into stage 2 at a rate F (kg/hr) and concentration z (wt fraction NaCl). The concentrations of brine in the two stages are x_1 and x_2 (wt fraction NaCl).

The brine liquid (L_2) from stage 2 is pumped into stage 1. The liquid from stage 1 is the concentrated product (L_1). The liquid holdups in the two stages are W_1 and W_2 (kg) and are assumed constant: W_1 is controlled by L_2, and W_2 is held by F. Production rate is established by flow-controlling L_1. The composition of brine in stage 1 is controlled by manipulating steam flow rate V_0.

Vapor holdup is negligible. Assume that both evaporators are perfectly mixed. Assume also that the dynamics of the reboilers and the condenser are instantaneous. Feed flow rate F is constant.

(*a*) Derive the nonlinear ODE dynamic mathematical model of this simplified process. (*Hint:* Your model should consist of two ODEs.)
(*b*) Convert the model to two linear ODEs.
(*c*) Using the following steady-state values of parameters, show that the linear model is

$$\frac{dx_1}{dt} = -1.5x_1 + 3.25x_2 + (0.513 \times 10^{-3})V$$

$$\frac{dx_2}{dt} = -3.25x_2 + 5z + (0.154 \times 10^{-3})V$$

$$F = 10{,}000 \text{ kg/hr}$$
$$\bar{z} = 0.20$$
$$\bar{V} = 3500 \text{ kg/hr}$$
$$W_1 = W_2 = 2000 \text{ kg}$$

(*d*) Starting with the equations given above for the linear model, derive the openloop transfer function that relates x_1 to V.
(*e*) If a composition transmitter is used with a span of 0.5 wt fraction and a control valve on the steam can pass 10,000 kg/hr when wide open, draw a root locus plot for the system if a proportional controller is used.

8.30. A process has the following openloop transfer function relating the controlled variable Y and the manipulated variable M.

$$\frac{Y_{(s)}}{M_{(s)}} = \frac{0.2e^{-7.4s}}{s}$$

The deadtime function e^{-Ds} can be approximated by a first-order Pade approximation.

$$e^{-Ds} \simeq \frac{1 - (D/2)s}{1 + (D/2)s}$$

(*a*) Using this approximation and assuming a proportional controller, what is the closedloop characteristic equation?
(*b*) Sketch a root locus plot.
(*c*) Determine the ultimate gain and ultimate frequency.
(*d*) If the controller gain is set equal to 0.675, what are the closedloop damping coefficient and closedloop time constant?

8.31. A process has an openloop transfer function $G_{M(s)}$ relating controlled and manipulated variables that has a steady-state gain of 3 and two identical first-order lags in series with 5-minute time constants.

(*a*) Using a proportional controller, derive the closedloop characteristic equation. Sketch a root locus plot. Solve for the value of controller gain that gives a closedloop damping coefficient of 0.3. What is the closedloop time constant using this controller gain?

(*b*) Using a proportional-integral controller with reset time set equal to 5 minutes, repeat all the parts of (*a*).

8.32. A satellite tracking antenna has an angle θ (radians) with the horizontal axis that must change as the satellite travels across the sky. To accomplish this, a feedback controller $G_{C(s)}$ is used. The controller's setpoint is the desired angle θ^{set}, and its output is the drive motor torque on the antenna T_C. Wind also exerts a torque on the antenna T_D. The equation of motion of the antenna is

$$J\ddot{\theta} + D\dot{\theta} = T_C + T_D$$

where $\dot{\theta}$ = time derivative of θ
$\ddot{\theta}$ = time derivative of $\dot{\theta}$
J = moment of inertia of the antenna
D = damping due to back emf of the DC motor driving the antenna

(*a*) Derive the openloop block diagram of the antenna. The inputs are T_C and T_D. The output is θ.

(*b*) The feedback controller has the transfer function

$$G_{C(s)} = \frac{T_{C(s)}}{E_{(s)}} = \frac{10s + 1}{s + 1}$$

where $E = \theta^{set} - \theta$. The numerical values are $D = 0.5$ and $J = 5$. What are the closedloop damping coefficient and the closedloop time constant of the system?

8.33. The temperature of water in a tank is controlled by adding a stream of hot water into the tank. A stream of cold water is also added to the tank. The transfer function between tank temperature T and hot-water flow rate F_H is

$$G_{M(s)} = \frac{T_{(s)}}{F_{H(s)}} = \frac{84.2}{0.98s + 1} \quad (°\text{F/gpm})$$

Time is in minutes. The temperature transmitter has a range of 50–200°F and has a dynamic response that can be approximated by a 0.5-min first-order lag. The hot-water control valve has linear installed characteristics and passes 4 gpm when wide open. Its dynamic response is a 10-sec first-order lag.

Calculate the ultimate gain and ultimate frequency of a proportional temperature controller.

8.34. The openloop transfer function $G_{M(s)}$ of a sterilizer relating the controlled variable temperature $T_{(s)}$ and the manipulated variable steam flow rate $F_{S(s)}$ is a gain $K_p = 2$ (with units of mA/mA when transmitter and valve gains are included), a first-order lag with time constant $\tau_o = 1$ minute, and an integrator in series.

(*a*) Calculate the gain K_c of a proportional feedback controller that gives a closedloop damping coefficient equal to 0.707.

(*b*) Calculate the closedloop time constant when this value of gain is used.

(*c*) Sketch a root locus plot for this system when a P-only controller is used.

(*d*) Sketch several root locus plots for this process when a PI controller is used for several values of reset τ_I, starting with very large values and then reducing reset toward $\tau_I = 1$ minute.

(*e*) Calculate the gain K_c of a proportional-derivative (PD) feedback controller that gives a closedloop damping coefficient equal to 0.707 when used to control this process.

$$G_{C(s)} = K_c \frac{s+1}{\frac{1}{5}s+1}$$

(*f*) Compare the closedloop time constant obtained in part (*e*) using a PD controller with that determined in part (*b*) using a P-only controller.

8.35. A process has an openloop transfer function $G_{M(s)}$ relating the controlled variable $Y_{(s)}$ to the manipulated variable $M_{(s)}$ that consists of a steady-state gain of -2 mol%/gpm and two first-order lags in series with time constants of 2 and 10 minutes. A composition transmitter with a gain of 0.1 mA/mol% is used. The control valve gain is 25 gpm/mA.

(*a*) What are the poles and zeros of the openloop process? Is it openloop stable?

(*b*) Solve for the openloop response of the process to a unit step change in the manipulated variable; i.e., solve analytically for $y_{(t)}$.

(*c*) If a proportional feedback controller is used with a value of $K_c = 3.8$, derive the closedloop transfer function relating the output signal from the composition transmitter (PV) to the controller setpoint signal (SP).

(*d*) What are the poles and zeros of the closedloop system? Is it closedloop stable? Is it overdamped or underdamped?

(*e*) Solve for the closedloop response of the PV signal for a unit step change in the SP signal.

8.36. The openloop transfer function between the controlled variable (tray 4 temperature) and the manipulated variable (steam flow rate) is

$$G_{M(s)} = \frac{T_{4(s)}}{F_{S(s)}} = \frac{3 \times 10^{-3}}{(10s+1)(2s+1)} \qquad (^\circ\text{F/lb steam/hr})$$

Time constants are in minutes. The steam control valve has linear installed characteristics and passes 50,000 lb/hr when wide open. The span of the temperature transmitter is 200°F. The temperature sensor has a dynamic first-order lag of 30 seconds.

(*a*) Calculate the ultimate gain and ultimate frequency if a proportional controller is used. Sketch a root locus plot.

(*b*) Calculate the ultimate gain and ultimate frequency if a PI controller is used with a reset time of 1 minute. Sketch a root locus plot.

8.37. A control system is needed to maintain the desired "roll attitude" of a space station. The openloop transfer function relating controlled variable θ (angle of rotation from the horizontal, in degrees) to the manipulated variable M (fuel flow to the thruster jets on the outside of the space station shell, in kg/sec) is a gain (5°/kg/sec) in series with two integrators. This transfer function is derived from an angular force balance on the space station. A roll-attitude sensor/transmitter has a span of 90°. The maximum flow rate through the control valve to the thrusters is 2 kg/sec.

(*a*) Can a proportional controller yield a closedloop-stable system? Explain your answer using a root locus plot.

(*b*) Sketch a root locus plot to show how a PD controller can be used to give a closedloop stable system.

$$G_{C(s)} = K_c \frac{\tau_D s + 1}{0.1\tau_D s + 1}$$

(*c*) If a τ_D of 10 seconds is used, calculate the value of controller gain K_c that gives an effective closedloop damping coefficient of 0.3. [Hint: For an underdamped system, the roots of a cubic characteristic equation must have one real root ($s_1 = \alpha_1$) and two complex-conjugate roots ($s_2 = \alpha_2 + i\omega_2$ and $s_3 = \alpha_2 - i\omega_2$). This means that the characteristic equation is

$$(s - \alpha_1)(s - \alpha_2 + i\omega_2)(s - \alpha_2 - i\omega_2) = 0$$

Thus, there are three unknowns: α_1, α_2, and ω_2. So you need three equations.]

8.38. A process has the following openloop transfer function relating controlled and manipulated variables.

$$G_{M(s)} = \frac{x_{P(s)}}{R_{(s)}} = \frac{(-5.339 \times 10^{-4})(s + 1)}{(0.3838s + 1)(1.121s + 1)} \quad \text{(mole fraction } A\text{/mol/min)}$$

The maximum flow rate of recycle R through a control valve is 200 mol/min, and the control valve has linear installed characteristics. The composition transmitter measuring x_P has a span of 0.20 mole fraction component A and has a dynamic first-order lag of 1 minute.

(*a*) What is the closedloop characteristic equation of the system if a proportional controller is used?

(*b*) Find the value of controller gain K_c that gives a closedloop damping coefficient of 0.3.

(*c*) What is the closedloop time constant?

CHAPTER 9

Laplace-Domain Analysis of Advanced Control Systems

In the last chapter we used Laplace-domain techniques to study the dynamics and stability of simple closedloop control systems. In this chapter we apply these same methods to more complex systems: cascade control, feedforward control, openloop-unstable processes, and processes with inverse response. We also discuss an alternative way to look at controller design that is called "model-based" control.

The tools used in this chapter are those developed in Chapters 7 and 8. We use transfer functions to design feedforward controllers or to develop the characteristic equation of the system and to find the location of its roots in the s plane.

9.1 CASCADE CONTROL

Cascade control was discussed qualitatively in Section 4.2. It employs two control loops; the secondary (or "slave") loop receives its setpoint from the primary (or "master") loop. Cascade control is used to improve load rejection and performance by decreasing closedloop time constants.

We can apply cascade control to two types of process structures. If the manipulated variable affects one variable, which in turn affects a second controlled variable, the structure leads to *series* cascade control. If the manipulated variable affects both variables directly, the structure leads to *parallel* cascade.

9.1.1 Series Cascade

Figure 9.1*a* shows an openloop process in which two transfer functions G_1 and G_2 are connected in series. The manipulated variable M enters G_1 and produces a change in Y_1. The Y_1 variable then enters G_2 and changes Y_2.

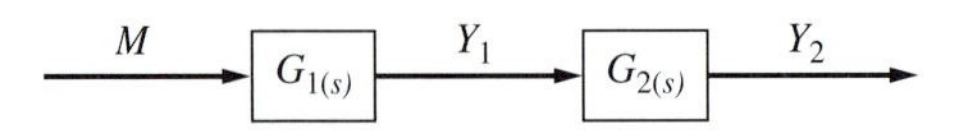

(*a*) Openloop process

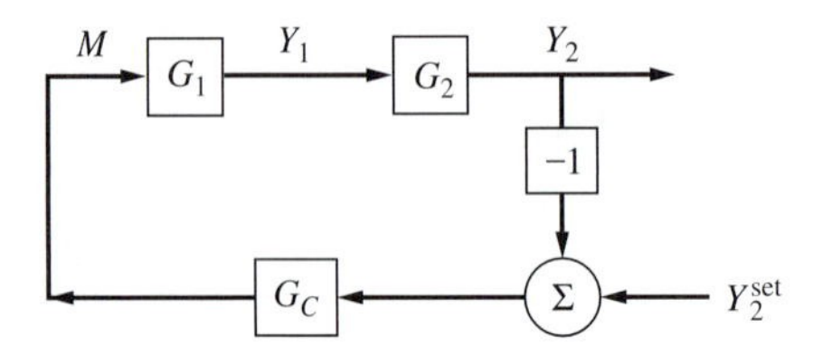

(*b*) Conventional feedback control

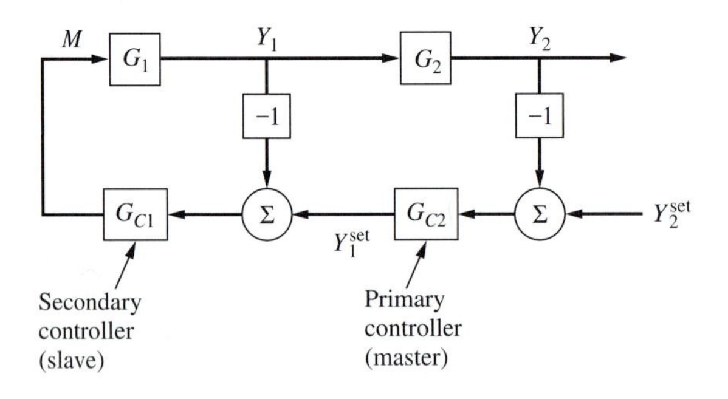

(*c*) Series cascade

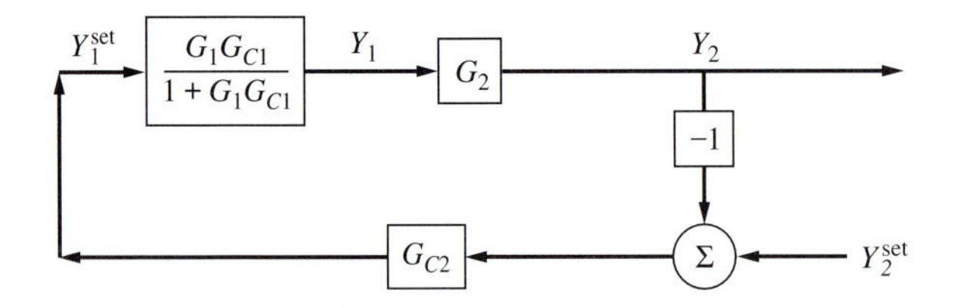

(*d*) Reduced block diagram

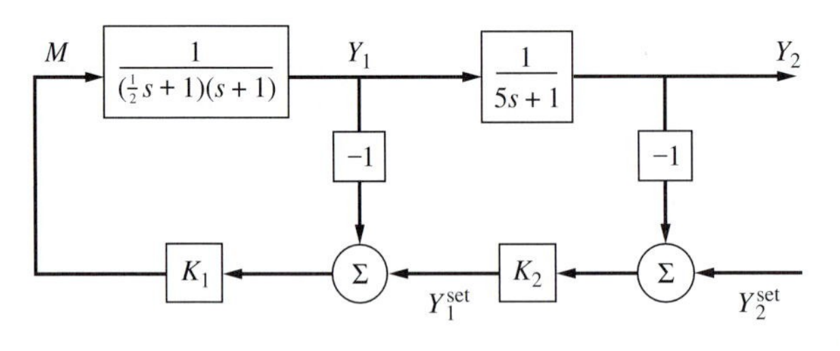

(*e*) Example 9.1

FIGURE 9.1
Series cascade. (*a*) Openloop process. (*b*) Conventional feedback control. (*c*) Series cascade. (*d*) Reduced block diagram. (*e*) Example 9.1.

Figure 9.1*b* shows the conventional feedback control system, where a single controller senses the controlled variable Y_2 and changes the manipulated variable M. The closedloop characteristic equation for this system was developed in Chapter 8.

$$1 + G_{1(s)} G_{2(s)} G_{C(s)} = 0 \tag{9.1}$$

Figure 9.1*c* shows a series cascade system. There are now two controllers. The secondary controller G_{C1} adjusts M to control the secondary variable Y_1. The setpoint signal Y_1^{set} to the G_{C1} controller comes from the primary controller; i.e., the output of the primary controller G_{C2} is the setpoint for the G_{C1} controller. The G_{C2} controller setpoint is Y_2^{set}.

The closedloop characteristic equation for this system is *not* the same as that given in Eq. (9.1). To derive it, let us first look at the secondary loop by itself. From the analysis presented in Chapter 8, the equation that describes this closedloop system is

$$Y_1 = \frac{G_1 G_{C1}}{1 + G_1 G_{C1}} Y_1^{set} \tag{9.2}$$

So to design the secondary controller G_{C1} we use the closedloop characteristic equation

$$1 + G_1 G_{C1} = 0 \tag{9.3}$$

Next we look at the controlled output variable Y_2. Figure 9.1*d* shows the reduced block diagram of the system in the conventional form. We can deduce the closedloop characteristic equation of this system by inspection.

$$1 + G_2 G_{C2} \left(\frac{G_1 G_{C1}}{1 + G_1 G_{C1}} \right) = 0 \tag{9.4}$$

However, let us derive it rigorously.

$$Y_2 = G_2 Y_1 \tag{9.5}$$

Substituting for Y_1 from Eq. (9.2) gives

$$Y_2 = G_2 \frac{G_1 G_{C1}}{1 + G_1 G_{C1}} Y_1^{set} \tag{9.6}$$

But Y_1^{set} is the output from the G_{C2} controller.

$$Y_1^{set} = G_{C2}(Y_2^{set} - Y_2) \tag{9.7}$$

Combining Eqs. (9.6) and (9.7) gives

$$Y_2 = G_2 \left(\frac{G_1 G_{C1}}{1 + G_1 G_{C1}} \right) G_{C2}(Y_2^{set} - Y_2)$$

$$Y_2 \left[1 + G_2 G_{C2} \left(\frac{G_1 G_{C1}}{1 + G_1 G_{C1}} \right) \right] = G_2 G_{C2} \left(\frac{G_1 G_{C1}}{1 + G_1 G_{C1}} \right) Y_2^{set}$$

Rearranging gives

$$Y_2 = \frac{G_2 G_{C2}\left(\dfrac{G_1 G_{C1}}{1 + G_1 G_{C1}}\right)}{\left[1 + G_2 G_{C2}\left(\dfrac{G_1 G_{C1}}{1 + G_1 G_{C1}}\right)\right]} Y_2^{set} \tag{9.8}$$

So Eq. (9.4) gives the closedloop characteristic equation of this series cascade system. A little additional rearrangement leads to a completely equivalent form:

$$Y_2 = \frac{G_1 G_2 G_{C1} G_{C1}}{1 + G_1 G_{C1}(1 + G_2 G_{C2})} Y_2^{set} \tag{9.9}$$

An alternative and equivalent closedloop characteristic equation is

$$1 + G_1 G_{C1}(1 + G_2 G_{C2}) = 0 \tag{9.10}$$

The roots of this equation dictate the dynamics of the series cascade system. Note that both of the openloop transfer functions are involved as well as both of the controllers. Equation (9.4) is a little more convenient to use than Eq. (9.10) because we can make conventional root locus plots, varying the gain of the G_{C2} controller, after the parameters of the G_{C1} controller have been specified.

The tuning procedure for a cascade control system is to tune the secondary controller first and then tune the primary controller with the secondary controller on automatic. As for the types of controller used, we often use a proportional controller in the secondary loop. Since it has only one tuning parameter, it is easy to tune. There is no need for integral action in the secondary controller because we don't care if there is offset in this loop. If we use a PI primary controller, the offset in the primary loop will be eliminated, which is our control objective.

EXAMPLE 9.1. Consider the process with a series cascade control system sketched in Fig. 9.1*e*. A typical example is a secondary loop in which the flow rate of condensate from a flooded reboiler is the manipulated variable M, the secondary variable is the flow rate of steam to the reboiler, and the primary variable is the temperature in a distillation column. We assume that the secondary controller G_{C1} and the primary controller G_{C2} are both proportional only.

$$G_{C1} = K_1 \qquad G_{C2} = K_2$$

In this example

$$G_1 = \frac{1}{(\frac{1}{2}s + 1)(s + 1)} \qquad G_2 = \frac{1}{5s + 1}$$

Conventional control. First we look at a conventional single proportional controller (K_c) that manipulates M to control Y_2^{set}. The closedloop characteristic equation is

$$1 + \frac{1}{(\frac{1}{2}s + 1)(s + 1)(5s + 1)} K_c = 0 \tag{9.11}$$

$$\tfrac{5}{2}s^3 + 8s^2 + \tfrac{13}{2}s + 1 + K_c = 0 \tag{9.12}$$

To solve for the ultimate gain and ultimate frequency, we substitute $i\omega$ for s.

$$
\begin{aligned}
-i\tfrac{5}{2}\omega^3 - 8\omega^2 + i\tfrac{13}{2}\omega + 1 + K_c &= 0 \\
(-8\omega^2 + 1 + K_c) + i(\tfrac{13}{2}\omega - \tfrac{5}{2}\omega^3) &= 0 + i0
\end{aligned}
\tag{9.13}
$$

Solving the two equations simultaneously for the two unknowns gives

$$K_u = \tfrac{99}{5} \qquad \text{and} \qquad \omega_u = \sqrt{\tfrac{13}{5}}$$

Designing the secondary (slave) loop. We pick a closedloop damping coefficient specification for the secondary loop of 0.707 and calculate the required value of K_1. The closedloop characteristic equation for the slave loop is

$$1 + K_1 \frac{1}{(\frac{1}{2}s + 1)(s + 1)} = 0 = \tfrac{1}{2}s^2 + \tfrac{3}{2}s + 1 + K_1 \tag{9.14}$$

Solving for the closedloop roots gives

$$s = -\tfrac{3}{2} \pm i\tfrac{1}{2}\sqrt{8K_1 - 1} \tag{9.15}$$

To have a damping coefficient of 0.707, the roots must lie on a radial line whose angle with the real axis is arccos(0.707) = 45°. See Fig. 9.2*a*. On this line the real and imaginary parts of the roots are equal. So for a closedloop damping coefficient of 0.707

$$\tfrac{3}{2} = \tfrac{1}{2}\sqrt{8K_1 - 1} \quad \Rightarrow \quad K_1 = \tfrac{5}{4} \tag{9.16}$$

Now the closedloop relationship between Y_1 and Y_1^{set} is

$$Y_1 = \frac{G_1 G_{C1}}{1 + G_1 G_{C1}} Y_1^{set} = \frac{\dfrac{1}{(\frac{1}{2}s + 1)(s + 1)}\left(\dfrac{5}{4}\right)}{1 + \dfrac{1}{(\frac{1}{2}s + 1)(s + 1)}\left(\dfrac{5}{4}\right)} Y_1^{set} \tag{9.17}$$

$$Y_1 = \frac{\frac{5}{2}}{s^2 + 3s + \frac{9}{2}} Y_1^{set} \tag{9.18}$$

Designing the primary (master) loop. The closedloop characteristic equation for the master loop is

$$1 + G_2 G_{C2}\left(\frac{G_1 G_{C1}}{1 + G_1 G_{C1}}\right) = 1 + \left(\frac{K_2}{5s + 1}\right)\left(\frac{\frac{5}{2}}{s^2 + 3s + \frac{9}{2}}\right) = 0 \tag{9.19}$$

$$5s^3 + 16s^2 + \tfrac{51}{2}s + \tfrac{9}{2} + \tfrac{5}{2}K_2 = 0 \tag{9.20}$$

Solving for the ultimate gain K_u and ultimate frequency ω_u by substituting $i\omega$ for s gives

$$K_u = 30.8 \qquad \omega_u = \sqrt{5.1} = 2.26$$

It is useful to compare these values with those found for a single conventional control loop, $K_u = 19.8$ and $\omega_u = 1.61$. We can see that cascade control results in higher controller gain and a smaller closedloop time constant (the reciprocal of the frequency). Therefore, the system will show faster response with cascade control than with a single loop. Figure 9.2*b* gives a root locus plot for the primary controller with the secondary controller gain set at $\frac{5}{4}$. Two of the loci start at the complex poles $s = -\frac{3}{2} \pm i\frac{3}{2}$ that come from the closedloop secondary loop. The other curve starts at the pole $s = -\frac{1}{5}$. ■

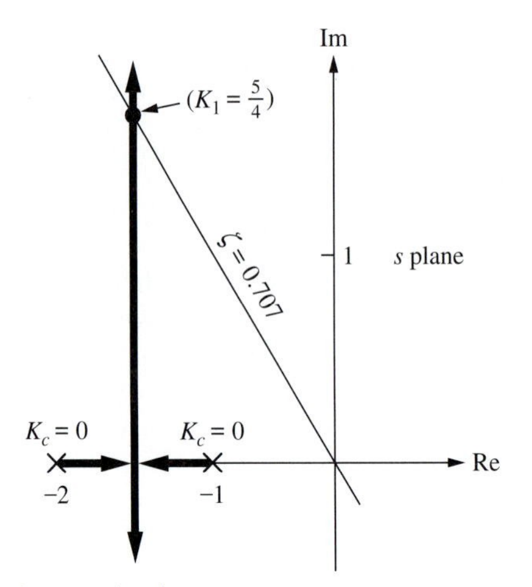

(a) Root locus for secondary loop

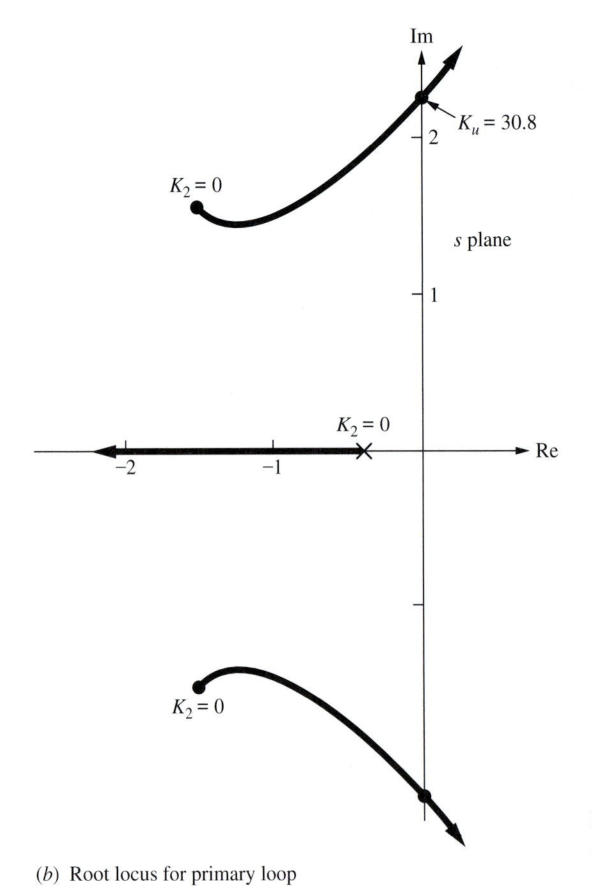

(b) Root locus for primary loop

FIGURE 9.2
(a) Root locus for secondary loop.
(b) Root locus for primary loop.

9.1.2 Parallel Cascade

Figure 9.3*a* shows a process where the manipulated variable affects the two controlled variables Y_1 and Y_2 in parallel. An important example is in distillation column control where reflux flow affects both distillate composition and a tray temperature. The process has a parallel structure, and this leads to a parallel cascade control system.

If only a single controller G_{C2} is used to control Y_2 by manipulating M, the closedloop characteristic equation is the conventional

$$1 + G_{2(s)}G_{C2(s)} = 0 \tag{9.21}$$

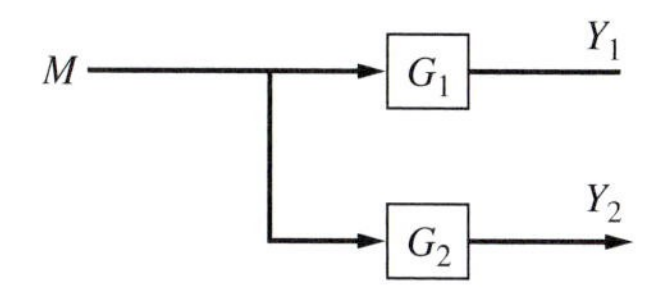

(*a*) Openloop process

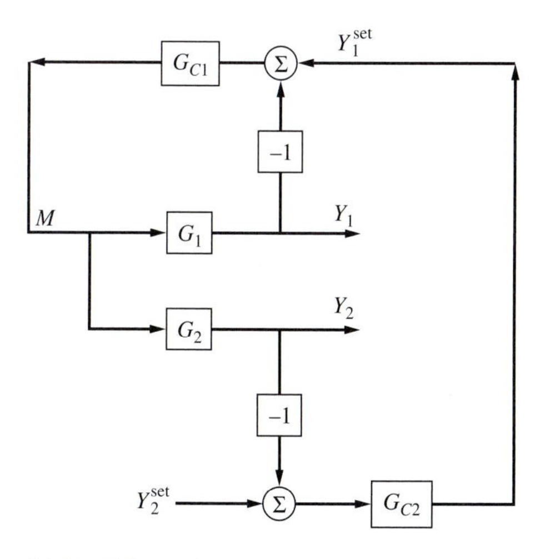

(*b*) Parallel cascade process

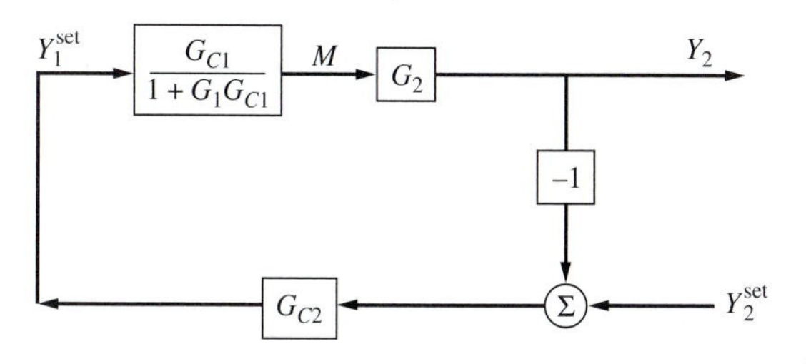

(*c*) Reduced block diagram

FIGURE 9.3
Parallel cascade. (*a*) Openloop process. (*b*) Parallel cascade control. (*c*) Reduced block diagram.

If, however, a cascade control system is used, as sketched in Fig. 9.3*b*, the closedloop characteristic equation is not that given in Eq. (9.21). To derive it, let us start with the secondary loop.

$$Y_1 = G_1 M = G_1 G_{C1}(Y_1^{set} - Y_1) \tag{9.22}$$

$$Y_1 = \frac{G_1 G_{C1}}{1 + G_1 G_{C1}} Y_1^{set} \tag{9.23}$$

Combining Eqs. (9.22) and (9.23) gives the closedloop relationship between M and Y_1^{set}.

$$M = \frac{1}{G_1} Y_1 = \frac{1}{G_1} \frac{G_1 G_{C1}}{1 + G_1 G_{C1}} Y_1^{set} = \frac{G_{C1}}{1 + G_1 G_{C1}} Y_1^{set} \tag{9.24}$$

Now we solve for the closedloop transfer function for the primary loop with the secondary loop on automatic. Figure 9.3*c* shows the simplified block diagram. By inspection we can see that the closedloop characteristic equation is

$$1 + G_2 G_{C2}\left(\frac{G_{C1}}{1 + G_1 G_{C1}}\right) = 0 \tag{9.25}$$

Note the difference between the series cascade [Eq. (9.4)] and the parallel cascade [Eq. (9.25)] characteristic equations.

9.2 FEEDFORWARD CONTROL

Most of the control systems we have discussed, simulated, and designed thus far in this book have been feedback control devices. A deviation of an output variable from a setpoint is detected. This error signal is fed into a feedback controller, which changes the manipulated variable. The controller makes no use of any information about the source, magnitude, or direction of the disturbance that has caused the output variable to change.

The basic notion of feedforward control is to detect disturbances as they enter the process and make adjustments in manipulated variables so that output variables are held constant. We do not wait until the disturbance has worked its way through the process and has upset everything to produce an error signal. If a disturbance can be detected as it enters the process, it makes sense to take immediate action to compensate for its effect on the process.

Feedforward control systems have gained wide acceptance in chemical engineering in the past three decades. They have demonstrated their ability to improve control, sometimes quite spectacularly. The dynamic responses of processes that have poor dynamics from a feedback control standpoint (high-order systems or systems with large deadtimes or inverse response) can often be greatly improved by using feedforward control. Distillation columns are one of the most common applications of feedforward control. We illustrate this improvement in this section by comparing the responses of systems using feedforward control with systems using conventional feedback control when load disturbances occur.

Feedforward control is probably used more in chemical engineering systems than in any other field of engineering. Our systems are often slow-moving, nonlinear, and multivariable, and contain appreciable deadtime. All these characteristics make life miserable for feedback controllers. Feedforward controllers can handle all these with relative ease as long as the disturbances can be measured and the dynamics of the process are known.

9.2.1 Linear Feedforward Control

A block diagram of a simple openloop process is sketched in Fig. 9.4*a*. The load disturbance $L_{(s)}$ and the manipulated variable $M_{(s)}$ affect the controlled variable $Y_{(s)}$. A conventional feedback control system is shown in Fig. 9.4*b*. The error signal $E_{(s)}$ is fed into a feedback controller $G_{C(s)}$ that changes the manipulated variable $M_{(s)}$.

Figure 9.4*c* shows the feedforward control system. The load disturbance $L_{(s)}$ still enters the process through the $G_{L(s)}$ process transfer function. The load disturbance is also fed into a feedforward control device that has a transfer function $G_{F(s)}$. The feedforward controller detects changes in the load $L_{(s)}$ and adjusts the manipulated variable $M_{(s)}$.

Thus, the transfer function of a feedforward controller is a relationship between a manipulated variable and a disturbance variable (usually a load change).

$$G_{F(s)} = \left(\frac{M}{L}\right)_{(s)} = \left(\frac{\text{manipulated variable}}{\text{disturbance}}\right)_{Y \text{ constant}} \tag{9.26}$$

To design a feedforward controller, that is, to find $G_{F(s)}$, we must know both $G_{L(s)}$ and $G_{M(s)}$. The objective of most feedforward controllers is to hold the controlled variable constant at its steady-state value. Therefore, the change or perturbation in $Y_{(s)}$ should be zero. The output $Y_{(s)}$ is given by the equation

$$Y_{(s)} = G_{L(s)}L_{(s)} + G_{M(s)}M_{(s)} \tag{9.27}$$

Setting $Y_{(s)}$ equal to zero and solving for the relationship between $M_{(s)}$ and $L_{(s)}$ give the feedforward controller transfer function.

$$\left(\frac{M_{(s)}}{L_{(s)}}\right)_{(Y=0)} \equiv G_{F(s)} = \left(\frac{-G_L}{G_M}\right)_{(s)} \tag{9.28}$$

EXAMPLE 9.2. Suppose we have a distillation column with the process transfer functions $G_{M(s)}$ and $G_{L(s)}$ relating bottoms composition x_B to steam flow rate F_s and to feed flow rate F_L.

$$\begin{aligned} \left(\frac{x_B}{F_s}\right)_{(s)} &= G_{M(s)} = \frac{K_M}{\tau_M s + 1} \\ \left(\frac{x_B}{F_L}\right)_{(s)} &= G_{L(s)} = \frac{K_L}{\tau_L s + 1} \end{aligned} \tag{9.29}$$

All these variables are perturbations from steady state. These transfer functions could have been derived from a mathematical model of the column or found experimentally.

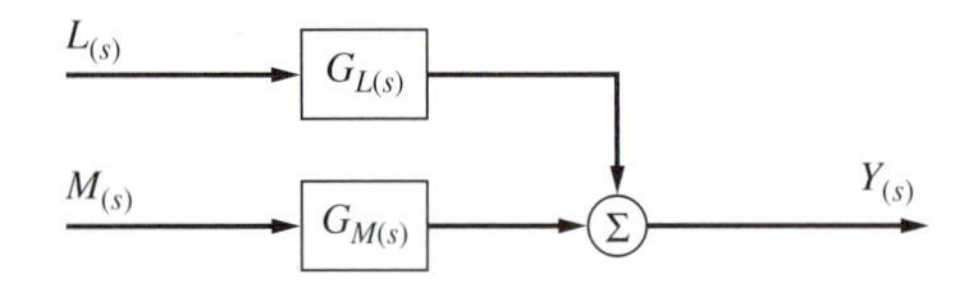

(*a*) Openloop

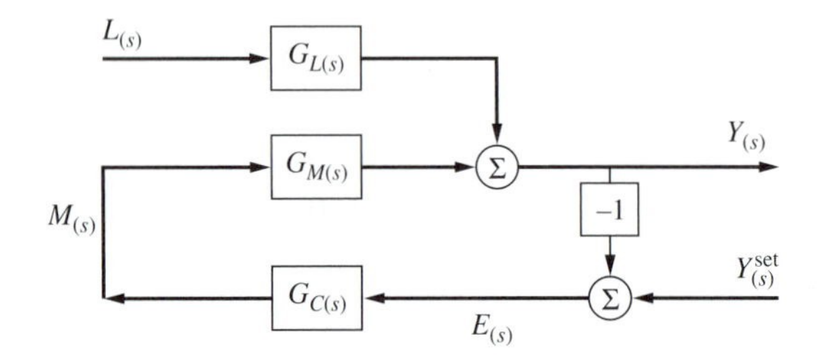

(*b*) Feedback control

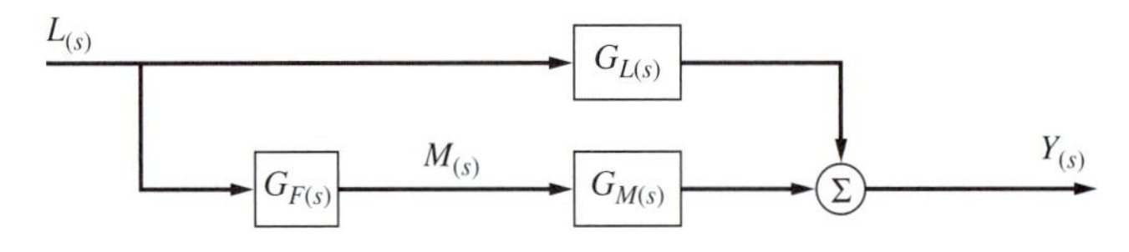

(*c*) Feedforward control

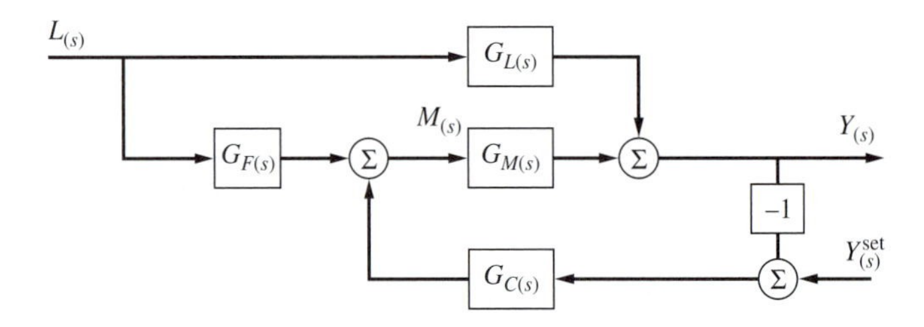

(*d*) Combined feedforward/feedback control

FIGURE 9.4
Block diagrams. (*a*) Openloop. (*b*) Feedback control. (*c*) Feedforward control. (*d*) Combined feedforward/feedback control.

We want to use a feedforward controller $G_{F(s)}$ to make adjustments in steam flow to the reboiler whenever the feed rate to the column changes, so that bottoms composition is held constant. The feedforward controller design equation [Eq. (9.28)] gives

$$G_{F(s)} = \left(\frac{-G_L}{G_M}\right)_{(s)} = \frac{-K_L/(\tau_L s + 1)}{K_M/(\tau_M s + 1)} = \frac{-K_L}{K_M}\frac{\tau_M s + 1}{\tau_L s + 1} \tag{19.30}$$

The feedforward controller contains a steady-state gain and dynamic terms. For this system the dynamic element is a first-order lead-lag. The unit step response of this lead-lag is an initial change to a value that is $(-K_L/K_M)(\tau_M/\tau_L)$, followed by an exponential rise or decay to the final steady-state value $-K_L/K_M$. ■

The advantage of feedforward control over feedback control is that perfect control can, in theory, be achieved. A disturbance produces no error in the controlled output variable if the feedforward controller is perfect. The disadvantages of feedforward control are:

1. The disturbance must be detected. If we cannot measure it, we cannot use feedforward control. This is one reason feedforward control for throughput changes is commonly used, whereas feedforward control for feed composition disturbances is only occasionally used. The former requires a flow measurement device, which is usually available. The latter requires a composition analyzer, which is often not available.
2. We must know how the disturbance and manipulated variables affect the process. The transfer functions $G_{L(s)}$ and $G_{M(s)}$ must be known, at least approximately. One of the nice features of feedforward control is that even crude, inexact feedforward controllers can be quite effective in reducing the upset caused by a disturbance.

In practice, many feedforward control systems are implemented by using ratio control systems, as discussed in Chapter 4. Most feedforward control systems are installed as combined feedforward-feedback systems. The feedforward controller takes care of the large and frequent measurable disturbances. The feedback controller takes care of any errors that come through the process because of inaccuracies in the feedforward controller as well as other unmeasured disturbances. Figure 9.4*d* shows the block diagram of a simple linear combined feedforward-feedback system. The manipulated variable is changed by both the feedforward controller and the feedback controller.

For linear systems the addition of the feedforward controller has *no* effect on the closedloop stability of the system. The denominators of the closedloop transfer functions are unchanged.

With feedback control:

$$Y_{(s)} = \frac{G_{L(s)}}{1 + G_{M(s)}G_{C(s)}} L_{(s)} + \frac{G_{M(s)}G_{C(s)}}{1 + G_{M(s)}G_{C(s)}} Y_{(s)}^{\text{set}} \tag{9.31}$$

With feedforward-feedback control:

$$Y_{(s)} = \frac{G_{L(s)} + G_{F(s)}G_{M(s)}}{1 + G_{M(s)}G_{C(s)}} L_{(s)} + \frac{G_{M(s)}G_{C(s)}}{1 + G_{M(s)}G_{C(s)}} Y_{(s)}^{\text{set}} \tag{9.32}$$

In a nonlinear system the addition of a feedforward controller often permits tighter tuning of the feedback controller because it reduces the magnitude of the disturbances that the feedback controller must cope with.

Figure 9.5*a* shows a typical implementation of a feedforward controller. A distillation column provides the specific example. Steam flow to the reboiler is ratioed to the feed flow rate. The feedforward controller gain is set in the ratio device. The dynamic elements of the feedforward controller are provided by the lead-lag unit.

Figure 9.5*b* shows a combined feedforward-feedback system where the feedback signal is added to the feedforward signal in a summing device. Figure 9.5*c* shows another combined system where the feedback signal is used to change the

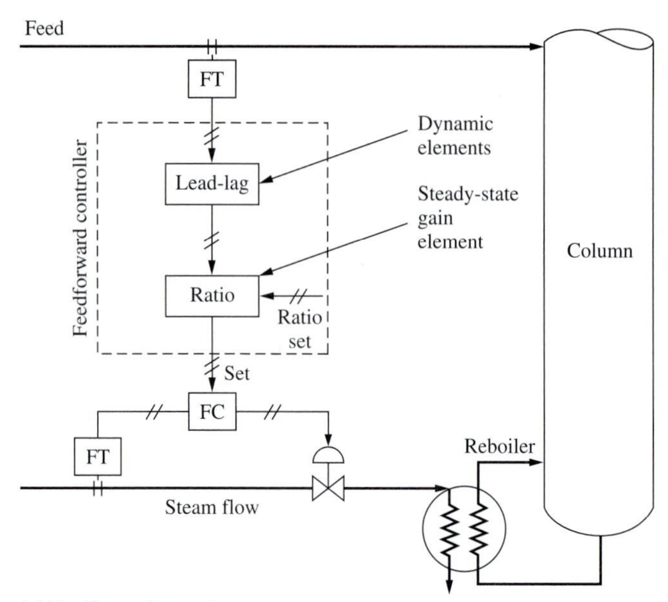

(*a*) Feedforward control

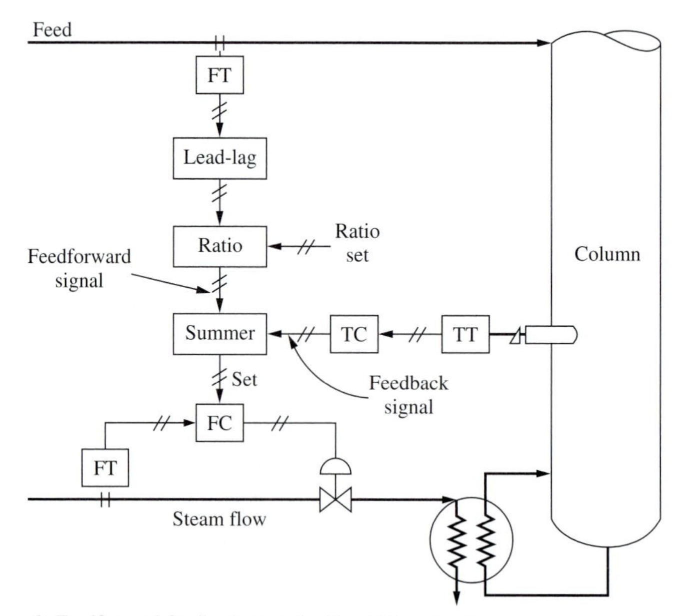

(*b*) Feedforward-feedback control with additive signals

FIGURE 9.5
Feedforward systems.

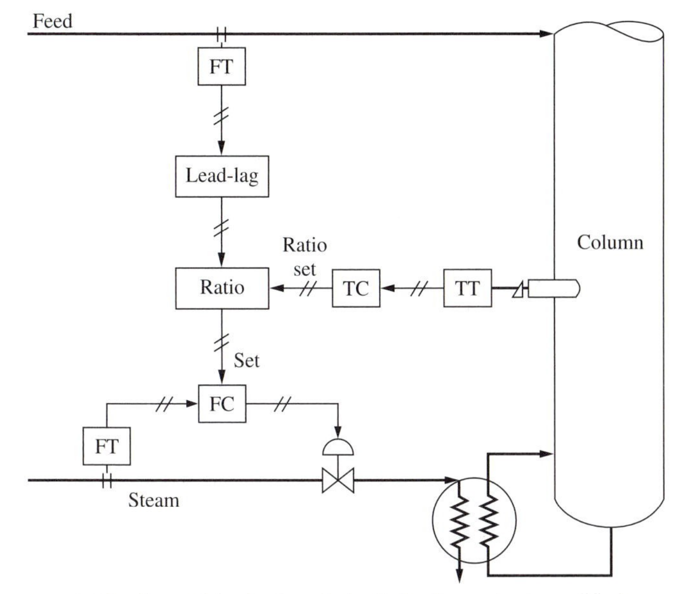

(c) Feedforward-feedback control with feedforward gain modified

FIGURE 9.5 (CONTINUED)
Feedforward systems.

feedforward controller gain in the ratio device. Figure 9.6 shows a combined feedforward-feedback control system for a distillation column where feed rate disturbances are detected and both steam flow and reflux flow are changed to hold constant both overhead and bottoms compositions. Two feedforward controllers are required.

Figure 9.7 shows some typical results of using feedforward control. A first-order lag is used in the feedforward controller so that the change in the manipulated variable is not instantaneous. The feedforward action is not perfect because the dynamics are not perfect, but there is a significant improvement over feedback control alone.

It is not always possible to achieve perfect feedforward control. If the $G_{M(s)}$ transfer function has a deadtime that is larger than the deadtime in the $G_{L(s)}$ transfer function, the feedforward controller will be physically unrealizable because it requires predictive action. Also, if the $G_{M(s)}$ transfer function is of higher order than the $G_{L(s)}$ transfer function, the feedforward controller will be physically unrealizable [see Eq. (9.28)].

9.2.2 Nonlinear Feedforward Control

There are no inherent linear limitations in feedforward control. Nonlinear feedforward controllers can be designed for nonlinear systems. The concepts are illustrated in Example 9.3.

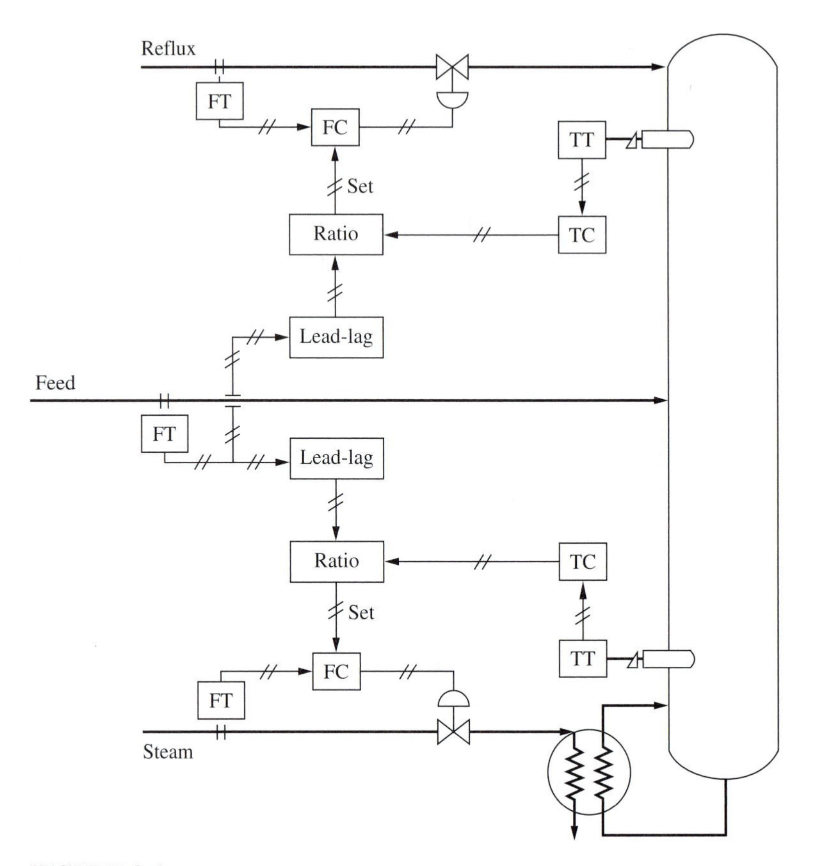

FIGURE 9.6
Combined feedforward-feedback system with two controlled variables.

EXAMPLE 9.3. The nonlinear ODEs describing the constant-holdup nonisothermal CSTR system are

$$\frac{dC_A}{dt} = \frac{F}{V}(C_{A0} - C_A) - C_A \alpha e^{-E/RT} \tag{9.33}$$

$$\frac{dT}{dt} = \frac{F}{V}(T_0 - T) - \left(\frac{\lambda}{\rho C_p}\right) C_A \alpha e^{-E/RT} - \left(\frac{UA}{C_p V \rho}\right)(T - T_J) \tag{9.34}$$

Let us choose a feedforward control system that holds both reactor temperature T and reactor concentration C_A constant at their steady-state values, $\overline{T}$ and $\overline{C}_A$. The feed flow rate F and the jacket temperature T_J are the manipulated variables. Disturbances are feed concentration C_{A0} and feed temperature T_0.

Noting that we are dealing with total variables now and not perturbations, the feedforward control objectives are

$$C_{A(t)} = \overline{C}_A \quad \text{and} \quad T_{(t)} = \overline{T} \tag{9.35}$$

Substituting these into Eqs. (9.33) and (9.34) gives

$$\frac{d\overline{C}_A}{dt} = 0 = \frac{F_{(t)}}{V}(C_{A0(t)} - \overline{C}_A) - \overline{C}_A \overline{k} \tag{9.36}$$

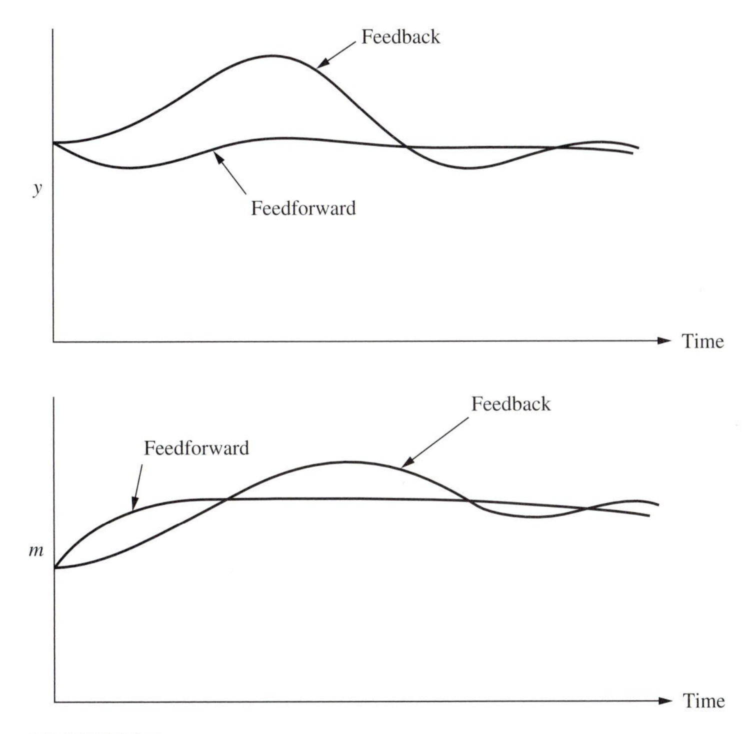

FIGURE 9.7
Feedforward control performance for load disturbance.

$$\frac{d\overline{T}}{dt} = 0 = \frac{F_{(t)}}{V}(T_{0(t)} - \overline{T}) - \left(\frac{\lambda}{\rho C_p}\right)\overline{C}_A\overline{k} - \left(\frac{UA}{C_p V \rho}\right)(\overline{T} - T_{J(t)}) \tag{9.37}$$

Rearranging Eq. (9.36) to find $F_{(t)}$, the manipulated variable, in terms of the disturbance $C_{A0(t)}$ gives the nonlinear feedforward controller relating the load variable C_{A0} to the manipulated variable F.

$$F_{(t)} = \frac{\overline{C}_A\overline{k}V}{C_{A0(t)} - \overline{C}_A} \tag{9.38}$$

The relationship is hyperbolic, as shown in Fig. 9.8. Feed rate must be decreased as feed concentration increases. This increases the holdup time, with constant volume, so that the additional reactant is consumed. Equation (9.38) tells us that feed flow rate does *not* have to be changed when feed temperature T_0 changes.

Substituting Eq. (9.38) into Eq. (9.37) and solving for the other manipulated variable T_J give

$$T_{J(t)} = \overline{T} + \frac{\overline{C}_A\overline{k}V}{UA}\left[\lambda + \frac{C_p(\overline{T} - T_{0(t)})}{C_{A0(t)} - \overline{C}_A}\right] \tag{9.39}$$

This is a second nonlinear feedforward relationship that shows how cooling-jacket temperature $T_{J(t)}$ must be changed as both feed concentration $C_{A0(t)}$ and feed temperature $T_{0(t)}$ change. Notice that the relationship between T_J and C_{A0} is nonlinear, but the relationship between T_J and T_0 is linear. ■

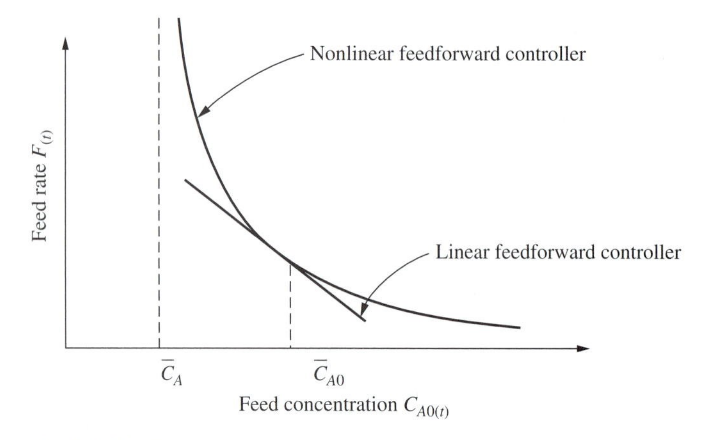

FIGURE 9.8
Nonlinear relationship between feed rate and feed concentration.

The preceding nonlinear feedforward controller equations were found analytically. In more complex systems, analytical methods become too complex, and numerical techniques must be used to find the required nonlinear changes in manipulated variables. The nonlinear steady-state changes can be found by using the nonlinear algebraic equations describing the process. The dynamic portion can often be approximated by linearizing around various steady states.

9.3 OPENLOOP-UNSTABLE PROCESSES

We remarked earlier in this book that one of the most interesting processes that chemical engineers have to control is the exothermic chemical reactor. This process can be openloop unstable.

Openloop instability means that reactor temperature will take off when there is no feedback control of cooling rate. It is easy to visualize qualitatively how this can occur. The reaction rate increases as the temperature climbs and more heat is given off. This heats the reactor to an even higher temperature, at which the reaction rate is still faster and even more heat is generated.

There is also an openloop-unstable mechanical system: the *inverted pendulum.* This is the problem of balancing a stick on the palm of your hand. You must keep moving your hand to keep the stick vertical. If you put your brain on manual and hold your hand still, the stick topples over. So the process is openloop unstable. If you think balancing an inverted pendulum is tough, try controlling a double inverted pendulum (two sticks on top of each other). You can see this done using a feedback controller at the French Science Museum in Paris.

We explore the effects of openloop instability quantitatively in the s plane. We discuss linear systems in which instability means that the reactor temperature theoretically goes to infinity. Because any real reactor system is nonlinear, reactor temperature will not increase without bounds. When the concentration of reactant begins to drop, the reaction rate eventually slows down. However, before it gets to that point the reactor may have blown a rupture disk or melted down! Nevertheless, linear techniques are very useful in looking at stability near some operating level. Mathematically, if the system is openloop unstable, its openloop transfer function $G_{M(s)}$ has at least one pole in the RHP.

9.3.1 Simple Systems

As a simple example, let us look at just the energy equation of the nonisothermal CSTR process of Example 7.6. We neglect any changes in C_A for the moment.

$$\frac{dT}{dt} = a_{22}T + a_{26}T_J + \cdots \tag{9.40}$$

Laplace transforming gives

$$\begin{aligned}(s - a_{22})T_{(s)} &= a_{26}T_{J(s)} + \cdots \\ T_{(s)} &= \frac{a_{26}}{s - a_{22}}T_{J(s)} + \cdots\end{aligned} \tag{9.41}$$

Thus, the stability of the system depends on the location of the pole a_{22}. If this pole is positive, the system is openloop unstable. The value of a_{22} is given in Eqs. (7.82).

$$a_{22} = \frac{-\lambda\bar{k}E\bar{C}_A}{\rho C_p R\bar{T}^2} - \frac{\bar{F}}{V} - \frac{UA}{V\rho C_p} \tag{9.42}$$

For the system to be openloop stable, $a_{22} < 0$.

$$\begin{aligned}\frac{-\lambda\bar{k}E\bar{C}_A}{\rho C_p R\bar{T}^2} - \frac{\bar{F}}{V} - \frac{UA}{V\rho C_p} &< 0 \\ \frac{-\lambda\bar{k}E\bar{C}_A}{\rho C_p R\bar{T}^2} &< \frac{\bar{F}}{V} + \frac{UA}{V\rho C_p}\end{aligned} \tag{9.43}$$

The left side of Eq. (9.43) represents the heat generation due to reaction. The right side represents heat removal due to sensible heat and the heat transfer to the jacket. Thus, our simple linear analysis tells us that the heat removal capacity must be greater than the heat generation if the system is to be stable. The actual stability requirement for the nonisothermal CSTR system is a little more complex than Eq. (9.43) because the concentration C_A does change.

A. First-order openloop-unstable process

Suppose we have a first-order process with the openloop transfer function

$$G_{M(s)} = \frac{K_p}{\tau_o s - 1} \tag{9.44}$$

Note that this is *not* a first-order lag because of the negative sign in the denominator. The system has an openloop pole in the RHP at $s = +1/\tau_o$. The unit step response of this system is an exponential that goes off to infinity as time increases.

Can we make the system stable by using feedback control? That is, can an openloop-unstable process be made closedloop stable by appropriate design of the feedback controller? Let us try a proportional controller: $G_{C(s)} = K_c$. The closedloop characteristic equation is

$$1 + G_{M(s)}G_{C(s)} = 1 + \frac{K_p}{\tau_o s - 1}K_c = 0$$
$$s = \frac{1 - K_c K_p}{\tau_o} \tag{9.45}$$

There is a single closedloop root. The root locus plot is given in Fig. 9.9*a*. It starts at the openloop pole in the RHP. The system is closedloop unstable for small values of controller gain. When the controller gain equals $1/K_p$, the closedloop root is located right at the origin. For gains greater than this, the root is in the LHP, so the system is closedloop stable.

Thus, in this system there is a *minimum* stable gain. Some of the systems studied up to now have had maximum values of gain $K_{\max}$ (or ultimate gain K_u) *beyond* which the system is closedloop unstable. Now we have a case that has a minimum gain $K_{\min}$ *below* which the system is closedloop unstable.

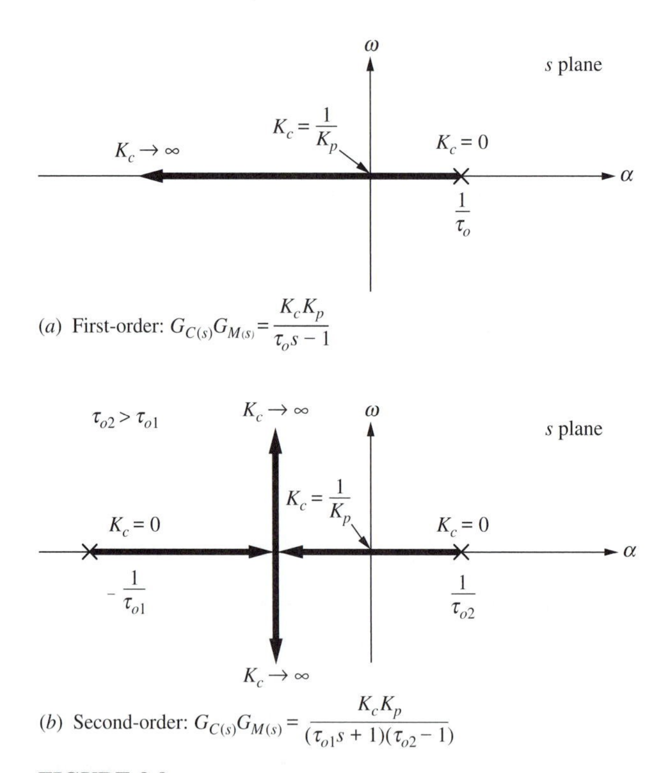

FIGURE 9.9
Root locus curves for openloop unstable processes (positive poles).

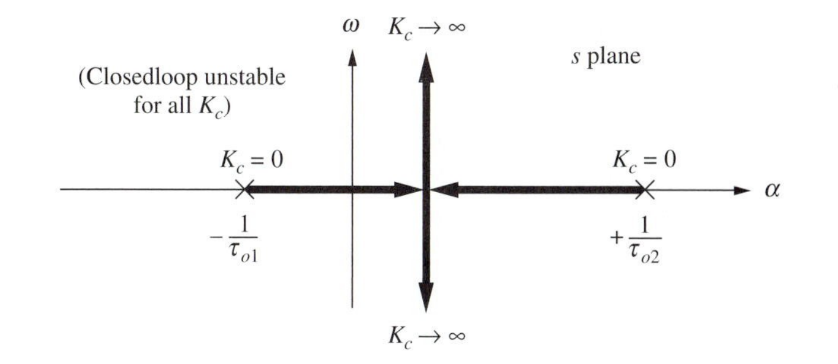

(c) $\tau_{o1} > \tau_{o2}$

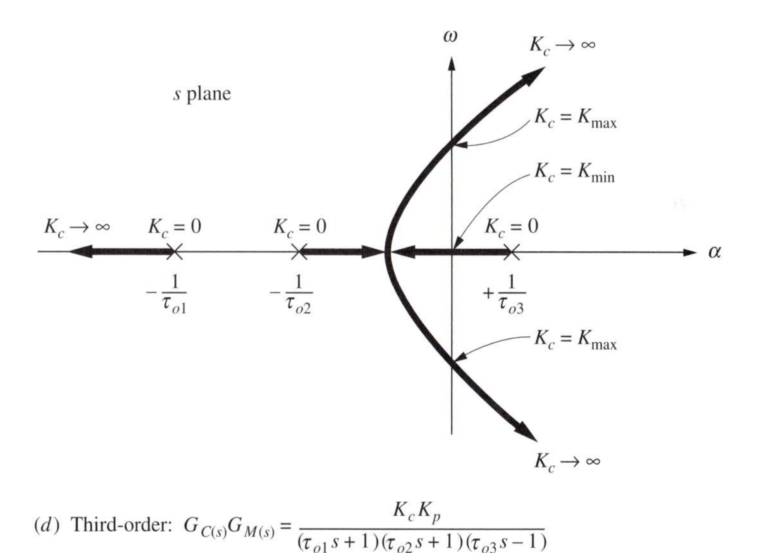

(d) Third-order: $G_{C(s)}G_{M(s)} = \dfrac{K_c K_p}{(\tau_{o1}s+1)(\tau_{o2}s+1)(\tau_{o3}s-1)}$

FIGURE 9.9 (CONTINUED)
Root locus curves for openloop unstable processes (positive poles).

B. Second-order openloop-unstable process

Consider the process given in Eq. (9.44) with a first-order lag added.

$$G_{M(s)} = \frac{K_p}{(\tau_{o1}s + 1)(\tau_{o2}s - 1)} \tag{9.46}$$

One of the roots of the openloop characteristic equation lies in the RHP at $s = +1/\tau_{o2}$.

Can we make this system closedloop stable? A proportional feedback controller gives a closedloop characteristic equation:

$$\begin{gathered} 1 + G_{M(s)}G_{C(s)} = 1 + \frac{K_p}{(\tau_{o1}s + 1)(\tau_{o2}s - 1)}K_c = 0 \\ \tau_{o1}\tau_{o2}s^2 + (\tau_{o2} - \tau_{o1})s + K_cK_p - 1 = 0 \end{gathered} \tag{9.47}$$

Two conditions must be satisfied if there are to be no positive roots of this closedloop characteristic equation:

$$\tau_{o2} > \tau_{o1} \quad \text{and} \quad K_c > \frac{1}{K_p} \tag{9.48}$$

Therefore, if $\tau_{o2} < \tau_{o1}$ a proportional controller cannot make the system closedloop stable. A controller with derivative action might be able to stabilize the system. Figures 9.9*b* and *c* give the root locus plots for the two cases $\tau_{o2} > \tau_{o1}$ and $\tau_{o2} < \tau_{o1}$. In the latter case there is always at least one closedloop root in the RHP, so the system is always unstable.

C. Third-order openloop-unstable process

If an additional lag is added to the system and a proportional controller is used, the closedloop characteristic equation becomes

$$1 + G_{M(s)}G_{C(s)} = 1 + \frac{K_p}{(\tau_{o1}s + 1)(\tau_{o2}s + 1)(\tau_{o3}s - 1)}K_c = 0 \tag{9.49}$$

Figure 9.9*d* gives a sketch of a typical root locus plot for this type of system. We now have a case of *conditional stability*. Below $K_{\min}$ the system is closedloop unstable. Above $K_{\max}$ the system is again closedloop unstable. A range of stable values of controller gain exists between these limits:

$$K_{\min} < K_c < K_{\max} \tag{9.50}$$

Clearly, the closer the values of $K_{\max}$ and $K_{\min}$ are to each other, the less controllable the system will be.

EXAMPLE 9.4. The transfer function relating process temperature T to cooling-water flow rate F_w in an openloop-unstable chemical reactor is

$$G_{M(s)} = \frac{-0.7\ (°\text{F/gpm})}{(s + 1)(\tau s - 1)} \tag{9.51}$$

where the time constants 1 and τ are in minutes. The temperature measurement has a dynamic first-order lag of 30 seconds. The range of the analog electronic (4 to 20 mA) temperature transmitter is 200 to 400°F. The control valve on the cooling water has linear installed characteristics and passes 500 gpm when wide open. The temperature controller is proportional.

(*a*) What is the closedloop characteristic equation of the system?

We must include the 0.5-minute lag of the temperature transmitter and the gains for both the transmitter and the valve.

$$1 + G_{M(s)}G_{T(s)}G_{V(s)}G_{C(s)} = 0$$

$$1 + \left[\frac{-0.7\ (°\text{F/gpm})}{(s + 1)(\tau s - 1)}\right]\left[\frac{16/200\ (\text{mA}/°\text{F})}{0.5s + 1}\right]\left[-500/16\ (\text{gpm/mA})\right][K_c] = 0$$

Note that the gain of the controller is chosen to be positive (reverse acting), so the controller output decreases as temperature increases, which increases cooling-water flow through the AC valve (this makes the gain of the control valve negative).

$$(0.5\tau)s^3 + (1.5\tau - 0.5)s^2 + (\tau - 1.5)s + (1.75K_c - 1) = 0 \tag{9.52}$$

(*b*) What is the minimum value of controller gain, K_{min}, that gives a closedloop-stable system?

Letting $s = i\omega$ in Eq. (9.52) gives two equations in two unknowns: K_c and ω. From the real part:

$$0.5\omega^2 - 1.5\omega^2\tau + 1.75K_c - 1 = 0 \tag{9.53}$$

From the imaginary part:

$$\omega\tau - 1.5\omega - 0.5\tau\omega^3 = 0 \tag{9.54}$$

There are two solutions for Eq. (9.54):

$$\omega = 0 \quad \text{and} \quad \omega = \sqrt{\frac{\tau - 1.5}{0.5\tau}} \tag{9.55}$$

Using $\omega = 0$ gives the minimum value of gain.

$$K_{min} = \frac{1}{1.75}$$

(*c*) Derive a relationship between τ (the positive pole) and the maximum closedloop stable gain, K_{max}.

Using the second value of ω in Eq. (9.55) gives K_{max}.

$$K_{max} = \frac{1.5 - 4.5\tau + 3\tau^2}{1.75\tau} \tag{9.56}$$

(*d*) Calculate K_{max} when $\tau = 5$ minutes and 10 minutes.

$$\text{For } \tau = 5, \qquad K_{max} = 6.17$$

$$\text{For } \tau = 10, \qquad K_{max} = 14.7$$

Note that this result shows that the smaller the value of τ (i.e., the closer the positive pole is to the value of the negative poles: $s = -1$ and -2), the more difficult it is to stabilize the system.

(*e*) At what value of τ will a proportional-only controller be unable to stabilize the system?

When $K_{max} = K_{min}$ the system will always be unstable.

$$\frac{1.5 - 4.5\tau + 3\tau^2}{1.75\tau} = \frac{1}{1.75} \quad \Rightarrow \quad \tau = 1.5 \text{ minutes}$$

Note that there are actually two values of τ that satisfy the equation above, but the limiting one is the larger of the two.

9.3.2 Effects of Lags

The systems explored in the preceding section illustrate a very important point about the control of openloop-unstable systems: The control of these systems becomes more difficult as the order of the system is increased and as the magnitudes of the first-order lags increase. Our examples demonstrated this quantitatively. For this reason, it is vital to design a reactor control system with very fast measurement dynamics and very fast heat removal dynamics. If the thermal lags in the temperature sensor and in the cooling jacket are not small, it may not be possible to stabilize the reactor with feedback control. Bare-bulb thermocouples and oversized cooling-water valves are often used to improve controllability.

9.3.3 PD Control

Up to this point we have looked at using proportional controllers on openloop-unstable systems. Controllability can often be improved by using derivative action in the controller. An example illustrates the point.

EXAMPLE 9.5. Let us take the same third-order process analyzed in Example 9.4. For $\tau = 5$ minutes and a proportional controller, the ultimate gain was 6.17 and the ultimate frequency was 1.18 rad/min.

Now we use a PD controller with τ_D set equal to 0.5 minutes (just to make the algebra work out nicely; this is not necessarily the optimal value of τ_D). The closedloop characteristic equation becomes

$$1 + \left[\frac{1.75}{(s+1)(5s-1)(0.5s+1)}\right] K_c \frac{\tau_D s + 1}{0.1\tau_D s + 1} = 0 \tag{9.57}$$

$$0.25s^3 + 5.2s^2 + 3.95s + 1.75K_c - 1 = 0$$

Solving for the ultimate gain and frequency gives $K_u = 47.5$ and $\omega_u = 3.97$. Comparing these with the results for P control shows a significant increase in gain and reduction in closedloop time constant.

9.3.4 Effect of Reactor Scale-up on Controllability

One of the classical problems in scaling up a jacketed reactor is the decrease in the ratio of heat transfer area to reactor volume as size is increased. This has a profound effect on the controllability of the system. Table 9.1 gives some results that quantify the effects for reactors varying from 5 gallons (typical pilot plant size) to 5000 gallons. Table 9.2 gives parameter values that are held constant as the reactor is scaled up.

TABLE 9.1
Effect of scale-up on controllability

Reactor volume (gal)	5	500	5000
Feed rate (lb_m/hr)	27.8	2780	27,800
Heat transfer (10^6 Btu/hr)	0.0028	0.28	2.8
Reactor height (ft)	1.504	6.98	15.04
Reactor diameter (ft)	0.752	3.49	7.52
Heat transfer area (ft^2)	3.99	86.15	400
Cooling-water flow (gpm)	0.086	11.58	240
Jacket temperature (°F)	135.3	118.3	93.3
Controller gains			
Max	169	100	144
Min	1.37	1.95	4.41
Ratio	124	51	33

TABLE 9.2
Reactor parameters

Reactor holdup time	1.2 h
Jacket holdup time	0.077 h
Overall heat transfer coefficient	150 Btu/h ft^2 °F
Heat capacity of products and feeds	0.75 Btu/lb$_m$ °F
Heat capacity of cooling water	1.0 Btu/lb$_m$ °F
Density of products and feeds	50 lb$_m$/ft^3
Density of cooling water	62.3 lb$_m$/ft^3
Inlet cooling-water temperature	70°F
Temperature measurement lag	30 s
Feed concentration	0.50 lb-mol A/ft^3
Feed temperature	70°F
Reactor temperature	140°F
Preexponential factor	7.08×10^{10} h^{-1}
Activation energy	30,000 Btu/lb-mol
Heat of reaction	−30,000 Btu/lb-mol
Steady-state concentration	0.245 lb-mol A/ft^3
Specific reaction rate	0.8672 h^{-1}
Temperature transmitter span	100°F
Cooling-water valve maximum flow rate	Twice normal design flow rate

Notice that the temperature difference between the cooling jacket and the reactor must be increased as the size of the reactor increases. The flow rate of cooling water also increases rapidly as reactor size increases.

The ratio of K_{max} to K_{min}, which is a measure of the controllability of the system, decreases from 124 for a 5-gallon reactor to 33 for a 5000-gallon reactor.

9.4
PROCESSES WITH INVERSE RESPONSE

Another interesting type of process is one that exhibits *inverse response*. This phenomenon, which occurs in a number of real systems, is sketched in Fig. 9.10*b*. The response of the output variable $y_{(t)}$ begins in the direction opposite of where it finishes. Thus, the process starts out in the wrong direction. You can imagine what this sort of behavior would do to a poor feedback controller in such a loop. We show quantitatively how inverse response degrades control loop performance.

An important example of a physical process that shows inverse response is the base of a distillation column with the reaction of bottoms composition and base level to a change in vapor boilup. In a binary distillation column, we know that an increase in vapor boilup V must drive more low-boiling material up the column and therefore decrease the mole fraction of light component in the bottoms x_B. However, the tray hydraulics can produce some unexpected results. When the vapor rate through a tray is increased, it tends to (1) back up more liquid in the downcomer to overcome the increase in pressure drop through the tray and (2) reduce the density of the liquid and vapor froth on the active part of the tray. The first effect momentarily reduces the liquid flow rates through the column while the liquid holdup in the downcomer is

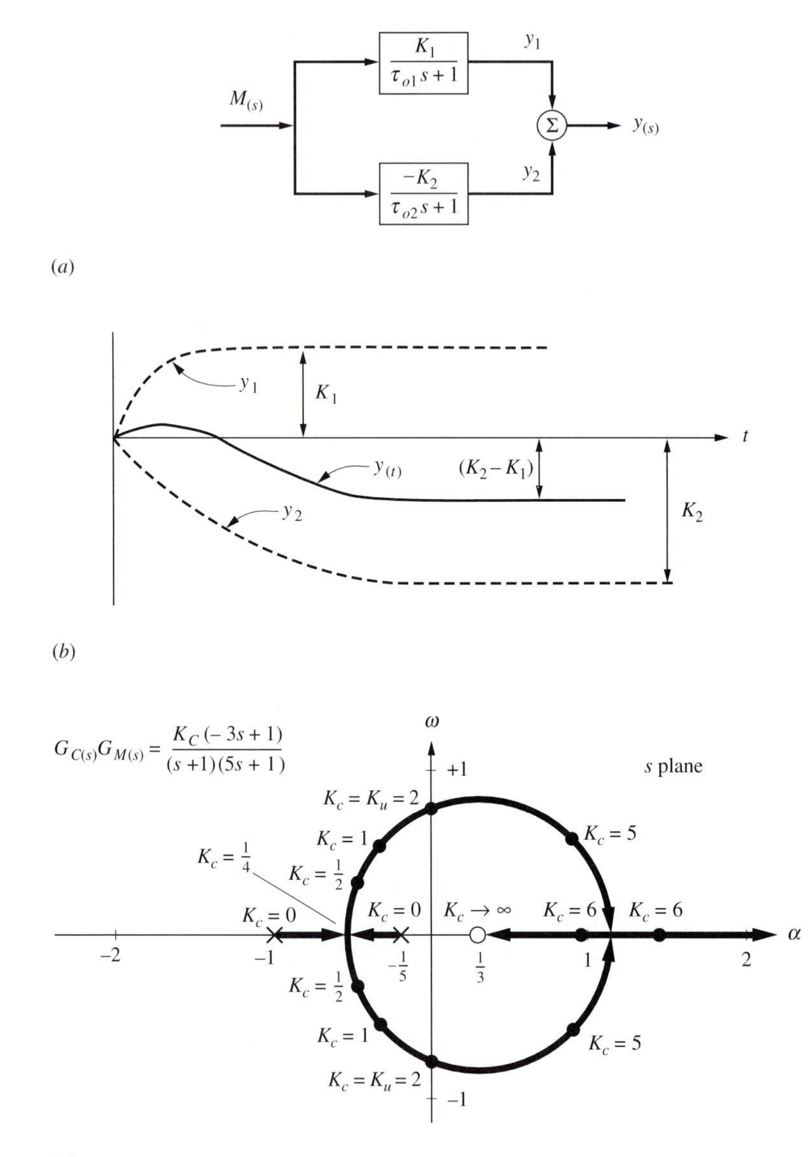

FIGURE 9.10
Process with inverse response. (*a*) Block diagram. (*b*) Step response. (*c*) Root locus plot.

building up. The second effect tends to momentarily increase the liquid rates since there is more height over the weir.

Which of these two opposing effects dominates depends on the tray design and operating level. The pressure drops through valve trays change little with vapor rates unless the valves are completely lifted. Therefore, the second effect is sometimes larger than the first. If this occurs, an increase in vapor boilup produces a transient increase in liquid rates down the column. This increase in liquid rates carries material that is richer in light component into the reboiler and momentarily increases x_B. Eventually, of course, the liquid rates will return to normal when the liquid inventory on the trays has dropped to the new steady-state levels. Then the effect of the increase in vapor boilup will drive x_B down. Thus, the vapor-liquid hydraulics can produce inverse response in the effect of V on x_B (and also on the liquid holdup in the base).

Mathematically, inverse response can be represented by a system that has a transfer function with a *positive zero,* a zero in the RHP. Consider the system sketched in Fig. 9.10*a*. There are two parallel first-order lags with gains of opposite sign. The transfer function for the overall system is

$$\frac{Y_{(s)}}{M_{(s)}} = \frac{K_1}{\tau_{o1}s + 1} - \frac{K_2}{\tau_{o2}s + 1} = \frac{(K_1\tau_{o2} - K_2\tau_{o1})s - (K_2 - K_1)}{(\tau_{o1}s + 1)(\tau_{o2}s + 1)} \tag{9.58}$$

If the K's and τ_o's are such that

$$\frac{\tau_{o2}}{\tau_{o1}} > \frac{K_2}{K_1} > 1$$

the system will show inverse response, as sketched in Fig. 9.10*b*. Eq. (9.58) can be rearranged as

$$\frac{Y_{(s)}}{M_{(s)}} = -(K_2 - K_1)\frac{-\left(\dfrac{K_1\tau_{o2} - K_2\tau_{o1}}{K_2 - K_1}\right)s + 1}{(\tau_{o1}s + 1)(\tau_{o2}s + 1)} \tag{9.59}$$

Thus, the system has a positive zero at

$$s = \frac{K_2 - K_1}{K_1\tau_{o2} - K_2\tau_{o1}}$$

Keep in mind that the positive zero does not make the system *openloop* unstable. Stability depends on the poles of the transfer function, not on the zeros. Positive zeros in a system do, however, affect *closedloop* stability, as the following example illustrates.

EXAMPLE 9.6. Let us take the same system used in Example 8.7 and add a positive zero at $s = +\frac{1}{3}$.

$$G_{M(s)} = \frac{-3s + 1}{(s + 1)(5s + 1)} \tag{9.60}$$

With a proportional feedback controller the closedloop characteristic equation is

$$1 + G_{M(s)}G_{C(s)} = 1 + \frac{-3s + 1}{(s + 1)(5s + 1)}K_c \tag{9.61}$$

$$5s^2 + (6 - 3K_c)s + 1 + K_c = 0$$

The root locus curves are shown in Fig. 9.10*c*. The loci start at the poles of the openloop transfer function: $s = -1$ and $s = -\frac{1}{5}$. Since the loci must end at the zeros of the openloop transfer function ($s = +\frac{1}{3}$), the curves swing over into the RHP. Therefore, the system is closedloop unstable for gains greater than 2.

Remember that in Example 8.8 adding a lead or a negative zero made the closedloop system more stable. In this example we have shown that adding a positive zero has the reverse effect. ■

9.5 MODEL-BASED CONTROL

Up to this point we have generally chosen a type of controller (P, PI, or PID) and determined the tuning constants that gave some desired performance (closedloop damping coefficient). We have used a model of the process to calculate the controller settings, but the structure of the model has not been explicitly involved in the controller design.

There are several alternative controller design methods that make more explicit use of a process model. We discuss two of these here.

9.5.1 Direct Synthesis

In direct synthesis the desired closedloop response for a given input is specified. Then, with the model of the process known, the required form and tuning of the feedback controller are back-calculated. These steps can be clarified with a simple example.

EXAMPLE 9.7. Suppose we have a process with the openloop transfer function

$$G_{M(s)} = \frac{K_p}{\tau_o s + 1} \tag{9.62}$$

where K_p and τ_o are the openloop gain and time constant. Let us assume that we want to specify the closedloop servo transfer function to be

$$\frac{Y_{(s)}}{Y^{set}_{(s)}} = \frac{1}{\tau_c s + 1} \tag{9.63}$$

That is, we want the process to respond to a step change in setpoint as a first-order process with a closedloop time constant τ_c. The steady-state gain between the controlled variable and the setpoint is specified as unity, so there will be no offset.

Now, knowing the process model and having specified the desired closedloop servo transfer function, we can solve for the feedback controller transfer function $G_{C(s)}$. We define the closedloop servo transfer function as $S_{(s)}$.

$$S_{(s)} \equiv \frac{Y_{(s)}}{Y^{set}_{(s)}} = \frac{G_{M(s)}G_{C(s)}}{1 + G_{M(s)}G_{C(s)}} \tag{9.64}$$

Equation (9.64) contains only one unknown (i.e., the feedback controller transfer function $G_{C(s)}$). Solving for $G_{C(s)}$ in terms of the known values of $G_{M(s)}$ and $S_{(s)}$ gives

$$G_{C(s)} = \frac{S_{(s)}}{(1 - S_{(s)})G_{M(s)}} \tag{9.65}$$

Equation (9.65) is a general solution for any process and for any desired closedloop servo transfer function. Plugging in the values for $G_{M(s)}$ and $S_{(s)}$ for the specific example gives

$$G_{C(s)} = \frac{\dfrac{1}{\tau_c s + 1}}{\left(1 - \dfrac{1}{\tau_c s + 1}\right)\dfrac{K_p}{\tau_o s + 1}} = \frac{\tau_o s + 1}{K_p \tau_c s} \tag{9.66}$$

Equation (9.66) can be rearranged to look just like a PI controller if K_c is set equal to $\tau_o/\tau_c K_p$ and the reset time τ_I is set equal to τ_o.

$$G_{C(s)} = K_c \frac{\tau_I s + 1}{\tau_I s} = \left(\frac{\tau_o}{\tau_c K_p}\right)\frac{\tau_o s + 1}{\tau_o s} \tag{9.67}$$

Thus, we find that the appropriate structure for the controller is PI, and we have solved analytically for the gain and reset time in terms of the parameters of the process model and the desired closedloop response.

Before we leave this example, it is important to make sure that you understand the limitations of the method. Suppose the process openloop transfer function also contained a deadtime.

$$G_{M(s)} = \frac{K_p e^{-Ds}}{\tau_o s + 1} \tag{9.68}$$

Using this $G_{M(s)}$ in Eq. (9.65) gives a new feedback controller:

$$G_{C(s)} = \frac{(\tau_o s + 1)e^{+Ds}}{K_p \tau_c s} \tag{9.69}$$

This controller is *not* physically realizable. The negative deadtime implies that we can change the output of the device D minutes before the input changes, which is impossible.

This case illustrates that the desired closedloop relationship *cannot* be chosen arbitrarily. You cannot make a jumbo jet behave like a jet fighter, or a garbage truck drive like a Ferrari! We must select the desired response so that the controller is physically realizable. In this case all we need to do is modify the specified closedloop servo transfer function $S_{(s)}$ to include the deadtime.

$$S_{(s)} = \frac{e^{-Ds}}{\tau_c s + 1} \tag{9.70}$$

Using this $S_{(s)}$ in Eq. (9.65) gives exactly the same $G_{C(s)}$ as found in Eq. (9.66), which is physically realizable.

As an additional case, suppose we had a second-order process transfer function.

$$G_{M(s)} = \frac{K_p}{(\tau_o s + 1)^2}$$

Specifying the original closedloop servo transfer function [Eq. (9.63)] and solving for the feedback controller using Eq. (9.65) gives

$$G_{C(s)} = \frac{(\tau_o s + 1)^2}{K_p \tau_c s} \tag{9.71}$$

Again, this controller is physically unrealizable because the order of the numerator is greater than the order of the denominator. We would have to modify our specified $S_{(s)}$ to make this controller realizable. ■

This type of controller design has been around for many years. The "pole placement" methods used in aerospace systems employ the same basic idea: the controller is designed to position the poles of the closedloop transfer function at the desired location in the s plane. This is exactly what we do when we specify the closedloop time constant in Eq. (9.63).

9.5.2 Internal Model Control

Garcia and Morari (*Ind. Eng. Chem. Process Des. Dev. 21:* 308, 1982) have used a similar approach in developing "internal model control" (IMC). The method gives the control engineer a different perspective on the controller design problem. The basic idea of IMC is to use a model of the openloop process $G_{M(s)}$ transfer function in such a way that the selection of the specified closedloop response yields a physically realizable feedback controller.

Figure 9.11 gives the IMC structure. The model of the process $\tilde{G}_{M(s)}$ is run in parallel with the actual process. The output of the model $\tilde{Y}$ is subtracted from the actual output of the process Y, and this signal is fed back into the controller $G_{IMC(s)}$. If our model is perfect ($\tilde{G}_M = G_M$), this signal is the effect of load disturbance on the output (since we have subtracted the effect of the manipulated variable M). Thus, we are "inferring" the load disturbance without having to measure it. This signal is

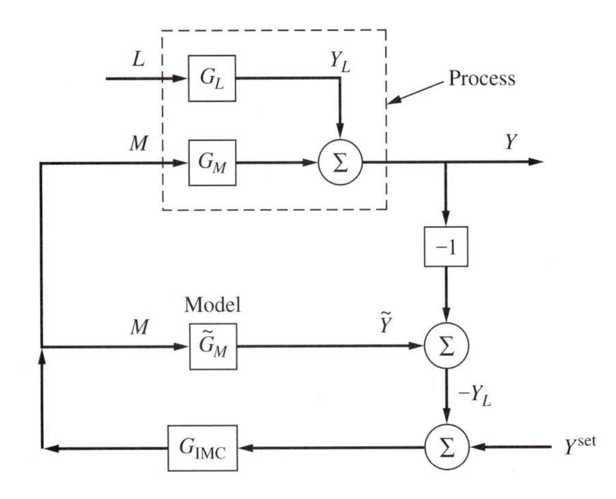

(*a*) Basic structure

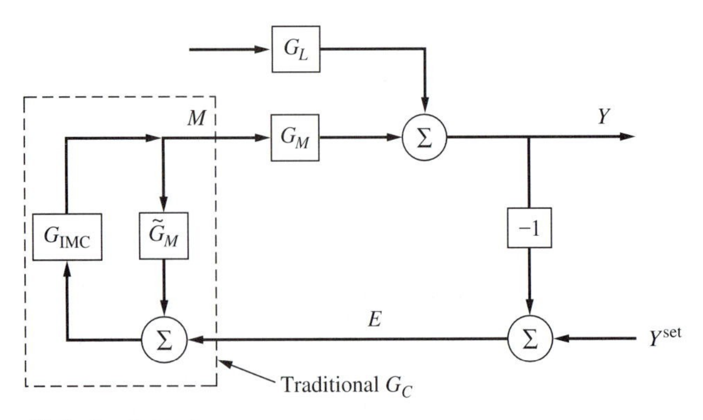

(*b*) Reduced structure

FIGURE 9.11
IMC.

Y_L, the output of thc process load transfer function, and is equal to $G_{L(s)}L_{(s)}$. We know from our studies of feedforward control [Eq. (9.28)] that if we change the manipulated variable $M_{(s)}$ by the relationship

$$M_{(s)} = \left(\frac{-G_L}{G_M}\right)_{(s)} L_{(s)} \tag{9.72}$$

we get perfect control of the the output $Y_{(s)}$. This tells us that if we could set the controller

$$G_{\text{IMC}(s)} = \frac{1}{G_{M(s)}} \tag{9.73}$$

we would get perfect control for load disturbances. In addition, this choice of $G_{\text{IMC}(s)}$ gives perfect control for setpoint disturbances: the total transfer function between the setpoint Y^{set} and Y is simply unity. Thus, the ideal controller is the inverse of the plant. We use this notion again in Chapter 13 when we consider multivariable processes.

However, there are two practical problems with this ideal choice of the feedback controller $G_{\text{IMC}(s)}$. First, it assumes that the model is perfect. More important, it assumes that the inverse of the plant model $G_{M(s)}$ is physically realizable. This is almost *never* true since most plants have deadtime or numerator polynomials that are of lower order than denominator polynomials.

So if we cannot attain perfect control, what do we do? From the IMC perspective, we simply break up the controller transfer function $G_{\text{IMC}(s)}$ into two parts. The first part is the inverse of $G_{M(s)}$. The second part, which Garcia and Morari call a "filter," is chosen to make the total $G_{\text{IMC}(s)}$ physically realizable. As we will show, this second part turns out to be the closedloop servo transfer function that we defined as $S_{(s)}$ in Eq. (9.64).

Referring to Fig. 9.11 and assuming that $G_M = \tilde{G}_M$, we see that

$$Y = G_L L + G_M M = Y_L + G_M G_{\text{IMC}(s)}(Y^{\text{set}} - Y_L) \tag{9.74}$$

$$Y = (G_M G_{\text{IMC}(s)})Y^{\text{set}} + (1 - G_M G_{\text{IMC}(s)})Y_L \tag{9.75}$$

Now if the controller transfer function $G_{\text{IMC}(s)}$ is selected as

$$G_{\text{IMC}(s)} = \frac{1}{G_{M(s)}} S_{(s)} \tag{9.76}$$

Eq. (9.75) becomes

$$Y_{(s)} = S_{(s)} Y_{(s)}^{\text{set}} + (1 - S_{(s)}) G_{L(s)} L_{(s)} \tag{9.77}$$

So the closedloop servo transfer function $S_{(s)}$ must be chosen such that $G_{\text{IMC}(s)}$ is physically realizable.

EXAMPLE 9.8. Let's take the process studied in Example 9.7. The process openloop transfer function is

$$G_{M(s)} = \frac{K_p}{\tau_o s + 1}$$

We want to design $G_{IMC(s)}$ using Eq. (9.76).

$$G_{IMC(s)} = \frac{1}{G_{M(s)}} S_{(s)} = \frac{1}{\dfrac{K_p}{\tau_o s + 1}} S_{(s)} = \frac{\tau_o s + 1}{K_p} S_{(s)} \tag{9.78}$$

Now the logical choice of $S_{(s)}$ that will make $G_{IMC(s)}$ physically realizable is the same as that chosen in Eq. (9.63).

$$S_{(s)} = \frac{Y_{(s)}}{Y_{(s)}^{set}} = \frac{1}{\tau_c s + 1}$$

So the IMC controller becomes a PD controller.

$$G_{IMC(s)} = \frac{\tau_o s + 1}{K_p(\tau_c s + 1)} \tag{9.79}$$

■

The IMC structure is an alternative way of looking at controller design. The model of the process is clearly indicated in the block diagram. The tuning of the $G_{IMC(s)}$ controller reduces to selecting a reasonable closedloop servo transfer function.

The reduced block diagram for the IMC structure shows that there is a precise relationship between the traditional feedback controller $G_{C(s)}$ and the $G_{IMC(s)}$ controller used in IMC.

$$\frac{M_{(s)}}{E_{(s)}} = G_{C(s)} = \frac{G_{IMC(s)}}{1 - G_{IMC(s)}\tilde{G}_{M(s)}} \tag{9.80}$$

The negative sign in the denominator of Eq. (9.80) comes from the positive feedback in the internal loop in the controller. Applying this equation to Example 9.8 gives

$$G_{C(s)} = \frac{G_{IMC(s)}}{1 - G_{IMC(s)}\tilde{G}_{M(s)}} = \frac{\dfrac{\tau_o s + 1}{K_p(\tau_c s + 1)}}{1 - \dfrac{\tau_o s + 1}{K_p(\tau_c s + 1)} \dfrac{K_p}{\tau_o s + 1}} = \frac{\tau_o s + 1}{K_p \tau_c s} \tag{9.81}$$

This is exactly the same result (a PI controller) that we found in Eq. (9.66).

EXAMPLE 9.9. Apply the IMC design to the process with the openloop transfer function

$$G_{M(s)} = \frac{K_p e^{-Ds}}{\tau_o s + 1} \tag{9.82}$$

Using Eq. (9.76) and substituting Eq. (9.82) give

$$G_{IMC(s)} = \frac{1}{G_{M(s)}} S_{(s)} = \frac{\tau_o s + 1}{K_p e^{-Ds}} S_{(s)}$$

Clearly the best way to select the closedloop servo transfer function $S_{(s)}$ to make $G_{IMC(s)}$ physically realizable is

$$S_{(s)} = \frac{e^{-Ds}}{\tau_c s + 1} \tag{9.83}$$

The response of Y to a step change in setpoint will be a deadtime of D minutes followed by an exponential rise. The IMC controller becomes a PD controller.

$$G_{\mathrm{IMC}(s)} = \frac{\tau_o s + 1}{K_p e^{-Ds}} \frac{e^{-Ds}}{\tau_c s + 1} = \frac{\tau_o s + 1}{K_p(\tau_c s + 1)} \tag{9.84}$$

It should be noted that the equivalent conventional controller $G_{C(s)}$ does *not* have the standard P, PI, or PID form.

$$G_{C(s)} = \frac{G_{\mathrm{IMC}(s)}}{1 - G_{\mathrm{IMC}(s)}\tilde{G}_{M(s)}} = \frac{\dfrac{\tau_o s + 1}{K_p(\tau_c s + 1)}}{1 - \dfrac{\tau_o s + 1}{K_p(\tau_c s + 1)} \dfrac{K_p e^{-Ds}}{\tau_o s + 1}} \tag{9.85}$$

$$= \frac{\tau_o s + 1}{K_p(\tau_c s + 1 - e^{-Ds})}$$

This controller has a uniquely new transfer function. ■

Maurath, Mellichamp, and Seborg (*IEC Res.* 27:956, 1988) give guidelines for selecting parameter values in IMC designs.

One final comment should be made about model-based control before we leave the subject. These model-based controllers depend quite strongly on the validity of the model, particularly its dynamic fidelity. If we have a poor model or if the plant parameters change, the performance of a model-based controller is usually seriously affected. Model-based controllers are less "robust" than the more conventional PI controllers. This lack of robustness can be a problem in the single-input, single-output (SISO) loops that we have been examining. It is an even more serious problem in multivariable systems, as we discuss in Chapters 12 and 13.

9.6 CONCLUSION

The material covered in this chapter should have convinced you of the usefulness of the "Russian" (Laplace domain) language. It permits us to look at fairly complex processes in a nice, compact way. We will find in the next two chapters that "Chinese" (frequency response) is even more useful for analyzing more realistically complex processes.

PROBLEMS

9.1. The load and manipulated-variable transfer functions of a process are

$$\frac{Y_{(s)}}{M_{(s)}} = G_{M(s)} = \frac{1}{(s+1)(5s+1)}$$

$$\frac{Y_{(s)}}{L_{(s)}} = G_{L(s)} = \frac{2}{(s+1)(5s+1)(\frac{1}{2}s+1)}$$

Derive the feedforward controller transfer function that will keep the process output $Y_{(s)}$ constant with load changes $L_{(s)}$.

9.2. Repeat Problem 9.1 with

$$G_{M(s)} = \frac{\frac{1}{2}s + 1}{(s+1)(5s+1)}$$

9.3. The transfer functions of a binary distillation column between distillate composition x_D and feed rate F, reflux rate R, and feed composition z are

$$\frac{x_D}{F} = \frac{K_F e^{-D_F s}}{(\tau_F s + 1)^2} \qquad \frac{x_D}{z} = \frac{K_z e^{-D_z s}}{(\tau_z s + 1)^2} \qquad \frac{x_D}{R} = \frac{K_R e^{-D_R s}}{\tau_R s + 1}$$

Find the feedforward controller transfer functions that will keep x_D constant, by manipulating R, despite changes in z and F. For what parameter values are these feedforward controllers physically realizable?

9.4. Greg Shinskey has suggested that the steady-state distillate and bottoms compositions in a binary distillation column can be approximately related by

$$\frac{x_D/(1 - x_D)}{x_B/(1 - x_B)} = \text{SF}$$

where SF is a separation factor. At total reflux it is equal to α^{N_T+1}, where α is the relative volatility and N_T is the number of theoretical trays. Assuming SF is a constant, derive the nonlinear steady-state relationship showing how distillate drawoff rate D must be manipulated, as feed rate F and feed composition z vary, in order to hold distillate composition x_D constant. Sketch this relationship for several values of SF and x_D.

9.5. Make root locus plots of first- and second-order openloop-unstable processes with PI feedback controllers.

9.6. Find the closedloop stability requirements for a third-order openloop-unstable process with a proportional controller:

$$G_{C(s)} G_{M(s)} = \frac{K_c K_p}{(\tau_{o1}s + 1)(\tau_{o2}s + 1)(\tau_{o3}s - 1)}$$

9.7. Find the value of feedback controller gain K_c that gives a closedloop system with a damping coefficient of 0.707 for a second-order openloop-unstable process with $\tau_{o2} > \tau_{o1}$:

$$G_{C(s)} G_{M(s)} = \frac{K_p K_c}{(\tau_{o1}s + 1)(\tau_{o2}s - 1)}$$

9.8. What is the ultimate gain and period of the system with a positive zero:

$$G_{M(s)} = \frac{-3s + 1}{(s+1)(5s+1)}$$

(*a*) With a proportional controller?
(*b*) With a PI controller for $\tau_I = 2$?

9.9. (*a*) Sketch the root locus plot of a system with openloop transfer function

$$G_{C(s)}G_{M(s)} = \frac{K_c}{(s+1)(s+5)(s-0.5)}$$

(*b*) For what values of gain K_c is the system closedloop stable?

9.10. Design a feedforward controller for the two-heated-tank process considered in Example 8.1. The load disturbance is inlet feed temperature T_0.

9.11. Modify your feedforward controller design of Problem 9.10 so that it can handle both feed temperature and feed flow rate changes and uses a feedback temperature controller to trim up the steam flow.

9.12. A "valve position controller" is used to minimize operating pressure in a distillation column. Assume that the openloop process transfer function between column pressure and cooling-water flow $G_M = P/F_w$ is known.
(*a*) Sketch a block diagram of the closedloop system.
(*b*) What is the closedloop characteristic equation of the system?

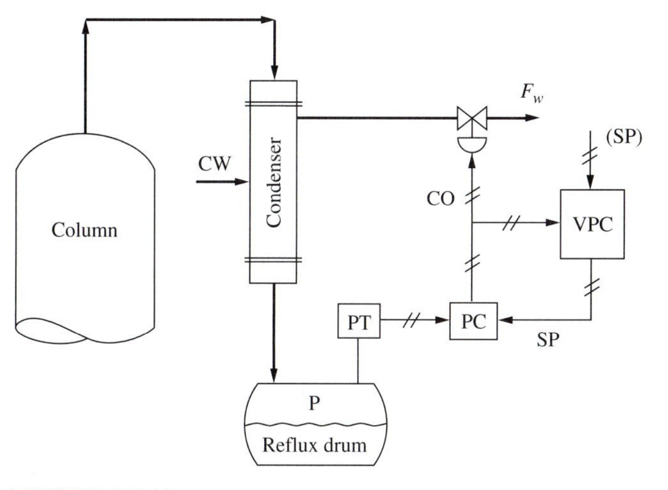

FIGURE P9.12

9.13. A proportional-only controller is used to control the liquid level in a tank by manipulating the outflow. It has been proposed that the steady-state offset of the proportional-only controller could be eliminated by using a combined feedforward-feedback system.

The flow rate into the tank is measured. The flow signal is sent through a first-order lag with time constant τ_F. The output of the lag is added to the output of the level controller. The sum of these two signals sets the outflow rate. Assume that the flow rate follows the setpoint signal to the flow controller exactly.
(*a*) Derive the closedloop transfer function between liquid level h and inflow rate F_0.
(*b*) Show that there is no steady-state offset of level from the setpoint for step changes in inflow rate.

9.14. A process has a positive zero:

$$G_M = \frac{-3s + 1}{(s + 1)(5s + 1)}$$

When a proportional-only feedback controller is used, the ultimate gain is 2. Outline your procedure for finding the optimal value of τ_D if a proportional-derivative controller is used. The optimum τ_D will give the maximum value for the ultimate gain.

9.15. Draw a block diagram of a process that has two manipulated variable inputs M_1 and M_2 that affect the output Y. A feedback controller G_{C1} is used to control Y by manipulating M_1 since the transfer function between M_1 and Y (G_{M1}) has a small time constant and small deadtime.

However, since M_1 is more expensive than M_2, we wish to minimize the long-term steady-state use of M_1. Therefore, a "valve position controller" G_{C2} is used to slowly drive M_1 to its setpoint M_1^{set}. What is the closedloop characteristic equation of the system?

9.16. Make root locus plots for the two processes given below, and calculate the ultimate gains and the gains that give closedloop damping coefficients of 0.707 for both processes.

$$(a)\ G_M G_C = \frac{K_c}{(10s + 1)(50s + 1)}$$

$$(b)\ G_M G_C = \frac{K_c(-\tau s + 1)}{(10s + 1)(50s + 1)} \quad \text{with } \tau = 6$$

For part (*b*) derive your expression for the ultimate gain as a general function of τ.

9.17. An openloop-unstable process has a transfer function containing a positive pole at $+1/\tau$ and a negative pole at $-1/a\tau$. Its steady-state gain is unity. If a proportional-only controller is used, what is the value of a that gives a closedloop damping coefficient of 0.5 when the controller gain is 10 times the minimum gain?

9.18. The "Smith predictor" for deadtime compensation is a feedback controller that has been modified by feeding back the controller output into the controller input through a trans-

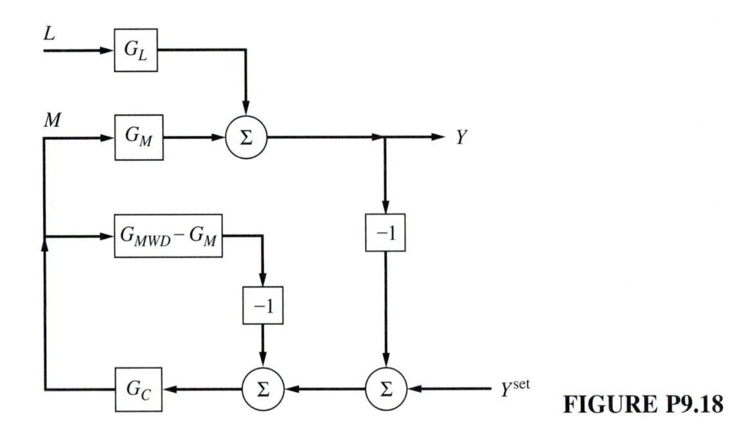

FIGURE P9.18

fer function $G_{S(s)} = G_{MWD} - G_M$. The transfer function G_{MWD} is the portion of the process openloop transfer function G_M that does not contain the deadtime D. What is the closedloop characteristic equation of the system?

9.19. We want to analyze the floating pressure "valve position control" (VPC) system proposed by Shinskey. Let the transfer function between cooling water flow and column pressure be

$$G_{M(s)} = \frac{0.1 \text{ psi/gpm}}{s(s+1)}$$

The span of the pressure transmitter is 50 psi. The control valve has linear installed characteristics and passes 600 gpm when wide open. A proportional-only pressure controller is used.

$$G_{C1} = K_1$$

An integral-only valve position controller is used.

$$G_{C2} = \frac{K_2}{s}$$

(*a*) Draw a block diagram of this VPC closedloop system.
(*b*) What is the closedloop characteristic equation of this system?
(*c*) Considering only the pressure control loop, determine the value of the pressure controller gain K_1 that gives a closedloop damping coefficient of 0.707 for the pressure control loop. Sketch a root locus plot for the pressure control loop.
(*d*) Using the value of K_1 found above and with the pressure controller on automatic, determine the VPC tuning constant (K_2) that makes the entire VPC–pressure controller system critically damped, i.e., that gives a closedloop damping coefficient of unity. Sketch a root locus plot for the VPC controller with the pressure controller on automatic.

9.20. Two openloop transfer functions $G_{M1(s)}$ and $G_{M2(s)}$ are connected in parallel. They have the same input $M_{(s)}$ but each has its own output, $Y_{1(s)}$ and $Y_{2(s)}$, respectively. In the closedloop system, a proportional controller K_1 is installed to control Y_1 by changing M.

However, a cascade system is used where another proportional controller K_2 is used to control Y_2 by changing the setpoint of the K_1 controller. Thus we have a parallel cascade system.
(*a*) Draw a block diagram of the system.
(*b*) Derive the closedloop transfer function between Y_2 and Y_2^{set}.
(*c*) What is the closedloop characteristic equation?
Suppose the openloop process transfer functions are

$$G_{M1(s)} = \frac{1}{(s+1)^2} \qquad G_{M2(s)} = \frac{1}{5s+1}$$

(*d*) Determine the value of K_1 that gives a closedloop damping coefficient of 0.5 in the secondary loop.
(*e*) Determine the ultimate gain in the primary K_2 loop when the secondary loop gain is $K_1 = 3$.

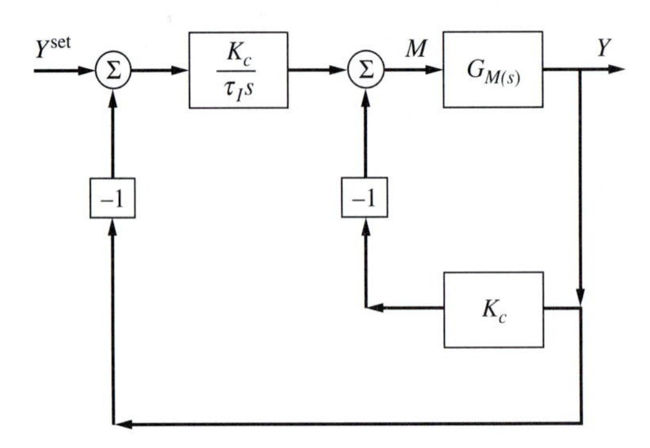

FIGURE P9.21

9.21. The openloop transfer function of a process is

$$G_{M(s)} = \frac{K_p}{\tau_o s + 1}$$

(*a*) If a conventional PI controller is used with gain K_c and reset time τ_I, derive the closedloop servo transfer function between the output $Y_{(s)}$ and the setpoint $Y^{set}_{(s)}$.

(*b*) A block diagram of a *nonconventional* PI controller is sketched in the accompanying figure. There are two feedback loops, one using proportional action on the process output and the other using integral action on the error signal. The element in the proportional loop is K_c, and the element in the integral loop is $K_c/(\tau_I s)$. Derive the closedloop servo transfer function between the output $Y_{(s)}$ and the setpoint $Y^{set}_{(s)}$ when this control structure is used.

(*c*) Compare the closedloop characteristic equations of the two types of controllers and the zeros of the two closedloop transfer functions.

PART THREE

Frequency-Domain Dynamics and Control

Our language lessons are coming along nicely. You should be fairly fluent in both English (differential equations) and Russian (Laplace transforms) by this time. We have found that only a small vocabulary is needed to handle our controller design problems. You must know the meaning and the pronunciation of only nine "words" in the two languages: first-order lag, first-order lead, deadtime, gain, second-order underdamped lag, integrator, derivative, positive zero, and positive pole.

We have found that dynamics are more conveniently handled in the Russian transfer function language than in the English ODE language. However, the manipulation of the algebraic equations becomes more difficult as the system becomes more complex and higher in order. If the system is Nth order, an Nth-order polynomial in s must be factored into its N roots. For N greater than 2, we usually abandon analytical methods and turn to numerical root-solving techniques, which are conveniently available in many commercial software packages such as MATLAB. Also, deadtime in the transfer function cannot be handled easily by the Russian Laplace-domain methods.

To overcome these problems, we must learn another language: Chinese. This is what we call the frequency-domain methods. These methods are a little more removed from our mother tongue of English and a little more abstract. But they are extremely powerful and very useful in dealing with realistically complex processes. Basically, this is because the manipulation of transfer functions becomes a problem of combining complex numbers numerically (addition, multiplication, etc.). This is easily done on a digital computer.

In Chapter 10 we learn this new "Chinese" language (including several dialects: Bode, Nyquist, and Nichols plots). In Chapter 11 we use frequency-domain methods to design closedloop feedback control systems.

As with Russian, you must learn only a limited Chinese vocabulary. The learning of 2000 to 3000 Chinese characters is *not* required (thank goodness)! Only nine Chinese words must be learned in each of the three dialects. It takes a little practice to get the hang of handling the complex numbers in various coordinate systems, but the effort is well worth it.

CHAPTER 10

Frequency-Domain Dynamics

10.1 DEFINITION

For most processes the *frequency response* is defined as the steady-state behavior of the system when forced by a sinusoidal input. Suppose the input $u_{(t)}$ to the process is a sine wave $u_{s(t)}$ of amplitude $\overline{U}$ and frequency ω as shown in Fig. 10.1.

$$u_{s(t)} = \overline{U}\sin(\omega t) \tag{10.1}$$

The period of one complete cycle is T units of time. Frequency is expressed in a variety of units. The electrical engineers usually use units of hertz (cycles per second):

$$\omega\ (\text{hertz}) = \frac{1}{T} \tag{10.2}$$

We chemical engineers find it more convenient to use frequency units of radians per time:

$$\omega\ (\text{radians/time}) = \frac{2\pi}{T} \tag{10.3}$$

The reason for this preference will become clear later in this chapter. Be very careful that you use frequency units that are consistent in terms of both time (seconds, minutes, or hours) and angles (cycles or radians). A very common error is to be off by a factor of 2π because of the radians/cycles variation or to be off by a factor of 60 because of mixing minutes and hours.

In a linear system, if the input is a sine wave with frequency ω, the output is also a sine wave with the same frequency. The output has, however, a different amplitude and "lags" (falls behind) or "leads" (rises ahead of) the input. Figure 10.2*a* shows the output $y_{s(t)}$ lagging the input $u_{s(t)}$ by T_y units of time. Figure 10.2*b* shows the output leading the input by T_y. The *phase angle* θ is defined as the angular difference

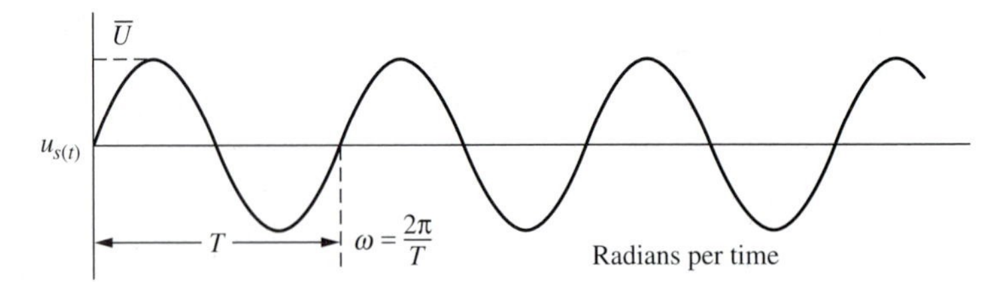

FIGURE 10.1
Sine wave input.

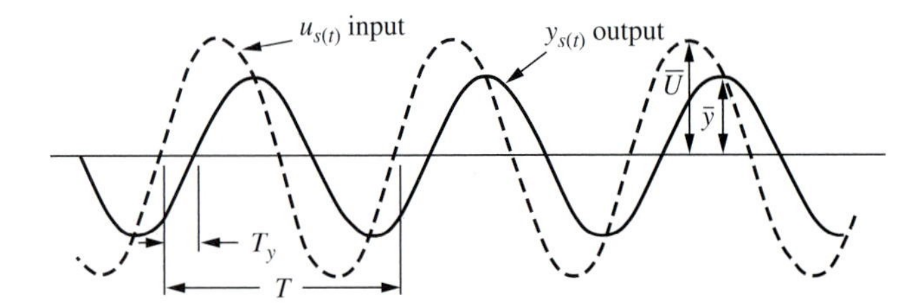

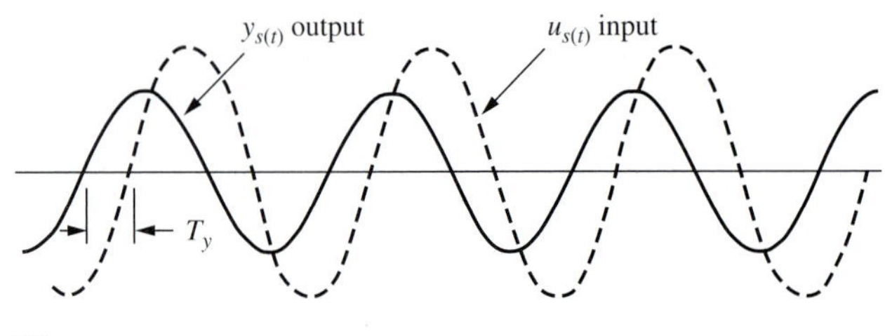

FIGURE 10.2
Sinusoidal input-output. (*a*) Output lags. (*b*) Output leads.

between the input and the output. In equation form,

$$y_{s(t)} = \bar{y}\sin(\omega t + \theta) \tag{10.4}$$

where $y_{s(t)}$ = output resulting from the sine wave input of frequency ω
$\bar{y}$ = maximum amplitude of the output y_s
θ = phase angle, in radians

If the output lags the input, θ is negative. If the output leads the input, θ is positive.

$$\theta = \frac{T_y}{T}2\pi \quad \text{(radians)} \qquad \theta = \frac{T_y}{T}360 \quad \text{(degrees)} \tag{10.5}$$

The magnitude ratio MR is defined as the ratio of the maximum amplitude of the output over the maximum amplitude of the input:

$$\text{MR} \equiv \frac{\bar{y}}{\bar{U}} \tag{10.6}$$

For a given process, both the phase angle θ and the magnitude ratio MR will change if frequency ω is changed. We must find out how θ and MR vary as ω covers a range from zero to infinity. Then we will know the system's *frequency response* (its Chinese translation). Different processes have different dependence of MR and θ on ω. Since each process is unique, the frequency-response curves are like fingerprints. By merely looking at curves of MR and θ, we can tell the kind of system (its order) and the values of parameters (time constants, steady-state gain, and damping coefficient).

There are a number of ways to obtain the frequency response of a process. Experimental methods, discussed in Chapter 16, are used when a mathematical model of the system is not available. If equations can be developed that adequately describe the system, the frequency response can be obtained directly from the system transfer function.

10.2 BASIC THEOREM

We will show that the frequency response of a system can be found simply by substituting $i\omega$ for s in the system transfer function $G_{(s)}$. Making the substitution $s = i\omega$ gives a complex number $G_{(i\omega)}$ with the following features:

1. A magnitude $|G_{(i\omega)}|$ that is the same as the magnitude ratio MR that would be obtained by forcing the system with a sine wave input of frequency ω.
2. A phase angle or *argument,* arg $G_{(i\omega)}$, that is equal to the phase angle θ that would be obtained from forcing the system with a sine wave of frequency ω.

$$|G_{(i\omega)}| = MR_{(\omega)} \tag{10.7}$$

$$\arg G_{(i\omega)} = \theta_{(\omega)} \tag{10.8}$$

$G_{(i\omega)}$ is a complex number, so it can be represented in terms of a real part and an imaginary part:

$$G_{(i\omega)} = \mathrm{Re}[G_{(i\omega)}] + i\,\mathrm{Im}[G_{(i\omega)}] \tag{10.9}$$

In polar form, the complex number $G_{(i\omega)}$ is represented as

$$G_{(i\omega)} = |G_{(i\omega)}|\, e^{i \arg G_{(i\omega)}} \tag{10.10}$$

where

$$|G_{(i\omega)}| = \text{absolute value of } G_{(i\omega)} = \sqrt{(\mathrm{Re}[G_{(i\omega)}])^2 + (\mathrm{Im}[G_{(i\omega)}])^2} \tag{10.11}$$

$$\arg G_{(i\omega)} = \text{argument of } G_{(i\omega)} = \arctan\left(\frac{\mathrm{Im}[G_{(i\omega)}]}{\mathrm{Re}[G_{(i\omega)}]}\right) \tag{10.12}$$

This very remarkable result [Eqs. (10.7) and (10.8)] permits us to go from the Laplace domain to the frequency domain with ease.

$$\text{Russian Laplace domain } G_{(s)} \quad \overset{s=i\omega}{\Longrightarrow} \quad \text{Chinese frequency domain } G_{(i\omega)}$$

Before we prove that this simple substitution is valid, let's illustrate its application in a specific example.

EXAMPLE 10.1. Suppose we want to find the frequency response of a first-order process with the transfer function

$$G_{(s)} = \frac{K_p}{\tau_o s + 1} \tag{10.13}$$

Substituting $s = i\omega$ gives

$$G_{(i\omega)} = \frac{K_p}{1 + i\omega\tau_o} \tag{10.14}$$

Multiplying numerator and denominator by the complex conjugate of the denominator gives

$$G_{(i\omega)} = \frac{K_p}{1 + i\omega\tau_o}\frac{1 - i\omega\tau_o}{1 - i\omega\tau_o} = \frac{K_p(1 - i\omega\tau_o)}{1 + \omega^2\tau_o^2} = \left[\frac{K_p}{1 + \omega^2\tau_o^2}\right] + i\left[\frac{-K_p\omega\tau_o}{1 + \omega^2\tau_o^2}\right] \tag{10.15}$$

Therefore

$$\mathrm{Re}[G_{(i\omega)}] = \frac{K_p}{1 + \omega^2\tau_o^2} \tag{10.16}$$

$$\mathrm{Im}[G_{(i\omega)}] = \frac{-K_p\omega\tau_o}{1 + \omega^2\tau_o^2} \tag{10.17}$$

Therefore

$$\mathrm{MR} = |G_{(i\omega)}| = \sqrt{\left(\frac{K_p}{1 + \omega^2\tau_o^2}\right)^2 + \left(\frac{-K_p\omega\tau_o}{1 + \omega^2\tau_o^2}\right)^2} = \frac{K_p}{\sqrt{1 + \omega^2\tau_o^2}} \tag{10.18}$$

$$\theta = \arg G_{(i\omega)} = \arctan\left[\frac{\dfrac{-K_p\omega\tau_o}{1 + \omega^2\tau_o^2}}{\dfrac{K_p}{1 + \omega^2\tau_o^2}}\right] = \arctan(-\omega\tau_o) \tag{10.19}$$

Notice that both MR and θ vary with frequency ω. ■

Now let us prove that this simple substitution $s = i\omega$ really works. Let $G_{(s)}$ be the transfer function of any arbitrary *N*th-order system. The only restriction we place on the system is that it is stable. If it were unstable and we forced it with a sine wave input, the output would go off to infinity. So we cannot experimentally get the frequency response of an unstable system. This does *not* mean that we cannot use frequency-domain methods for openloop-unstable systems. We return to this subject in Chapter 11.

If the system is initially at rest (all derivatives equal zero) and we start to force it with a sine wave $u_{s(t)}$, the output $y_{(t)}$ will go through some transient period as shown in Fig. 10.3 and then settle down to a steady sinusoidal oscillation. In the Laplace domain, the output is by definition

$$Y_{(s)} = G_{(s)}U_{(s)} \tag{10.20}$$

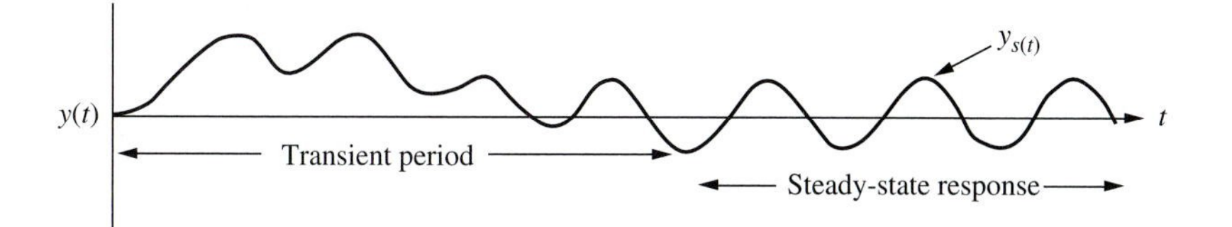

FIGURE 10.3
Response of a system to a sine wave input.

For the sine wave input $u_{(t)} = \overline{U}\sin(\omega t)$. Laplace transforming,

$$U_{(s)} = \overline{U}\frac{\omega}{s^2 + \omega^2} \tag{10.21}$$

Therefore, the output with this sine wave input is

$$Y_{(s)} = G_{(s)}U_{(s)} = G_{(s)}\frac{\overline{U}\omega}{s^2 + \omega^2}$$

$G_{(s)}$ is a ratio of polynomials in s that can be factored into poles and zeros.

$$G_{(s)} = \frac{(s - z_1)(s - z_2)\cdots(s - z_M)}{(s - p_1)(s - p_2)\cdots(s - p_N)}$$

$$\begin{aligned} Y_{(s)} &= \frac{(s - z_1)(s - z_2)\cdots(s - z_M)}{(s - p_1)(s - p_2)\cdots(s - p_N)}\left(\frac{\overline{U}\omega}{s^2 + \omega^2}\right) \\ &= \frac{(s - z_1)(s - z_2)\cdots(s - z_M)}{(s - p_1)(s - p_2)\cdots(s - p_N)}\frac{\overline{U}\omega}{(s + i\omega)(s - i\omega)} \end{aligned} \tag{10.22}$$

Partial fractions expansion gives

$$Y_{(s)} = \frac{A}{s + i\omega} + \frac{B}{s - i\omega} + \frac{C}{s - p_1} + \cdots + \frac{W}{s - p_N} \tag{10.23}$$

where

$$A = \lim_{s \to -i\omega}[(s + i\omega)Y_{(s)}] = \lim_{s \to -i\omega}\left[\frac{\omega\overline{U}G_{(s)}}{s - i\omega}\right] = -\frac{\overline{U}}{2i}G_{(-i\omega)}$$

$$B = \lim_{s \to i\omega}[(s - i\omega)Y_{(s)}] = \lim_{s \to i\omega}\left[\frac{\omega\overline{U}G_{(s)}}{s + i\omega}\right] = \frac{\overline{U}}{2i}G_{(i\omega)}$$

$$C = \lim_{s \to p_1}[(s - p_1)Y_{(s)}]$$

Substituting into Eq. (10.23) and inverting to the time domain,

$$y_{(t)} = \left(\frac{-\overline{U}}{2i}G_{(-i\omega)}\right)e^{-i\omega t} + \left(\frac{\overline{U}}{2i}G_{(i\omega)}\right)e^{i\omega t} + \sum_{j=1}^{N} c_j e^{p_j t} \tag{10.24}$$

Now we are interested only in the steady-state response after the initial transients have died out and the system has settled into a sustained oscillation. As time goes to infinity, all the exponential terms in the summation shown in Eq. (10.24) decay to zero. The system is stable, so all the poles p_j must be negative. The steady-state output with a sine wave input, which we called $y_{s(t)}$, is

$$y_{s(t)} = \frac{\overline{U}}{2i}[G_{(i\omega)}e^{i\omega t} - G_{(-i\omega)}e^{-i\omega t}] \tag{10.25}$$

The $G_{(i\omega)}$ and $G_{(-i\omega)}$ terms are complex numbers and can be put into polar form:

$$\begin{aligned} G_{(i\omega)} &= \left|G_{(i\omega)}\right| e^{i \arg G_{(i\omega)}} \\ G_{(-i\omega)} &= \left|G_{(-i\omega)}\right| e^{i \arg G_{(-i\omega)}} = \left|G_{(i\omega)}\right| e^{-i \arg G_{(i\omega)}} \end{aligned} \tag{10.26}$$

Equation (10.25) becomes

$$y_{s(t)} = \overline{U}\left|G_{(i\omega)}\right| \left[\frac{e^{i(\omega t + \arg G_{(i\omega)})} - e^{-i(\omega t + \arg G_{(i\omega)})}}{2i}\right]$$

Therefore,

$$\frac{y_{s(t)}}{\overline{U}} = \left|G_{(i\omega)}\right| \sin(\omega t + \arg G_{(i\omega)}) \tag{10.27}$$

We have proved what we set out to prove: (1) the magnitude ratio MR is the absolute value of $G_{(s)}$ with s set equal to $i\omega$, and (2) the phase angle is the argument of $G_{(s)}$ with s set equal to $i\omega$.

10.3 REPRESENTATION

Three different kinds of plots are commonly used to show how magnitude ratio (absolute magnitude) and phase angle (argument) vary with frequency ω. They are called *Nyquist, Bode* (pronounced "Bow-dee"), and *Nichols* plots. After defining each of them, we show what some common transfer functions look like in the three different plots.

10.3.1 Nyquist Plots

A Nyquist plot (also called a *polar plot* or a *G plane plot*) is generated by plotting the complex number $G_{(i\omega)}$ in a two-dimensional diagram whose ordinate is the imaginary part of $G_{(i\omega)}$ and whose abscissa is the real part of $G_{(i\omega)}$. The real and imaginary parts of $G_{(i\omega)}$ at a specific frequency ω_1 define a point in this coordinate system. As shown in Fig. 10.4*a*, either rectangular (real versus imaginary) or polar (absolute magnitude versus phase angle) can be used to locate the point. As frequency is varied continuously from zero to infinity, a curve is formed in the G plane, as shown in Fig. 10.4*b*. Frequency is a parameter along this curve. The shape and location of the curve are unique characteristics of the system. Let us show what the Nyquist plots of some simple transfer functions look like.

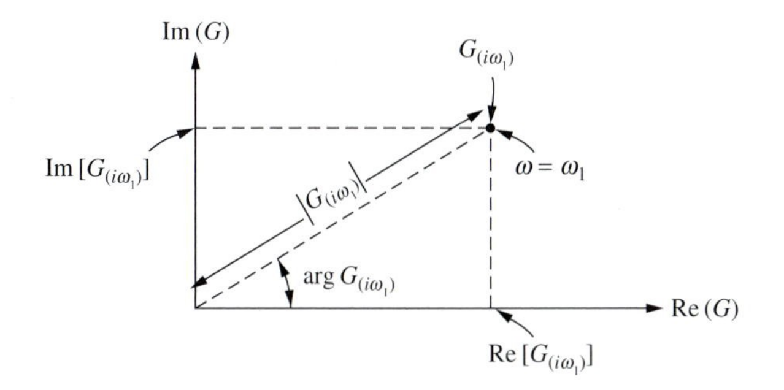

(a)

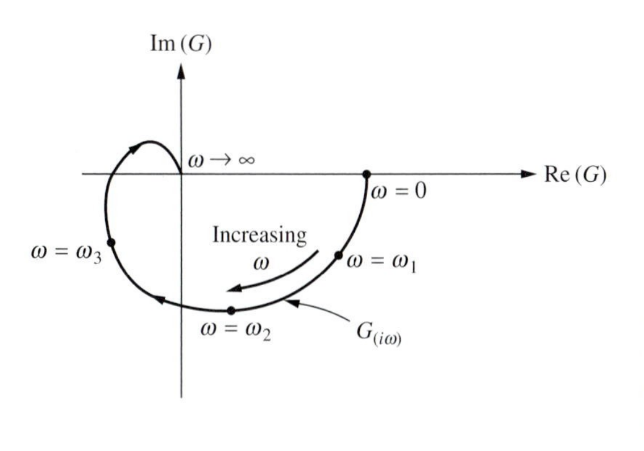

(b)

FIGURE 10.4
Nyquist plots in the G plane.
(a) Single point $G_{(i\omega_1)}$.
(b) Complete curve $G_{(i\omega)}$.

A. First-order lag

$$G_{(s)} = \frac{K_p}{\tau_o s + 1}$$

We developed $G_{(i\omega)}$ for this transfer function in Example 10.1 [Eqs. (10.18) and (10.19)].

$$|G_{(i\omega)}| = \frac{K_p}{\sqrt{1 + \omega^2 \tau_o^2}} \qquad \arg G_{(i\omega)} = \arctan(-\omega \tau_o) \tag{10.28}$$

When frequency is zero, $|G|$ is equal to K_p and $\arg G$ is equal to zero. So the Nyquist plot starts ($\omega = 0$) on the positive real axis at $\text{Re}[G] = K_p$.

When frequency is equal to the reciprocal of the time constant ($\omega = 1/\tau_o$),

$$|G_{(i\omega)}| = \frac{K_p}{\sqrt{1 + (1/\tau_o)^2 \tau_o^2}} = \frac{K_p}{\sqrt{2}}$$

$$\arg G_{(i\omega)} = \arctan[-(1/\tau_o)\tau_o] = -45^\circ = -\frac{\pi}{4} \text{ radians}$$

This illustrates why we use frequency in radians per time: there is a convenient relationship between the time constant of the process and frequency if these units are used.

As frequency goes to infinity, $|G_{(i\omega)}|$ goes to zero and arg $G_{(i\omega)}$ goes to $-90°$ or $-\pi/2$ radians. All of these points are shown in Fig. 10.5. The complete Nyquist plot is a semicircle. This is a unique curve. Anytime you see it, you know you are dealing with a first-order lag. The effect of changing the gain K_p is also shown in Fig. 10.5. The magnitude of each point is changed, but the phase angle is not affected.

B. First-order lead

$$G_{(s)} = \tau_z s + 1 \quad \Rightarrow \quad G_{(i\omega)} = 1 + i\omega\tau_z$$

The real part is constant at +1. The imaginary part increases directly with frequency.

$$|G_{(i\omega)}| = \sqrt{1 + \omega^2\tau_z^2} \qquad \arg G_{(i\omega)} = \arctan(\omega\tau_z) \tag{10.29}$$

When $\omega = 0$, $\arg G = 0$ and $|G| = 1$. As ω goes to infinity, $|G|$ becomes infinite and $\arg G$ goes to $+90°$ or $+\pi/2$ radians. The Nyquist plot is shown in Fig. 10.6.

C. Deadtime

$$G_{(s)} = e^{-Ds} \quad \Rightarrow \quad G_{(i\omega)} = e^{-i\omega D}$$

This is a complex number with magnitude of 1 and argument equal to $-\omega D$.

$$|G_{(i\omega)}| = 1 \qquad \arg G_{(i\omega)} = -\omega D \tag{10.30}$$

Deadtime changes the phase angle but has no effect on the magnitude; the magnitude is unity at all frequencies. The Nyquist plot is shown in Fig. 10.7. The curve moves around the unit circle as ω increases.

D. Deadtime and first-order lag

Combining the transfer functions for deadtime and first-order lag gives

$$G_{(s)} = \frac{K_p e^{-Ds}}{\tau_o s + 1}$$

$$G_{(i\omega)} = \frac{K_p e^{-Di\omega}}{1 + i\omega\tau_o} = \left(\frac{K_p}{\sqrt{1 + \omega^2\tau_o^2}} e^{i\arctan(-\omega\tau_o)}\right) e^{-i\omega D}$$

$$= \frac{K_p}{\sqrt{1 + \omega^2\tau_o^2}} e^{i[\arctan(-\omega\tau_o) - D\omega]}$$

Therefore,

$$|G_{(i\omega)}| = \frac{K_p}{\sqrt{1 + \omega^2\tau_o^2}} \qquad \arg G_{(i\omega)} = \arctan(-\omega\tau_o) - D\omega \tag{10.31}$$

Note that the magnitude is exactly the same as for the first-order lag alone. Phase angle is decreased by the deadtime contribution. Figure 10.8 shows that the Nyquist plot is a spiral that wraps around the origin as it shrinks in magnitude.

This example illustrates a very important property of complex numbers. The magnitude of the product of two complex numbers is the *product* of their magnitudes. The argument of the product of two complex numbers is the *sum* of their arguments.

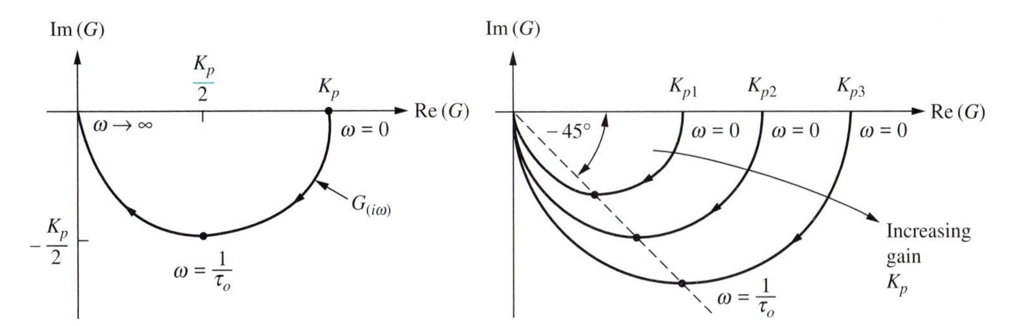

FIGURE 10.5
Nyquist plot of first-order lag.

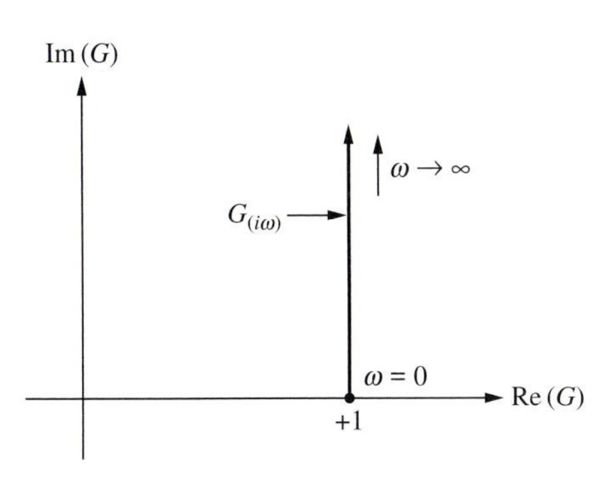

FIGURE 10.6
Nyquist plot for first-order lead.

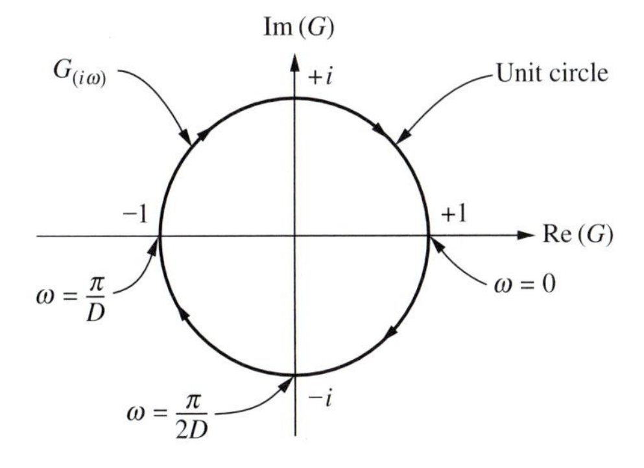

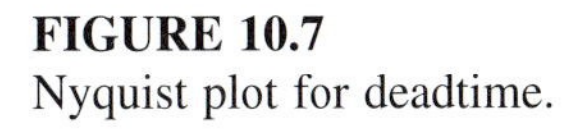

FIGURE 10.7
Nyquist plot for deadtime.

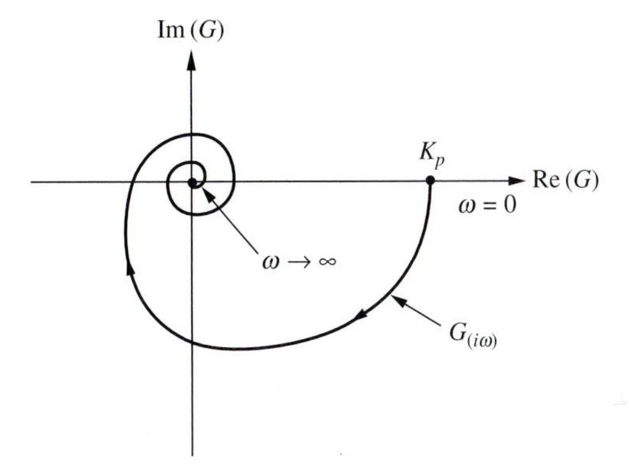

FIGURE 10.8
Nyquist plot for deadtime with first-order lag.

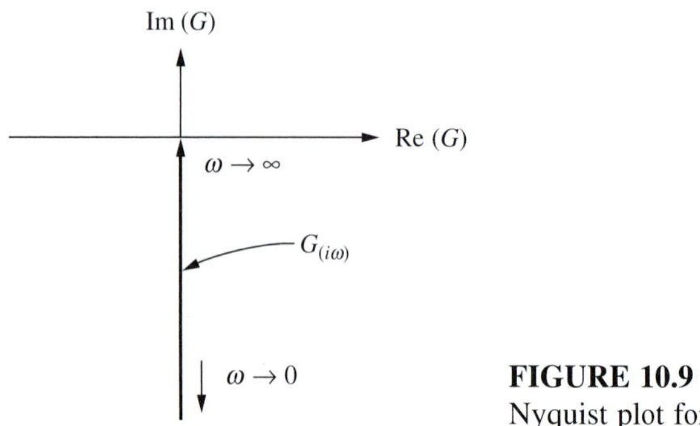

FIGURE 10.9
Nyquist plot for an integrator.

E. Integrator

The transfer function for a pure integrator is $G_{(s)} = 1/s$. Going into the frequency domain by substituting $s = i\omega$ gives

$$G_{(i\omega)} = \frac{1}{i\omega} = -\frac{1}{\omega}i$$

$G_{(i\omega)}$ is a pure imaginary number (its real part is zero), lying on the imaginary axis. It starts at minus infinity when ω is zero and goes to the origin as $\omega \rightarrow \infty$. The Nyquist plot is sketched in Fig. 10.9.

$$\begin{aligned} \left|G_{(i\omega)}\right| &= \frac{1}{\omega} \\ \arg G_{(i\omega)} &= \arctan\left(\frac{-1/\omega}{0}\right) = -90^\circ = -\frac{\pi}{2} \text{ radians} \end{aligned} \tag{10.32}$$

F. Integrator and first-order lag

Combining the transfer functions for an integrator and first-order lag gives

$$\begin{aligned} G_{(s)} &= \frac{K_p}{s(\tau_o s + 1)} \\ G_{(i\omega)} &= \frac{K_p}{-\omega^2\tau_o + i\omega} = \frac{-K_p\tau_o\omega - K_p i}{\omega(\omega^2\tau_o^2 + 1)} \\ \left|G_{(i\omega)}\right| &= \frac{K_p}{\omega\sqrt{1 + \omega^2\tau_o^2}} \\ \arg G_{(i\omega)} &= \arctan\left(\frac{-1}{-\omega\tau_o}\right) = -\frac{\pi}{2} + \arctan(-\omega\tau_o) \end{aligned} \tag{10.33}$$

The Nyquist curve is shown in Fig. 10.10. Note that the results given in Eq. (10.33) could have been derived by combining the magnitudes and arguments of an integrator [Eq. (10.32)] and a first-order lag [Eq. (10.28)].

G. Second-order underdamped system

The second-order underdamped system is probably the most important transfer function that we need to translate into the frequency domain. Since we often design

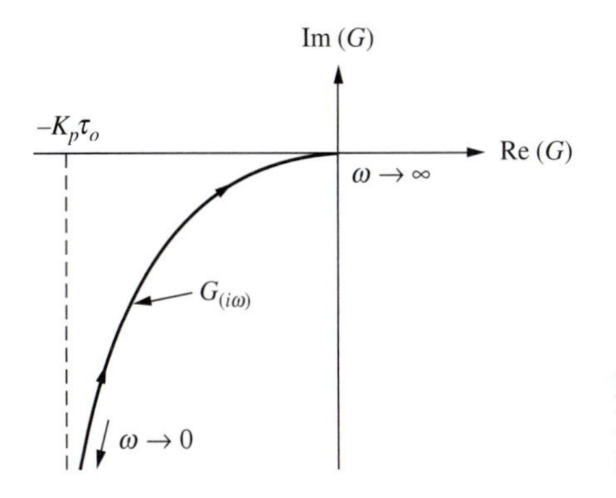

FIGURE 10.10
Nyquist plot of an integrator and first-order lag.

for a desired closedloop damping coefficient, we need to know what the Nyquist plot of such a system looks like. Most of the underdamped systems that we have in chemical engineering are closedloop systems (process with a controller).

$$G_{(s)} = \frac{K_p}{\tau_o^2 s^2 + 2\zeta\tau_o s + 1}$$

$$G_{(i\omega)} = \frac{K_p}{(1 - \tau_o^2\omega^2) + i(2\zeta\tau_o\omega)} = \frac{K_p(1 - \tau_o^2\omega^2) - iK_p(2\zeta\tau_o\omega)}{(1 - \tau_o^2\omega^2)^2 + (2\zeta\tau_o\omega)^2}$$

$$|G_{(i\omega)}| = \frac{K_p}{\sqrt{(1 - \tau_o^2\omega^2)^2 + (2\zeta\tau_o\omega)^2}} \tag{10.34}$$

$$\arg G_{(i\omega)} = \arctan\left(\frac{-2\zeta\tau_o\omega}{1 - \tau_o^2\omega^2}\right) \tag{10.35}$$

Figure 10.11 shows the Nyquist plot. It starts at K_p on the positive real axis. It intersects the imaginary axis ($\arg G = -\pi/2$) when $\omega = 1/\tau_o$. At this point, $|G| = K_p/2\zeta$. Therefore, the smaller the damping coefficient, the farther out on the negative imaginary axis the curve will cross. This shape is unique to an underdamped system. Anytime you see a hump in the curve, you know the damping coefficient must be less than unity. As ω goes to infinity, the magnitude goes to zero and the phase angle goes to $-\pi$ radians ($-180°$).

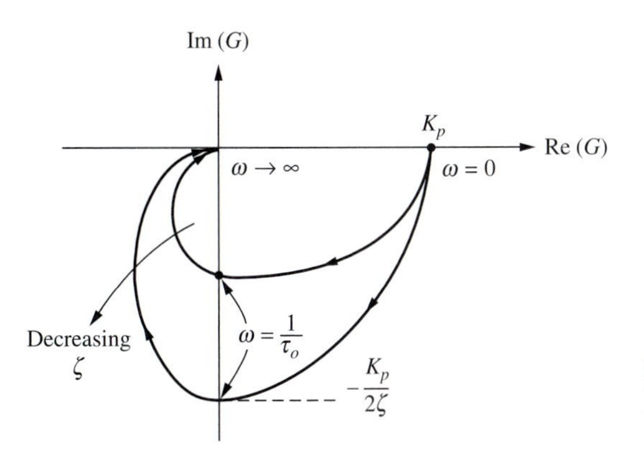

FIGURE 10.11
Nyquist plot of a second-order system.

The results of the preceding examples show that adding lags (poles) to the transfer function moves the Nyquist plot clockwise around the origin in the G plane. Adding leads (zeros) moves it counterclockwise. We will return to this generalization in the next chapter when we start designing controllers that shift these curves in the desired way.

10.3.2 Bode Plots

The Nyquist plot presents all the frequency information in a compact, one-curve form. Bode plots require two curves to be plotted instead of one. This increase in the number of plots is well worth the trouble because complex transfer functions can be handled much more easily using Bode plots. The two curves show how magnitude ratio and phase angle (argument) vary with frequency.

Phase angle is usually plotted against the log of frequency, using semilog graph paper as illustrated in Fig. 10.12. The magnitude ratio is sometimes plotted against the log of frequency on a log-log plot. However, usually it is more convenient to convert magnitude to *log modulus,* defined by the equation

$$L \equiv \text{log modulus} \equiv 20 \log_{10} |G_{(i\omega)}| \tag{10.36}$$

Then semilog graph paper can be used to plot both phase angle and log modulus versus the log of frequency, as shown in Fig. 10.12. There are very practical reasons for using these kinds of graphs, as you will find out shortly.

The units of log modulus are decibels (dB), a term originally used in communications engineering to indicate the ratio of two values of power. The scale in Fig. 10.13 is convenient to use to convert back and forth from magnitude to decibels.

Let us look at the Bode plots of some common transfer functions. We calculated the magnitudes and phase angles for most of them in the previous section. The job now is to plot them in this new coordinate system.

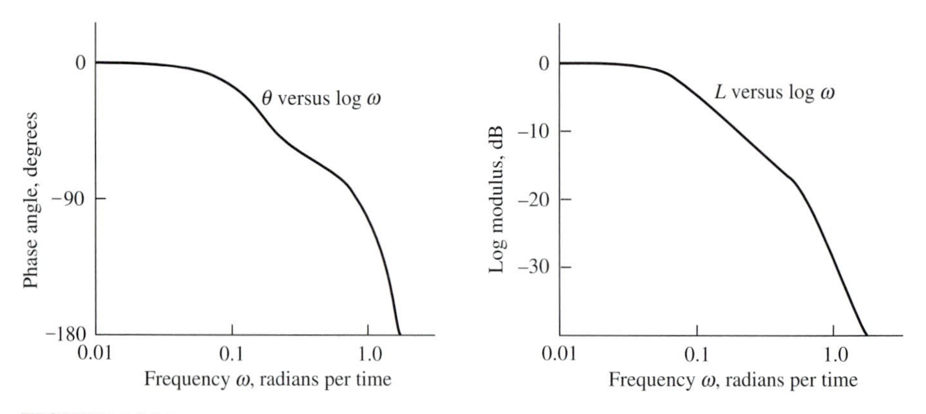

FIGURE 10.12
Bode plots of phase angle and log modulus versus the logarithm of frequency.

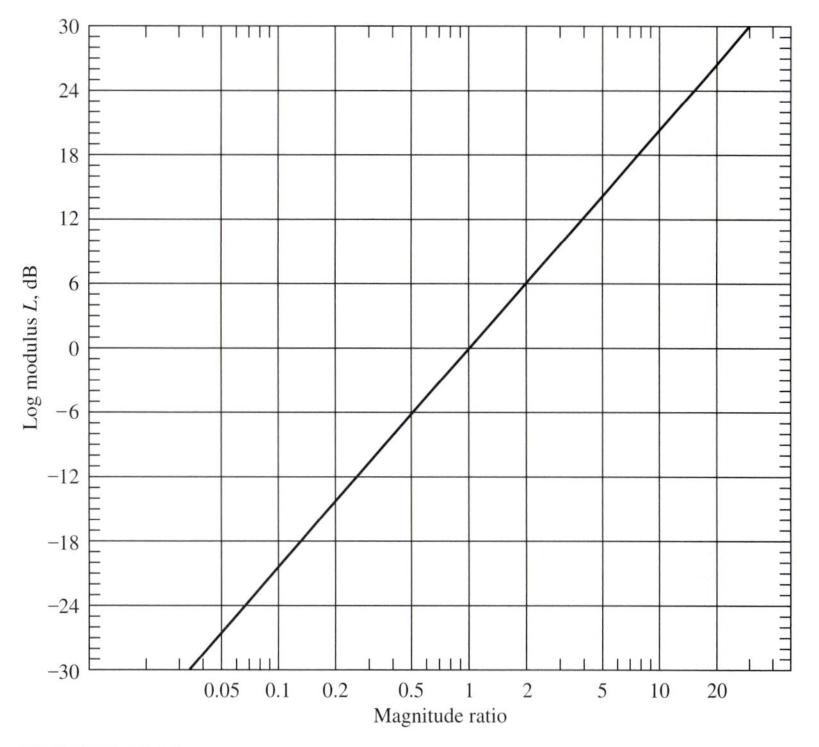

FIGURE 10.13
Conversion between magnitude ratio and log modulus.

A. Gain

If $G_{(s)}$ is just a constant K_p, $|G_{(i\omega)}| = K_p$ and phase angle $= \arg G_{(i\omega)} = 0$. Neither magnitude nor phase angle varies with frequency. The log modulus is

$$L = 20 \log_{10} |K_p| \tag{10.37}$$

Both the phase angle and log modulus curves are horizontal lines on a Bode plot, as shown in Fig. 10.14.

If K_p is less than 1, L is negative. If K_p is unity, L is zero. If K_p is greater than 1, L is positive. Increasing K_p by a factor of 10 (a "decade") increases L by a factor of

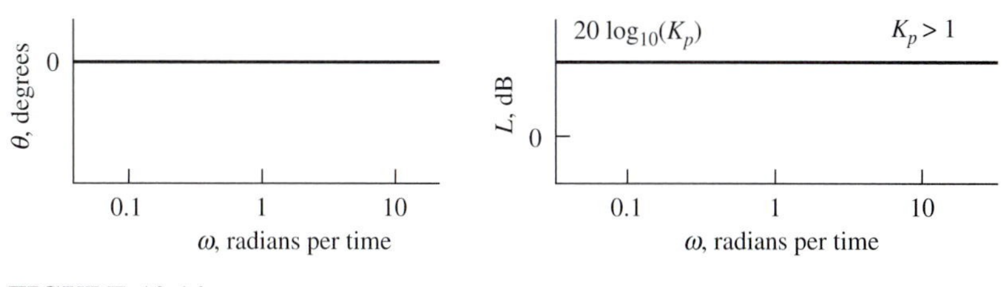

FIGURE 10.14
Bode plots of gain.

20 dB. Increasing K_p moves the L curve up in the Bode plot. Decreasing K_p moves the L curve down.

B. First-order lag

$$G_{(s)} = \frac{1}{\tau_o s + 1}$$

From Eq. (10.28) (with $K_p = 1$)

$$\left|G_{(i\omega)}\right| = \frac{1}{\sqrt{1 + \omega^2\tau_o^2}} \qquad \arg G_{(i\omega)} = \arctan(-\omega\tau_o) \tag{10.38}$$

$$L = 20\log_{10}\frac{1}{\sqrt{1 + \omega^2\tau_o^2}} = -10\log_{10}(1 + \omega^2\tau_o^2) \tag{10.39}$$

The Bode plots are shown in Fig. 10.15. One of the most convenient features of Bode plots is that the L curves can be easily sketched by considering the low- and high-frequency asymptotes. As ω goes to zero, L goes to zero. As ω becomes very large, Eq. (10.39) reduces to

$$\lim_{\omega\to\infty} L = -10\log_{10}(\omega^2\tau_o^2) = -20\log_{10}(\omega) - 20\log_{10}(\tau_o) \tag{10.40}$$

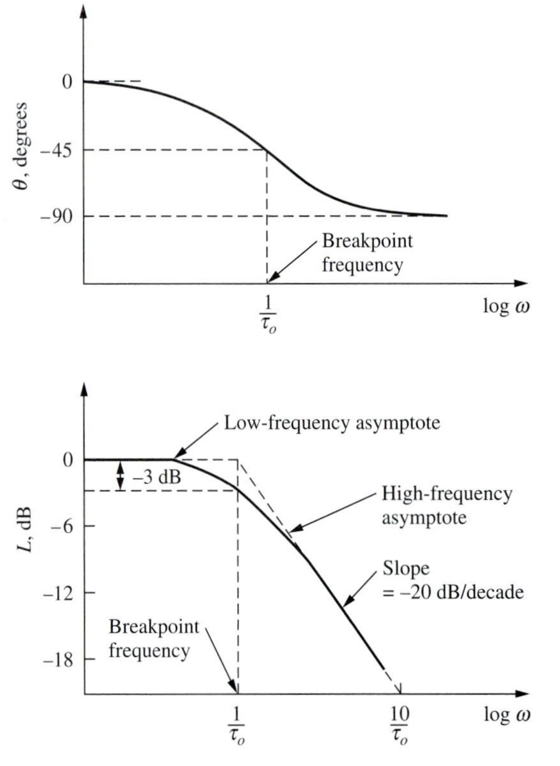

FIGURE 10.15
Bode plots for first-order lag.

This is the equation of a straight line of L versus log ω. It has a slope of -20. L will decrease 20 dB as log ω increases by 1 (or as ω increases by 10, a decade). Therefore the slope of the high-frequency asymptote is -20 dB/decade.

The high-frequency asymptote intersects the $L = 0$ line at $\omega = 1/\tau_o$. This is called the *breakpoint frequency*. The log modulus is "flat" (horizontal) out to this point and then begins to drop off.

Thus, the L curve can be easily sketched by drawing a line with slope of -20 dB/decade from the breakpoint frequency on the $L = 0$ line. Notice also that the phase angle is $-45°$ at the breakpoint frequency, which is the reciprocal of the time constant. The L curve has a value of -3 dB at the breakpoint frequency, as shown in Fig. 10.15.

C. First-order lead

$$G_{(s)} = \tau_z s + 1 \qquad G_{(i\omega)} = 1 + i\omega\tau_z$$

From Eq. (10.29),

$$\begin{aligned} \left|G_{(i\omega)}\right| &= \sqrt{1 + \omega^2\tau_z^2} \qquad \arg G_{(i\omega)} = \arctan(\omega\tau_z) \\ L &= 20\log_{10}\sqrt{1 + \omega^2\tau_z^2} \end{aligned} \tag{10.41}$$

These curves are shown in Fig. 10.16. The high-frequency asymptote has a slope of $+20$ dB/decade. The breakpoint frequency is $1/\tau_z$. The phase angle goes from zero

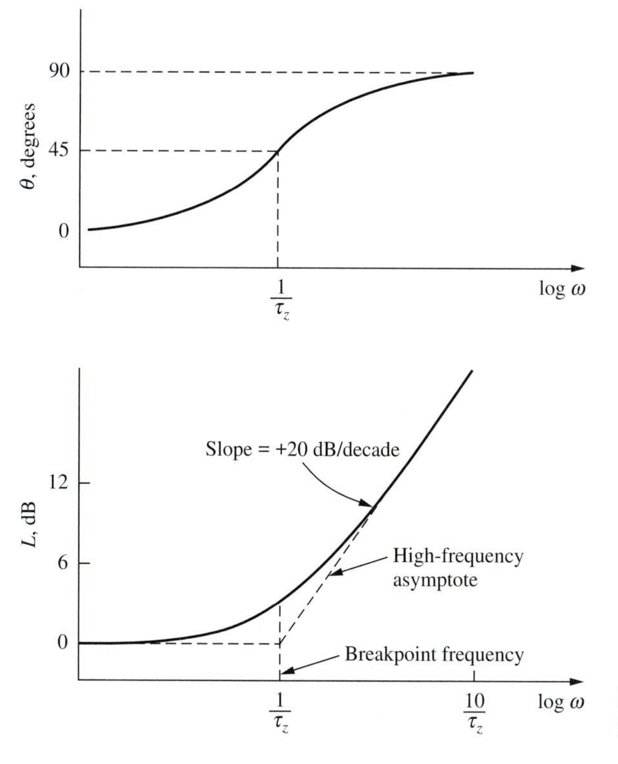

FIGURE 10.16
Bode plots for first-order lead.

to $+90°$ and is $+45°$ at $\omega = 1/\tau_z$. Thus, a lead contributes positive phase angle. A lag contributes negative phase angle. A gain doesn't change the phase angle.

D. Deadtime

$$G_{(i\omega)} = e^{-iD\omega}$$

$$L = 20\log_{10}\left|G_{(i\omega)}\right| = 20\log_{10}(1) = 0 \tag{10.42}$$

$$\arg G_{(i\omega)} = -\omega D \tag{10.43}$$

As shown in Fig. 10.17, the deadtime transfer function has a flat $L = 0$ dB curve for all frequencies, but the phase angle drops off to minus infinity. The phase angle is down to $-180°$ when the frequency is π/D. So the bigger the deadtime, the lower the frequency at which the phase angle drops off rapidly.

E. *n*th power of *s*

The general category of s^n plots includes a differentiator ($n = +1$) and an integrator ($n = -1$).

$$\begin{aligned} G_{(s)} &= s^n \quad n = \pm 1, \pm 2, \ldots \qquad G_{(i\omega)} = \omega^n i^n \\ L &= 20\log_{10}(\omega^n) = 20n\log_{10}\omega \end{aligned} \tag{10.44}$$

The L curve is a straight line in the (L–$\log\omega$) plane of the Bode plot with a slope of $20n$. See Fig. 10.18.

$$\arg G_{(i\omega)} = \arctan\left[\frac{\operatorname{Im}(G)}{\operatorname{Re}(G)}\right]$$

If n is odd, $G_{(i\omega)}$ is a pure imaginary number and the phase angle is the arctan of infinity ($\theta = \pm 90°, \pm 270°, \ldots$). If n is even, $G_{(i\omega)}$ is a real number and the phase angle is the arctan of zero ($\theta = \pm 0°, \pm 180°, \ldots$). Therefore, the phase angle changes by $90°$ or $\pi/2$ radians for each successive integer value of n.

$$\arg G_{(i\omega)} = n\left(\frac{\pi}{2}\right) \text{ radians} \tag{10.45}$$

The important specific values of n are:

1. $n = 1$: This is the transfer function of an ideal derivative.

$$G_{(i\omega)} = i\omega \quad \Rightarrow \quad L = 20\log_{10}\omega \tag{10.46}$$

This is a straight line with a slope of +20 dB/decade.

$$\theta = \arctan\left(\frac{\omega}{0}\right) = +90° \tag{10.47}$$

So an ideal derivative *increases* phase angle by 90°.

2. $n = -1$: This is the transfer function of an integrator.

$$G_{(i\omega)} = \frac{1}{i\omega} = \left(-\frac{1}{\omega}\right)i \quad \Rightarrow \quad L = -20\log_{10}\omega \tag{10.48}$$

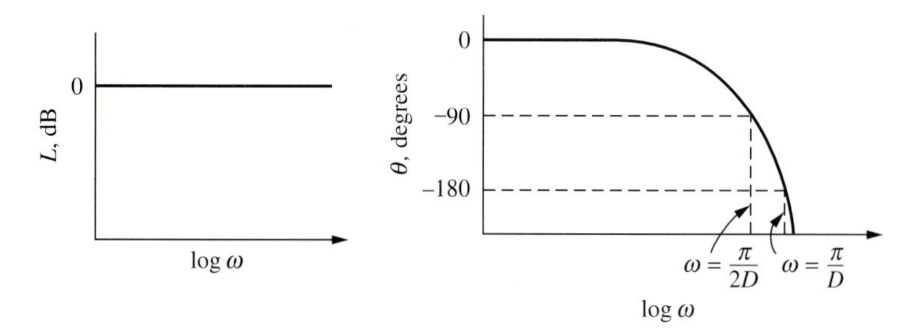

FIGURE 10.17
Bode plots for deadtime.

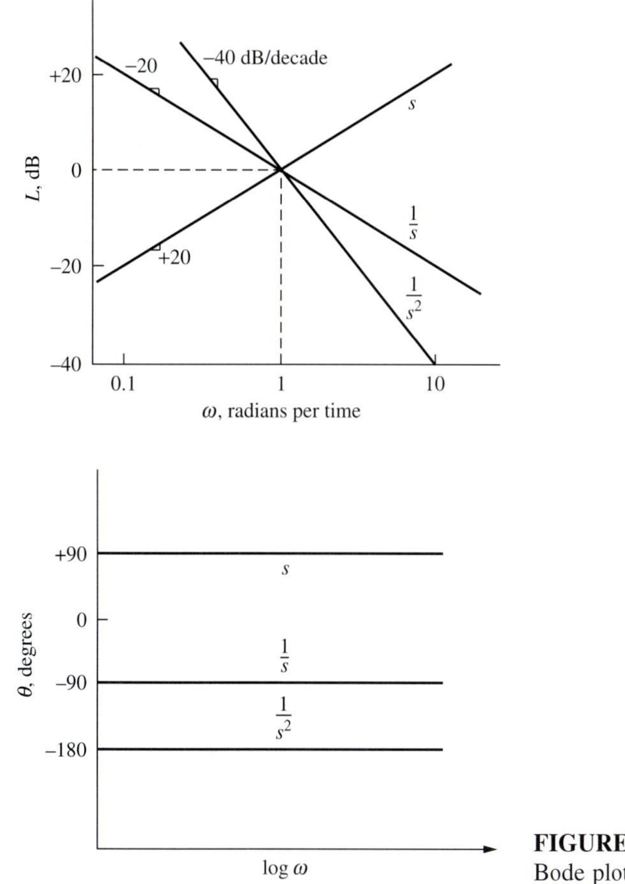

FIGURE 10.18
Bode plots for s^n.

This is a straight line with a slope of −20 dB/decade.

$$\arg G_{(i\omega)} = \arctan\left(\frac{-1/\omega}{0}\right) = -90° \tag{10.49}$$

Thus, an integrator *decreases* phase angle by 90°.

3. $n = -2$: Two integrators in series would produce a straight-line L curve with a slope of −40 dB/decade and a straight-line θ curve at −180°.

F. Second-order underdamped lag

Equations (10.34) and (10.35) give (with $K_p = 1$)

$$|G_{(i\omega)}| = \frac{1}{\sqrt{(1-\tau_o^2\omega^2)^2 + (2\zeta\tau_o\omega)^2}} \qquad \arg G_{(i\omega)} = \arctan\left(\frac{-2\zeta\tau_o\omega}{1-\tau_o^2\omega^2}\right)$$

$$L = 20\log_{10}\left[\frac{1}{\sqrt{(1-\tau_o^2\omega^2)^2 + (2\zeta\tau_o\omega)^2}}\right] \tag{10.50}$$

Figure 10.19 shows the Bode plots for several values of damping coefficient ζ. The breakpoint frequency is the reciprocal of the time constant. The high-frequency asymptote has a slope of −40 dB/decade.

$$\lim_{\omega\to\infty} L = 20\log_{10}\left(\frac{1}{\tau_o^2\omega^2}\right) = -40\log_{10}(\omega\tau_o)$$

Note the unique shape of the log modulus curves in Fig. 10.19. The lower the damping coefficient, the higher the peak in the L curve. A damping coefficient of about 0.4 gives a peak of about +2 dB. We use this property extensively in tuning feedback controllers. We adjust the controller gain to give a maximum peak of +2 dB in the log modulus curve for the closedloop servo transfer function Y/Y^{set}.

G. General transfer functions in series

The historical reason for the widespread use of Bode plots is that, before the use of computers, they made it possible to handle complex processes fairly easily using graphical techniques. A complex transfer function can be broken down into its simple elements: leads, lags, gains, deadtimes, etc. Then each of these is plotted on the same Bode plots. Finally, the total complex transfer function is obtained by adding the individual log modulus curves and the individual phase curves at each value of frequency.

Consider a general transfer function $G_{(s)}$ that can be broken up into two simple transfer functions $G_{1(s)}$ and $G_{2(s)}$:

$$G_{(s)} = G_{1(s)}G_{2(s)}$$

In the frequency domain

$$G_{(i\omega)} = G_{1(i\omega)}G_{2(i\omega)} \tag{10.51}$$

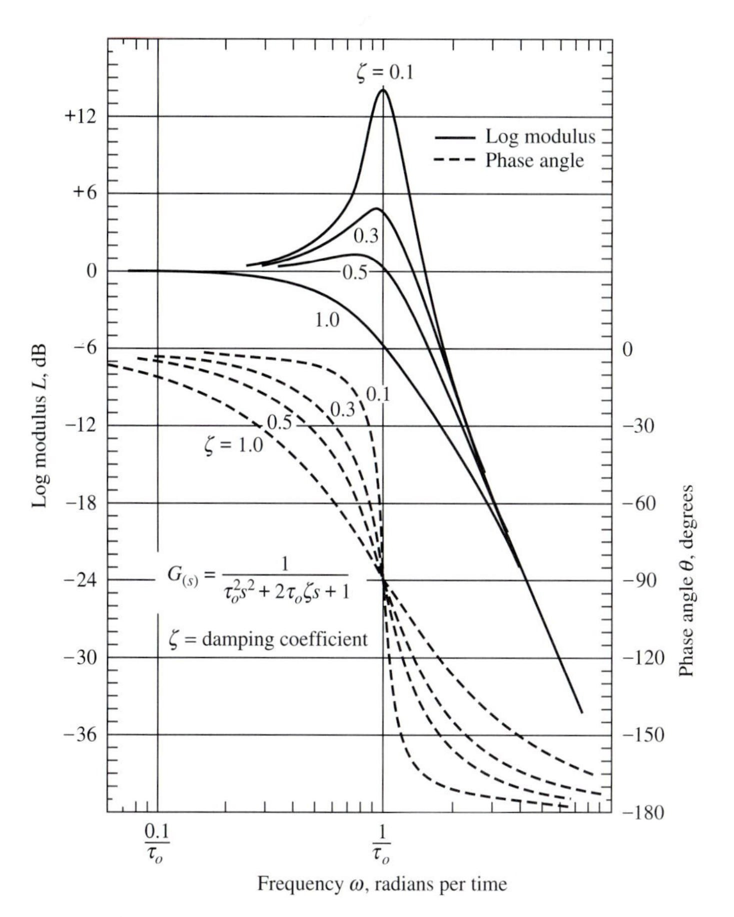

FIGURE 10.19
Second-order system Bode plots.

Each of the G's is a complex number and can be expressed in polar form:

$$
\begin{aligned}
G_{1(i\omega)} &= \left|G_{1(i\omega)}\right| e^{i \arg G_{1(i\omega)}} \\
G_{2(i\omega)} &= \left|G_{2(i\omega)}\right| e^{i \arg G_{2(i\omega)}} \\
G_{(i\omega)} &= \left|G_{(i\omega)}\right| e^{i \arg G_{(i\omega)}}
\end{aligned}
$$

Combining,

$$
\begin{aligned}
G_{(i\omega)} &= \left|G_{1(i\omega)}\right| e^{i \arg G_{1(i\omega)}} \left|G_{2(i\omega)}\right| e^{i \arg G_{2(i\omega)}} \\
\left|G_{(i\omega)}\right| e^{i \arg G_{(i\omega)}} &= \left|G_{1(i\omega)}\right| \left|G_{2(i\omega)}\right| e^{i[\arg G_{1(i\omega)} + \arg G_{2(i\omega)}]}
\end{aligned}
\tag{10.52}
$$

Taking the logarithm of both sides gives

$$
\ln|G| + i \arg G = \ln|G_1| + \ln|G_2| + i(\arg G_1 + \arg G_2)
$$

Therefore, the log modulus curves and phase angle curves of the individual components are simply added at each value of frequency to get the total L and θ curves for the complex transfer function.

$$20\log_{10}|G| = 20\log_{10}|G_1| + 20\log_{10}|G_2| \tag{10.53}$$

$$\arg G = \arg G_1 + \arg G_2 \tag{10.54}$$

EXAMPLE 10.2. Consider the transfer function $G_{(s)}$:

$$G_{(s)} = \frac{1}{(\tau_{o1}s + 1)(\tau_{o2}s + 1)}$$

Bode plots of the individual transfer functions G_1 and G_2 are sketched in Fig. 10.20 and added to give $G_{(i\omega)}$.

$$G_{1(s)} = \frac{1}{\tau_{o1}s + 1} \qquad G_{2(s)} = \frac{1}{\tau_{o2}s + 1}$$

Note that the total phase angle drops down to $-180°$ and the slope of the high-frequency asymptote of the log modulus line is -40 dB/decade, since the process is net second order. ■

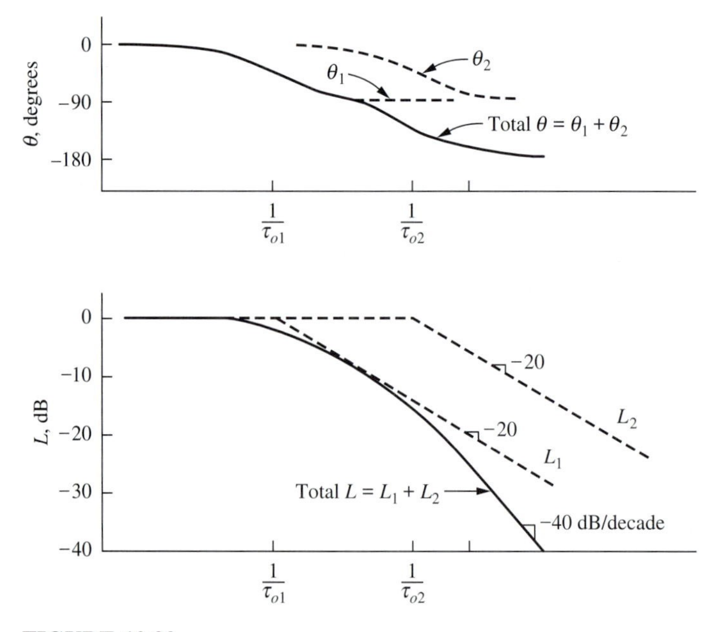

FIGURE 10.20
Bode plots for two lags.

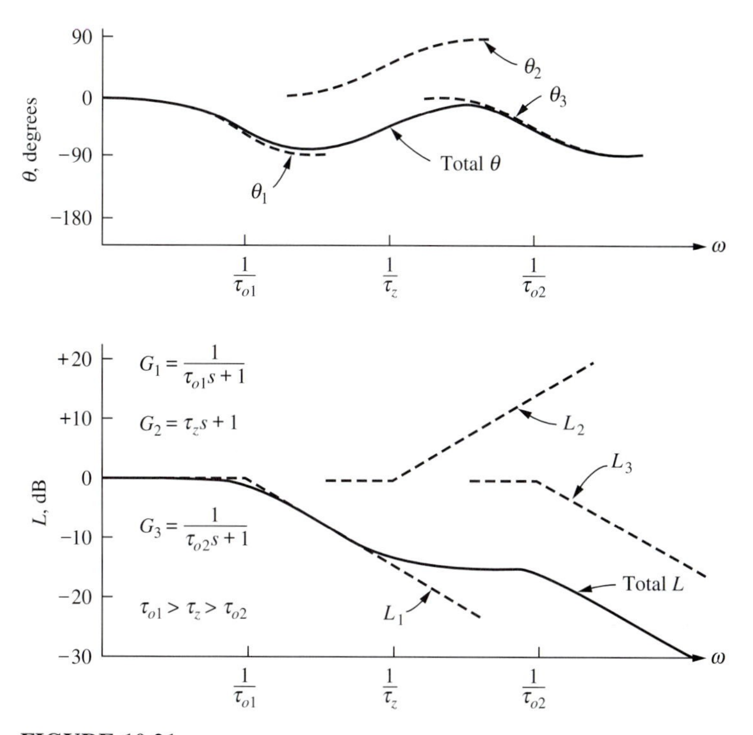

FIGURE 10.21
Bode plots for two lags and one lead.

EXAMPLE 10.3. Bode plots for the transfer function

$$G_{(s)} = \frac{\tau_z s + 1}{(\tau_{o1} s + 1)(\tau_{o2} s + 1)}$$

are sketched in Fig. 10.21. Note that the phase angle goes to $-90°$ and the slope of the log modulus line is -20 dB/decade at high frequencies because the system is net first order (the order of the numerator polynomial M is 1 and the order of the denominator polynomial N is 2). ■

EXAMPLE 10.4. Figure 10.22 gives Bode plots for a transfer containing a first-order lag and deadtime.

$$G_{(s)} = \frac{e^{-Ds}}{s + 1}$$

Several different values of deadtime D are shown. The L curve is the same for all values of D. Only the phase-angle curve is changed as D changes. The larger the value of D, the lower the frequency at which the phase angle drops to $-180°$. As we learn in Chapter 11, the lower the phase angle curve, the poorer the control. We show quantitatively in Chapter 11 how increasing deadtime degrades feedback control performance. You have already seen this effect in our simulation studies in Chapter 1. ■

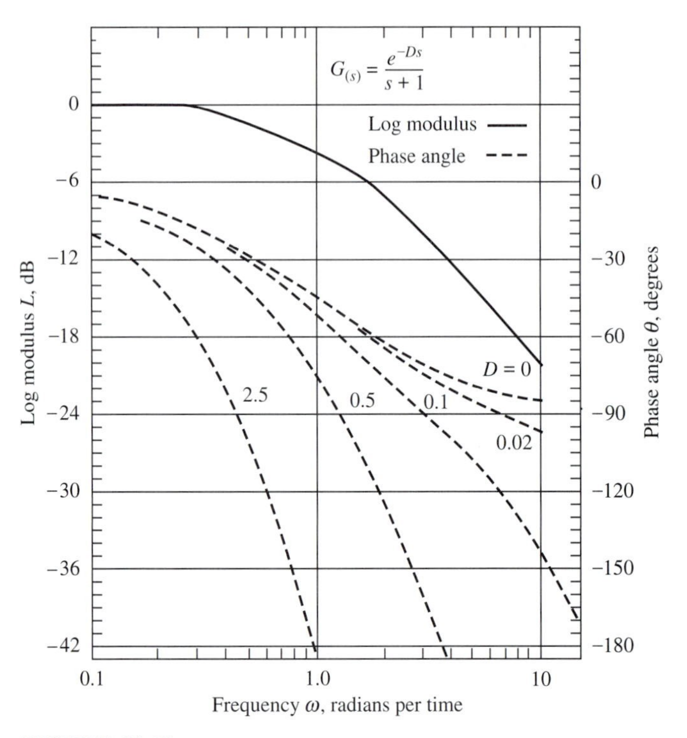

FIGURE 10.22
Bode plots for lag and deadtime.

10.3.3 Nichols Plots

The final plot that we need to learn how to make is called a Nichols plot. It is a single curve in a coordinate system with phase angle as the abscissa and log modulus as the ordinate. Frequency is a parameter along the curve. Figure 10.23 gives Nichols plots of some simple transfer functions.

At this point you may be asking why we need another type of plot. After all, the Nyquist and the Bode plots are simply different ways to plot complex numbers. As we will see in Chapter 11, all of these plots (Nichols, Bode, and Nyquist) are very useful for designing control systems. Each has its own application, so we have to learn all three of them. Keep in mind, however, that the main workhorse of our Chinese language is the Bode plot. We usually make it first since it is easy to construct from its individual simple elements. Then we use the Bode plot to sketch the Nyquist and Nichols plots.

10.4 COMPUTER PLOTTING

Up to this point we have stressed graphical methods for quickly sketching Bode plots. Since the combining of simple transfer functions is just numerical manipulation

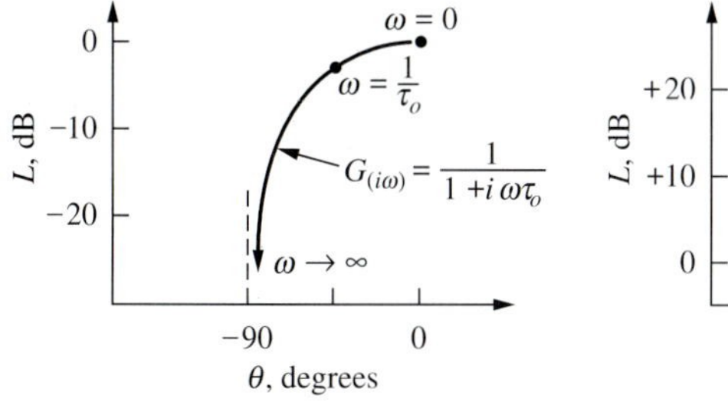

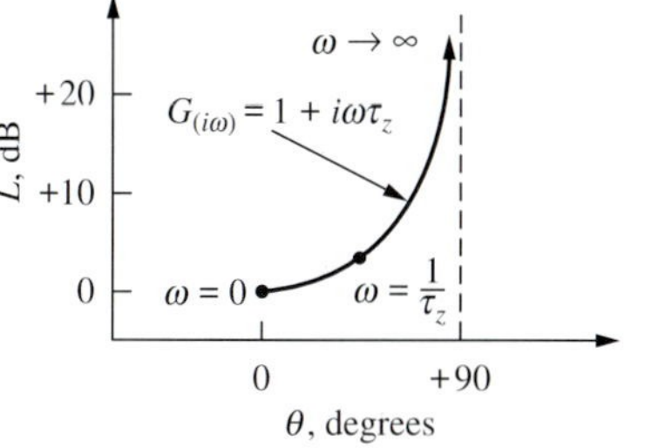

(b)

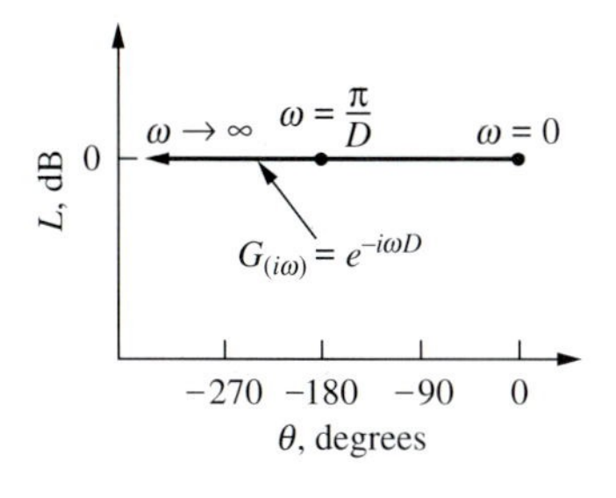

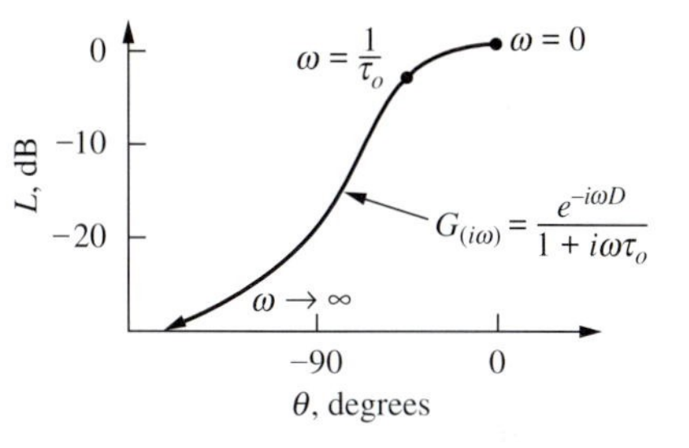

(d)

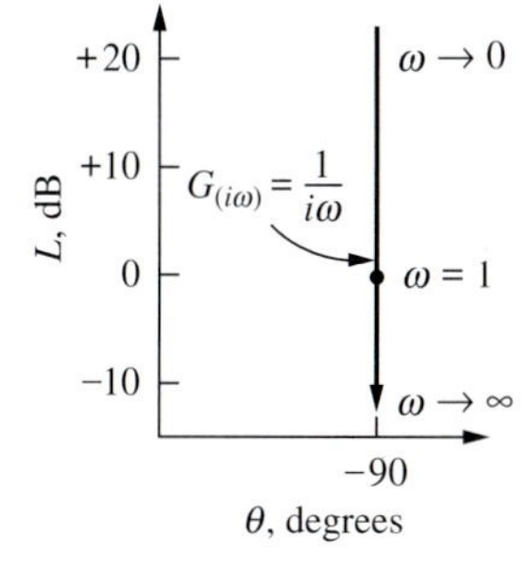

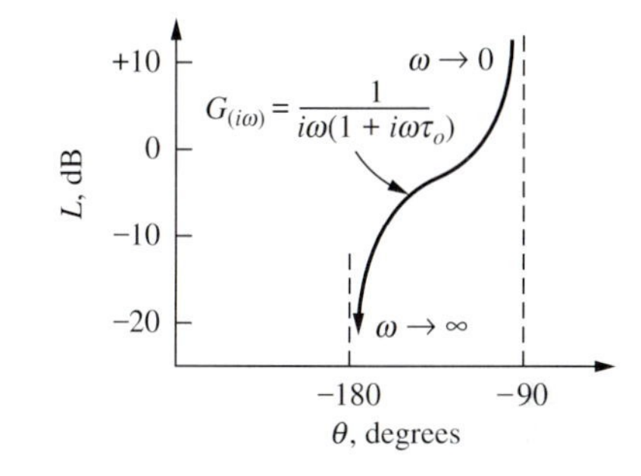

(e)

(f)

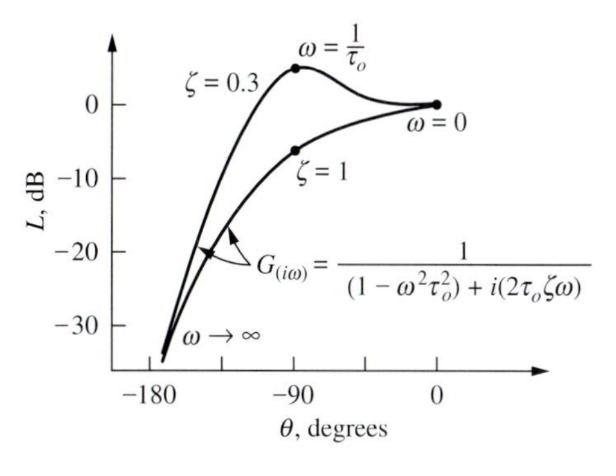

(g)

FIGURE 10.23
Nichols plots. (*a*) First-order lag. (*b*) First-order lead. (*c*) Deadtime. (*d*) Deadtime and lag. (*e*) Integrator. (*f*) Integrator and lag. (*g*) Second-order underdamped lag.

of complex numbers, a digital computer can easily generate any of the desired forms of the complex number: real and imaginary parts, magnitude, log modulus, and phase angle. We give two approaches for generating these plots on a computer: using FORTRAN and using MATLAB.

10.4.1 FORTRAN Programs for Plotting Frequency Response

Table 10.1 gives a FORTRAN program that calculates the frequency response of several simple systems. Figure 10.24 gives Bode plots for four different transfer functions. The variables must be declared *COMPLEX* at the beginning of the program. Note in the program how the phase angles are calculated by summing the

TABLE 10.1
FORTRAN program for frequency response

```
c DIGITAL PROGRAM TO CALCULATE FREQUENCY RESPONSE
      DIMENSION G(4),DB(4),DEG(4)
      COMPLEX G
      OPEN(7,FILE='FREQ.DAT')
      WRITE(7,1)
      WRITE(7,2)
    1 FORMAT(1X,'FREQ REG1 IMG1 REG2 IMG2 REG3 IMG3 REG4
     +IMG4')
    2 FORMAT(1X,'DBG1 DEGG1 DBG2 DEGG2 DBG3 DEGG3 DBG4
     +DEGG4')
C W  IS FREQUENCY IN RADIANS PER MINUTE
      W=0.01
      DW=10.**(0.2)
      DO 100 I=1,11
C G1 IS FIRST ORDER LAG WITH TAU = 1
      G(1)=1./CMPLX(1.,1.0*W)
C G2 IS TWO FIRST-ORDER LAGS WITH TAU'S = 1
      G(2)=G(1)*G(1)
C G3 IS TWO FIRST-ORDER LAGS WITH TAU'S = 1 AND 10
      G(3)=G(1)/CMPLX(1.,10.*W)
C G4 IS SECOND-ORDER UNDERDAMPED LAG WITH TAU=1 AND ZETA=0.3
      G(4)=1./CMPLX(1.-W**2,2.*W*0.3)
      DO 10 J=1,4
   10 DB(J)=20.*ALOG10(CABS(G(J)))
      DEG(1)=ATAN(-W)*180./3.1416
      DEG(2)=DEG(1)+DEG(1)
      DEG(3)=DEG(1)+ATAN(-10.*W)*180./3.1416
      DEG(4)=ATAN2(-2.*W*0.3,1.-W**2)*180./3.1416
      WRITE(7,4)W,G
      WRITE(7,3)(DB(J),DEG(J),J=1,4)
    4 FORMAT(1X,9F7.3)
    3 FORMAT(8X,8F7.1)
  100 W=W*DW
      STOP
      END
```

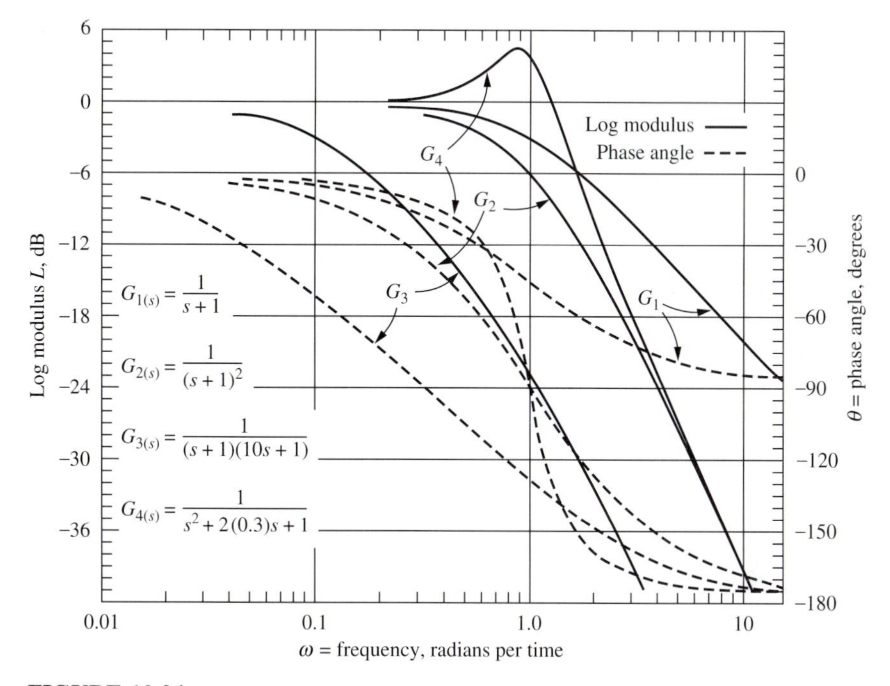

FIGURE 10.24
Bode plots of several systems.

arguments of the individual components. This is particularly useful when there is a deadtime element in the transfer function. If you try to calculate the phase angle from the final total complex number, the FORTRAN subroutines ATAN and ATAN2 cannot determine what quadrant you are in. ATAN has only one argument and therefore can track the complex number only in the first or fourth quadrant. So phase angles between $-90°$ and $-180°$ will be reported as $+90°$ to $0°$. The subroutine ATAN2, since it has two arguments (the imaginary and the real part of the number), can accurately track the phase angle between $+180°$ and $-180°$, but not beyond. Getting the phase angle by summing the angles of the components eliminates all these problems.

A complex number G always has two parts: real and imaginary. These parts can be specified using the statement

$$G{=}CMPLX(X,Y)$$

where G = complex number
X = real part of G
Y = imaginary part of G

Complex numbers can be added ($G{=}G1{+}G2$), multiplied ($G{=}G1{*}G2$), and divided ($G{=}G1/G2$). The magnitude of a complex number can be found by using the statement

$$XX{=}CABS(G)$$

where XX = the real number that is the magnitude of G. Knowing the complex number G, we can find its real and imaginary parts by using the statements

$$X{=}REAL(G) \qquad Y{=}AIMAG(G)$$

A deadtime element $G_{(s)} = e^{-Ds}$

$$G_{(i\omega)} = e^{-iD\omega} = \cos(D\omega) - i\sin(D\omega) \tag{10.55}$$

can be calculated using the statement

$$G{=}CMPLX(COS(D{*}W),-SIN(D{*}W))$$

where D = deadtime
W = frequency, in radians per minute in the FORTRAN program

The program in Table 10.1 illustrates the use of some of these complex FORTRAN statements.

10.4.2 MATLAB Program for Plotting Frequency Response

Table 10.2 gives a MATLAB program that generates Nyquist, Bode, and Nichols plots for the three-heated-tank process. Figure 10.25 gives the plots. The *num* and *den* polynomials are defined in the same way as in the root locus plots in Chapter 8. The frequency range of interest is specified by a *logspace* function from $\omega = 0.01$ to 10 radians/time. The magnitudes and phase angles of $G_{(i\omega)}$ are found by using the *[mag,phase,w]=bode(num,den,w)* statement. The ultimate gain and frequency are found by searching through the vector of phase angles until the $-180°$ point is crossed.

The real and the imaginary parts of $G_{(i\omega)}$ for making the Nyquist plot can be found in two different ways. The easiest is to use the *[greal,gimag,w]=nyquist(num, den,w)* statement. Alternatively, the real part can be calculated from the product of the magnitude and the cosine of the phase angle (in radians) at each frequency. This term-by-term multiplication is accomplished in MATLAB by using the .* operation.

Handling deadtime in MATLAB is not at all obvious. Larry Ricker (private communication, 1993) suggested a method for accomplishing it using the *polyval* function. As illustrated in the program given in Table 10.3, the numerator and denominator polynomials are evaluated at each frequency point. Then these polynomials are divided at each frequency point by using the ./ division operation. Finally, each of the resulting complex numbers is multiplied by the corresponding deadtime *exp(−d*w)* at that frequency by the .* multiplication operation.

Another problem encountered in systems with deadtime is large phase angles. As the curves wrap around the origin for higher frequencies, it becomes difficult to track the phase angle. MATLAB has a convenient solution to this problem: the *unwrap* command. As illustrated in Table 10.3, the phase angles are first calculated for each frequency by using a "for" loop to run through all the frequency points. Then the *unwrap(radians)* command is used to avoid the jumps in phase angle.

TABLE 10.2
MATLAB program for frequency response plots

```
% Program "tempbode.m" uses Matlab to plot Bode, Nyquist
% and Nichols plots for three heated tank process
%
% Form numerator and denominator polynomials
num=1.333;
den=conv([0.1 1],[0.1 1]);
den=conv(den,[0.1 1]);
% Specify frequency range from 0.1 to 100 radians/hour (600 points)
w=logspace(-1,2,600);
%
% Calculate magnitudes and phase angles at all frequencies
% using the "bode" function
%
[mag,phase,w]=bode(num,den,w);
db=20*log10(mag);
%
% Calculate ultimate gain and frequency
n=1;
while phase(n)>=-180;n=n+1;end
kult=1/mag(n);
wult=w(n);
%
% Plot Bode plot
%
clf
axis('normal')
subplot(211)
semilogx(w,db)
title('Bode Plot for Three Heated Tank Process')
xlabel('Frequency (radians/hr)')
ylabel('Log Modulus (dB)')
grid
text(2,-10,['Ku=',num2str(kult)])
text(2,-20,['wu=',num2str(wult)])
subplot(212)
semilogx(w,phase)
xlabel('Frequency (radians/hr)')
ylabel('Phase Angle (degrees)')
grid
pause
print -dps pfig1025.ps
%
% Make Nichols plot
%
clf
plot(phase,db)
title('Nichol Plot for Three Heated Tank Process')
xlabel('Phase Angle (degrees)')
ylabel('Log Modulus (dB)')
grid
pause
```

TABLE 10.2 (CONTINUED)
MATLAB program for frequency response plots

```
% Alternatively you can use "nichols" command
%     [mag,phase,w]=nichols(num,den,w);
print -dps -append pfig1025
%
% Make Nyquist plot
%
% Using the "nyquist" command
[greal,gimag,w]=nyquist(num,den,w);
% Alternatively you can calculate the real and imaginary parts
%     from the magnitudes and phase angles
%radians=phase*3.1416/180;
%greal=mag.*cos(radians);
%gimag=mag.*sin(radians);
clf
axis('square')
plot(greal,gimag)
grid
title('Nyquist Plot for Three Heated Tank Process')
xlabel('Real(G)')
ylabel('Imag(G)')
pause
print -dps -append pfig1025
```

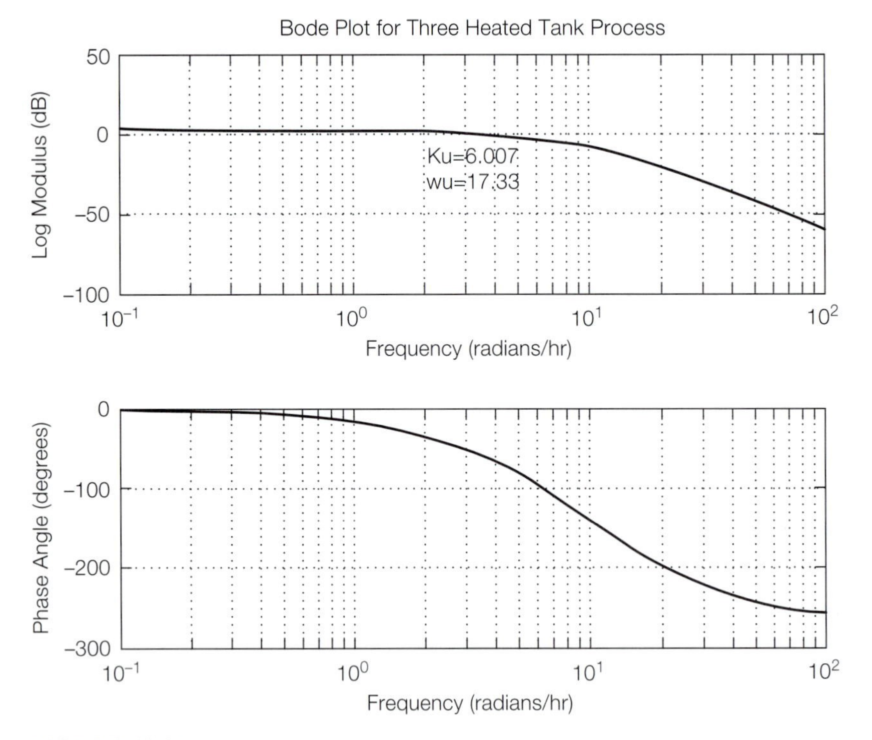

FIGURE 10.25
Frequency response plots for three-heated-tank process.

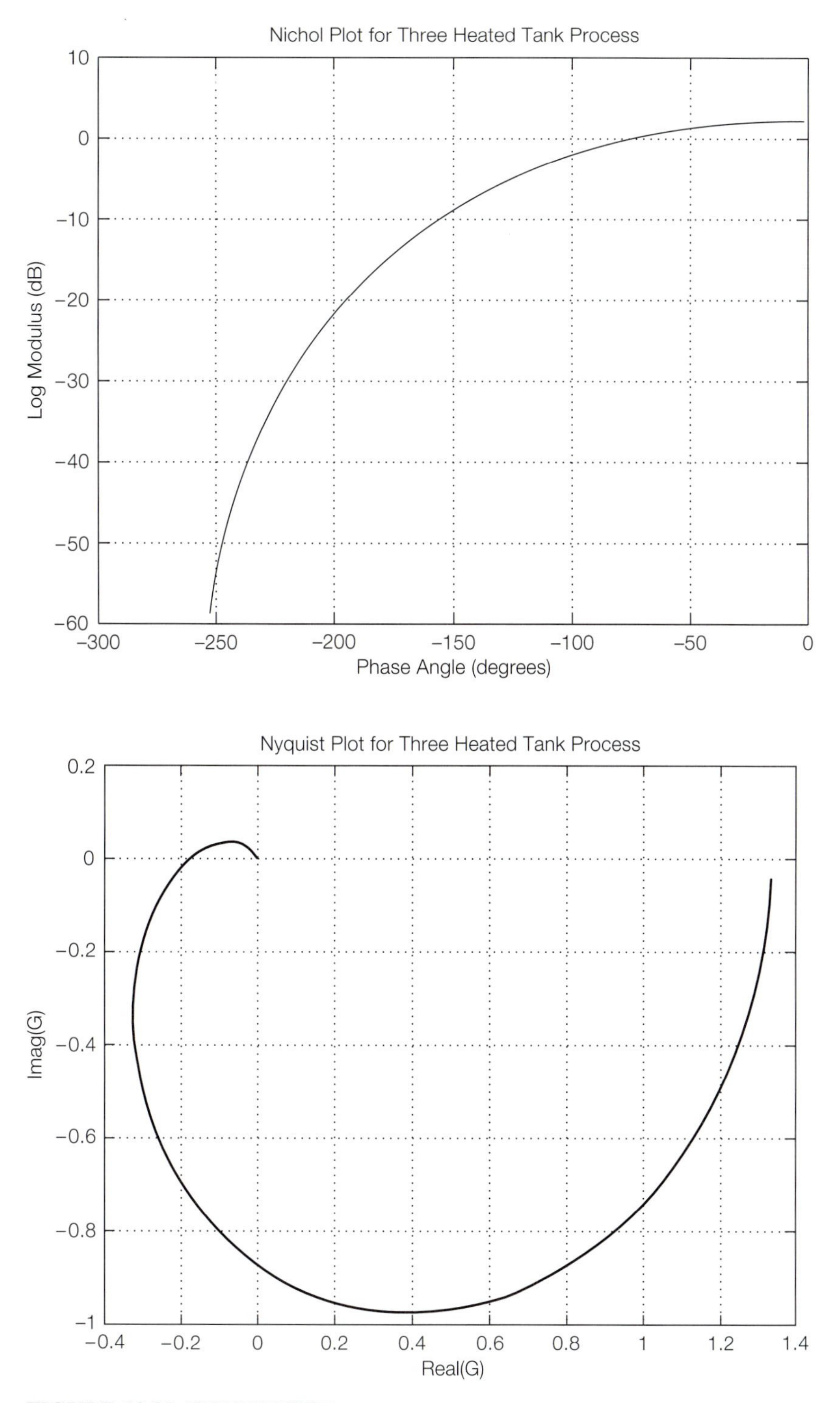

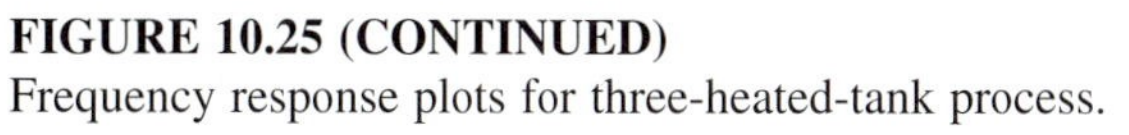

FIGURE 10.25 (CONTINUED)
Frequency response plots for three-heated-tank process.

TABLE 10.3
MATLAB program for deadtime Bode plots

```
% Program "deadtime.m"
% Calculate frequency response of process with deadtime
% using the Larry Ricker method.
%
% Process is a first-order lag with time constant tau=10 minute,
% a steady-state gain of kp=1 and a d=5 minute deadtime.
%
tau=10; kp=1; d=5;
num=1;
den=[10 1];
% Specify frequency range from 0.01 to 1 radians/minute (400 points)
w=logspace(-2,0,400);
% Define complex variable "s"
s=w*sqrt(-1);
%
% Evaluate numerator and denominator polynomials at all frequencies
% Note the "./" operator which does term by term division
%
g=polyval(num,s) ./ polyval(den,s);
%
% Multiple by deadtime
% Note the ".*" operator which does term by term multiplication
%
gdead=g .* exp(-d*s);
%
% Calculate log modulus
%
db=20*log10(abs(gdead));
%
% Calculate phase angles
%
nw=length(w);
for n=1:nw
radians(n)=atan2(imag(gdead(n)),real(gdead(n)));
end
% Use "unwrap" operator to remove 360 degree jumps in phase angles
phase=180*(unwrap(radians))/3.1416;
%
% Plot Bode plot
clf
axis('normal')
subplot(211)
semilogx(w,db)
title('Bode Plot for Deadtime Process')
xlabel('Frequency (radians/min)')
ylabel('Log Modulus (dB)')
text(0.02,-10,'Tau=10')
text(0.02,-15,'Deadtime=5')
grid
```

TABLE 10.3 (CONTINUED)
MATLAB program for deadtime Bode plots

```
subplot(212)
semilogx(w,phase)
xlabel('Frequency (radians/min)')
ylabel('Phase Angle (degrees)')
grid
pause
print -dps pfig1026.ps
```

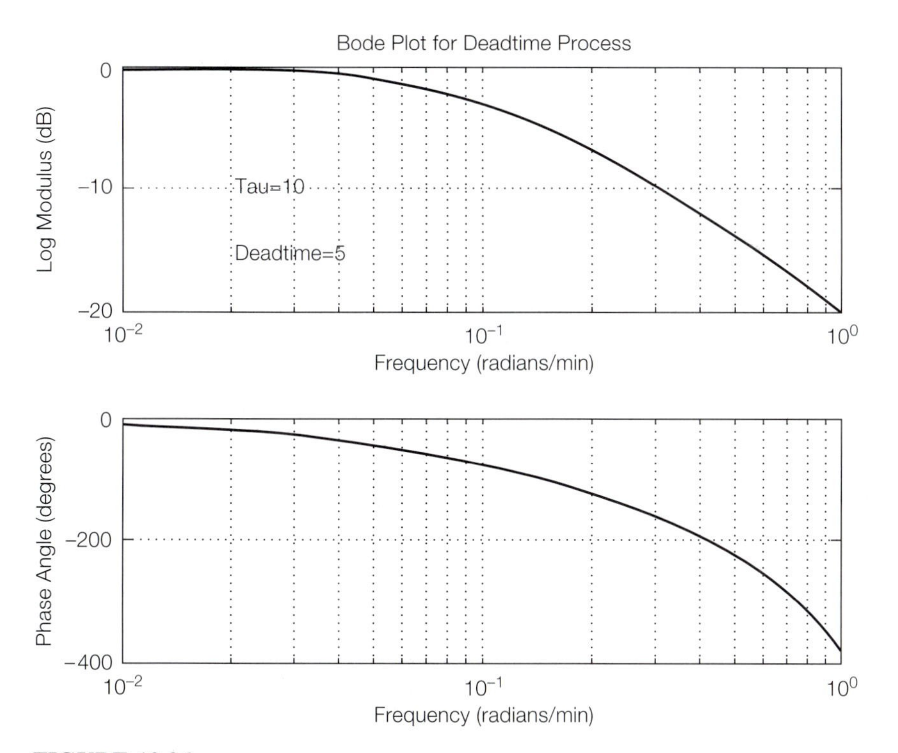

FIGURE 10.26

Figure 10.26 gives the resulting plots for a first-order lag and deadtime process.

$$G_{(i\omega)} = \frac{e^{-5s}}{10s + 1} \tag{10.56}$$

10.5 CONCLUSION

We have laid the foundation for our adventure into the China mainland. We've learned the language, and we have learned some useful graphical and computer software tools for working with it. In the next chapter we apply all these to the problem of designing controllers for simple SISO systems. In later chapters we use these frequency-domain methods to tackle some very complex and important problems: multivariable systems and system identification.

PROBLEMS

10.1. Sketch Nyquist, Bode, and Nichols plots for the following transfer functions:

(*a*) $G_{(s)} = \dfrac{1}{(s+1)^3}$

(*b*) $G_{(s)} = \dfrac{1}{(s+1)(10s+1)(100s+1)}$

(*c*) $G_{(s)} = \dfrac{1}{s^2(s+1)}$

(*d*) $G_{(s)} = \dfrac{\tau s+1}{(\tau/6)s+1}$

(*e*) $G_{(s)} = \dfrac{s}{2s+1}$

(*f*) $G_{(s)} = \dfrac{1}{(10s+1)(s^2+s+1)}$

10.2. Draw Bode plots for the transfer functions:
(*a*) $G_{(s)} = 0.5$
(*b*) $G_{(s)} = 5.0$

10.3. Sketch Nyquist, Bode, and Nichols plots for the proportional-integral feedback controller $G_{C(s)}$:

$$G_{C(s)} = K_c\left(1 + \frac{1}{\tau_I s}\right)$$

10.4. Sketch Nyquist, Bode, and Nichols plots for a system with the transfer function

$$G_{(s)} = \frac{-3s+1}{(s+1)(5s+1)}$$

10.5. Draw the Bode plots of the transfer function

$$G_{(s)} = \frac{7.5(s+0.2)}{s(s+1)^3}$$

10.6. Write a digital computer program that gives the real and imaginary parts, log modulus, and phase angle for the transfer functions:

(*a*) $G_{(s)} = G_{C(s)}\dfrac{P_{11}P_{22} - P_{12}P_{21}}{P_{22}}$

where

$$G_{C(s)} = 8\left(1 + \frac{1}{400s}\right)$$

$$P_{11(s)} = \frac{1}{(1 + 167s)(1 + s)(1 + 0.1s)^4}$$

$$P_{12(s)} = \frac{0.85}{(1 + 83s)(1 + s)^2}$$

$$P_{21(s)} = \frac{0.85}{(1 + 167s)(1 + 0.5s)^3(1 + s)}$$

$$P_{22(s)} = \frac{1}{(1 + 167s)(1 + s)^2}$$

(*b*) $G_{(s)} = \dfrac{e^{-0.1s}}{s + 1 + e^{-0.1s}}$

10.7. Draw Bode, Nyquist, and Nichols plots for the transfer functions:

(*a*) $A_{L(s)} = \dfrac{G_{(s)}}{1 + G_{C(s)}G_{(s)}}$

(*b*) $A_{s(s)} = \dfrac{G_{C(s)}G_{(s)}}{1 + G_{C(s)}G_{(s)}}$

where

$$G_{C(s)} = K_c\left(1 + \frac{1}{\tau_I s}\right) \qquad K_c = 6 \qquad \tau_I = 6$$

$$G_{(s)} = \frac{1}{\tau_o s + 1} \qquad \tau_o = 10$$

10.8. Draw the Bode plot of

$$G_{(s)} = \frac{1 - e^{-Ds}}{s}$$

10.9. A process is forced by sinusoidal input $u_{s(t)}$. The output is a sine wave $y_{s(t)}$. If these two signals are connected to an x-y recorder, we get a Lissajous plot. Time is the parameter along the curve, which repeats itself with each cycle. The shape of the curve will change if the frequency is changed and will be different for different kinds of processes.

(*a*) How can the magnitude ratio MR and phase angle θ be found from this curve?

(*b*) Sketch Lissajous curves for the following systems:

(*i*) $G_{(s)} = K_p$

(*ii*) $G_{(s)} = \dfrac{1}{s}$ at $\omega = 1$ radian/time

(*iii*) $G_{(s)} = \dfrac{1}{\tau_o s + 1}$ at $\omega = \dfrac{1}{\tau_o}$ radians/time

CHAPTER 11

Frequency-Domain Analysis of Closedloop Systems

The design of feedback controllers in the frequency domain is the subject of this chapter. The Chinese language that we learned in Chapter 10 is used to tune controllers. Frequency-domain methods have the significant advantage of being easy to use for high-order systems and for systems with deadtime.

We show in Section 11.1 that *closedloop stability* can be determined from the frequency response plot of the total *openloop* transfer function of the system (process openloop transfer function *and* feedback controller $G_{M(i\omega)}G_{C(i\omega)}$). This means that a Bode plot of $G_{M(i\omega)}G_{C(i\omega)}$ is all we need. As you remember from Chapter 10, the total frequency response curve of a complex system is easily obtained on a Bode plot by splitting the system into its simple elements, plotting each of these, and merely adding log moduli and phase angles together. Therefore, the graphical generation of the required $G_{M(i\omega)}G_{C(i\omega)}$ curve is relatively easy. Of course, all this algebraic manipulation of complex numbers can be even more easily performed on a digital computer.

11.1 NYQUIST STABILITY CRITERION

The Nyquist stability criterion is a method for determining the stability of systems in the frequency domain. It is almost always applied to closedloop systems. A working, but not completely general, statement of the Nyquist stability criterion is:

> *If a polar plot of the total openloop transfer function of the system $G_{M(i\omega)}G_{C(i\omega)}$ wraps around the $(-1, 0)$ point in the G_MG_C plane as frequency ω goes from zero to infinity, the system is* closedloop *unstable.*

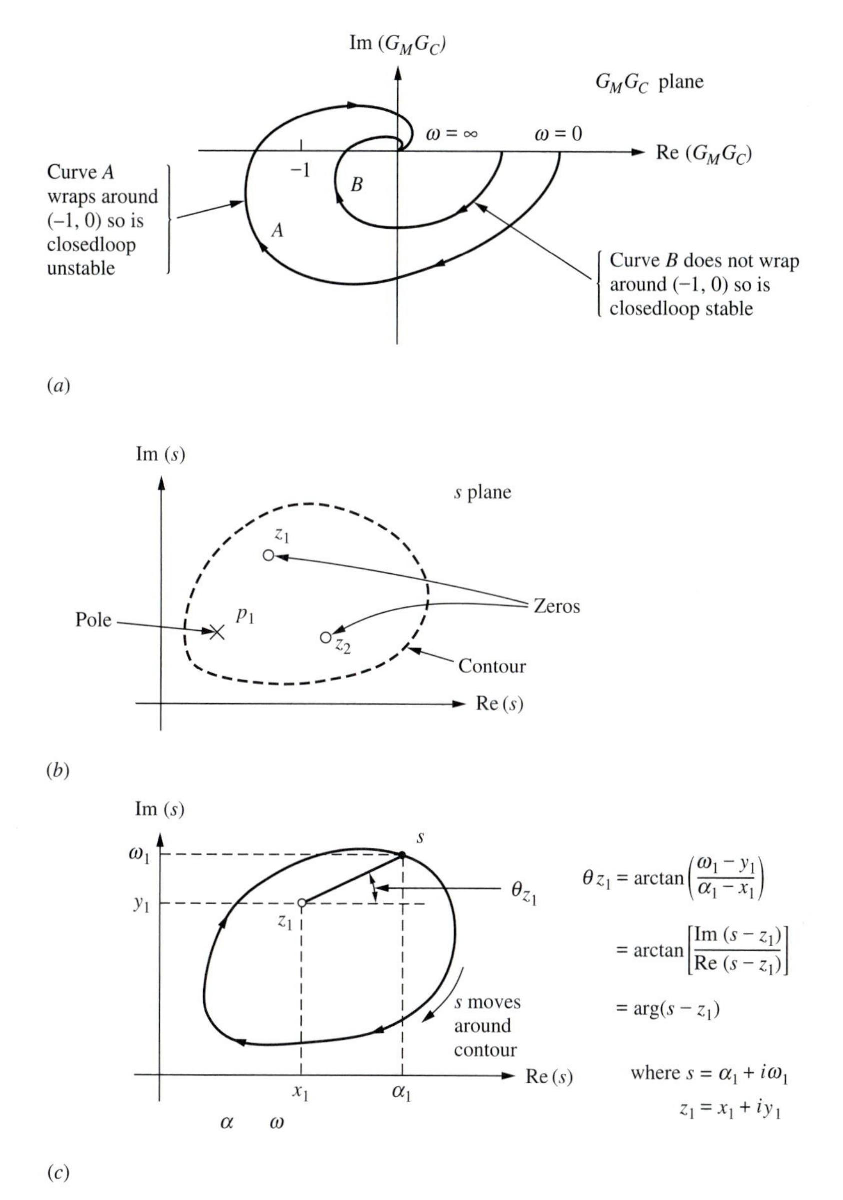

FIGURE 11.1
(*a*) Polar plots showing closedloop stability or instability. (*b*) *s* plane location of zeros and poles. (*c*) Argument of $(s - z_1)$.

The two polar plots sketched in Fig. 11.1*a* show that system *A* is closedloop unstable whereas system *B* is closedloop stable.

On the surface, the Nyquist stability criterion is quite remarkable. We are able to deduce something about the stability of the *closedloop* system by making a frequency response plot of the *openloop* system! And the encirclement of the mystical, magical $(-1, 0)$ point somehow tells us that the system is closedloop unstable. This all looks like blue smoke and mirrors. However, as we will prove, it all goes back to finding out if there are any roots of the closedloop characteristic equation in the RHP (positive real roots).

11.1.1 Proof

A. Complex variable theorem

The Nyquist stability criterion is derived from a theorem of complex variables.

If a complex function $F_{(s)}$ has Z zeros and P poles inside a certain area of the s plane, the number N of encirclements of the origin that a mapping of a closed contour around the area makes in the F plane is equal to Z − P.

$$Z - P = N \tag{11.1}$$

Consider the hypothetical function $F_{(s)}$ of Eq. (11.2) with two zeros at $s = z_1$ and $s = z_2$ and one pole at $s = p_1$.

$$F_{(s)} = \frac{(s - z_1)(s - z_2)}{s - p_1} \tag{11.2}$$

The locations of the zeros and the pole are sketched in the s plane in Fig. 11.1*b*.

The argument of $F_{(s)}$ is

$$\arg F_{(s)} = \arg\left[\frac{(s - z_1)(s - z_2)}{s - p_1}\right] \tag{11.3}$$
$$\arg F_{(s)} = \arg(s - z_1) + \arg(s - z_2) - \arg(s - p_1)$$

Remember, the argument of the product of two complex numbers z_1 and z_2 is the sum of the arguments.

$$z_1 z_2 = (r_1 e^{i\theta_1})(r_2 e^{i\theta_2}) = r_1 r_2 e^{i(\theta_1 + \theta_2)}$$
$$\arg(z_1 z_2) = \theta_1 + \theta_2$$

And the argument of the quotient of two complex numbers is the difference between the arguments.

$$\frac{z_1}{z_2} = \frac{r_1 e^{i\theta_1}}{r_2 e^{i\theta_2}} = \frac{r_1}{r_2} e^{i(\theta_1 - \theta_2)}$$
$$\arg\left(\frac{z_1}{z_2}\right) = \theta_1 - \theta_2$$

Let us pick an arbitrary point s on the contour and draw a line from the zero z_1 to this point (see Fig. 11.1*c*). The angle between this line and the horizontal, θ_{z_1}, is equal to the argument of $(s - z_1)$. Now let the point s move completely around the contour. The angle θ_{z_1} or $\arg(s - z_1)$ will increase by 2π radians. Therefore, $\arg F_{(s)}$ will *increase* by 2π radians for each zero inside the contour.

A similar development shows that arg $F_{(s)}$ *decreases* by 2π for each pole inside the contour because of the negative sign in Eq. (11.3). Two zeros and one pole mean that arg $F_{(s)}$ must show a net increase of $+2\pi$. Thus, a plot of $F_{(s)}$ in the complex F plane (real part of $F_{(s)}$ versus imaginary part of $F_{(s)}$) must encircle the origin once as s goes completely around the contour.

In this system $Z = 2$ and $P = 1$, and we have found that $N = Z - P = 2 - 1 = 1$. Generalizing to a system with Z zeros and P poles gives the desired theorem [Eq. 11.1].

If any of the zeros or poles are repeated, of order M, they contribute $2\pi M$ radians. Thus, Z is the number of zeros inside the contour with Mth-order zeros counted M times. And P is the number of poles inside the contour with Nth-order poles counted N times.

B. Application of theorem to closedloop stability

To check the stability of a system, we are interested in the roots or zeros of the characteristic equation. If any of them lies in the right half of the s plane, the system is unstable. For a closedloop system, the characteristic equation is

$$1 + G_{M(s)}G_{C(s)} = 0 \tag{11.4}$$

So for a closedloop system, the function we are interested in is

$$F_{(s)} = 1 + G_{M(s)}G_{C(s)} \tag{11.5}$$

If this function has any zeros in the RHP, the closedloop system is unstable.

If we pick a contour that goes completely around the right half of the s plane and plot $1 + G_{M(s)}G_{C(s)}$, Eq. (11.1) tells us that the number of encirclements of the origin in this $(1 + G_M G_C)$ plane will be equal to the difference between the zeros and poles of $1 + G_M G_C$ that lie in the RHP. Figure 11.2 shows a case where there are two zeros in the RHP and no poles. There are two encirclements of the origin in the $(1 + G_M G_C)$ plane.

We are familiar with making plots of complex functions like $G_{M(i\omega)}G_{C(i\omega)}$ in the $G_M G_C$ plane. It is therefore easier (but more confusing unless you are careful to keep track of the "apples" and the "oranges") to use the $G_M G_C$ plane instead of the $(1 + G_M G_C)$ plane. The origin in the $(1 + G_M G_C)$ plane maps into the $(-1, 0)$ point in the $G_M G_C$ plane since the real part of every point is moved to the left one unit. We therefore look at encirclements of the $(-1, 0)$ point in the $G_M G_C$ plane, instead of encirclements of the origin in the $(1 + G_M G_C)$ plane.

After we map the contour into the $G_M G_C$ plane and count the number N of encirclements of the $(-1, 0)$ point, we know the difference between the number of zeros Z and the number of poles P that lie in the RHP. We want to find out if there are any *zeros* of the function $F_{(s)} = 1 + G_{M(s)}G_{C(s)}$ in the RHP. Therefore, we must find the number of poles of $F_{(s)}$ in the RHP before we can determine the number of zeros.

$$Z = N + P \tag{11.6}$$

The poles of the function $F_{(s)} = 1 + G_{M(s)}G_{C(s)}$ are the same as the poles of $G_{M(s)}G_{C(s)}$. If the process is openloop stable, there are no poles of $G_{M(s)}G_{C(s)}$ in the RHP. An openloop-stable process means that $P = 0$. Therefore, the number N of

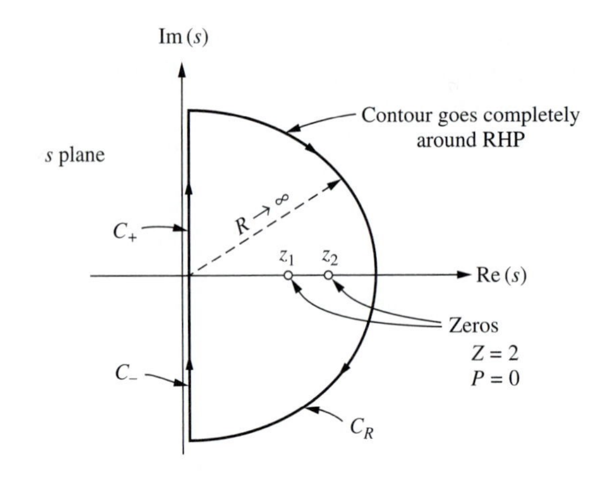

(a)

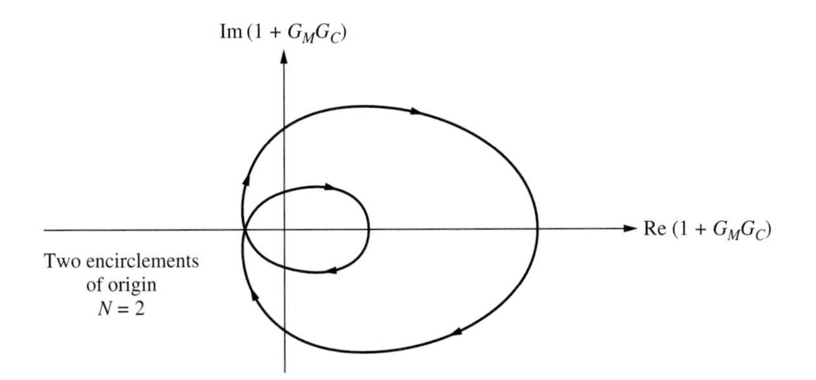

(b)

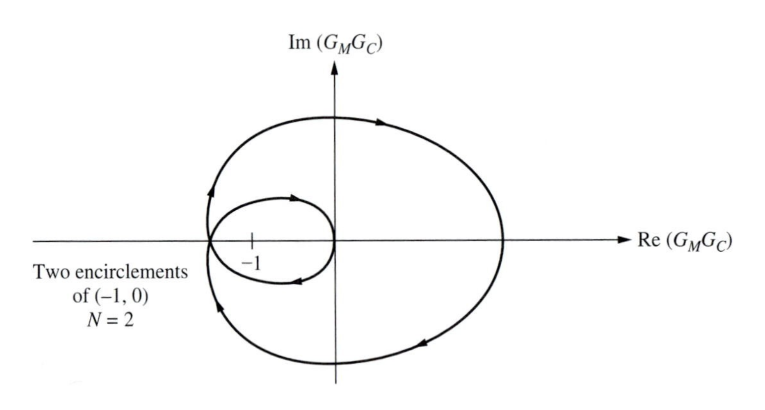

(c)

FIGURE 11.2
(a) s plane area of interest. (b) $(1 + G_{M(s)}G_{C(s)})$ plane. (c) $G_{M(s)}G_{C(s)}$ plane.

encirclements of the $(-1, 0)$ point is equal to the number of zeros of $1 + G_{M(s)}G_{C(s)}$ in the RHP for an openloop-stable process. Any encirclement means the closedloop system is unstable.

If the process is openloop *unstable*, $G_{M(s)}$ has one or more poles in the RHP, so $F_{(s)} = 1 + G_{M(s)}G_{C(s)}$ also has one or more poles in the RHP. We can find out how many poles there are by solving for the roots of the *openloop* characteristic equation (the denominator of $G_{M(s)}$). Once the number of poles P is known, the number of zeros can be found from Eq. (11.6).

11.1.2 Examples

Let us illustrate the mapping of the contour that goes around the entire right half of the s plane using some examples.

EXAMPLE 11.1. Consider the three-CSTR process

$$G_{M(s)} = \frac{\frac{1}{8}}{(s+1)^3}$$

With a proportional feedback controller, the total openloop transfer function (process and controller) is

$$G_{M(s)}G_{C(s)} = \frac{\frac{1}{8}K_c}{(s+1)^3} \tag{11.7}$$

This system is openloop stable (the three poles are all in the left half of the s plane), so $P = 0$.

The contour around the entire RHP is shown in Fig. 11.2*a*. Let us split it up into three parts: C_+, the path up the positive imaginary axis from the origin to $+\infty$; C_R, the path around the infinitely large semicircle; and C_-, the path back up the negative imaginary axis from $-\infty$ to the origin.

C_+ contour. On the C_+ contour the variable s is a pure imaginary number. Thus, $s = i\omega$ as ω goes from 0 to $+\infty$. Substituting $i\omega$ for s in the total openloop system transfer function gives

$$G_{M(i\omega)}G_{C(i\omega)} = \frac{\frac{1}{8}K_c}{(i\omega+1)^3} \tag{11.8}$$

We now let ω take on values from 0 to $+\infty$ and plot the real and imaginary parts of $G_{M(i\omega)}G_{C(i\omega)}$. This, of course, is just a polar plot of $G_{M(s)}G_{C(s)}$, as sketched in Fig 11.3*a*. The plot starts ($\omega = 0$) at $\frac{1}{8}K_c$ on the positive real axis. It ends at the origin, as ω goes to infinity, with a phase angle of $-270°$.

C_R contour. On the C_R contour,

$$s = Re^{i\theta} \tag{11.9}$$

R will go to infinity and θ will take on values from $+\pi/2$ through 0 to $-\pi/2$ radians. Substituting Eq. (11.9) into $G_{M(s)}G_{C(s)}$ gives

$$G_{M(s)}G_{C(s)} = \frac{\frac{1}{8}K_c}{(Re^{i\theta}+1)^3} \tag{11.10}$$

As R becomes large, the $+1$ term in the denominator can be neglected.

$$\lim_{R\to\infty} G_M G_C = \lim_{R\to\infty}\left(\frac{K_c}{8R^3}e^{-3\theta i}\right) \tag{11.11}$$

The magnitude of $G_M G_C$ goes to zero as R goes to infinity. Thus, the infinitely large semicircle in the s plane maps into a point (the origin) in the $G_M G_C$ plane (Fig. 11.3*b*). The argument of $G_M G_C$ goes from $-3\pi/2$ through 0 to $+3\pi/2$ radians.

C_- contour. On the C_- contour s is again equal to $i\omega$, but now ω takes on values from $-\infty$ to 0. The $G_M G_C$ on this path is just the complex conjugate of the path with positive values of ω. See Fig. 11.3*c*.

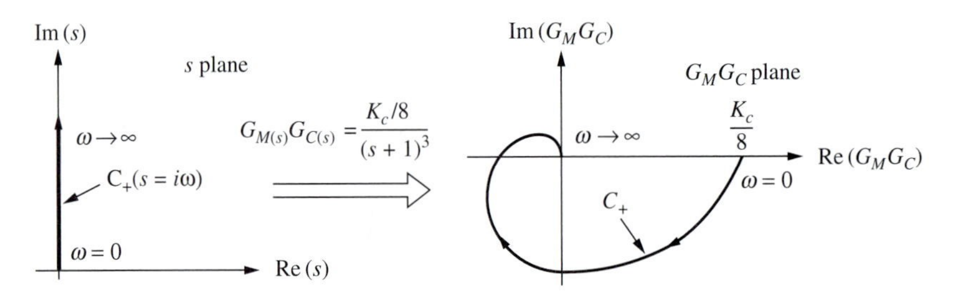

(*a*) Contour C_+

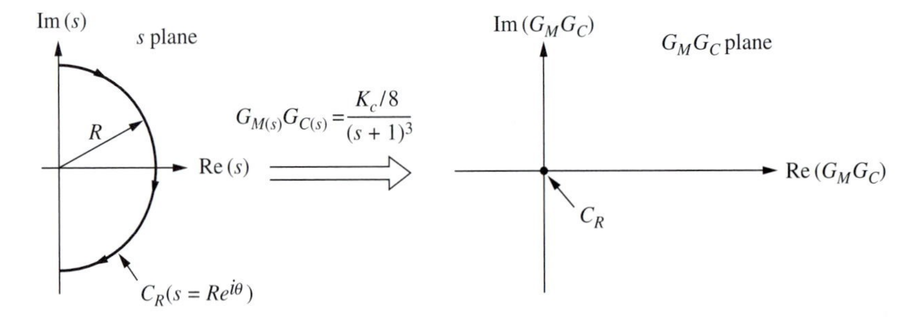

(*b*) Contour C_R

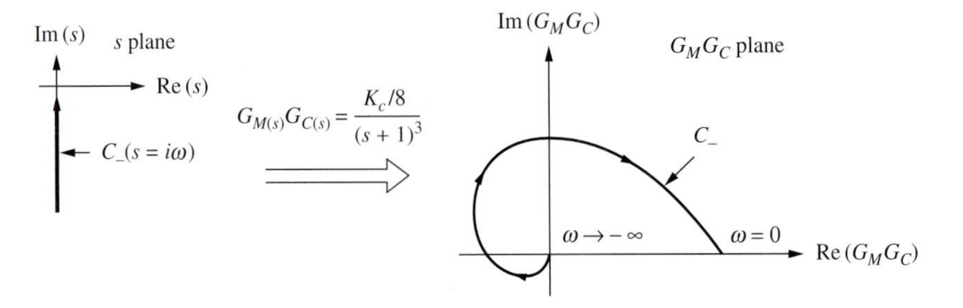

(*c*) Contour C_-

FIGURE 11.3
Nyquist plots of three-CSTR system with proportional controller.

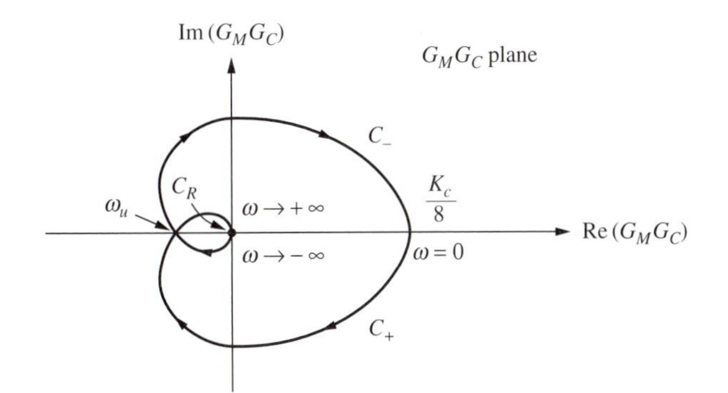

(*d*) Complete contour

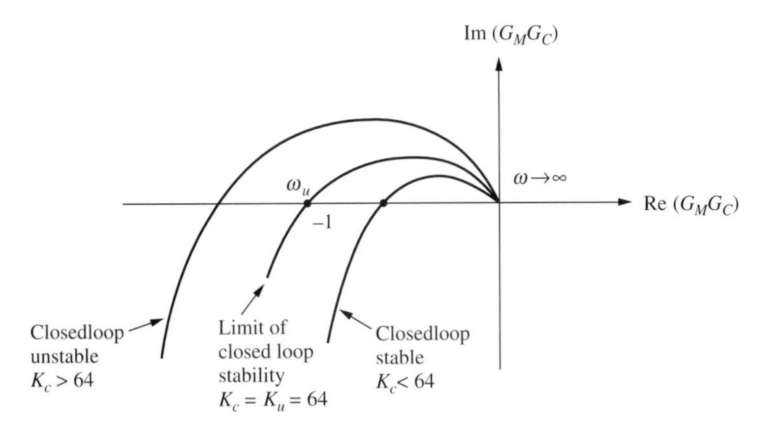

(*e*) Intersections on negative real axis

FIGURE 11.3 (CONTINUED)
Nyquist plots of three-CSTR system with proportional controller.

The complete contour is shown in Fig. 11.3*d*. The bigger the value of K_c, the farther out on the positive real axis the $G_M G_C$ plot starts, and the farther out on the negative real axis is the intersection with the $G_M G_C$ plot.

If the $G_M G_C$ plot crosses the negative real axis beyond (to the left of) the critical $(-1, 0)$ point, the system is closedloop unstable. There would then be two encirclements of the $(-1, 0)$ point, and therefore $N = 2$. Since we know P is zero, there must be two zeros in the RHP.

If the $G_M G_C$ plot crosses the negative real axis between the origin and the $(-1, 0)$ point, the system is closedloop stable. Now $N = 0$, and therefore $Z = N = 0$. There are no zeros of the closedloop characteristic equation in the RHP.

There is some critical value of gain K_c at which the $G_M G_C$ plot goes right through the $(-1, 0)$ point. This is the limit of closedloop stability. See Fig. 11.3*e*. The value of K_c at this limit should be the ultimate gain K_u that we dealt with before in making root locus plots of this system. We found in Chapter 8 that $K_u = 64$ and $\omega_u = \sqrt{3}$. Let us see if the frequency-domain Nyquist stability criterion studied in this chapter gives the same results.

At the limit of closedloop stability

$$G_{M(i\omega)}G_{C(i\omega)} = -1 + i0 \tag{11.12}$$

$$\left[\frac{\frac{1}{8}K_c}{s^3 + 3s^2 + 3s + 1}\right]_{s=i\omega} = \frac{\frac{1}{8}K_c}{(1 - 3\omega^2) + i(3\omega - \omega^3)}$$

$$= \frac{(\frac{1}{8}K_c)(1 - 3\omega^2)}{(1 - 3\omega^2)^2 + (3\omega - \omega^3)^2}$$

$$+ i\frac{(\frac{1}{8}K_c)(\omega^3 - 3\omega)}{(1 - 3\omega^2)^2 + (3\omega - \omega^3)^2} \tag{11.13}$$

Equating the imaginary part of the preceding equation to zero gives

$$\frac{(\frac{1}{8}K_c)(\omega^3 - 3\omega)}{(1 - 3\omega^2)^2 + (3\omega - \omega^3)^2} = 0$$

$$\omega = \sqrt{3} = \omega_u$$

This is exactly what we found from our root locus plot. This is the value of frequency at the intersection of the $G_M G_C$ plot with the negative real axis.

Equating the real part of Eq. (11.13) to -1 gives

$$\frac{(\frac{1}{8}K_c)(1 - 3\omega^2)}{(1 - 3\omega^2)^2 + (3\omega - \omega^3)^2} = -1$$

$$\frac{(\frac{1}{8}K_c)[1 - 3(3)]}{[1 - 3(3)]^2 + (3\sqrt{3} - 3\sqrt{3})^2} = -1$$

$$\frac{-K_c}{64} = -1 \quad \Rightarrow \quad K_c = 64 = K_u$$

This is the same ultimate gain that we found from the root locus plot.

Remember also that for gains greater than the ultimate gain, the root locus plot showed two roots of the closedloop characteristic equation in the RHP. This is exactly the result we get from the Nyquist stability criterion ($N = 2 = Z$). ■

EXAMPLE 11.2. The system of Example 8.5 is second order.

$$G_{M(s)}G_{C(s)} = \frac{K_c}{(s + 1)(5s + 1)} \tag{11.14}$$

It has two poles, both in the LHP: $s = -1$ and $s = -\frac{1}{5}$. Thus, the number of poles of $G_M G_C$ in the RHP is zero: $P = 0$. Let us break up the contour around the entire RHP into the same three parts used in the previous example.

C_+ contour. $s = i\omega$ as ω goes from 0 to $+\infty$. This is just the polar plot of $G_{M(i\omega)}G_{C(i\omega)}$. See Fig. 11.4*a*.

C_R contour. $s = Re^{i\theta}$ as $R \to \infty$ and θ goes from $\pi/2$ to $-\pi/2$.

$$G_{M(s)}G_{C(s)} = \frac{K_c}{(Re^{i\theta} + 1)(5Re^{i\theta} + 1)}$$

$$\lim_{R\to\infty} G_{M(s)}G_{C(s)} = \lim_{R\to\infty}\left[\frac{K_c}{5R^2}e^{-2\theta i}\right] = 0 \tag{11.15}$$

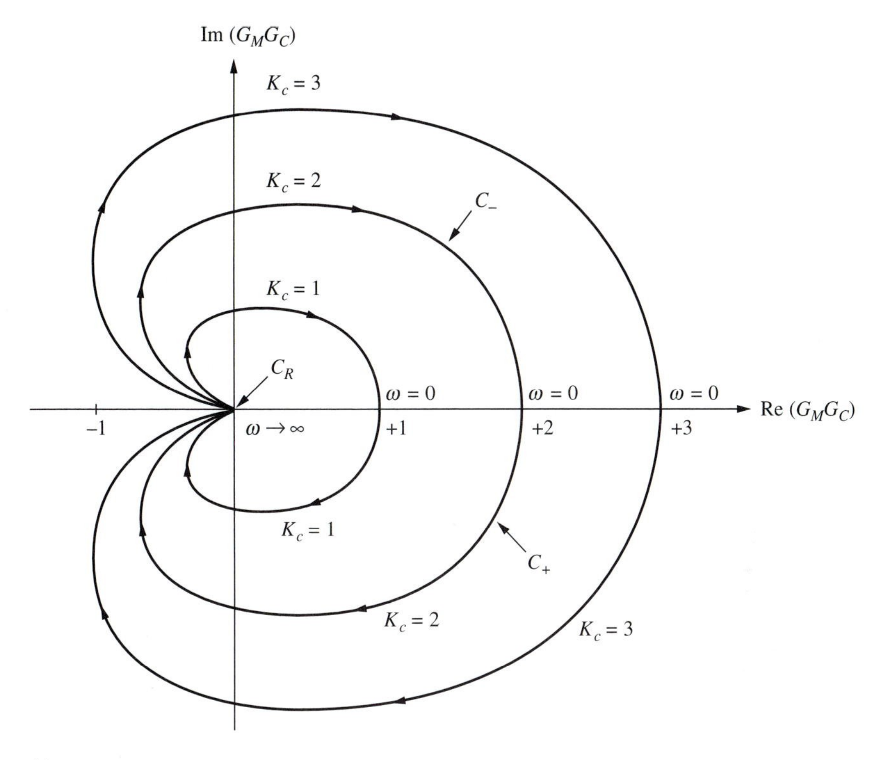

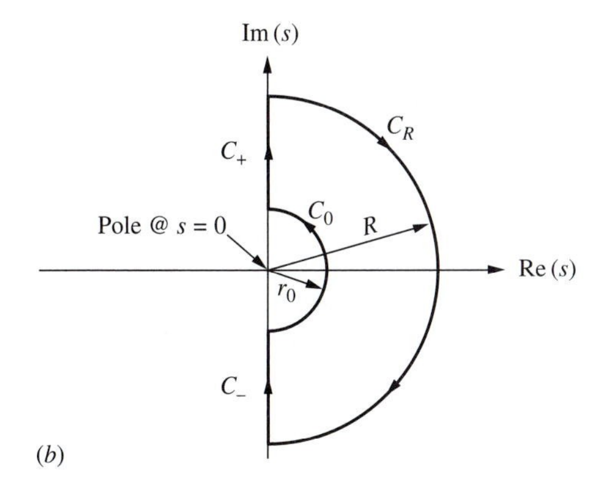

FIGURE 11.4
(*a*) Nyquist plot of the second-order system. (*b*) s plane contour to avoid pole at origin.

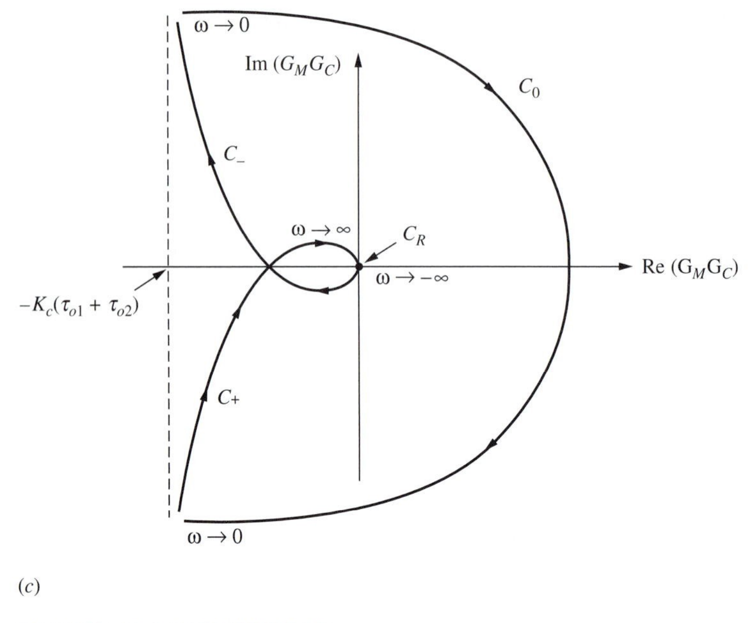

FIGURE 11.4 (CONTINUED)
(*c*) Nyquist plot of system with integrator.

Thus, the infinite semicircle in the *s* plane again maps into the origin in the $G_M G_C$ plane. This happens for all transfer functions where the order of the denominator is greater than the order of the numerator.

C_- contour. $s = i\omega$ as ω goes from $-\infty$ to 0. The $G_{M(i\omega)}G_{C(i\omega)}$ curve for negative values of ω is the reflection over the real axis of the curve for positive values of ω. So we really don't need to plot the C_- contour. The C_+ contour gives us all the information we need.

The complete Nyquist plot is shown in Fig. 11.4*a* for several values of gain K_c. Notice that the curves will *never* encircle the $(-1, 0)$ point, even as the gain is made infinitely large. This means that this second-order system can never be closedloop unstable. This is exactly what our root locus curves showed in Chapter 8.

As the gain is increased, the $G_M G_C$ curve gets closer and closer to the $(-1, 0)$ point. Later in this chapter we use the closeness to the $(-1, 0)$ point as a specification for designing controllers. ■

EXAMPLE 11.3. If the openloop transfer function of the system has poles that lie on the imaginary axis, the *s* plane contour must be modified slightly to exclude these poles. A system with an integrator is a common example.

$$G_{M(s)}G_{C(s)} = \frac{K_c}{s(\tau_{o1}s + 1)(\tau_{o2}s + 1)} \tag{11.16}$$

This system has a pole at the origin. We pick a contour in the *s* plane that goes counterclockwise around the origin, excluding the pole from the area enclosed by the contour. As shown in Fig. 11.4*b*, the contour C_0 is a semicircle of radius r_0. And r_0 is made to approach zero.

C_+ contour. $s = i\omega$ as ω goes from r_0 to R, with r_0 going to 0 and R going to $+\infty$.

$$\begin{aligned} G_{M(i\omega)}G_{C(i\omega)} &= \frac{K_c}{i\omega(1 + i\omega\tau_{o1})(1 + i\omega\tau_{o2})} \\ &= \frac{-K_c\omega(\tau_{o1} + \tau_{o2}) - iK_c(1 - \tau_{o1}\tau_{o2}\omega^2)}{\omega^2(\tau_{o1} + \tau_{o2})^2 + \omega(1 - \tau_{o1}\tau_{o2}\omega^2)^2} \end{aligned} \tag{11.17}$$

The polar plot is shown in Fig. 11.4c.

C_R contour. $s = Re^{i\theta}$.

$$\begin{aligned} G_{M(s)}G_{C(s)} &= \frac{K_c}{Re^{i\theta}(\tau_{o1}Re^{i\theta} + 1)(\tau_{o2}Re^{i\theta} + 1)} \\ \lim_{R\to\infty}[G_{M(s)}G_{C(s)}] &= \lim_{R\to\infty}\left(\frac{K_c}{R^3\tau_{o1}\tau_{o2}}e^{-3\theta i}\right) = 0 \end{aligned} \tag{11.18}$$

The C_R contour maps into the origin in the $G_M G_C$ plane.

C_- contour. The $G_M G_C$ curve is the reflection over the real axis of the $G_M G_C$ curve for the C_+ contour.

C_0 contour. On this small semicircular contour

$$s = r_0 e^{i\theta} \tag{11.19}$$

The radius r_0 goes to zero and θ goes from $-\pi/2$ through 0 to $+\pi/2$ radians. The system transfer function becomes

$$G_{M(s)}G_{C(s)} = \frac{K_c}{r_0 e^{i\theta}(\tau_{o1}r_0e^{i\theta} + 1)(\tau_{o2}r_0e^{i\theta} + 1)} \tag{11.20}$$

As r_0 gets very small, the $\tau_{o1}r_0e^{i\theta}$ and $\tau_{o2}r_0e^{i\theta}$ terms become negligible compared with unity.

$$\lim_{r_0\to 0}\left(G_{M(s)}G_{C(s)}\right) = \lim_{r_0\to 0}\left(\frac{K_c}{r_0e^{i\theta}}\right) = \lim_{r_0\to 0}\left(\frac{K_c}{r_0}e^{-i\theta}\right) \tag{11.21}$$

Thus, the C_0 contour maps into a semicircle in the $G_M G_C$ plane with a radius that goes to infinity and a phase angle that goes from $+\pi/2$ through 0 to $-\pi/2$. See Fig. 11.4c. The Nyquist plot does not encircle the $(-1, 0)$ point if the polar plot of $G_{M(i\omega)}G_{C(i\omega)}$ crosses the negative real axis inside the unit circle. The system would then be closedloop *stable.*

The maximum value of K_c for which the system is still closedloop stable can be found by setting the real part of $G_{M(i\omega)}G_{C(i\omega)}$ equal to -1 and the imaginary part equal to 0. The results are

$$K_u = \frac{\tau_{o1} + \tau_{o2}}{\tau_{o1}\tau_{o2}} \qquad \omega_u = \frac{1}{\sqrt{\tau_{o1}\tau_{o2}}} \tag{11.22}$$

■

As we have seen in the three examples, the C_+ contour usually is the only one that we need to map into the $G_M G_C$ plane. Therefore, from now on we make only polar (or Bode or Nichols) plots of $G_{M(i\omega)}G_{C(i\omega)}$.

EXAMPLE 11.4. Figure 11.5a shows the polar plot of an interesting system that has conditional stability. The system openloop transfer function has the form

$$G_{M(s)}G_{C(s)} = \frac{K_c(\tau_{z1}s + 1)}{(\tau_{o1}s + 1)(\tau_{o2}s + 1)(\tau_{o3}s + 1)(\tau_{o4}s + 1)} \tag{11.23}$$

If the controller gain K_c is such that the $(-1, 0)$ point is in the stable region indicated in Fig. 11.5a, the system is closedloop stable. Let us define three values of controller gain:

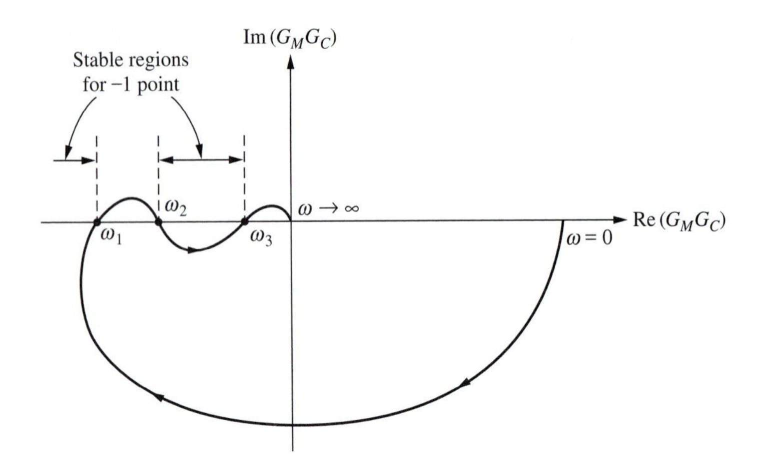

(*a*) Nyquist plot

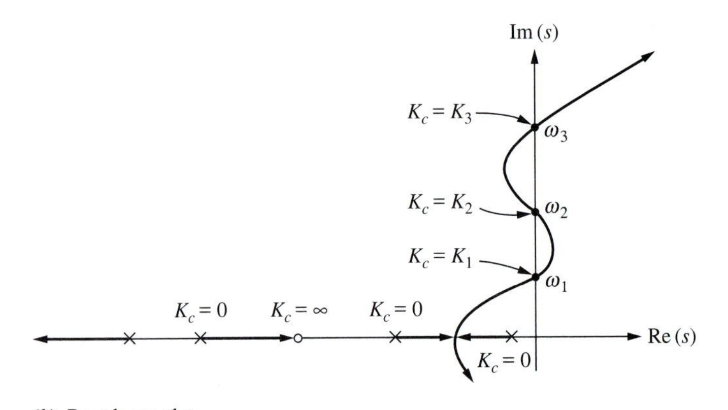

(*b*) Root locus plot

FIGURE 11.5
System with conditional stability.

$$
\begin{aligned}
K_1 &= \text{value of } K_c \text{ when } \left|G_{M(i\omega_1)} G_{C(i\omega_1)}\right| = 1 \\
K_2 &= \text{value of } K_c \text{ when } \left|G_{M(i\omega_2)} G_{C(i\omega_2)}\right| = 1 \\
K_3 &= \text{value of } K_c \text{ when } \left|G_{M(i\omega_3)} G_{C(i\omega_3)}\right| = 1
\end{aligned}
$$

The system is closedloop stable for two ranges of feedback controller gain:

$$K_c < K_1 \quad \text{and} \quad K_2 < K_c < K_3 \tag{11.24}$$

This conditional stability is shown on a root locus plot for this system in Fig. 11.5*b*. ■

11.1.3 Representation

In Chapter 10 we presented three different kinds of graphs that were used to represent the frequency response of a system: Nyquist, Bode, and Nichols plots. The Nyquist stability criterion was developed in the previous section for Nyquist or polar plots.

The critical point for closedloop stability was shown to be the $(-1, 0)$ point on the Nyquist plot.

Naturally we also can show closedloop stability or instability on Bode and Nichols plots. The $(-1, 0)$ point has a phase angle of $-180°$ and a magnitude of unity or a log modulus of 0 decibels. The stability limit on Bode and Nichols plots

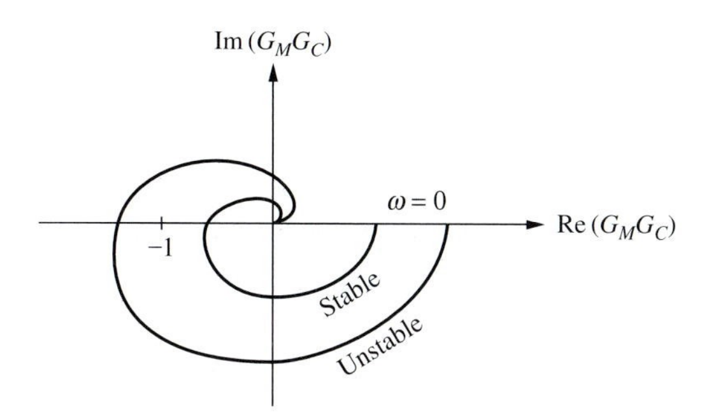

(*a*) Nyquist plot

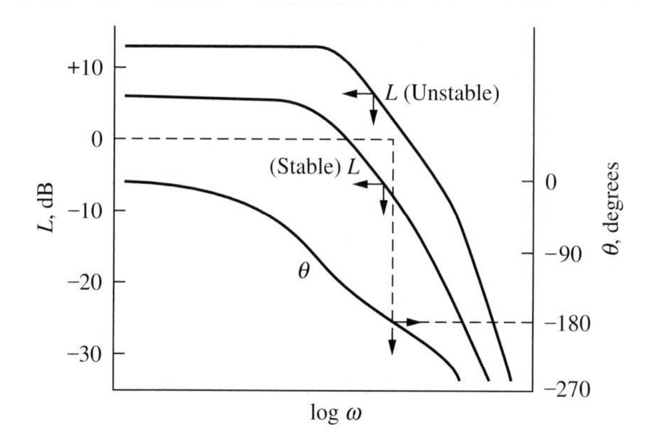

(*b*) Bode plot

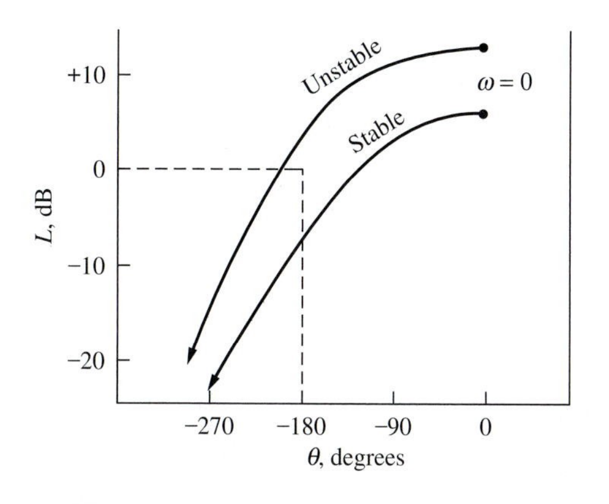

(*c*) Nichols plot

FIGURE 11.6
Stable and unstable closedloop systems in Nyquist, Bode, and Nichols plots.

is therefore the (0 dB, $-180°$) point. At the limit of closedloop stability

$$L = 0 \text{ dB} \quad \text{and} \quad \theta = -180° \tag{11.25}$$

The system is closedloop stable if

$$L < 0 \text{ dB} \quad \text{at } \theta = -180°$$

$$\theta > -180° \quad \text{at } L = 0 \text{ dB}$$

Figure 11.6 illustrates stable and unstable closedloop systems on the three types of plots.

Keep in mind that we are talking about *closedloop* stability and that we are studying it by making frequency response plots of the total *openloop* system transfer function. These log modulus and phase angle plots are for the openloop system. So we could use the terminology L_0 and θ_0 for our Bode and Nichols plots of the openloop $G_M G_C$ frequency response plots.

We consider openloop-stable systems most of the time. We show how to deal with openloop-unstable processes in Section 11.4.

11.2 CLOSEDLOOP SPECIFICATIONS IN THE FREQUENCY DOMAIN

There are two basic types of specifications commonly used in the frequency domain. The first type, *phase margin* and *gain margin*, specifies how near the *openloop* $G_{M(i\omega)} G_{C(i\omega)}$ polar plot is to the critical $(-1, 0)$ point. The second type, *maximum closedloop log modulus*, specifies the height of the resonant peak on the log modulus Bode plot of the *closedloop* servo transfer function. So keep the apples and the oranges straight. We make *openloop* transfer function plots and look at the $(-1, 0)$ point. We make *closedloop* servo transfer function plots and look at the peak in the log modulus curve (indicating an underdamped system). But in both cases we are concerned with *closedloop* stability.

These specifications are easy to use, as we show with some examples in Section 11.4. They can be related qualitatively to time-domain specifications such as damping coefficient.

11.2.1 Phase Margin

Phase margin (PM) is defined as the angle between the negative real axis and a radial line drawn from the origin to the point where the $G_M G_C$ curve intersects the unit circle. See Fig. 11.7. The definition is more compact in equation form.

$$\text{PM} = 180° + (\arg G_M G_C)_{|G_M G_C| = 1} \tag{11.26}$$

If the $G_M G_C$ polar plot goes through the $(-1, 0)$ point, the phase margin is zero. If the $G_M G_C$ polar plot crosses the negative real axis to the right of the $(-1, 0)$ point, the

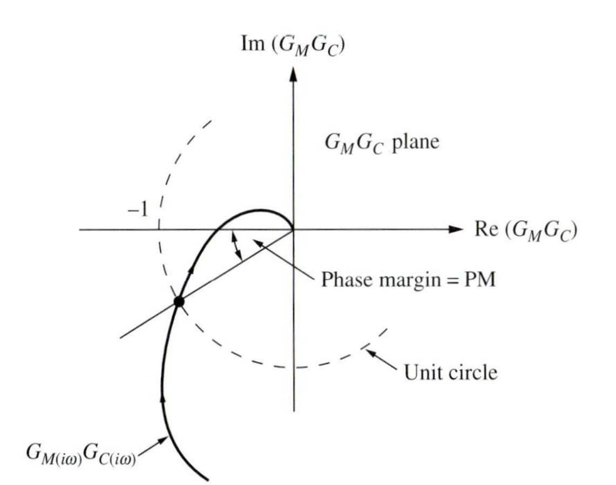

(a) Nyquist plot

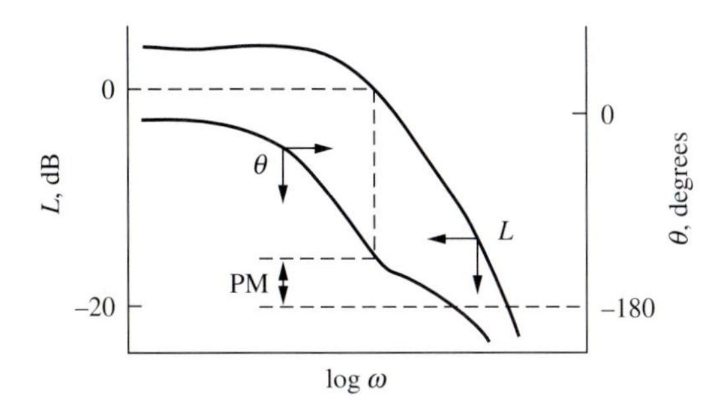

(b) Bode plot

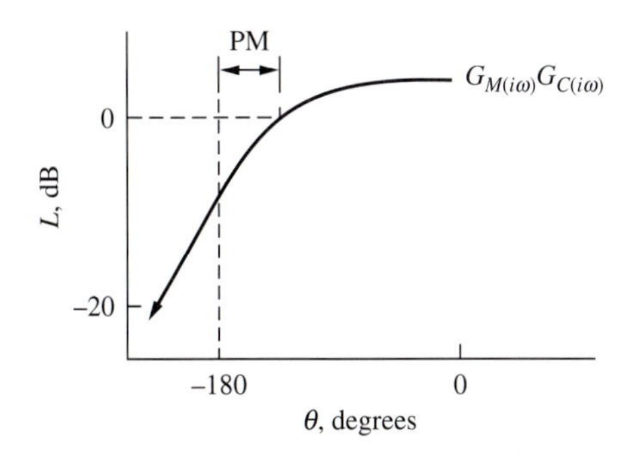

(c) Nichols plot

FIGURE 11.7
Phase margin.

phase margin is some positive angle. The bigger the phase margin, the more stable is the closedloop system. A negative phase margin means an unstable closedloop system.

Phase margins of around 45° are often used. Figure 11.7 shows how phase margin is found on Bode and Nichols plots.

11.2.2 Gain Margin

Gain margin (GM) is defined as the reciprocal of the intersection of the $G_M G_C$ polar plot on the negative real axis.

$$\text{GM} = \frac{1}{|G_M G_C|_{\arg G_M G_C = -180°}} \tag{11.27}$$

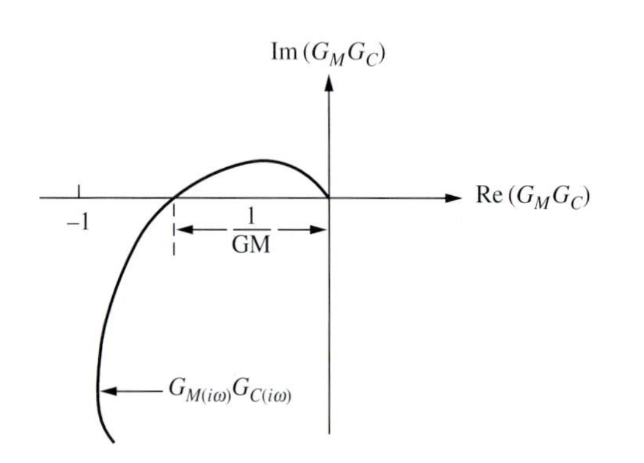

(*a*) Nyquist plot

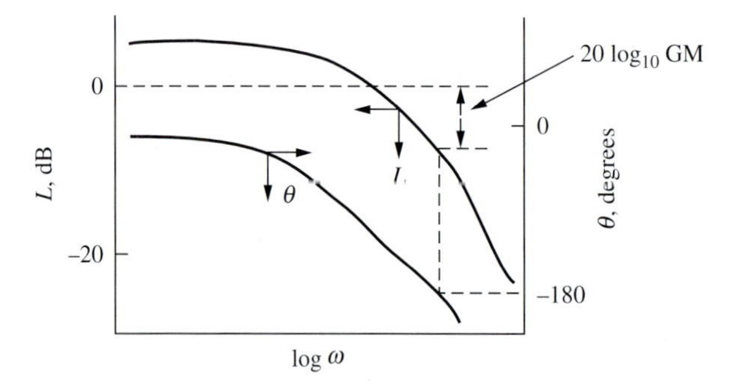

(*b*) Bode plot

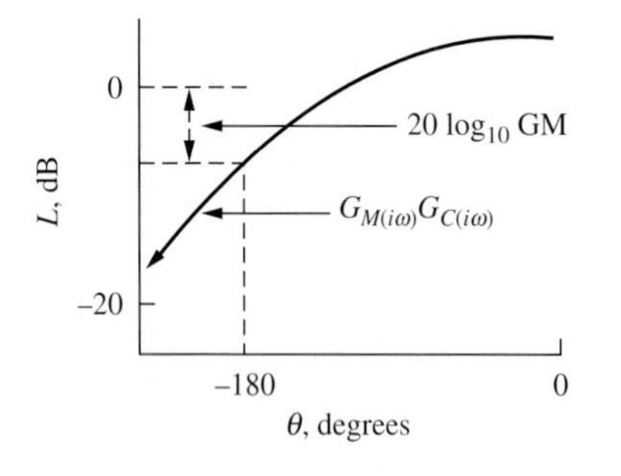

(*c*) Nichols plot

FIGURE 11.8
Gain margin.

Figure 11.8 shows gain margins on Nyquist, Bode, and Nichols plots. Gain margins are sometimes reported in decibels.

If the $G_M G_C$ curve goes through the critical $(-1, 0)$ point, the gain margin is unity (0 dB). If the $G_M G_C$ curve crosses the negative real axis between the origin and -1, the gain margin is greater than 1. Therefore, the bigger the gain margin, the more stable the system is, i.e., the farther away from -1 the curve crosses the real axis. Gain margins of around 2 are often used.

A system must be third or higher order (or have deadtime) to have a meaningful gain margin. Polar plots of first- and second-order systems do not intersect the negative real axis.

11.2.3 Maximum Closedloop Log Modulus ($L_c^{\max}$)

The most useful frequency-domain specification is the maximum closedloop log modulus. The phase margin and gain margin specifications can sometimes give poor results when the shape of the frequency response curve is unusual.

For example, consider the Nyquist plot of a process sketched in Fig. 11.9*a*, where the shape of the $G_M G_C$ curve gives a good phase margin but the curve still passes very close to the $(-1, 0)$ point. The damping coefficient of this system would be quite low. This type of $G_M G_C$ curve is commonly encountered when the process has a large deadtime. Figure 11.9*b* shows a $G_M G_C$ curve that has a good gain margin but passes too close to the $(-1, 0)$ point. These two cases illustrate that using phase or gain margins does not necessarily give the desired degree of damping. This is because each of these criteria measures the closeness of the $G_M G_C$ curve to the $(-1, 0)$ point at only one particular spot.

The maximum closedloop log modulus does not have this problem since it directly measures the closeness of the $G_M G_C$ curve to the $(-1, 0)$ point at all frequencies. The closedloop log modulus refers to the closedloop servo transfer function:

$$\frac{Y_{(s)}}{Y_{(s)}^{\text{set}}} = \frac{G_{M(s)} G_{C(s)}}{1 + G_{M(s)} G_{C(s)}} \tag{11.28}$$

The feedback controller is designed to give a maximum resonant peak or hump in the closedloop log modulus plot.

All the Nyquist, Bode, and Nichols plots discussed in previous sections have been for *openloop* system transfer functions $G_{M(i\omega)} G_{C(i\omega)}$. Frequency response plots can be made for any type of system, openloop or closedloop. The two closedloop transfer functions that we derived in Chapter 8 show how the output $Y_{(s)}$ is affected in a closedloop system by a setpoint input $Y_{(s)}^{\text{set}}$ and by a load $L_{(s)}$. Equation (11.28) gives the closedloop servo transfer function. Equation (11.29) gives the closedloop load transfer function.

$$\frac{Y_{(s)}}{L_{(s)}} = \frac{G_{L(s)}}{1 + G_{M(s)} G_{C(s)}} \tag{11.29}$$

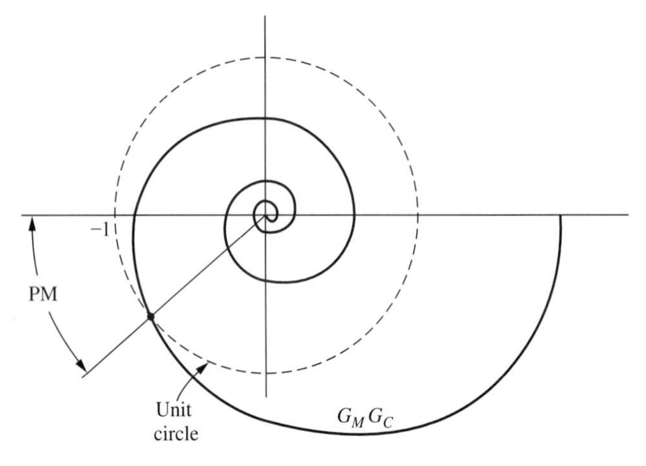

(*a*) Phase margin doesn't work

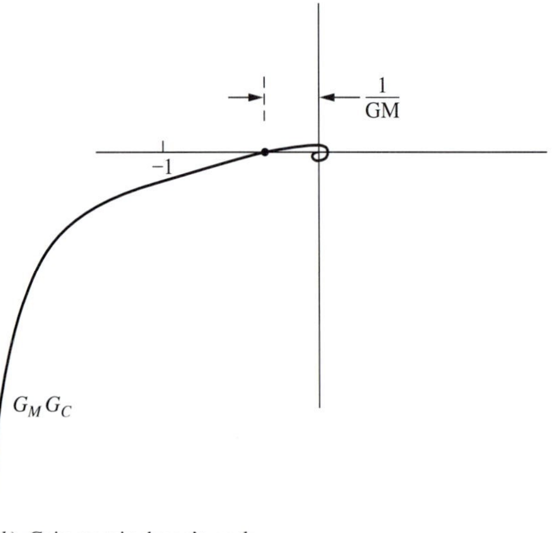

(*b*) Gain margin doesn't work

FIGURE 11.9
Nyquist plots.

Typical log modulus Bode plots of these two closedloop transfer functions are shown in Fig. 11.10*a*. If it were possible to achieve perfect or ideal control, the two ideal closedloop transfer functions would be

$$\frac{Y_{(s)}}{L_{(s)}} = 0 \quad \text{and} \quad \frac{Y_{(s)}}{Y^{\text{set}}_{(s)}} = 1 \tag{11.30}$$

Equation (11.30) says that we want the output to track the setpoint perfectly for all frequencies, and we want the output to be unaffected by the load disturbance for all frequencies. Log modulus curves for these ideal (but unattainable) closedloop systems are shown in Fig. 11.10*b*.

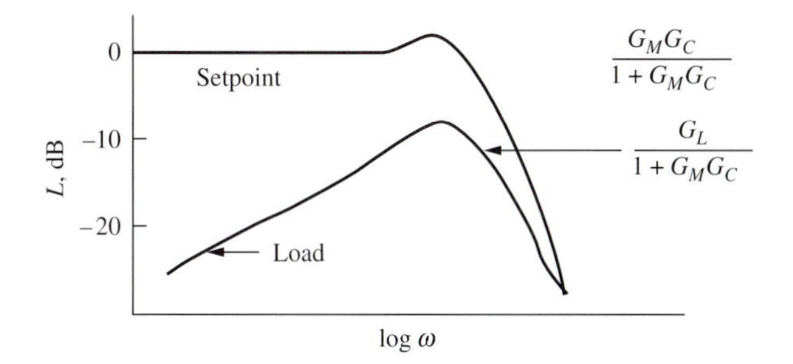

(*a*) Load and setpoint closedloop transfer functions

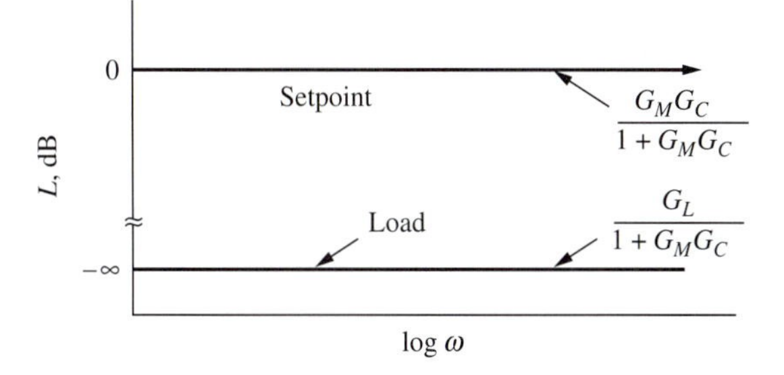

(*b*) Ideal load and setpoint closedloop transfer functions

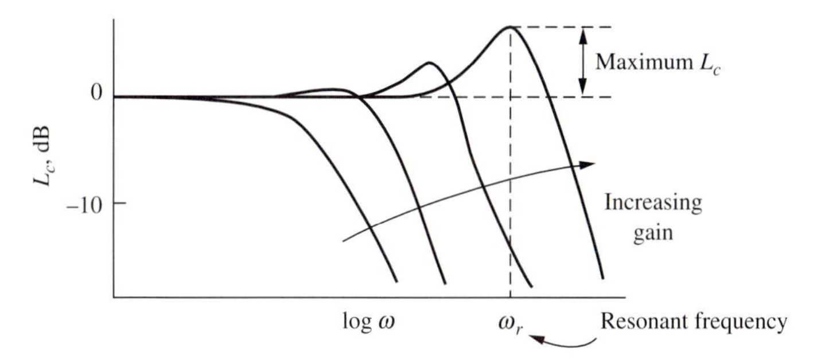

(*c*) Typical setpoint closedloop transfer functions

FIGURE 11.10
Closedloop log modulus curves.

In most systems, the closedloop servo log modulus curves move out to higher frequencies as the gain of the feedback controller is increased. The system has a "wider bandwidth," as the mechanical and electrical engineers say. This is desirable since it means a faster closedloop system. Remember, the breakpoint frequency is the reciprocal of the closedloop time constant.

But the height of the resonant peak also increases as the controller gain is increased. This means that the closedloop system becomes more underdamped. The effects of increasing controller gain are sketched in Fig. 11.10*c*.

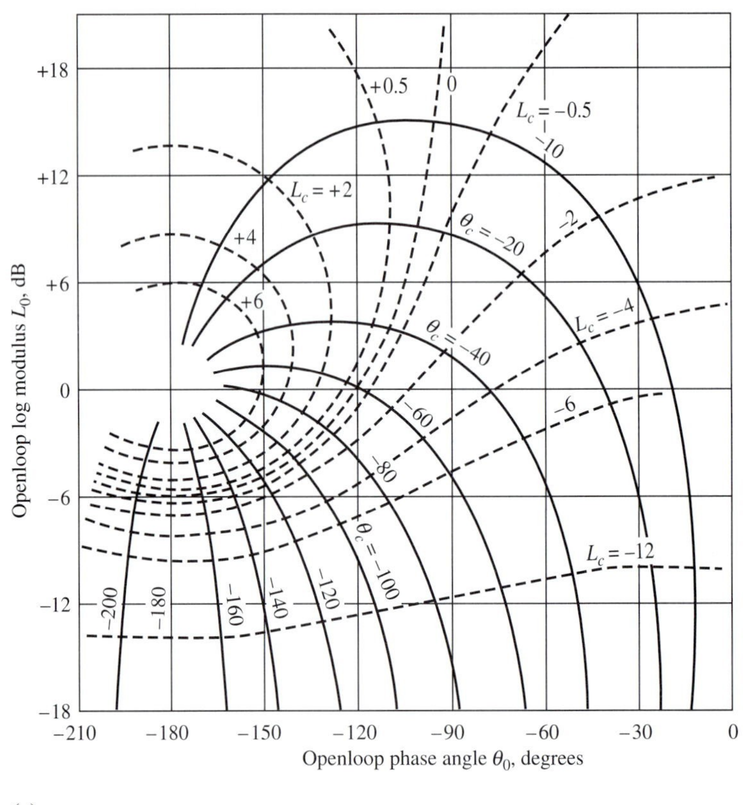

(*a*)

FIGURE 11.11
(*a*) Nichols chart.

A commonly used maximum closedloop log modulus specification is +2 dB. The controller parameters are adjusted to give a maximum peak in the closedloop servo log modulus curve of +2 dB. This corresponds to a magnitude ratio of 1.3 and is approximately equivalent to an underdamped system with a damping coefficient of 0.4.

Both the openloop and the closedloop frequency response curves can be easily generated on a digital computer by using the complex variables and functions in FORTRAN discussed in Chapter 10 or by using MATLAB software. The frequency response curves for the closedloop servo transfer function can also be fairly easily found graphically by using a Nichols chart. This chart was developed many years ago, before computers were available, and was widely used because it greatly facilitated the conversion of openloop frequency response to closedloop frequency response.

A Nichols chart is a graph that shows what the closedloop log modulus L_c and closedloop phase angle θ_c are for any given openloop log modulus L_0 and openloop phase angle θ_0. See Fig. 11.11*a*. The graph is completely general and can

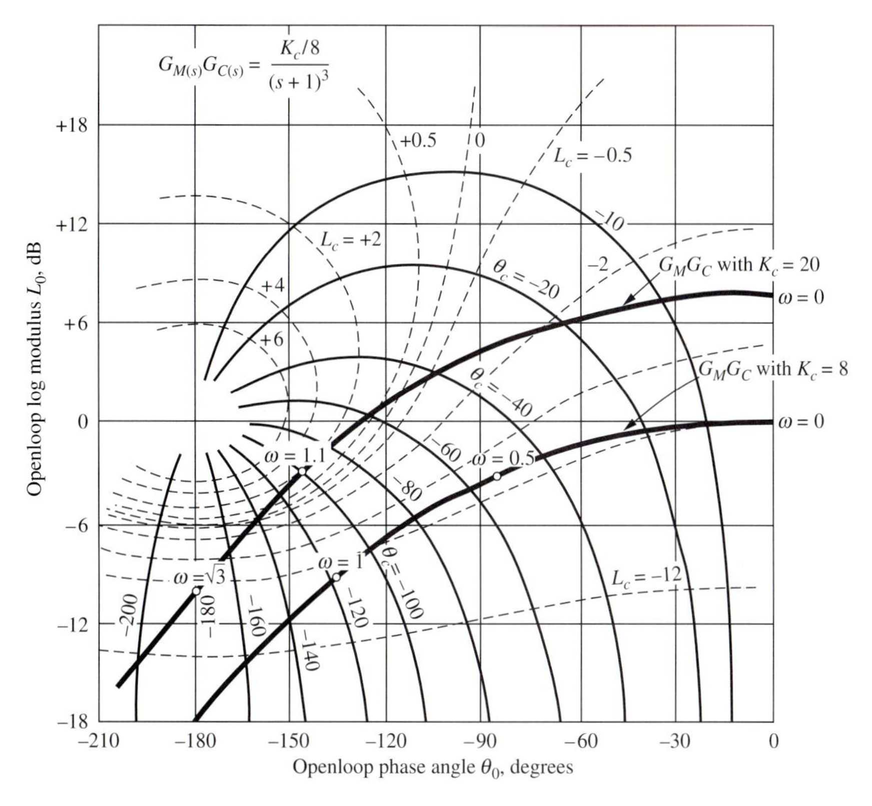

(*b*)

FIGURE 11.11 (CONTINUED)
(*b*) Nichols chart with a three-CSTR system openloop $G_{M(i\omega)}G_{C(i\omega)}$ plotted.

be used for any SISO system. To prove this, let us choose an arbitrary openloop $G_{M(i\omega)}G_{C(i\omega)}$. In polar form, the openloop complex function is

$$G_{M(i\omega)}G_{C(i\omega)} = r_0 e^{i\theta_0} \tag{11.31}$$

where r_0 = magnitude of the openloop complex function at frequency ω
θ_0 = argument of the openloop complex function at frequency ω

The closedloop servo transfer function is

$$\frac{G_{M(s)}G_{C(s)}}{1 + G_{M(s)}G_{C(s)}} = \frac{r_0 e^{i\theta_0}}{1 + r_0 e^{i\theta_0}} \tag{11.32}$$

Putting this complex function into polar form gives

$$r_c e^{i\theta_c} = \frac{r_0 e^{i\theta_0}}{1 + r_0 e^{i\theta_0}} \tag{11.33}$$

where r_c = magnitude of the closedloop complex function at frequency ω
θ_c = argument of the closedloop complex function at frequency ω

Equation (11.33) can be rearranged to get r_c and θ_c as explicit functions of r_0 and θ_0.

$$r_c = \frac{r_0}{\sqrt{1 + 2r_0 \cos\theta_0 + r_0^2}} \tag{11.34}$$

$$\theta_c = \arctan\left(\frac{\sin\theta_0}{r_0 \cos\theta_0}\right) \tag{11.35}$$

Thus, for any arbitrary system with the given openloop parameters θ_0 and L_0, Eqs. (11.34) and (11.35) give the closedloop parameters θ_c and L_c. The Nichols chart is a plot of these relationships.

The graphical procedure for using a Nichols chart is first to construct the openloop $G_M G_C$ Bode plots. Then we draw an openloop Nichols plot of $G_{M(i\omega)}G_{C(i\omega)}$. Finally we sketch this openloop curve of L_0 versus θ_0 onto a Nichols chart. At each point on this curve (which corresponds to a certain value of frequency), the values of the closedloop log modulus L_c can be read off.

Figure 11.11*b* is a Nichols chart with two $G_M G_C$ curves plotted on it. They are from the three-CSTR system with a proportional controller.

$$G_{M(s)}G_{C(s)} = \frac{\frac{1}{8}K_c}{(s+1)^3} \tag{11.36}$$

The two curves have two different values of controller gain: $K_c = 8$ and $K_c = 20$. The openloop Bode plots of $G_M G_C$ and the closedloop Bode plots of $G_{M(s)}G_{C(s)}/(1 + G_{M(s)}G_{C(s)})$, with $K_c = 20$, are given in Fig. 11.12.

The lines of constant closedloop log modulus L_c are part of the Nichols chart. If we are designing a closedloop system for an L_c^{max} specification, we merely have to adjust the controller type and settings so that the openloop $G_M G_C$ curve is tangent to the desired L_c line on the Nichols chart. For example, the $G_M G_C$ curve in Fig. 11.11*b* with $K_c = 20$ is just tangent to the +2 dB L_c line of the Nichols chart. The value of frequency at the point of tangency, 1.1 rad/min, is the closedloop resonant frequency ω_r. The peak in the log modulus plot is clearly seen in the closedloop curves given in Fig. 11.12.

There are two aspects of using the maximum closedloop log modulus specification that you should be aware of. First, the L_c curves can display multiple peaks because the Nyquist plot of some complex processes can approach the $(-1, 0)$ point at several frequency points along its trajectory. We are always looking for the highest peak, so make sure you cover the entire frequency range of interest when you plot the L_c curve. This multiple-peak phenomenon can be a particularly confusing problem when you are using a computer program to determine the controller gain that gives a desired L_c^{max}. Keep in mind that multiple peaks can occur in some systems.

The second thing that you should be alert to is that a plot of $G_{M(s)}G_{C(s)}/(1 + G_{M(s)}G_{C(s)})$ tells you only how close you are to the $(-1, 0)$ point. As you make the gain bigger and bigger, approaching the ultimate gain, the peak in the curve increases. At the ultimate gain the peak height is infinite. However, if you continue to make the same plot for gains greater than the ultimate gain, the peak height will

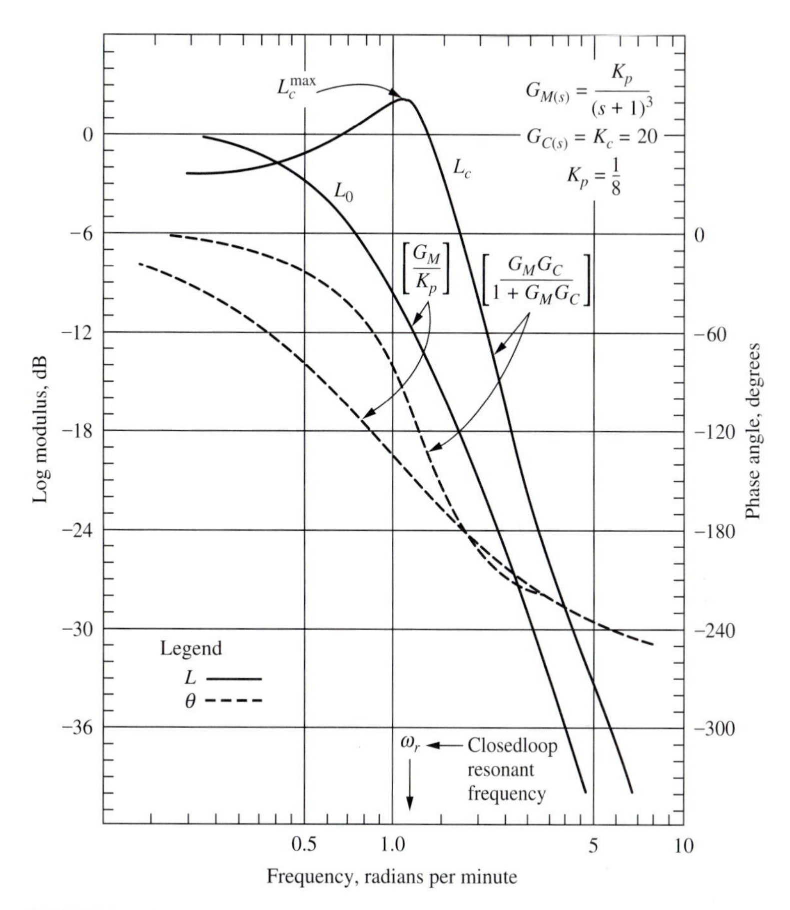

FIGURE 11.12
Openloop and closedloop Bode plots for a three-CSTR system.

decrease. This is because the function we are plotting measures only the closeness of the $G_M G_C$ curve to the $(-1, 0)$ point. After the $(-1, 0)$ point has been encircled, increasing the controller gain moves the $G_M G_C$ curve further away from the $(-1, 0)$ point on the other side. So be sure to check that the $G_M G_C$ curve does not encircle the $(-1, 0)$ point.

11.3 FREQUENCY RESPONSE OF FEEDBACK CONTROLLERS

Before we give some examples of the design of feedback controllers in the frequency domain, it would be wise to show what the common P, PI, and PID controllers look like in the frequency domain. This forms the $G_{C(i\omega)}$ that we combine with the process $G_{M(i\omega)}$ to get the total openloop Bode plots of $G_{M(i\omega)} G_{C(i\omega)}$.

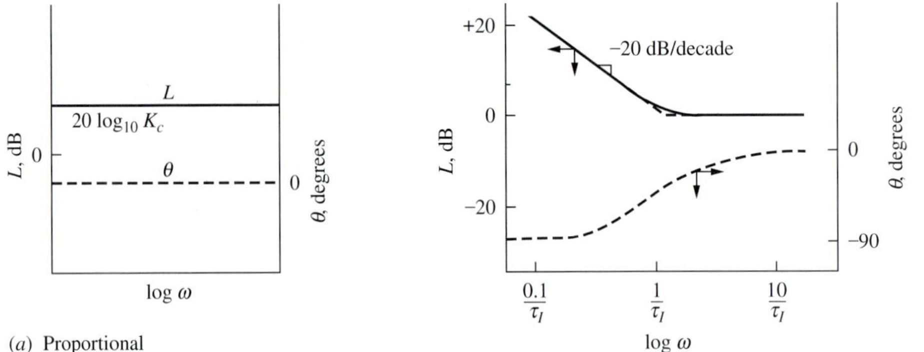

(a) Proportional

(b) Proportional-integral

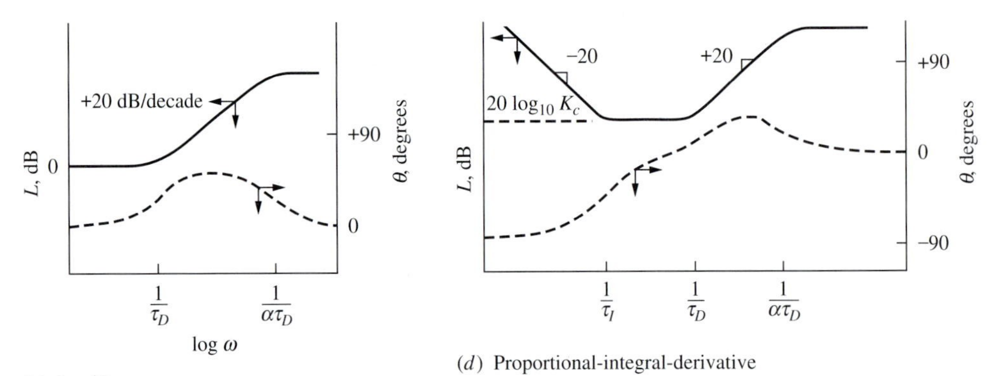

(c) Lead-lag

(d) Proportional-integral-derivative

FIGURE 11.13
Bode plots of controllers.

11.3.1 Proportional Controller (P)

The transfer function of a P controller is $G_{C(s)} = K_c$. Substituting $s = i\omega$ gives

$$G_{C(i\omega)} = K_c \tag{11.37}$$

A proportional controller merely multiplies the magnitude of $G_{M(i\omega)}$ at every frequency by a constant K_c. On a Bode plot, this means that a proportional controller raises the log modulus curve by $20 \log_{10} K_c$ decibels but has no effect on the phase angle curve. See Fig. 11.13*a*.

11.3.2 Proportional-Integral Controller (PI)

The transfer function of a PI controller is

$$G_{C(s)} = K_c\left(1 + \frac{1}{\tau_I s}\right) = K_c\left(\frac{\tau_I s + 1}{\tau_I s}\right)$$

$$G_{C(i\omega)} = K_c \left(\frac{\tau_I i\omega + 1}{\tau_I i\omega} \right) \tag{11.38}$$

The Bode plot of this combination of an integrator and a first-order lead is shown in Fig. 11.13*b*. At low frequencies, a PI controller amplifies magnitudes and contributes $-90°$ of phase angle lag. This loss of phase angle is undesirable from a dynamic standpoint since it moves the $G_M G_C$ polar plot closer to the $(-1, 0)$ point.

11.3.3 Proportional-Integral-Derivative Controller (PID)

$$G_{C(s)} = K_c \left(\frac{\tau_I s + 1}{\tau_I s} \right) \left(\frac{\tau_D s + 1}{\alpha \tau_D s + 1} \right) \tag{11.39}$$

The Bode plot for the lead-lag element is sketched in Fig. 11.13*c*. It contributes positive phase angle advance over a range of frequencies between $1/\tau_D$ and $1/\alpha\tau_D$.

The lead-lag element can move the $G_M G_C$ curve away from the $(-1, 0)$ point and improve stability. When the derivative setting on a PID controller is tuned, the location of the phase angle advance is shifted so that it occurs near the critical $(-1, 0)$ point.

11.4 EXAMPLES

11.4.1 Three-CSTR Process

The openloop transfer function for a three-CSTR process is

$$G_{M(s)} = \frac{\frac{1}{8}}{(s + 1)^3}$$

Before we design controllers in the frequency domain, it might be interesting to see what the frequency-domain indicators of closedloop performance turn out to be when the Ziegler-Nichols settings are used on this system. Table 11.1 shows the phase

TABLE 11.1

Frequency-domain indicators that result from Ziegler-Nichols settings and 0.316 damping coefficient settings

	Ziegler-Nichols			0.316 damping coefficient			
	P	PI	PID	P	PI	PID	PID
K_c	32	29.1	37.6	17	13	30	17
τ_I	—	3.03	1.82	—	3.03	1.82	1.82
τ_D	—	—	0.45	—	—	0.9	0.45
Phase margin (°)	28	13	22	64	52	38	41
Gain margin	2	1.6	7	3.8	3.5	10	15
L_c^{max} (dB)	6.9	13	8.3	0.5	1.9	3.8	3.2
Resonant frequency ω_r (rad/min)	1.3	1.3	1.6	1.0	0.8	1.6	1.0

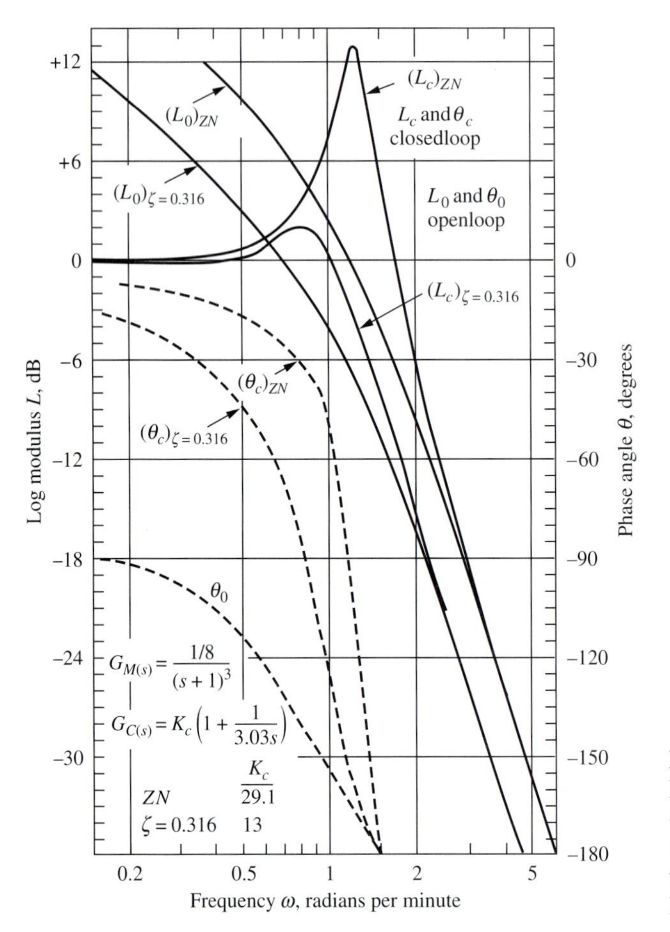

FIGURE 11.14
Bode plots of openloop $G_M G_C$ and closedloop $G_M G_C/(1 + G_M G_C)$ for a three-CSTR system and PI controllers.

and gain margins and the maximum closedloop log moduli that the Ziegler-Nichols settings give. Also shown in Table 11.1 are the results when the settings for a damping coefficient of 0.316 are used.

The Ziegler-Nichols settings give quite small phase and gain margins and large maximum closedloop log moduli. The $\zeta = 0.316$ settings are more conservative. Figure 11.14 shows the closedloop and openloop Bode plots for the PI controllers with the two different settings.

Now we are ready to find the controller settings required to give various frequency-domain specifications with P, PI, and PID controllers.

A. Proportional controller

Gain margin. Suppose we want to find the value of feedback controller gain K_c that gives a gain margin GM $= 2$. We must find the value of K_c that makes the Nyquist plot of

$$G_{M(s)}G_{C(s)} = \frac{\frac{1}{8}K_c}{(s+1)^3}$$

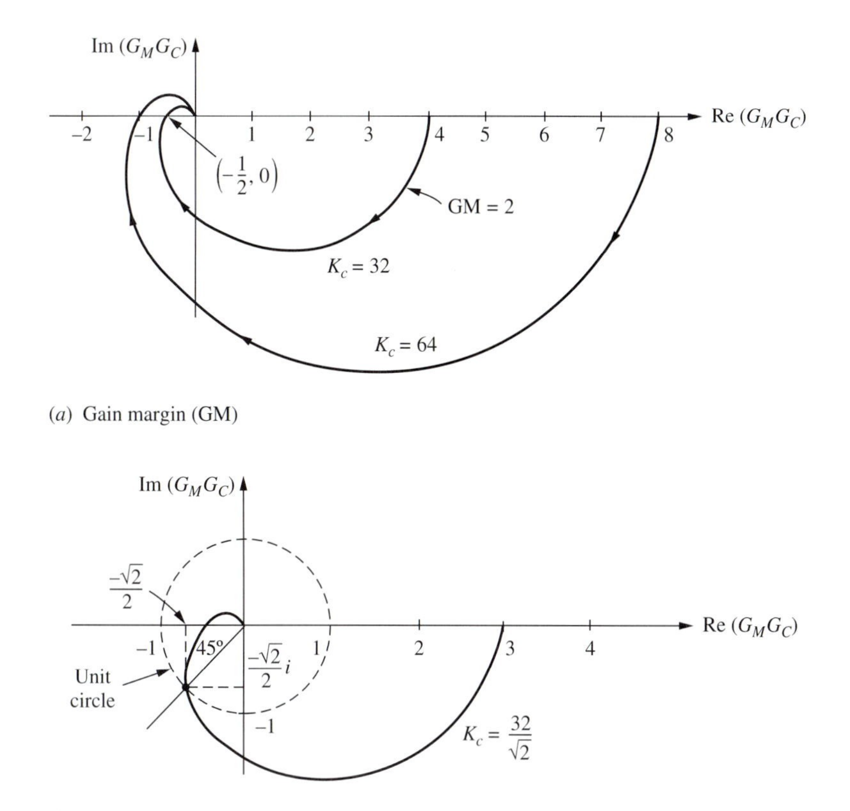

FIGURE 11.15
Nyquist plots for a three-CSTR system with proportional controllers.

cross the negative real axis at $(-0.5, 0)$. As shown in Fig. 11.15*a*, the ultimate gain is 64. Thus, a gain of 32 will reduce the magnitude of each point by one-half and make the $G_M G_C$ polar curve pass through the $(-0.5, 0)$ point.

Figure 11.16 shows the same result in Bode plot form. When the phase angle is $-180°$ (at frequency $\omega_u = \sqrt{3}$), the magnitude must be 0.5 or the log modulus must be -6 dB. Thus, the log modulus curve must be raised $+12$ dB (gain 4) above its position when the controller gain is 8. Therefore, the total gain must be 32 for GM $= 2$. Notice that this is the Ziegler-Nichols setting.

Phase margin. To get a 45° phase margin, we must find the value of K_c that makes the Nyquist plot pass through the unit circle when the phase angle is $-135°$, as shown in Fig. 11.15*b*. The real and imaginary parts of $G_M G_C$ must both be equal to $-\frac{1}{2}\sqrt{2}$ at this point on the unit circle. Solving the two simultaneous equations gives

$$K_c = \frac{32}{\sqrt{2}} = 22.6 \quad \text{and} \quad \omega = 1 \text{ rad/min}$$

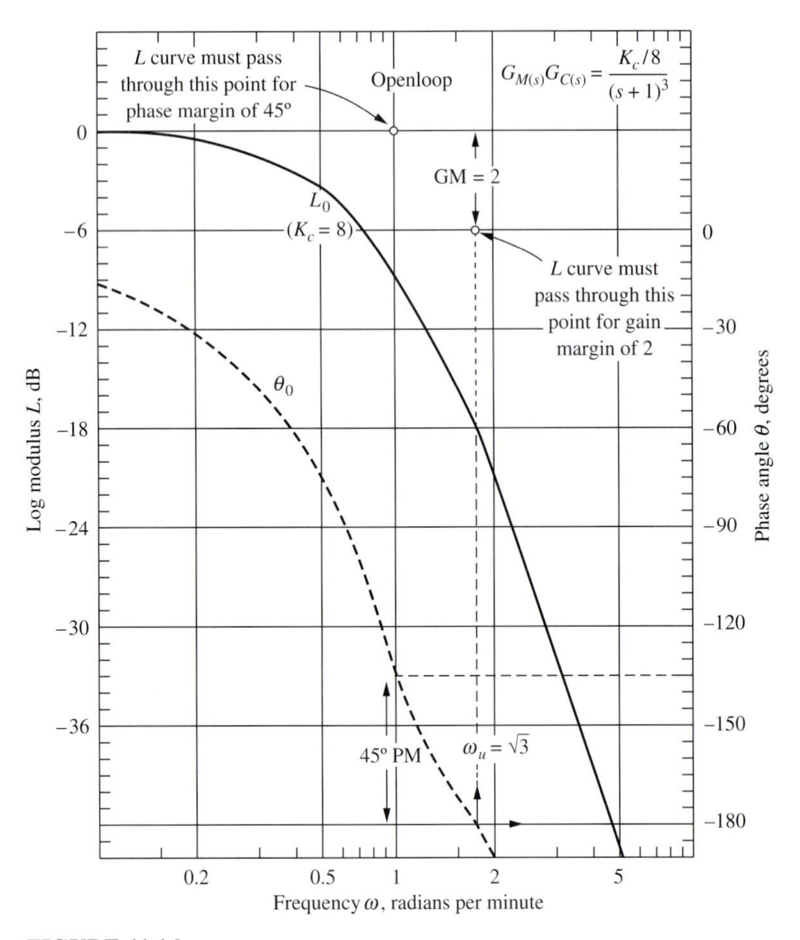

FIGURE 11.16

Bode plots of a three-CSTR system with proportional controller.

On a Bode plot (Fig. 11.16), the log modulus curve of $G_M G_C$ must pass through the 0-dB point when the phase angle curve is at $-135°$. This occurs at $\omega = 1$ rad/min. The log modulus curve for $K_c = 8$ must be raised +9 dB (gain 2.82). Therefore, the controller gain must be $(8)(2.82) = 22.6$.

Notice that this gain is lower than that needed to give a gain margin of 2. The gain margin with $K_c = 22.6$ can be easily found from the Bode plot. When the phase angle is $-180°$, the log modulus is -18 dB (for $K_c = 8$). If the gain of 22.6 is used, the log modulus is raised +9 dB. The log modulus is now -9 dB at the $-180°$ frequency, giving a gain margin of 2.82.

Maximum closedloop log modulus. We have already designed (in Section 11.2.3) a proportional controller that gave $L_c^{\max}$ of +2 dB. Figure 11.11*b* gives a Nichols chart with the $G_M G_C$ curve for this system. A gain of 20 makes the openloop $G_M G_C$ curve tangent to the +2-dB L_c curve on the Nichols chart.

From the three preceding cases we can conclude that, for this third-order system with three equal first-order lags, the +2-dB L_c^{max} specification is the most conservative, the 45° PM is next, and the 2 GM gives the controller gain that is closest to instability. These results are typical, but different types of processes can give different results.

B. Proportional-integral controllers

A PI controller has two adjustable parameters, and therefore we should, theoretically, be able to set two frequency-domain specifications and find the values of τ_I and K_c that satisfy them. We cannot make this choice of specifications completely arbitrary. For example, we cannot achieve a 45° phase margin and a gain margin of 2 with a PI controller in this three-CSTR system. A PI controller cannot reshape the Nyquist plot to make it pass through both the $(-\frac{1}{2}\sqrt{2}, -\frac{1}{2}\sqrt{2})$ point and the $(-0.5, 0)$ point because of the loss of phase angle at low frequencies.

Let us design a PI controller for a +2-dB L_c^{max} specification. For proportional controllers, all we have to do is find the value of K_c that makes the $G_M G_C$ curve on a Nichols chart tangent to the +2-dB L_c line. For a PI controller there are two parameters to find. Design procedures and guides have been developed over the years for finding the values of τ_I.

One easy approach is to use the Ziegler-Nichols value or the Tyreus-Luyben value for reset time. Integral action is utilized only to eliminate steady-state offset, so it is not too critical what value is used as long as it is reasonable, i.e., about the same magnitude as the process time constant.

In the three-CSTR example the ultimate gain and ultimate frequency are $K_u = 64$ and $\omega_u = \sqrt{3}$. The Ziegler-Nichols value for τ_I for a PI controller is 3.03 minutes, and the Tyreus-Luyben value is 3.63 minutes. Figure 11.17 gives Bode plots of the openloop and closedloop system with a PI controller using the two reset times. The controller gains that give +2-dB maximum closedloop log modulus for the two reset times are $K_c = 12.97$ with ZN reset and 17.32 for TL reset. Note that the bandwidth using the TL reset is a little wider (resonant frequency is 1.02 rad/min compared to 0.848 rad/min for ZN reset), so the closedloop time constant is smaller.

C. Proportional-integral-derivative controllers

PID controllers provide three adjustable parameters. We should theoretically be able to satisfy three specifications. A practical design procedure that we have used with good success for many years is outlined below.

1. Set τ_I equal to the Ziegler-Nichols value or the Tyreus-Luyben value.
2. Pick a value of τ_D and find the value of K_c that gives +2-dB L_c^{max}.
3. Repeat step 2 for a whole range of τ_D values, with a new value of K_c calculated at each new value of τ_D such that an $L_c^{max} = +2$ dB is achieved.
4. Select the value of τ_D that gives the maximum value of K_c. This τ_D^{opt} gives the largest gain and therefore the smallest closedloop time constant for a specified closedloop damping coefficient (as inferred from the L_c^{max} specification). Figure 11.18 gives the plot of K_c versus τ_D for the three-CSTR process. The optimal value for the derivative time is 1.0 minutes, giving a controller gain of 25.

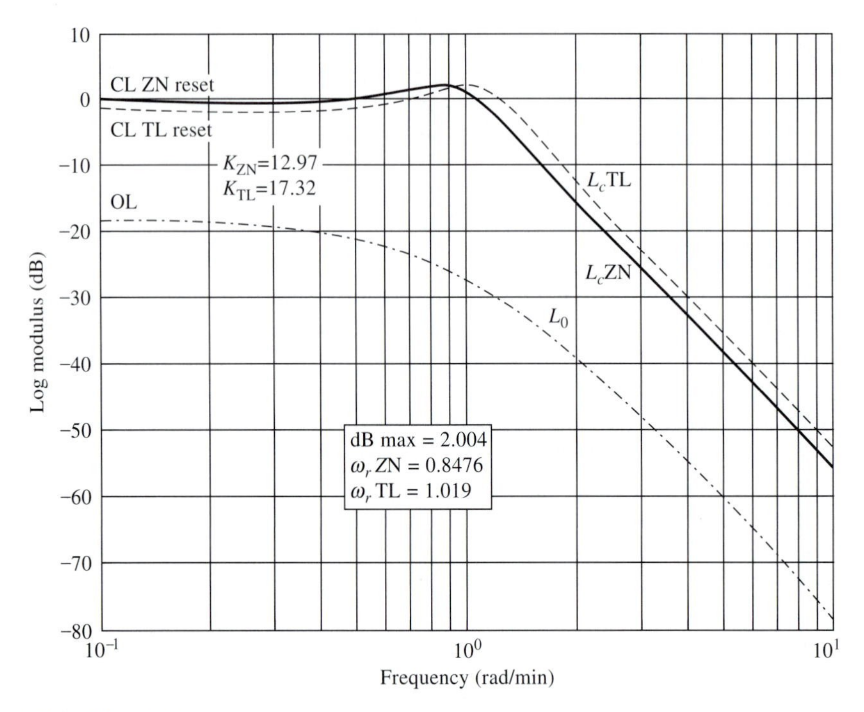

FIGURE 11.17
Three-CSTR process with PI control.

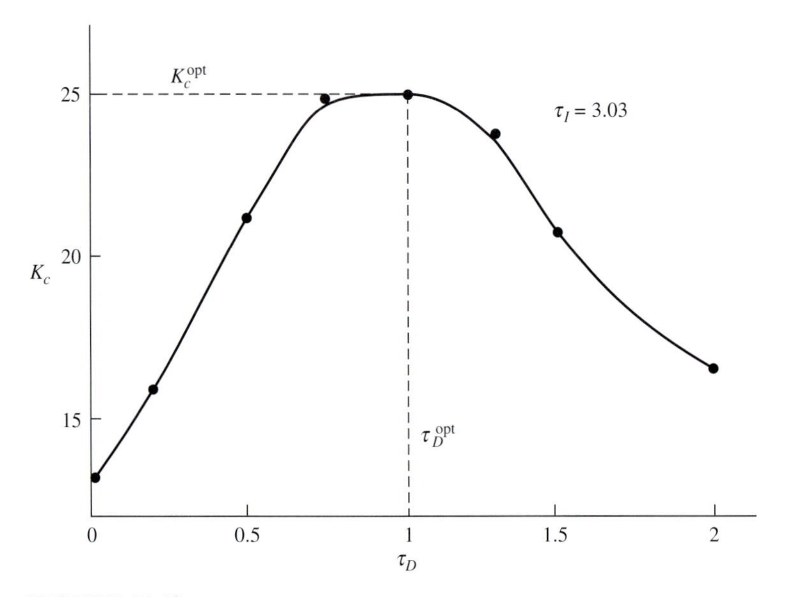

FIGURE 11.18
Optimal τ_D.

This procedure leads to the controller settings $K_c = 25$, $\tau_I = 3.03$ min, and $\tau_D = 1$ min. The gain and phase margins with these settings are 6.3 and 48°, respectively. The Ziegler-Nichols settings are $K_c = 37.6$, $\tau_I = 1.82$ min, and $\tau_D = 0.45$ min. The maximum closedloop log modulus for the ZN settings is +8.3 dB (Table 11.1), which is too underdamped for most chemical engineering systems.

11.4.2 First-Order Lag with Deadtime

Many chemical engineering systems can be modeled by a transfer function involving a first-order lag with deadtime. Let us consider a typical transfer function:

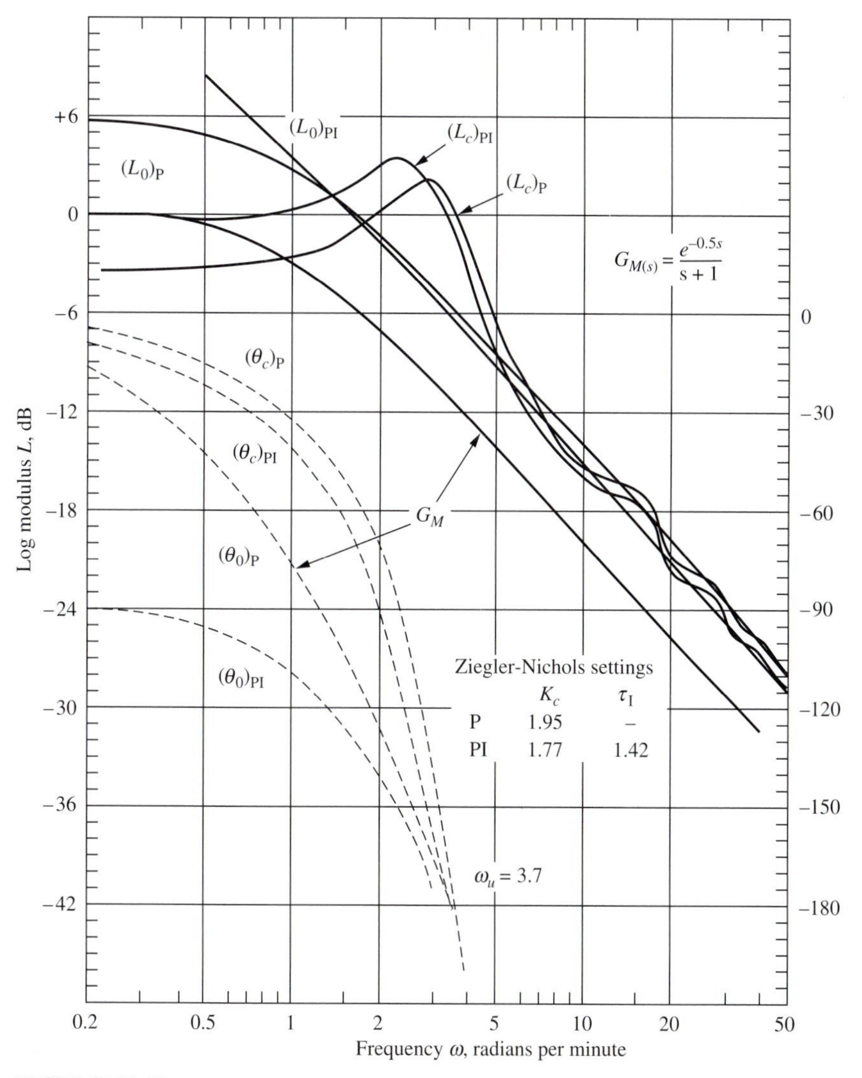

FIGURE 11.19
Openloop and closedloop plots for deadtime-with-lag process.

$$G_{M(s)} = \frac{K_p e^{-Ds}}{\tau_o s + 1} \tag{11.40}$$

We look at several values of deadtime D. For all cases the values of K_p and τ_o are unity. Other values of τ_o simply modify the frequency and time scales. Other values of K_p modify the controller gain.

The Bode plot of $G_{M(i\omega)}$ is given in Fig. 11.19 for $D = 0.5$. The ultimate gain is 3.9 (11.6 dB), and the ultimate frequency is 3.7 rad/min. The ZN controller settings for P and PI controllers and the corresponding phase and gain margins and log moduli are shown in Table 11.2 for several values of deadtime D. Also shown are the K_c values for a proportional controller that gives +2-dB maximum closedloop log modulus.

Notice that the ZN settings give very large phase margins for large deadtimes. This illustrates that the phase margin criterion would result in poor control for large-deadtime processes. The ZN settings also give L_c^{max} values that are too large when the deadtime is small, but too small when the deadtime is large. The $L_c^{max} = +2$ dB specification gives reliable controller settings for all values of deadtime.

Figure 11.19 shows the closedloop servo transfer function Bode plots for P and PI controllers with the ZN settings for a deadtime of 0.5 minutes. The effect of the deadtime on the first-order lag is to drop the phase angle below $-180°$. The system can be made closedloop unstable if the gain is high enough. Since there is always some deadtime in any real system, all real processes can be made closedloop unstable by making the feedback controller gain high enough.

TABLE 11.2
Settings for first-order lag process with deadtime ($K_p = \tau_o = 1$)

	Deadtime D		
	0.1	**0.5**	**2**
ZN (P)			
K_c	8.18	1.90	0.760
GM	2.0	2.0	2.0
PM (°)	51	71	180
L_c^{max} (dB)	2.8	1.6	0.34
ZN (PI)			
K_c	7.43	1.73	0.690
τ_I	0.321	1.42	4.58
GM	1.9	2.0	2.1
PM (°)	30	53	98
L_c^{max} (dB)	6.0	2.9	0.43
+2 dB tuning (P)			
K_c	7.70	1.95	0.833
GM	2.1	2.0	1.8
PM (°)	54	73	180
L_c^{max} (dB)	2.0	2.0	2.0

Notice in Fig. 11.19 that the L_c curve for the P controller does not approach 0 dB at low frequencies. This shows that there is a steady-state offset with a proportional controller. The L_c curve for the PI controller does go to 0 dB at low frequencies because the integrator drives the closedloop servo transfer function to unity (i.e., no offset).

11.4.3 Openloop-Unstable Processes

The Nyquist stability criterion can be used for openloop-unstable processes, but we have to use the complete, rigorous version with P (the number of poles of the closedloop characteristic equation in the RHP) no longer equal to zero.

Consider the simple openloop-unstable process

$$G_{M(s)} = \frac{K_p}{\tau_o s - 1} \tag{11.41}$$

We found in Chapter 9, using Laplace-domain root locus plots, that we could make this system closedloop stable by using a proportional controller with a gain K_c greater than $1/K_p$. Let us see if the Nyquist stability criterion leads us to the same conclusion. It certainly should if it is any good, because a table in Chinese must be a table in Russian!

First of all, we know immediately that the openloop system transfer function $G_{M(s)}G_{C(s)}$ has one pole (at $s = +1/\tau_o$) in the RHP. Therefore, the closedloop characteristic equation

$$1 + G_{M(s)}G_{C(s)} = 0$$

must also have one pole in the RHP, so $P = 1$.

On the C_+ contour up the imaginary axis, $s = i\omega$. We must make a polar plot of $G_{M(i\omega)}G_{C(i\omega)}$.

$$G_{M(i\omega)}G_{C(i\omega)} = \frac{K_c K_p}{\tau_o i\omega - 1} = \frac{K_c K_p(-1 - i\omega\tau_o)}{i + \omega^2\tau_o^2} \tag{11.42}$$

Figure 11.20 shows that the curve starts ($\omega = 0$) at $-K_c K_p$ on the negative real axis where the phase angle is $-180°$. It ends at the origin, coming in with an angle of $-90°$. The C_R contour maps into the origin. The C_- contour is the reflection of the C_+ contour over the real axis.

If $K_c > 1/K_p$, the $(-1, 0)$ point is encircled. But the encirclement is in a *counterclockwise* direction! You recall that all the curves considered up to now have encircled the $(-1, 0)$ point in a clockwise direction. A clockwise encirclement is a positive N. A counterclockwise encirclement is a negative N. Therefore, $N = -1$ for this example if $K_c > 1/K_p$. The number of zeros of the closedloop characteristic equation in the RHP is then

$$Z = P + N = 1 + (-1) = 0$$

Thus, the system is closedloop stable if $K_c > 1/K_p$. This is exactly the conclusion we reached using root locus methods. So the Chinese frequency-domain conclusions are the same as the Russian Laplace-domain conclusions.

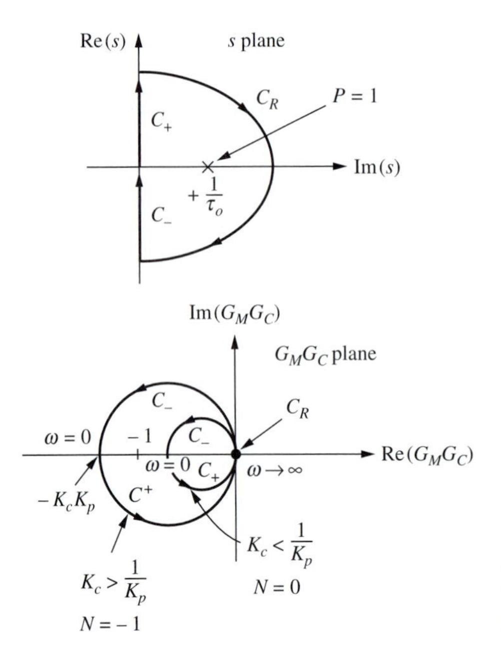

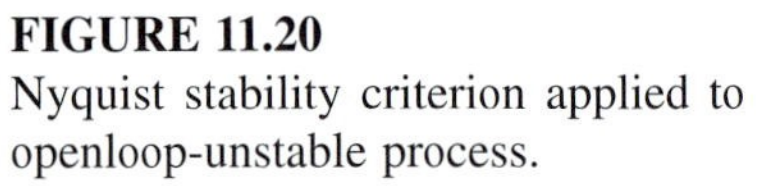

FIGURE 11.20
Nyquist stability criterion applied to openloop-unstable process.

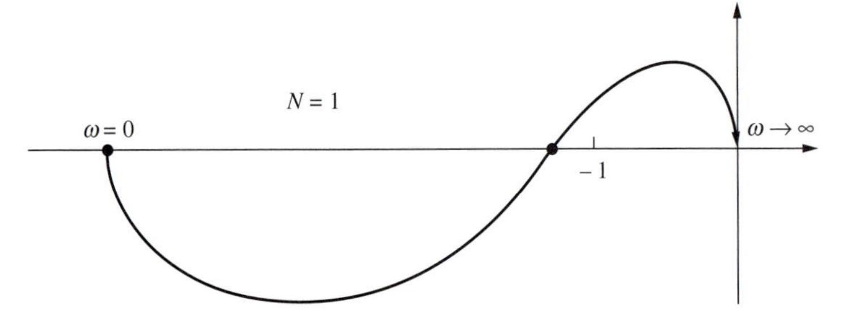

(*a*) $K_c > K_{max}$

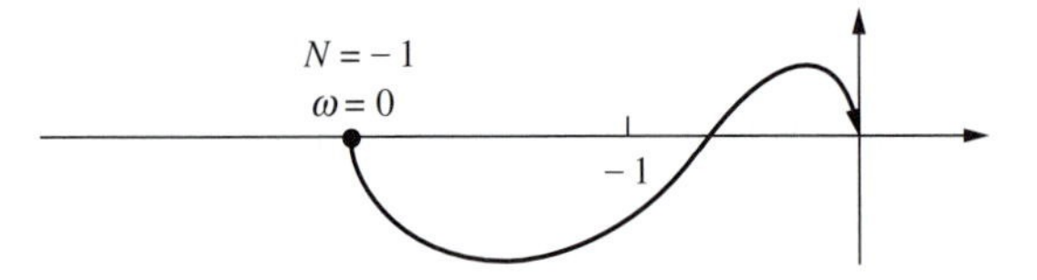

(*b*) $K_{min} < K_c < K_{max}$

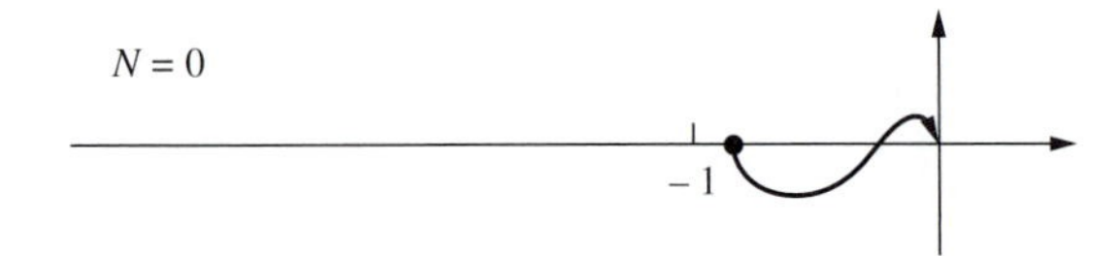

(*c*) $K_c < K_{min}$

FIGURE 11.21
Nyquist plots for openloop-unstable process.

If $K_c < 1/K_p$, the $(-1, 0)$ point is not encircled and $N = 0$. The number of zeros of the closedloop characteristic equation is

$$Z = P + N = 1 + 0 = 1$$

The closedloop system has one zero in the RHP and is unstable if $K_c < 1/K_p$.

Figure 11.21 gives Nyquist plots for higher-order systems. For the third-order system, conditional stability can occur: the closedloop system is stable for controller gains between K_{min} and K_{max}. For gains greater than K_{max} there is one positive encirclement of the $(-1, 0)$ point, so $N = 1$ and $Z = P + N = 1 + 1 = 2$. The system is closedloop unstable. For gains less than K_{min} there is *no* encirclement, and $N = 0$. This makes $Z = 1$, and the system is again closedloop unstable. But for gains between K_{min} and K_{max} there is one negative encirclement of the $(-1, 0)$ point, so $N = -1$ and $Z = P + N = 1 - 1 = 0$. The system is closedloop stable since its closedloop characteristic equation has no zeros in the right half of the s plane.

11.5 USE OF MATLAB FOR FREQUENCY RESPONSE PLOTS

The MATLAB Control Toolbox makes it quite easy to generate frequency response plots for both openloop and closedloop systems. Table 11.3 gives a MATLAB program that generates Bode plots and Nichols plots. The process example is the three-heated-tank system with the openloop transfer function

$$G_{M(s)}G_{V(s)}G_{T(s)} = \frac{1.333}{(0.1s + 1)^3} \tag{11.43}$$

The program goes through the following steps:

1. The numerator and denominator polynomials are formed for the openloop transfer function.
2. The ultimate gain and frequency are calculated.
3. The Ziegler-Nichols and Tyreus-Luyben settings for PI controllers are calculated.
4. The numerator and denominator polynomials are formed for the total openloop transfer functions: the product of the G_M and G_C transfer functions. Three controllers are studied: a proportional controller with $K_c = K_u/2$, a PI controller with ZN settings, and a PI controller with TL settings.
5. The magnitudes and phase angles of the total openloop transfer functions are calculated using the *bode* function.
6. The numerator and denominator polynomials of the closedloop servo transfer functions are formed using the *[numcl,dencl]= cloop(numol,denol,−1)* command. This command converts an openloop transfer function into a closedloop transfer function, assuming negative unity feedback. A block diagram of a unity feedback system is given in Fig. 11.22.
7. The magnitudes and phase angles of the closedloop servo transfer functions are calculated for all three controllers.

TABLE 11.3
MATLAB program for openloop and closedloop frequency response plots

```
% Program "tempfreq.m" uses Matlab to analyze the three-heated-tank process.
% (1) Calculate ultimate gain and frequency
% (2) Make Bode plot of total openloop transfer function
% (process and control) for P and PI controllers
% with ZN and Tl settings.
% (3) Make Nichols plots of three controllers
%
% Form openloop process transfer function numerator and denominator
num=1.333;
den=conv([0.1 1],[0.1 1]);
den=conv(den,[0.1 1]);
%
% Specify frequency values
w=logspace(0,2,600);
%
% Calculate magnitudes and phase angles for all frequencies
[mag,phase,w]=bode(num,den,w);
%
% Calculate ultimate gain and frequency
n=1;
while phase(n)>=-180;n=n+1;end
kult=1/mag(n);
wult=w(n);
% Calculate Ziegler-Nichols and Tyreus-Luyben PI controller settings
pult=2*3.1416/wult;
kzn=kult/2.2;
tzn=pult/1.2;
ktl=kult/3.2;
ttl=2.2*pult;
%
% Form transfer functions for process and PI controllers
nzn=kzn*conv(num,[tzn 1]);
dzn=conv(den,[tzn 0]);
ntl=ktl*conv(num,[ttl 1]);
dtl=conv(den,[ttl 0]);
% Form transfer functions for process and P controller (Kc=Kult/2)
np=kult*num/2;
%*******************************
% Calculate magnitudes and phase angles
%
% For P control
[mag,phase,w]=bode(np,den,w);
db=20*log10(mag);
%
% Form unity-feedback closedloop transfer functions
[ncl,dcl]=cloop(np,den,-1);
[mcl,pcl,w]=bode(ncl,dcl,w);
cldb=20*log10(mcl);
```

TABLE 11.3 (CONTINUED)

MATLAB program for openloop and closedloop frequency response plots

```
%
% Pick up maximum CL peak and resonant frequency
[dbmax,nmax]=max(cldb);
wr=w(nmax);
%
% Plot openloop and closedloop log modulus
clf
semilogx(w,db,'-',w,cldb,'--')
xlabel('Frequency (radians/hr)')
ylabel('Log Modulus (dB)')
grid
title('Three Heated Tank Process; P control')
text(wr,15,['Lcmax=',num2str(dbmax)])
text(wr,10,['wr=',num2str(wr)])
text(2,-10,['Ku=',num2str(kult)])
text(2,-15,['wu=',num2str(wult)])
legend('Openloop','Closedloop')
pause
print -dps pfig1123.ps
%
%*****************************
% PI control: ZN and Tl
%
% Form closedloop transfer functions
[nclzn,dclzn]=cloop(nzn,dzn,-1);
[mclzn,pclzn,w]=bode(nclzn,dclzn,w);
[mzn,pzn,w]=bode(nzn,dzn,w);
clzn=20*log10(mclzn);
% Pick up maximum CL peak
[dbznmax,nmax]=max(clzn);
wrzn=w(nmax);
%
% Form closedloop transfer functions
[ncltl,dcltl]=cloop(ntl,dtl,-1);
[mcltl,pcltl,w]=bode(ncltl,dcltl,w);
[mtl,ptl,w]=bode(ntl,dtl,w);
cltl=20*log10(mcltl);
% Pick up maximum CL peak
[dbtlmax,nmax]=max(cltl);
wrtl=w(nmax);
%
% Plot closedloop log modulus for ZN and Tl settings
semilogx(w,clzn,'-',w,cltl,'--')
xlabel('Frequency (radians/hr)')
ylabel('CL Log Modulus (dB)')
title('Three Heated Tank Process; Closedloop Lc with PI control')
grid
legend('PI ZN','PI TL')
text(wrzn,15,['ZNmax=',num2str(dbznmax)])
```

TABLE 11.3 (CONTINUED)
MATLAB program for openloop and closedloop frequency response plots

```
text(wrzn,10,['ZNwr=',num2str(wrzn)])
text(wrtl,-20,['TLmax=',num2str(dbtlmax)])
text(wrtl,-30,['TLwr=',num2str(wrtl)])
text(2,-10,['KZN=',num2str(kzn)])
text(2,-15,['TZN=',num2str(tzn)])
text(2,-20,['KTL=',num2str(ktl)])
text(2,-25,['TTL=',num2str(ttl)])
pause
print -dps pfig1124
%
%****************************************
%
% Make Nichols plots
%
dbzn=20*log10(mzn);
dbtl=20*log10(mtl);
plot(pzn,dbzn,'-',ptl,dbtl,'.',phase,db,'--')
title('Three Heated Tank Process; Nichols Plots')
xlabel('Open-Loop Phase (deg)')
ylabel('Open-Loop Gain (db)')
ngrid
legend('PI ZN','PI TL','P')
pause
print -dps pfig1125
```

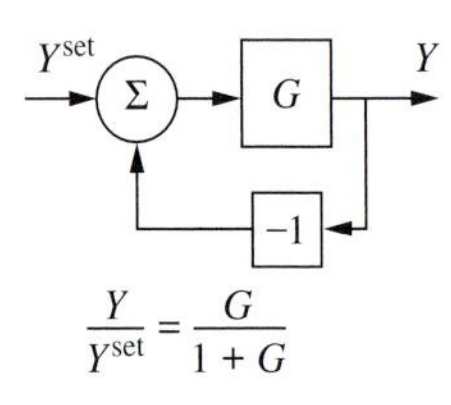

FIGURE 11.22
Unity feedback loop.

8. The peak in the closedloop log modulus curve (*dbmax*) and the frequency at which it occurs (*wr*, the resonant frequency of the closedloop system) are calculated using the *[dbmax,nmax]=max(cldb)* command, which selects the largest value in a vector of numbers *cldb* and gives the row *nmax* where it is located.
9. Bode plots are generated (Figs. 11.23 and 11.24) using the *semilogx* plotting command, which gives a logarithmic scale on the abscissa.
10. Nichols plots are generated (Fig. 11.25). The *ngrid* command draws the Nichols chart curves of closedloop log modulus L_c.

We again find that the Ziegler-Nichols settings give large peaks in the L_c curve (+13 dB for the PI controller). The Tyreus-Luyben settings give a more conservative +3.5 dB.

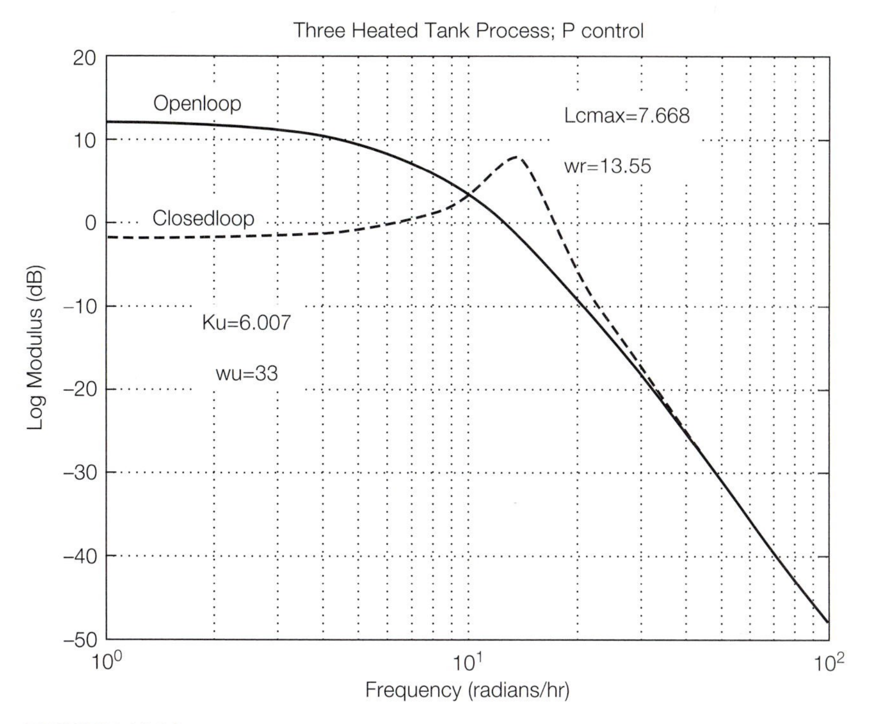

FIGURE 11.23

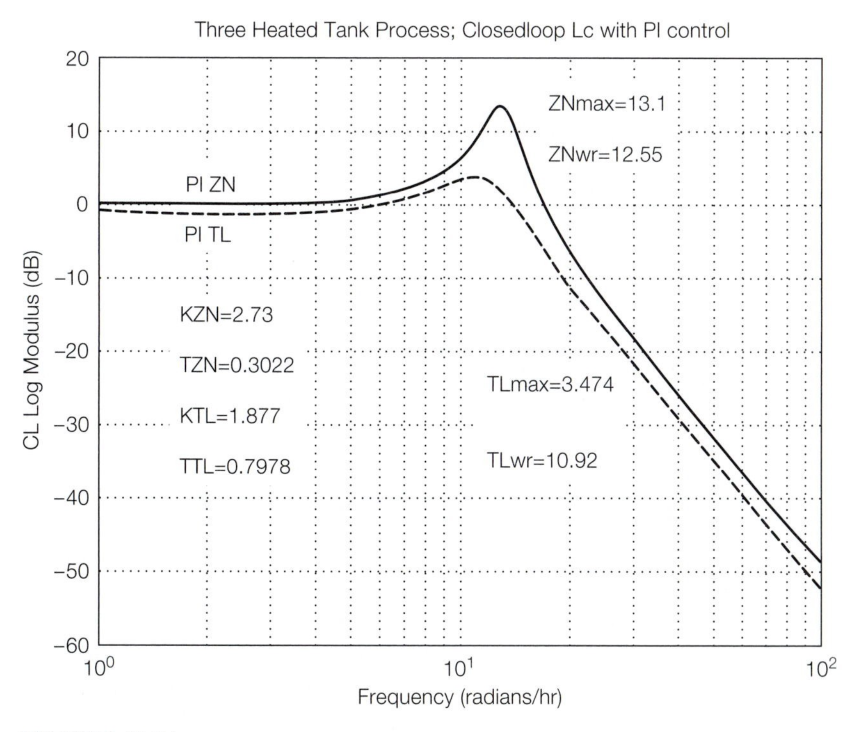

FIGURE 11.24

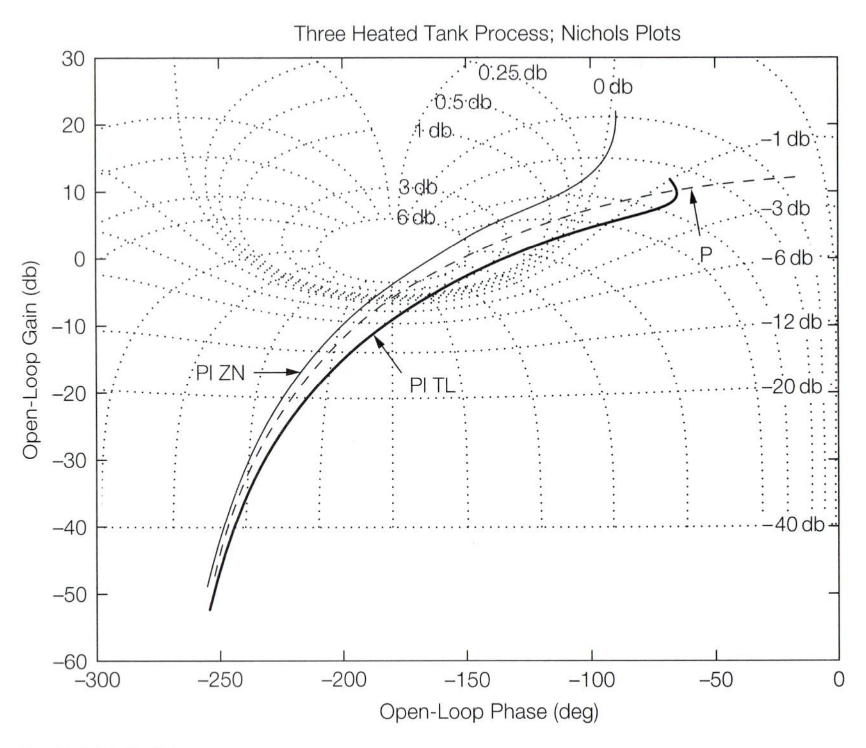

FIGURE 11.25

11.6 CAPACITY-BASED METHOD FOR QUANTIFYING CONTROLLABILITY

Back in Chapter 5 we discussed a method for quantitatively incorporating the economics of control into conventional steady-state design economics. For a given process and control system, the idea is to specify upper and lower limits on product quality and determine what fraction of the time the process is within these specification limits. The capacity of the plant is equal to this fraction times the product flow rate. Periods of off-specification production must be handled by reworking or disposing of the material, which carries some economic cost. The steady-state economic calculations are made using the on-specification capacity.

In Chapter 5 we illustrated how this capacity factor can be obtained from simulation studies. These can be time consuming, and the screening of a large number of alternative flowsheets becomes intractable. In addition, simulation studies require the specification of a time sequence of disturbances. At the conceptual design stage we seldom know exactly what type of disturbances the plant will experience. We may be able to estimate what the disturbance variables are and their likely magnitudes, but we rarely can estimate the frequency content of the disturbances, i.e., how rapidly they change.

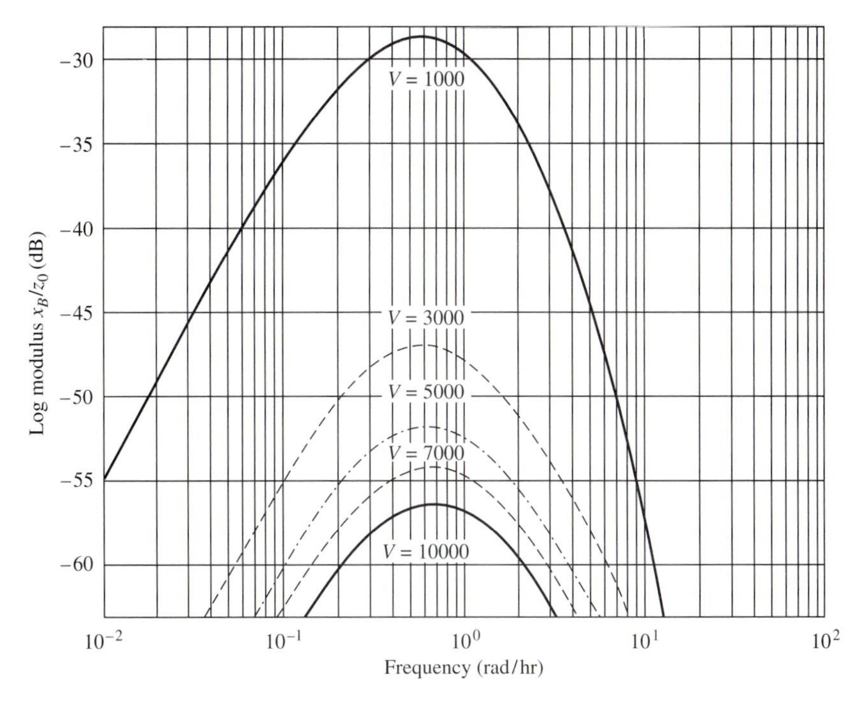

FIGURE 11.26
Closedloop regulator transfer function.

Frequency-domain methods can greatly simplify the procedure. The idea is to use a linear model of the process, design a control system, and calculate the magnitude ratios of the closedloop *regulator* or *load* transfer functions over a frequency range. Figure 11.26 illustrates the typical shape of these curves. The example is the reactor-stripper process with recycle discussed in Chapter 5. The measure of product quality is the bottoms composition x_B in the stripper. The load disturbance is a change in the fresh feed composition z_0. The closedloop regulator transfer function is $x_{B(s)}/z_{0(s)}$.

At low frequencies the PI controller eliminates steady-state offset, so the $x_{B(s)}/z_{0(s)}$ ratio goes to zero and the log modulus goes to minus infinity. At high frequencies the process itself filters out the disturbances. There is a maximum in the curve at some frequency. A disturbance entering the closedloop system at this frequency gives the largest variation in the output. Therefore a "worst-case" situation is to have a sine wave load disturbance at this frequency with a maximum expected amplitude. Figure 11.26 clearly shows the significant effect of reactor holdup V_R on controllability. As V_R increases, the control of x_B becomes better and better.

The time-domain response of the linear system to this sinusoidal input can be calculated analytically, and the fraction of time that the output variable is outside a given specification range can be easily computed. Then the profitability is determined using this fraction of on-specification product and combining the conventional capital, energy, and raw material costs.

The calculations are done very rapidly on a computer since they involve only frequency-domain (complex number) manipulations. Dynamic simulations are not required. This allows the investigation of a large number of alternative process flowsheets. It does require the determination of the transfer functions for each design. Several commercial software packages provide the capability of calculating these functions numerically using state-space methods discussed in Chapter 12.

Another significant advantage is that the method eliminates the need to specify the shapes of the disturbances. The worst-case frequencies are determined as part of the method.

11.7 CONCLUSION

In this chapter the power and usefulness of frequency response methods should have become apparent to you. Very complex processes (high order and deadtimes) can be handled with ease. In Chapter 13 we apply these methods to multivariable processes with equal success. In Chapter 16 we show that frequency-domain methods are invaluable for process identification (experimentally determining the dynamics of a system from plant data).

Chinese (frequency response) is a very powerful language. It permits us to quantitatively analyze quite complex processes.

PROBLEMS

11.1. (*a*) Make Bode, Nyquist, and Nichols plots of the system with $K_c = 1$:

$$G_{C(s)}G_{M(s)} = \frac{K_c}{(s+1)(5s+1)(\frac{1}{2}s+1)}$$

(*b*) Find the value of gain K_c that gives a phase margin of 45°. What is the gain margin?
(*c*) Find the value of gain K_c that gives a gain margin of 2. What is the phase margin?
(*d*) Find the value of gain K_c that gives a maximum closedloop log modulus of +2 dB. What are the gain and phase margins with this value of gain?
(*e*) Find the Ziegler-Nichols settings for this process, and calculate the gain and phase margins and maximum closedloop log moduli that these settings give for P, PI, and PID controllers.

11.2. (*a*) Make Bode, Nyquist, and Nichols plots of the system with $K_c = 1$:

$$G_{C(s)}G_{M(s)} = \frac{K_c(\frac{1}{2}s+1)}{(s+1)(5s+1)}$$

(*b*) Find the value of gain K_c that gives a phase margin of 45°. What is the maximum closedloop log modulus with this value of gain?
(*c*) Find the value of gain K_c that gives a maximum closedloop log modulus of +2 dB. What is the phase margin with this value of gain?

11.3. (*a*) Make Bode, Nyquist, and Nichols plots of the system with $K_c = 1$:

$$G_{M(s)}G_{C(s)} = \frac{K_c(-3s+1)}{(s+1)(5s+1)}$$

(*b*) Find the ultimate gain and frequency.
(*c*) Find the value of K_c that gives a phase margin of 45°.
(*d*) Find the value of K_c that gives a gain margin of 2.
(*e*) Find the value of K_c that gives a maximum closedloop log modulus of +2 dB.

11.4. Repeat Problem 11.3 for the system

$$G_{M(s)}G_{C(s)} = K_c\left(1+\frac{1}{2s}\right)\frac{-3s+1}{(s+1)(5s+1)}$$

11.5. How would you use the "$Z - P = N$" theorem to develop a test for openloop stability?

11.6. A process has G_M and G_L openloop transfer functions that are first-order lags and gains: τ_M, τ_L, K_M, and K_L. Assume τ_M is twice τ_L. Sketch the log modulus Bode plot for the closedloop load transfer function when:
(*a*) A proportional-only feedback controller is used with $K_c K_M = 8$.
(*b*) A PI controller is used, with $\tau_I = \tau_M$ and the same gain as above.

11.7. (*a*) Sketch Bode, Nichols, and Nyquist plots of the closedloop servo and closedloop load transfer functions of the process

$$G_{L(s)} = G_{M(s)} = \frac{1}{10s+1} \qquad G_{C(s)} = 6\left(1+\frac{1}{6s}\right)$$

(*b*) Calculate the phase margin and maximum closedloop log modulus for the system.

11.8. Using a first-order Pade approximation of deadtime, find the ultimate gain and frequency of the system

$$G_{M(s)}G_{C(s)} = \frac{K_c e^{-0.5s}}{s+1}$$

Compare your answers with Section 11.4.2.

11.9. (*a*) Draw Bode, Nyquist, and Nichols plots of the system

$$G_{M(s)}G_{C(s)} = \frac{K_c}{(s+1)(s+5)(s-0.5)}$$

(*b*) Use the Nyquist stability criterion to find the values of K_c for which the system is closedloop stable.

11.10. (*a*) Make Nyquist and Bode plots of the openloop transfer function

$$G_{M(s)}G_{C(s)} = \frac{K_c}{s^2(s+1)}$$

(*b*) Is this system closedloop stable? Will using a PI controller stabilize it?
(*c*) Will a lead-lag element used as a feedback controller provide enough phase angle advance to meet a 45° phase margin specification? Will two lead-lags in series be enough?

(*d*) Use two elements and find the values of τ_D and K_c that give a 45° phase margin. What is the gain margin?

$$G_{C(s)} = K_c \left[\frac{\tau_D s + 1}{(\tau_D/20)s + 1} \right]^2$$

11.11. Find the largest value of deadtime D that can be tolerated in a process

$$G_{M(s)} = e^{-Ds}/s$$

and still achieve a 45° phase margin with a feedback controller having a reset time constant $\tau_I = 1$ minute. Find the value of gain K_c that gives the 45° of phase margin with the value of deadtime found above.

11.12. A process consists of two transfer functions in series. The first, G_{M1}, relates the manipulated variable M to the variable x_1 and is a steady-state gain of 1 and two first-order lags in series with equal time constants of 1 minute.

$$G_{M1(s)} = \frac{1}{(s+1)^2}$$

The second, G_{M2}, relates x_1 to the controlled variable x_2 and is a steady-state gain of 1 and a first-order lag with a time constant of 5 minutes.

$$G_{M2(s)} = \frac{1}{5s+1}$$

If a single proportional controller is used to control x_2 by manipulating M, determine the gain that gives a phase margin of 45°. What is the maximum closedloop log modulus when this gain is used?

11.13. Suppose we want to use a cascade control system in the process considered in Problem 11.12. The secondary or slave loop will control x_1 by manipulating M. The primary or master loop will control x_2 by changing the setpoint x_1^{set} of the secondary controller.

(*a*) Design a proportional secondary controller (K_1) that gives a phase margin of 45° for the secondary loop.

(*b*) Using a value of gain for the secondary loop of $K_1 = 6.82$, design the master proportional controller (K_2) that gives a phase margin of 45° for the primary loop.

(*c*) What is the maximum closedloop log modulus for the primary loop when this value of gain is used?

11.14. A process has the following transfer function:

$$G_M = \frac{1}{(\tau_o s + 1)^2}$$

It is controlled using a PI controller with τ_I set equal to τ_o.

(*a*) Sketch a root locus plot and calculate the controller gain that gives a closedloop damping coefficient of 0.4.

(*b*) Sketch Bode, Nyquist, and Nichols plots of $G_M G_C$.

(*c*) Calculate analytically the gain that gives a phase margin of 45° and check your answer graphically.

(*d*) Determine the values of the maximum closedloop log modulus for the two values of gains from (*a*) and (*c*).

11.15. An openloop-unstable, second-order process has one positive pole at $+1/\tau_1$ and one negative pole at $-1/\tau_2$. If a proportional controller is used and if $\tau_1 < \tau_2$, show by using a root locus plot and then by using the Nyquist stability criterion that the system is always unstable.

11.16. A process has the openloop transfer function $G_M = K_p/[s(\tau_o s + 1)]$. A proportional-only controller is used. Calculate analytically the closedloop damping coefficient that is equivalent to a phase margin of 45°. What is the maximum closedloop log modulus when the controller gain that gives a 45° phase margin is used?

11.17. A process has the openloop transfer function

$$G_{M(s)} = \frac{K_p e^{-Ds}}{\tau_o s + 1}$$

where $K_p = 1, \tau_o = 1, D = 0.3$.

(*a*) Draw a Bode plot for the openloop system.

(*b*) What is the ultimate gain and ultimate frequency of this system?

(*c*) Using Ziegler-Nichols settings, draw a Bode plot for $G_M G_C$ when a PI controller is used on this process.

11.18. Cold, 70°F liquid is fed at a rate of 250 gpm into a 500-gallon, perfectly mixed tank. The tank is heated by steam, which condenses in a jacket surrounding the vessel. The heat of condensation of the steam is 950 Btu/lb. The liquid in the tank is heated to 180°F, under steady-state conditions, and continuously withdrawn from the tank to maintain a constant level. Heat capacity of the liquid is 0.9 Btu/lb °F; density is 8.33 lb/gal.

The control valve on the steam has linear installed characteristics and passes 500 lb/min when wide open. An electronic temperature transmitter (range 50–250°F) is used. A temperature measurement lag of 10 seconds and a heat transfer lag of 30 seconds can be assumed. A proportional-only temperature controller is used.

(*a*) Derive a mathematical model of the system.

(*b*) Derive the openloop transfer functions between the output variable temperature (T) and the two input variables steam flow rate (F_s) and liquid inlet temperature (T_0).

(*c*) Sketch a root locus plot for the closedloop system.

(*d*) What are the ultimate gain and ultimate frequency ω_u?

(*e*) If a PI controller is used with $\tau_I = 5/\omega_u$, what value of controller gain gives a maximum closedloop modulus of +2 dB?

(*f*) What are the gain margin and phase margin with the controller settings of part (*e*)?

11.19. Prepare a plot of closedloop damping coefficient versus phase margin for a process with openloop transfer function

$$G_M G_C = \frac{K_c}{(\tau_o s + 1)^2}$$

11.20. A process has the following transfer function relating controlled and manipulated variables:

$$G_M = \frac{1}{(s + 1)^2}$$

If a proportional-only controller is used, what value of controller gain will give a maximum closedloop log modulus of +2 dB?

(*a*) Solve this problem analytically.
(*b*) Check your answer graphically using Bode and Nichols plots.
(*c*) Draw a root locus plot and determine the closedloop damping coefficient for the value of gain found in (*a*) and (*b*).

11.21. A process has openloop process transfer function

$$G_M = \frac{1}{(s-1)(\frac{1}{10}s+1)^2}$$

(*a*) Plot Bode, Nyquist, and Nichols plots for this system.
(*b*) If a proportional-only controller is used, over what range of controller gains K_c will the system be closedloop stable? Use frequency-domain methods to determine your answer, but confirm with a root locus plot.
(*c*) What value of gain gives the smallest maximum closedloop log modulus?

11.22. Derive an analytical relationship between openloop maximum log modulus and damping coefficient for a second-order underdamped openloop system with a gain of unity. Show that a damping coefficient of 0.4 corresponds to a maximum log modulus of +2.7 dB.

11.23. A first-order lag process with a time constant of 1 minute and a steady-state gain of $5°\text{F}/10^3$ lb/hr is to be controlled with a PI feedback controller.

(*a*) Sketch Bode plots of $G_M G_C$ when the reset time τ_I is very much smaller than 1 minute and when it is much larger than 1 minute.
(*b*) Find the biggest value of reset time τ_I for which a phase margin of 45° is feasible.

11.24. A process has an openloop transfer function that is approximately a pure deadtime of D minutes. A proportional-derivative controller is to be used with a value of α equal to 0.1. What is the optimal value of the derivative time constant τ_D? Note that part of this problem involves defining what is meant by "optimal."

11.25. A process has the following openloop transfer function:

$$G_M = \frac{0.5}{(10s+1)(50s+1)}$$

A proportional-only feedback controller is used.

(*a*) Make a root locus plot for the closedloop system.
(*b*) Calculate the value of controller gain that will give a closedloop damping coefficient of 0.707.
(*c*) Using this value of gain, what is the phase margin?
(*d*) What is the maximum closedloop log modulus?

11.26. The following frequency response data were obtained by pulse-testing a closedloop system that contained a proportional-only controller with a proportional band of 25. Controller setpoint was pulsed, and the process variable signal was recorded as the output signal.

(*a*) What is the openloop frequency response of the process?
(*b*) What is the openloop process transfer function?

ω (rad/min)	Log modulus (dB)	Phase angle (°)
0.1	2.50	−2
0.4	2.54	−6
0.8	2.65	−12
1.0	2.74	−16
2.0	3.39	−34
4.0	3.90	−90
6.3	− 2.31	−151
8	− 7.1	−172
10	−12.1	−187
16	−22.3	−211
20	−27.6	−221
40	−44.5	−244
80	−64.6	−256

11.27. Gas at 100°C flows through two pressurized cylindrical vessels in series at a rate of 1000 kg/hr. The first tank is 2 meters in diameter and 5 meters high. The second is 3 meters in diameter and 8 meters high. The molecular weight of the gas is 30.

The first tank operates at 2000 kPa at the initial steady state. There is a pressure drop between the vessels that varies linearly with gas flow rate F_1. This pressure drop is 100 kPa when the flow rate is 1000 kg/hr.

$$P_1 - P_2 = K_1 F_1$$

Assume the perfect gas law can be used [$R = 8.314$ (kPa m^3)/(kg-mol K)]. Assume the pressure transmitter range is 1800–2000 kPa and that the valve has linear installed characteristics with a maximum flow rate of 2000 kg/hr.

(*a*) Derive a mathematical model for the system.
(*b*) Determine the openloop transfer function between P_1 and the two inputs F_0 and F_2.
(*c*) Assuming a proportional-only controller is used to manipulate F_2 to control P_1, make a root locus plot and calculate the controller gain that gives a closedloop damping coefficient of 0.3.
(*d*) Draw Bode, Nyquist, and Nichols plots for the openloop process transfer function P_1/F_2.
(*e*) Calculate the controller gain that gives 45° of phase margin.

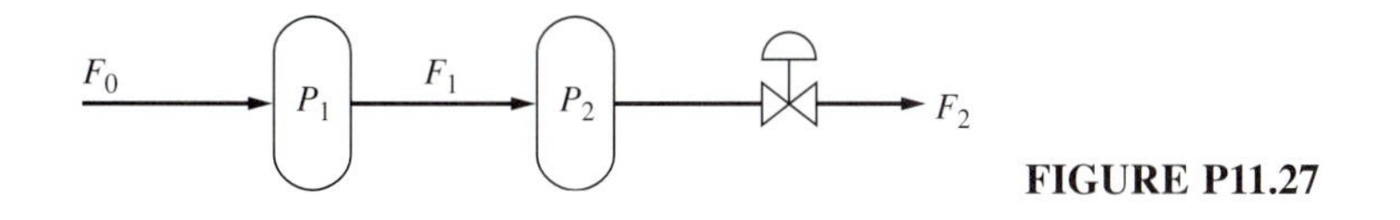

FIGURE P11.27

11.28. A process has the following openloop transfer function G_M relating controlled to manipulated variables:

$$G_{M(s)} = \frac{e^{-Ds}}{\tau_o s - 1}$$

(*a*) Sketch Bode, Nyquist, and Nichols plots for the system.
(*b*) If a proportional feedback controller is used, what is the largest ratio of D/τ_o for which a phase margin of 45° can be obtained?

11.29. A process has a transfer function $G_M = 1/[s(\tau_o s+1)]$. It is controlled by a PI controller with reset time τ_I. Sketch the $G_M G_C$ plot of phase angle versus frequency, the Nyquist plot of $G_M G_C$, and a root locus plot for the case where:
(*a*) $\tau_o > \tau_I$
(*b*) $\tau_o < \tau_I$

11.30. A process has an openloop transfer function that contains a positive pole at $+1/\tau_o$, a negative pole at $-10/\tau_o$, and a gain of unity. If a proportional-only controller is used, find the two values of controller gain that give a maximum closedloop log modulus of +2 dB.
(*a*) Do this problem graphically.
(*b*) Solve it analytically.

11.31. A two-pressurized-tank process has the openloop transfer function

$$G_{M(s)} = \frac{K_p}{(\tau_o s + 1)s}$$

If a PI controller is used, find the smallest value of the ratio of the reset time τ_I to the process time constant τ_o for which a maximum closedloop log modulus of +2 dB is attainable.

11.32. A process with an openloop transfer function consisting of a steady-state gain, deadtime, and first-order lag is to be controlled by a PI controller. The deadtime (D) is one-fifth the magnitude of the time constant (τ_o).

Sketch Bode, Nyquist, and Nichols plots of the total openloop transfer function ($G_M G_C$) when:
(*a*) $\tau_I > 5\tau_o$
(*b*) $\tau_I = \tau_o$
(*c*) $\tau_I = 2D/\pi$

11.33. Write two computer programs, one in FORTRAN and one in MATLAB, that calculate the feedback controller gain K_c that gives a maximum closedloop log modulus of +2 dB for a process with the openloop transfer function

$$G_M = \frac{K_p e^{-Ds}}{\tau_o s + 1}$$

and a PI controller with the reset time set equal to $2D/\pi$.

11.34. A process has the openloop transfer function

$$G_M = \frac{1}{(s-1)(0.1s+1)}$$

Sketch phase angle plots for $G_M G_C$ and root locus plots when:
(*a*) A proportional controller is used.
(*b*) A PI controller is used with:
  (*i*) $\tau_I \gg 0.1$
  (*ii*) $\tau_I \ll 0.1$
  (*iii*) $\tau_I = 0.1$

How would you find the maximum value of τ_I for which a 45° phase margin is attainable?

11.35. The frequency response Bode plot of the output of a closedloop system for setpoint changes, using a proportional controller with a gain of 10, shows the following features:

- The low-frequency asymptote on the log modulus plot is -0.828 dB.
- The breakpoint frequency is 11 rad/min.
- The slope of the high-frequency asymptote is -20 dB/decade.
- The phase angle goes to $-90°$ as frequency becomes large.

Calculate the form and the constants in the openloop transfer function relating the controlled and the manipulated variables.

11.36. A process has the following transfer function:

$$G_{M(s)} = \frac{24.85(5s+1)e^{-5s}}{(2.93s+1)(16.95s+1)}$$

(*a*) Make a Bode plot of the openloop system.
(*b*) Determine the ultimate gain and frequency if a proportional controller is used.
(*c*) What value of gain for a proportional controller gives a maximum closedloop log modulus of $+2$ dB?
(*d*) If a PI controller is used with Ziegler-Nichols settings, what are the phase margin, gain margin, and maximum closedloop log modulus?

11.37. The block diagram of a DC motor is shown.
(*a*) Show that the openloop transfer function between the manipulated variable M (voltage to the armature) and the controlled variable Y (angular position of the motor) is

$$G_M = \frac{16.13}{s(0.129s+1)}$$

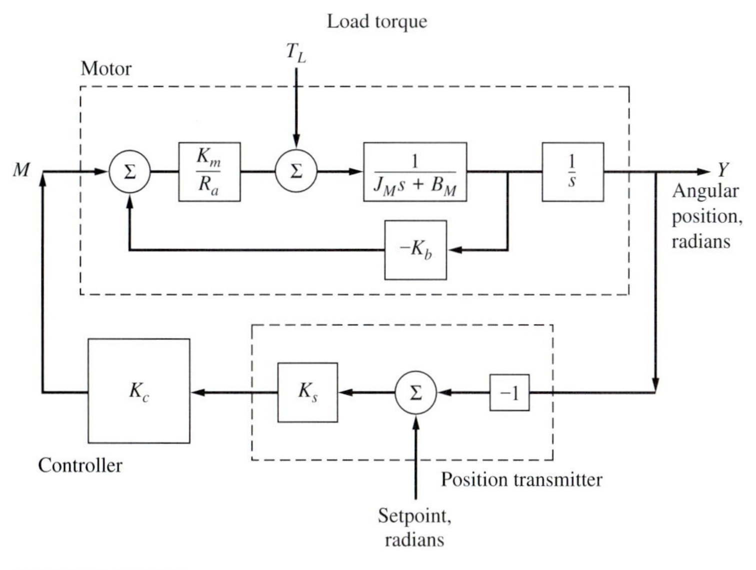

FIGURE P11.37

(*b*) Draw a root locus plot for this sytem if a proportional feedback controller is used.
(*c*) What value of gain gives a closedloop damping coefficient of 0.707?
(*d*) What value of gain gives a phase margin of 45°?
(*e*) What is the maximum closedloop log modulus using this gain?

R_a = armature resistance = 1 Ω
K_m = motor torque constant = 10 oz-in/A
K_b = back emf constant = 0.052 V/rad/sec
J_M = motor inertia = 0.08 oz-in/rad/sec^2
B_M = motor viscous friction = 0.1 oz-in/rad/sec
K_s = transmitter gain = 1 V/rad
K_c = feedback controller gain (V/V)

11.38. A process has the following openloop transfer function:

$$G_M = \frac{1}{(s+1)(2s+1)}$$

(*a*) Sketch the root locus diagram if a proportional feedback controller is used. Show on the root locus plot the range of K_c values for which the system is overdamped. For what K_c values is the system underdamped? What value of K_c yields a system with a damping coefficient of 0.707?
(*b*) Suppose a PI controller is used with reset time equal to 1 minute. Sketch the root locus diagram and identify the range of K_c values that give an underdamped response.
(*c*) Sketch the root locus diagram if the integral time τ_I is much greater than 1 minute. Repeat the sketch for τ_I values that are much smaller than 1 minute. Make clear on your sketches whether the system can become unstable.
(*d*) Consider the case where $\tau_I = \frac{1}{3}$ minute. Sketch the Bode diagram. It is sufficient to build the overall sketch from the component transfer functions. Label all asymptotes, slopes, and breakpoint frequencies.
(*e*) Sketch the Nyquist plot for case (*d*). Is the closedloop system stable for any K_c?
(*f*) Give the characteristic equation for the closedloop system with a PI controller and arbitrary reset time. Is there a range of τ_I values for which the system is always stable regardless of K_c value?

11.39. Consider the process

$$G_{M(s)} = \frac{e^{-3s}}{(s+1)^2(2s+1)}$$

(*a*) Use frequency-domain methods to solve analytically for the ultimate gain and ultimate frequency of a proportional controller.
(*b*) Suppose a PID controller is used with a reset time of 10.73 minutes and a derivative time of 0.15 minutes. The α constant in the derivative unit denominator is 0.1. Use graphical frequency-domain methods to find the value of controller gain that gives a maximum closedloop log modulus of +2 dB.

11.40. A process has an openloop transfer function relating controlled and manipulated variables that is a steady-state gain of unity and two first-order lags in series. The two time constants are 1 and 5 minutes. A proportional feedback controller is used.
(*a*) Develop an analytical relationship between the closedloop time constant τ_c and the closedloop damping coefficient ζ_c.

(*b*) If a closedloop damping coefficient of 0.3 is desired, what value of controller gain is required and what is the closedloop time constant?
(*c*) Repeat parts (*a*) and (*b*) using a proportional-derivative (PD) controller with $\tau_D = 1$ and $\alpha = 0.1$.
(*d*) Using a proportional controller, what value of controller gain gives a phase margin of 45°?
(*e*) What is the maximum closedloop log modulus when the controller gain is 10? When it is 20?

11.41. The openloop process transfer function relating the controlled and manipulated variables is

$$G_{M(s)} = \frac{0.488e^{-0.2s}}{(11.5s + 1)(0.167s + 1)(0.083s + 1)}$$

(*a*) Draw Bode plots for the openloop log modulus and openloop phase angle.
(*b*) Find graphically the ultimate gain and ultimate frequency.
(*c*) Determine the value of controller gain that gives 45° of phase margin using a proportional controller.
(*d*) Determine the value of controller gain that gives a +2-dB maximum closedloop log modulus.

11.42. A process has one openloop pole located at $+1/\tau$ and one openloop pole located at $-1/a\tau$. The steady-state gain of the openloop process is unity. A proportional controller is used. Using frequency-domain methods:
(*a*) Show that the parameter a must be less than unity for the system to be closedloop stable.
(*b*) Show that the minimum controller gain is unity.
(*c*) Find the largest value of a that will still permit a phase margin of 45° to be achieved.

11.43. The openloop transfer function $G_{M(s)}$ of a sterilizer relating the controlled variable temperature $T_{(s)}$ and the manipulated variable steam flow rate $F_{S(s)}$ is a gain $K_p = 2$ (with units of mA/mA when transmitter and valve gains have been included), a first-order lag with time constant $\tau_o = 1$ minute, and an integrator in series.

The value of gain for a proportional-only controller that gives 45° of phase margin is $K_c = 0.707$.
(*a*) Calculate the closedloop damping coefficient of the system when this gain is used.
(*b*) Sketch several Bode plots of the openloop phase angle for this process when a PI controller is used for several values of reset τ_I, starting with very large values and then reducing reset toward $\tau_I = 1$ minute.
(*c*) Calculate the gain K_c of a proportional-derivative (PD) feedback controller that gives a phase margin of 45° when used to control this process.

$$G_{C(s)} = K_c \frac{s + 1}{\frac{1}{5}s + 1}$$

(*d*) Sketch a Nichols plot and use a Nichols chart to estimate the maximum closedloop log modulus when the PD controller is used with the gain calculated in part (*c*).

11.44. We can often capture the feedback-control-relevant dynamics of a process by assuming that the openloop process transfer function $G_{M(s)}$ relating the controlled and the manipulated variables is a simple gain, a deadtime, and a pure integrator in series.

(*a*) Sketch Bode and Nyquist plots for this process for a gain $K_p = 1$ and a deadtime $D = 0.5$ minutes.
(*b*) Calculate the ultimate frequency ω_u and ultimate gain K_u *analytically* for arbitrary values of gain and deadtime, and confirm your results *graphically* using the Bode plot from part (*a*) for the specific numerical values given.
(*c*) Calculate analytically and graphically the value of controller gain that gives a phase margin of 45°.
(*d*) Use a Nichols chart to determine the maximum closedloop log modulus if the gain calculated in part (*c*) is used.
(*e*) Calculate the TLC settings for a PI controller ($K_c = K_u/3.2$; $\tau_I = 2.2P_u$) and generate a Bode plot for $G_M G_C$ when these controller constants are used.
(*f*) What are the phase margin, gain margin, and maximum closedloop log modulus when this PI controller is used?

11.45. A process has the following openloop transfer function between the controlled variable Y and manipulated variable M:

$$\frac{Y}{M} = G_{M(s)} = \frac{K_p e^{-Ds}}{\tau_o s - 1}$$

(*a*) If a proportional analog controller is used, calculate the ultimate gain and ultimate frequency for the numerical values $K_p = 2$, $\tau_o = 10$, and $D = 1$.
(*b*) Sketch Nyquist and Bode plots of $G_{M(i\omega)}$. Calculate the value of controller gain that gives a phase margin of 45°.
(*c*) Use a Nichols chart to determine the maximum closedloop log modulus if the controller gain is 3.19.

11.46. A process has an openloop transfer function that is a double integrator.

$$G_{M(s)} = \frac{1}{s^2}$$

(*a*) Using a root locus plot, show that a proportional feedback controller cannot produce a closedloop stable system.
(*b*) Using frequency-domain methods, show the same result as in (*a*).

11.47. The openloop transfer function for a process is

$$G_{M(s)} = \frac{K_0}{\tau_o s + 1}$$

(*a*) If a proportional-only controller is used, calculate the closedloop servo transfer function Y/Y^{set}, expressing the closedloop gain K_{cl} and closedloop time constant τ_{cl} in terms of the openloop gain K_0, the openloop time constant τ_o, and the controller gain K_c. Sketch a closedloop log modulus plot for several values of controller gain.
(*b*) Repeat part (*a*) using a proportional-integral feedback controller with reset time $\tau_I = \tau_o$.

11.48. The openloop transfer function $G_{M(s)}$ of a process relating the controlled variable $Y_{(s)}$ and the manipulated variable $M_{(s)}$ is a gain $K_p = 3$ (with units of mA/mA when transmitter and valve gains have been included) and two first-order lags in series with time constants $\tau_1 = 2$ min and $\tau_2 = 0.4$ min.

(*a*) If a proportional controller is used, sketch a root locus plot.
(*b*) Calculate the controller gain that gives a closedloop damping coefficient of 0.3.
(*c*) What is the closedloop time constant when this gain is used?
(*d*) Make a Bode plot of the openloop system.
(*e*) Using a controller gain of 6.3, calculate the phase margin analytically and graphically.

11.49. A deadtime element ($D = 0.1$ min) is added in series with the lags in the process considered in Problem 11.48.

(*a*) Make a Bode plot of the openloop system with a proportional controller and $K_c = 1$.
(*b*) Determine graphically the ultimate frequency and the ultimate gain.
(*c*) Determine graphically the phase margin if a controller gain of 6.3 is used.
(*d*) Determine graphically the maximum closedloop log modulus if a controller gain of 6.3 is used.
(*e*) Calculate the Tyreus-Luyben settings for a PI controller.
(*f*) Determine graphically the phase and gain margins when these settings are used.

PART FOUR

Multivariable Processes

Perhaps the area of process control that has changed the most drastically in the last two decades is multivariable control. This change was driven by the increasing occurrence of highly complex and interacting processes. Such processes arise from the design of plants that are subject to rigid product quality specifications, are more energy efficient, have more material integration, and have better environmental performance. The tenfold increase in energy prices in the 1970s spurred activity to make chemical and petroleum processes more efficient. The result has been an increasing number of plants with complex interconnections of both material flows and energy exchange. The control engineer must be able to design control systems that give effective control in this multivariable environment. Multivariable systems contain more than one controlled variable and manipulated variable (the type of system we have studied so far).

We need to learn a little bit of yet another language! In previous chapters we have found the perspectives of time (English), Laplace (Russian), and frequency (Chinese) to be useful. Now we must learn some matrix methods and their use in the "state-space" approach to control systems design. Let's call this state-space methodology the "Greek" language.

The next two chapters are devoted to this subject. Chapter 12 summarizes some useful matrix notation and discusses stability and interaction in multivariable systems. Chapter 13 presents a practical procedure for designing conventional multiloop SISO controllers (the diagonal control structure).

It should be emphasized that the area of multivariable control is still in an early stage of development. Many active research programs are under way around the world to study this problem, and every year brings many new developments. The methods and procedures presented in this book should be viewed as a summary of some of the practical tools developed so far. Improved methods will undoubtedly grow from current and future research.

CHAPTER 12

Matrix Representation and Analysis

12.1 MATRIX REPRESENTATION

Many books have been written on matrix notation and linear algebra. Their elegance has great appeal to many mathematically inclined individuals. Many hard-nosed engineers, however, are interested not so much in elegance as in useful tools to solve real problems. We attempt in this chapter to weed out most of the chaff, blue smoke, and mirrors. Only those aspects that we have found to have useful engineering applications are summarized here. For a more extensive treatment, the readable book *Control System Design: An Introduction to State-Space Methods* by Bernard Friedland (1986, McGraw-Hill, New York) is recommended.

We use the symbolism of a double underline ($\underline{\underline{A}}$) for a matrix and a single underline ($\underline{x}$) for a vector, i.e., a matrix with only one column. This helps us keep track of which quantities are matrices, which are vectors, and which are scalar terms.

We assume that you have had some exposure to matrices so that the standard matrix operations are familiar to you. All you need to remember is how the inverse of a matrix, the determinant of a matrix, and the transpose of a matrix are calculated and how to add, subtract, and multiply matrices.

12.1.1 Matrix Properties

A host of matrix properties have been studied by the mathematicians. We discuss only the notions of "eigenvalues" and "singular values" since these are valuable in our design methods for multivariable systems. *Eigenvalues* is simply another name for the roots of the characteristic equation of the system. *Singular values* give us a measure of the size of the matrix and an indication of how close it is to being "singular." A matrix is singular if its determinant is zero. Since the determinant appears in the denominator when the inverse of a matrix is taken [Eq. (12.1)], the inverse will not exist if the matrix is singular.

$$[\underline{\underline{A}}]^{-1} = \frac{[\text{Cofactor}[\underline{\underline{A}}])]^{\text{T}}}{\text{Det}\,\underline{\underline{A}}} \tag{12.1}$$

A. Eigenvalues

The eigenvalues of a square $N \times N$ matrix are the N roots of the scalar equation

$$\boxed{\text{Det}[\lambda \underline{\underline{I}} - \underline{\underline{A}}] = 0} \tag{12.2}$$

where λ is a scalar quantity. Since there are N eigenvalues, it is convenient to define the vector $\underline{\lambda}$, of length N, that consists of the eigenvalues: $\lambda_1, \lambda_2, \lambda_3, \ldots, \lambda_N$.

$$\underline{\lambda} = \begin{bmatrix} \lambda_1 \\ \lambda_2 \\ \lambda_3 \\ \vdots \\ \lambda_N \end{bmatrix} \tag{12.3}$$

We use the notation that the expression $\underline{\lambda}_{[\underline{\underline{A}}]}$ means the vector of eigenvalues of the matrix $\underline{\underline{A}}$. Thus, the eigenvalues of a matrix $[\underline{\underline{I}} + \underline{\underline{G_M}}\,\underline{\underline{G_C}}]$ is written $\underline{\lambda}_{[\underline{\underline{I}}+\underline{\underline{G_M}}\,\underline{\underline{G_C}}]}$.

EXAMPLE 12.1. Calculate the eigenvalues of the following matrix $\underline{\underline{A}}$.

$$\underline{\underline{A}} = \begin{bmatrix} -2 & 0 \\ 2 & -4 \end{bmatrix} \tag{12.4}$$

$$\text{Det}[\lambda \underline{\underline{I}} - \underline{\underline{A}}] = 0$$

$$\lambda \underline{\underline{I}} - \underline{\underline{A}} = \lambda \begin{bmatrix} 1 & 0 \\ 0 & 1 \end{bmatrix} - \begin{bmatrix} -2 & 0 \\ 2 & -4 \end{bmatrix} = \begin{bmatrix} (\lambda + 2) & 0 \\ -2 & (\lambda + 4) \end{bmatrix}$$

$$\text{Det}[\lambda \underline{\underline{I}} - \underline{\underline{A}}] = \text{Det}\begin{bmatrix} (\lambda + 2) & 0 \\ -2 & (\lambda + 4) \end{bmatrix} = (\lambda + 2)(\lambda + 4) - (0)(-2)$$

$$\lambda^2 + 6\lambda + 8 = 0 = (\lambda + 2)(\lambda + 4)$$

The roots of this equation are $\lambda = -2$ and $\lambda = -4$. Therefore, the eigenvalues of the matrix given in Eq. (12.4) are $\lambda_1 = -2$ and $\lambda_2 = -4$.

$$\underline{\lambda}_{[\underline{\underline{A}}]} = \begin{bmatrix} -2 \\ -4 \end{bmatrix} \tag{12.5}$$

■

It might be useful at this point to provide some motivation for defining such seemingly abstract quantities as eigenvalues. Consider a system of N linear ordinary differential equations that model a chemical process.

$$\frac{d\underline{x}}{dt} = \underline{\underline{A}}\,\underline{x} + \underline{\underline{B}}\,\underline{u} \tag{12.6}$$

where $\underline{x}$ = vector of the N *state variables* of the system
$\underline{\underline{A}}$ = $N \times N$ matrix of constants
$\underline{\underline{B}}$ = a different $N \times M$ matrix of constants
$\underline{u}$ = vector of the M input variables of the system

We show in Section 12.1.3 that the eigenvalues of the $\underline{\underline{A}}$ matrix are the roots of the characteristic equation of the system. Thus, the eigenvalues tell us whether the system is stable or unstable, fast or slow, overdamped or underdamped. They are essential for the analysis of dynamic systems.

B. Singular values

The singular values of a matrix are a measure of how close the matrix is to being "singular," i.e., to having a determinant that is zero. A matrix that is $N \times N$ has N singular values. We use the symbol σ_i for a singular value. The largest magnitude σ_i is called the maximum singular value, and the notation $\sigma^{\max}$ is used. The smallest magnitude σ_i is called the minimum singular value ($\sigma^{\min}$). The ratio of the maximum and minimum singular values is called the "condition number."

The N singular values of a *real* $N \times N$ matrix (i.e., all elements of the matrix are real numbers) are defined as the square root of the eigenvalues of the matrix formed by multiplying the original matrix by its transpose.

$$\sigma_{i[\underline{\underline{A}}]} = \sqrt{\lambda_{i[\underline{\underline{A}}^{T}\underline{\underline{A}}]}} \qquad i = 1, 2, \ldots, N \tag{12.7}$$

EXAMPLE 12.2. Find the singular values of the $\underline{\underline{A}}$ matrix from Example 12.1.

$$\underline{\underline{A}} = \begin{bmatrix} -2 & 0 \\ 2 & -4 \end{bmatrix} \qquad \underline{\underline{A}}^{T} = \begin{bmatrix} -2 & 2 \\ 0 & -4 \end{bmatrix}$$

$$\underline{\underline{A}}^{T}\underline{\underline{A}} = \begin{bmatrix} -2 & 2 \\ 0 & -4 \end{bmatrix}\begin{bmatrix} -2 & 0 \\ 2 & -4 \end{bmatrix} = \begin{bmatrix} 8 & -8 \\ -8 & 16 \end{bmatrix} \tag{12.8}$$

To get the eigenvalues of this matrix we use Eq. (12.2).

$$\text{Det}[\lambda\underline{\underline{I}} - \underline{\underline{A}}^{T}\underline{\underline{A}}] = 0$$

$$\text{Det}\begin{bmatrix} (\lambda - 8) & 8 \\ 8 & (\lambda - 16) \end{bmatrix} = 0 = (\lambda - 8)(\lambda - 16) - 64$$

$$\lambda^2 - 24\lambda + 128 - 64 = 0 = \lambda^2 - 24\lambda + 64$$

$$\lambda_1 = 20.94 \qquad \lambda_2 = 3.06$$

$$\sigma_1 = \sqrt{20.94} = 4.58 \qquad \sigma_2 = \sqrt{3.06} = 1.75 \tag{12.9}$$

Neither of these singular values is small, so the matrix is not close to being singular. The determinant of $\underline{\underline{A}}$ is 8, so it is indeed not singular. ■

EXAMPLE 12.3. Calculate the singular values of the matrix

$$\underline{\underline{A}} = \begin{bmatrix} -1 & 1 \\ 1 & -1 \end{bmatrix}$$

Note that the determinant of this matrix is zero, so it is singular.

$$\underline{\underline{A}}^{\mathrm{T}} = \begin{bmatrix} -1 & 1 \\ 1 & -1 \end{bmatrix}$$

$$\underline{\underline{A}}^{\mathrm{T}}\underline{\underline{A}} = \begin{bmatrix} -1 & 1 \\ 1 & -1 \end{bmatrix}\begin{bmatrix} -1 & 1 \\ 1 & -1 \end{bmatrix} = \begin{bmatrix} 2 & -2 \\ -2 & 2 \end{bmatrix}$$

$$\mathrm{Det}[\lambda \underline{\underline{I}} - \underline{\underline{A}}^{\mathrm{T}}\underline{\underline{A}}] = 0$$

$$\mathrm{Det}\begin{bmatrix} (\lambda - 2) & 2 \\ 2 & (\lambda - 2) \end{bmatrix} = 0 = (\lambda - 2)(\lambda - 2) - 4$$

$$\lambda^2 - 4\lambda + 4 - 4 = \lambda(\lambda - 4) = 0$$

$$\lambda_1 = 0 \quad \lambda_2 = 4 \qquad \sigma_1 = 0 \quad \sigma_2 = 2$$

The singular value of zero tells us that the matrix is singular. ■

The singular values of a *complex* matrix are similar to those of a real matrix. The only difference is that we use the conjugate transpose.

$$\sigma_{i[\underline{\underline{A}}]} = \sqrt{\lambda_{i[\underline{\underline{A}}^{\mathrm{CT}}\underline{\underline{A}}]}} \qquad i = 1, 2, \ldots, N \tag{12.10}$$

First we calculate the conjugate transpose (the transpose of the matrix with all of the signs of the imaginary parts changed). Then we multiply $\underline{\underline{A}}$ by it. Then we calculate the eigenvalues. These can be found using Eq. (12.2) for simple systems. In more realistic problems we use MATLAB or the IMSL subroutine EIGCC. Note that the product of a complex matrix with its conjugate transpose gives a complex matrix (called a "hermitian" matrix) that has real elements on the diagonal and has real eigenvalues. Thus, all the singular values of a complex matrix are real numbers.

EXAMPLE 12.4. Calculate the singular values of the complex matrix

$$\underline{\underline{A}} = \begin{bmatrix} (1 + i) & (1 + i) \\ (2 + i) & (1 + i) \end{bmatrix} \tag{12.11}$$

$$\underline{\underline{A}}^{\mathrm{CT}} = \begin{bmatrix} (1 - i) & (2 - i) \\ (1 - i) & (1 - i) \end{bmatrix}$$

$$\underline{\underline{A}}^{\mathrm{CT}}\underline{\underline{A}} = \begin{bmatrix} (1 - i) & (2 - i) \\ (1 - i) & (1 - i) \end{bmatrix}\begin{bmatrix} (1 + i) & (1 + i) \\ (2 + i) & (1 + i) \end{bmatrix} = \begin{bmatrix} 7 & (5 + i) \\ (5 - i) & 4 \end{bmatrix}$$

$$\mathrm{Det}[\lambda \underline{\underline{I}} - \underline{\underline{A}}^{\mathrm{CT}}\underline{\underline{A}}] = 0$$

$$\mathrm{Det}\left[\lambda \begin{bmatrix} 1 & 0 \\ 0 & 1 \end{bmatrix} - \begin{bmatrix} 7 & (5 + i) \\ (5 - i) & 4 \end{bmatrix}\right] = 0$$

$$\text{Det}\begin{bmatrix} (\lambda - 7) & (-5 - i) \\ (-5 + i) & (\lambda - 4) \end{bmatrix} = 0$$

$$\lambda^2 - 11\lambda + 28 - 25 - 1 = \lambda^2 - 11\lambda + 2 = 0$$

$$\lambda_1 = 10.63 \quad \lambda_2 = 0.185 \qquad \sigma_1 = 3.29 \quad \sigma_2 = 0.430$$

Note that the singular values are real. ■

12.1.2 Transfer Function Representation

A. Openloop system

Let us first consider an openloop process with N controlled variables, N manipulated variables, and one load disturbance. The system can be described in the Laplace domain by N equations that give the transfer functions showing how all of the manipulated variables and the load disturbance affect each of the controlled variables through their appropriate transfer functions.

$$\begin{aligned} Y_1 &= G_{M_{11}} m_1 + G_{M_{12}} m_2 + \cdots + G_{M_{1N}} m_N + G_{L_1} L \\ Y_2 &= G_{M_{21}} m_1 + G_{M_{22}} m_2 + \cdots + G_{M_{2N}} m_N + G_{L_2} L \\ &\vdots \\ Y_N &= G_{M_{N1}} m_1 + G_{M_{N2}} m_2 + \cdots + G_{M_{NN}} m_N + G_{L_N} L \end{aligned} \tag{12.12}$$

All the variables are in the Laplace domain, as are all of the transfer functions. This set of N equations is very conveniently represented by one matrix equation.

$$\underline{Y} = \underline{\underline{G_{M(s)}}}\,\underline{m_{(s)}} + \underline{G_{L(s)}} L_{(s)} \tag{12.13}$$

where $\underline{Y}$ = vector of N controlled variables
$\underline{\underline{G_M}}$ = $N \times N$ matrix of process openloop transfer functions relating the controlled variables and the manipulated variables
$\underline{m}$ = vector of N manipulated variables
$\underline{G_L}$ = vector of process openloop transfer functions relating the controlled variables and the load disturbance
$L_{(s)}$ = load disturbance

These relationships are shown pictorially in Fig. 12.1. We use only one load variable in this development to keep things as simple as possible. Clearly, there could be several load disturbances, which would just appear as additional terms to Eqs. (12.12). Then $L_{(s)}$ in Eq. (12.13) becomes a vector, and G_L becomes a matrix with N rows and as many columns as there are load disturbances. Since the effects of each of the

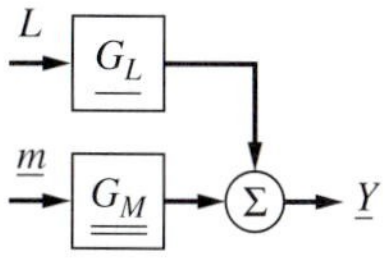

FIGURE 12.1

load disturbances can be considered one at a time, we do it that way to simplify the mathematics. Note that the effects of each of the manipulated variables can also be considered one at a time if we were looking only at the *openloop* system or if we were considering controlling only one variable. However, when we go to a multivariable *closedloop* system, the effects of all manipulated variables must be considered simultaneously.

B. Closedloop system

Figure 12.2 gives the matrix block diagram description of the openloop system with a feedback control system added. The $\underline{\underline{I}}$ matrix is the identity matrix. The $\underline{\underline{G_{C(s)}}}$ matrix contains the feedback controllers. Most industrial processes use conventional single-input, single-output (SISO) feedback controllers. One controller is used in each loop to regulate one controlled variable by changing one manipulated variable. In this case the $\underline{\underline{G_{C(s)}}}$ matrix has only diagonal elements. All the off-diagonal elements are zero.

$$\underline{\underline{G_{C(s)}}} = \begin{bmatrix} G_{C1} & 0 & \cdots & 0 \\ 0 & G_{C2} & 0 & \cdots \\ \cdots & \cdots & \cdots & \cdots \\ 0 & \cdots & 0 & G_{CN} \end{bmatrix} \tag{12.14}$$

where $G_{C1}, G_{C2}, \ldots, G_{CN}$ are the individual controllers in each of the N loops. We call this multiloop SISO system a "diagonal controller" structure. It is important to recognize right from the beginning that having multiple SISO controllers does *not* mean that we can tune each controller independently. As we will soon see, the dynamics and stability of this multivariable closedloop process depend on the settings of all controllers.

Controller structures that are not diagonal but have elements in all positions in the $\underline{\underline{G_{C(s)}}}$ matrix are called multivariable controllers.

$$\underline{\underline{G_{C(s)}}} = \begin{bmatrix} G_{C11} & G_{C12} & \cdots & G_{C1N} \\ G_{C21} & G_{C22} & \cdots & G_{C2N} \\ \cdots & \cdots & \cdots & \cdots \\ G_{CN1} & G_{CN2} & \cdots & G_{CNN} \end{bmatrix} \tag{12.15}$$

The feedback controller matrix gives the transfer functions between the manipulated variables and the errors.

$$\underline{m} = \underline{\underline{G_{C(s)}}}\,\underline{E} \tag{12.16}$$

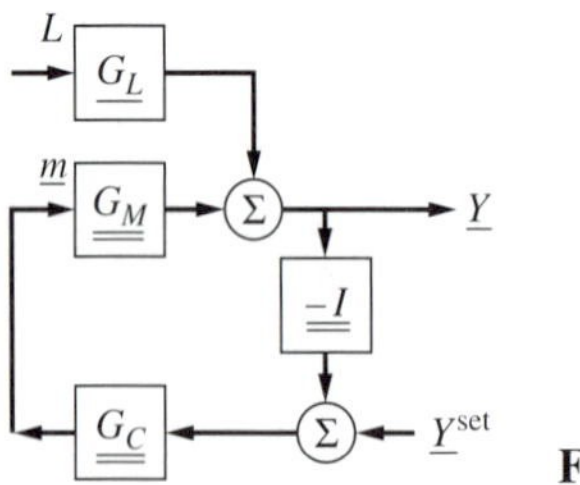

FIGURE 12.2

Since the errors are the differences between setpoints and controlled variables,

$$\underline{m} = \underline{\underline{G_{C(s)}}}[\underline{Y}^{\text{set}} - \underline{Y}] \tag{12.17}$$

Substituting for $\underline{m}$ in Eq. (12.13) gives

$$\underline{Y} = \underline{\underline{G_{M(s)}}}\,\underline{\underline{G_{C(s)}}}[\underline{Y}^{\text{set}} - \underline{Y}] + \underline{G_{L(s)}}L_{(s)} \tag{12.18}$$

Bringing all the terms with $\underline{Y}$ to the left side gives

$$\underline{\underline{I}}\,\underline{Y} + \underline{\underline{G_{M(s)}}}\,\underline{\underline{G_{C(s)}}}\,\underline{Y} = \underline{\underline{G_{M(s)}}}\,\underline{\underline{G_{C(s)}}}\,\underline{Y}^{\text{set}} + \underline{G_{L(s)}}L_{(s)} \tag{12.19}$$

$$[\underline{\underline{I}} + \underline{\underline{G_{M(s)}}}\,\underline{\underline{G_{C(s)}}}]\underline{Y} = \underline{\underline{G_{M(s)}}}\,\underline{\underline{G_{C(s)}}}\,\underline{Y}^{\text{set}} + \underline{G_{L(s)}}L_{(s)} \tag{12.20}$$

$$\begin{aligned}\underline{Y} = &\left[[\underline{\underline{I}} + \underline{\underline{G_{M(s)}}}\,\underline{\underline{G_{C(s)}}}]^{-1}\underline{\underline{G_{M(s)}}}\,\underline{\underline{G_{C(s)}}}\right]\underline{Y}^{\text{set}} \\ &+ \left[[\underline{\underline{I}} + \underline{\underline{G_{M(s)}}}\,\underline{\underline{G_{C(s)}}}]^{-1}\underline{G_{L(s)}}\right]L_{(s)}\end{aligned} \tag{12.21}$$

Equation (12.21) gives the effects of setpoint and load changes on the controlled variables in the closedloop multivariable environment. The matrix (of order $N \times N$) multiplying the vector of setpoints is the closedloop servo transfer function matrix. The matrix ($N \times 1$) multiplying the load disturbance is the closedloop regulator transfer function vector.

It is clear that this matrix equation is very similar to the scalar equation describing a closedloop system derived back in Chapter 8 for SISO systems.

$$Y = \left[\frac{G_M G_C}{1 + G_M G_C}\right] Y^{\text{set}} + \left[\frac{G_L}{1 + G_M G_C}\right] L \tag{12.22}$$

Now we have matrix inverses to worry about, but the structure is essentially the same.

12.1.3 State Variables

The "states" of a dynamic system are simply the variables that appear in the time differential. The time-domain differential equation description of multivariable systems can be used instead of Laplace-domain transfer functions. Naturally, the two are related, and we derive these relationships below. State variables are very popular in electrical and mechanical engineering control problems, which tend to be of lower order (fewer differential equations) than chemical engineering control problems. Transfer function representation is more useful in practical process control problems because the matrices are of lower order than would be required by a state variable representation. For example, a distillation column can be represented by a 2×2 transfer function matrix. The number of state variables of the column might be 200.

State variables appear naturally in the differential equations describing chemical engineering systems because our mathematical models are based on a number of first-order differential equations: component balances, energy equations, etc. If there are N such equations, they can be linearized (if necessary) and written in matrix form

$$\frac{d\underline{x}}{dt} = \underline{\underline{A}}\,\underline{x} + \underline{\underline{B}}\,\underline{m} + \underline{D}L \tag{12.23}$$

where $\underline{x}$ = vector of the N state variables of the system.

EXAMPLE 12.5. The irreversible chemical reaction A → B takes place in two perfectly mixed reactors connected in series, as shown in Fig. 12.3. The reaction rate is proportional to the concentration of reactant. Let x_1 be the concentration of reactant A in the first tank and x_2 the concentration in the second tank. The concentration of reactant in the feed is x_0. The feed flow rate is F. Both x_0 and F can be manipulated. Assume the specific reaction rates k_1 and k_2 in each tank are constant (isothermal operation). Assume constant volumes V_1 and V_2.

The component balances for the system are

$$\begin{aligned} V_1\frac{dx_1}{dt} &= F(x_0 - x_1) - k_1 V_1 x_1 \\ V_2\frac{dx_2}{dt} &= F(x_1 - x_2) - k_2 V_2 x_2 \end{aligned} \tag{12.24}$$

Linearizing around the initial steady state gives two linear ODEs.

$$\begin{aligned} V_1\frac{dx_1}{dt} &= -(\overline{F} + k_1 V_1)x_1 + (\overline{F})x_0 + (\overline{x}_0 - \overline{x}_1)F \\ V_2\frac{dx_2}{dt} &= (\overline{F})x_1 - (\overline{F} + k_2 V_2)x_2 + (\overline{x}_1 - \overline{x}_2)F \end{aligned} \tag{12.25}$$

These two equations in matrix form are

$$\frac{d}{dt}\begin{bmatrix} x_1 \\ x_2 \end{bmatrix} = \begin{bmatrix} \left(-k_1 - \dfrac{\overline{F}}{V_1}\right) & 0 \\ \left(\dfrac{\overline{F}}{V_2}\right) & \left(-k_2 - \dfrac{\overline{F}}{V_2}\right) \end{bmatrix}\begin{bmatrix} x_1 \\ x_2 \end{bmatrix} + \begin{bmatrix} \left(\dfrac{\overline{F}}{V_1}\right) & \left(\dfrac{\overline{x}_0 - \overline{x}_1}{V_1}\right) \\ 0 & \left(\dfrac{\overline{x}_1 - \overline{x}_2}{V_2}\right) \end{bmatrix}\begin{bmatrix} x_0 \\ F \end{bmatrix} \tag{12.26}$$

The state variables are the two concentrations. The feed concentration x_0 and the feed flow rate F are the manipulated variables. To take a specific numerical case, let $k_1 = 1\ \text{min}^{-1}$, $k_2 = 2\ \text{min}^{-1}$, $V_1 = 100\ \text{ft}^3$, and $V_2 = 50\ \text{ft}^3$. The initial steady-state conditions are $\overline{F} = 100\ \text{ft}^3/\text{min}$, $\overline{x}_0 = 0.5$ mol A/ft^3, $\overline{x}_1 = 0.25$ mol A/ft^3, and $\overline{x}_2 = 0.125$ mol A/ft^3. This gives the $\underline{\underline{A}}$ matrix that we used in Example 12.1.

$$\underline{\underline{A}} = \begin{bmatrix} -2 & 0 \\ 2 & -4 \end{bmatrix}$$

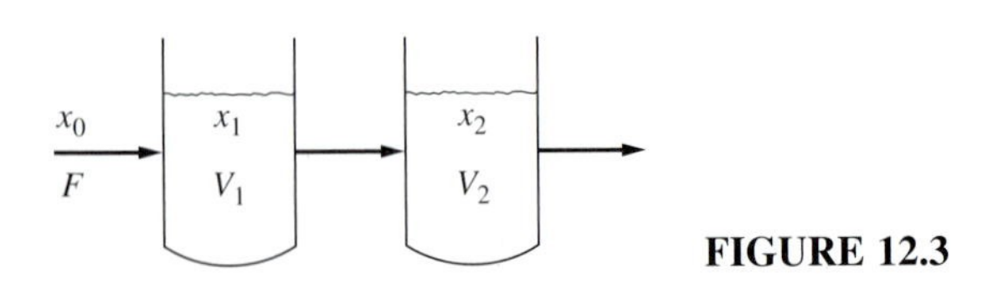

FIGURE 12.3

The $\underline{\underline{B}}$ matrix is

$$\underline{\underline{B}} = \begin{bmatrix} 1 & 0.0025 \\ 0 & 0.0025 \end{bmatrix}$$ ■

State variable representation can be transformed into transfer function representation by Laplace-transforming the set of N linear ordinary differential equations [Eq. (12.23)].

$$\frac{d\underline{x}}{dt} = \underline{\underline{A}}\,\underline{x} + \underline{\underline{B}}\,\underline{m} + \underline{D}L$$

$$s\underline{x}_{(s)} = \underline{\underline{A}}\,\underline{x}_{(s)} + \underline{\underline{B}}\,\underline{m}_{(s)} + \underline{D}L_{(s)}$$

$$[s\underline{\underline{I}} - \underline{\underline{A}}]\underline{x}_{(s)} = \underline{\underline{B}}\,\underline{m}_{(s)} + \underline{D}L_{(s)}$$

$$\underline{x}_{(s)} = \left[[s\underline{\underline{I}} - \underline{\underline{A}}]^{-1}\underline{\underline{B}}\right]\underline{m}_{(s)} + \left[[s\underline{\underline{I}} - \underline{\underline{A}}]^{-1}\underline{D}\right]L_{(s)} \tag{12.27}$$

Comparing this with Eq. (12.13) and considering the case where the controlled variables Y are the same as the state variables, we see how the transfer function matrix $\underline{\underline{G_{M(s)}}}$ and transfer function vector $\underline{G_{L(s)}}$ are related to the $\underline{\underline{A}}$ and $\underline{\underline{B}}$ matrices and to the $\underline{D}$ vector.

EXAMPLE 12.6. Determine the transfer function matrix $\underline{\underline{G_{M(s)}}}$ for the system described in Example 12.5.

$$\frac{d}{dt}\begin{bmatrix} x_1 \\ x_2 \end{bmatrix} = \begin{bmatrix} -2 & 0 \\ 2 & -4 \end{bmatrix}\begin{bmatrix} x_1 \\ x_2 \end{bmatrix} + \begin{bmatrix} 1 & 0.0025 \\ 0 & 0.0025 \end{bmatrix}\begin{bmatrix} x_0 \\ F \end{bmatrix}$$

$$\underline{\underline{G_{M(s)}}} = [s\underline{\underline{I}} - \underline{\underline{A}}]^{-1}\underline{\underline{B}} = \left[s\begin{bmatrix} 1 & 0 \\ 0 & 1 \end{bmatrix} - \begin{bmatrix} -2 & 0 \\ 2 & -4 \end{bmatrix}\right]^{-1}\begin{bmatrix} 1 & 0.0025 \\ 0 & 0.0025 \end{bmatrix}$$

$$= \begin{bmatrix} (s+2) & 0 \\ -2 & (s+4) \end{bmatrix}^{-1}\begin{bmatrix} 1 & 0.0025 \\ 0 & 0.0025 \end{bmatrix}$$

$$= \frac{\begin{bmatrix} (s+4) & 0 \\ 2 & (s+2) \end{bmatrix}}{(s+2)(s+4)}\begin{bmatrix} 1 & 0.0025 \\ 0 & 0.0025 \end{bmatrix}$$

$$\begin{bmatrix} x_1 \\ x_2 \end{bmatrix} = \underline{\underline{G_{M(s)}}}\begin{bmatrix} x_0 \\ F \end{bmatrix} = \begin{bmatrix} \dfrac{1}{s+2} & \dfrac{0.0025}{s+2} \\ \dfrac{2}{(s+2)(s+4)} & \dfrac{0.0025}{s+2} \end{bmatrix}\begin{bmatrix} x_0 \\ F \end{bmatrix} \tag{12.28}$$ ■

The system considered in the preceding example has a characteristic equation that is the denominator of the transfer function set equal to zero. This is true, of course, for any system. Since the system is uncontrolled, the openloop characteristic equation is [using Eq. (12.28)]

$$(s+2)(s+4) = 0$$

The roots of the openloop characteristic equation are $s = -2$ and $s = -4$. These are exactly the values we calculated for the eigenvalues of the $\underline{\underline{A}}$ matrix of this system (see Example 12.1)!

> *The eigenvalues of the $\underline{\underline{A}}$ matrix are equal to the roots of the characteristic equation of the system.*

The $\underline{\underline{A}}$ matrix is the matrix that multiplies the $\underline{x}$ vector when the differential equations are in the standard form $d\underline{x}/dt = \underline{\underline{A}}\,\underline{x}$.

We have considered openloop systems up to this point, but the mathematics apply to *any* system, openloop or closedloop. Suppose an openloop system is described by

$$\frac{d\underline{x}}{dt} = \underline{\underline{A}}\,\underline{x} + \underline{\underline{B}}\,\underline{m} + \underline{D}L \tag{12.29}$$

The eigenvalues of the $\underline{\underline{A}}$ matrix, $\lambda_{[\underline{\underline{A}}]}$, are the *openloop eigenvalues* and are equal to the roots of the openloop characteristic equation. To help us keep straight on what are "apples" versus "oranges," we call the openloop eigenvalues λ_{OL}.

Now suppose a feedback controller is added to the system. The manipulated variables $\underline{m}$ are set by the feedback controller. To keep things as simple as possible, let us make two assumptions that are not very good but permit us to illustrate an important point. We assume that the feedback controller matrix $\underline{\underline{G_{C(s)}}}$ consists of just constants (gains) $\underline{\underline{K}}$, and we assume that there are as many manipulated variables $\underline{m}$ as state variables $\underline{x}$.

$$\underline{m} = \underline{\underline{K}}[\underline{x}^{set} - \underline{x}] \tag{12.30}$$

Substituting into Eq. (12.29) gives

$$\frac{d\underline{x}}{dt} = \underline{\underline{A}}\,\underline{x} + \underline{\underline{B}}\,\underline{\underline{K}}[\underline{x}^{set} - \underline{x}] + \underline{D}L$$

Rearranging to put the differential equations in the standard form gives

$$\frac{d\underline{x}}{dt} = [\underline{\underline{A}} - \underline{\underline{B}}\,\underline{\underline{K}}]\underline{x} + \underline{\underline{B}}\,\underline{\underline{K}}\,\underline{x}^{set} + \underline{D}L \tag{12.31}$$

This equation describes the closedloop system. Let us define the matrix that multiplies $\underline{x}$ as the "closedloop $\underline{\underline{A}}$" matrix and use the symbol $\underline{\underline{A_{CL}}}$.

$$\frac{d\underline{x}}{dt} = \underline{\underline{A_{CL}}}\,\underline{x} + \underline{\underline{B}}\,\underline{\underline{K}}\,\underline{x}^{set} + \underline{D}L \tag{12.32}$$

Thus, the characteristic matrix for this closedloop system is the $\underline{\underline{A_{CL}}}$ matrix. Its eigenvalues will be the *closedloop eigenvalues,* and they will be the roots of the closedloop characteristic equation.

The purpose of the preceding discussion is to contrast openloop eigenvalues and closedloop eigenvalues. We must use the appropriate eigenvalue for the system we are studying. The "Greek" state-space language uses the term *eigenvalue* instead of the Russian-language, Laplace transfer function term *root of the characteristic equation.* But whatever the language, they are exactly the same thing. So we have openloop and closedloop eigenvalues, or we have roots of the openloop and closed-loop characteristic equations.

EXAMPLE 12.7. The openloop eigenvalues for the two-reactor system studied in Example 12.6 were $\lambda_{OL} = -2, -4$. Calculate the closedloop eigenvalues if two proportional controllers are used. G_{C1} manipulates x_0 to control x_1, and G_{C2} manipulates F to control x_2.

$$\underline{\underline{A_{CL}}} = \underline{\underline{A}} - \underline{\underline{B}}\,\underline{\underline{K}} = \begin{bmatrix} -2 & 0 \\ 2 & -4 \end{bmatrix} - \begin{bmatrix} 1 & 0.0025 \\ 0 & 0.0025 \end{bmatrix} \begin{bmatrix} K_1 & 0 \\ 0 & K_2 \end{bmatrix}$$
$$= \begin{bmatrix} (-2 - K_1) & -0.0025K_2 \\ 2 & (-4 - 0.0025K_2) \end{bmatrix} \tag{12.33}$$

Using Eq. (12.2) to solve for the eigenvalues of this matrix gives the closedloop eigenvalues.

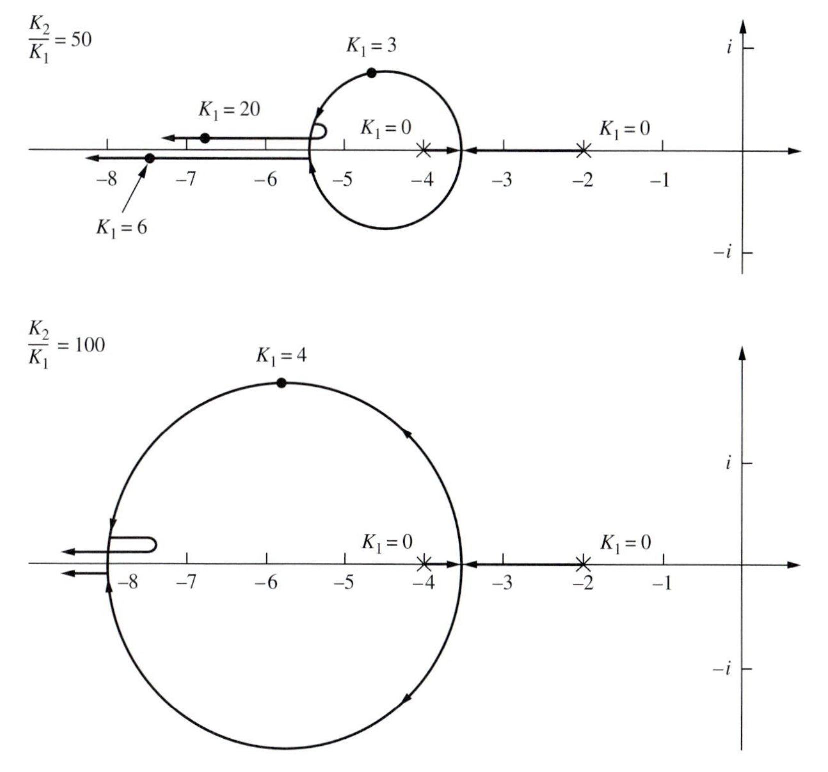

FIGURE 12.4
Closedloop eigenvalues.

$$\text{Det}[\lambda_{CL}\underline{\underline{I}} - \underline{\underline{A_{CL}}}] = 0 \qquad (12.34)$$

$$\text{Det}\left[\lambda_{CL}\begin{bmatrix}1 & 0\\ 0 & 1\end{bmatrix} - \begin{bmatrix}(-2-K_1) & -0.0025K_2\\ 2 & (-4-0.0025K_2)\end{bmatrix}\right] = 0$$

$$\text{Det}\begin{bmatrix}(\lambda_{CL}+2+K_1) & 0.0025K_2\\ -2 & (\lambda_{CL}+4+0.0025K_2)\end{bmatrix} = 0$$

$$\lambda_{CL}^2 + \lambda_{CL}(6 + K_1 + 0.0025K_2) + (8 + 4K_1 + 0.01K_2 + 0.0025K_1K_2) = 0 \qquad (12.35)$$

For $K_1 = 1$ and $K_2 = 100$, the closedloop eigenvalues are $\lambda_{CL} = -3.62 \pm i0.331$. For $K_1 = 5$ and $K_2 = 500$, $\lambda_{CL} = -6.12 \pm i1.32$. Figure 12.4 is a plot of the closedloop eigenvalues as a function of the two controller gains. Note that this is *not* a traditional SISO root locus plot, so some of the traditional rules do not apply. Both gains are changing along the curves. The shapes of the curves are quite unusual. For example, the two loci both run out the negative real axis as the gains become large. ■

12.2 STABILITY

12.2.1 Closedloop Characteristic Equation

Remember that the inverse of a matrix has the determinant of the matrix in the denominator of each element. Therefore, the denominators of all of the transfer functions in Eq. (12.21) contain $\text{Det}[\underline{\underline{I}} + \underline{\underline{G_{M(s)}}}\,\underline{\underline{G_{C(s)}}}]$. Now we know that the characteristic equation of any system is the denominator set equal to zero. Therefore, the closedloop characteristic equation of the multivariable system with feedback controllers is the simple *scalar* equation

$$\boxed{\text{Det}[\underline{\underline{I}} + \underline{\underline{G_{M(s)}}}\,\underline{\underline{G_{C(s)}}}] = 0} \qquad (12.36)$$

We use this multivariable closedloop characteristic equation in Chapter 13 to design controllers in a multivariable process.

EXAMPLE 12.8. Determine the closedloop characteristic equation for the system whose openloop transfer function matrix was derived in Example 12.6. Use a diagonal controller structure (two SISO controllers) that are proportional only.

$$\underline{\underline{G_{C(s)}}} = \begin{bmatrix}K_1 & 0\\ 0 & K_2\end{bmatrix} \qquad \underline{\underline{G_{M(s)}}} = \begin{bmatrix}\dfrac{1}{s+2} & \dfrac{0.0025}{s+2}\\[2ex] \dfrac{2}{(s+2)(s+4)} & \dfrac{0.0025}{s+2}\end{bmatrix}$$

$$\underline{\underline{G_{M(s)}}}\,\underline{\underline{G_{C(s)}}} = \begin{bmatrix}\dfrac{K_1}{s+2} & \dfrac{0.0025K_2}{s+2}\\[2ex] \dfrac{2K_1}{(s+2)(s+4)} & \dfrac{0.0025K_2}{s+2}\end{bmatrix}$$

$$\underline{\underline{I}} + \underline{\underline{G_{M(s)}}}\,\underline{\underline{G_{C(s)}}} = \begin{bmatrix} \left(1 + \dfrac{K_1}{s+2}\right) & \dfrac{0.0025K_2}{s+2} \\ \dfrac{2K_1}{(s+2)(s+4)} & \left(1 + \dfrac{0.0025K_2}{s+2}\right) \end{bmatrix}$$

$$\text{Det}[\underline{\underline{I}} + \underline{\underline{G_{M(s)}}}\,\underline{\underline{G_{C(s)}}}] = 0$$

$$\left(1 + \frac{K_1}{s+2}\right)\left(1 + \frac{0.0025K_2}{s+2}\right) - \left(\frac{0.0025K_2}{s+2}\right)\left(\frac{2K_1}{(s+2)(s+4)}\right) = 0 \qquad (12.37)$$

$$1 + \frac{K_1}{s+2} + \frac{0.0025K_2}{s+2} + \frac{0.0025K_1K_2}{(s+2)^2} - \frac{0.005K_1K_2}{(s+2)^2(s+4)} = 0$$

$$1 + \frac{K_1}{s+2} + \frac{0.0025K_2}{s+2} + \frac{0.0025K_1K_2}{(s+2)^2}\left(1 - \frac{2}{s+4}\right) = 0$$

$$1 + \frac{K_1}{s+2} + \frac{0.0025K_2}{s+2} + \frac{0.0025K_1K_2}{(s+2)^2}\left(\frac{s+4-2}{s+4}\right) = 0$$

$$1 + \frac{K_1}{s+2} + \frac{0.0025K_2}{s+2} + \frac{0.0025K_1K_2}{(s+2)(s+4)} = 0$$

$$s^2 + 6s + 8 + K_1(s+4) + 0.0025K_2(s+4) + 0.0025K_1K_2 = 0$$

$$s^2 + s(6 + K_1 + 0.0025K_2) + (8 + 4K_1 + 0.01K_2 + 0.0025K_1K_2) = 0 \qquad (12.38)$$

Note that this is exactly the same characteristic equation that we found using the transfer function notation [see Eq. (12.35)]. Remember that these values of s are the roots of the closedloop characteristic equation. ■

EXAMPLE 12.9. Determine the closedloop characteristic equation for a 2×2 process with a diagonal feedback controller.

$$\underline{\underline{G_{M(s)}}} = \begin{bmatrix} G_{M11} & G_{M12} \\ G_{M21} & G_{M22} \end{bmatrix} \qquad \underline{\underline{G_{C(s)}}} = \begin{bmatrix} G_{C1} & 0 \\ 0 & G_{C2} \end{bmatrix} \qquad (12.39)$$

$$\text{Det}[\underline{\underline{I}} + \underline{\underline{G_{M(s)}}}\,\underline{\underline{G_{C(s)}}}] = \text{Det}\left[\begin{bmatrix} 1 & 0 \\ 0 & 1 \end{bmatrix} + \begin{bmatrix} G_{M11} & G_{M12} \\ G_{M21} & G_{M22} \end{bmatrix}\begin{bmatrix} G_{C1} & 0 \\ 0 & G_{C2} \end{bmatrix}\right] = 0$$

$$\text{Det}\left[\begin{bmatrix} 1 & 0 \\ 0 & 1 \end{bmatrix} + \begin{bmatrix} G_{C1}G_{M11} & G_{C2}G_{M12} \\ G_{C1}G_{M21} & G_{C2}G_{M22} \end{bmatrix}\right] = 0$$

$$\text{Det}\begin{bmatrix} (1 + G_{C1}G_{M11}) & (G_{C2}G_{M12}) \\ (G_{C1}G_{M21}) & (1 + G_{C2}G_{M22}) \end{bmatrix} = 0 \qquad (12.40)$$

$$(1 + G_{C1}G_{M11})(1 + G_{C2}G_{M22}) - G_{C2}G_{M12}G_{C1}G_{M21} = 0$$

$$1 + G_{C1}G_{M11} + G_{C2}G_{M22} + G_{C1}G_{C2}(G_{M11}G_{M22} - G_{M12}G_{M21}) = 0$$

Notice that the closedloop characteristic equation depends on the tuning of *both* feedback controllers. ■

12.2.2 Multivariable Nyquist Plot

The Nyquist stability criterion developed in Chapter 11 can be directly applied to multivariable processes. As you should recall, the procedure is based on a complex variable theorem that says that the difference between the number of zeros and poles of a function inside a closed contour can be found by plotting the function and looking at the number of times it encircles the origin. We can use this theorem to find out if the closedloop characteristic equation has any roots or zeros in the right half of the s plane. The s variable follows a closed contour that completely surrounds the entire right half of the s plane. Since the closedloop characteristic equation is given in Eq. (12.36), the function of interest is

$$F_{(s)} = \text{Det}[\underline{\underline{I}} + \underline{\underline{G_{M(s)}}}\,\underline{\underline{G_{C(s)}}}] \tag{12.41}$$

The contour of $F_{(s)}$ is plotted in the F plane. The number of encirclements of the origin made by this plot is equal to the difference between the number of zeros and the number of poles of $F_{(s)}$ in the right half of the s plane.

If the process is openloop stable, none of the transfer functions in $\underline{\underline{G_{M(s)}}}$ has any poles in the right half of the s plane. And the feedback controllers in $\underline{\underline{G_{C(s)}}}$ are always chosen to be openloop stable (P, PI, or PID action), so $\underline{\underline{G_{C(s)}}}$ has no poles in the right half of the s plane. Clearly, the poles of $F_{(s)}$ are the poles of $\underline{\underline{G_{M(s)}}}\,\underline{\underline{G_{C(s)}}}$. Thus, if the process is openloop stable, the $F_{(s)}$ function has *no* poles in the right half of the s plane. So the number of encirclements of the origin made by the $F_{(s)}$ function is equal to the number of zeros in the right half of the s plane.

Thus the Nyquist stability criterion for a multivariable openloop-stable process is:

> *If a plot of* $\text{Det}[\underline{\underline{I}} + \underline{\underline{G_{M(i\omega)}}}\,\underline{\underline{G_{C(i\omega)}}}]$ *encircles the origin, the system is closedloop unstable!*

Remember that this is a simple scalar curve in the F plane, which varies with frequency ω.

The usual way to use the Nyquist stability criterion in scalar SISO systems is *not* to plot $1 + G_{M(i\omega)}G_{C(i\omega)}$ and look at encirclements of the origin. Instead we simply plot just $G_{M(i\omega)}G_{C(i\omega)}$ and look at encirclements of the $(-1, 0)$ point. To use a similar plot in multivariable systems we define a function $W_{(i\omega)}$ as follows:

$$W_{(i\omega)} = -1 + \text{Det}[\underline{\underline{I}} + \underline{\underline{G_{M(i\omega)}}}\,\underline{\underline{G_{C(i\omega)}}}] \tag{12.42}$$

Then the number of encirclements of the $(-1, 0)$ point made by $W_{(i\omega)}$ as ω varies from 0 to ∞ gives the number of zeros of the closedloop characteristic equation in the right half of the s plane.

EXAMPLE 12.10. The Wood and Berry (*Chem. Eng. Sci. 28:*1707, 1973) distillation column is a 2×2 system with the following openloop process transfer functions:

$$\underline{\underline{G_{M(s)}}} = \begin{bmatrix} G_{M11} & G_{M12} \\ G_{M21} & G_{M22} \end{bmatrix} = \begin{bmatrix} \dfrac{12.8e^{-s}}{16.7s + 1} & \dfrac{-18.9e^{-3s}}{21s + 1} \\ \dfrac{6.6e^{-7s}}{10.9s + 1} & \dfrac{-19.4e^{-3s}}{14.4s + 1} \end{bmatrix} \tag{12.43}$$

The process is openloop stable with no poles in the right half of the s plane. The authors used a diagonal controller structure with PI controllers and found, by empirical tuning, the following settings: $K_{c1} = 0.20$, $K_{c2} = -0.04$, $\tau_{I1} = 4.44$, and $\tau_{I2} = 2.67$. The feedback controller matrix was

$$\underline{\underline{G_{C(s)}}} = \begin{bmatrix} \dfrac{K_{c1}(\tau_{I1}s + 1)}{\tau_{I1}s} & 0 \\ 0 & \dfrac{K_{c2}(\tau_{I2}s + 1)}{\tau_{I2}s} \end{bmatrix} \tag{12.44}$$

Table 12.1 gives a MATLAB program that generates a W plot for the Wood and Berry column. After the four transfer functions are formed for the process and the two transfer functions are formed for the controllers, they are evaluated at each frequency using the *polyval* command. The identity matrix is formed by using the *eye(size(g))* command. Then the W function is calculated at each frequency using the *wnyquist(nw)=−1+det(eye(size(g))+g*gc);* command. This calculation is a good example of how easy it is to handle complex matrix calculations in MATLAB.

Figure 12.5 gives the W plane plots when the empirical settings are used and when the Ziegler-Nichols (ZN) settings for each individual controller are used ($K_{c1} = 0.960$, $K_{c2} = -0.19$, $\tau_{I1} = 3.25$, and $\tau_{I2} = 9.2$). The curve with the empirical settings does not encircle the $(-1, 0)$ point, and therefore the system is closedloop stable. Figure 12.6 gives the response of the system to a unit step change in x_1^{set}, verifying that the multivariable system is indeed closedloop stable.

The W plane curve using the ZN settings gets very close to the $(-1, 0)$ point, indicating that the system is closedloop unstable with these settings. This example illustrates that tuning each loop independently with the other loops on manual does not necessarily give a stable system when all loops are on automatic.

Note that the W plots with PI controllers start on the negative real axis. This is due to the two integrators, one in each controller, which give 180° of phase angle lag at low frequencies. As shown in Eq. (12.40), the product of the G_{C1} and G_{C2} controllers appears in the closedloop characteristic equation. ■

TABLE 12.1
W curves for 2 × 2 Wood and Berry column

```
% Program "wnyquist.m"
% Plots W curves for 2x2 Wood and Berry column
%
% Define transfer functions without deadtimes
numg11=12.8;
deng11=[16.7 1];
numg12=-18.9;
deng12=[21 1];
numg21=6.6;
deng21=[10.9 1];
numg22=-19.4;
deng22=[14.4 1];
d=[1 3
  7 3];
% Give ZN settings
kczn=[0.96 0
      0 -0.19];
```

TABLE 12.1 (CONTINUED)
W curves for 2 × 2 Wood and Berry column

```
resetzn=[3.25 0
         0 9.28];
% Give Empirical settings
kcemp=[0.2 0
       0 -0.04];
resetemp=[4.44 0
          0 2.67];
% Set frequencies
i=sqrt(-1);
w=logspace(-1,1,200);
s=i*w;
%
% Use Ziegler-Nichols settings
% Form controller transfer function
numgczn11=kczn(1,1)*[resetzn(1,1) 1];
dengczn11=[resetzn(1,1) 0];
numgczn22=kczn(2,2)*[resetzn(2,2) 1];
dengczn22=[resetzn(2,2) 0];
%
% Use empirical settings
% Form controller transfer function
numgcemp11=kcemp(1,1)*[resetemp(1,1) 1];
dengcemp11=[resetemp(1,1) 0];
numgcemp22=kcemp(2,2)*[resetemp(2,2) 1];
dengcemp22=[resetemp(2,2) 0];
%
% Loop to vary frequency
nwtot=length(w);
for nw=1:nwtot
wn=w(nw);
% Process g's
g(1,1)=polyval(numg11,s(nw)) / polyval(deng11,s(nw));
g(1,1)=g(1,1)*exp(-d(1,1)*s(nw));
g(1,2)=polyval(numg12,s(nw)) / polyval(deng12,s(nw));
g(1,2)=g(1,2)*exp(-d(1,2)*s(nw));
g(2,1)=polyval(numg21,s(nw)) / polyval(deng21,s(nw));
g(2,1)=g(2,1)*exp(-d(2,1)*s(nw));
g(2,2)=polyval(numg22,s(nw)) / polyval(deng22,s(nw));
g(2,2)=g(2,2)*exp(-d(2,2)*s(nw));
% Controller gc's
% Ziegler-Nichols
gczn(1,1)=polyval(numgczn11,s(nw)) / polyval(dengczn11,s(nw));
gczn(1,2)=0;
gczn(2,1)=0;
gczn(2,2)=polyval(numgczn22,s(nw)) / polyval(dengczn22,s(nw));
% Empirical
gcemp(1,1)=polyval(numgcemp11,s(nw)) / polyval(dengcemp11,s(nw));
gcemp(1,2)=0;
gcemp(2,1)=0;
gcemp(2,2)=polyval(numgcemp22,s(nw)) / polyval(dengcemp22,s(nw));
```

TABLE 12.1 (CONTINUED)
W curves for 2 × 2 Wood and Berry column

```
% Calculate wzn and wemp function
% "eye" operation forms an identity matrix
wzn(nw)=-1+det(eye(size(g))+g*gczn);
wemp(nw)=-1+det(eye(size(g))+g*gcemp);
end
% End of frequency loop
%
%
% Plot W function
clf
axis('equal');
plot(real(wzn),imag(wzn),'-',real(wemp),imag(wemp),'--')
axis([-2 1 -2 1]);
xlabel('Real(W)')
ylabel('Imag(W)')
text(-1.5,-1,'KZN=0.96/-0.19')
text(-1.5,-1.2,'ResetZN=3.25/9.28')
text(0,-1,'Kemp=0.2/-0.04')
text(0,-1.2,'Resetemp=4.44/2.67')
legend('ZN Settings','Empirical')
grid
pause
print -dps pfig125.ps
```

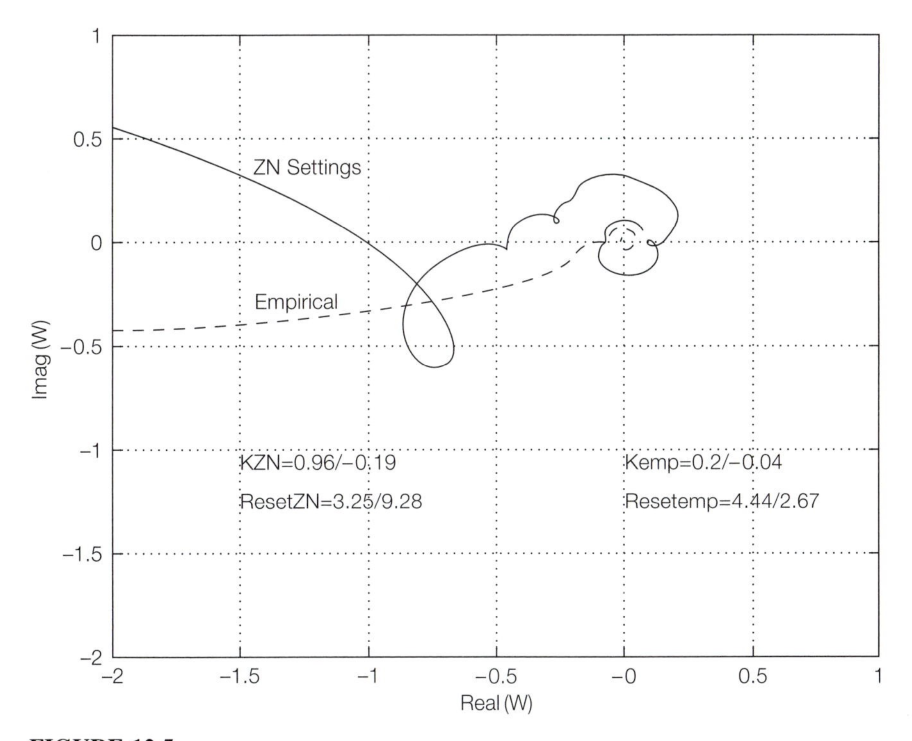

FIGURE 12.5
W function for Wood-Berry column with ZN and empirical tuning.

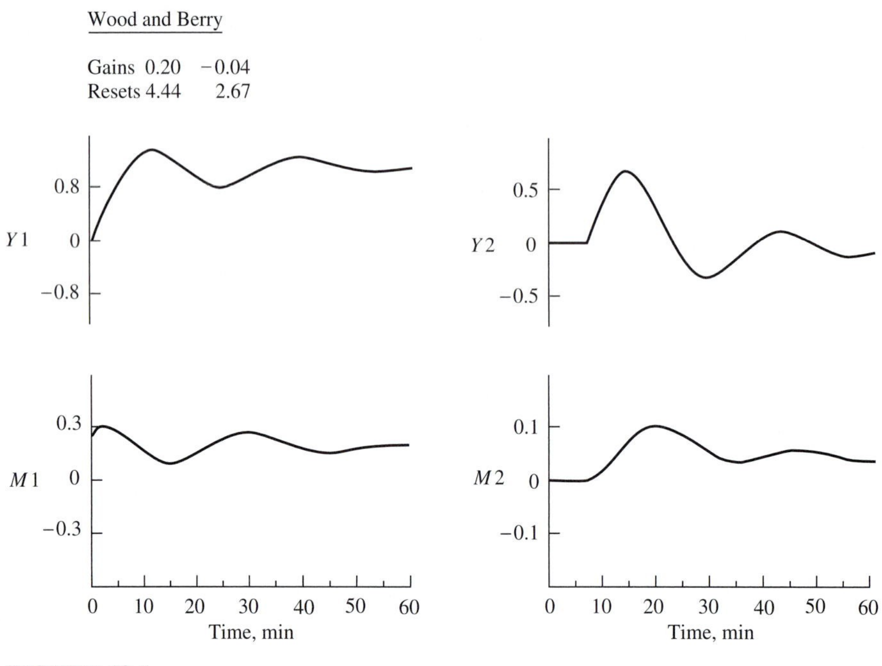

FIGURE 12.6

12.2.3 Niederlinski Index

A fairly useful stability analysis method is the Niederlinski index. It can eliminate unworkable pairings of variables at an early stage in the design. The settings of the controllers do not have to be known, but it applies only when integral action is used in all loops. It utilizes only the steady-state gains of the process transfer function matrix.

The method is a "necessary but not sufficient condition" for stability of a closed-loop system with integral action. If the index is negative, the system *will be* unstable for any controller settings (this is called "integral instability"). If the index is positive, the system may or may not be stable. Further analysis is necessary.

$$\text{Niederlinski index} = \text{NI} = \frac{\text{Det}[\underline{\underline{K_P}}]}{\prod_{j=1}^{N} K_{Pjj}} \tag{12.45}$$

where $\underline{\underline{K_P}} = \underline{\underline{G_{M(0)}}}$ = matrix of steady-state gains from the process openloop $\underline{\underline{G_M}}$ transfer function

$K_{P_{jj}}$ = diagonal elements in steady-state gain matrix

EXAMPLE 12.11. Calculate the Niederlinski index for the Wood and Berry column.

$$\underline{\underline{K_P}} = \underline{\underline{G_{M(0)}}} = \begin{bmatrix} 12.8 & -18.9 \\ 6.6 & -19.4 \end{bmatrix} \tag{12.46}$$

$$\text{NI} = \frac{\text{Det}[\underline{\underline{K_P}}]}{\prod_{j=1}^{N} K_{P_{jj}}} = \frac{(12.8)(-19.4) - (-18.9)(6.6)}{(12.8)(-19.4)} = 0.498 \tag{12.47}$$

Since the NI is positive, the closedloop system with the specified pairing *may* be stable.

Notice that pairing assumes that distillate composition x_D is controlled by reflux R and that bottoms composition x_B is controlled by vapor boilup V.

$$\begin{bmatrix} x_D \\ x_B \end{bmatrix} = \begin{bmatrix} 12.8 & -18.9 \\ 6.6 & -19.4 \end{bmatrix} \begin{bmatrix} R \\ V \end{bmatrix} \tag{12.48}$$

If the pairing had been reversed, the steady-state gain matrix would be

$$\begin{bmatrix} x_D \\ x_B \end{bmatrix} = \begin{bmatrix} -18.9 & 12.8 \\ -19.4 & 6.6 \end{bmatrix} \begin{bmatrix} V \\ R \end{bmatrix} \tag{12.49}$$

and the NI for this pairing would be

$$\text{NI} = \frac{\text{Det}[\underline{\underline{K_P}}]}{\prod_{j=1}^{N} K_{P_{jj}}} = \frac{(-18.9)(6.6) - (12.8)(-19.4)}{(-18.9)(6.6)} = -0.991$$

Therefore, the pairing of x_D with V and x_B with R gives a closedloop system that is "integrally unstable" for any controller tuning. ■

12.3 INTERACTION

Interaction among control loops in a multivariable system has been the subject of much research over the last 30 years. All of this work is based on the premise that interaction is undesirable. This is true for setpoint disturbances. We would like to change a setpoint in one loop without affecting the other loops. And if the loops do not interact, each individual loop can be tuned by itself, and the whole system should be stable if each individual loop is stable.

Unfortunately, much of this interaction analysis work has clouded the issue of how to design an effective control system for a multivariable process. In most process control applications the problem is not setpoint response but load response. We want a system that holds the process at the desired values in the face of load disturbances. Interaction is therefore not necessarily bad; in fact, in some systems it helps in rejecting the effects of load disturbances. Niederlinski (*AIChE Journal 17*:1261, 1971) showed in an early paper that the use of decouplers made the load rejection worse.

Therefore, the following discussions of the relative gain array (RGA) and decoupling are quite brief. We include them not because they are all that useful, but because they are part of the history of multivariable control. You should be aware of what they are and what their limitations are so that when you see them being misapplied (which, unfortunately, occurs quite often) you can be knowledgeably skeptical of the conclusions drawn.

12.3.1 Relative Gain Array

Undoubtedly the most discussed method for studying interaction is the RGA. It was proposed by Bristol (*IEEE Trans. Autom. Control AC-II:* 133, 1966) and has been extensively applied (and, in our opinion, often misapplied) by many workers. Detailed discussions are presented by Shinskey (*Process Control Systems,* 1967, McGraw-Hill, New York) and McAvoy (*Interaction Analysis,* 1983, Instr. Soc. America, Research Triangle Park, NC). The RGA has the advantage of being easy to calculate and requires only steady-state gain information.

A. Definition

The RGA is a matrix of numbers. The ijth element in the array is called β_{ij}. It is the ratio of the steady-state gain between the ith controlled variable and the jth manipulated variable when all other manipulated variables are constant, divided by the steady-state gain between the same two variables when all other controlled variables are constant.

$$\beta_{ij} = \frac{[Y_i/m_j]_{\overline{m}_k}}{[Y_i/m_j]_{\overline{Y}_k}} \tag{12.50}$$

For example, suppose we have a 2×2 system with the steady-state gains $K_{p_{ij}}$.

$$\begin{aligned} Y_1 &= K_{p_{11}} m_1 + K_{p_{12}} m_2 \\ Y_2 &= K_{p_{21}} m_1 + K_{p_{22}} m_2 \end{aligned} \tag{12.51}$$

For this system, the gain between Y_1 and m_1 when m_2 is constant is

$$[Y_1/m_1]_{\overline{m}_2} = K_{p_{11}}$$

The gain between Y_1 and m_1 when Y_2 is constant ($Y_2 = 0$) is found from solving the equations

$$\begin{aligned} Y_1 &= K_{p_{11}} m_1 + K_{p_{12}} m_2 \\ 0 &= K_{p_{21}} m_1 + K_{p_{22}} m_2 \end{aligned} \tag{12.52}$$

$$Y_1 = K_{p_{11}} m_1 + K_{p_{12}}[-K_{p_{21}} m_1/K_{p_{22}}]$$

$$Y_1 = \left[\frac{K_{p_{11}}K_{p_{22}} - K_{p_{12}}K_{p_{21}}}{K_{p_{22}}}\right] m_1 \tag{12.53}$$

$$[Y_1/m_1]_{\overline{Y}_2} = \left[\frac{K_{p_{11}}K_{p_{22}} - K_{p_{12}}K_{p_{21}}}{K_{p_{22}}}\right] \tag{12.54}$$

Therefore, the β_{11} term in the RGA is

$$\beta_{11} = \frac{K_{p_{11}}K_{p_{22}}}{K_{p_{11}}K_{p_{22}} - K_{p_{12}}K_{p_{21}}} \tag{12.55}$$

$$\beta_{11} = \frac{1}{1 - \dfrac{K_{p_{12}} K_{p_{21}}}{K_{p_{11}} K_{p_{22}}}} \tag{12.56}$$

EXAMPLE 12.12. Calculate the β_{11} element of the RGA for the Wood and Berry column.

$$\underline{\underline{K_P}} = \underline{\underline{G_{M(0)}}} = \begin{bmatrix} 12.8 & -18.9 \\ 6.6 & -19.4 \end{bmatrix}$$

$$\beta_{11} = \frac{1}{1 - \dfrac{K_{p_{12}} K_{p_{21}}}{K_{p_{11}} K_{p_{22}}}} = \frac{1}{1 - \dfrac{(-18.9)(6.6)}{(12.8)(-19.4)}} = 2.01$$

■

Equation (12.56) applies to only a 2×2 system. The elements of the RGA can be calculated for a system of any size by using the following equation.

$$\beta_{ij} = (ij\text{th element of } \underline{\underline{K_P}})(ij\text{th element of } [\underline{\underline{K_P}}^{-1}]^{\mathrm{T}}) \tag{12.57}$$

Note that Eq. (12.57) does *not* say that we take the ijth element of the product of the $\underline{\underline{K_P}}$ and $[\underline{\underline{K_P}}^{-1}]^{\mathrm{T}}$ matrices.

EXAMPLE 12.13. Use Eq. (12.57) to calculate all of the elements of the Wood and Berry column.

$$\underline{\underline{K_P}} = \begin{bmatrix} 12.8 & -18.9 \\ 6.6 & -19.4 \end{bmatrix}$$

$$\underline{\underline{K_P}}^{-1} = \frac{\begin{bmatrix} -19.4 & 18.9 \\ -6.6 & 12.8 \end{bmatrix}}{-123.58} \qquad [\underline{\underline{K_P}}^{-1}]^{\mathrm{T}} = \frac{\begin{bmatrix} -19.4 & -6.6 \\ 18.9 & 12.8 \end{bmatrix}}{-123.58}$$

$$\beta_{11} = (12.8)\left(\frac{-19.4}{-123.58}\right) = 2.01$$

This is the same result we obtained using Eq. (12.56).

$$\beta_{12} = (-18.9)\left(\frac{-6.6}{-123.58}\right) = -1.01$$

$$\beta_{21} = (6.6)\left(\frac{18.9}{-123.58}\right) = -1.01$$

$$\beta_{22} = (-19.4)\left(\frac{12.8}{-123.58}\right) = 2.01$$

$$\underline{\underline{\mathrm{RGA}}} = \begin{bmatrix} \beta_{11} & \beta_{12} \\ \beta_{21} & \beta_{22} \end{bmatrix} = \begin{bmatrix} 2.01 & -1.01 \\ -1.01 & 2.01 \end{bmatrix} \tag{12.58}$$

Note that the sum of the elements in each row is 1. The sum of the elements in each column is also 1. This property holds for any RGA, so in the 2×2 case we only have to calculate one element. ■

EXAMPLE 12.14. Calculate the RGA for the 3×3 system studied by Ogunnaike and Ray (*AIChE Journal 25:*1043, 1979).

TABLE 12.2
MATLAB program to calculate RGA for Ogunnaike-Ray column

```
% Program "rga.m"
% Calculates rga for OR column
% Give steady-state gain matrix
k=[0.66 -0.61 -0.0049
  1.11 -2.36 -0.012
  -34.68 46.2 0.87];
% Calculate matrix inverse
kinvers=inv(k);
% Take transpose
kintran=kinvers';
%
% Do term-by-term multiplication using ".*" operator
rga=k .* kintran
```

$$\underline{\underline{G_{M(s)}}} = \begin{bmatrix} \dfrac{0.66e^{-2.6s}}{6.7s+1} & \dfrac{-0.61e^{-3.5s}}{8.64s+1} & \dfrac{-0.0049e^{-s}}{9.06s+1} \\ \dfrac{1.11e^{-6.5s}}{3.25s+1} & \dfrac{-2.36e^{-3s}}{5s+1} & \dfrac{-0.012e^{-1.2s}}{7.09s+1} \\ \dfrac{-34.68e^{-9.2s}}{8.15s+1} & \dfrac{46.2e^{-9.4s}}{10.9s+1} & \dfrac{0.87(11.61s+1)e^{-s}}{(3.89s+1)(18.8s+1)} \end{bmatrix} \tag{12.59}$$

$$\underline{\underline{K_P}} = \begin{bmatrix} 0.66 & -0.61 & -0.0049 \\ 1.11 & -2.36 & -0.012 \\ -34.68 & 46.2 & 0.87 \end{bmatrix}$$

Table 12.2 gives a MATLAB program using Eq. (12.57) to calculate the elements of the 3×3 RGA matrix.

$$\underline{\underline{RGA}} = \begin{bmatrix} 1.96 & -0.66 & -0.30 \\ -0.67 & 1.89 & -0.22 \\ -0.29 & -0.23 & 1.52 \end{bmatrix} \tag{12.60}$$

Note that the sums of the elements in all rows and all columns are 1. ■

B. Uses and limitations

The elements in the RGA can be numbers that vary from very large negative values to very large positive values. If the RGA is close to 1, there should be little effect on the control loop by closing the other loops in the multivariable system. Therefore, there should be less interaction, so the proponents of the RGA claim that variables should be paired so that they have RGA elements near 1. Numbers around 0.5 indicate interaction. Numbers that are very large indicate interaction. Numbers that are negative indicate that the sign of the controller may have to be different when other loops are on automatic.

As pointed out earlier, the problem with pairings to avoid interaction is that interaction is not necessarily a bad thing. Therefore, the use of the RGA in deciding how to pair variables is not an effective tool for process control applications. Likewise, the use of the RGA in deciding what control structure (choice of manipulated and controlled variables) is best is not effective. What is important is the ability of the control system to keep the process at setpoint in the face of load disturbances. Thus, load rejection is the most important criterion for deciding what variables to pair and what controller structure is best.

The RGA is useful for avoiding poor pairings. If the diagonal element in the RGA is negative, the system may show integral instability—the same situation that we discussed in the context of the Niederlinski index. Very large values of the RGA indicate that the system can be quite sensitive to changes in the parameter values.

12.3.2 Decoupling

Some of the earliest work in multivariable control involved the use of decouplers to remove the interaction between the loops. Figure 12.7 gives the basic structure of the system. The decoupling matrix $\underline{\underline{D}}_{(s)}$ is chosen such that each loop does not affect the other. Figure 12.8 shows the details of a 2×2 system. The decoupling element D_{ij} can be selected in a number of ways. One of the most straightforward is to set $D_{11} = D_{22} = 1$ and design the D_{12} and D_{21} elements so that they cancel (in a feedforward way) the effect of each manipulated variable in the other loop. For example, suppose Y_1 is not at its setpoint but Y_2 is. The G_{C1} controller changes m_1 to drive Y_1 back to Y_1^{set}. But the change in m_1 disturbs Y_2 through the G_{M21} transfer function.

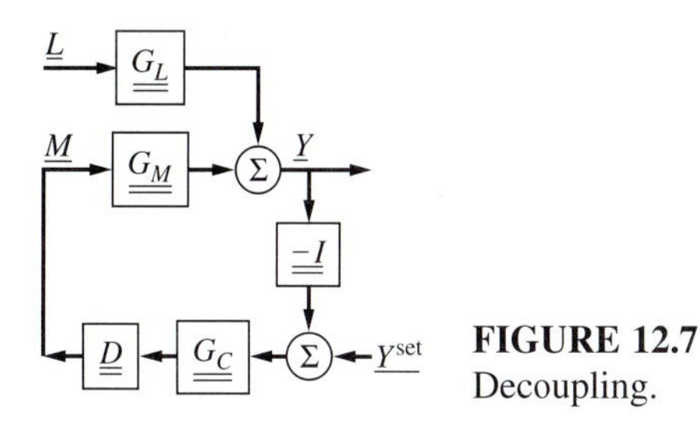

FIGURE 12.7
Decoupling.

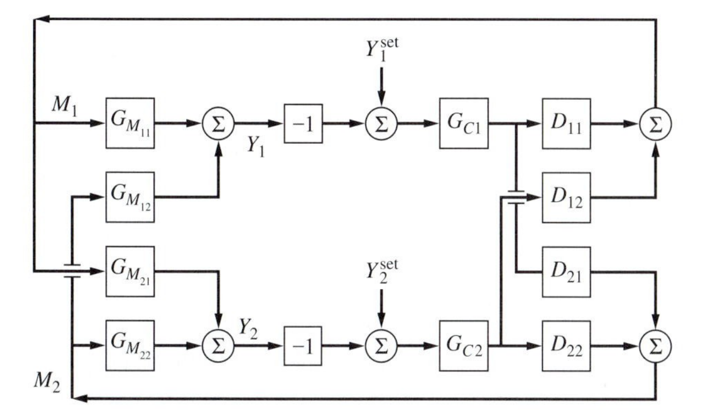

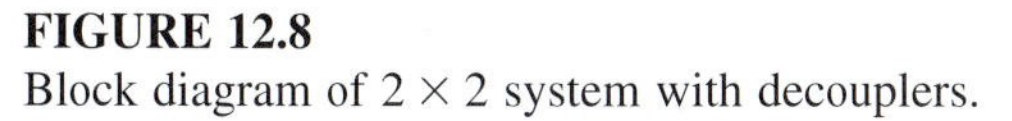

FIGURE 12.8
Block diagram of 2×2 system with decouplers.

If, however, the D_{21} decoupler element is set equal to $(-G_{M21}/G_{M22})$, there is a change in m_2 that comes through the G_{M22} transfer function and cancels out the effect of the change in m_1 on Y_2.

$$D_{21} = \frac{-G_{M21}}{G_{M22}} \tag{12.61}$$

Using the same arguments for the other loop, the D_{12} decoupler could be set equal to

$$D_{12} = \frac{-G_{M12}}{G_{M11}} \tag{12.62}$$

This "simplified decoupling" splits the two loops so that they can be independently tuned. Note, however, that the closedloop characteristic equations for the two loops are *not* $1 + G_{M11}G_{C1} = 0$ and $1 + G_{M22}G_{C2} = 0$. The presence of the decouplers changes the closedloop characteristic equations to

$$1 + G_{C1}\frac{G_{M11}G_{M22} - G_{M12}G_{M21}}{G_{M22}} = 0 \tag{12.63}$$

$$1 + G_{C2}\frac{G_{M11}G_{M22} - G_{M12}G_{M21}}{G_{M11}} = 0 \tag{12.64}$$

Other choices of decouplers are also possible. However, since decoupling may degrade the load rejection capability of the system, the use of decouplers is not recommended except in those cases where setpoint changes are the major disturbances.

12.4 CONCLUSION

The notation used for multivariable systems was reviewed in this chapter, and some important concepts were developed. The most important topic is the derivation of the characteristic equation for a closedloop multivariable process, which permits us to determine if the system is stable or unstable.

PROBLEMS

12.1. Wardle and Wood (*Inst. Chem. Eng. Symp. Ser. 32:*1, 1969) give the following transfer function matrix for an industrial distillation column:

$$\underline{\underline{G_M}} = \begin{bmatrix} \dfrac{0.126e^{-6s}}{60s+1} & \dfrac{-0.101e^{-12s}}{(48s+1)(45s+1)} \\[2ex] \dfrac{0.094e^{-8s}}{38s+1} & \dfrac{-0.12e^{-8s}}{35s+1} \end{bmatrix}$$

The empirical PI controller settings reported were

$$K_c = 18, -24 \qquad \tau_I = 19, 24$$

(*a*) Use a multivariable Nyquist plot to see if the system is closedloop stable.
(*b*) Calculate the values of the RGA and the Niederlinski index.

12.2. A distillation column has the following transfer function matrix:

$$\underline{\underline{G_M}} = \begin{bmatrix} \dfrac{34}{(54s+1)(0.5s+1)^2} & \dfrac{-44.7}{(114s+1)(0.5s+1)^2} \\ \dfrac{31.6}{(78s+1)(0.5s+1)^2} & \dfrac{-45.2}{(42s+1)(0.5s+1)^2} \end{bmatrix}$$

Empirical PI diagonal controller settings are:

$$K_c = 1.6, -1.6 \qquad \tau_I = 20, 9 \quad \text{minutes}$$

(*a*) Check the closedloop stability of the system using a multivariable Nyquist plot.
(*b*) Calculate values of the RGA and the Niederlinski index.

12.3. A distillation column is described by the following linear ODEs:

$$\frac{dx_D}{dt} = -4.74x_D + 5.99x_B + 0.708R - 0.472V$$

$$\frac{dx_B}{dt} = 10.84x_D - 18.24x_B + 1.28R - 1.92V + 4z$$

(*a*) Use state-variable matrix methods to derive the openloop transfer function matrix.
(*b*) What are the openloop eigenvalues of the system?
(*c*) If the openloop steady-state gain matrix is

$$\underline{\underline{K_p}} = \begin{bmatrix} 0.958 & -0.936 \\ 0.6390 & -0.661 \end{bmatrix}$$

calculate the RGA and the Niederlinski index.

12.4. A 2×2 process has the openloop transfer function matrix

$$\underline{\underline{G_M}} = \frac{1}{(s+1)} \begin{bmatrix} 2 & -1 \\ 2 & 2 \end{bmatrix}$$

A diagonal proportional feedback controller is used with both gains set equal to K_c. Time is in minutes.
(*a*) What is the openloop time constant of the system?
(*b*) Calculate the closedloop eigenvalues as functions of K_c.
(*c*) What value of K_c will give a closedloop time constant of 0.1 minute?

12.5. Air and water streams are fed into a pressurized tank through two control valves. Air flows out of the top of the tank through a restriction, and water flows out the bottom through another restriction. The linearized equations describing the system are

$$\frac{dP}{dt} = -0.8P + 0.5F_a + 0.1F_w$$

$$\frac{dh}{dt} = -0.4P - 0.1h + 0.5F_w$$

where P = pressure
h = liquid height
F_a = air flow rate into tank
F_w = water flow rate into tank

Use state variable methods to calculate:
(*a*) The openloop eigenvalues and the openloop transfer function matrix.
(*b*) The closedloop eigenvalues if two proportional SISO controllers are used with gains of 5 (for the pressure controller manipulating air flow) and 2 (for the level controller manipulating water flow).
(*c*) Calculate the RGA and the Niederlinski index for this system.

12.6. A 2 × 2 process has the openloop transfer function matrix

$$\underline{\underline{G_M}} = \frac{1}{(s+1)}\begin{bmatrix} 2 & 1 \\ 1 & 2 \end{bmatrix}$$

A diagonal proportional feedback controller is used with both gains set equal to K_c. Time is in minutes.
(*a*) What is the openloop eigenvalue of the system?
(*b*) What value of K_c will give a minimum closedloop eigenvalue of -10?

12.7. Calculate the RGA, Niederlinski index, minimum singular value, and condition number of the following matrix of steady-state gains.

$$\begin{bmatrix} -37.7 & 0.647 \\ -43.8 & -1.71 \end{bmatrix}$$

12.8. An openloop system is described by the following two differential equations:

$$5\frac{dx_1}{dt} + x_1 = K_p m$$

$$\frac{dx_2}{dt} + x_2 = x_1$$

(*a*) Calculate the openloop eigenvalues of the system.
(*b*) Suppose we use a multivariable proportional controller of the type

$$m = K_{c1}x_1 + K_{c2}x_2$$

Derive the closedloop characteristic equation as a polynomial in λ in terms of the process gain K_p and the controller gains K_{c1} and K_{c2}.
(*c*) What values must the controller gains have to position the closedloop eigenvalues at -2?
(*d*) Repeat (*c*) to position them at -5.

12.9. A distillation column has the following transfer function matrix relating controlled variables $x_{D2(2)}$ and $x_{B2(3)}$ with manipulated variables R_2 and Q_{R2}:

$$\underline{\underline{G_M}} = \begin{bmatrix} \dfrac{1.6e^{-1.3s}}{(13s+1)(3s+1)} & \dfrac{-1.2e^{-1.05s}}{(15.5s+1)(3s+1)} \\ \dfrac{-7.5e^{-2.3s}}{(37.3s+1)(2s+1)} & \dfrac{23.1e^{-s}}{(42s+1)(2s+1)} \end{bmatrix}$$

(*a*) Calculate values of the RGA and the minimum singular value.
(*b*) Calculate the Niederlinski index for the two possible pairings:
 (*i*) $x_{D2(2)}-R_2$ and $x_{B2(3)}-Q_{R2}$
 (*ii*) $x_{D2(2)}-Q_{R2}$ and $x_{B2(3)}-R_2$

12.10. The dynamics of a Patriot missile launcher located somewhere in Saudi Arabia to shoot down incoming SCUD missiles are given by the following openloop transfer function between θ (the angle between the horizon and the missile direction) and M (the power to the motor that positions the launcher):

$$\frac{\theta}{M} = G_{M(s)} = \frac{1}{s(s+1)}$$

(*a*) Convert the openloop Patriot missile launcher from Laplace-domain transfer function form to state-space form

$$\underline{\dot{x}} = \underline{\underline{A}}\,\underline{x} + \underline{B}m$$

where $x_1 = \theta$ and $x_2 = \dot{\theta}$.
(*b*) What are the openloop eigenvalues λ_0 of the system?
(*c*) Suppose a state feedback controller is used: $m = K_1\theta + K_2\dot{\theta}$. Calculate the values of the controller gains that will position the closedloop eigenvalues at $\lambda_c = -5$ in the complex plane.

12.11. A process has an openloop transfer function relating controlled and manipulated variables that is a steady-state gain K_p and two first-order lags in series (τ_1 and τ_2).
(*a*) Convert this process to openloop state-space form. What are the openloop eigenvalues?
(*b*) Calculate the gains required in a state feedback controller that will position both closedloop eigenvalues at τ_{CL}. Your answer should be general equations in terms of the openloop parameters.

12.12. A multivariable process is described by two ODEs:

$$2\frac{dx_1}{dt} + x_1 = 0.5x_2 + 2m_1 + 0.5m_2$$

$$10\frac{dx_2}{dt} + x_2 = 1.5x_1 + m_1 + 3m_2$$

(*a*) What is the characteristic equation of the system?
(*b*) What are the eigenvalues of the system?
(*c*) Put into state variable form and calculate the $\underline{\underline{A}}$ and $\underline{\underline{B}}$ matrices.
(*d*) Calculate the $\underline{\underline{G_{M(s)}}}$ transfer function matrix.
(*e*) Calculate the $\overline{\text{RGA}}$ and the Niederlinski index.

CHAPTER 13

Design of Controllers for Multivariable Processes

In the last chapter we developed some mathematical tools and some methods of analyzing multivariable closedloop systems. This chapter studies the development of control structures for these processes. Because of their widespread use in real industrial applications, conventional diagonal control structures are discussed. These systems, which are also called *decentralized* control, consist of multiloop SISO controllers with one controlled variable paired with one manipulated variable. The major idea in this chapter is that these SISO controllers should be tuned simultaneously, with the interactions in the process taken into account.

13.1 PROBLEM DEFINITION

Most industrial control systems use the multiloop SISO diagonal control structure. It is the most simple and understandable structure. Operators and plant engineers can use it and modify it when necessary. It does not require an expert in applied mathematics to design and maintain it. In addition, the performance of these diagonal controller structures is usually quite adequate for process control applications. In fact, there has been little quantitative unbiased data showing that the performances of the more sophisticated controller structures are really any better! The slight improvement is seldom worth the price of the additional complexity and engineering cost of implementation and maintenance.

A number of critical questions must be answered in developing a control system for a plant. What should be controlled? What should be manipulated? How should the controlled and manipulated variables be paired in a multivariable plant? How do we tune the controllers?

The procedure discussed in this chapter provides a practical approach to answering these questions. It was developed to provide a workable, stable, simple

SISO system with only a modest amount of engineering effort. The resulting diagonal controller can then serve as a realistic benchmark, against which the more complex multivariable controller structures can be compared.

The limitations of the procedure should be pointed out. It does not apply to openloop-unstable systems. It also does not work well when the time constants of the transfer functions are quite different, i.e., some parts much faster than others. The fast and slow sections should be designed separately in such a case.

The procedure has been tested primarily on realistic distillation column models. This choice was deliberate because most industrial processes have similar gain, deadtime, and lag transfer functions. Undoubtedly, some pathological transfer functions can be found that the procedure cannot handle. But we are interested in a practical engineering tool, not elegant, rigorous, all-inclusive mathematical theorems.

The steps in the procedure are summarized below. Each step is discussed in more detail in later sections of this chapter.

1. *Select controlled variables.* Use primarily engineering judgment based on process understanding.
2. *Select manipulated variables.* Find the set of manipulated variables that gives the largest minimum singular value of the steady-state gain matrix.
3. *Eliminate unworkable variable pairings.* Eliminate pairings with negative Niederlinski indices or that have obvious poor dynamic relationships.
4. *Find the best pairing from the remaining sets.*
   a. Tune all combinations using BLT tuning.
   b. Select the pairing that gives the lowest-magnitude closedloop regulator transfer function.

13.2 SELECTION OF CONTROLLED VARIABLES

13.2.1 Engineering Judgment

Engineering judgment is the principal tool for deciding what variables to control. A good understanding of the process leads in most cases to a logical choice of what needs to be controlled. Considerations such as economics, safety, constraints, and the availability and reliability of sensors must be factored into this decision. We must control inventories (liquid levels and gas pressures), product qualities, and production rate.

For example, in a distillation column we are usually interested in controlling the purity of the distillate and bottoms product streams. In chemical reactors, heat exchangers, and furnaces the usual controlled variable is temperature. In most cases these choices are fairly obvious. It should be remembered that controlled variables need not be simple, directly measured variables. They can also be computed from a number of sensor inputs. Common examples are heat removal rates, mass flow rates, and ratios of flow rates.

However, sometimes selection of the appropriate controlled variable is not so easy. For example, in a distillation column it is frequently difficult and expensive to measure product compositions directly with sensors such as gas chromatographs. Instead, temperatures on various trays are controlled. The selection of the best control tray to use requires a considerable amount of knowledge about the column, its operation, and its performance. Varying amounts of non-key components in the feed can significantly affect the best choice of control trays.

13.2.2 Singular Value Decomposition

The use of singular value decomposition (SVD), introduced into chemical engineering by Moore and Downs (*Proc. JACC,* Paper WP-7C, 1981), gives some guidance on the question of what variables to control. They used SVD to select the best tray temperatures in a distillation column. SVD involves expressing the matrix of plant transfer function steady-state gains $\underline{\underline{K_p}}$ as the product of three matrices: a $\underline{\underline{U}}$ matrix, a diagonal $\underline{\underline{\Sigma}}$ matrix, and a $\underline{\underline{V}}^{\mathrm{T}}$ matrix.

$$\underline{\underline{K_p}} = \underline{\underline{U}}\,\underline{\underline{\Sigma}}\,\underline{\underline{V}}^{\mathrm{T}} \tag{13.1}$$

The diagonal $\underline{\underline{\Sigma}}$ matrix contains as its elements the singular values of the $\underline{\underline{K_p}}$ matrix. The biggest elements in each column of the $\underline{\underline{U}}$ matrix indicate which outputs of the process are the most sensitive. Thus, SVD can be used to help select which tray temperatures in a distillation column should be controlled. The following example from the Moore and Downs paper illustrates the procedure.

EXAMPLE 13.1. A nine-tray distillation column separating isopropanol and water has the following steady-state gains between tray temperatures and the manipulated variables reflux R and heat input Q.

Tray number	$\Delta T_n/\Delta R$	$\Delta T_n/\Delta Q$
9	−0.00773271	0.0134723
8	−0.2399404	0.2378752
7	−2.5041590	2.4223120
6	−5.9972530	5.7837800
5	−1.6773120	1.6581630
4	0.0217166	0.0259478
3	0.1976678	−0.1586702
2	0.1289912	−0.1068900
1	0.0646059	−0.0538632

The entries in this table are the elements in the steady-state gain matrix of the column $\underline{\underline{K_p}}$, which has nine rows and two columns.

$$\begin{bmatrix} \Delta T_9 \\ \Delta T_8 \\ \vdots \\ \Delta T_1 \end{bmatrix} = \underline{\underline{K_p}} \begin{bmatrix} R \\ Q \end{bmatrix} \tag{13.2}$$

Now $\underline{\underline{K_p}}$ is decomposed into the product of three matrices.

$$\underline{\underline{K_p}} = \underline{\underline{U}}\,\underline{\underline{\Sigma}}\,\underline{\underline{V}}^{\mathrm{T}} \tag{13.3}$$

$$\underline{\underline{U}} = \begin{bmatrix} -0.0015968 & -0.0828981 \\ -0.0361514 & -0.0835548 \\ -0.3728142 & -0.0391486 \\ -0.8915611 & 0.1473784 \\ -0.2523673 & -0.6482796 \\ -0.0002581 & -0.6482796 \\ 0.0270092 & -0.4463671 \\ 0.0178741 & -0.2450451 \\ 0.0089766 & -0.1182182 \end{bmatrix} \tag{13.4}$$

$$\underline{\underline{\Sigma}} = \begin{bmatrix} 9.3452 & 0 \\ 0 & 0.052061 \end{bmatrix} \tag{13.5}$$

$$\underline{\underline{V}}^{\mathrm{T}} = \begin{bmatrix} 0.7191619 & -0.6948426 \\ -0.6948426 & -0.7191619 \end{bmatrix} \tag{13.6}$$

The largest element in the first column in $\underline{\underline{U}}$ is -0.8915611, which corresponds to tray 6. Therefore, SVD would suggest the control of tray 6 temperature. ■

The software to do the SVD calculations is readily available (*Computer Methods for Mathematical Computations* by Forsythe, Malcolm, and Moler, 1977, Prentice Hall, Englewood Cliffs, NJ), and they can be easily performed using MATLAB.

13.3 SELECTION OF MANIPULATED VARIABLES

Once the controlled variables have been specified, the control structure depends only on the choice of manipulated variables. For a given process, selecting different manipulated variables will produce different control structure alternatives. These *control* structures are independent of the *controller* structure, i.e., pairing of variables in a diagonal multiloop SISO structure or one multivariable controller.

For example, in a distillation column the manipulated variables could be the flow rates of reflux and vapor boilup (R, V) to control distillate and bottoms compositions. This choice gives one possible control structure. Alternatively, we could choose to manipulate the flow rates of distillate and vapor boilup (D, V). This yields another control structure for the same basic distillation process.

The set of manipulated variables that gives the *largest* minimum singular value $\sigma^{\min}$ of the steady-state gain matrix is the best. In Chapter 9 we showed that a perfect controller would be the inverse of the plant. Therefore, we want a plant that can be easily inverted (not close to singular). Since the minimum singular value measures how close the gain matrix is from being singular, the choice of manipulated variables should be based on finding the set that gives a plant with a large minimum singular value.

The difference in these values must be fairly large to be meaningful. If one set of manipulated variables has a $\sigma^{\min}$ of 10 and another set has a $\sigma^{\min}$ of 1, you can conclude that the first set is better. However, if the two sets give numbers that are close, such as 10 and 8, there is probably little difference, and the two sets of manipulated variables may be equally effective.

This selection of control structure is independent of variable pairing and controller tuning. The minimum singular value is a measure of the inherent ability of the process (with the specified choice of manipulated variables) to handle disturbances, changes in operating conditions, etc.

The problem of the effect of scaling on singular values is handled by expressing the gains of all the plant transfer functions in dimensionless form. The gains with engineering units are divided by transmitter spans and multiplied by valve gains. This yields the dimensionless gain that the controller sees and has to cope with.

13.4 ELIMINATION OF POOR PAIRINGS

The Niederlinski index (Section 12.2.4) can be used to eliminate some of the pairings. Negative values of this index mean unstable pairings, independent of controller tuning. As illustrated in Example 12.12, pairing x_D with V and x_B with R in the Wood and Berry column gives a negative Niederlinski index, so this pairing should not be used.

For a 2×2 system, a negative Niederlinski index is equivalent to pairing on a negative RGA element. For example, in Example 12.13 the β_{11} RGA element is positive (2.01). This says that the x_D to R pairing is okay. However, the β_{12} element is negative (−1.01), telling us that the x_D to V pairing is not okay.

Probably the most important method for eliminating poor pairings is the use of a little common sense. We know that more lags or bigger deadtimes in a control loop lead to poorer performance. We also know that the manipulated variable should be able to cause a significant change in the controlled variable; i.e., we need a big stick. In addition, we know that the manipulated variable should affect the controlled variable quickly.

Therefore, we should pair variables that are related through low-order transfer functions having large steady-state gains, small time constants, and small deadtimes. A number of dynamically poor pairings can be eliminated by inspection.

For example, in distillation we generally do not attempt to control a temperature or composition in the base of the column by manipulating reflux. There is typically a liquid hydraulic lag of 6 seconds per tray, so a change in reflux to the top of a 50-tray column does not change the liquid flow at the bottom of the column for about 5 minutes. The dynamic performance of this loop is poor, so we do not pair bottoms composition with reflux no matter what the RGA tells us to do. On the other hand, the vapor boilup affects all sections of the column quite quickly, so it can be paired with a controlled variable at the top of the column with no dynamic problem.

13.5 BLT TUNING

One of the major questions in multivariable control is how to tune controllers in a diagonal multiloop SISO system. If PI controllers are used, there are $2N$ tuning parameters to be selected. The gains and reset times must be specified so that the overall system is stable and gives acceptable load responses. Once a consistent and rational tuning procedure is available, the pairing problem can be attacked.

The tuning procedure discussed in this section (called BLT, biggest log-modulus tuning) provides such a standard tuning methodology. It satisfies the objective of arriving at reasonable controller settings with only a small amount of engineering and computational effort. We do not claim that the method produces the best possible results or that some other tuning or controller structure will not give superior performance. However, the method is easy to use, is easily understandable by control engineers, and leads to settings that compare very favorably with the empirical settings found by the exhaustive and expensive trial-and-error tuning methods used in many studies.

The method should be viewed in the same light as the classical SISO Ziegler-Nichols method. It gives reasonable settings that provide a starting point for further tuning and a benchmark for comparative studies.

BLT tuning involves the following four steps:

1. Calculate the Ziegler-Nichols settings for each individual loop. The ultimate gain and ultimate frequency ω_u of each diagonal transfer function $G_{jj(s)}$ are calculated in the classical SISO way. To do this numerically, a value of frequency ω is guessed. The phase angle is calculated, and the frequency is varied to find the point where the Nyquist plot of $G_{jj(i\omega)}$ crosses the negative real axis (phase angle is $-180°$). The frequency where this occurs is ω_u. The reciprocal of the real part of $G_{jj(i\omega)}$ is the ultimate gain.
2. A detuning factor F is assumed. F should always be greater than 1. Typical values are between 1.5 and 4. The gains of *all* feedback controllers K_{ci} are calculated by *dividing* the Ziegler-Nichols gains K_{ZNi} by the factor F.

$$K_{ci} = \frac{K_{ZNi}}{F} \tag{13.7}$$

where $K_{ZNi} = K_{ui}/2.2$. Then all feedback controller reset times τ_{Ii} are calculated by multiplying the Ziegler-Nichols reset times τ_{ZN} by the same factor F.

$$\tau_{Ii} = \tau_{ZNi} F \tag{13.8}$$

where

$$\tau_{ZNi} = \frac{2\pi}{1.2\omega_{ui}} \tag{13.9}$$

The F factor can be considered as a detuning factor that is applied to all loops. The larger the value of F, the more stable the system is, but the more sluggish are the setpoint and load responses. The method yields settings that give a reasonable compromise between stability (robustness) and performance (quickness) in multivariable systems.

3. Based on the guessed value of F and the resulting controller settings, a multivariable Nyquist plot of the scalar function $W_{(i\omega)} = -1 + \text{Det}[\underline{\underline{I}} + \underline{\underline{G_{M(i\omega)}}}\,\underline{\underline{G_{C(i\omega)}}}]$ is made. [See Section 12.2.2, Eq. (12.42).] The closer this contour is to the $(-1, 0)$ point, the closer the system is to instability. The quantity $W/(1 + W)$ is similar to the closedloop servo transfer function for a SISO loop $G_M G_C/(1 + G_M G_C)$. Therefore, based on intuition and empirical grounds, we define a multivariable closedloop log modulus L_{cm}.

$$L_{cm} = 20 \log_{10} \left| \frac{W}{1 + W} \right| \tag{13.10}$$

The peak in the plot of L_{cm} over the entire frequency range is the biggest log modulus $L_{cm}^{\max}$.

4. The F factor is varied until $L_{cm}^{\max}$ is equal to $2N$, where N is the order of the system. For $N = 1$, the SISO case, we get the familiar $+2$ dB maximum closedloop log modulus criterion. For a 2×2 system, a $+4$ dB value of $L_{cm}^{\max}$ is used; for a 3×3, $+6$ dB; and so forth. This empirically determined criterion has been tested on a large number of cases and gives reasonable performance that is a little on the conservative side.

This tuning method should be viewed as giving preliminary controller settings that can be used as a benchmark for comparative studies. The procedure guarantees that the system is stable with all controllers on automatic and also that each individual loop is stable if all others are on manual (the F factor is limited to values greater than 1, so the settings are always more conservative than the Ziegler-Nichols values). Thus, a portion of the integrity question is automatically answered. However, further checks of stability would have to be made for other combinations of manual and automatic operation.

The method weights each loop equally; i.e., all loops are equally detuned. If it is important to keep tighter control of some variables than others, the method can be easily modified by using different weighting factors for different controlled variables. The less important loop could be detuned more than the more important loop.

Table 13.1 gives a MATLAB program that calculates the BLT tuning for the Wood and Berry column, and Fig. 13.1 shows the $W_{(i\omega)}$ plot with the BLT settings. The program forms the numerator and denominator polynomials for all four openloop process transfer functions. Then a "while" loop is used to converge on the value of f that gives *dbmax*=4. At each value of f, the controller constants are calculated and the controller numerator and denominator polynomials are formed. Then the process and controller transfer functions are calculated using the *polyval* operation over a range of frequencies (using a "for" loop). The *wnyquist* is calculated at each frequency. The peak in the log modulus curve is picked off using the *max*(*dbcl*) operation. Simple interval halving is used to change the guessed value of f at each pass through the loop.

The resulting controller settings are compared with the empirical setting in Table 13.2. The BLT settings usually have larger gains and larger reset times than the

TABLE 13.1
MATLAB program for BLT tuning

```
% Program "bltwb.m"
% does BLT tuning for the 2x2 Wood and Berry column
%
% Define transfer functions without deadtimes
numg11=12.8;
deng11=[16.7 1];
numg12=-18.9;
deng12=[21 1];
numg21=6.6;
deng21=[10.9 1];
numg22=-19.4;
deng22=[14.4 1];
d=[1 3
   7 3];
% Give ZN settings
kczn=[0.96  0
      0 -0.19];
resetzn=[3.25 0
         0  9.28];
i=sqrt(-1);
w=logspace(-1,0.8,200);
s=i*w;
f=2.54;
df=0.01;
loop=0;
flagm=-1;
flagp=-1;
dbmax=-100;
%
% Main loop to vary f in BLT tuning
%
while abs(dbmax-4)>0.05
kc=kczn/f;
reset=resetzn*f;
% Form controller transfer function
numgc11=kc(1,1)*[reset(1,1) 1];
dengc11=[reset(1,1) 0];
numgc22=kc(2,2)*[reset(2,2) 1];
dengc22=[reset(2,2) 0];
%
% Inside loop to vary frequency
nwtot=length(w);
for nw=1:nwtot
wn=w(nw);
% Process g's
g(1,1)=polyval(numg11,s(nw)) / polyval(deng11,s(nw));
g(1,1)=g(1,1)*exp(-d(1,1)*s(nw));
g(1,2)=polyval(numg12,s(nw)) / polyval(deng12,s(nw));
g(1,2)=g(1,2)*exp(-d(1,2)*s(nw));
g(2,1)=polyval(numg21,s(nw)) / polyval(deng21,s(nw));
g(2,1)=g(2,1)*exp(-d(2,1)*s(nw));
g(2,2)=polyval(numg22,s(nw)) / polyval(deng22,s(nw));
g(2,2)=g(2,2)*exp(-d(2,2)*s(nw));
```

TABLE 13.1 (CONTINUED)
MATLAB program for BLT tuning

```
% Controller gc's
gc(1,1)=polyval(numgc11,s(nw)) / polyval(dengc11,s(nw));
gc(1,2)=0;
gc(2,1)=0;
gc(2,2)=polyval(numgc22,s(nw)) / polyval(dengc22,s(nw));
% Calculate w function
% "eye" operation forms an identity matrix
wnyquist(nw)=-1+det(eye(size(g))+g*gc);
% Calculate lc function
lc(nw)=wnyquist(nw)/(1+wnyquist(nw));
dbcl(nw)=20*log10(abs(lc(nw)));
%
% End of inside loop sweeping through frequencies
end
% Pick off peak in closedloop log modulus

[dbmax,nmax]=max(dbcl);
wmax=w(nmax);
%
loop=loop+1;
if loop>10,break,end
%
% Test if +4 dB and reguess f factor
%
if dbmax>4
if flagp>0,df=df/2;end
flagm=1;
f=f+df;
else
if flagm>0,df=df/2;end
flagp=1;
f=f-df;
if f<1,f=1;end
end
% End of "while" loop to find correct f factor
end
%
% Plot W function
clf
axis('equal');
plot(real(wnyquist),imag(wnyquist))
axis([-2 1 -2 1]);
xlabel('Real(W)')
ylabel('Imag(W)')
grid
text(-1.8,.4,['f=',num2str(f)])
text(-1.8,.2,['Kc=',num2str(kc(1,1)),'/',num2str(kc(2,2))]);
text(-1.8,0,['Reset=',num2str(reset(1,1)),'/',num2str(reset(2,2))]);
pause
print -dps pfig131.p
```

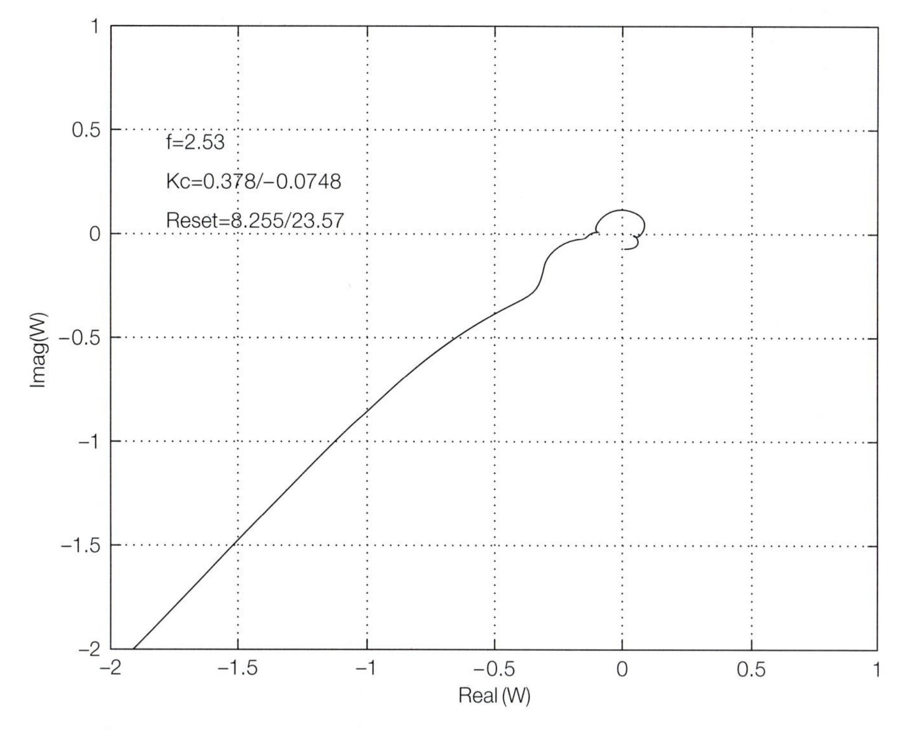

FIGURE 13.1
W function for WB column with BLT tuning.

TABLE 13.2
BLT, Ziegler-Nichols, and empirical controller tuning

	Wood and Berry	Ogunnaike and Ray
	Empirical	
K_c	0.2/−0.04	1.2/−0.15/0.6
τ_I	4.44/2.67	5/10/4
	Z-N	
K_c	0.96/−0.19	3.24/−0.63/5.66
τ_I	3.25/9.2	7.62/8.36/3.08
	BLT	
L_{cm}^{max} (dB)	+4	+6
F factor	2.55	2.15
K_c	0.375/−0.075	1.51/−0.295/2.63
τ_I	8.29/23.6	16.4/18/6.61

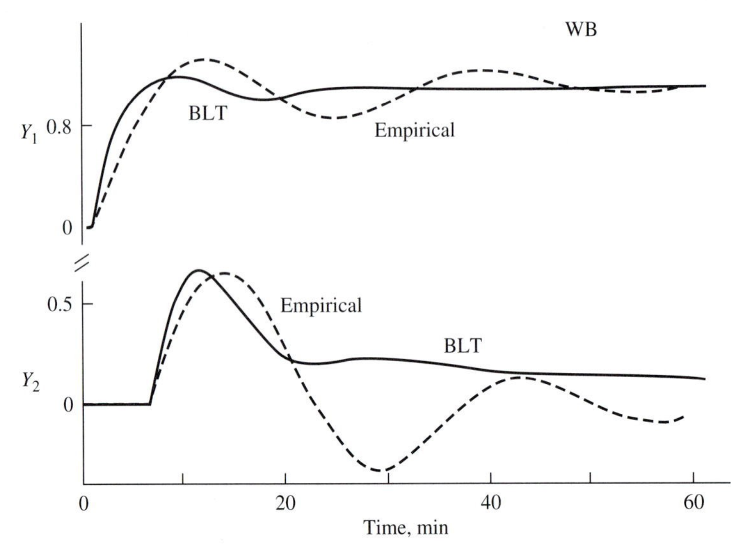

FIGURE 13.2
Wood and Berry column: BLT and empirical tuning.

empirical. Time responses using the two sets of controller tuning parameters are compared in Fig. 13.2.

Results for the 3×3 Ogunnaike and Ray column are given in Table 13.2 and in Fig. 13.3. The "+4" and "+6" refer to the value of $L_{cm}^{\max}$ used. Both of these cases illustrate that the BLT procedure gives reasonable controller settings.

As noted previously, if the process transfer functions have greatly differing time constants, the BLT procedure does not work well; it tends to give a response that is too oscillatory. The problem can be handled by breaking up the system into fast and slow sections and applying BLT to each subsection.

The BLT procedure was applied with PI controllers. The method can be extended to include derivative action (PID controllers) by using two detuning factors: F detunes the ZN reset and gain values, and F_D detunes the ZN derivative value. The optimum value of F_D is that which gives the minimum value of F and still satisfies the $+2N$ maximum closedloop log modulus criterion (see the paper by Monica, Yu, and Luyben in *IEC Res.* *27:*969, 1988).

13.6 LOAD REJECTION PERFORMANCE

In most chemical processes the principal control problem is load rejection. We want a control system that keeps the controlled variables at or near their setpoints in the face of load disturbances. Thus, the closedloop regulator transfer function is the most important.

The ideal closedloop relationship between the controlled variable and the load is zero. Of course, this can never be achieved, but the smaller the magnitude of the

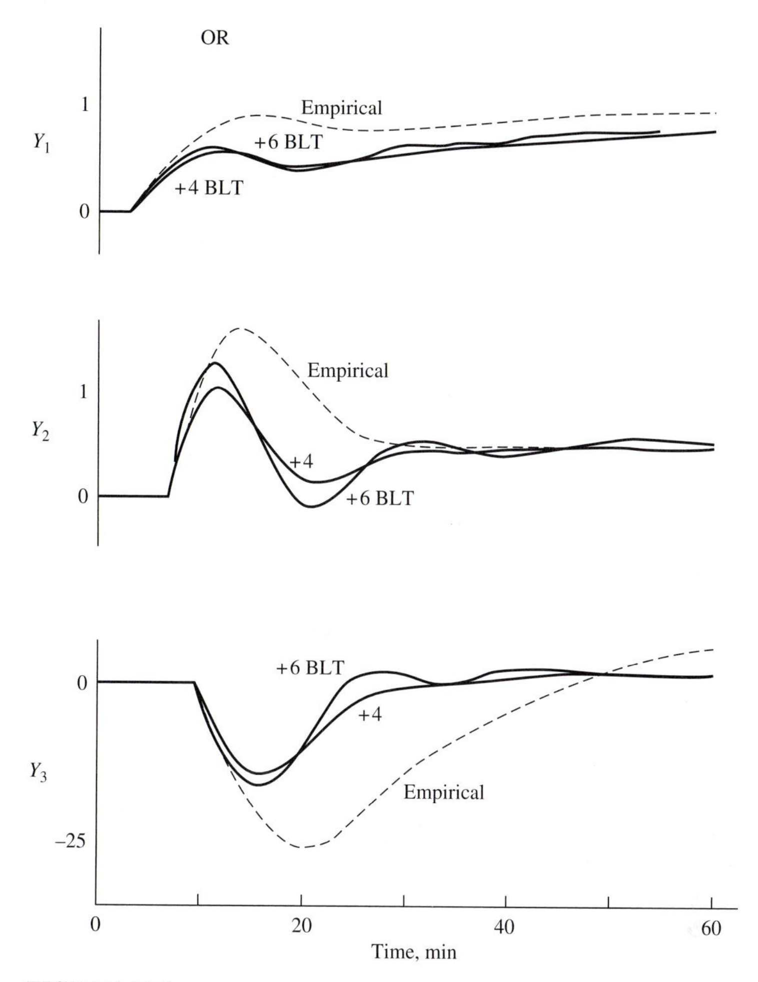

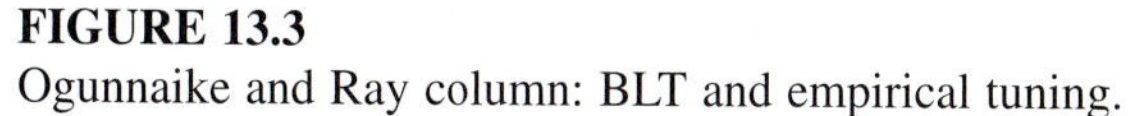

FIGURE 13.3
Ogunnaike and Ray column: BLT and empirical tuning.

closedloop regulator transfer function, the better the control. Thus, a rational criterion for selecting the best pairing of variables is to choose the one that gives the smallest peaks in a plot of the elements of the closedloop regulator transfer function matrix.

The closedloop relationships for a multivariable process were derived in Chapter 12 [Eq. (12.21)].

$$\underline{Y} = \left[[\underline{\underline{I}} + \underline{\underline{G_{M(s)}}}\,\underline{\underline{G_{C(s)}}}]^{-1} \underline{\underline{G_{M(s)}}}\,\underline{\underline{G_{C(s)}}} \right] \underline{Y}^{\text{set}} + \left[[\underline{\underline{I}} + \underline{\underline{G_{M(s)}}}\,\underline{\underline{G_{C(s)}}}]^{-1} \underline{G_{L(s)}} \right] L_{(s)} \tag{13.11}$$

For a 3×3 system there are three elements in the $\underline{x}$ vector, so three curves are plotted for each of the three elements in the vector $[[\underline{\underline{I}} + \underline{\underline{G_{M(i\omega)}}}\,\underline{\underline{G_{C(i\omega)}}}]^{-1} \quad \underline{G_{L(i\omega)}}]$ for the specified load variable L. Table 13.3 gives a MATLAB program that calculates these curves for the 3×3 Ogunnaike and Ray column. Figure 13.4 plots the log modulus of the three closedloop regulator transfer functions.

TABLE 13.3
Load rejection for the Ogunnaike-Ray column

```
% Program "lreject.m"
% calculates closedloop regulator transfer functions
% for Ogunnaike and Ray 3x3 column
%
% Define GM transfer functions without deadtimes
numg11=0.66;
deng11=[6.7 1];
numg12=-0.61;
deng12=[8.64 1];
numg13=-0.0049;
deng13=[9.06 1];

numg21=1.11;
deng21=[3.25 1];
numg22=-2.36;
deng22=[5 1];
numg23=-0.012;
deng23=[7.09 1];

numg31=-34.68;
deng31=[8.15 1];
numg32=46.2;
deng32=[10.9 1];
numg33=0.87*[11.61 1];
deng33=conv([3.89 1],[18.8 1]);

d=[2.6 3.5 1
   6.5 3   1.2
   9.2 9.4 1];
%
% Give empirical settings
kc=[1.2 0 0
    0 -0.15 0
    0 0 0.6];
reset=[5 0  0
       0 10 0
       0 0  4];
%
% Form GL transfer functions
numgL1=0.14;
dengL1=[6.2 1];
numgL2=0.53;
dengL2=[6.9 1];
numgL3=-11.54;
dengL3=[7.01 1];
% GL deadtimes
dgL=[12 10.5 0.6];
%
i=sqrt(-1);
w=logspace(-2,1,90);
s=i*w;
```

TABLE 13.3 (CONTINUED)
Load rejection for the Ogunnaike-Ray column

```
% Form controller transfer function
numgc1=kc(1,1)*[reset(1,1) 1];
dengc1=[reset(1,1) 0];
numgc2=kc(2,2)*[reset(2,2) 1];
dengc2=[reset(2,2) 0];
numgc3=kc(3,3)*[reset(3,3) 1];
dengc3=[reset(3,3) 0];
%
% Loop to vary frequency
nwtot=length(w);
for nw=1:nwtot
wn=w(nw);

% Process GM tranfer functions
g(1,1)=polyval(numg11,s(nw)) / polyval(deng11,s(nw));
g(1,1)=g(1,1)*exp(-d(1,1)*s(nw));
g(1,2)=polyval(numg12,s(nw)) / polyval(deng12,s(nw));
g(1,2)=g(1,2)*exp(-d(1,2)*s(nw));
g(1,3)=polyval(numg13,s(nw)) / polyval(deng13,s(nw));
g(1,3)=g(1,3)*exp(-d(1,3)*s(nw));

g(2,1)=polyval(numg21,s(nw)) / polyval(deng21,s(nw));
g(2,1)=g(2,1)*exp(-d(2,1)*s(nw));
g(2,2)=polyval(numg22,s(nw)) / polyval(deng22,s(nw));
g(2,2)=g(2,2)*exp(-d(2,2)*s(nw));
g(2,3)=polyval(numg23,s(nw)) / polyval(deng23,s(nw));
g(2,3)=g(2,3)*exp(-d(2,3)*s(nw));

g(3,1)=polyval(numg31,s(nw)) / polyval(deng31,s(nw));
g(3,1)=g(3,1)*exp(-d(3,1)*s(nw));
g(3,2)=polyval(numg32,s(nw)) / polyval(deng32,s(nw));
g(3,2)=g(3,2)*exp(-d(3,2)*s(nw));
g(3,3)=polyval(numg33,s(nw)) / polyval(deng33,s(nw));
g(3,3)=g(3,3)*exp(-d(3,3)*s(nw));
%
% Calculate GL transfer functions
gL1=polyval(numgL1,s(nw)) / polyval(dengL1,s(nw));
gL1=gL1*exp(-dgL(1)*s(nw));
gL2=polyval(numgL2,s(nw)) / polyval(dengL2,s(nw));
gL2=gL2*exp(-dgL(2)*s(nw));
gL3=polyval(numgL3,s(nw)) / polyval(dengL3,s(nw));
gL3=gL3*exp(-dgL(3)*s(nw));
% Form vector of GL transfer functions
gL=[gL1 gL2 gL3];
% Controller gc's
gc(1,1)=polyval(numgc1,s(nw)) / polyval(dengc 1,s(nw));
gc(1,2)=0;
gc(1,3)=0;
gc(2,1)=0;
gc(2,2)=polyval(numgc2,s(nw)) / polyval(dengc2,s(nw));
gc(2,3)=0;
gc(3,1)=0;
gc(3,2)=0;
gc(3,3)=polyval(numgc3,s(nw)) / polyval(dengc3,s(nw));
```

TABLE 13.3 (CONTINUED)
Load rejection for the Ogunnaike-Ray column

```
% Calculate closedloop regulator function
%
clreg=inv(eye(size(g))+g*gc)*gL;
dbcl1(nw)=20*log10(abs(clreg(1)));
dbcl2(nw)=20*log10(abs(clreg(2)));
dbcl3(nw)=20*log10(abs(clreg(3)));

%
end
% End of frequency loop
%
%
% Plot dbcl function
clf
semilogx(w,dbcl1,'-',w,dbcl2,'--',w,dbcl3,'-.')
xlabel('Frequency (radians/minute)')
ylabel('Closedloop Regulator (dB)')
grid
legend('Y1/L','Y2/L','Y3/L')

print -dps pfig134.ps
```

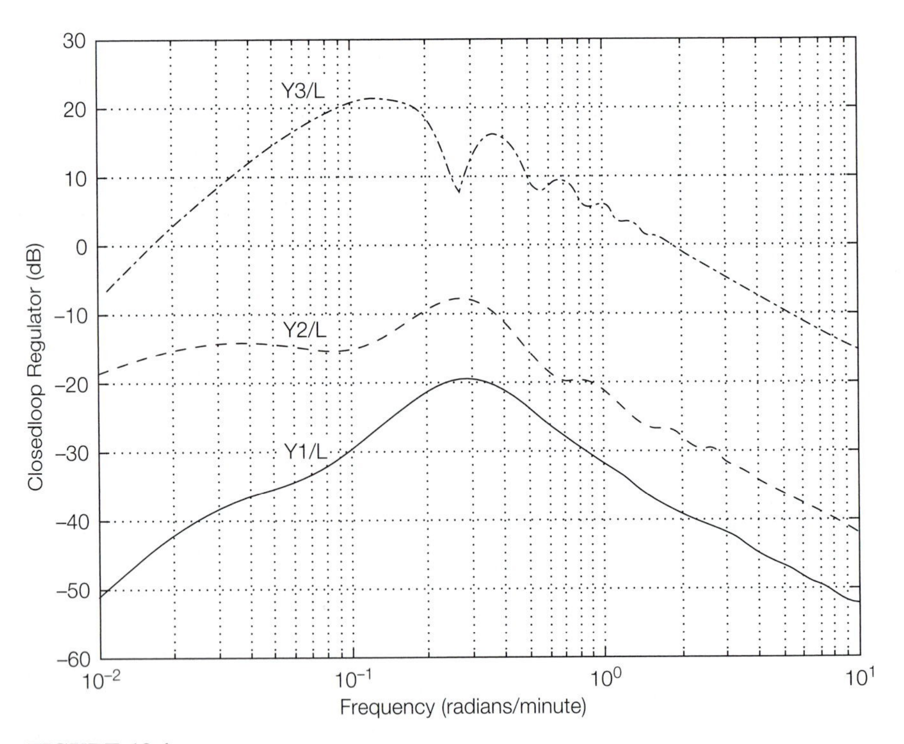

FIGURE 13.4
Load rejection for Ogunnaike-Ray column.

The set of variable pairings that gives the smallest peaks in these closedloop regulator transfer functions should be selected since it provides a control structure with the best load rejection.

13.7 MODEL PREDICTIVE CONTROL

In the past decade, a great deal of activity in industry and in academia has focused on the use of process models to develop new types of multivariable controllers. Numerous papers have been published over the last several years on multivariable control systems generically called model predictive control (MPC). MPC applications have been reported for the most part in the petroleum industry, on units such as fluid catalytic crackers, hydrocrackers, and petroleum fractionating towers. Such operations are characterized by being multivariable in nature, having many constraints, and processing large volumes of material. One of the most popular commercial applications of MPC is dynamic matrix control (DMC). However, MPC has yet to demonstrate any advantages over conventional strategies on a number of important processes, particularly in the chemical industry. A particularly insightful test of MPC has recently been reported by N. L. Ricker and J. H. Lee (*Comput. Chem. Eng. 19*:961, 1995), in which they state that MPC is no panacea.

We present here only some of the basic ideas of model predictive control. A thorough treatment of this subject would be quite extensive and is available in other, more advanced textbooks (see Ogunnaike and Ray, *Process Dynamics, Modeling, and Control,* 1994, Oxford University Press, New York, pp. 991–1032; or Seborg, Edgar, and Mellichamp, *Process Dynamics and Control,* 1989, Wiley, New York, pp. 649–667). Our objective in this book is to strip away everything but the essentials. A chemical engineer who is thoroughly grounded in the essence of chemical process control will be capable of understanding the features of model predictive control should the need arise.

The basic idea of MPC is to use a process model (either linear or nonlinear) to calculate the best changes in the manipulated variables that will achieve a specified desired result in the controlled variables. At each point in time the output variables are measured. Then the optimization procedure calculates the moves in the manipulated variables for several time steps into the future. The first of these changes is made and has a certain effect on the controlled variables. At the next time step the new values of the controlled variables are measured and incorporated into the optimization problem, which is re-solved to obtain new manipulated variable values.

Suppose we have a model that relates the process outputs $y_{(t)}$ to the manipulated variables $u_{(t)}$:

$$y_{(t)} = f[u_{(t)}] \tag{13.12}$$

We also have desired values for the output variables $y_{(t)}^{\text{set}}$. Then we can formulate an optimization problem to minimize an objective function J, which is typically the square of the difference between the setpoints and the controlled variables over a specified time period P.

$$\min_{u_{(t)}} J = [y_{(t)}^{\text{set}} - y_{(t)}]^2 \quad \text{with } t \text{ over period } P \tag{13.13}$$

We solve this optimization problem for the manipulated inputs as functions of time that provide the smallest value of the specified objective function. The model is used to predict what the output variables will be in the future ($y_{(t)}^{\text{predict}}$). Since the model is not perfect and does not contain information about all unmeasured disturbances, the MPC algorithm must account for this by updating the model prediction information at each time step from the measured output values.

The simple objective function shown in Eq. (13.13) does not consider other factors such as the magnitudes of the changes in manipulated variables at each point in time. In many processes rapid and large swings in the manipulated variables are undesirable. Therefore, other forms of the objective function can be used. We can penalize changes in the manipulated variables by adding to the objective function a term that weights the changes in $u_{(t)}$.

$$\min_{u_t} J = [y_{(t)}^{\text{set}} - y_{(t)}]^2 + W u_{(t)}^2 \tag{13.14}$$

The manipulated and controlled variables have physical constraints that must be considered in the optimization. The values of the outputs predicted by the model can be used to avoid hitting constraints. In conventional control these constraints are handled by overrides, whereas in MPC they are incorporated into the optimization algorithm.

MPC is basically just an alternative way to look at the process control problem. We hope this brief summary conveys a little of what is behind this approach and helps to remove a little of the mystery from the method.

13.8 CONCLUSION

This chapter has presented a simple procedure for developing control structures for multivariable processes. The method has been applied to many real industrial plants with good success. It requires a modest amount of engineering time to obtain the transfer functions of the process (see Chapter 16 for more discussion on this subject). The calculations are easily accomplished on a small computer.

PROBLEMS

13.1. Calculate the BLT settings for the Wardle and Wood column (see Problem 12.1).

13.2. Determine the BLT settings for the 2×2 multivariable system given in Problem 12.2.

13.3. Alatiqi (*IEC Proc. Des. Dev.* 25:762, 1986) presented the transfer functions for a 4×4 multivariable complex distillation column with sidestream stripper for separating a ternary mixture into three products. There are four controlled variables: purities of the three product streams (x_{D1}, x_{S2}, and x_{B3}) and a temperature difference ΔT to minimize energy consumption. There are four manipulated variables: reflux R, heat input to the reboiler Q_R, heat input to the stripper reboiler Q_S, and flow rate of feed to the stripper L_S. The 4×4 matrix of openloop transfer functions relating controlled and manipulated variables is

$$\begin{bmatrix} \dfrac{4.09e^{-1.3s}}{(33s+1)(8.3s+1)} & \dfrac{-6.36e^{-1.2s}}{(31.6s+1)(20s+1)} & \dfrac{-0.25e^{-1.4s}}{21s+1} & \dfrac{-0.49e^{-6s}}{(22s+1)^2} \\ \dfrac{-4.17e^{-5s}}{45s+1} & \dfrac{6.93e^{-1.02s}}{44.6s+1} & \dfrac{-0.05e^{-6s}}{(34.5s+1)^2} & \dfrac{1.53e^{-3.8s}}{48s+1} \\ \dfrac{1.73e^{-18s}}{(13s+1)^2} & \dfrac{5.11e^{-12s}}{(13.3s+1)^2} & \dfrac{4.61e^{-1.01s}}{18.5s+1} & \dfrac{-5.49e^{-1.5s}}{15s+1} \\ \dfrac{-11.2e^{-2.6s}}{(43s+1)(6.5s+1)} & \dfrac{14(10s+1)e^{-0.02s}}{(45s+1)(17.4s^2+3s+1)} & \dfrac{0.1e^{-0.05s}}{(31.6s+1)(5s+1)} & \dfrac{4.49e^{-0.6s}}{(48s+1)(6.3s+1)} \end{bmatrix}$$

Calculate the BLT settings.

13.4. Both the temperature T and the liquid height H in a tank are controlled simultaneously by manipulating the flow rates of hot water F_H and cold water F_C into the vessel. Liquid leaves the vessel through a hand valve, so the flow rate out of the vessel depends on the square root of the height of liquid in the tank: $F = K\sqrt{h}$. The inlet temperature of the cold water is 70°F and for the hot water it is 160°F. The diameter of the tank is 12 inches. The temperature transmitter range is 50–200°F. The level transmitter range is 0–16 inches of water. The hot and cold water control valves have linear installed characteristics and can pass 3 gpm and 4 gpm, respectively, when wide open.

At the initial steady state, the height of liquid in the tank is 8 inches and the flow rates of hot and cold water are each 2 gpm. A stream of cold water can be added to the tank at a rate of F_L (gpm) as a load disturbance. F_L is initially zero.

The temperature transmitter has a dynamic response that can be approximated by a 30-second first-order lag. The level transmitter has a 15-second first-order lag. Both control valves have 10-second first-order lags.

(*a*) Develop a nonlinear mathematical model of the process. Linearize to obtain a linear model in terms of perturbation variables that has the form

$$\frac{dh}{dt} = a_{11}h + a_{12}T + b_{11}F_C + b_{12}F_H + d_1F_L$$

$$\frac{dT}{dt} = a_{21}h + a_{22}T + b_{21}F_C + b_{22}F_H + d_2F_L$$

(*b*) Laplace-transform and rearrange to obtain the openloop transfer functions between the controlled variables (h and T) and the manipulated variables (F_C and F_H).

(*c*) Determine the ultimate gain and ultimate frequency of each individual control loop using the diagonal elements of the openloop transfer function matrix. Calculate the Ziegler-Nichols settings for each controller assuming the use of PI controllers.

(*d*) Simulate the nonlinear system. Test each controller individually (with the other on manual) for small setpoint changes. Then put both loops on automatic and make a load disturbance in F_L.

(*e*) Calculate the BLT settings for the multivariable system and test these settings on the nonlinear model for a load disturbance in F_L.

PART FIVE

Sampled-Data Systems

All the control systems we have studied in previous parts of this book used continuous analog devices. All control signals were continuously generated by transmitters, fed continuously to analog controllers, and sent continuously to control valves. The development of digital control computers and chromatographic composition analyzers has resulted in a large number of control systems that have discontinuous, intermittent components. The operational nature of both these devices means that their input and output signals are discrete.

A distributed control system (DCS) uses digital computers to service a number of control loops. At a given instant in time, the computer looks at one loop, checking the value of the controlled variable and computing a new signal to send to the control valve. The controller output signal for this loop is then held constant as the computer moves on to look at all the other loops. The controller output signal is changed only at discrete moments in time.

To analyze systems with discontinuous control elements we need to learn another new "language." The mathematical tool of *z transformation* is used to design control systems for discrete systems. As we show in the next chapter, z transforms are to sampled-data systems what Laplace transforms are to continuous systems. The mathematics in the z domain and in the Laplace domain are very similar. We have to learn how to translate our small list of words from English and Russian into the language of z transforms, which we call German.

In Chapter 14 we define mathematically the sampling process, derive the z transforms of common functions (learn our German vocabulary), develop transfer functions in the z domain, and discuss stability. Design of digital controllers is studied in Chapter 15 using root locus and frequency response methods in the z plane. We use practically all the stability analysis and controller design techniques that we introduced in the Laplace and frequency domains, now applying them in the z domain for sampled-data systems.

CHAPTER 14

Sampling, z Transforms, and Stability

14.1 INTRODUCTION

14.1.1 Definition

Sampled-data systems have signals that are discontinuous or discrete. Figure 14.1 shows a continuous analog signal or function $f_{(t)}$ being fed into a sampler. Every T_s minutes the sampler closes for a brief instant. The output of the sampler $f_{s(t)}$ is therefore an intermittent series of pulses. Between sampling times, the sampler output is zero. At the instant of sampling the output of the sampler is equal to the input function.

$$\begin{aligned} f_{s(t)} &= f_{(nT_s)} \qquad \text{for } t = nT_s \\ f_{s(t)} &= 0 \qquad \text{for } t \neq nT_s \end{aligned} \tag{14.1}$$

14.1.2 Occurrence of Sampled-Data Systems in Chemical Engineering

Chromatographs and digital control computers are the principal components that produce sampled-data systems in chemical engineering processes. Figure 14.2 shows a typical chromatograph system. The process output variable $y_{(t)}$ is sampled every T_s minutes. The sample is injected into a chromatographic column that has a retention time of D_c minutes, which is essentially a pure time delay or deadtime. The sampling period T_s is usually set equal to the chromatograph cycle time D_c. The detector on the output of the column produces a signal that can be related to composition. The "peak picker" (or an area integrator) converts the detector signal into a composition signal. The maximum value or peak on the chromatograph curve can sometimes be used directly, but usually the areas under the curves are integrated and converted to a composition signal. This signal is generated only every T_s minutes. It is fed into a digital computer, which serves as a feedback controller. The output of the computer is fed into a device called a *hold* that clamps the signal until the next sampling period;

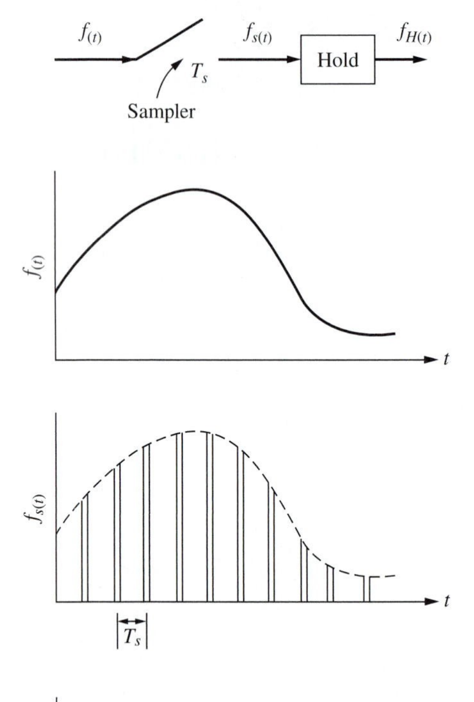

FIGURE 14.1
Sampled signals.

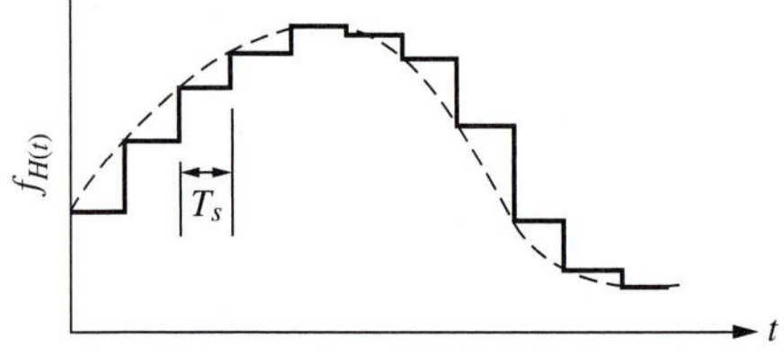

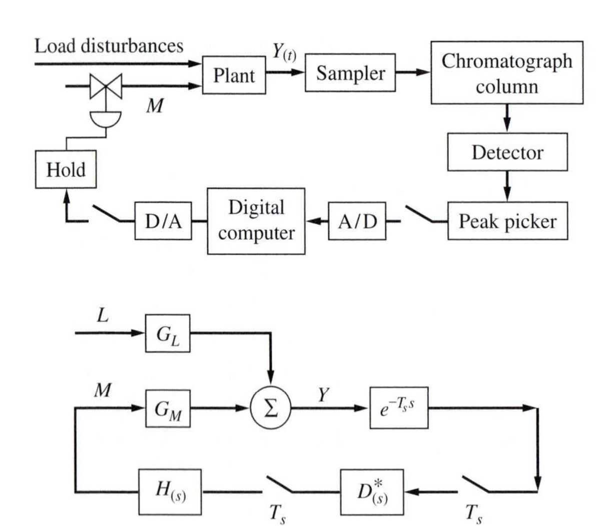

FIGURE 14.2
Chromatograph loop.

i.e., the output of the hold is maintained at a constant value over the sampling period. The hold converts the discrete signal, which is a series of pulses, into a continuous signal that is a stairstep function. The equivalent block diagram of this system is shown at the bottom of Fig. 14.2. The transfer function of the hold is $H_{(s)}$. The transfer function of the computer controller is $D^*_{(s)}$.

Figure 14.3 shows a digital control computer. The process output variables y_1, $y_2, \ldots, y_N$ are sensed and converted to continuous analog signals by transmitters $T_1, T_2, \ldots, T_N$. These data signals enter the digital computer through a multiplexed analog-to-digital (A/D) converter. The feedback control calculations are done in the

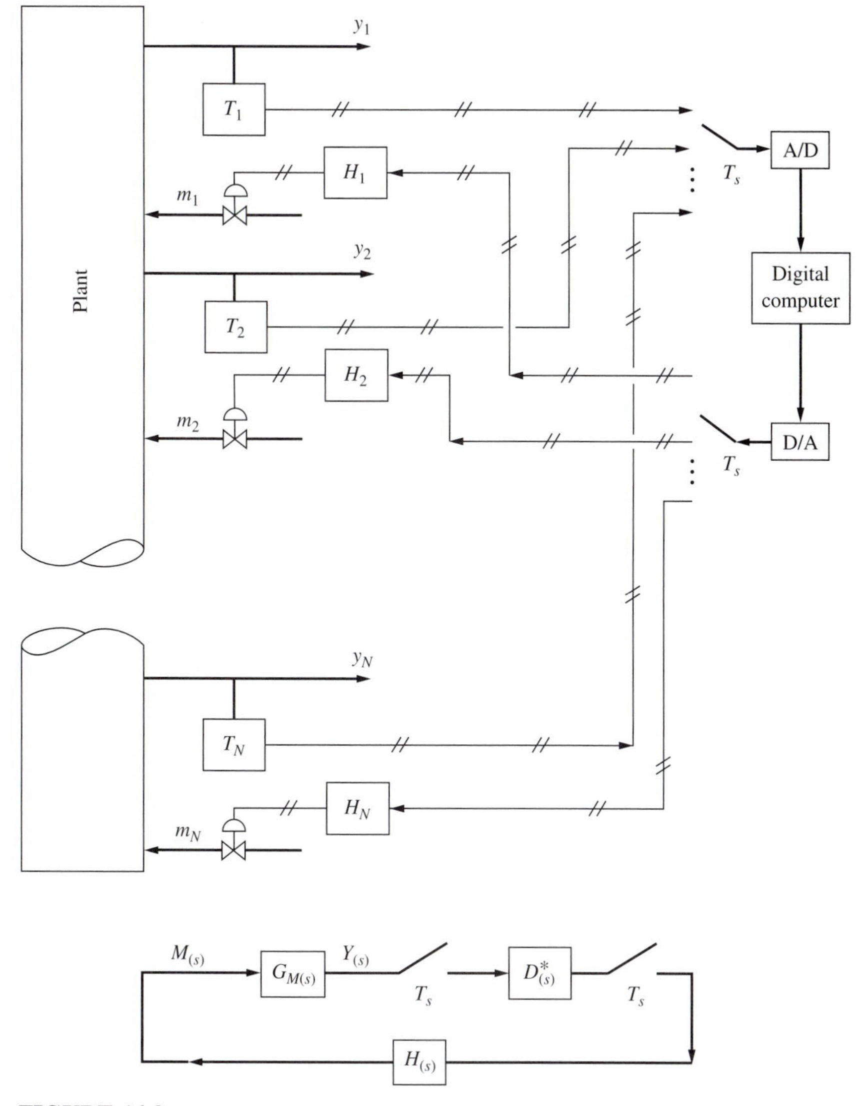

FIGURE 14.3
Computer control.

computer using some algorithm, and the calculated controller output signals are sent to holds associated with each control valve through a multiplexed digital-to-analog (D/A) converter. A block diagram of one loop is shown in the bottom of Fig. 14.3.

The sampling rate of these digital control computers can vary from several times a second to only several times an hour. The dynamics of the process dictate the sampling time required. The faster the process, the smaller the sampling period T_s must be for good control. One of the important questions that we explore in this chapter and the following one is what the sampling rate should be for a given process. For a given number of loops, the smaller the value of T_s specified, the faster the computer and the input/output equipment must be. This increases the cost of the digital hardware.

14.2 IMPULSE SAMPLER

A real sampler (Fig. 14.1) is closed for a finite period of time. This time of closure is usually small compared with the sampling period T_S. Therefore, the real sampler can be closely approximated by an *impulse sampler.* An impulse sampler is a device that converts a continuous input signal to a sequence of impulses or delta functions. Remember, these are *impulses,* not pulses. The height of each of these impulses is infinite. The width of each is zero. The area or "strength" of the impulse is equal to the magnitude of the input function at the sampling instant.

$$\int_{-nT_s}^{+nT_s} f^*_{(t)}\,dt = f_{(nT_s)} \tag{14.2}$$

If the units of $f_{(t)}$ are, for example, kilograms, the units of $f^*_{(t)}$ are kilograms/minute.

The impulse sampler is, of course, a mathematical fiction; it is not physically realizable. But the behavior of a real sampler-and-hold circuit is practically identical to that of the idealized impulse sampler-and-hold circuit. The impulse sampler is used in the analysis of sampled-data systems and in the design of sampled-data controllers because it greatly simplifies the calculations.

Let us now define an infinite sequence of unit impulses $\delta_{(t)}$ or Dirac delta functions whose strengths are all equal to unity. One unit impulse occurs at every sampling time. We call this series of unit impulses, shown in Fig. 14.4, the function $I_{(t)}$.

$$\begin{aligned} I_{(t)} &= \delta_{(t)} + \delta_{(t-T_s)} + \delta_{(t-2T_s)} + \delta_{(t-3T_s)} + \cdots \\ I_{(t)} &\equiv \sum_{n=0}^{\infty} \delta_{(t-nT_s)} \end{aligned} \tag{14.3}$$

Thus, the sequence of impulses $f^*_{(t)}$ that comes out of an impulse sampler can be expressed as

$$\begin{aligned} f^*_{(t)} &= f_{(t)} I_{(t)} = f_{(0)}\delta_{(t)} + f_{(T_s)}\delta_{(t-T_s)} + f_{(2T_s)}\delta_{(t-2T_s)} + \cdots \\ &= \sum_{n=0}^{\infty} f_{(nT_s)}\delta_{(t-nT_s)} \end{aligned} \tag{14.4}$$

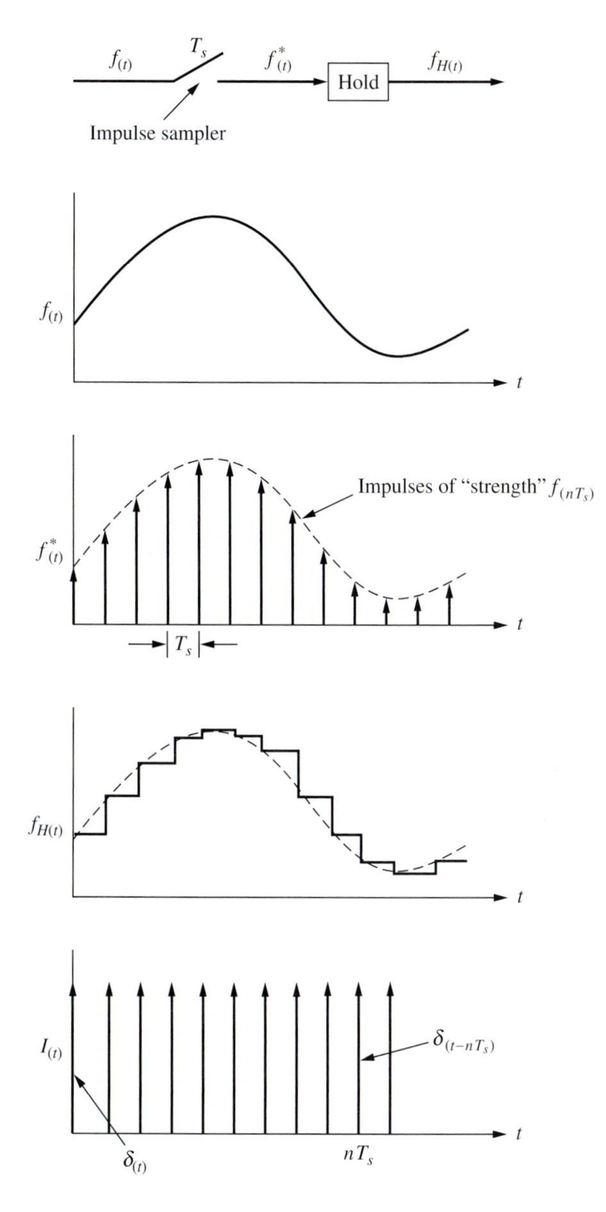

FIGURE 14.4
Impulse sampler.

Laplace-transforming Eq. (14.4) gives

$$\mathscr{L}[f^*_{(t)}] = \mathscr{L}\left[\sum_{n=0}^{\infty} f_{(nT_s)}\delta_{(t-nT_s)}\right] = \sum_{n=0}^{\infty} f_{(nT_s)}\mathscr{L}[\delta_{(t-nT_s)}]$$

$$= \sum_{n=0}^{\infty} f_{(nT_s)}e^{-nT_s s}\mathscr{L}[\delta_{(t)}]$$

$$F^*_{(s)} \equiv \sum_{n=0}^{\infty} f_{(nT_s)}e^{-nT_s s} \tag{14.5}$$

Equation (14.4) expresses the sequence of impulses that exits from an impulse sampler in the time domain. Equation (14.5) gives the sequence in the Laplace domain. Substituting $i\omega$ for s gives the impulse sequence in the frequency domain.

$$F^*_{(i\omega)} \equiv \sum_{n=0}^{\infty} f_{(nT_s)} e^{-in\omega T_s} \tag{14.6}$$

The sequence of impulses $f^*_{(t)}$ can be represented in an alternative manner. The $I_{(t)}$ function is a periodic function (see Fig. 14.4) with period T_s and a frequency ω_s in radians per minute.

$$\omega_s = \frac{2\pi}{T_s} \tag{14.7}$$

Since $I_{(t)}$ is periodic, it can be represented as a complex Fourier series:

$$I_{(t)} = \sum_{n=-\infty}^{n=+\infty} C_n e^{in\omega_s t} \tag{14.8}$$

where

$$C_n = \frac{1}{T_s} \int_{-T_s/2}^{+T_s/2} I_{(t)} e^{-in\omega_s t}\, dt \tag{14.9}$$

Over the interval from $-T_s/2$ to $+T_s/2$ the function $I_{(t)}$ is just $\delta_{(t)}$. Therefore, Eq. (14.9) becomes

$$C_n = \frac{1}{T_s} \int_{-T_s/2}^{+T_s/2} \delta_{(t)} e^{-in\omega_s t}\, dt = \frac{1}{T_s} \left[e^{-in\omega_s t}\right]_{t=0} = \frac{1}{T_s} \tag{14.10}$$

Remember, multiplying a function $f_{(t)}$ by the Dirac delta function and integrating give $f_{(0)}$. Therefore, $I_{(t)}$ becomes

$$I_{(t)} = \frac{1}{T_s} \sum_{n=-\infty}^{n=+\infty} e^{in\omega_s t}$$

The sequence of impulses $f^*_{(t)}$ can be expressed as a doubly infinite series:

$$f^*_{(t)} = f_{(t)} I_{(t)} = \frac{1}{T_s} \sum_{n=-\infty}^{n=+\infty} f_{(t)} e^{in\omega_s t} \tag{14.11}$$

Remember that the Laplace transformation of a function multiplied by an exponential e^{at} is simply the Laplace transform of the function with $(s - a)$ substituted for s.

$$\mathcal{L}[f_{(t)} e^{at}] = \int_0^{\infty} f_{(t)} e^{at} e^{-st}\, dt = \int_0^{\infty} f_{(t)} e^{-(s-a)t}\, dt \equiv F_{(s-a)}$$

So Laplace-transforming Eq. (14.11) gives

$$F^*_{(s)} = \frac{1}{T_s} \sum_{n=-\infty}^{n=+\infty} F_{(s-in\omega_s)} = \frac{1}{T_s} \sum_{n=-\infty}^{n=+\infty} F_{(s+in\omega_s)} \tag{14.12}$$

Substituting $i\omega$ for s gives

$$F^*_{(i\omega)} = \frac{1}{T_s} \sum_{n=-\infty}^{n=+\infty} F_{[i(\omega + n\omega_s)]} \tag{14.13}$$

Equation (14.4) is completely equivalent to Eq. (14.11) in the time domain. Equation (14.5) is equivalent to Eq. (14.12) in the Laplace domain. Equation (14.6) is equivalent to Eq. (14.13) in the frequency domain. We use these alternative forms of representation in several ways later.

14.3 BASIC SAMPLING THEOREM

A very important theorem of sampled-data systems is:

To obtain dynamic information about a plant from a signal that contains components out to a frequency $\omega_{\max}$, the sampling frequency ω_s must be set at a rate greater than twice $\omega_{\max}$.

$$\omega_s > 2\omega_{\max} \tag{14.14}$$

EXAMPLE 14.1. Suppose we have a signal that has components out to 100 rad/min. We must set the sampling frequency at a rate greater than 200 rad/min.

$$\omega_s > 200 \text{ rad/min}$$

$$T_s = \frac{2\pi}{\omega_s} = \frac{2\pi}{200} = 0.0314 \text{ min}$$

■

This basic sampling theorem has profound implications. It says that any high-frequency components in the signal (for example, 60-cycle-per-second electrical noise) can necessitate very fast sampling, even if the basic process is quite slow. It is therefore always recommended that signals be analog filtered *before* they are sampled. This eliminates the unimportant high-frequency components. Trying to filter the data after it has been sampled using a digital filter does not work.

To prove the sampling theorem, let us consider a continuous $f_{(t)}$ that is a sine wave with a frequency ω_0 and an amplitude A_0.

$$f_{(t)} = A_0 \sin(\omega_0 t) \tag{14.15}$$

$$f_{(t)} = A_0 \frac{e^{i\omega_0 t} - e^{-i\omega_0 t}}{2i} \tag{14.16}$$

Suppose we sample this $f_{(t)}$ with an impulse sampler. The sequence of impulses $f^*_{(t)}$ coming out of the impulse sampler will be, according to Eq. (14.11),

$$f^*_{(t)} = \frac{1}{T_s} \sum_{n=-\infty}^{n=+\infty} f_{(t)} e^{in\omega_s t} = \frac{1}{T_s} \sum_{n=-\infty}^{n=+\infty} A_0 \left(\frac{e^{i\omega_0 t} - e^{-i\omega_0 t}}{2i} \right) e^{in\omega_s t}$$

$$= \frac{A_0}{2iT_s} \sum_{n=-\infty}^{n=+\infty} \left(e^{i(\omega_0 + n\omega_s)t} - e^{-i(\omega_0 - n\omega_s)t} \right)$$

Now we write out a few of the terms, grouping some of the positive and negative n terms.

$$f^*_{(t)} = \frac{A_0}{T_s}\left(\frac{e^{i\omega_0 t} - e^{-i\omega_0 t}}{2i} + \frac{e^{i(\omega_0+\omega_s)t} - e^{-i(\omega_0+\omega_s)t}}{2i} + \frac{e^{i(\omega_0+2\omega_s)t} - e^{-i(\omega_0+2\omega_s)t}}{2i}\right.$$
$$\left. + \frac{e^{i(\omega_0-\omega_s)t} - e^{-i(\omega_0-\omega_s)t}}{2i} + \frac{e^{i(\omega_0-2\omega_s)t} - e^{-i(\omega_0-2\omega_s)t}}{2i} + \cdots\right)$$

$$f^*_{(t)} = \frac{A_0}{T_s}\{\sin(\omega_0 t) + \sin[(\omega_0+\omega_s)t] + \sin[(\omega_0+2\omega_s)t] + \sin[(\omega_0-\omega_s)t] + \sin[(\omega_0-2\omega_s)t] + \cdots\} \tag{14.17}$$

Thus, the sampled function $f^*_{(t)}$ contains a primary component at frequency ω_0 plus an infinite number of complementary components at frequencies $\omega_0 + \omega_s$, $\omega_0 + 2\omega_s, \ldots, \omega_0 - \omega_s, \omega_0 - 2\omega_s, \ldots$. The amplitude of each component is the amplitude of the original sine wave $f_{(t)}$ attenuated by $1/T_s$. The sampling process produces a signal with components at frequencies that are multiples of the sampling frequency plus the original frequency of the continuous signal before sampling. Figure 14.5*a* illustrates this in terms of the frequency spectrum of the signal. This is referred to by electrical engineers as "aliasing."

Now suppose we have a continuous function $f_{(t)}$ that contains components over a range of frequencies. Figure 14.5*b* shows its frequency spectrum $f_{(\omega)}$. If this signal is sent through an impulse sampler, the output $f^*_{(t)}$ has a frequency spectrum $f^*_{(\omega)}$ as shown in Fig. 14.5*b*. If the sampling rate or sampling frequency ω_s is high, there is no overlap between the primary and complementary components. Therefore, $f^*_{(t)}$ can be filtered to remove all the high-frequency complementary components, leaving just the primary component. This can then be related to the original continuous function. Therefore, if the sampling frequency is greater than twice the highest frequency in the original signal, the original signal can be determined from the sampled signal.

If, however, the sampling frequency is less than twice the highest frequency in the original signal, the primary and complementary components overlap. Then the sampled signal cannot be filtered to recover the original signal, and the sampled signal predicts incorrectly the steady-state gain and the dynamic components of the original signal.

Figure 14.5*b* shows that $f^*_{(\omega)}$ is a periodic function of frequency ω. Its period is ω_s.

$$f^*_{(\omega)} = f^*_{(\omega+\omega_s)} = f^*_{(\omega+2\omega_s)} = \cdots \tag{14.18}$$

This equation can also be written

$$f^*_{(i\omega)} = f^*_{(i\omega+i\omega_s)} = f^*_{(i\omega+i2\omega_s)} = \cdots \tag{14.19}$$

Going into the Laplace domain by substituting s for $i\omega$ gives

$$F^*_{(s)} = F^*_{(s+i\omega_s)} = F^*_{(s+i2\omega_s)} = \cdots \tag{14.20}$$

Thus $F^*_{(s)}$ is a periodic function of s with period $i\omega_s$. We use this periodicity property to develop pulse transfer functions in Section 14.5.

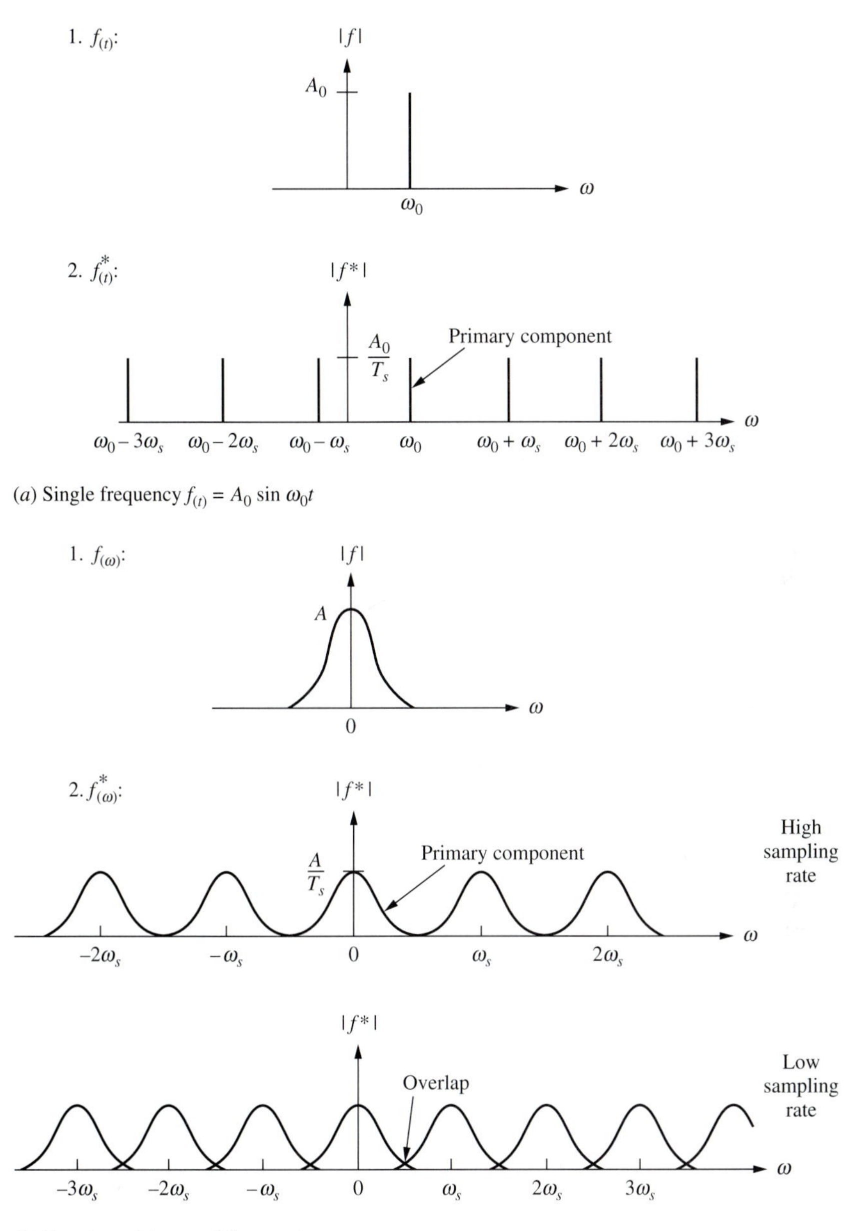

(a) Single frequency $f_{(t)} = A_0 \sin \omega_0 t$

(b) Function with several frequencies

FIGURE 14.5
Frequency spectrum of continuous and sampled signals.

14.4 z TRANSFORMATION

14.4.1 Definition

Sequences of impulses, such as the output of an impulse sampler, can be z transformed. For a specified sampling period T_s, the z transformation of an impulse-sampled signal $f^*_{(t)}$ is defined by the equation

$$\mathscr{Z}[f^*_{(t)}] \equiv f_{(0)} + f_{(T_s)}z^{-1} + f_{(2T_s)}z^{-2} + f_{(3T_s)}z^{-3} + \cdots + f_{(nT_s)}z^{-n} + \cdots \quad (14.21)$$

The notation $\mathscr{Z}[]$ means the z transformation operation. The $f_{(nT_s)}$ values are the magnitudes of the continuous function $f_{(t)}$ (before impulse sampling) at the sampling periods. We use the notation that the z transform of $f^*_{(t)}$ is $F_{(z)}$.

$$\boxed{\mathscr{Z}[f^*_{(t)}] \equiv F_{(z)} = \sum_{n=0}^{\infty} f_{(nT_s)}z^{-n}} \quad (14.22)$$

The z variable can be considered an "ordering" variable whose exponent represents the position of the impulse in the infinite sequence $f^*_{(t)}$.

Comparing Eqs. (14.5) and (14.22), we can see that the s and z variables are related by

$$\boxed{z = e^{T_s s}} \quad (14.23)$$

We make frequent use of this very important relationship between these two complex variables.

Keep in mind the concept that we always take z transforms of impulse-sampled signals, *not* continuous functions. We also use the notation

$$\mathscr{Z}[F^*_{(s)}] = \mathscr{Z}[f^*_{(t)}] \equiv F_{(z)} \quad (14.24)$$

This means exactly the same thing as Eq. (14.22). We can go directly from the time domain $f^*_{(t)}$ to the z domain. Or we can go from the time domain $f^*_{(t)}$ to the Laplace domain $F^*_{(s)}$ and on to the z domain $F_{(z)}$.

14.4.2 Derivation of z Transforms of Common Functions

Just as we did in learning Russian (Laplace transforms), we need to develop a small German vocabulary of z transforms.

A. Step function

$$f_{(t)} = Ku_{n(t)}$$

Passing the step function through an impulse sampler gives $f^*_{(t)} = Ku_{n(t)}I_{(t)}$, where $I_{(t)}$ is the sequence of unit impulses defined in Eq. (14.3). Using the definition of z transformation [Eq. (14.22)] gives

$$\mathscr{Z}[f^*_{(t)}] = \sum_{n=0}^{\infty} f_{(nT_s)} z^{-n} = f_{(0)} + f_{(T_s)} z^{-1} + f_{(2T_s)} z^{-2} + \cdots$$
$$= K + Kz^{-1} + Kz^{-2} + Kz^{-3} + \cdots$$
$$= K(1 + z^{-1} + z^{-2} + z^{-3} + \cdots)$$
$$= K\frac{1}{1 - z^{-1}}$$

provided $|z^{-1}| < 1$. This requirement is analogous to the requirement in Laplace transformation that s be large enough that the integral converges. Since $z^{-1} = e^{-T_s s}$, s must be large enough to keep the exponential less than 1.

The z transform of the impulse-sampled step function is

$$\mathscr{Z}[Ku_{n(t)}I_{(t)}] = K\frac{z}{z-1} \tag{14.25}$$

B. Ramp function

$$f_{(t)} = Kt \quad \Rightarrow \quad f^*_{(t)} = KtI_{(t)}$$

$$\mathscr{Z}[f^*_{(t)}] = \sum_{n=0}^{\infty} f_{(nT_s)} z^{-n} = f_{(0)} + f_{(T_s)} z^{-1} + f_{(2T_s)} z^{-2} + \cdots$$
$$= 0 + KT_s z^{-1} + 2KT_s z^{-2} + 3KT_s z^{-3} + \cdots$$
$$= KT_s z^{-1}(1 + 2z^{-1} + 3z^{-2} + \cdots) = \frac{KT_s z^{-1}}{(1 - z^{-1})^2}$$

for $|z^{-1}| < 1$. The z transform of the impulse-sampled ramp function is

$$\mathscr{Z}[KtI_{(t)}] = \frac{KT_s z}{(z-1)^2} \tag{14.26}$$

Notice the similarity between the Laplace domain and the z domain. The Laplace transformation of a constant (K) is K/s and of a ramp (Kt) is K/s^2. The z transformation of a constant is $Kz/(z-1)$ and of a ramp is $KT_s z/(z-1)^2$. Thus, the s in the denominator of a Laplace transformation and the $(z-1)$ in the denominator of a z transformation behave somewhat similarly.

You should now be able to guess the z transformation of t^2. We know there will be an s^3 term in the denominator of the Laplace transformation of this function. So we can extrapolate our results to predict that there will be a $(z-1)^3$ in the denominator of the z transformation.

We find later in this chapter that a $(z-1)$ in the denominator of a transfer function in the z domain means that there is an integrator in the system, just as the presence of an s in the denominator in the Laplace domain tells us there is an integrator.

C. Exponential

$$f_{(t)} = Ke^{-at}$$

$$\begin{aligned} F_{(z)} &= \sum_{n=0}^{\infty} (Ke^{-anT_s}) z^{-n} \\ &= K[1 + (e^{-aT_s}z^{-1}) + (e^{-aT_s}z^{-1})^2 + (e^{-aT_s}z^{-1})^3 + \cdots] \\ &= K\frac{1}{1 - e^{-aT_s}z^{-1}} \quad \text{for } \left|e^{-aT_s}z^{-1}\right| < 1 \end{aligned}$$

The z transform of the impulse-sampled exponential function is

$$\mathscr{Z}[Ke^{-at}I_{(t)}] = \frac{Kz}{z - e^{-aT_s}} \tag{14.27}$$

Remember that the Laplace transformation of the exponential was $K/(s + a)$. So the $(s + a)$ term in the denominator of a Laplace transformation is similar to the $(z - e^{-aT_s})$ term in a z transformation. Both indicate an exponential function. In the s plane we have a pole at $s = -a$. In the z plane we find later in this chapter that we have a pole at $z = e^{-aT_s}$. So we can immediately conclude that poles on the negative real axis in the s plane "map" (to use the complex-variable term) onto the positive real axis between 0 and +1.

D. Exponential multiplied by time

In the Laplace domain we found that repeated roots $1/(s + a)^2$ occur when we have the exponential multiplied by time. We can guess that similar repeated roots should occur in the z domain. Let us consider a very general function:

$$f_{(t)} = \frac{K}{p!} t^p e^{-at} \tag{14.28}$$

This function can be expressed in the alternative form

$$f_{(t)} = (-1)^p \frac{K}{p!} \frac{\partial^p (e^{-at})}{\partial a^p} \tag{14.29}$$

The z transformation of this function after impulse sampling is

$$\begin{aligned} F_{(z)} &= \sum_{n=0}^{\infty} (-1)^p \frac{K}{p!} \frac{\partial^p (e^{-anT_s})}{\partial a^p} z^{-n} \\ &= (-1)^p \frac{K}{p!} \frac{\partial^p}{\partial a^p} \left[\sum_{n=0}^{\infty} \left(z^{-1} e^{-aT_s} \right)^n \right] \\ &= (-1)^p \frac{K}{p!} \frac{\partial^p}{\partial a^p} \left(\frac{z}{z - e^{-aT_s}} \right) \end{aligned} \tag{14.30}$$

EXAMPLE 14.2. Take the case where $p = 1$.

$$\mathscr{Z}[Kte^{-at}I_{(t)}] = -K\frac{\partial}{\partial a}\left(\frac{z}{z - e^{-aT_s}}\right) = \frac{KT_s e^{-aT_s} z}{(z - e^{-aT_s})^2} \tag{14.31}$$

So we get a repeated root in the z plane, just as we did in the s plane. ■

E. Sine

$$f_{(t)} = \sin(\omega t)$$

$$F_{(z)} = \sum_{n=0}^{\infty}\left(\frac{e^{in\omega T_s} - e^{-in\omega T_s}}{2i}\right)z^{-n}$$

$$= \frac{1}{2i}\left(\frac{1}{1 - e^{i\omega T_s}z^{-1}} - \frac{1}{1 - e^{-i\omega T_s}z^{-1}}\right) \tag{14.32}$$

$$= \frac{1}{2i}\left(\frac{e^{i\omega T_s}z^{-1} - e^{-i\omega T_s}z^{-1}}{1 + z^{-2} - e^{i\omega T_s}z^{-1} - e^{-i\omega T_s}z^{-1}}\right)$$

$$= \frac{1}{2i}\left(\frac{z^{-1}(2i)\sin(\omega T_s)}{1 + z^{-2} - 2z^{-1}\cos(\omega T_s)}\right) = \frac{z\sin(\omega T_s)}{z^2 + 1 - 2z\cos(\omega T_s)}$$

F. Unit impulse function

By definition, the z transformation of an impulse-sampled function is

$$F_{(z)} = f_{(0)} + f_{(T_s)}z^{-1} + f_{(2T_s)}z^{-2} + \cdots$$

If $f_{(t)}$ is a unit impulse, putting it through an impulse sampler should give an $f^*_{(t)}$ that is still just a unit impulse $\delta_{(t)}$. But Eq. (14.4) says that

$$f^*_{(t)} = f_{(0)}\delta_{(t)} + f_{(T_s)}\delta_{(t-T_s)} + f_{(2T_s)}\delta_{(t-2T_s)} + \cdots$$

But if $f^*_{(t)}$ must be equal to just $\delta_{(t)}$, the term $f_{(0)}$ in the equation above must be equal to 1 and all the other terms $f_{(T_s)}, f_{(2T_s)}, \ldots$ must be equal to zero. Therefore, the z transformation of the unit impulse is unity.

$$\mathcal{Z}[\delta_{(t)}] = 1 \tag{14.33}$$

14.4.3 Effect of Deadtime

Deadtime in a sampled-data system is very easily handled, particularly if the deadtime D is an integer multiple of the sampling period T_s. Let us assume that

$$D = kT_s \tag{14.34}$$

where k is an integer. Consider an arbitrary function $f_{(t-D)}$. The original function $f_{(t)}$ before the time delay is assumed to be zero for time less than zero. Running the delayed function through an impulse sampler and z transforming give

$$\mathcal{Z}[f^*_{(t-D)}] = \sum_{n=0}^{\infty} f_{(nT_s-kT_s)}z^{-n}$$

Now we let $x \equiv n - k$.

$$\mathcal{Z}[f^*_{(t-D)}] = \sum_{x=-k}^{\infty} f_{(xT_s)}z^{-x-k} = \left[\sum_{x=0}^{\infty} f_{(xT_s)}z^{-x}\right]z^{-k}$$

since $f_{(xT_s)} = 0$ for $x < 0$. The term in the brackets is just the z transform of $f^*_{(t)}$ since x is a dummy variable of summation.

$$\boxed{\mathscr{Z}[f^*_{(t-D)}] = F_{(z)} z^{-k}} \tag{14.35}$$

Therefore, the deadtime transfer function in the z domain is z^{-k}.

14.4.4 z Transform Theorems

Just as for Laplace transforms, there are several useful theorems for z transforms.

A. Linearity

The linearity property is easily proved from the definition of z transformation.

$$\mathscr{Z}[f^*_{1(t)} + f^*_{2(t)}] = \mathscr{Z}[f^*_{1(t)}] + \mathscr{Z}[f^*_{2(t)}] \tag{14.36}$$

B. Scale change

$$\mathscr{Z}[e^{-at} f^*_{(t)}] = F_{(ze^{aT_s})} = F_{(z_1)} = \mathscr{Z}_1[f^*_{(t)}] \tag{14.37}$$

The notation $\mathscr{Z}_1[]$ means z transforming using the z_1 variable where $z_1 \equiv ze^{aT_s}$.
This theorem is proved by going back to the definition of z transformation.

$$\mathscr{Z}[e^{-at} f^*_{(t)}] = \sum_{n=0}^{\infty} e^{-anT_s} f_{(nT_s)} z^{-n} = \sum_{n=0}^{\infty} f_{(nT_s)} (ze^{aT_s})^{-n}$$

Now substitute $z_1 = ze^{aT_s}$ into the equation above.

$$\mathscr{Z}[e^{-at} f^*_{(t)}] = \sum_{n=0}^{\infty} f_{(nT_s)} (z_1)^{-n} = F_{(z_1)}$$

EXAMPLE 14.3. Suppose we want to take the z transformation of the function $f^*_{(t)} = Kte^{-at} I_{(t)}$. Using Eqs. (14.26) and (14.37) gives

$$\mathscr{Z}[Kte^{-at} I_{(t)}] = \mathscr{Z}_1[KtI_{(t)}] = \frac{KT_s z_1}{(z_1 - 1)^2}$$

Substituting $z_1 = ze^{aT_s}$ gives

$$\mathscr{Z}[Kte^{-at} I_{(t)}] = \frac{KT_s ze^{aT_s}}{(ze^{aT_s} - 1)^2} = \frac{KT_s ze^{-aT_s}}{(z - e^{-aT_s})^2} \tag{14.38}$$

This is exactly what we found in Example 14.2. ■

C. Final-value theorem

$$\lim_{t \to \infty} f_{(t)} = \lim_{z \to 1} \left(\frac{z-1}{z} F_{(z)} \right) \tag{14.39}$$

To prove this theorem, let $f_{(t)}$ be the step response of an arbitrary, openloop-stable Nth-order system:

$$f_{(t)} = Ku_{n(t)} + \sum_{i=1}^{N} K_i e^{-a_i t}$$

The steady-state value of $f_{(t)}$ or the limit of $f_{(t)}$ as time goes to infinity is K. Running $f_{(t)}$ through an impulse sampler and z transforming give

$$\mathfrak{Z}[f^*_{(t)}] = \mathfrak{Z}[Ku_{n(t)}I_{(t)}] + \mathfrak{Z}\left[\sum_{i=1}^{N} K_i e^{-a_i t} I_{(t)}\right]$$

$$F_{(z)} = \frac{Kz}{z-1} + \sum_{i=1}^{N} K_i \frac{z}{z - e^{-a_i T_s}}$$

Multiplying both sides by $(z-1)/z$ and letting $z \to 1$ give

$$\lim_{z \to 1}\left(\frac{z-1}{z} F_{(z)}\right) = K = \lim_{t \to \infty} f_{(t)}$$

D. Initial-value theorem

$$\lim_{t \to 0} f_{(t)} = \lim_{z \to \infty} F_{(z)} \tag{14.40}$$

The definition of the z transform of $f^*_{(t)}$ is

$$F_{(z)} = f_{(0)} + f_{(T_s)} z^{-1} + f_{(2T_s)} z^{-2} + \cdots$$

Letting z go to infinity (for $|z^{-1}| < 1$) in this equation gives $f_{(0)}$, which is the limit of $f_{(t)}$ as $t \to 0$.

14.4.5 Inversion

We sometimes want to invert from the z domain back to the time domain. The inversion gives the values of the function $f_{(t)}$ only at the sampling instants.

$$\mathfrak{Z}^{-1}[F_{(z)}] = f_{(nT_s)} \quad \text{for } n = 0, 1, 2, 3, \ldots \tag{14.41}$$

The z transformation of an impulse-sampled function is unique; i.e., there is only one $F_{(z)}$ that is the z transform of a given $f^*_{(t)}$. The inverse z transform of any $F_{(z)}$ is also unique; i.e., there is only one $f_{(nT_s)}$ that corresponds to a given $F_{(z)}$.

However, keep in mind that more than one continuous function $f_{(t)}$ gives the same impulse-sampled function $f^*_{(t)}$. The sampled function $f^*_{(t)}$ contains information about the original continuous function $f_{(t)}$ only at the sampling times. This nonuniqueness between $f^*_{(t)}$ (and $F_{(z)}$) and $f_{(t)}$ is illustrated in Fig. 14.6. Both continuous functions $f_{1(t)}$ and $f_{2(t)}$ pass through the same points at the sampling times but are different in between the sampling instants. They would have exactly the same z transformation.

There are several ways to invert z transforms.

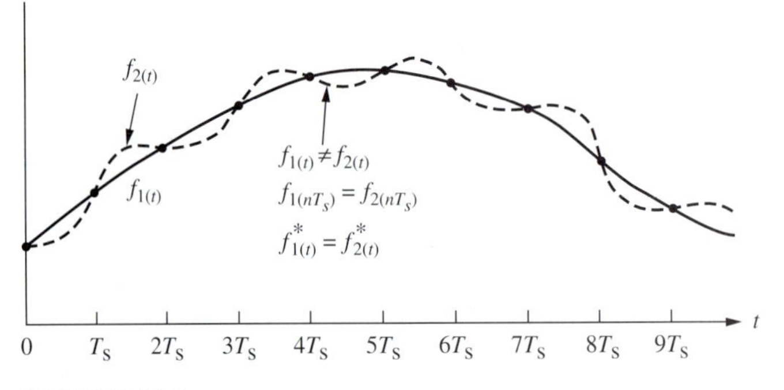

FIGURE 14.6
Continuous functions with identical values at sampling time.

A. Partial-fractions expansion

The classical mathematical method for inverting a z transform is to use the linearity theorem [Eq. (14.36)]. We expand the function $F_{(z)}$ into a sum of simple terms and invert each individually. This is completely analogous to Laplace transformation inversion. Let $F_{(z)}$ be a ratio of polynomials in z, Mth-order in the numerator and Nth-order in the denominator. We factor the denominator into its N roots: $p_1, p_2, p_3, \ldots, p_N$.

$$F_{(z)} = \frac{Z_{(z)}}{(z - p_1)(z - p_2)(z - p_3)\cdots(z - p_N)} \tag{14.42}$$

where $Z_{(z)} = M$th-order numerator polynomial. Each root p_i can be expressed in terms of the sampling period:

$$p_i = e^{-a_i T_s} \tag{14.43}$$

Using partial-fractions expansion, Eq. (14.42) becomes

$$F_{(z)} = \frac{Az}{z - p_1} + \frac{Bz}{z - p_2} + \frac{Cz}{z - p_3} + \cdots + \frac{Wz}{z - p_N} \tag{14.44}$$

$$= \frac{Az}{z - e^{-a_1 T_s}} + \frac{Bz}{z - e^{-a_2 T_s}} + \frac{Cz}{z - e^{-a_3 T_s}} + \cdots + \frac{Wz}{z - e^{-a_N T_s}} \tag{14.45}$$

The coefficients $A, B, C, \ldots, W$ are found and $F_{(z)}$ is inverted term by term to give

$$\mathscr{Z}^{-1}[F_{(z)}] = f_{(nT_s)} = Ae^{-a_1 nT_s} + Be^{-a_2 nT_s} + \cdots + We^{-a_N nT_s} \tag{14.46}$$

EXAMPLE 14.4. We show in Example 14.8 that the closedloop response to a unit step change in setpoint with a sampled-data proportional controller and a first-order process is

$$Y_{(z)} = \frac{K_c K_p (1 - b) z}{(z - 1)[z - b + K_c K_p (1 - b)]} \tag{14.47}$$

TABLE 14.1
Results for Example 14.4

t	n	$y_{(nT_s)}$ $K_c = 4.5$	$y_{(nT_s)}$ $K_c = 12$
0	0	0	0
0.2	1	0.8159	2.176
0.4	2	0.8184	−0.8098
0.6	3	0.8184	3.254
0.8	4	0.8184	−2.310

where $b \equiv e^{T_s/\tau_o}$
K_c = feedback controller gain
K_p = process steady-state gain
τ_o = process time constant

For the numerical values of $K_p = \tau_o = 1$, $K_c = 4.5$, and $T_s = 0.2$, $Y_{(z)}$ becomes

$$Y_{(z)} = \frac{0.8159z}{(z-1)(z-0.003019)} \tag{14.48}$$

Expanding in partial fractions gives

$$Y_{(z)} = \frac{0.8159z}{z-1} - \frac{0.8159z}{z-0.003019} \tag{14.49}$$

The pole at 0.003019 can be expressed as

$$0.003019 = e^{-5.803} = e^{-aT_s}$$

The value of the term aT_s is 5.803.

$$Y_{(z)} = \frac{0.8159z}{z-1} - \frac{0.8159z}{z-e^{-5.803}} \tag{14.50}$$

Inverting each of the terms above by inspection gives

$$y_{(nT_s)} = y_{(0.2n)} = 0.8159 - 0.8159e^{-naT_s} = 0.8159(1-e^{-5.803n}) \tag{14.51}$$

Table 14.1 gives the calculated results of $y_{(nT_s)}$ as a function of time. ■

B. Long division

An interesting z transform inversion technique is simple long division of the numerator by the denominator of $F_{(z)}$. The ease with which z transforms can be inverted with this technique is one of the reasons z transforms are often used.

By definition,

$$F_{(z)} = f_{(0)} + f_{(T_s)}z^{-1} + f_{(2T_s)}z^{-2} + f_{(3T_s)}z^{-3} + \cdots$$

If we can get $F_{(z)}$ in terms of an infinite series of powers of z^{-1}, the coefficients in front of all the terms give the values of $f_{(nT_s)}$. The infinite series is obtained by merely dividing the numerator of $F_{(z)}$ by the denominator of $F_{(z)}$.

$$F_{(z)} = \frac{Z_{(z)}}{P_{(z)}} = f_{(0)} + f_{(T_s)}z^{-1} + f_{(2T_s)}z^{-2} + f_{(3T_s)}z^{-3} + \cdots \tag{14.52}$$

where $Z_{(z)}$ and $P_{(z)}$ are polynomials in z. The method is easily understood by looking at a specific example.

EXAMPLE 14.5. The function considered in Example 14.4 is

$$Y_{(z)} = \frac{0.8159z}{z^2 - 1.003019z + 0.003019}$$

Long division gives

$$\begin{array}{r|l}
 & 0.8159z^{-1} + 0.8184z^{-2} + 0.8184z^{-3} + \\
\hline
z^2 - 1.003019z + 0.003019 & 0.8159z \\
 & \underline{0.8159z - 0.8184 + 0.0025z^{-1}} \\
 & \qquad 0.8184 - 0.0025z^{-1} \\
 & \qquad \underline{0.8184 - 0.8209z^{-1} + 0.0025z^{-2}} \\
 & \qquad\qquad 0.8184z^{-1} - 0.0025z^{-2} \\
 & \qquad\qquad \cdots\cdots\cdots\cdots
\end{array}$$

Therefore

$$\begin{aligned}
f_{(0)} &= 0 \\
f_{(T_s)} &= f_{(0.2)} = 0.8159 \\
f_{(2T_s)} &= f_{(0.4)} = 0.8184 \\
f_{(3T_s)} &= f_{(0.6)} = 0.8184 \\
&\vdots
\end{aligned}$$

These are, of course, exactly the same results we found by partial-fractions expansion in Example 14.4. ■

EXAMPLE 14.6. If the value of K_c in Example 14.4 is changed to 12, we show later in this chapter that $Y_{(z)}$ becomes

$$Y_{(z)} = \frac{2.1756z}{z^2 + 0.3722z - 1.357}$$

Inverting by long division gives

$$Y_{(z)} = 2.1756z^{-1} - 0.8098z^{-2} + 3.254z^{-3} - 2.310z^{-4} + \cdots \tag{14.53}$$

The system is unstable for this value of gain ($K_c = 12$), as we show later in this chapter. Notice that this example demonstrates that a first-order process controlled by a sampled-data proportional controller can be made closedloop unstable if the gain is high enough. With the use of an analog controller, the first-order process can *never* be closedloop unstable. Thus, there is a very important difference between continuous and discrete closedloop systems. Analog continuous controllers have an inherent advantage over discrete sampled-data controllers because they know what the output is doing at all points in time. The discrete controller knows only what the output is at the sampling times. ■

Inversion of z transforms by long division is very easily accomplished numerically by a digital computer. The FORTRAN subroutine LONGD given in Table 14.2 performs this long division. The output variable Y is calculated for NT sampling

TABLE 14.2
Long division subroutine

```
c program "longdtest.f"
c     does long-division using subroutine longd
        dimension a(10),b(10),y(100)
c Case for kc=12

        a0=0.
        a(1)=2.1756
        a(2)=0.
        b(1)=0.3722
        b(2)=-1.357
        n=2
        m=1
        nt=6
        y0=0.
        call longd(a0,a,b,y0,y,n,m,nt)

        do 10 k=1,nt
        write(6,1)k,y(k)
     10 continue
      1 format(' n=',i2,' y=',f10.5)
        stop
        end
c
        subroutine longd(a0,a,b,y0,y,n,m,nt)
        dimension a(10),b(10),y(100),d(10)
        nmax=n
        if(m.gt.n) nmax=m
        do 10 k=1,nmax
        d(k)=a(k)
        if(k.gt.m) d(k)=0.
     10 continue
        d(nmax+1)=0.
        if(a0.eq.0.)go to 30
        y0=a0
        do 20 k=1,nmax
     20 d(k)=d(k)-y0*b(k)
        y(1)=d(1)
        go to 40
     30 y0=0.
        y(1)=a(1)
     40 do 100 j=2,nt
        do 50 k=1,nmax
     50 d(k)=d(k+1)-y(j-1)*b(k)
    100 y(j)=d(1)
        return
        end
```

TABLE 14.3
MATLAB program for inversion

```
% Program "testdimpulse.m"
% Inverts z transform from Example 14.6 (Kc=12)
%
% Form numerator and denominator polynomials
num=[2.1756 0];
den=[1 0.3722 -1.357];
% Specify number of sampling periods and sampling period Ts
ntotal=5;
ts=0.2;
%
% Use "dimpulse" command to invert F(z)
[y,x]=dimpulse(num,den,ntotal);
%
% Calculate time
npts=length(y);
points=[0:1:(npts-1)];
t=ts*points;
clf
plot(t,y)
pause
```

times, given the coefficients $A0, A(1), A(2), \ldots, A(M)$ of the numerator and the coefficients $B(1), B(2), \ldots, B(N)$ of the denominator.

$$\begin{aligned} Y_{(z)} &= Y0 + Y(1)z^{-1} + Y(2)z^{-2} + Y(3)z^{-3} + \cdots \\ &= \frac{A0 + A(1)z^{-1} + A(2)z^{-2} + \cdots + A(M)z^{-M}}{1 + B(1)z^{-1} + B(2)z^{-2} + \cdots + B(N)z^{-N}} \end{aligned} \tag{14.54}$$

C. Use of MATLAB for inversion

Now that we have discussed the classical inversion methods, we are ready to see how inversion of z transforms can be easily accomplished using MATLAB software. Specific numerical values of parameters must be specified. Table 14.3 gives a program that solves for the values of the output at the sampling periods for the $Y_{(z)}$ considered in Example 14.6. First the numerator and denominator polynomials are formed. The number of sampling periods (*ntotal*) is specified, and the sampling period is set. Then the *[y,x]=dimpulse(num,den,ntotal)* command is used to generate the output sequence $y_{(nT_s)}$ at each value of n. The results are the same as those obtained by long division.

14.5 PULSE TRANSFER FUNCTIONS

We know how to find the z transformations of functions. Let us now turn to the problem of expressing input/output transfer function relationships in the z domain. Figure 14.7 shows a system with samplers on the input and on the output of the

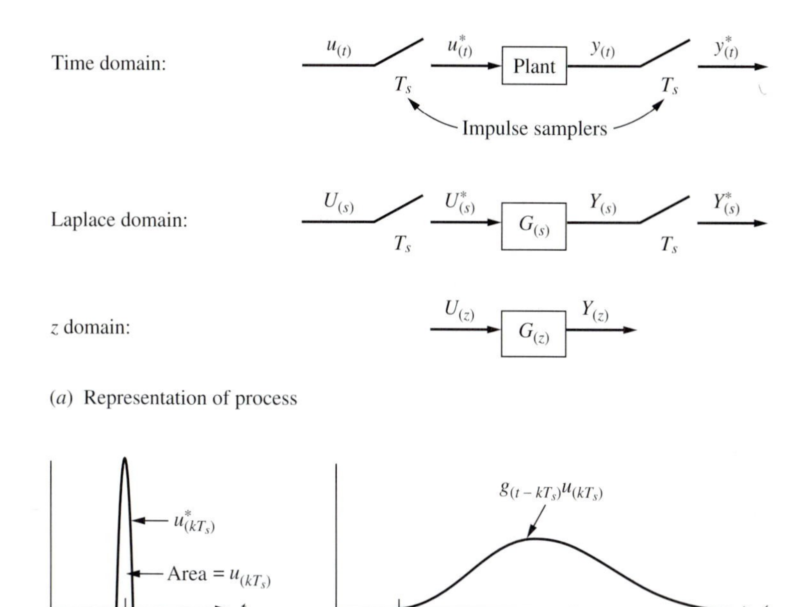

FIGURE 14.7
Pulse transfer functions.

process. Time-, Laplace-, and z-domain representations are shown. $G_{(z)}$ is called a *pulse transfer function.* It is defined below.

A sequence of impulses $u^*_{(t)}$ comes out of the impulse sampler on the input of the process. Each of these impulses produces a response from the process. Consider the kth impulse $u^*_{(kT_s)}$. Its area or strength is $u_{(kT_s)}$. Its effect on the continuous output of the plant $y_{(t)}$ is

$$y_{k(t)} = g_{(t-kT_s)}u_{(kT_s)} \tag{14.55}$$

where $y_{k(t)}$ = response of the process to the kth impulse
$g_{(t)}$ = unit impulse response of the process $= \mathcal{L}^{-1}[G_{(s)}]$

Figure 14.7 shows these functions.

The system is linear, so the total output $y_{(t)}$ is the sum of all the y_k's.

$$y_{(t)} = \sum_{k=0}^{\infty} y_{k(t)} = \sum_{k=0}^{\infty} g_{(t-kT_s)}u_{(kT_s)} \tag{14.56}$$

At the sampling times, the value of $y_{(t)}$ is $y_{(nT_s)}$.

$$y_{(nT_s)} = \sum_{k=0}^{\infty} g_{(nT_s-kT_s)}u_{(kT_s)} \tag{14.57}$$

The continuous function $y_{(t)}$ coming out of the process is then impulse sampled, producing a sequence of impulses $y^*_{(t)}$. If we z-transform $y^*_{(t)}$, we get

$$\mathscr{Z}[y^*_{(t)}] = \sum_{n=0}^{\infty} y_{(nT_s)} z^{-n} = Y_{(z)}$$

$$Y_{(z)} = \sum_{n=0}^{\infty} \left(\sum_{k=0}^{\infty} g_{(nT_s - kT_s)} u_{(kT_s)} \right) z^{-n} \tag{14.58}$$

Letting $p = n - k$ and remembering that $g_{(t)} = 0$ for $t < 0$ give

$$\begin{aligned} Y_{(z)} &= \sum_{p=0}^{\infty} \sum_{k=0}^{\infty} g_{(pT_s)} u_{(kT_s)} z^{-(p+k)} \\ &= \left(\sum_{p=0}^{\infty} g_{(pT_s)} z^{-p} \right) \left(\sum_{k=0}^{\infty} u_{(kT_s)} z^{-k} \right) \end{aligned} \tag{14.59}$$

$$Y_{(z)} = G_{(z)} U_{(z)} \tag{14.60}$$

The pulse transfer function $G_{(z)}$ is defined as the first term on the right-hand side of Eq. (14.59).

$$G_{(z)} \equiv \sum_{p=0}^{\infty} g_{(pT_s)} z^{-p} \tag{14.61}$$

Defining $G_{(z)}$ in this way permits us to use transfer functions in the z domain [Eq. (14.60)] just as we use transfer functions in the Laplace domain. $G_{(z)}$ is the z transform of the impulse-sampled response $g^*_{(t)}$ of the process to a unit impulse function $\delta_{(t)}$. In z-transforming functions, we used the notation $\mathscr{Z}[f^*_{(t)}] = \mathscr{Z}[F^*_{(s)}] = F_{(z)}$. In handling pulse transfer functions, we use similar notation.

$$\mathscr{Z}[g^*_{(t)}] = \mathscr{Z}[G^*_{(s)}] = G_{(z)} \tag{14.62}$$

where $G^*_{(s)}$ is the Laplace transform of the impulse-sampled response $g^*_{(t)}$ of the process to a unit impulse input.

$$G^*_{(s)} = \mathscr{L}[g^*_{(t)}] \tag{14.63}$$

$G^*_{(s)}$ can also be expressed, using Eq. (14.12), as

$$G^*_{(s)} = \frac{1}{T_s} \sum_{n=-\infty}^{n=+\infty} G_{(s+in\omega_s)} \tag{14.64}$$

We show how these pulse transfer functions are applied to openloop and closedloop systems in Section 14.7.

14.6 HOLD DEVICES

A hold device is always needed in a sampled-data process control system. The *zero-order* hold converts the sequence of impulses of an impulse-sampled function $f^*_{(t)}$ to

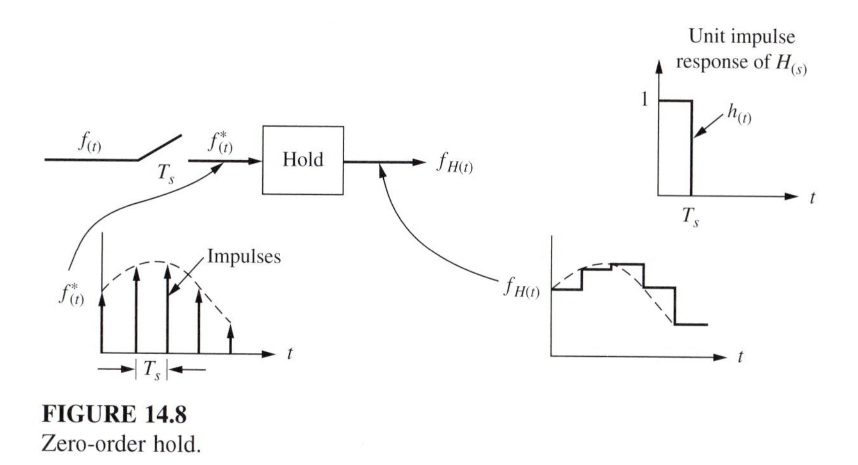

FIGURE 14.8
Zero-order hold.

a continuous stairstep function $f_{H(t)}$. The hold must convert an impulse $f^*_{(t)}$ of area or strength $f_{(nT_s)}$ at time $t = nT_s$ to a square *pulse* (not an impulse) of height $f_{(nT_s)}$ and width T_s. See Fig. 14.8. Let the unit *impulse* response of the hold be defined as $h_{(t)}$. If the hold does what we want it to do (i.e., convert an impulse to a step up and then a step down after T_s minutes), its unit impulse response must be

$$h_{(t)} = u_{n(t)} - u_{n(t-T_s)} \tag{14.65}$$

where $u_{n(t)}$ is the unit step function. Therefore the Laplace-domain transfer function $H_{(s)}$ of a zero-order hold is

$$H_{(s)} = \mathscr{L}[h_{(t)}] = \mathscr{L}[u_{(t)} - u_{(t-T_s)}] = \frac{1}{s} - \frac{e^{-T_s s}}{s}$$

$$\boxed{H_{(s)} = \frac{1 - e^{-T_s s}}{s}} \tag{14.66}$$

14.7 OPENLOOP AND CLOSEDLOOP SYSTEMS

We are now ready to use the concepts of impulse-sampled functions, pulse transfer functions, and holds to study the dynamics of sampled-data systems. Consider the sampled-data system shown in Fig. 14.9a in the Laplace domain. The input enters through an impulse sampler. The continuous output of the process $Y_{(s)}$ is

$$Y_{(s)} = G_{(s)} U^*_{(s)} \tag{14.67}$$

$Y_{(s)}$ is then impulse sampled to give $Y^*_{(s)}$. Equation (14.13) says that $Y^*_{(s)}$ is

$$Y^*_{(s)} = \frac{1}{T_s} \sum_{n=-\infty}^{n=+\infty} Y_{(s+in\omega_s)}$$

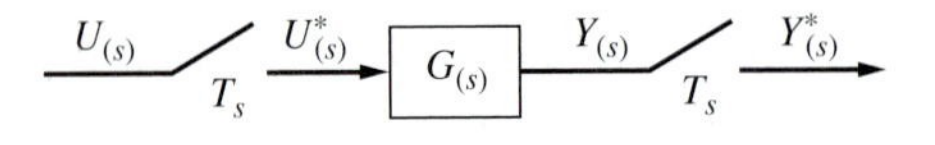

(*a*) Single element

(*b*) Series elements with intermediate sampler

(*c*) Series elements that are continuous

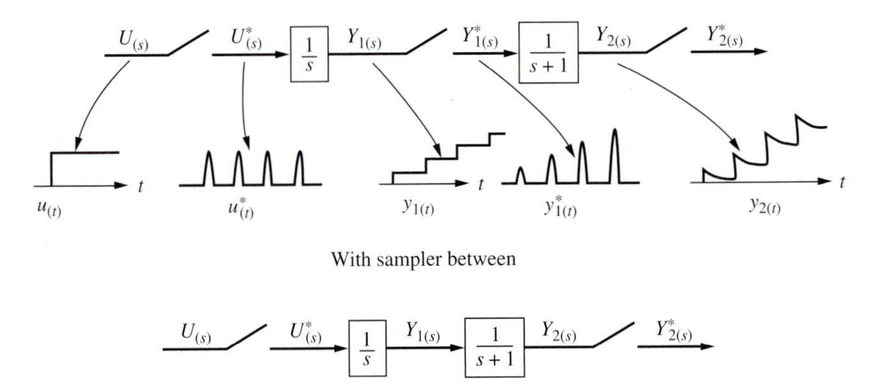

With sampler between

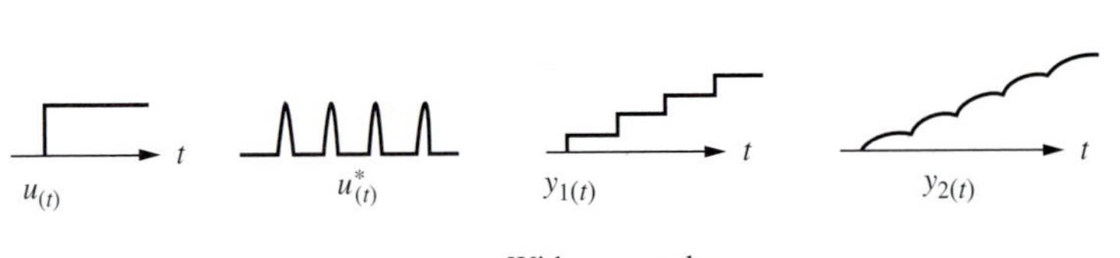

Without sampler

(*d*) System of Example 14.7

FIGURE 14.9
Openloop sampled-data systems.

Substituting for $Y_{(s+in\omega_s)}$, using Eq. (14.67), gives

$$Y^*_{(s)} = \frac{1}{T_s} \sum_{n=-\infty}^{n=+\infty} G_{(s+in\omega_s)} U^*_{(s+in\omega_s)} \tag{14.68}$$

We showed [Eq. (14.20)] that the Laplace transform of an impulse-sampled function is periodic.

$$U^*_{(s)} = U^*_{(s+i\omega_s)} = U^*_{(s-i\omega_s)} = U^*_{(s+i2\omega_s)} = \cdots \tag{14.69}$$

Therefore, the $U^*_{(s+in\omega_s)}$ terms can be factored out of the summation in Eq. (14.68) to give

$$Y^*_{(s)} = \left(\frac{1}{T_s}\sum_{n=-\infty}^{n=+\infty} G_{(s+in\omega_s)}\right)U^*_{(s)}$$

The term in the parentheses is $G^*_{(s)}$ according to Eq. (14.64), and therefore the output of the process in the Laplace domain is

$$Y^*_{(s)} = G^*_{(s)}U^*_{(s)} \tag{14.70}$$

By z-transforming this equation, using Eq. (14.61), the output in the z domain is

$$Y_{(z)} = G_{(z)}U_{(z)} \tag{14.71}$$

Now consider the system shown in Fig. 14.9*b*, where there are two elements separated by a sampler. The continuous output $Y_{1(s)}$ is

$$Y_{1(s)} = G_{1(s)}U^*_{(s)}$$

When $Y_{1(s)}$ goes through the impulse sampler it becomes $Y^*_{1(s)}$, which can be expressed [see Eq. (14.69)] as

$$Y^*_{1(s)} = \frac{1}{T_s}\sum_{n=-\infty}^{n=+\infty} G_{1(s+in\omega_s)}U^*_{(s+in\omega_s)} = G^*_{1(s)}U^*_{(s)} \tag{14.72}$$

The continuous function $Y_{2(s)}$ is

$$Y_{2(s)} = G_{2(s)}Y^*_{1(s)}$$

The impulse-sampled $Y^*_{2(s)}$ is

$$\begin{aligned} Y^*_{2(s)} &= G^*_{2(s)}Y^*_{1(s)} = G^*_{2(s)}(G^*_{1(s)}U^*_{(s)}) \\ &= G^*_{1(s)}G^*_{2(s)}U^*_{(s)} \end{aligned} \tag{14.73}$$

In the z domain, this equation becomes

$$Y_{2(z)} = G_{1(z)}G_{2(z)}U_{(z)} \tag{14.74}$$

Thus, the overall transfer function of the process can be expressed as a product of the two individual pulse transfer functions if there is an impulse sampler between the elements.

Consider now the system shown in Fig. 14.9*c*, where the two continuous elements $G_{1(s)}$ and $G_{2(s)}$ do *not* have a sampler between them. The continuous output $Y_{2(s)}$ is

$$Y_{2(s)} = G_{2(s)}Y_{1(s)} = G_{2(s)}G_{1(s)}U^*_{(s)} \tag{14.75}$$

Sampling the output gives

$$Y^*_{2(s)} = \frac{1}{T_s}\sum_{n=-\infty}^{n=+\infty} G_{1(s+in\omega_s)}G_{2(s+in\omega_s)}U^*_{(s+in\omega_s)}$$

$$= \left(\frac{1}{T_s}\sum_{n=-\infty}^{n=+\infty} G_{1(s+in\omega_s)}G_{2(s+in\omega_s)}\right)U^*_{(s)} \tag{14.76}$$

The term in parentheses is the Laplace transformation of the impulse-sampled response of the *total combined* process to a unit impulse input. We call this $(G_1G_2)^*_{(s)}$ in the Laplace domain and $(G_1G_2)_{(z)}$ in the z domain.

$$Y^*_{2(s)} = (G_1G_2)^*_{(s)}U^*_{(s)} \tag{14.77}$$

$$Y_{2(z)} = (G_1G_2)_{(z)}U_{(z)} \tag{14.78}$$

Equation (14.78) looks somewhat like Eq. (14.74), but it is not at all the same. The two processes are physically different: one has a sampler between the G_1 and G_2 elements, and the other does not.

$$\begin{aligned}(G_1G_2)^*_{(s)} &\neq G^*_{1(s)}G^*_{2(s)}\\ (G_1G_2)_{(z)} &\neq G_{1(z)}G_{2(z)}\end{aligned} \tag{14.79}$$

Let us take a specific example to illustrate the difference between these two systems.

EXAMPLE 14.7. Suppose the system has two elements as shown in Fig. 14.9*d*.

$$G_{1(s)} = \frac{1}{s} \qquad G_{2(s)} = \frac{1}{s+1} \tag{14.80}$$

With an impulse sampler between the elements, the overall system transfer function is, from Eq. (14.74),

$$\begin{aligned}G_{1(z)}G_{2(z)} &= \mathscr{Z}[g^*_{1(t)}]\mathscr{Z}[g^*_{2(t)}] = \mathscr{Z}\left[I_{(t)}\mathscr{L}^{-1}\left(\frac{1}{s}\right)\right]\mathscr{Z}\left[I_{(t)}\mathscr{L}^{-1}\left(\frac{1}{s+1}\right)\right]\\ &= \mathscr{Z}[I_{(t)}u_{n(t)}]\mathscr{Z}[I_{(t)}e^{-t}] = \left(\frac{z}{z-1}\right)\left(\frac{z}{z-e^{-T_s}}\right)\end{aligned} \tag{14.81}$$

In the preceding calculation, we went through the time domain, getting $g_{(t)}$ by inverting $G_{(s)}$ and then z-transforming $g^*_{(t)}$. The operation can be represented more concisely by going directly from the Laplace domain to the z domain.

$$G_{1(z)}G_{2(z)} = \mathscr{Z}[G^*_{1(s)}]\mathscr{Z}[G^*_{2(s)}] = \mathscr{Z}\left[\frac{1}{s}\right]\mathscr{Z}\left[\frac{1}{s+1}\right] \tag{14.82}$$

This equation is a shorthand expression for Eq. (14.81). The inversion to the impulse response $g_{(t)}$ and the impulse sampling to get $g^*_{(t)}$ is implied in the notation $\mathscr{Z}[1/s]$ and $\mathscr{Z}[1/(s+1)]$.

$$G_{1(z)}G_{2(z)} = \mathscr{Z}\left[\frac{1}{s}\right]\mathscr{Z}\left[\frac{1}{s+1}\right] = \left(\frac{z}{z-1}\right)\left(\frac{z}{z-e^{-T_s}}\right) \tag{14.83}$$

The responses of $y^*_{(t)}$, $y_{1(t)}$, and $y_{2(t)}$ to a unit step change in $u_{(t)}$ are sketched in Fig. 14.7*d*.

Without a sampler between the elements, the overall system transfer function is [Eq. (14.78)]

$$
\begin{aligned}
(G_1G_2)_{(z)} &= \mathcal{Z}\left[I_{(t)}\mathcal{L}^{-1}\left(\frac{1}{s(s+1)}\right)\right] = \mathcal{Z}\left[I_{(t)}\mathcal{L}^{-1}\left(\frac{1}{s} - \frac{1}{s+1}\right)\right] \\
&= \mathcal{Z}[I_{(t)}u_{n(t)} - I_{(t)}e^{-t}] = \frac{z}{z-1} - \frac{z}{z-e^{-T_s}} \\
&= \frac{z(z-e^{-T_s})}{(z-1)(z-e^{-T_s})}
\end{aligned}
\tag{14.84}
$$

Using the shorthand notation,

$$
\begin{aligned}
(G_1G_2)_{(z)} &= \mathcal{Z}\left[\frac{1}{s(s+1)}\right] = \mathcal{Z}\left[\frac{1}{s} - \frac{1}{s+1}\right] = \frac{z}{z-1} - \frac{z}{z-e^{-T_s}} \\
&= \frac{z(z-e^{-T_s})}{(z-1)(z-e^{-T_s})}
\end{aligned}
$$

From now on we use the shorthand, Laplace-domain notation, but keep in mind what is implied in its use.

Notice that Eq. (14.83) is not equal to Eq. (14.84). The responses of the two systems' $y_{2(t)}$ are not the same, as shown in Fig. 14.9*d*, because the systems are physically different. ■

Now let us look at a closedloop system with a sampled-data digital controller as shown in Fig. 14.10. The equations describing the system are

$$Y_{(s)} = G_{L(s)}L_{(s)} + H_{(s)}G_{M(s)}M^*_{(s)} \tag{14.85}$$

$$M_{(s)} = D^*_{(s)}(Y^{\text{set}*}_{(s)} - Y^*_{(s)}) \tag{14.86}$$

Sampling $Y_{(s)}$ and $M_{(s)}$ gives

$$
\begin{aligned}
Y^*_{(s)} &= [G_L L]^*_{(s)} + [HG_M]^*_{(s)}M^*_{(s)} \\
M^*_{(s)} &= D^*_{(s)}Y^{\text{set}*}_{(s)} - D^*_{(s)}Y^*_{(s)}
\end{aligned}
$$

z transforming and combining give

$$
\begin{aligned}
Y_{(z)} &= (G_L L)_{(z)} + (HG_M)_{(z)}(D_{(z)}Y^{\text{set}}_{(z)} - D_{(z)}Y_{(z)}) \\
&= \frac{(G_L L)_{(z)} + (HG_M)_{(z)}D_{(z)}Y^{\text{set}}_{(z)}}{1 + (HG_M)_{(z)}D_{(z)}}
\end{aligned}
\tag{14.87}
$$

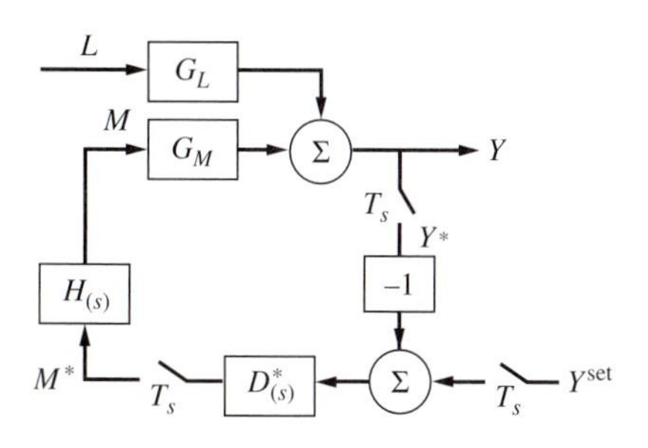

FIGURE 14.10
Closedloop sampled-data block diagram.

In this system we obtain an explicit input/output transfer function relationship between $Y_{(z)}$ and $Y^{set}_{(z)}$:

$$\frac{Y_{(z)}}{Y^{set}_{(z)}} = \frac{(HG_M)_{(z)}D_{(z)}}{1 + (HG_M)_{(z)}D_{(z)}} \tag{14.88}$$

EXAMPLE 14.8. Consider a first-order process with the transfer function

$$G_{M(s)} = \frac{K_p}{\tau_o s + 1} \tag{14.89}$$

A zero-order hold and a proportional controller are used.

$$D_{(z)} = K_c \tag{14.90}$$

$$H_{(s)} = \frac{1 - e^{-T_s s}}{s} \tag{14.91}$$

We want to find the response of the closedloop system to a unit step change in setpoint.

$$Y^{set}_{(s)} = \frac{1}{s} \tag{14.92}$$

We need to find $(HG_M)_{(z)}$ and $Y^{set}_{(z)}$ to plug into Eq. (14.88). There is no load disturbance, so L is equal to zero.

$$\begin{aligned}\mathscr{Z}[HG_M] &= \mathscr{Z}\left[\frac{1 - e^{-T_s s}}{s}\frac{K_p}{\tau_o s + 1}\right] \\ &= K_p \mathscr{Z}\left[\frac{1 - e^{-T_s s}}{s(\tau_o s + 1)}\right] \\ &= K_p(1 - z^{-1})\mathscr{Z}\left[\frac{1}{s(\tau_o s + 1)}\right]\end{aligned}$$

since

$$\begin{aligned}\mathscr{Z}[(1 - e^{-T_s s})F^*_{(s)}] &= \mathscr{Z}[F^*_{(s)}] - \mathscr{Z}[e^{-T_s s}F^*_{(s)}] \\ &= F_{(z)} - z^{-1}F_{(z)} = (1 - z^{-1})F_{(z)}\end{aligned}$$

$$\begin{aligned}\mathscr{Z}[HG_M] &= K_p\left[1 - \frac{1}{z}\right]\mathscr{Z}\left[\frac{1}{s} - \frac{1}{s + 1/\tau_o}\right] \\ &= K_p\left[\frac{z-1}{z}\right]\left[\frac{z}{z-1} - \frac{z}{z-b}\right] \\ &= \frac{K_p(1-b)}{z-b}\end{aligned} \tag{14.93}$$

where

$$b \equiv e^{-T_s/\tau_o} \tag{14.94}$$

Since $Y^{set}_{(s)} = 1/s$, z transforming gives

$$Y^{set}_{(z)} = \frac{z}{z-1} \tag{14.95}$$

Therefore, $Y_{(z)}$ becomes

$$Y_{(z)} = \frac{(HG_M)_{(z)}D_{(z)}}{1+(HG_M)_{(z)}D_{(z)}}Y^{\text{set}}_{(z)} = \frac{\dfrac{K_p(1-b)}{z-b}K_c}{1+\dfrac{K_cK_p(1-b)}{z-b}}Y^{\text{set}}_{(z)}$$

$$= \frac{K_cK_p(1-b)}{[z-b+K_cK_p(1-b)]}Y^{\text{set}}_{(z)} \tag{14.96}$$

Combining Eqs. (14.95) and (14.96) gives

$$Y_{(z)} = \frac{K_cK_p(1-b)z}{z^2 - z[1+b-K_cK_p(1-b)] + [b-K_cK_p(1-b)]} \tag{14.97}$$

Let us take a specific numerical case where $\tau_o = K_p = 1$, $T_s = 0.2$, and $K_c = 4.5$. The value of b is $e^{-T_s/\tau_o} = e^{-0.2/1} = 0.8187$. Equation (14.97) gives

$$Y_{(z)} = \frac{0.8159z}{z^2 - 1.00302z + 0.003019} \tag{14.98}$$

This is the function we used in Example 14.4. ■

In the preceding example we derived the expression for $Y_{(z)}$ analytically for a step change in $Y^{\text{set}}_{(z)}$. An alternative approach is to use MATLAB to calculate the step response of the closedloop servo transfer function

$$\frac{Y_{(z)}}{Y^{\text{set}}_{(z)}} = \frac{(HG_M)_{(z)}D_{(z)}}{1+(HG_M)_{(z)}D_{(z)}}$$

Equation (14.96) shows that for the process considered in the example

$$\frac{Y_{(z)}}{Y^{\text{set}}_{(z)}} = \frac{K_cK_p(1-b)}{[z-b+K_cK_p(1-b)]} \tag{14.99}$$

Table 14.4 gives a MATLAB program that calculates the step response using the *[y,x]=dstep(num,den,n)* command. The numerator and the denominator of the

TABLE 14.4
MATLAB program for step response

```
% Program "testdstep.m"
%  gets setpoint step response for closedloop process
%  with sampled-data P controller with Kc=4.5, 6.57, and 12
%  sampling period = Ts = 0.2
% Openloop continuous process G(s)=Kp/(To*s+1)
% Closedloop sampled-data Y/Yset with zero-order hold
%        Y/Yset = Kc*Kp*(1-b)/[z-b + Kc*Kp*(1-b)]
%
% Closedloop M/Yset = Kc*(Yset-Y)/Yset
%
%        M/Yset = Kc*(z-b)/[z-b + Kc*Kp*(1-b)]
%
%    where b=exp(-Ts/To)
% Parameter Values: To=Kp=1
```

TABLE 14.4 (CONTINUED)
MATLAB program for step response

```
ntotal=5;
ts=0.2;
b=exp(-ts);

% Kc=4.5
kc=4.5;
num1=kc*(1-b);
den1=[1 (-b+kc*(1-b))];
% Calculate Y step response
[y1,x]=dstep(num1,den1,ntotal);
% Calculate M step response
numm1=kc*[z-b];
[m1,x]=dstep(numm1,den1,ntotal);

% Kc=6.57
kc=6.57;
num2=kc*(1-b);
den2=[1 (-b+kc*(1-b))];
% Calculate Y step response
[y2,x]=dstep(num2,den2,ntotal);
% Calculate M step response
numm2=kc*[z-b];
[m2,x]=dstep(numm2,den2,ntotal);

% Kc=12
kc=12;
num3=kc*(1-b);
den3=[1 (-b+kc*(1-b))];
% Calculate Y step response
[y3,x]=dstep(num3,den3,ntotal);
% Calculate M step response
numm3=kc*[z-b];
[m3,x]=dstep(numm3,den3,ntotal);
%
% Convert number of points to time scale
npts=length(y1);
points=[0:1:(npts-1)]';
t=ts*points;
%
% Convert M plots for zero-order holds
[t1,mm1]=stairs(t,m1);
[t2,mm2]=stairs(t,m2);
[t3,mm3]=stairs(t,m3);

clf
orient tall
subplot(211)
plot(t,y1,'+',t,y2,'x',t,y3,'o')
xlabel('Time (minutes)')
ylabel('Output Y')
grid
legend('Kc=4.5','Kc=6.57','Kc=12')
```

TABLE 14.4 (CONTINUED)
MATLAB program for step response

```
subplot(212)
plot(t1,mm1,':',t2,mm2,'--',t3,mm3,'-')

xlabel('Time (minutes)')
ylabel('Manipulated Variable M')
grid
legend('Kc=4.5','Kc=6.57','Kc=12')

print -dps pfig1411.ps
```

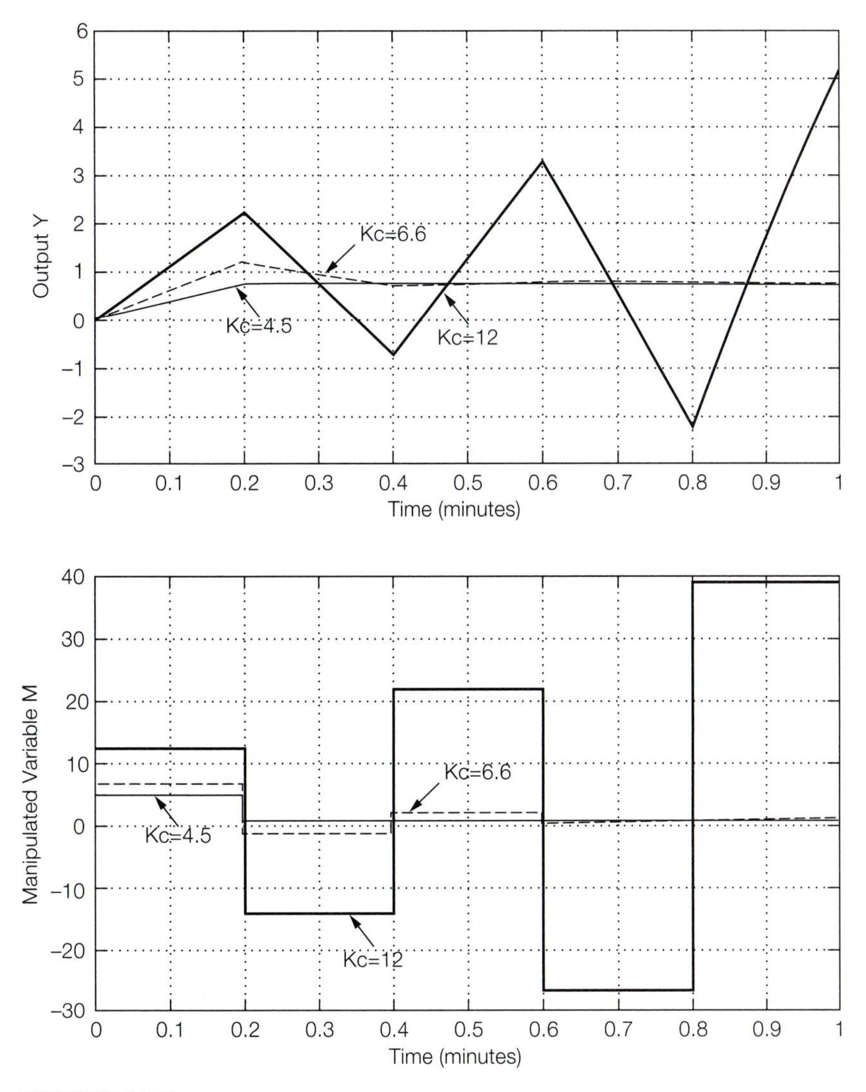

FIGURE 14.11
Sampled-data control: $T_s = 0.2$.

TABLE 14.5
MATLAB program to derive pulse transfer functions

```
% Program "convert.m"
%   illustrated conversion from continuous to discrete transfer function
%   First example is GM(s)=1/(s+1), with sampling period Ts=0.2
%
% form numerator and denominator polynomials for continuous process
num=1;
den=[1  1];
printsys(num,den,'s')
pause
ts=0.2;
%
% Find sampled-data transfer function using "c2dm" command
[numd,dend]=c2dm(num,den,ts,'zoh');
printsys(numd,dend,'z')
pause
%
% Second example is GM(s)=2.315/[(s+1)(5s+1)], with Ts=0.5
num2=2.315;
den2=conv([1  1],[5  1]);
printsys(num2,den2,'s')
pause
ts2=0.5;
[numd2,dend2]=c2dm(num2,den2,ts2,'zoh');
printsys(numd2,dend2,'z')
roots(numd2)
roots(dend2)
pause
```

TABLE 14.6
MATLAB program to obtain closedloop transfer function from openloop transfer function

```
% Program "discretefb.m"
%   Calculates closedloop transfer functions
%   from openloop pulse transfer functions using "cloop" command
% Process is G(s)=1/(s+1)
% Sampled-data controller is proportional with Kc=4.5
kc=4.5;
num=1;
den=[1  1];
ts=0.2;
%
% Convert continuous to discrete using zero-order hold
%
[numd,dend]=c2dm(num,den,ts,'zoh');
printsys(numd,dend,'z')
pause
%
% Calculate unity-feedback closedloop discrete transfer function
[ncl,dcl]=cloop(numd*kc,dend,-1);
printsys(ncl,dcl,'z')
```

closedloop servo transfer function are formed using Eq. (14.99) for three different values of controller gain. Results are plotted in Fig. 14.11. Both the output Y and the manipulated variable M are plotted. The *stairs* command produces the stairstep functions of the manipulated variables coming from the zero-order hold. The values of Y are plotted only at the sampling points. These are what the computer sees. The continuous output of the process $y_{(t)}$ is a series of exponential responses between the sampling points.

Instead of analytically deriving the z-domain pulse transfer functions, we can use MATLAB to convert from $G_{M(s)}$ to $HG_{M(z)}$ with the *[numd,dend]=c2dm(num,den, ts,'zoh')* command. Table 14.5 gives a program that illustrates the procedure for two different processes: a first-order lag and a second-order lag.

Instead of deriving the closedloop transfer function analytically, we can have MATLAB do it for us by using the *[numcl,dencl]=cloop(numol,denol,−1)* command as shown in Table 14.6.

14.8 STABILITY IN THE z PLANE

The stability of any system is determined by the location of the roots of its characteristic equation (or the poles of its transfer function). The characteristic equation of a continuous system is a polynomial in the complex variable s. If all the roots of this polynomial are in the left half of the s plane, the system is stable. For a continuous closedloop system, all the roots of $1 + G_{M(s)}G_{C(s)}$ must lie in the left half of the s plane. Thus, the region of stability in continuous systems is the left half of the s plane.

The stability of a sampled-data system is determined by the location of the roots of a characteristic equation that is a polynomial in the complex variable z. This characteristic equation is the denominator of the system transfer function set equal to zero. The roots of this polynomial (the poles of the system transfer function) are plotted in the z plane. The ordinate is the imaginary part of z, and the abscissa is the real part of z.

The region of stability in the z plane can be found directly from the region of stability in the s plane by using the basic relationship between the complex variables s and z [Eq. (14.23)].

$$z = e^{T_s s} \tag{14.100}$$

Figure 14.12 shows the s plane. Let the real part of s be α and the imaginary part of s be ω.

$$s = \alpha + i\omega \tag{14.101}$$

The stability region in the s plane is where α, the real part of s, is negative. Substituting Eq. (14.101) into Eq. (14.100) gives

$$z = e^{T_s(\alpha + i\omega)} = (e^{\alpha T_s})e^{i\omega T_s} \tag{14.102}$$

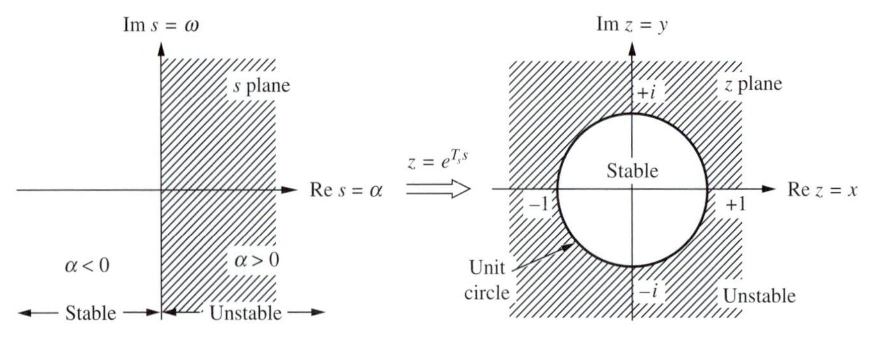

FIGURE 14.12
Stability regions in the s plane and in the z plane.

The absolute value of z, $|z|$, is $e^{\alpha T_s}$. When α is negative, $|z|$ is less than 1. When α is positive, $|z|$ is greater than 1. Therefore, the left half of the s plane maps into the inside of the unit circle in the z plane, as shown in Fig. 14.12.

A sampled-data system is stable if all the roots of its characteristic equation (the poles of its transfer function) lie inside the unit circle in the z plane.

First let's consider an openloop system with the openloop transfer function

$$HG_{M(z)} = \frac{(z - z_1)(z - z_2)\cdots(z - z_M)}{(z - p_1)(z - p_2)\cdots(z - p_N)} \tag{14.103}$$

The stability of this openloop system depends on the values of the poles of the openloop transfer function. If all the p_i lie inside the unit circle, the system is openloop stable.

The more important problem is closedloop stability. The equation describing the closedloop digital control system of Section 14.7 is

$$Y_{(z)} = \frac{(G_L L)_{(z)} + (HG_M)_{(z)} D_{(z)} Y_{(z)}^{\text{set}}}{1 + (HG_M)_{(z)} D_{(z)}} \tag{14.104}$$

The closedloop stability of this system depends on the location of the roots of the characteristic equation:

$$1 + (HG_M)_{(z)} D_{(z)} = 0 \tag{14.105}$$

If all the roots lie inside the unit circle, the system is closedloop stable.

EXAMPLE 14.9. Consider a first-order process with a zero-order hold and proportional sampled-data controller.

$$G_{M(s)} = \frac{K_p}{\tau_o s + 1}$$

As we developed in the previous section, the openloop pulse transfer function for this process is

$$(HG_M)_{(z)} = \mathcal{Z}\left[\frac{1 - e^{-T_s s}}{s} \frac{K_p}{\tau_o s + 1}\right] = \frac{K_p(1 - b)}{z - b} \tag{14.106}$$

where $b \equiv e^{-T_s/\tau_o}$. (14.107)

The *openloop* characteristic equation is

$$z - b = 0$$

The root of the openloop characteristic equation is b. Since b is less than 1, this root lies inside the unit circle and the system is *openloop* stable.

The *closedloop* characteristic equation for this system is

$$1 + (HG_M)_{(z)} D_{(z)} = 1 + \frac{K_c K_p (1 - b)}{z - b} = 0 \tag{14.108}$$

since $D_{(z)} = K_c$. Solving for the closedloop root gives

$$z = b - K_c K_p (1 - b) \tag{14.109}$$

There is a single root. It lies on the real axis in the z plane, and its location depends on the value of the feedback controller gain K_c. When the feedback controller gain is zero (the openloop system), the root lies at $z = b$. As K_c is increased, the closedloop root moves to the left along the real axis in the z plane. We return to this example in the next chapter. ■

14.9 CONCLUSION

In this chapter we have learned some of the mathematics and notation of z transforms and developed transfer function descriptions of sampled-data systems. The usefulness of MATLAB in obtaining transfer functions and step responses has been illustrated. The stability region is the interior of the unit circle in the z plane.

PROBLEMS

14.1. Derive the z transforms of the following functions:

(*a*) $f^*_{(t)} = I_{(t)} t^2$

(*b*) $f^*_{(t)} = I_{(t)} t^2 e^{-at}$

(*c*) $f^*_{(t)} = I_{(t)} \cos(\omega t)$

(*d*) $f^*_{(t)} = I_{(t)} e^{-\zeta t/\tau_o}\left[\cos\left(\frac{\sqrt{1-\zeta^2}}{\tau_o} t\right) + \frac{\zeta}{\sqrt{1-\zeta^2}} \sin\left(\frac{\sqrt{1-\zeta^2}}{\tau_o} t\right)\right]$

(*e*) $f^*_{(t)} = I_{(t)} K_p \tau_o \left(\frac{t}{\tau_o} - 1 + e^{-t/\tau_o}\right)$

(*f*) $f^*_{(t)} = I_{(t)}e^{-a(t-kTs)}$ where k is an integer

(*g*) $f^*_{(t)} = I_{(t)}\dfrac{K}{\tau_{o1} - \tau_{o2}}[e^{-t/\tau_{o1}} - e^{-t/\tau_{o2}}]$

14.2. Find the pulse transfer functions in the z domain $(HG_M)_{(z)}$ for the following systems ($H_{(s)}$ is zero-order hold):

(*a*) $G_{M(s)} = \dfrac{K_p}{(\tau_{o1}s + 1)(\tau_{o2}s + 1)}$

(*b*) $G_{M(s)} = \dfrac{K_p e^{-kT_s s}}{(\tau_{o1}s + 1)(\tau_{o2}s + 1)}$

14.3. Find $y_{(nT_s)}$ for a unit step input in $u_{(t)}$ for the system in part (*a*) of Problem 14.2 by partial-fractions expansion and by long division. Use the following numerical values of parameters:

$$K_p = \tau_{o2} = 1 \qquad \tau_{o1} = 5 \qquad T_s = 0.5$$

14.4. Repeat Problem 14.3 for part (*b*) of Problem 14.2. Use $k = 3$.

14.5. Use the subroutine LONGD given in Table 14.2 to find the response of the closedloop system of Example 14.4 to a unit step load disturbance. Use values of $\tau_o = K_p = 1$:
(*a*) With $T_s = 0.2$ and $K_c = 2, 4, 6, 8, 10, 12$
(*b*) With $T_s = 0.4$ and $K_c = 2, 4, 6, 8, 10, 12$
(*c*) With $T_s = 0.6$ and $K_c = 2, 4, 6, 8, 10, 12$
What do you conclude about the effect of sampling time on stability from these results?

14.6. Find the outputs $y_{2(nT_s)}$ of the two systems of Example 14.7 for a unit step input in $u_{(t)}$. Use partial-fractions expansion and long division.

14.7. Repeat Problem 14.6 for a ramp input in $u_{(t)}$.

14.8. A distillation column has an approximate transfer function between overhead composition x_D and reflux flow rate R of

$$G_{M(s)} = \frac{x_{D(s)}}{R_{(s)}} = \frac{0.0092}{(5s + 1)^2}\,\frac{\text{mole fraction}}{\text{mol/min}}$$

A chromatograph must be used to detect x_D. A sampled-data P controller is used with a gain of 1000. Calculate the response of x_D to a unit step change in setpoint for different chromatograph cycle times D_c (5, 10, and 20 minutes). The sampling period T_s is set equal to the chromatograph cycle time.

14.9. A tubular chemical reactor's response to a change in feed concentration is found to be essentially a pure deadtime D with attenuation K_p. A computer monitors the outlet concentration $C_{AL(t)}$ and changes the feed concentration $C_{AO(t)}$, through a zero-order hold, using proportional action. The sampling period T_s can be adjusted to an integer multiple of D. Calculate the response of C_{AL} for a unit step change in setpoint C_{AL}^{set} for $D/T_s = 1$ and $D/T_s = 2$:
(*a*) With $K_c = 1/K_p$
(*b*) With $K_c = 1/2K_p$

CHAPTER 15

Stability Analysis of Sampled-Data Systems

We developed the mathematical tool of z transformation in the last chapter. Now we are ready to apply it to analyze the dynamics of sampled-data systems. Our primary task is to design sampled-data feedback controllers for these systems. We explore the very important impact of sampling period T_s on these designs.

The first two sampled-data controller design methods use conventional root locus and frequency response methods, which are completely analogous to the techniques in continuous systems. Instead of looking at the s plane, however, we look at the z plane. The third sampled-data controller design method is similar to the direct synthesis method discussed in Chapter 9.

15.1 ROOT LOCUS DESIGN METHODS

With continuous systems we make root locus plots in the s plane. Controller gain is varied from zero to infinity, and the roots of the closedloop characteristic equation are plotted. Time constants, damping coefficients, and stability can be easily determined from the positions of the roots in the s plane. The limit of stability is the imaginary axis. Lines of constant closedloop damping coefficient are radial straight lines from the origin. The closedloop time constant is the reciprocal of the distance from the origin.

With sampled-data systems root locus plots can be made in the z plane in almost exactly the same way. Controller gain is varied from zero to infinity, and the roots of the closedloop characteristic equation $1 + HG_{M(z)}D_{(z)} = 0$ are plotted. When the roots lie inside the unit circle, the system is closedloop stable. When the roots lie outside the unit circle, the system is closedloop unstable.

In continuous systems lines of constant damping coefficient ζ in the s plane are radial lines from the origin, as sketched in Fig. 15.1.

$$\zeta = \cos\theta \tag{15.1}$$

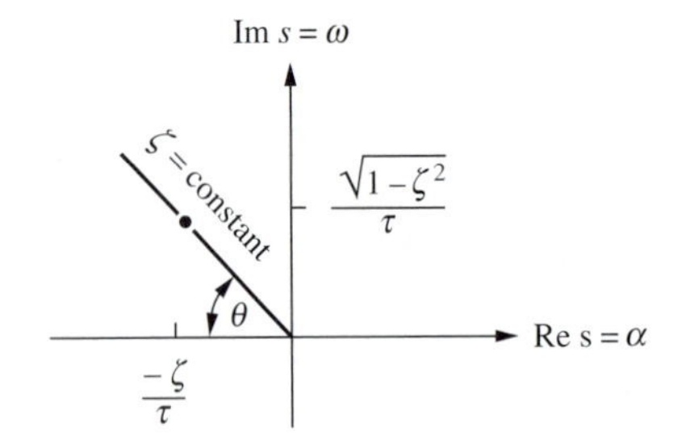

(*a*) In the *s* plane

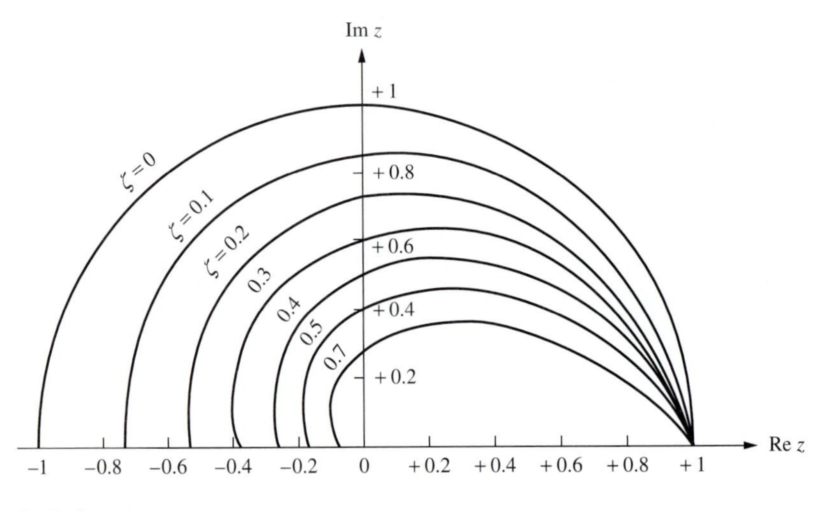

(*b*) In the *z* plane

FIGURE 15.1
Lines of constant damping coefficient.

where θ is the angle between the radial line and the negative real axis. Along a line of constant ζ in the s plane, the tangent of θ is

$$\tan\theta = \frac{\omega}{\alpha} = \frac{\left(\sqrt{1-\zeta^2}\right)/\tau}{-\zeta/\tau} = \frac{\sqrt{1-\zeta^2}}{-\zeta} \tag{15.2}$$

Using Eq. (15.2), the real part of s, α, can be expressed in terms of the imaginary part of s, ω, and the damping coefficient ζ.

$$\alpha = \frac{-\zeta\omega}{\sqrt{1-\zeta^2}} \tag{15.3}$$

These lines of constant damping coefficient can be mapped into the z plane. The z variable along a line of constant damping coefficient is

$$z = e^{T_s s} = e^{T_s(\alpha+i\omega)} = e^{\alpha T_s} e^{i\omega T_s} = \exp\left(-\frac{\zeta\omega T_s}{\sqrt{1-\zeta^2}}\right) e^{i\omega T_s} \tag{15.4}$$

Lines of constant damping coefficient in the z plane can be generated by picking a value of ζ and varying ω in Eq. (15.4) from 0 to $\omega_s/2$. See Section 15.2 for a discussion of the required range of ω.

Figure 15.1*b* shows these curves in the z plane. On the unit circle, the damping coefficient is zero. On the positive real axis, the damping coefficient is greater than 1. At the origin, the damping coefficient is exactly unity.

Notice the very significant result that the damping coefficient is less than 1 on the *negative* real axis. This means that in sampled-data systems a real root can give underdamped response. This can never happen in a continuous system; the roots must be complex to give underdamped response.

So we can design sampled-data controllers for a desired closedloop damping coefficient by adjusting the controller gain to position the roots on the desired damping line. Some examples illustrate the method and point out the differences and the similarities between continuous systems and sampled-data systems.

EXAMPLE 15.1. Let us make a root locus plot for the first-order system considered in Example 14.10.

$$G_{M(s)} = \frac{K_p}{\tau_o s + 1} \quad \Rightarrow \quad (HG_M)_{(z)} = \frac{K_p(1-b)}{z-b}$$

Using a proportional sampled-data controller, the closedloop characteristic equation for this system is

$$1 + (HG_M)_{(z)}D_{(z)} = 1 + \frac{K_c K_p(1-b)}{z-b} = 0$$

The closedloop root is

$$z = b - K_c K_p(1-b) \tag{15.5}$$

Figure 15.2 shows the root locus plot. It starts ($K_c = 0$) at $z = b$ on the positive real axis inside the unit circle, so it is openloop stable. As K_c is increased, the closedloop root moves to the left.

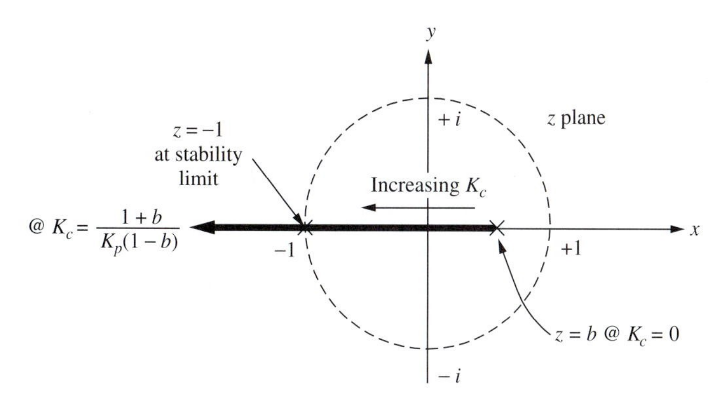

FIGURE 15.2
z plane location of roots (Example 15.1).

At the origin (where $z = 0$) the damping coefficient is unity. Solving Eq. (15.5) for the value of controller gain that gives this "critically damped" system yields

$$0 = b - K_c K_p(1 - b) \quad \Rightarrow \quad (K_c)_{\zeta=1} = \frac{b}{K_p(1 - b)} \tag{15.6}$$

For gains less than this, the system is overdamped. For gains greater than this, the system is underdamped! This is distinctly different from continuous systems, where a first-order continuous system can never be underdamped.

If we wanted to design for a closedloop damping of 0.3, we would use Eq. (15.4) with $\zeta = 0.3$ and with $\omega T_s = \pi$ because we are on the negative real axis, where the argument of z is π. Equation (15.4) shows that the argument of z on a line of constant damping coefficient is ωT_s.

$$\arg z = \omega T_s = \pi \tag{15.7}$$

$$z = \exp\left(-\frac{0.3\pi}{\sqrt{1 - (0.3)^2}}\right) e^{i\pi} = (0.372)(\cos \pi + i \sin \pi) = -0.372 \tag{15.8}$$

So positioning the closedloop root on the negative real axis at -0.372 gives a closedloop system with a damping coefficient of 0.3. Solving for the required gain gives

$$-0.372 = b - K_c K_p(1 - b) \quad \Rightarrow \quad (K_c)_{\zeta=0.3} = \frac{b + 0.372}{K_p(1 - b)} \tag{15.9}$$

The system reaches the limit of closedloop stability when the root crosses the unit circle at $z = -1$. The value of controller gain at this limit is the ultimate gain K_u.

$$-1 = b - K_u K_p(1 - b) \quad \Rightarrow \quad K_u = \frac{1 + b}{K_p(1 - b)} \tag{15.10}$$

Let us take some numerical values to show the effect of changing the sampling period. Let $K_p = \tau_o = 1$. Table 15.1 gives values for the critical gain ($\zeta = 1$), the gain that gives $\zeta = 0.3$, and the ultimate gain for different sampling periods. Note that the gains decrease as T_s increases, showing that control gets worse as the sampling period gets bigger. Remember from Example 14.10 with $T_s = 0.2$ that a K_c of 12 gave a closedloop-unstable response. Table 15.1 shows that $K_u = 10$ for this T_s. ■

TABLE 15.1
Effect of sampling period

T_s	b	$(K_c)_{\zeta=1}$	$(K_c)_{\zeta=0.3}$	K_u
First-order lag				
0.1	0.905	9.51	13.4	20.0
0.2	0.819	4.52	6.57	10.0
0.5	0.606	1.54	2.49	4.08
1.0	0.368	0.582	1.17	2.16
First-order lag with deadtime $D = T_s$				
0.1		2.15		10.5
0.2		0.925		5.52
0.5		0.234		2.54
1.0		0.054		1.58

This example demonstrates several extremely important facts about sampled-data control. This simple first-order system, which could never be made closed-loop unstable in a *continuous* control system, *can* become closedloop unstable in a *sampled-data* system. This is an extremely important difference between continuous control and sampled-data control. It points out that continuous control is almost always *better* than sampled-data control!

It is logical to ask at this point why we use sampled-data, digital-computer control if it is inherently worse than continuous analog control. Computer control offers a number of advantages that outweigh its theoretical dynamic disadvantages: cost; ease of configuring and tuning control loops; ease of maintenance; reliability; data acquisition capability; built-in alarms; ability to handle nonstandard, complex, and nonlinear control algorithms; etc. But keep in mind that from a purely technical control performance standpoint, an analog controller can do a better job than a sampled-data controller in most applications. The process of sampling results in some loss of information: we don't know what is going on in between the sampling periods. Thus, there is an inherent degradation of dynamic performance.

EXAMPLE 15.2. Consider the first-order lag process with a deadtime of one sampling period.

$$G_{M(s)} = \frac{K_p e^{-T_s s}}{\tau_o s + 1} \quad \Rightarrow \quad (HG_M)_{(z)} = \frac{K_p(1-b)}{z(z-b)} \tag{15.11}$$

The addition of the one-sampling-period deadtime increases the order of the denominator polynomial to 2. Using a proportional sampled-data controller gives the closedloop characteristic equation

$$\begin{aligned} 1 + (HG_M)_{(z)} D_{(z)} &= 1 + \frac{K_c K_p (1-b)}{z(z-b)} = 0 \\ z^2 - bz + K_c K_p (1-b) &= 0 \end{aligned} \tag{15.12}$$

There are two root loci, as shown in Fig. 15.3*a* for the numerical values $K_p = \tau_o = 1$ and $T_s = 0.2$ ($b = 0.819$).

$$z = \frac{b \pm \sqrt{b^2 - 4K_c K_p (1-b)}}{2} \tag{15.13}$$

The paths start ($K_c = 0$) at $z = 0$ and $z = b$. They come together on the positive real axis at $z = b/2$ when

$$K_c = \frac{b^2}{4K_p(1-b)} = 0.925$$

This value of controller gain gives a critically damped system. For larger controller gains, the system is underdamped. The complex roots are

$$z = \frac{b}{2} \pm i \frac{1}{2} \sqrt{4K_c K_p (1-b) - b^2} \tag{15.14}$$

The value of gain that gives a closedloop damping coefficient of 0.3 can be found by using Eqs. (15.4) and (15.14) with $\zeta = 0.3$.

$$\exp\left(- \frac{0.3 \omega T_s}{\sqrt{1 - (0.3)^2}} \right) e^{i \omega T_s} = \frac{b}{2} \pm i \frac{1}{2} \sqrt{4K_c K_p (1-b) - b^2}$$

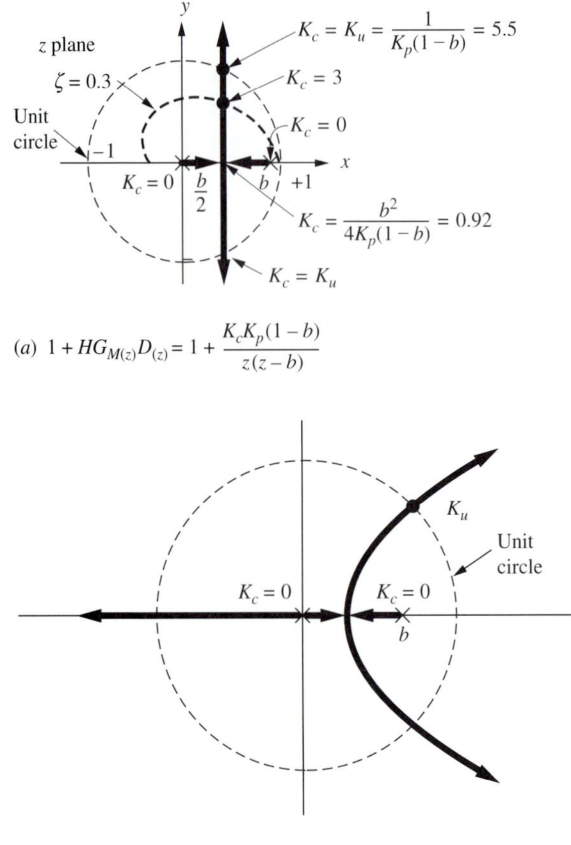

(b) $HG_{M(z)} = \dfrac{K_p(1-b)}{z^2(z-b)}$

FIGURE 15.3
Root locus plots in z plane.

$$e^{-0.3145\omega T_s}[\cos(\omega T_s) + i\sin(\omega T_s)] = \frac{b}{2} \pm i\frac{1}{2}\sqrt{4K_cK_p(1-b) - b^2}$$

Equating the real and imaginary parts of the left and right sides of the above equation gives two equations in the two unknowns, ω and K_c, for a given sampling period T_s.

The maximum K_c for which the closedloop system is stable occurs when the paths cross the unit circle. At that point the magnitude of z, $|z|$, is unity:

$$|z| = \sqrt{\left(\frac{b}{2}\right)^2 + \left[\frac{\sqrt{4K_uK_p(1-b) - b^2}}{2}\right]^2}$$

$$1 = \tfrac{1}{2}\sqrt{4K_uK_p(1-b)} \tag{15.15}$$

$$K_u = \frac{1}{K_p(1-b)}$$

Table 15.1 shows numerical values for different sampling periods. The addition of the deadtime to the lag reduces the gains. ■

EXAMPLE 15.3. Adding a deadtime to the first-order lag process that is equal to *two* sampling periods gives a third-order system in the z plane.

$$1 + (HG_M)_{(z)} D_{(z)} = 1 + \frac{K_c K_p (1-b)}{z^2(z-b)} = 0$$
$$z^3 - bz^2 + K_c K_p (1-b) = 0 \tag{15.16}$$

Figure 15.3*b* shows the root locus plot in the z plane. There are now three loci. The ultimate gain can occur on either the real-root path (at $z = -1$) or on the complex-conjugate-roots path. We can solve for these two values of controller gain and see which is smaller.

At $z = -1$. Using Eq. (15.16) gives

$$-1 - b + K_u K_p (1-b) = 0 \quad \Rightarrow \quad K_u = \frac{1+b}{K_p(1-b)} \tag{15.17}$$

For the numerical case $K_p = \tau_o = 1$ and $T_s = 0.5$, the result is 4.09.

On the complex-conjugate path. At the unit circle $z = e^{i\theta}$. Substituting into Eq. (15.16) gives

$$e^{i3\theta} - be^{i2\theta} + K_u K_p (1-b) = 0$$
$$\cos(3\theta) + i\sin(3\theta) - b[\cos(2\theta) + i\sin(2\theta)] + K_u K_p(1-b) = 0$$
$$\{\cos(3\theta) - b\cos(2\theta) + K_u K_p(1-b)\} + i\{\sin(3\theta) - b\sin(2\theta)\} = 0 + i0$$

Equating real and imaginary parts of both sides of the equation above gives two equations and two unknowns: θ and K_u.

$$\cos(3\theta) - b\cos(2\theta) + K_u K_p(1-b) = 0 \tag{15.18}$$

$$\sin(3\theta) - b\sin(2\theta) = 0 \tag{15.19}$$

Equation (15.19) can be solved for θ. For the numerical case $K_p = \tau_o = 1$ and $T_s = 0.5$, the result is $\theta = 0.83$ radians. Then Eq. (15.18) can be solved for K_u (= 1.88 for the numerical case). Since this is smaller than that calculated from Eq. (15.17), the ultimate gain is 1.88. Note that this is lower than the ultimate gain for the process with the smaller deadtime. ■

EXAMPLE 15.4. Now let's look at a second-order system.

$$G_{M(s)} = \frac{K_p}{(\tau_{o1}s + 1)(\tau_{o2}s + 1)} \tag{15.20}$$

Using a zero-order hold gives an openloop transfer function

$$HG_{M(z)} = \frac{K_p a_0 (z - z_1)}{(z - p_1)(z - p_2)} \tag{15.21}$$

where

$$p_1 = e^{-T_s/\tau_{o1}} \quad \text{and} \quad p_2 = e^{-T_s/\tau_{o2}} \tag{15.22}$$

$$a_0 = 1 + \frac{p_2\tau_{o2} - p_1\tau_{o1}}{\tau_{o1} - \tau_{o2}} \tag{15.23}$$

$$z_1 = \frac{p_1 p_2(\tau_{o2} - \tau_{o1}) + p_2\tau_{o1} - p_1\tau_{o2}}{\tau_{o1} - \tau_{o2} + p_2\tau_{o2} - p_1\tau_{o1}} \tag{15.24}$$

As a specific numerical example, consider the two-heated-tank process from Example 9.1 with the openloop transfer function

$$G_{M(s)} = \frac{2.315}{(s+1)(5s+1)}$$

Using a zero-order hold and a sampling period $T_s = 0.5$ minutes gives

$$HG_{M(z)} = \frac{2.315(0.0209)(z+0.8133)}{(z-0.607)(z-0.905)} \tag{15.25}$$

Several important features should be noted. The first-order process considered in Example 15.1 gives a pulse transfer function that is also first order; i.e., the denominator of the transfer function is first order in z. The second-order process considered in this example gives a sampled-data pulse transfer function that has a second-order denominator polynomial. These results can be generalized to an Nth-order system. The order of s in the denominator of the continuous transfer function is the same as the order of z in the denominator of the corresponding sampled-data transfer function.

However, note that in the case of the second-order system, the sampled-data transfer function has a zero, whereas the continuous transfer function does not. So in this respect the analogy between continuous and sampled-data transfer functions does not hold.

The closedloop characteristic equation for the second-order system with a proportional sampled-data controller is

$$1 + HG_{M(z)}D_{(z)} = 1 + \frac{K_c K_p a_0(z-z_1)}{(z-p_1)(z-p_2)} = 0$$

For the numerical example

$$1 + \frac{0.0479K_c(z+0.8133)}{(z-0.607)(z-0.905)} = 0 \tag{15.26}$$

$$z^2 + z(0.0479K_c - 1.512) + 0.549 + 0.039K_c = 0 \tag{15.27}$$

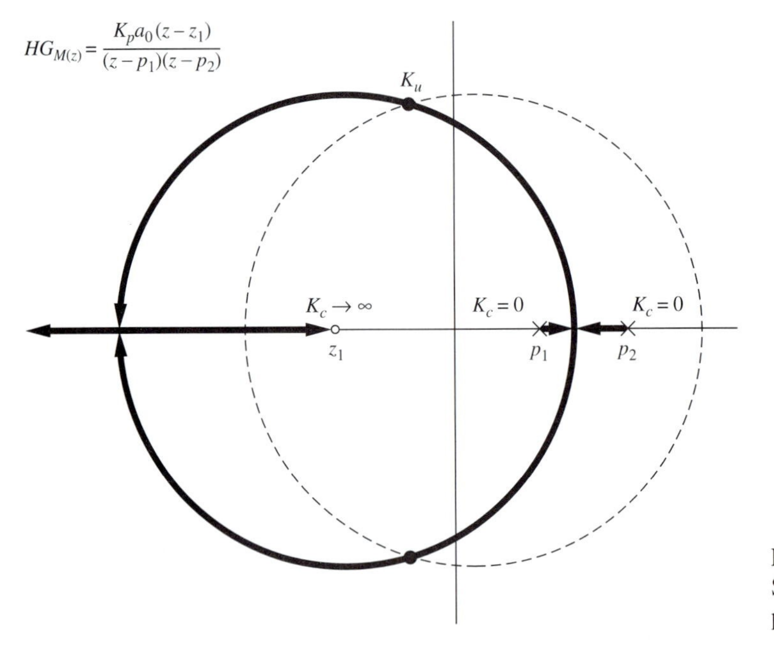

FIGURE 15.4
Second-order process.

To find the ultimate gain for the second-order system, we set $|z| = 1$ and solve for K_u.

$$z = \frac{-(0.0479K_u - 1.512)}{2} \pm i\frac{\sqrt{4(0.549 + 0.039K_u) - (0.0479K_u - 1.512)^2}}{2}$$

$$1 = \sqrt{\left(\frac{0.0479K_u - 1.512}{2}\right)^2 + \left(\frac{\sqrt{4(0.549 + 0.039K_u) - (0.0479K_u - 1.512)^2}}{2}\right)^2}$$

$$K_u = 11.6$$

This is the ultimate gain if the sampling period is 0.5 minutes. If $T_s = 2$ minutes is used, the openloop pulse transfer function becomes

$$HG_{M(z)} = \frac{0.4535(z + 0.455)}{(z - 0.135)(z - 0.67)} \tag{15.28}$$

The location of the zero is closer to the origin. This reduces the radius of the circular part of the root locus plot (see Fig. 15.4) and reduces the ultimate gain to 4.45. ■

15.2 FREQUENCY-DOMAIN DESIGN TECHNIQUES

Sampled-data control systems can be designed in the frequency domain by using the same techniques that we employed for continuous systems. The Nyquist stability criterion is applied to the appropriate closedloop characteristic equation to find the number of zeros outside the unit circle.

15.2.1 Nyquist Stability Criterion

The closedloop characteristic equation of a sampled-data system is

$$1 + HG_{M(z)}D_{(z)} = 0$$

We want to find out if the function $F_{(z)} \equiv 1 + HG_{M(z)}D_{(z)}$ has any zeros or roots outside the unit circle in the z plane. If it does, the system is closedloop unstable.

We can apply the $Z - P = N$ theorem of Chapter 11 to this new problem. We pick a contour that goes completely around the area in the z plane that is outside the unit circle, as shown in Fig. 15.5. We then plot $HG_{M(z)}D_{(z)}$ in the HG_MD plane and look at the encirclements of the $(-1, 0)$ point to find N.

If the number of poles of $HG_{M(z)}D_{(z)}$ outside the unit circle is known, the number of zeros outside the unit circle can be calculated from

$$Z = N + P \tag{15.29}$$

If the system is openloop stable, there will be no poles of $HG_{M(z)}D_{(z)}$ outside the unit circle, and $P = 0$. Thus, the Nyquist stability criterion can be applied directly to sampled-data systems.

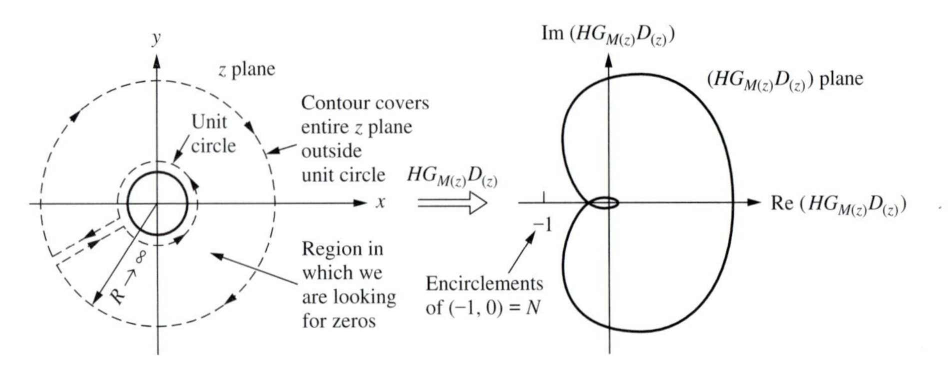

FIGURE 15.5
Nyquist stability criterion.

15.2.2 Rigorous Method

The relationship between z and s is

$$z = e^{T_s s}$$

Going from the Laplace domain to the frequency domain by substituting $s = i\omega$ gives

$$z = e^{i\omega T_s} \tag{15.30}$$

To make a Nyquist plot, we merely substitute $e^{i\omega T_s}$ for z in the $HG_{M(z)}D_{(z)}$ function and make a polar plot as frequency ω goes from 0 to $\omega_s/2$, where

$$\omega_s = \frac{2\pi}{T_s} \tag{15.31}$$

The reason we have to cover only this frequency range is demonstrated in the next example.

EXAMPLE 15.5. Let's consider the first-order lag process with a proportional sampled-data controller.

$$HG_{M(z)}D_{(z)} = \frac{K_c K_p(1-b)}{z-b} \tag{15.32}$$

Note that this function has *no* poles outside the unit circle, so $P = 0$.

We must let z move around a closed contour in the z plane that completely encloses the area outside the unit circle. Figure 15.6*a* shows such a contour. The C_+ contour starts at $z = +1$ and moves along the top of the unit circle to $z = -1$. The C_{out} contour goes from $z = -1$ to $-\infty$. The C_R contour is a circle of infinite radius going all the way around the z plane. Finally, the C_{in} and C_- contours get us back to our starting point at $z = +1$.

On the C_+ contour $z = e^{i\theta}$ since the magnitude of z is unity on the unit circle. The angle θ goes from 0 to $+\pi$. Now, from Eq. (15.30),

$$z = e^{i\omega T_s}$$

Therefore, on the C_+ contour

$$\omega T_s = \theta$$

As θ goes from 0 to $+\pi$, ω must go from 0 to π/T_s. Using Eq. (15.31) gives

$$\frac{\pi}{T_s} = \frac{\pi}{\left(\dfrac{2\pi}{\omega_s}\right)} = \omega_s/2$$

Therefore, moving around the top of the unit circle is equivalent to letting frequency ω vary from 0 to $\omega_s/2$.

Substituting $z = e^{i\omega T_s}$ into Eq. (15.32) gives

$$HG_{M(i\omega)}D_{(i\omega)} = \frac{K_c K_p(1-b)}{e^{i\omega T_s} - b} \tag{15.33}$$

Figure 15.6*b* shows the Nyquist plot. At $\omega = 0$ (where $z = +1$), the plot starts on the positive real axis at $K_c K_p$. When $\omega = \omega_s/2$ (where $z = -1$)

$$HG_{M(i\omega_s/2)}D_{(i\omega_s/2)} = \frac{K_c K_p(1-b)}{-1-b} = -\frac{K_c K_p(1-b)}{1+b} \tag{15.34}$$

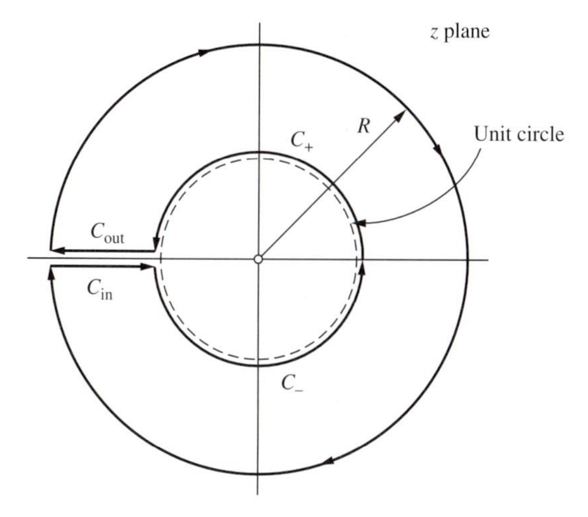

(*a*) *z*-plane contours

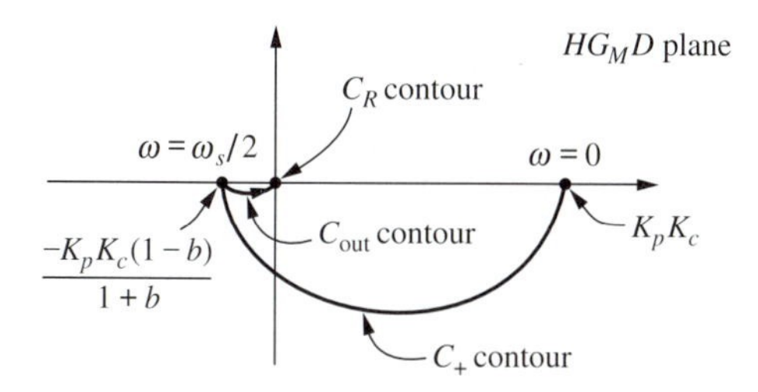

(*b*) First-order lag

FIGURE 15.6
Frequency-domain methods.

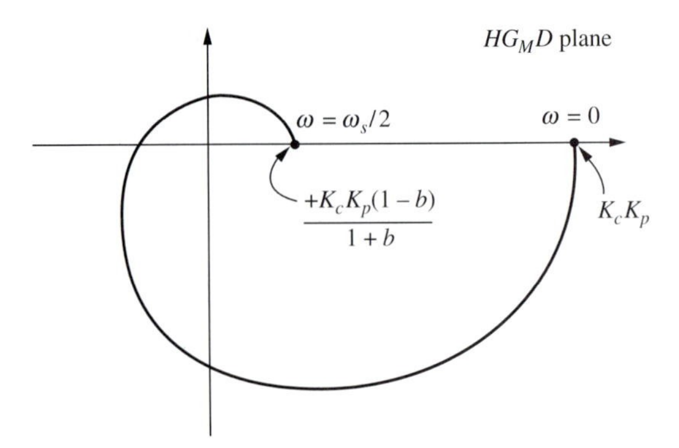

(c) First-order lag with deadtime ($D = T_s$)

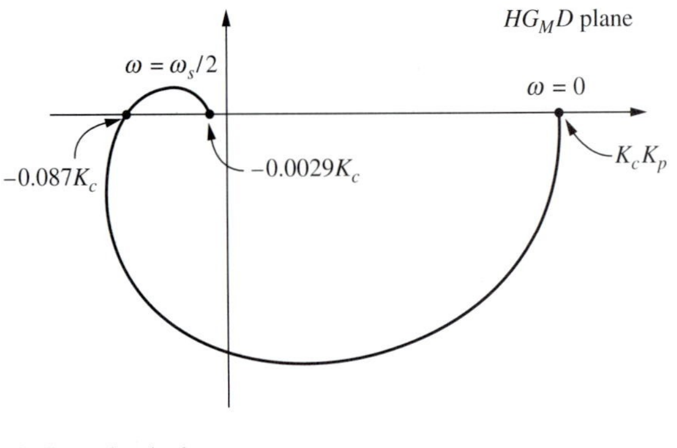

(d) Second-order lag

FIGURE 15.6 (CONTINUED)
Frequency-domain methods.

Thus, the Nyquist plot of the sampled-data system does not end at the origin as the Nyquist plot of the continuous system does. It ends on the negative real axis at the value given in Eq. (15.34).

As we show later after completing the contours, this means that if the controller gain is made big enough, the Nyquist plot *does* encircle the $(-1, 0)$ point. If $N = 1, Z = 1$ for this system since $P = 0$. Thus, there is one zero or root of the closedloop characteristic equation outside the unit circle.

The limiting gain occurs when the $HG_{M(i\omega)}D_{(i\omega)}$ curve ends right at -1. From Eq. (15.34)

$$-1 = -\frac{K_u K_p(1-b)}{1+b} \quad \Rightarrow \quad K_u = \frac{1+b}{K_p(1-b)}$$

This is exactly what our root locus analysis showed.

On the C_{out} contour, $z = re^{i\pi} = -r$ as r goes from 1 to ∞. Substituting $-r$ for z in Eq. (15.32) gives

$$HG_{M(z)}D_{(z)} = \frac{K_c K_p(1-b)}{-r-b} \tag{15.35}$$

As r goes from 1 to ∞, $HG_{M(z)}D_{(z)}$ goes from $-K_c K_p(1-b)/(1+b)$ to 0. See Fig. 15.6b.

On the C_R contour, $z = Re^{i\theta}$ where $R \to \infty$ and θ goes from π through 0 to $-\pi$. Substituting into Eq. (15.32) gives

$$HG_{M(z)}D_{(z)} = \frac{K_cK_p(1-b)}{Re^{i\theta}-b} \tag{15.36}$$

As $R \to \infty$, $HG_{M(z)}D_{(z)} \to 0$. Therefore, the infinite circle in the z plane maps into the origin in the $HG_{M(z)}D_{(z)}$ plane.

The C_{in} contour is just the reverse of the C_{out} contour, going from the origin out along the negative real axis. The C_- contour is just the reflection of the C_+ contour over the real axis. So, just as in a continuous system, we only have to plot the C_+ contour. If it goes around the $(-1, 0)$ point, this sampled-data system is closedloop unstable since $P = 0$.

The farther the curve is from the $(-1, 0)$ point, the more stable the system. We can use exactly the same frequency-domain specifications we used for continuous systems: phase margin, gain margin, and maximum closedloop log modulus. The last is obtained by plotting the function $HG_{M(i\omega)}D_{(i\omega)}/(1 + HG_{M(i\omega)}D_{(i\omega)})$. For this process (with $\tau_o = K_p = 1$) with a proportional sampled-data controller and a sampling period of 0.5 minutes, the controller gain that gives a phase margin of 45° is $K_c = 3.43$. The controller gain that gives a +2-dB maximum closedloop log modulus is $K_c = 2.28$. The ultimate gain is $K_u = 4.08$. ■

EXAMPLE 15.6. If a deadtime of one sampling period is added to the process considered in the previous example,

$$HG_{M(z)}D_{(z)} = \frac{K_cK_p(1-b)}{z(z-b)} \tag{15.37}$$

We substitute $e^{i\omega T_s}$ for z in Eq. (15.37) and let ω go from 0 to $\omega_s/2$. At $\omega = 0$, where $z = +1$, the Nyquist plot starts at K_cK_p on the positive real axis. At $\omega = \omega_s/2$, where $z = -1$, the curve ends at

$$HG_{M(i\omega)}D_{(i\omega)} = \frac{K_cK_p(1-b)}{-1(-1-b)} = \frac{K_cK_p(1-b)}{1+b} \tag{15.38}$$

This is on the positive real axis. Figure 15.6*c* shows the complete curve in the HG_MD plane. At some frequency the curve crosses the negative real axis. This occurs when the real part of $HG_{M(i\omega)}D_{(i\omega)}$ is equal to -0.394 for the numerical case considered in the previous example. Thus the ultimate gain is

$$K_u = 1/0.394 = 2.54$$

Note that this is smaller than the ultimate gain for the process with no deadtime. The controller gain that gives a +2-dB maximum closedloop log modulus is $K_c = 1.36$.

The final phase angle (at $\omega_s/2$) for the first-order lag process with no deadtime was $-180°$. For the process with a deadtime of one sampling period, it was $-360°$. If we had a deadtime that was equal to two sampling periods, the final phase angle would be $-540°$. Every multiple of the sampling period subtracts 180° from the final phase angle.

Remember that in a continuous system, the presence of deadtime made the phase angle go to $-\infty$ as ω went to ∞. So the effect of deadtime on the Nyquist plots of sampled-data systems is different than its effect in continuous systems. ■

EXAMPLE 15.7. As our last example, let's consider the second-order two-heated-tank process studied in Example 15.4.

$$G_{M(s)} = \frac{2.315}{(s+1)(5s+1)}$$

Using a zero-order hold, a proportional sampled-data controller, and a sampling period of $T_s = 0.5$ minutes gives

$$D_{(z)}HG_{M(z)} = \frac{0.0479K_c(z + 0.8133)}{(z - 0.607)(z - 0.905)} \tag{15.39}$$

We substitute $e^{i\omega T_s}$ for z and let ω vary from 0 to $\omega_s/2$.

$$\omega_s = \frac{2\pi}{T_s} = \frac{2\pi}{0.5} = 4\pi$$

At $\omega = 0$, the Nyquist plot starts at $2.315K_c$. This is the same starting point that the continuous system would have. At $\omega = \omega_s/2 = 2\pi$, where $z = -1$, the Nyquist plot ends on the negative real axis at

$$\frac{0.0479K_c(-1 + 0.8133)}{(-1 - 0.607)(-1 - 0.905)} = -0.0029K_c$$

The entire curve is given in Fig. 15.6*d*. It crosses the negative real axis at $-0.087K_c$. So the ultimate gain is $K_u = 1/0.087 = 11.6$, which is the same result we obtained from the root locus analysis.

The controller gain that gives a phase margin of 45° is $K_c = 2.88$. The controller gain that gives a +2-dB maximum closedloop log modulus is $K_c = 2.68$. ■

15.2.3 Approximate Method

To generate the $HG_{M(i\omega)}$ Nyquist plots discussed above, the z transform of the appropriate transfer functions must first be obtained. Then $e^{i\omega T_s}$ is substituted for z, and ω is varied from 0 to $\omega_s/2$. There is an alternative method that is often more convenient to use, particularly in high-order systems. Equation (15.40) gives a doubly infinite series representation of $HG_{M(i\omega)}$.

$$HG_{M(i\omega)} = \frac{1}{T_s}\sum_{n=-\infty}^{+\infty} H_{(i\omega+in\omega_s)}G_{M(i\omega+in\omega_s)} \tag{15.40}$$

where $H_{(s)}$ and $G_{M(s)}$ are the transfer functions of the original continuous elements before z transforming.

If the series in Eq. (15.40) converges in a reasonable number of terms, we can approximate $HG_{M(i\omega)}$ with a few terms in the series. Usually two or three are all that are required.

$$HG_{M(i\omega)} = \frac{1}{T_s}\sum_{n=-\infty}^{+\infty} H_{(i\omega+in\omega_s)}G_{M(i\omega+in\omega_s)} \tag{15.41}$$

$$HG_{M(i\omega)} \simeq \frac{1}{T_s}[H_{(i\omega)}G_{M(i\omega)} + H_{(i\omega+i\omega_s)}G_{M(i\omega+i\omega_s)} + H_{(i\omega-i\omega_s)}G_{M(i\omega-i\omega_s)} + H_{(i\omega+i2\omega_s)}G_{M(i\omega+i2\omega_s)} + H_{(i\omega-i2\omega_s)}G_{M(i\omega-i2\omega_s)}] \tag{15.42}$$

This series approximation can be easily generated on a digital computer.

The big advantage of this method is that the analytical step of taking the z transformation is eliminated. You just deal with the original continuous transfer functions. For complex, high-order systems, this can eliminate a lot of messy algebra.

15.2.4 Use of MATLAB

The frequency response of sampled-data systems can be easily calculated using MATLAB software. Table 15.2 gives a program that generates a Nyquist plot for the first-order process with a deadtime of one sampling period.

After parameter values are specified ($K_c = 3$, $T_s = 0.2$, $\tau_o = 1$, and $K_p = 1$), the frequency range is specified from $\omega = 0.01$ to $\omega = \omega_s/2$. The vector of complex variables z for each frequency is calculated. Then the complex function $HG_{M(i\omega)}D_{(i\omega)}$ is calculated in two steps:

```
hgmd=kc*kp*(1-b)./(z-b);
hgmd=hgmd ./ z;
```

Term-by-term division is specified by the use of the "./" operator.

TABLE 15.2
MATLAB program for discrete frequency response

```
% Program "dfreq.m" generates Nyquist plot
%   for first-order process with deadtime = sampling period
%   using a sampled-data P controller and zero-order hold
% GM(s) = Kp*exp(-Ts*s)/(tauo*s+1)
%
% Give parameter values
kc=3;
ts=0.2;
tauo=1;
kp=1;
b=exp(-ts/tauo);
% Calculate sampling frequency
ws=2*pi/ts;
% Specify frequency range from 0.01 up to ws/2
w=[0.01:0.01:ws/2];
% Calculate vector of z values
i=sqrt(-1);
z=exp(i*w*ts);
%
% Calculate value of HGM(iw)*D(iw) at each frequency
%
hgmd=kc*kp*(1-b) ./ (z-b);
hgmd=hgmd ./ z;
%
clf
plot(hgmd)
grid
title('Polar Plot for HGM(iw)*D(iw)')
axis('square')
axis([-1  3.5  -2.5  2]);
xlabel('Real HGM*D')
ylabel('Imag HGM*D')
text(0.5,-0.5,'Ts=0.2, Kc=3')
pause
```

15.3 PHYSICAL REALIZABILITY

In a digital computer control system the feedback controller $D_{(z)}$ has a pulse transfer function. What we need is an equation or algorithm that can be programmed into the digital computer. At the sampling time for a given loop, the computer looks at the current process output $y_{(t)}$, compares it with a setpoint, and calculates a current value of the error $e_{(t)}$. This error plus some old values of the error and old values of the controller output or manipulated variable that have been stored in computer memory are then used to calculate a new value of the controller output $m_{(t)}$.

These algorithms are basically difference equations that relate the current value of m to the current value of e and old values of m and e. These difference equations can be derived from the pulse transfer function $D_{(z)}$.

Suppose the current moment in time is the nth sampling period $t = nT_s$. The current value of the error $e_{(t)}$ is $e_{(nT_s)}$. We will call this e_n. The value of $e_{(t)}$ at the previous sampling time was e_{n-1}. Other old values of error are e_{n-2}, e_{n-3}, etc. The value of the controller output $m_{(t)}$ that is computed at the current instant in time $t = nT_s$ is $m_{(nT_s)}$ or m_n. Old values are m_{n-1}, m_{n-2}, etc. Suppose we have the following difference equation:

$$\begin{aligned} m_n &= b_0 e_n + b_1 e_{n-1} + b_2 e_{n-2} + \cdots + b_M e_{n-M} \\ &\quad - a_1 m_{n-1} - a_2 m_{n-2} - a_3 m_{n-3} - \cdots - a_N m_{n-N} \end{aligned} \tag{15.43}$$

$$\begin{aligned} m_{(nT_s)} &= b_0 e_{(nT_s)} + b_1 e_{(nT_s-T_s)} + b_2 e_{(nT_s-2T_s)} + \cdots + b_M e_{(nT_s-MT_s)} \\ &\quad - a_1 m_{(nT_s-T_s)} - a_2 m_{(nT_s-2T_s)} - a_3 m_{(nT_s-3T_s)} - \cdots - a_N m_{(nT_s-NT_s)} \end{aligned} \tag{15.44}$$

Limiting t to some multiple of T_s,

$$\begin{aligned} m_{(t)} &= b_0 e_{(t)} + b_1 e_{(t-T_s)} + b_2 e_{(t-2T_s)} + \cdots + b_M e_{(t-MT_s)} \\ &\quad - a_1 m_{(t-T_s)} - a_2 m_{(t-2T_s)} - a_3 m_{(t-3T_s)} - \cdots - a_N m_{(t-NT_s)} \end{aligned} \tag{15.45}$$

If each of these time functions is impulse sampled and z transformed, Eq. (15.45) becomes

$$\begin{aligned} M_{(z)} &= b_0 E_{(z)} + b_1 z^{-1} E_{(z)} + b_2 z^{-2} E_{(z)} + b_3 z^{-3} E_{(z)} + \cdots \\ &\quad - a_1 z^{-1} M_{(z)} - a_2 z^{-2} M_{(z)} - \cdots - a_N z^{-N} M_{(z)} \end{aligned} \tag{15.46}$$

Putting this in terms of a pulse transfer function gives

$$D_{(z)} = \frac{M_{(z)}}{E_{(z)}} = \frac{b_0 + b_1 z^{-1} + b_2 z^{-2} + \cdots + b_M z^{-M}}{1 + a_1 z^{-1} + a_2 z^{-2} + \cdots + a_N z^{-N}} \tag{15.47}$$

A sampled-data controller is a ratio of polynomials in either positive or negative powers of z. It can be directly converted into a difference equation for programming into the computer.

Continuous transfer functions are physically realizable if the order of the polynomial in s of the numerator is less than or equal to the order of the polynomial in s of the denominator.

The physical realizability of pulsed transfer functions uses the basic criterion that the current output of a device (digital computer) cannot depend upon future information about the input. We cannot build a gadget that can predict the future.

If $D_{(z)}$ is expressed as a polynomial in negative powers of z, as in Eq. (15.47), the requirement for physical realizability is that there must be a "1" term in the denominator. If $D_{(z)}$ is expressed as a polynomial in positive powers of z, as shown in Eq. (15.48) below, the requirement for physical realizability is that the order of the numerator polynomial in z must be less than or equal to the order of the denominator polynomial in z. These two ways of expressing physical realizability are completely equivalent, but since the second is analogous to continuous transfer functions in s, it is probably used more often.

$$D_{(z)} = \frac{M_{(z)}}{E_{(z)}} = \frac{b_0 z^M + b_1 z^{M-1} + b_2 z^{M-2} + \cdots + b_M}{z^N + a_1 z^{N-1} + a_2 z^{N-2} + \cdots + a_N} \tag{15.48}$$

Multiplying numerator and denominator by z^{-N} and converting to difference equation form give the current value of the output m_n:

$$\begin{aligned} m_n = {} & b_0 e_{n+M-N} + b_1 e_{n+M-N-1} + \cdots + b_M e_{n-N} \\ & - a_1 m_{n-1} - a_2 m_{n-2} - \cdots - a_N m_{n-N} \end{aligned} \tag{15.49}$$

If the order of the numerator M is greater than the order of the denominator N in Eq. (15.48), the calculation of m_n requires future values of error. For example, if $M - N = 1$, Eq. (15.49) tells us that we need to know e_{n+1} or $e_{(t+T_s)}$ to calculate m_n or $m_{(t)}$. Since we do not know at time t what the error $e_{(t+T_s)}$ will be one sampling period in the future, this calculation is physically impossible.

15.4 MINIMAL-PROTOTYPE DESIGN

One of the most interesting and unique approaches to the design of sampled-data controllers is called minimal-prototype design. It is one of the earliest examples of model-based or direct-synthesis controllers.

The basic idea is to specify the desired response of the system to a particular type of disturbance and then, knowing the model of the process, back-calculate the controller required. There is no guarantee that the minimal-prototype controller is physically realizable for the given process and the specified response. Therefore, the specified response may have to be modified to make the controller realizable.

Let us consider the closedloop response of an arbitrary system with a sampled-data controller.

$$Y_{(z)} = \frac{D_{(z)} HG_{M(z)}}{1 + D_{(z)} HG_{M(z)}} Y_{(z)}^{\text{set}} + \frac{G_L L_{(z)}}{1 + D_{(z)} HG_{M(z)}} \tag{15.50}$$

If we consider for the moment only changes in setpoint,

$$\frac{Y_{(z)}}{Y_{(z)}^{\text{set}}} = \frac{D_{(z)} HG_{M(z)}}{1 + D_{(z)} HG_{M(z)}} \tag{15.51}$$

If we specify the form of the input $Y_{(z)}^{\text{set}}$ and the desired form of the output $Y_{(z)}$, and if the process and hold transfer functions are known, we can rearrange Eq. (15.51) to give the required controller designed for setpoint changes $D_{S(z)}$.

$$D_{S(z)} = \frac{Y_{(z)}}{HG_{M(z)}(Y_{(z)}^{\text{set}} - Y_{(z)})} \tag{15.52}$$

If all of the terms on the right side of this equation have been specified, the controller can be calculated.

EXAMPLE 15.9. The first-order lag process with a zero-order hold has a pulse transfer function

$$HG_{M(z)} = \frac{K_p(1-b)}{z-b} \tag{15.53}$$

where $b \equiv e^{-T_s/\tau_o}$. Suppose we want to derive a minimal-prototype controller for step changes in setpoint.

$$Y_{(z)}^{\text{set}} = \mathscr{Z}\left[\frac{1}{s}\right] = \frac{z}{z-1} \tag{15.54}$$

We know that it is impossible to have the output of the process respond instantaneously to the change in setpoint. Therefore, the best possible response that we could expect from the process would be to drive the output $Y_{(z)}$ up the setpoint in one sampling period. This is sketched in Fig. 15.7*a*. Remember, we are specifying only the values of the variables at the sampling times.

The output at $t = 0$ is zero. At $t = T_s$, the output should be 1 and should stay at 1 for all subsequent sampling times. Therefore, the desired $Y_{(z)}$ is

$$\begin{aligned} Y_{(z)} &\equiv y_{(0)} + y_{(T_s)}z^{-1} + y_{(2T_s)}z^{-2} + y_{(3T_s)}z^{-3} + \cdots \\ &= 0 + z^{-1} + z^{-2} + z^{-3} + \cdots \end{aligned} \tag{15.55}$$

$$Y_{(z)} = \frac{z^{-1}}{1-z^{-1}} = \frac{1}{z-1} \tag{15.56}$$

Plugging these specified functions for $Y_{(z)}$ and $Y_{(z)}^{\text{set}}$ into Eq. (15.52) gives

$$D_{S(z)} = \frac{Y_{(z)}}{HG_{M(z)}(Y_{(z)}^{\text{set}} - Y_{(z)})} = \frac{\dfrac{1}{z-1}}{(HG_{M(z)})\left(\dfrac{z}{z-1} - \dfrac{1}{z-1}\right)} \tag{15.57}$$

$$D_{S(z)} = \frac{1}{HG_{M(z)}(z-1)}$$

Now, for this first-order process, Eq. (15.53) gives $HG_{M(z)}$. Plugging this into Eq. (15.57) gives the minimal-prototype controller.

$$D_{S(z)} = \frac{1}{HG_{M(z)}(z-1)} = \frac{1}{\dfrac{K_p(1-b)}{z-b}(z-1)} = \frac{z-b}{K_p(1-b)(z-1)} \tag{15.58}$$

This sampled-data controller is physically realizable since the order of the polynomial in the numerator is equal to the order of the polynomial in the denominator. Therefore, the desired setpoint response is achievable for this process.

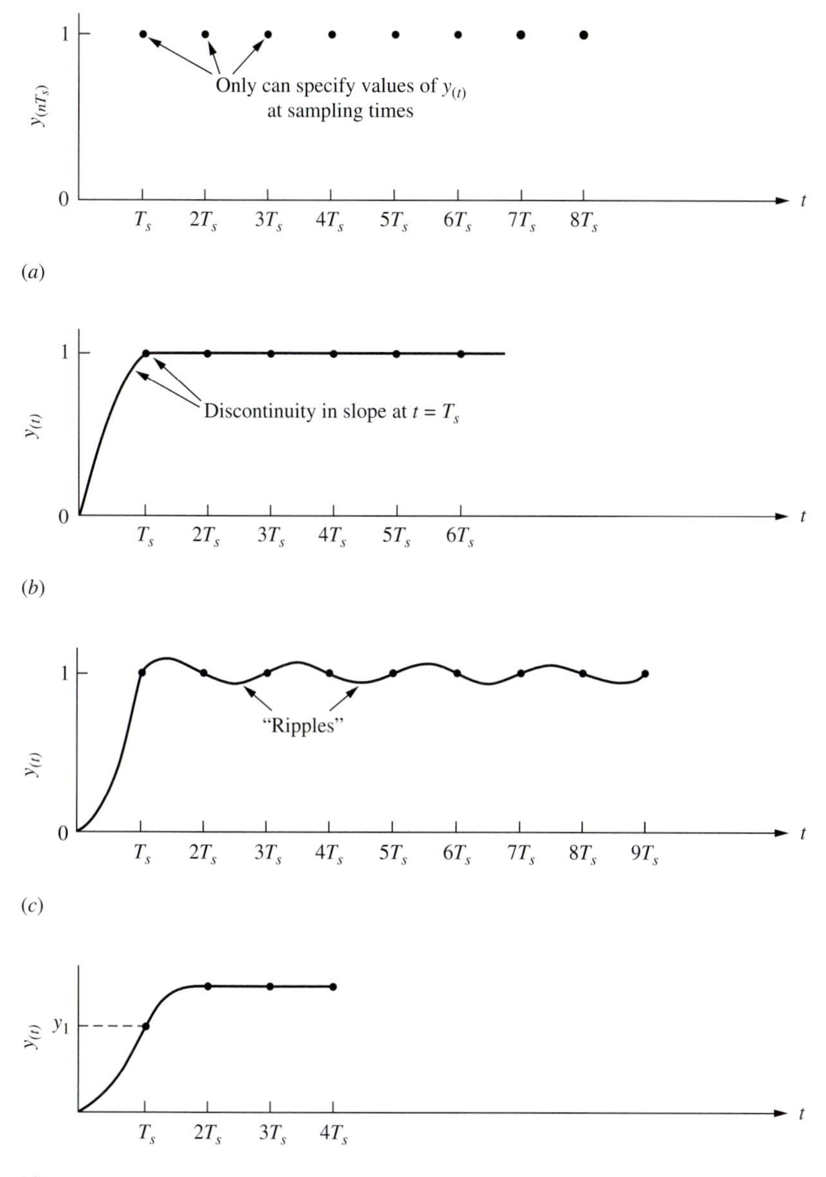

FIGURE 15.7
Minimal-prototype responses. (*a*) Desired response to unit step change in setpoint. (*b*) Response of first-order process. (*c*) Response of second-order system when driven to setpoint in one sampling period. (*d*) Modified response of second-order system to take two sampling periods to reach setpoint without rippling.

Before we leave this example, let's look at the closedloop characteristic equation of the system.

$$1 + D_{S(z)} HG_{M(z)} = 0$$

If we substitute Eqs. (15.58) and (15.53), we get

$$1 + \frac{z-b}{K_p(1-b)(z-1)} \frac{K_p(1-b)}{z-b} = 0$$

$$1 + \frac{1}{z-1} = 0 \quad \Rightarrow \quad z = 0$$

Thus, the closedloop root is located at the origin. This corresponds to a critically damped closedloop system ($\zeta = 1$). The specified response in the output was for no overshoot, so this damping coefficient is to be expected. ■

EXAMPLE 15.10. If we have a first-order lag process with a deadtime equal to one sampling period, the process transfer function becomes

$$HG_{M(z)} = \frac{K_p(1-b)}{z(z-b)} \tag{15.59}$$

Suppose we specified the same kind of response for a step change in setpoint as in Example 15.9: the output is driven to the setpoint in one sampling period. Substituting our new process transfer function into Eq. (15.58) gives

$$D_{S(z)} = \frac{1}{HG_{M(z)}(z-1)} = \frac{1}{\dfrac{K_p(1-b)}{z(z-b)}(z-1)} = \frac{z(z-b)}{K_p(1-b)(z-1)} \tag{15.60}$$

This controller is *not* physically realizable because the order of the numerator is higher (second) than the order of the denominator (first). Therefore, we cannot achieve the response specified. This result should really be no surprise. The deadtime does not let the output even begin to change during the first sampling period, and we cannot drive the output up to its setpoint instantaneously at $t = T_s$.

Let us back off on the specified output and allow two sampling periods to drive the output to the setpoint.

$$\begin{aligned} Y_{(z)} &\equiv y_{(0)} + y_{(T_s)} z^{-1} + y_{(2T_s)} z^{-2} + y_{(3T_s)} z^{-3} + \cdots \\ &= 0 + (0) z^{-1} + z^{-2} + z^{-3} + \cdots \end{aligned} \tag{15.61}$$

$$Y_{(z)} = \frac{z^{-2}}{1 - z^{-1}} = \frac{1}{z(z-1)} \tag{15.62}$$

Now the minimal-prototype controller for step changes in setpoint is

$$\begin{aligned} D_{S(z)} &= \frac{Y_{(z)}}{HG_{M(z)}(Y_{(z)}^{\text{set}} - Y_{(z)})} = \frac{\dfrac{1}{z(z-1)}}{\left[\dfrac{K_p(1-b)}{z(z-b)}\right]\left[\dfrac{z}{z-1} - \dfrac{1}{z(z-1)}\right]} \\ &= \frac{z(z-b)}{K_p(1-b)(z^2-1)} \end{aligned} \tag{15.63}$$

The controller is physically realizable since $N = 2$ and $M = 2$. Note that there are two poles, one at $z = +1$ and the other at $z = -1$, and there are two zeros (at $z = 0$ and $z = b$). ■

A first-order process can be driven to the setpoint in one sampling period and held right on the setpoint even between sampling periods. This is possible because we can change the slope of a first-order process response curve, as shown in Fig. 15.7*b*.

If the process is second or higher order, we are not able to make a discontinuous change in the slope of the response curve. Consequently, we would expect a second-order process to overshoot the setpoint if we forced it to reach the setpoint in one sampling period. The output would oscillate between sampling periods, and the manipulated variable would change at each sampling period. This is called *rippling* and is illustrated in Fig. 15.7*c*.

Rippling is undesirable since we do not want to keep wiggling the control valve. We may want to modify the specified output response to eliminate rippling. Allowing two sampling periods for the process to come up to the setpoint gives us two switches of the manipulated variable and should let us bring a second-order process up to the setpoint without rippling. This is illustrated in Example 15.11. In general, an Nth-order process must be given N sampling periods to come up to the setpoint if the response is to be completely ripple free.

Since we know only the values of the output $y_{(nT_s)}$ at the sampling times, we cannot use $Y_{(z)}$ to see if there are ripples. We can see what the manipulated variable $m_{(nT_s)}$ is doing at each sampling period. If it is changing, rippling is occurring. So we choose $Y_{(z)}$ such that $M_{(z)}$ does not ripple.

$$M_{(z)} = D_{(z)}E_{(z)} = D_{(z)}[Y_{(z)}^{\text{set}} - Y_{(z)}] \tag{15.64}$$

If the controller is designed for setpoint changes [Eq. (15.52)],

$$\begin{aligned} D_{(z)} &= \frac{Y_{(z)}}{HG_{M(z)}(Y_{(z)}^{\text{set}} - Y_{(z)})} \\ M_{(z)} &= \frac{Y_{(z)}}{HG_{M(z)}} \end{aligned} \tag{15.65}$$

Let's check the first-order system from Example 15.9.

$$HG_{M(z)} = \frac{K_p(1-b)}{z-b} \quad \text{and} \quad Y_{(z)} = \frac{1}{z-1}$$

$$M_{(z)} = \frac{Y_{(z)}}{HG_{M(z)}} = \frac{\dfrac{1}{z-1}}{\dfrac{K_p(1-b)}{z-b}} = \frac{z-b}{K_p(1-b)(z-1)} \tag{15.66}$$

Using long division to see the values of $m_{(nT_s)}$ gives

$$M_{(z)} = \frac{1}{K_p(1-b)} + \frac{1}{K_p}z^{-1} + \frac{1}{K_p}z^{-2} + \frac{1}{K_p}z^{-3} + \cdots \tag{15.67}$$

Thus, the manipulated variable holds constant after the first sampling period, indicating no rippling.

EXAMPLE 15.11. The second-order process considered in Example 15.4 has the following openloop transfer function:

$$G_{M(s)} = \frac{K_p}{(\tau_{o1}s + 1)(\tau_{o2}s + 1)} \tag{15.68}$$

Using a zero-order hold gives an openloop transfer function

$$HG_{M(z)} = \frac{K_p a_0(z - z_1)}{(z - p_1)(z - p_2)} \tag{15.69}$$

We want to design a minimal-prototype controller for a unit step setpoint change. The output is supposed to come up to the new setpoint in one sampling period. Substituting Eq. (15.69) into Eq. (15.52) gives

$$\begin{aligned} D_{S(z)} &= \frac{1}{HG_{M(z)}(z-1)} = \frac{1}{\dfrac{K_p a_0(z - z_1)}{(z - p_1)(z - p_2)}(z-1)} \\ &= \frac{(z - p_1)(z - p_2)}{K_p a_0(z - z_1)(z-1)} \end{aligned} \tag{15.70}$$

This controller is physically realizable. Therefore, minimal-prototype control should be attainable. But what about intersample rippling? Let us check the manipulated variable.

$$M_{(z)} = \frac{Y_{(z)}}{HG_{M(z)}} = \frac{\dfrac{1}{z-1}}{\dfrac{K_p a_0(z - z_1)}{(z - p_1)(z - p_2)}} = \frac{(z - p_1)(z - p_2)}{a_0(z-1)(z - z_1)} \tag{15.71}$$

Long division shows that the manipulated variable changes at each sampling period, so rippling occurs.

For a specific numerical case ($K_p = \tau_{o1} = 1$; $\tau_{o2} = 5$; $T_s = 0.2$), the parameter values are $p_1 = 0.8187$, $p_2 = 0.9608$, $a_0 = 0.0037$, and $z_1 = -0.923$.

$$M_{(z)} = 270 - 460z^{-1} + 427z^{-2} - 392z^{-3} + 364z^{-4} - 334z^{-5} + \cdots \tag{15.72}$$

This system exhibits rippling.

To prevent rippling, we modify our desired output response to give the system two sampling periods to come up to the setpoint. The value of $y_{(t)}$ at the first sampling period, the y_1 shown in Fig. 15.7*d*, is unspecified at this point. The output $Y_{(z)}$ is now

$$\begin{aligned} Y_{(z)} &= y_1 z^{-1} + z^{-2} + z^{-3} + \cdots \\ &= y_1 z^{-1} + z^{-2}(1 + z^{-1} + z^{-2} + z^{-3} + \cdots) \\ &= y_1 z^{-1} + z^{-2}\frac{1}{1 - z^{-1}} \end{aligned} \tag{15.73}$$

$$Y_{(z)} = \frac{y_1 z + 1 - y_1}{z(z-1)}$$

The setpoint disturbance is still a unit step:

$$Y_{(z)}^{\text{set}} = \frac{z}{z-1}$$

The new controller is

$$D_{(z)} = \frac{Y_{(z)}}{HG_{M(z)}(Y^{set}_{(z)} - Y_{(z)})} = \frac{\dfrac{y_1 z + 1 - y_1}{z(z-1)}}{\left[\dfrac{K_p a_0 (z - z_1)}{(z - p_1)(z - p_2)}\right]\left[\dfrac{z}{z-1} - \dfrac{y_1 z + 1 - y_1}{z(z-1)}\right]} \tag{15.74}$$

$$= \frac{(z - p_1)(z - p_2)(y_1 z + 1 - y_1)}{a_0 (z - z_1)(z - 1)(z + 1 - y_1)}$$

The manipulated variable is

$$M_{(z)} = \frac{Y_{(z)}}{HG_{M(z)}} = \frac{(z - p_1)(z - p_2)(y_1 z + 1 - y_1)}{K_p a_0 (z - z_1) z (z - 1)} \tag{15.75}$$

Rippling occurs whenever the denominator of $M_{(z)}$ contains any terms other than z or $z - 1$. Therefore, the $z - z_1$ term must be eliminated by picking y_1 such that the term $z - z_1$ cancels out.

$$\frac{1 - y_1}{y_1} = -z_1 \quad \Rightarrow \quad y_1 = \frac{1}{1 - z_1} \tag{15.76}$$

Then $M_{(z)}$ becomes

$$M_{(z)} = \frac{y_1 (z - p_1)(z - p_2)}{K_p a_0 z (z - 1)}$$
$$= \frac{y_1}{K_p a_0} + \frac{(1 - p_1 - p_2)}{K_p a_0} z^{-1} + z^{-2} + z^{-3} + z^{-4} + \cdots \tag{15.77}$$

Thus, there is no rippling. ■

15.5 CONCLUSION

The design of digital compensators was discussed in this chapter. The conventional root locus, frequency response, and direct synthesis methods used in continuous systems in the s plane can be directly extended to sampled-data systems in the z plane.

PROBLEMS

15.1. Find the maximum value of K_c for which a proportional sampled-data controller with zero-order hold is closedloop stable for the three-CSTR process

$$G_{M(s)} = \frac{\frac{1}{8}}{(s + 1)^3}$$

Use sampling times of 0.1 and 1 minute.

15.2. Repeat Problem 15.1 using a sampled-data PI controller.

$$D_{(z)} = \frac{K_c}{\alpha}\frac{z-\alpha}{z-1} \qquad \text{where } \alpha \equiv \frac{\tau_I}{\tau_I + T_s}$$

Use τ_I values of 0.5 and 2 minutes.

15.3. Make Nyquist plots for the process of Problem 15.1 and find the value of gain that gives the following specifications:

- Gain margin of 2
- Phase margin of 45°
- Maximum closedloop log modulus of +2 dB

15.4. Make a root locus plot of the system in Problem 15.1 and find the value of gain that gives a closedloop damping coefficient ζ equal to 0.3.

15.5. A distillation column has an approximate transfer function between distillate composition x_D and reflux flow rate R of

$$G_{M(s)} = \frac{0.0092}{(5s+1)^2}$$

Distillate composition is measured by a chromatograph with a deadtime equal to the sampling period. If a proportional sampled-data controller is used with a zero-order hold, calculate the ultimate gain for $T_s = 2$ and 10.

15.6. Grandpa McCoy has decided to open up a new Liquid Lightning plant in the California gold fields. He plans to stay in Kentucky, and he must direct operation of the plant using the pony express. It takes two days for a message to be carried in either direction, and a rider arrives each day.

The new Liquid Lightning reactor is a single, isothermal, constant-holdup CSTR in which the concentration of ethanol, C, is controlled by manual changes in the feed concentration, C_0. Ethanol undergoes an irreversible first-order reaction at a specific reaction rate $k = 0.25$/day. The volume of the reactor is 100 barrels, and the throughput is 25 barrels/day.

Grandpa will receive information from the plant every day telling him what the concentration C was two days earlier. He will send back instructions on how to change C_0. What is the largest change Grandpa can make in C_0 as a percentage of C without causing the concentration in the reactor to begin oscillating?

15.7. A process has the following transfer function relating the controlled and manipulated variables:

$$G_{M(s)} = \frac{-s+1}{s+1}$$

(*a*) If a zero-order hold and a proportional digital controller are used with sampling period T_s, determine the openloop pulse transfer function $HG_{M(z)}$.

(*b*) Calculate the value of controller gain that puts the system right at the limit of closedloop stability.

(*c*) Calculate the controller gain that gives a closedloop damping coefficient of 0.3.

15.8. A process controlled by a proportional digital controller with zero-order hold and sampling period $T_s = 0.25$ has the openloop pulse transfer function

$$HG_{M(z)} = \frac{-z + 1.2212}{z - 0.7788}$$

If a unit step change is made in the setpoint, calculate the closedloop response of the process at the sampling times if a controller gain of 0.722 is used.

15.9. Make a root locus plot for the process considered in Problem 15.7 in the z plane.

15.10. A first-order lag process with a zero-order hold is controlled by a proportional sampled-data controller.

(*a*) What value of gain gives a critically damped closedloop system?
(*b*) What is the gain margin when this value of gain is used?
(*c*) What is the steady-state error for a unit step change in setpoint when this value of gain is used?

15.11. A pressurized tank has the openloop transfer function between pressure in the first tank and gas flow from the second tank

$$G_{M(s)} = \frac{0.2386}{s(0.7137s + 1)}$$

If a zero-order hold and a proportional sampled-data controller are used and the sampling time is 1 minute:

(*a*) Make a root locus plot in the z plane.
(*b*) Find the value of controller gain that gives a damping coefficient of 0.3.
(*c*) Find the controller gain that gives 45° of phase margin.
(*d*) Find the gain that gives a maximum closedloop log modulus of +2 dB.

15.12. Repeat Problem 15.11 for a process with the openloop transfer function

$$G_{M(s)} = \frac{-s + 1}{(s + 1)^2}$$

15.13. For the process considered in Problem 15.11, generate Nyquist and Bode plots by the rigorous method and by the approximate method using several values of n.

15.14. A process has the following openloop transfer function relating the controlled and manipulated variables:

$$G_{M(s)} = \frac{-s + 1}{s + 1}$$

If a zero-order hold is used with sampling period T_s, the openloop pulse transfer function is

$$HG_{M(z)} = \frac{-z + 2 - b}{z - b}$$

Design a sampled-data minimal-prototype digital compensator for step changes in setpoint that does not ripple.

15.15. A process has openloop load and manipulated variable transfer functions

$$G_{M(s)} = \frac{e^{-T_s s}}{s+1} \qquad G_{L(s)} = \frac{e^{-2T_s s}}{s+1}$$

The digital compensator that gives minimal-prototype setpoint response for step changes is

$$D_{s(z)} = \frac{z(z-b)}{(z^2-1)(1-b)}$$

Determine the closedloop response of the system when this controller is used and a unit step change in the load input occurs.

15.16. An openloop-unstable, first-order process has the transfer function

$$G_{M(s)} = \frac{K_p}{\tau_o s - 1}$$

A discrete approximation of a PI controller is used.

$$D_{(z)} = \frac{K_c}{\alpha}\frac{z-\alpha}{z-1} \qquad \text{where } \alpha \equiv \frac{\tau_I}{\tau_I + T_s}$$

(*a*) Sketch a root locus plot for this system. Show the effect of changing the reset time τ_I from very large to very small values.

(*b*) Find the maximum value of controller gain (K_{max}) for which the system is closedloop stable as a general function of τ_o, τ_I, and T_s.

(*c*) Calculate the numerical value of K_{max} for the case $\tau_o = K_p = \tau_I = 1$ and $T_s = 0.25$.

15.17. Design a minimal-prototype sampled-data controller for the pressurized tank process considered in Problem 15.11 that will bring the pressure up to the setpoint in one sampling period for a unit step change in setpoint.

(*a*) Calculate how the manipulated variable changes with time to test for rippling.

(*b*) Repeat for the case when the controller brings the pressure up to the setpoint in two sampling periods without rippling.

15.18. Design a minimal-prototype sampled-data controller for a first-order system with a deadtime that is three sampling periods. The input is a unit step change in setpoint.

15.19. Design minimal-prototype controllers for step changes in setpoint and load for a process that is a pure integrator.

$$G_{M(s)} = G_{L(s)} = \frac{1}{s}$$

For setpoint changes, the process should be brought up to the setpoint in one sampling period. For load changes, the process should be driven back to its initial steady-state value in two sampling periods. Calculate the controller outputs for both controllers to check for rippling.

15.20. Design a minimal-prototype sampled-data controller for a process with an openloop process transfer function that is a pure deadtime.

$$G_{M(s)} = e^{-kT_s s}$$

where k is an integer. Design the controller for a unit step change in setpoint and the best possible response in the controlled variable.

15.21. A process has the following openloop transfer function relating the controlled variable Y and the manipulated variable M.

$$\frac{Y_{(s)}}{M_{(s)}} = \frac{1}{(s+1)^2}$$

(a) If a proportional analog controller is used, derive the relationships that show how the closedloop time constant τ_{CL} and the closedloop damping coefficient ζ_{CL} vary with controller gain K_c.

(b) If a proportional digital controller is used with a zero-order hold and sampling time T_s:

(i) Derive the openloop pulse transfer function $HG_{M(z)}$ for any sampling period and for a specific value of 0.5 minutes.

(ii) Make a root locus plot in the z plane.

(iii) Prove that the ultimate gain is 9.84.

15.22. The output of a process Y is affected by two inputs M_1 and M_2 through two transfer functions G_1 and G_2.

$$Y = G_1 M_1 + G_2 M_2$$

where

$$G_1 = \frac{1}{s+1} \quad \text{and} \quad G_2 = \frac{0.2}{5s+1}$$

Since the transfer function G_1 has a smaller time constant and larger gain, we want to use M_1 to control Y.

(a) Using a digital proportional controller D_1 and zero-order hold with sampling period $T_s = 0.5$, find the value of controller gain K_c that gives a closedloop damping coefficient of 0.5.

However, using manipulated variable M_1 is more expensive than using M_2 because M_2 is cheaper. Therefore, we want to use a "valve position controller" (VPC), a simple type of optimizing control, that will slowly change M_2 in such a way that only a small amount of M_1 is used under steady-state conditions. This is accomplished by using a second digital controller D_2 that has as its "process variable" signal the *output* signal from the D_1 controller (M_1^*) and has as its setpoint signal M_1^{set}, which is set at a small value. The output of the D_2 controller (M_2^*) is sent to a zero-order hold whose output is M_2.

(b) Draw a block diagram of this sampled-data VPC system.

(c) What is the closedloop characteristic equation of the entire system with the D_1 controller tuned using the gain determined in part (a)?

(d) Solve for the roots of the closedloop characteristic equation as functions of the gain K_2 of the D_2 controller.

15.23. A process has an openloop transfer function relating the controlled variable Y and the manipulated variable M that is a pure integrator with unity gain.

$$\frac{Y_{(s)}}{M_{(s)}} = \frac{1}{s}$$

(*a*) Sketch root locus plots in the s plane for analog P and PI controllers.
(*b*) Derive the openloop pulse transfer function for the sampled-data system if a zero-order hold is used with a sampling time T_s.
(*c*) Sketch the root locus plot in the z plane if a proportional sampled-data controller is used. What value of controller gain gives a critically damped closedloop system? What is the ultimate gain?
(*d*) Sketch the root locus plot in the z plane if the following sampled-data controller is used.

$$D_{(z)} = \frac{K_c}{\alpha} \frac{z - \alpha}{z - 1}$$

where $\alpha = \dfrac{\tau_I}{\tau_I + T_s}$

τ_I = reset time
K_c = controller gain

(*e*) Derive the relationship between the ultimate gain and the parameters τ_I and T_s.
(*f*) What is the maximum closedloop damping coefficient that can be achieved in this system? Your result should be an equation that gives ζ_{CL} as a function of α. For the numerical values $\tau_I = 1$ and $T_s = 0.2$, the maximum closedloop damping coefficient should be 0.298.

15.24. (*a*) Make a frequency response Nyquist plot for the pure integrator process with a proportional sampled-data controller.
(*b*) Use the Nyquist stability criterion to find the ultimate gain.
(*c*) What is the phase margin of the system if $K_c = 1/T_s$?

15.25. A pure integrating process with unity gain is controlled by a PI controller with reset $\tau_I = 1$ minute.
(*a*) If the controller is analog, what value of controller gain gives a closedloop damping coefficient of 0.3?
(*b*) Now suppose the controller is sampled-data and a zero-order hold is used.

$$D_{(z)} = \frac{K_c}{\alpha} \frac{z - \alpha}{z - 1}$$

where $\alpha = \tau_I/(\tau_I + T_s)$. Sketch a root locus plot in the z plane.
(*c*) What values of controller gain give a closedloop damping coefficient of 0.3, and what is the ultimate gain if the sampling period $T_s = 0.2$?
(*d*) Sketch Nyquist plots of $G_{M(i\omega)}G_{C(i\omega)}$ for the continuous system and of $HG_{M(i\omega)}D_{(i\omega)}$ for the sampled-data system.

15.26. A process has the following openloop transfer function relating controlled and manipulated variables:

$$G_{M(s)} = \frac{Y}{M} = \frac{2}{s(10s+1)}$$

where time is in minutes.

(*a*) If a sampled-data proportional controller is used with a zero-order hold and sampling period T_s, derive the openloop pulse transfer function of the process.

(*b*) Using a sampling period of 0.5 minutes, show that the openloop pulse transfer function is

$$HG_{M(z)} = \frac{0.02459(z+0.9835)}{(z-1)(z-0.9512)}$$

(*c*) Sketch a root locus plot.

(*d*) Calculate the ultimate gain and ultimate frequency of this sampled-data system.

15.27. Derive an analytical relationship that can be used to calculate the damping coefficient of a sampled-data system from known values x and y of the real and imaginary parts of the complex variable z for any location in the z plane. Check your result by showing that the damping coefficient is 0.3 when $z = -0.372 + i0$.

15.28. (*a*) Design a minimal-prototype controller for a pure integrator process with a zero-order hold and a sampled-data controller that, for step changes in setpoint, brings the process output up to the setpoint in one sampling period.

(*b*) If the process has an openloop transfer relating the output to the *load* input that is also a pure integrator with unity gain, find the output of the closedloop system for a step change in load when the controller derived in part (*a*) is used.

15.29. A process has the same transfer functions relating the controlled variable to the manipulated variable and to the load variable:

$$G_{M(s)} = G_{L(s)} = \frac{1}{(\tau_{o1}s+1)(\tau_{o2}s+1)}$$

When a zero-order hold is used in a sampled-data system with $T_s = 0.2$, $\tau_1 = 1$, and $\tau_2 = 5$, the pulse transfer function is

$$HG_{M(z)} = \frac{0.0037(z+0.923)}{(z-0.8187)(z-0.9608)}$$

Design a minimal-prototype sampled-data controller for a step change in load that will not give rippling.

15.30. A process has the following transfer functions relating load disturbance L and manipulated input M to the controlled variable Y.

$$G_{M(s)} = \frac{Y}{M} = \frac{K_M}{\tau_o s+1} \qquad G_{L(s)} = \frac{Y}{L} = \frac{K_L}{\tau_o s+1}$$

Design a minimal-prototype controller for a unit step change in load such that the maximum change in the manipulated variable M cannot exceed some specified maximum value M_{max}.

$$M_{max} = RK_L/K_M$$

R is a number greater than unity but less than $1 + b$, where $b = e^{-T_s/\tau_o}$ for sampling period T_s.

15.31. The openloop transfer function of a process is

$$G_{M(s)} = \frac{K_p e^{-T_s s}}{s}$$

(*a*) If an analog proportional controller is used, calculate the ultimate gain and ultimate frequency.
(*b*) If a digital proportional controller and zero-order hold are used, derive $HG_{M(z)}$.
(*c*) Sketch a root locus plot in the z plane and find the ultimate gain and ultimate frequency.
(*d*) Design a minimal-prototype digital compensator for step changes in setpoint. Sketch the time-domain curves for the output and the manipulated variable.
(*e*) If a digital compensator is used on this process that has the form

$$D_{(z)} = \frac{K_c(z - z_1)}{(z - p_1)}$$

where $z_1 = 1$ and $p_1 = 0$, sketch a root locus plot in the z plane and calculate the ultimate gain and ultimate frequency.

15.32. The process considered in Problem 11.45 is now controlled by a digital controller.
(*a*) Using a zero-order hold in the sampled-data system with sampling period T_s (where $D = T_s$), derive the pulse transfer function of the system.
(*b*) For a sampling period $T_s = 1$ and the numerical values of parameters given in Problem 11.46, calculate the ultimate gain and frequency if a proportional digital compensator is used.
(*c*) Calculate the value of the gain of a proportional digital compensator that gives a closedloop damping coefficient of 0.3.
(*d*) Sketch a Nyquist plot of the sampled-data system.
(*e*) Design a minimal-prototype digital compensator for a step change in setpoint.
(*f*) Sketch time-domain plots of the controlled and manipulated variables.

15.33. The openloop transfer function $G_{M(s)}$ of a process relating the controlled variable $Y_{(s)}$ and the manipulated variable $M_{(s)}$ is a gain K_p (with units of mA/mA when transmitter and valve gains are included) and a first-order lag with time constant τ_o. A sampled-data PI feedback controller $D_{(z)}$ is used with a sampling period T_s.

$$D_{(z)} = \frac{K_c(z - \alpha)}{\alpha(z - 1)}$$

where K_c = controller gain
$\alpha = \tau_I/(\tau_I + T_s)$
τ_I = reset time

(*a*) What is the closedloop characteristic equation?
(*b*) If $T_s = 0.5$ minutes and $\tau_I = 1$ minute, sketch a root locus plot.
(*c*) Calculate the ultimate gain in terms of α, K_p, T_s, and τ_o. Calculate a numerical value for K_u if $K_p = \tau_o = 1$.
(*d*) Calculate the value of controller gain that gives a closedloop damping coefficient of 0.3.
(*e*) Sketch a Nyquist plot of $HG_{M(i\omega)}D_{(i\omega)}$.

PART SIX

Process Identification

CHAPTER 16

Process Identification

The dynamic relationships discussed thus far in this book have been determined from mathematical models of the process. Equations based on fundamental physical and chemical laws were developed to describe the time-dependent behavior of the system. We assumed that the values of all parameters, such as holdups, reaction rates, and heat transfer coefficients, were known. Thus, the dynamic behavior was predicted on essentially a theoretical basis.

For a process already in operation, there is an alternative approach based on experimental dynamic data obtained from plant tests. The experimental approach is sometimes used when the process is thought to be too complex to model from first principles. More often, however, we use it to find the values of some model parameters that are unknown. Although many of the parameters can be calculated from steady-state plant data, some must be found from dynamic tests (e.g., holdups in nonreactive systems). Additionally, we employ dynamic plant experiments to confirm the predictions of a theoretical mathematical model. Verification is a critical step in a model's development and application.

In performing plant tests it is important to consider how much the process can be upset and how long testing can last. In devising tests we need to consider all of the disturbances and upsets that can potentially occur during the course of the test.

Experimental identification of process dynamics has been an active area of research for many years by workers in several areas of engineering. The literature is extensive, and entire books have been devoted to the subject. The theoretical aspects are covered in *System Identification,* by L. Ljung (1987, Prentice-Hall, Englewood Cliffs, NJ.) A user-friendly discussion of some of the practical aspects of identification is provided by R. C. McFarlane and D. E. Rivera in "Identification of Distillation Systems," Chapter 7 in *Practical Distillation Control* (1992, Van Nostrand Reinhold, New York).

Although many techniques have been proposed, we limit our discussion to the methods that are widely used in the chemical and petroleum industries. Only the identification of linear transfer function models is discussed. We illustrate the use of the MATLAB System Identification Toolbox.

16.1 FUNDAMENTAL CONCEPTS

16.1.1 Control-Relevant Identification

Whenever we want to identify a model of the process, we have to specify how we will use the information. In this book we are interested in control, so we want to obtain good models for designing controllers. To design feedback controllers, a model must be accurate over the frequency range near the $(-1, 0)$ point. Its fidelity at higher or lower frequencies is not important. This means that an accurate value of the steady-state gain (the $\omega = 0$ point) is *not* required.

Figure 16.1 illustrates the point. Suppose we have two models of a process, model A and model B, and we know exactly what the true transfer function of the process is. The step responses of the alternative models are compared with the response of the real process in Fig. 16.1*a*. Model A fits the real plant well near the final steady state, but its fidelity is poor during the initial part of the transient. Model B is just the opposite. Which model is better for use in designing a feedback controller?

Figure 16.1*b* gives the Nyquist plots of the plant and the two models. The model B curve is closer than the model A curve to the real process curve over the frequency range near the $(-1, 0)$ point. Therefore, model B should be used for feedback controller design.

However, for feedforward controller design, model A should be used because having the correct steady-state gain is more important in feedforward control than the correct dynamics.

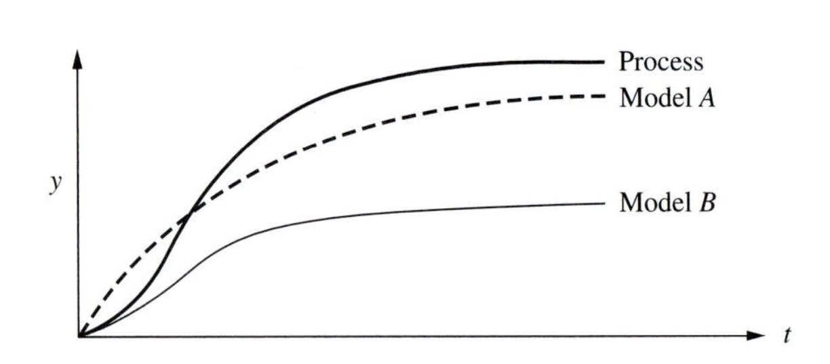

(*a*)

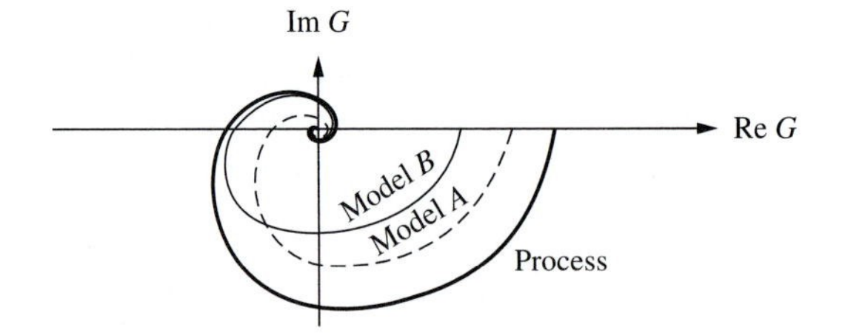

(*b*)

FIGURE 16.1
Control-relevant models.

16.1.2 Frequency Content of the Input Signal

Identification requires that the process be disturbed by some input signal. If the input signal has little frequency content (small magnitude) over some frequency range, the accuracy of the identified model is poor. This is because we are obtaining a transfer function that is a ratio of functions.

$$G_{M(i\omega)} = \frac{Y_{(i\omega)}}{U_{(i\omega)}} \tag{16.1}$$

If the input U does not provide enough excitation of the process over the important frequency range, the model fidelity is poor, particularly in processes with appreciable noise. This is why direct sine wave testing at a frequency near the ultimate frequency and relay feedback testing are such useful methods.

16.1.3 Model Order

Part of the identification problem is to determine the "best" order of the model. The higher the order, the more parameters there are to identify and the more difficult the estimation problem becomes. A "parsimonious" model (one with few parameters) is the easiest to identify. A large number of parameters gives high variance (a measure of the difference between the model predictions and the actual plant) and a poorly conditioned estimation problem (the matrix to be inverted in the numerical solution technique is nearly singular). These difficulties can be overcome by using a large number of data points, but this increases the duration of the plant testing, which is undesirable because it increases the likelihood of the plant being disturbed by other events.

A common way to determine the best model order is to use "model validation." The experimental data are separated into two sets. A specific number of parameters is assumed. The first set is used in the numerical calculations to identify a model. Then the predictions of this model are compared with the actual data from the second set (the variance is calculated). A different model order is assumed and the procedure is repeated. A plot of the variance of the model in the prediction of the second set of data versus the number of parameters is usually a curve that goes through a minimum. This is the best model order.

16.2 DIRECT METHODS

16.2.1 Time-Domain Fitting of Step Test Data

The most direct way of obtaining an empirical linear dynamic model of a process is to find the parameters (deadtime, time constant, and damping coefficient) that fit the experimentally obtained step response data. The process being identified is usually openloop, but experimental testing of closedloop systems is also possible.

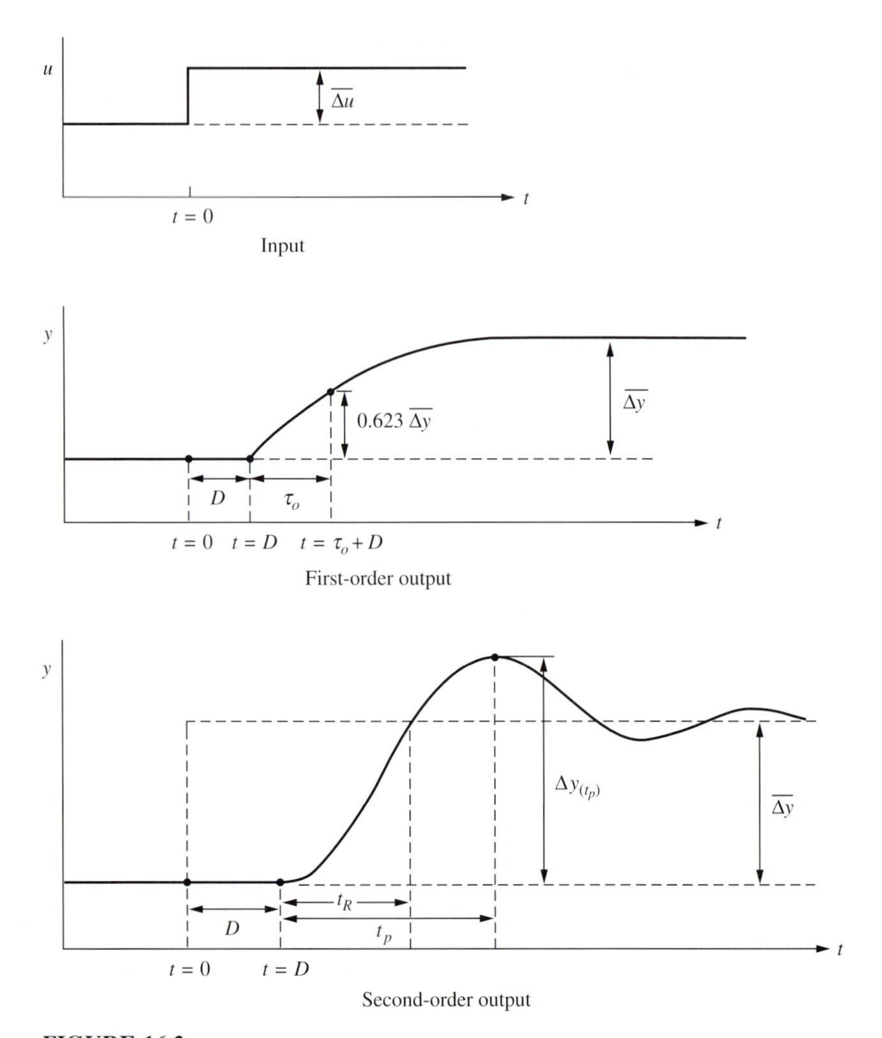

FIGURE 16.2
Step response.

We put in a step disturbance $u_{(t)}$ and record the output variable $y_{(t)}$ as a function of time, as illustrated in Fig. 16.2. The quick-and-dirty engineering approach is to simply look at the shape of the $y_{(t)}$ curve and find some approximate transfer function $G_{(s)}$ that would give the same type of step response.

Probably 80 percent of all chemical engineering openloop processes can be modeled by a gain, a deadtime, and one lag.

$$G_{(s)} = K_p \frac{e^{-Ds}}{\tau_o s + 1} \tag{16.2}$$

The steady-state gain K_p is easily obtained from the ratio of the final steady-state change in the output $\overline{\Delta y}$ over the size of the step input $\overline{\Delta u}$. The deadtime can be easily read from the $y_{(t)}$ curve. The time constant can be estimated from the time it takes the output $y_{(t)}$ to reach 62.3 percent of the final steady-state change.

Closedloop processes are usually tuned to be somewhat underdamped, so a second-order underdamped model must be used.

$$G_{(s)} = K_p \frac{e^{-Ds}}{\tau^2 s^2 + 2\tau\zeta s + 1} \tag{16.3}$$

As shown in Fig. 16.2, the steady-state gain and deadtime are obtained in the same way as with a first-order model. The damping coefficient ζ can be calculated from the "peak overshoot ratio," POR (see Problem 2.7), using Eq. (16.4).

$$\text{POR} = e^{-\pi \cot \phi} \tag{16.4}$$

$$\text{where POR} = \frac{\Delta y_{(t_p)} - \overline{\Delta y}}{\overline{\Delta y}} \tag{16.5}$$

$$\phi = \arccos \zeta \tag{16.6}$$

$\Delta y_{(t_p)}$ = change in $y_{(t)}$ at the peak overshoot

t_p = time to reach the peak overshoot (excluding the deadtime)

Then the time constant τ can be calculated from Eq. (16.7).

$$\frac{t_R}{\tau} = \frac{\pi - \phi}{\sin \phi} \tag{16.7}$$

where t_R is the time it takes the output to reach the final steady-state value for the first time (see Fig. 16.2).

These "eyeball" estimation methods are simple and easy to use. They can provide a rough model that is adequate for many engineering purposes. For example, an approximate model can be used to get preliminary values for controller settings.

However, these crude methods cannot provide a precise, higher-order model, and they are quite sensitive to nonlinearity. Most chemical engineering processes are fairly nonlinear. A step test drives the process away from the initial steady state, and the values of the parameters of a linear transfer function model may be significantly in error. If the magnitude of the step change could be made very small (sometimes as small as 10^{-4} to 10^{-6} percent of the normal value of the input), nonlinearity would not be a problem. But in most plant situations, such small changes give output responses that cannot be seen because of the normal noise in the signals. Thus, step testing has definite limitations for plant studies.

16.2.2 Direct Sine Wave Testing

The next level of dynamic testing is with direct sine waves. The input of the plant, which is usually a control valve position or a flow controller setpoint, is varied sinusoidally at a fixed frequency ω. After waiting for all transients to die out and for a steady oscillation in the output to be established, the amplitude ratio and phase angle are found by recording input and output data. The data point at this frequency is plotted on a Nyquist, Bode, or Nichols plot. See Fig. 16.3*a*. Then the frequency is changed to another value, and a new amplitude ratio and a new phase angle are

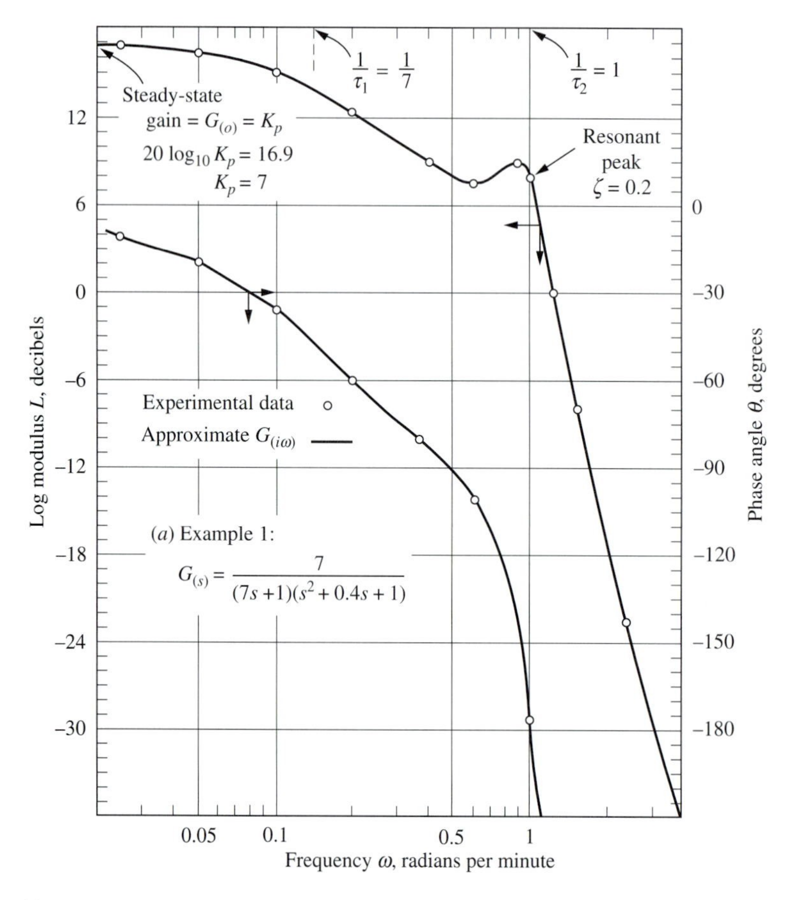

(a)

FIGURE 16.3
Fitting an approximate transfer function to experimental frequency response data.

determined. Thus, the complete frequency response curve is found experimentally by varying frequency over the range of interest. Once the $G_{(i\omega)}$ curves have been found, they can be used directly to examine the dynamics and stability of the system or to design controllers in the frequency domain (see Chapter 11).

If a transfer function model is desired, approximate transfer functions can be fit to the experimental $G_{(i\omega)}$ curves. First the log modulus Bode plot is used. The low-frequency asymptote gives the steady-state gain. The time constants can be found from the breakpoint frequency and the slope of the high-frequency asymptote. The damping coefficient can be found from the resonant peak.

Once the log modulus curve has been adequately fitted by an approximate transfer function $G^A_{(i\omega)}$, the phase angle of $G^A_{(i\omega)}$ is compared with the experimental phase angle curve. The difference is usually the contribution of deadtime. The procedure is illustrated in Fig. 16.3*b*.

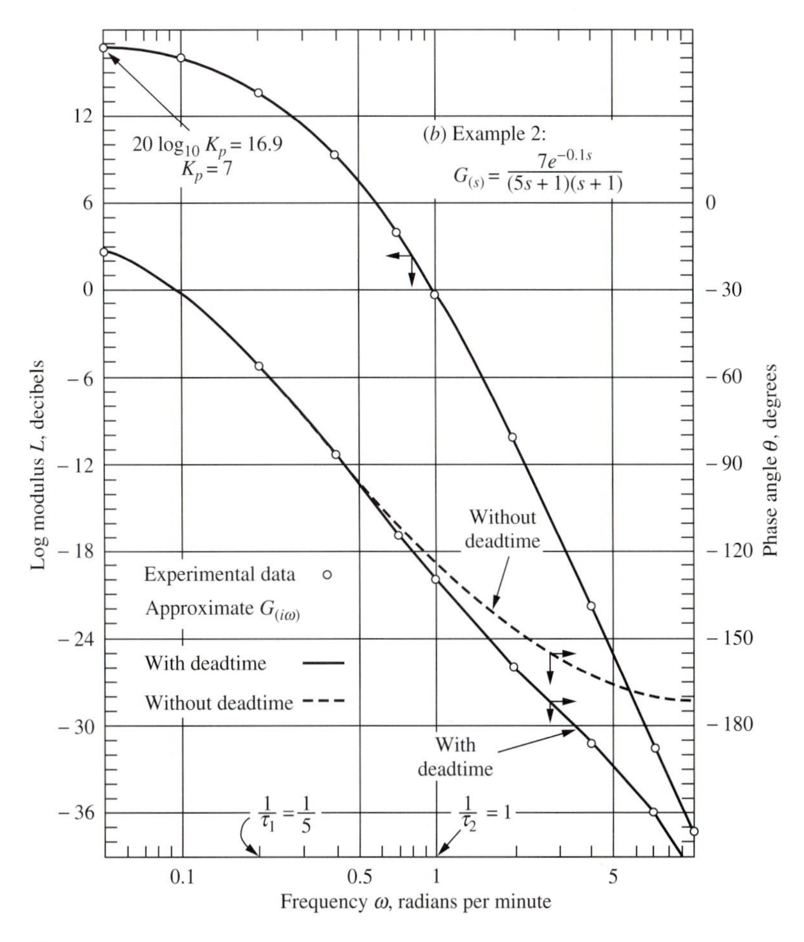

(b)

FIGURE 16.3 (CONTINUED)
Fitting an approximate transfer function to experimental frequency response data.

It is usually important to get an accurate fit of the model frequency response to the experimental frequency response near the critical region where the phase angle is between $-135°$ and $-180°$. It doesn't matter how well or how poorly the approximate transfer function fits the data once the phase angle has dropped below $-180°$. So the fitting of the approximate transfer function should weight heavily the differences between the model and the data over this frequency range.

Direct sine wave testing is an extremely useful way to obtain precise dynamic data. Damping coefficients, time constants, and system order can all be quite accurately found. Direct sine wave testing is particularly useful for processes with signals that are noisy. Since you are putting in a sine wave signal with a known frequency and the output signal has this same frequency, you can easily filter out all of the noise signals at other frequencies and obtain an output signal with a much higher signal-to-noise ratio.

The main disadvantage of direct sine wave testing is that it can be very time consuming when applied to typical chemical process equipment with large time constants. The steady-state oscillation must be established at each value of frequency. It can take days to generate the complete frequency response curves of a slow process. While the test is being conducted over this long period of time, other disturbances and changes in operating conditions can occur and affect the results of the test. Therefore, direct sine wave testing is only rarely used to get the complete frequency response. However, it can be very useful for obtaining accurate data at one or two important frequencies. For example, it can be used to get amplitude and phase angle data near the critical $-180°$ point.

16.3 PULSE TESTING

One useful and practical method for obtaining experimental dynamic data from many chemical engineering processes is pulse testing. It yields reasonably accurate frequency response curves and requires only a fraction of the time that direct sine wave testing takes.

An input pulse $u_{(t)}$ of fairly arbitrary shape is put into the process. This pulse starts and ends at the same value and is often just a square pulse (i.e., a step up at time zero and a step back to the original value at a later time t_u). See Fig. 16.4. The response of the output is recorded. It typically returns eventually to its original steady-state values. If $y_{(t)}$ and $u_{(t)}$ are perturbations from steady state, they start and end at zero.

The input and output functions are then Fourier transformed and divided to give the system transfer function in the frequency domain $G_{(i\omega)}$. These calculations can be done using the *spa* function in MATLAB. The vectors of input values (u) and output values (y) are combined into the vector z.

$$z=[y \quad u];$$

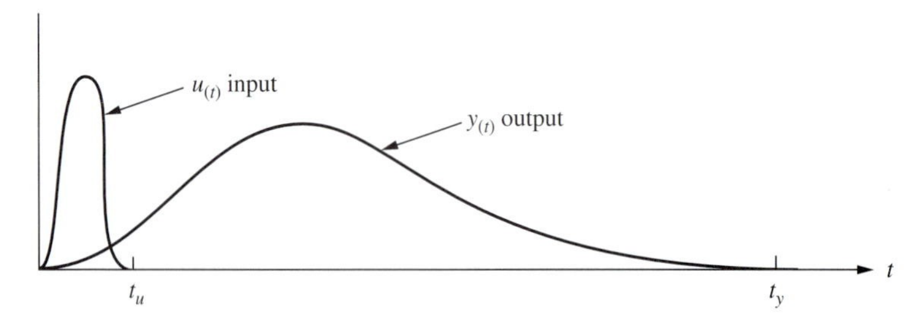

FIGURE 16.4
Pulse test input and output curves.

The sampling time (*ts*) and the frequency range of interest (*w*) are defined; for example: *w=[0.1:0.01:10];* Then *spa* is used to get the frequency response. The simple version is

$$g = spa(z);$$

We have found that using the default values for the arguments in the *spa* function often gives poor results, so the full version is recommended.

$$g = spa(z,M,w,[\,],ts);$$

The *[ ]* is a default value for the fourth parameter. The *M* parameter should be varied by starting with M = total number of data points and reducing it until reasonable results are obtained. For noise-free pulse test data, setting *M* equal to the number of data points works well. As the noise level increases, *M* should be reduced.

Bode plots can be generated by calculating the magnitude ratios and phase angles of *G* by using the *getff* command.

$$[w,mag,phase] = getff(g);$$

In theory, only one pulse input is required to generate the entire frequency response curve. In practice, several pulses are usually needed to establish the required size and duration of the input pulse. We need to keep the width of the pulse fairly small to prevent its "frequency content" from becoming too low at higher frequencies. A good rule of thumb is to keep the width of the pulse less than about half the smallest time constant of interest. If the dynamics of the process are completely unknown, it takes a few trials to establish a reasonable pulse width. If the width of the pulse is too small for a given pulse height, the system is disturbed very little, and it becomes difficult to separate the real output signal from noise and experimental error. The height of the pulse can be increased to "kick" the process more, but there is a limit here also.

We want to obtain an experimental linear dynamic model of the system in the form of $G_{(i\omega)}$. It must be a linear model since the notion of a transfer function applies only to a linear system. The process is usually nonlinear, and we are obtaining a model that is linearized around the steady-state operation level. If the height of the pulse is too high, we may drive the process out of the linear range. Therefore, pulses of various heights should be tried. It is also a good idea to make both positive and negative pulses in the input (increase and decrease). The computed $G_{(i\omega)}$'s should be identical if the region of linearity is not exceeded. For highly nonlinear processes, this is difficult to do. Therefore, pulse testing does not work very well with highly nonlinear processes.

Pulse testing also has problems in situations where load disturbances occur at the same time as the pulse is occurring. These other disturbances can affect the shape of the output response and produce poor results. The output of the process may not return to its original value because of load disturbances. We are trying to extract a lot of information from one pulse test, i.e., the whole frequency response curve. This is asking a lot from one experiment.

16.4 RELAY FEEDBACK IDENTIFICATION

If the purpose of our identification is to obtain information for feedback controller design, we really need data only in the frequency range where the phase angle approaches $-180°$. For example, we might use the ultimate frequency ω_u and the ultimate gain K_u to calculate the Ziegler-Nichols or Tyreus-Luyben settings. In many situations what we need is not an accurate frequency response curve over the entire frequency range, but only the ultimate gain and ultimate frequency.

16.4.1 Autotuning

Astrom and Hagglund (*Proceedings of the 1983 IFAC Conference,* San Francisco) suggested an "autotune" procedure that is a very attractive technique for determining the ultimate frequency and ultimate gain. We call this method ATV, for "autotune variation." The acronym also stands for "all-terrain vehicle" which makes it easy to remember and is not completely inappropriate since ATV does provide a useful tool for the rough and rocky road of process identification.

ATV is illustrated in Fig. 16.5. A relay of height h is inserted as a feedback controller. The manipulated variable m is increased by h above the steady-state value. When the controlled variable y crosses the setpoint, the relay reduces m to a value h below the steady-state value. The system responds to this "bang-bang" control by producing a limit cycle, provided the system phase angle drops below $-180°$, which is true for all real processes.

The period of the limit cycle is the ultimate period (P_u) for the transfer function relating the controlled variable y and the manipulated variable m. So the ultimate frequency is

$$\omega_u = \frac{2\pi}{P_u} \tag{16.8}$$

The ultimate gain of the same transfer function is given by

$$K_u = \frac{4h}{a\pi} \tag{16.9}$$

where h = height of the relay
a = amplitude of the primary harmonic of the output y

It should be noted that Eqs. (16.8) and (16.9) give approximate values for ω_u and K_u because the relay feedback introduces a nonlinearity into the system. However, for most systems, the approximation is close enough for engineering purposes.

Astrom's autotune method has several distinct advantages over openloop pulse testing:

1. There is no need for a priori knowledge of the system time constants. The method automatically results in a sustained oscillation at the critical frequency of the process. The only parameter that has to be specified is the height of the relay

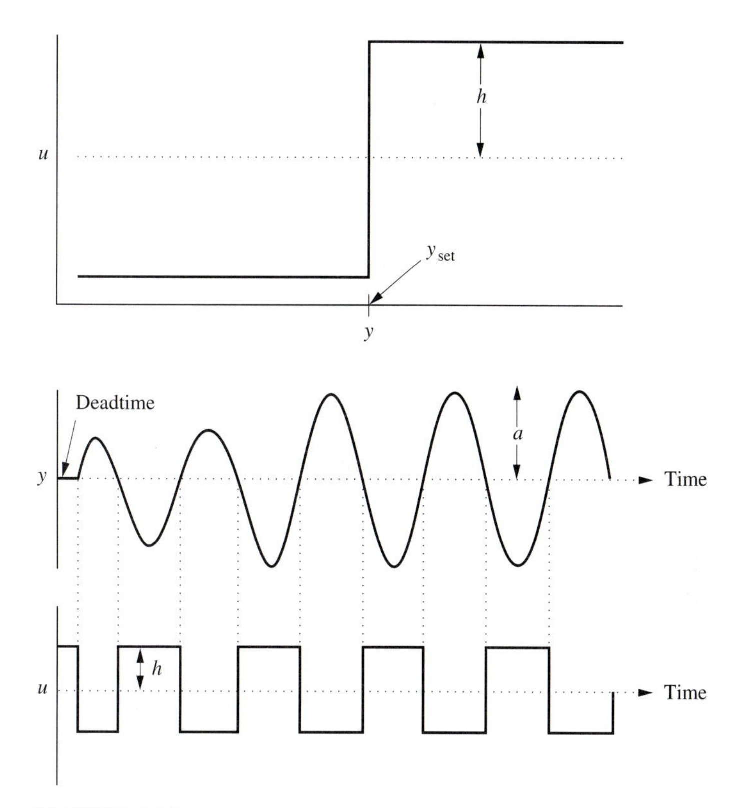

FIGURE 16.5

step. This would typically be set at 5 to 10 percent of the manipulated variable range.

2. ATV is a closedloop test, so the process will not drift away from the setpoint. This keeps the process in the linear region where we are trying to get transfer functions. This is precisely why the method works well on highly nonlinear processes. The process is never pushed very far away from the steady-state conditions.
3. Accurate information is obtained around the important frequency, i.e., near phase angles of $-180°$. In contrast, pulse testing tries to extract information for a range of frequencies. It is inherently less accurate than a method that concentrates on a specific frequency. Remember, however, that we do not have to specify the frequency. The relay feedback automatically finds it.

16.4.2 Approximate Transfer Functions

Once the relay feedback test has been performed and the ultimate gain and ultimate frequency have been determined, we may simply use the results to calculate controller setting. Alternatively, it is possible to use this information to calculate approximate transfer functions. The idea is to pick a very simple form for the transfer function and find the parameter values that fit the ATV results.

These simple approximate transfer function models are particularly useful in multivariable systems. For example, to use the BLT method discussed in Chapter 13, we need all the $N \times N$ transfer functions relating the N inputs to the N outputs. The ATV method provides a quick and fairly accurate way to obtain all these transfer functions.

The simplest possible form for a transfer function is a gain, an integrator, and a deadtime. We have used this model with good success on a variety of processes. It works well for feedback controller design because it does a good job in fitting the important frequency range near the $(-1, 0)$ point.

$$G_{(s)} = \frac{K_p e^{-Ds}}{s} \tag{16.10}$$

From the autotune test, the frequency (ω_u), the argument of $G_{(i\omega)}$ at this frequency ($-\pi$ radians), and the magnitude of $G_{(i\omega)}$ ($= 1/K_u$) are known. From Eq. (16.10)

$$\arg G_{(i\omega_u)} = -\pi = -\frac{\pi}{2} - D\omega_u \tag{16.11}$$

$$\left|G_{(i\omega)}\right| = \frac{1}{K_u} = \frac{K_p}{\omega_u} \tag{16.12}$$

EXAMPLE 16.1. Suppose an ATV test gives the results $K_u = 5.05$ and $\omega_u = 0.542$ rad/min. Equation (16.11) is solved for D, and Eq. (16.12) is solved for K_p.

$$D = \frac{\pi}{2\omega_u} \tag{16.13}$$

$$K_p = \frac{\omega_u}{K_u} \tag{16.14}$$

For the example, $D = 2.90$ minutes and $K_p = 0.107$. The approximate transfer function is

$$G_{(s)} = \frac{0.107 e^{-2.9s}}{s}$$

■

Let us emphasize once again that the important feature of the ATV method is to give transfer function models that fit the frequency response data very well near the ultimate frequency, which determines closedloop stability.

16.5 LEAST-SQUARES METHODS

Instead of converting the step or pulse responses of a system to frequency response curves, it is fairly easy to use classical least-squares methods to solve for the "best" values of model parameters that fit the time-domain data. Any type of input forcing function can be used: steps, pulses, or a sequence of positive and negative pulses. Figure 16.6 shows some typical input/output data from a process. The specific ex-

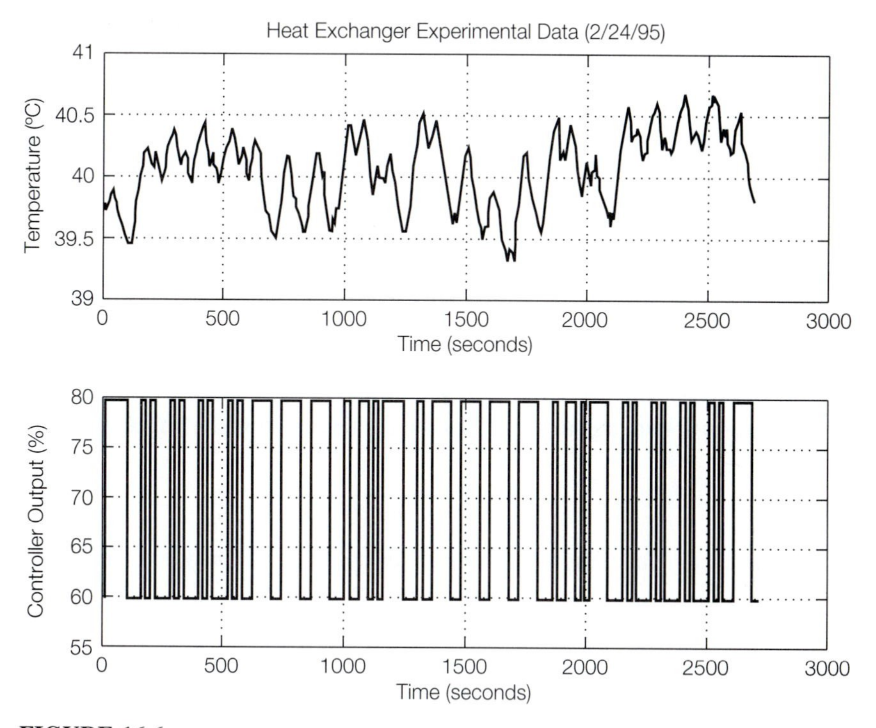

FIGURE 16.6

ample is a heat exchanger in which the manipulated variable is steam flow rate and the output variable is the temperature of the process stream leaving the exchanger.

A very popular sequence of inputs is the "pseudo-random binary sequence" (PRBS), which is illustrated in Fig. 16.6. It is easy to generate and has some attractive statistical properties (see *System Identification for Self-Adaptive Control,* by W. D. T. Davies, 1970, Wiley-Interscience, London).

Whatever the form of the input, the basic idea is to use a difference equation model for the process in which the current output y_n is related to previous values of the output ($y_{n-1}, y_{n-2}, \ldots$) and present and past values of the input ($u_n, u_{n-1}, \ldots$). In the simple model structures, the relationship is linear, so classical least-squares can be used to solve for the best values of the unknown coefficients. These difference equation models occur naturally in sampled-data systems (see Chapter 15) and can be easily converted to Laplace-domain transfer function models.

For example, consider a process with a continuous transfer function

$$\frac{Y_{(s)}}{U_{(s)}} = G_{(s)} = \frac{K_p e^{-Ds}}{\tau_o s + 1} \tag{16.15}$$

The pulse transfer function of this process with a zero-order hold is [see Chapter 14, Eq. (14.93)]

$$\frac{Y_{(z)}}{U_{(z)}} = HG_{(z)} = \frac{K_p(1-b)}{z^{nk}(z-b)} = \frac{K_p(1-b)z^{-1}}{z^{nk}(1-bz^{-1})} = \frac{z^{-nk}(b_0 + b_1 z^{-1})}{1 + a_1 z^{-1}} \tag{16.16}$$

where $nk = D/T_s$
T_s = sampling period
$b = \exp(-T_s/\tau_o)$
$b_0 = 0$
$b_1 = K_p(1 - b)$
$a_1 = -b$

The discrete transfer function has three parameters that need to be identified: nk, b_1, and a_1. Of course, if we are identifying an unknown plant, we do not know what the real order of the systems is. A crucial part of the identification problem is the determination of what model structure yields the best fit to the real plant. In addition to finding the deadtime (nk), we must find the number of a_k terms (na) and the number of b_k terms (nb) to use in the model.

A process with two lags in series has the continuous transfer function

$$G_{(s)} = \frac{K_p e^{-Ds}}{(\tau_1 s + 1)(\tau_2 s + 1)} \tag{16.17}$$

The discrete transfer function with a zero-order hold is [see Chapter 15, Eq. (15.21) for the definitions of a_0, z_1, p_1, and p_2 in terms of τ_1, τ_2, and T_s]

$$\begin{aligned} HG_{(z)} &= \frac{z^{-nk} K_p a_0 (z - z_1)}{(z - p_1)(z - p_2)} = \frac{z^{-nk} K_p a_0 (z^{-1} - z_1 z^{-2})}{1 - (p_1 + p_2) z^{-1} + p_1 p_2 z^{-2}} \\ &= \frac{z^{-nk}(b_1 z^{-1} + b_2 z^{-2})}{1 + a_1 z^{-1} + a_2 z^{-2}} \end{aligned} \tag{16.18}$$

where $b_0 = 0$
$b_1 = K_p a_0$
$b_2 = -K_p a_0 z_1$
$a_1 = -(p_1 + p_2)$
$a_2 = p_1 p_2$

Now there are five parameters to identify: nk, b_1, b_2, a_1, and a_2.

Let us assume a model of the general form

$$\begin{aligned} \overline{y}_n = {} & b_1 u_{n-1} + b_2 u_{n-2} + \cdots + b_{nb} u_{n-nb} \\ & - a_1 y_{n-1} - a_2 y_{n-2} - \cdots - a_{na} y_{n-na} \end{aligned} \tag{16.19}$$

where $\overline{y}_n$ = the predicted value of the current output of the process. The unknown parameters are $b_1, \ldots, b_{nb}, a_1, a_2, \ldots, a_{na}$. In the identification literature, models such as Eq. (16.19) are usually written

$$\overline{y}_{(t)} = \frac{B_{(z)}}{A_{(z)}} u_{(t-nk)} \tag{16.20}$$

where the deadtime ($D = nkT_s$) is factored out. Other types of models are often used in which disturbances and noise are included in the equations. These require other numerical solution methods, which are available in the MATLAB Identification Toolbox, as illustrated in the next section.

It is convenient to define a vector $\underline{\theta}$ that contains these unknown parameters.

$$\underline{\theta} = [b_1 \quad b_2 \quad \ldots \quad b_{nb} \quad a_1 \quad a_2 \quad \ldots \quad a_{na}]^T \tag{16.21}$$

There are $(nb + na)$ unknown parameters. Let us define NC as this total. So the vector has NC rows.

Now suppose we have NP data points that give values of the output y_n for known values of $y_{n-1}, y_{n-2}, \ldots, y_{n-N}, u_n, u_{n-1}, \ldots, u_{n-M}$. The data could be grouped as shown:

	y_n values	y_{n-1} values	...	y_{n-N} values	u_n values	u_{n-1} values	...	u_{n-M} values
1	$y_{1,n}$	$y_{1,n-1}$		$y_{1,n-N}$	$u_{1,n}$	$u_{1,n-1}$		$u_{1,n-M}$
2	$y_{2,n}$	$y_{2,n-1}$		$y_{2,n-N}$	$u_{2,n}$	$u_{2,n-1}$		$u_{2,n-M}$
⋮								
NP	$y_{NP,n}$	$y_{NP,n-1}$		$y_{NP,n-N}$	$u_{NP,n}$	$u_{NP,n-1}$		$u_{NP,n-M}$

Our objective is to minimize the sum of the squares of the differences between the actual measured data points $(y_{i,n})$ and those predicted by our model equation $(\overline{y}_{i,n})$.

$$J = \sum_{i=1}^{NP} (y_{i,n} - \overline{y}_{i,n})^2 \tag{16.22}$$

This is a least-squares problem that is solved by taking partial derivatives of J with respect to each of the unknown parameters (the NC elements of the $\underline{\theta}$ vector) and setting these partial derivatives equal to zero. This gives NC equations in NC unknowns. The solution is compactly written in matrix form:

$$\underline{\theta} = [\underline{\underline{A}}^T \underline{\underline{A}}]^{-1} \underline{\underline{A}}^T \underline{y} \tag{16.23}$$

where the $\underline{\underline{A}}$ matrix (with NP rows and NC columns) and the $\underline{y}$ vector (with NP rows) are defined to represent the data points

$$\underline{\underline{A}} = \begin{bmatrix} y_{1,n-1} & y_{1,n-2} & \cdots & y_{1,n-N} & u_{1,n} & \cdots & u_{1,n-M} \\ y_{2,n-1} & y_{2,n-2} & \cdots & y_{2,n-N} & u_{2,n} & \cdots & u_{2,n-M} \\ \cdots & \cdots & \cdots & \cdots & \cdots & \cdots & \cdots \\ y_{NP,n-1} & & & \cdots & & & u_{NP,n-M} \end{bmatrix} \tag{16.24}$$

$$\underline{y} \equiv [y_{1,n} \quad y_{2,n} \quad y_{3,n} \quad \cdots \quad y_{2,NP}]^T \tag{16.25}$$

Several commercial software packages are available that make it easy to perform this type of identification. We illustrate the use of MATLAB in the next section.

One important comment should be made at this point. These least-squares methods are getting the best fit in the time domain, and they tend to give models that are more accurate at low frequencies than near the ultimate frequency. So you should be careful in your testing to make sure that your input signal has a significant frequency content near the ultimate frequency.

16.6 USE OF THE MATLAB IDENTIFICATION TOOLBOX

We illustrate some of the simple MATLAB identification tools in this section. Table 16.1 gives a MATLAB program that analyzes some real experimental data from a tube-in-shell heat exchanger in the Interdisciplinary Controls Laboratory at Lehigh University. The output variable (y) is the temperature of the hot water leaving the heat exchanger (in degrees centigrade). The input variable (u) is the signal to the cold-water control valve (in percent of scale). The input test signal is a PRBS variation of the signal to the control valve. Figure 16.6 shows the raw experimental data. There are a total of 1351 data points. The sampling period is 2 seconds.

In the program in Table 16.1, the data are "detrended" (the *mean* is subtracted from each data point), and the *spa* function is used to calculate the frequency response of the transfer function relating y and u. Figure 16.7 gives results. The parameter M is set equal to 300.

Then a parametric model is calculated from the data by using the *arx* command to fit a first-order model to the data. The detrended input and output data are combined in z:

$$z = [-y \quad u];$$

The negative sign is used on y since the steady-state gain is negative for this process (an increase in cold-water flow causes a decrease in exit hot-water temperature). From step test data of the process, the deadtime is estimated to be 16 seconds. Therefore a value of $nk = 8$ is assumed. Various other values of nk can be explored to find the one that gives the minimum variance.

A first-order model is assumed, so *na* and *nb* are both set equal to 1. The command *th=arx[z,[na nb nk]);* calculates the values of $b_1 = 0.0016$ and $a_1 = -0.953$. The variance is given in the *th* matrix of results in the (1, 1) element (0.0022). Choosing values of nk greater or less than 8 gives larger variances. The time constant and steady-state gain can be calculated from the a_1 and b_1 values: $\tau_o = 113$ seconds and $K_p = -0.0779$°C/%. These results are similar to those found by fitting a first-order model to step test data from the process.

$$G_{(s)} = \frac{-0.0779e^{-16s}}{113s + 1} \tag{16.26}$$

The frequency response of this first-order model is compared in Fig. 16.7 with that calculated directly from the data. The last part of the program in Table 16.1 calculates the predicted output of the process *ycalc* by feeding the actual input values into the model. The model predictions are compared with the actual experimental data in Fig. 16.8.

If a second-order model is used, the parameters *na* and *nb* are changed to 2. The variance is reduced to 0.0019. Parameter values are

$$\begin{aligned} a_1 &= -0.6739 \\ a_2 &= -0.3077 \\ b_1 &= -0.00051 \\ b_2 &= 0.0025 \end{aligned}$$

TABLE 16.1
Use of MATLAB identification toolbox

```
% Program "prbstest.m" to use matlab functions on prbs data
load prbs3.dat
% Read variables
t=prbs3(:,1);
temp=prbs3(:,2);
cw=prbs3(:,9);
% Initialize time
t=t-t(1);
% Plot raw data
clf
subplot(211)
plot(t,temp)
grid
title('Heat Exchanger Experimental Data (2/24/95)')
xlabel('Time (seconds)')
ylabel('Temperature (C)')
subplot(212)
plot(t,cw)
grid
xlabel('Time (seconds)')
ylabel('Controller Output (%)')
pause
% Detrend data
y=detrend(temp);
u=detrend(cw);
%
% Define vector z
% Use negative y because transfer function gain is negative
z=[-y u];
%
% Set frequency
w=[0.001:0.001:0.15];
%
% Calculate frequency response from data
%
nu=length(u);
ts=2;
g=spa(z,300,w,[],ts);
[w,mag,phase]=getff(g);
db=20*log10(mag);
%
% Get first-order model
na=1;
nb=1;
nk=8;
z=[-y u];
th=arx(z,[na nb nk]);
ts=2;
th=sett(th,ts);
```

TABLE 16.1 (CONTINUED)
Use of MATLAB identification toolbox

```
%
% Calculate frequency response from model
gmodel=trf(th,1,w);
[w,magmodel,phasemodel]=getff(gmodel);
%
% Plot Bode plot
db=20*log10(mag);
dbmodel=20*log10(magmodel);
clf
subplot(211)
semilogx(w,db,'-',w,dbmodel,'--')
title('Frequency Response')
grid
legend('Data (M=300,D=16 sec)','First-Order Model')
xlabel('Frequency (radians/sec)')
ylabel('Log Modulus (dB)')
subplot(212)
semilogx(w,phase,'-',w,phasemodel,'--')
grid
legend('Data','Model')
xlabel('Frequency (radians/sec)')
ylabel('Phase Angle (degrees)')
pause
variance=th(1,1)
fpe=th(2,1)
a1=th(3,1)
b1=th(3,na+1)
pause
tauo=-ts/log(-a1)
kp=b1/(1+a1)
pause
%
% Recalculate model using +y to get correct sign in gain
z=[y u];
th=arx(z,[na nb nk]);
ts=2;
th=sett(th,ts);
%
% Calculate time response of model with input u
ycalc=idsim(u,th);
clf
plot(t,y,'-',t,ycalc,'--')
legend('Data','Model')
title('Data versus First-Order Model Predictions (D=16 sec.)')
xlabel('Time (seconds)')
ylabel('Output')
grid
```

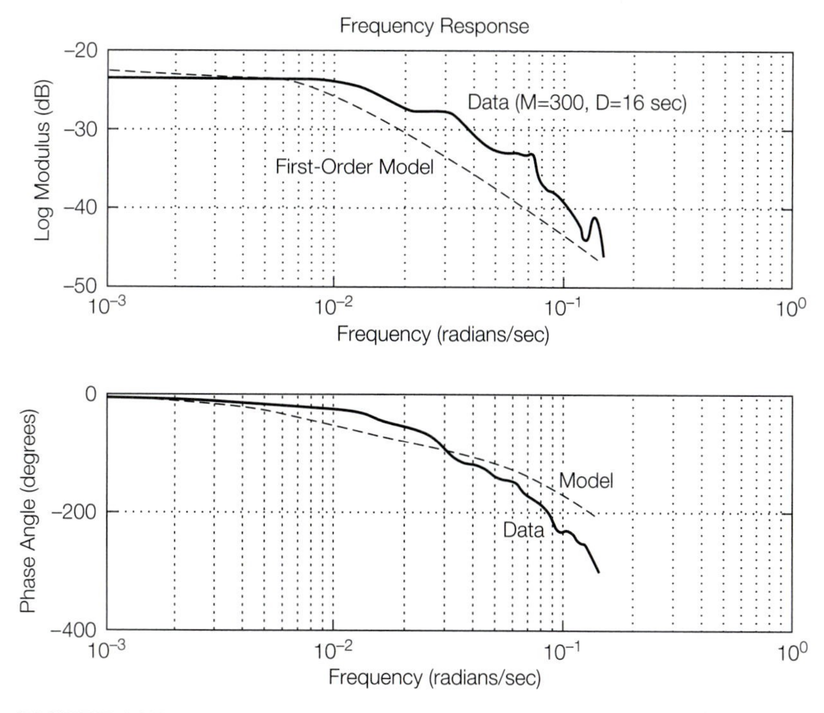

FIGURE 16.7

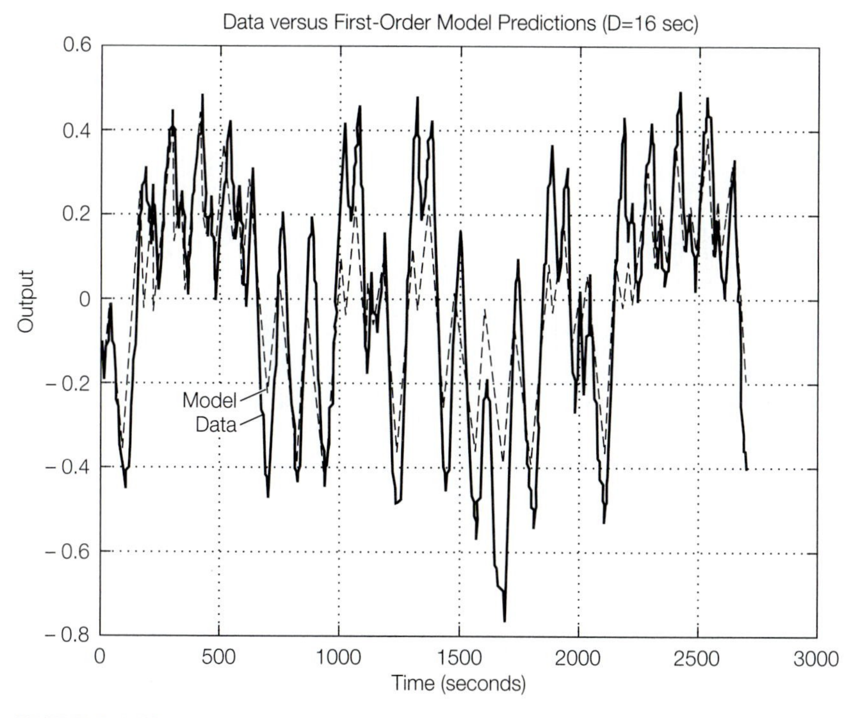

FIGURE 16.8

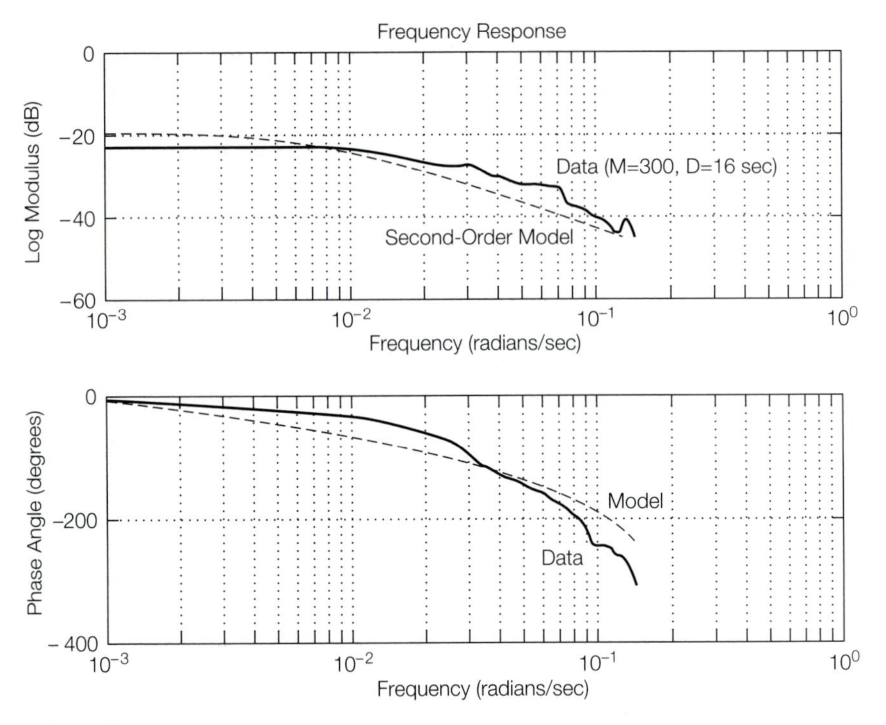

FIGURE 16.9

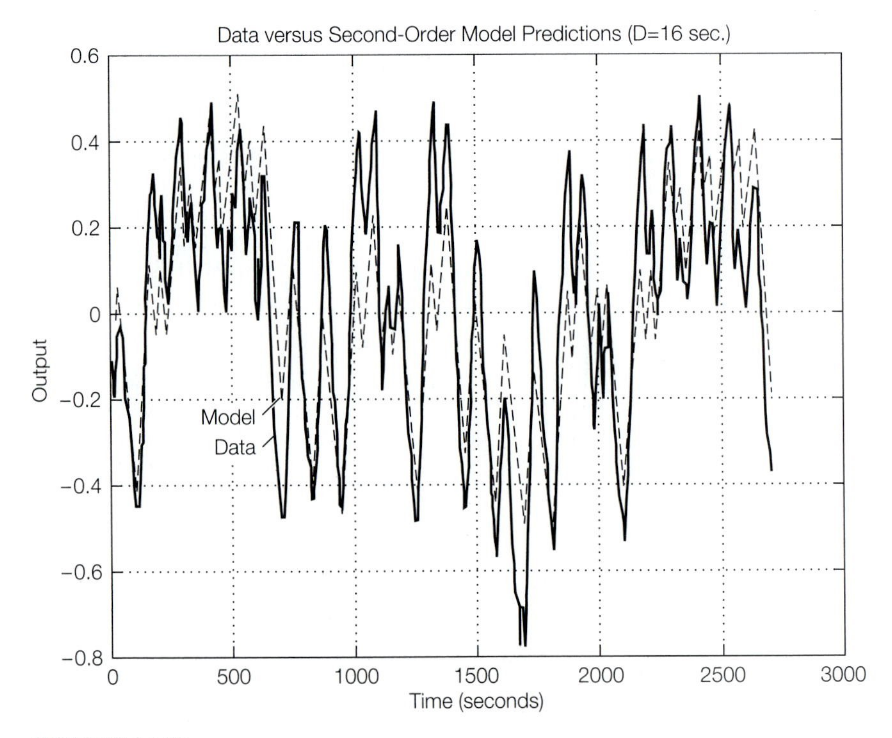

FIGURE 16.10

Figures 16.9 and 16.10 give frequency response and time-domain results for the second-order model.

Using the MATLAB Identification Toolbox and PRBS test signals is much more complex than using simple step tests or relay feedback tests. Much more data are required and the analysis is more difficult. However, the PRBS method is probably better in situations where the signals from the process are heavily corrupted by noise that cannot be simply filtered out.

16.7 CONCLUSION

This chapter has illustrated several identification methods that are used to determine dynamic parameters or models from experimental plant data. The simple and effective relay feedback test is a powerful tool for practical identification if the objective is the design of feedback controllers. The more complex and elegant statistical methods are currently popular with the theoreticians, but they require a very large amount of data (long test periods) and their effective use requires a high level of technical expertise. It is very easy to get completely inaccurate results from these sophisticated tests if the user is not aware of all the potential pitfalls (both fundamental and numerical).

It is almost never possible to take a huge quantity of data from the plant historian files and gain any understanding of how the process works, either at steady state or dynamically. Deliberate and well-conceived plant tests are usually required to gain useful information.

We strongly recommend the use of the simplest method that does the job. In most applications the tool of choice is the relay feedback test (ATV). It is quick, simple, and accurate, and it works to accomplish its goal. What more could be asked of a practical tool?

PROBLEMS

16.1 Using the simulation program given in Appendix A for the three-heated-tank process and a relay feedback test, determine the ultimate gain and ultimate frequency for the loop in which the temperature in the third tank T_3 is controlled by manipulating the heat input to the first tank Q_1. Compare these results with the theoretical values obtained in Example 8.8.

16.2 Perform a step test on the three-heated-tank process and fit a first-order lag plus deadtime model to the response curve. Calculate the ultimate gain and the ultimate frequency from the transfer function and compare with the results from Problem 16.1.

16.3 The following frequency response data were obtained from direct sine wave tests of a chemical plant. Fit an approximate transfer function $G_{(s)}$ to these data.

Frequency (rad/min)	Real part	Imaginary part	Log modulus (dB)	Phase angle (°)
0.01	6.964	−0.522	16.88	−4.2
0.02	6.859	−1.028	16.82	−8.5
0.04	6.467	−1.942	16.59	−16.7
0.08	5.254	−3.202	15.78	−31.3
0.10	4.568	−3.554	15.25	−37.9
0.20	2.096	−3.673	12.52	−60.2
0.40	0.324	−2.741	8.82	−83.2
0.63	−0.557	−2.234	7.49	−103
0.80	−1.462	−2.083	8.11	−125
1.00	−2.472	−0.104	7.87	−177
1.41	−0.282	0.547	−4.21	−243
2.00	−0.021	0.160	−15.81	−262
4.00	0.004	0.016	−35.5	−285
8.00	0.001	0.001	−53.9	−312

16.4 Disturb the three-heated-tank process with a PRBS input. Use the MATLAB Identification Toolbox to identify several alternative models, ranging from first to third order. Compare the ultimate frequencies and ultimate gains calculated from these models with the values obtained in Problem 16.1.

16.5 Simulate several first-order lag plus deadtime processes on a digital computer with a relay feedback. Compare the ultimate gains and frequencies obtained by the autotune method with the real values of ω_u and K_u obtained from the transfer functions.

APPENDIX A

TABLE A.1

```
c Program "tc3.f"
c
c P temperature control of three heated tanks in series
      dimension t(3)
      real kc(3)
      dimension dt(3)
      dimension tp(2000),t1p(3,2000),t2p(3,2000),t3p(3,2000)
      dimension q1p(3,2000)
      open(7,file='temp1.dat')
      open(8,file='temp2.dat')
      open(9,file='temp3.dat')
      data delta,tstop/.001,1./
      data to,f/70.,1000./
      data  kc/2.,4.,8./
      data dtprint,dtplot/0.05,.01/
c Make three runs with different controller settings
      do 1000 nc=1,3
      time=0.
      tprint=0.
      tplot=0.
      np=0
      do 10 ntank=1,3
   10 t(ntank)=150.
c controller calculation to get Q1
c All control signals are in 4-20 milliamperes
  100 continue
      pv=4.+(t(3)-50.)*16./200.
      if(pv.gt.20.)pv=20.
      if(pv.lt.4.)pv=4.
      e=12.-pv
      co=7.6+kc(nc)*e
```

TABLE A.1 (CONTINUED)

```
      if(co.gt.20.)co=20.
      if(co.lt.4.)co=4.
      q1=(co-4.)*10.e06/16.
c print and store for plotting
      if(time.lt.tprint)go to 30
      write(6,21)time,t,q1*1.e-06
   21 format(' time=',f6.2,' t=',3f7.2,'    q1=',3f7.2)
      tprint=tprint+dtprint
   30 if(time.lt.tplot)go to 40
      np=np+1
      tp(np)=time
      t1p(nc,np)=t(1)
      t2p(nc,np)=t(2)
      t3p(nc,np)=t(3)
      q1p(nc,np)=q1/1000000.
      tplot=tplot+dtplot
   40 continue
c evaluate all derivatives
      dt(1)=f*(to-t(1))/100.+   q1/(100.*50.*0.75)
      do 50 ntank=2,3
   50 dt(ntank)=f*(t(ntank-1)-t(ntank))/100.
c integrate ala Euler
      time=time+delta
      do 60 ntank=1,3
   60 t(ntank)=t(ntank)+dt(ntank)*delta
      if(time.lt.tstop)go to 100
c change controller settings
 1000 continue
c store data for plotting using MATLAB
      do 110 j=1,np
      write(7,111)tp(j),t1p(1,j),t2p(1,j),t3p(1,j),g1p(1,j)
      write(8,111)t1p(2,j),t2p(2,j),t3p(2,j),q1p(2,j)
  110 write(9,111)t1p(3,j),t2p(3,j),t3p(3,j),q1p(3,j)
  111 format(7(1x,f7.3))
      stop
      end
```

TABLE A.2

```
c Program "tc1dead.f"
c
c P temperature control of one heated tank with deadtime
        real kc(3),ndead
        dimension tp(2000),t1p(3,2000),q1p(3,2000),t1dead(1000)
        open(7,file='temp1.dat')
        open(8,file='temp2.dat')
        open(9,file='temp3.dat')
        data delta,tstop/.0001,0.25/
        data to,f/70.,1000./
        data kc/2.,4.,8./
        data dtprint,dtplot/0.05,.01/
c Disturbance is drop in To from 90 to 70
        to=70.
c Make three runs with different controller settings
        do 1000 nc=1,3
        time=0.
        tprint=0.
        tplot=0.
        np=0
        t1=150.
c Initialize t1dead array
        dead=0.01
        ndead=dead/delta
        if(ndead.gt.999)then
           write(6,*)' ndead too big'
           stop
           endif
        do 5 j=1,ndead
 5      t1dead(j)=t1
c controller calculation to get Q1
c All control signals are in 4-20 milliamperes
  100 continue
        pv=4.+(t1dead(ndead)-50.)*16./200.
        if(pv.gt.20.)pv=20.
        if(pv.lt.4.)pv=4.
        e=12.-pv
        co=7.6+kc(nc)*e
        if(co.gt.20.)co=20.
        if(co.lt.4.)co=4.
        q1=(co-4.)*10.e06/16.
c print and store for plotting
        if(time.lt.tprint)go to 30
        write(6,21)time,t1,q1*1.e-0.6
   21 format(' time=',f6.2,' t1=',f7.2,'    q1=',3f7.2)
        tprint=tprint+dtprint
   30 if(time.lt.tplot)go to 40
        np=np+1
        tp(np)=time
        t1p(nc,np)=t1
        q1p(nc,np)=q1/1000000.
        tplot=tplot+dtplot
```

TABLE A.2 (CONTINUED)

```
   40 continue
c evaluate derivatives
      dt1=f*(to-t1)/100.+   q1/(100.*50.*0.75)
c integrate ala Euler
      time=time+delta
      t1=t1+dt1*delta
c**********************
c Deadtime shift
      do 50 j=1,ndead
50    t1dead(ndead+1-j)=t1dead(ndead-j)
      t1dead(1)=t1
c**********************
      if(time.lt.tstop)go to 100
c change controller settings
 1000 continue
c store data for plotting using MATLAB
      do 110 j=1,np
      write(7,111)tp(j),t1p(1,j),q1p(1,j)
      write(8,111)t1p(2,j),q1p(2,j)
  110 write (9,111)t1p(3,j),q1p(3,j)
  111 format(3(1x,f7.3))
      stop
      end
```

NONLINEAR MODEL

The equations describing the nth jacketed, constant holdup, nonisothermal CSTR are

$$\frac{dz_n}{dt} = \left(\frac{F}{V_n}\right)z_{n-1} - \left(\frac{F}{V_n}\right)z_n - z_n k_0 e^{-E/R(T_n+460)}$$

where F = flow rate (lb-mol hr^{-1}) = 100
z_n = composition in nth stage (mole fraction A)
k_0 = preexponential factor (hr^{-1})
E = activation energy (Btu lb-mol^{-1}) = 30,000
R = 1.99 Btu lb-mol^{-1} °R^{-1}
T_n = reactor temperature (°F) = 140 at steady state

$$\frac{dT_n}{dt} = \left(\frac{F}{V_n}\right)T_{n-1} - \left(\frac{F}{V_n}\right)T_n - \frac{\lambda z_n}{Mc_p}k_0 e^{-E/R(T_n+460)} - \frac{UA_{Hn}}{c_p M V_n}(T_n - T_{Jn})$$

where T_n = temperature of nth reactor (°F)
λ = heat of reaction (Btu (lb-mol of A)$^{-1}$) = 30,000
M = molecular weight (lb lb-mol^{-1}) = 50
U = overall heat-transfer coefficient (Btu h^{-1} °F^{-1} ft^{-2})
A_{Hn} = heat-transfer area through reactor wall (ft^2) = $\pi D_n L_n$
D_n = reactor diameter (ft)
L_n = reactor height (ft) = $2D_n$
T_{Jn} = temperature in cooling jacket of nth reactor (°F)

$$\frac{dT_{Jn}}{dt} = \left(\frac{F_{Jn}}{V_{Jn}}\right)(T_{J0} - T_{Jn}) + \frac{UA_{Hn}}{c_J \rho_J V_{Jn}}(T_n - T_{Jn})$$

where F_{Jn} = coolant flow rate in jacket of nth stage (ft^3 h^{-1})
V_{Jn} = jacket volume in nth stage (ft^3) = $A_{Hn}(0.333)$ since a 4-in. jacket was assumed
T_{J0} = inlet coolant temperature (°F) = 70

APPENDIX B

Instrumentation Hardware

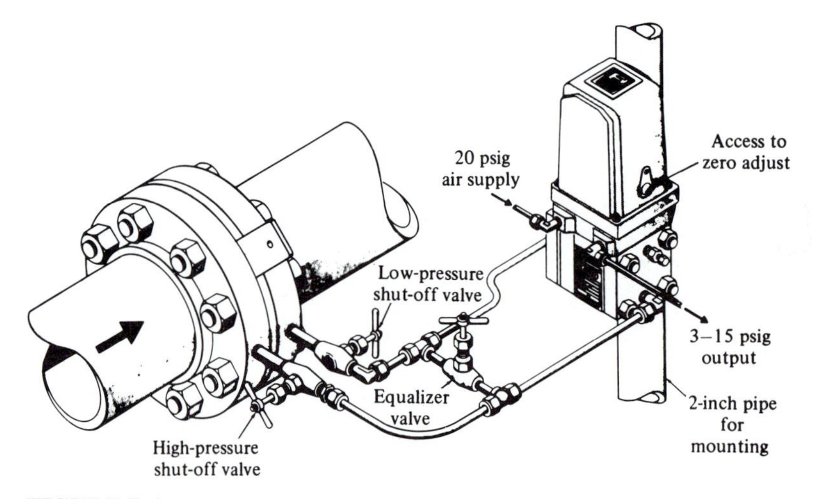

FIGURE B.1
Pneumatic differential pressure transmitter. Typical installation with orifice plate to sense flow rate. (*Courtesy of Fischer and Porter Company.*)

FIGURE B.2
Electronic differential-pressure transmitter. (*Courtesy of Honeywell.*)

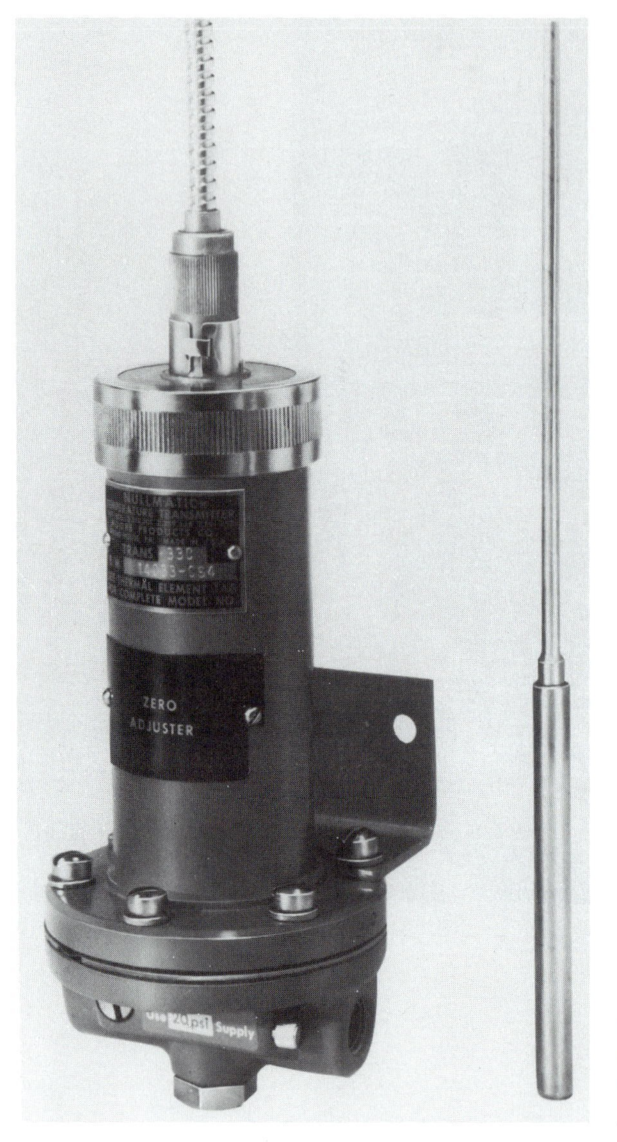

FIGURE B.3
Filled-bulb temperature transmitter. (*Courtesy of Moore Products.*)

FIGURE B.4
Control valve. (*Courtesy of Honeywell.*)

FIGURE B.5
Butterfly control valve with positioner. (*Courtesy of Foxboro.*)

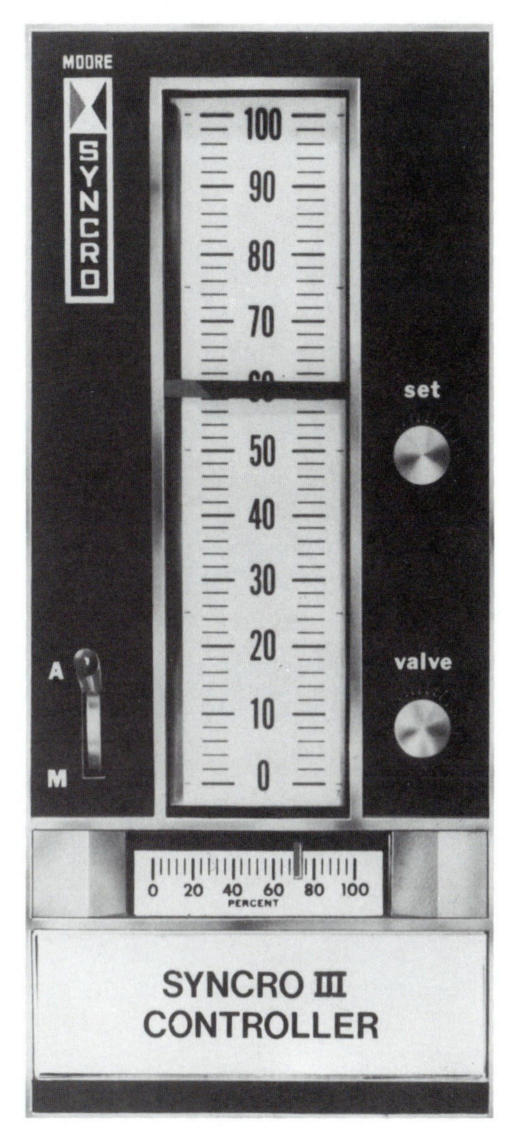

FIGURE B.6
Pneumatic control station. (*Courtesy of Moore Products.*)

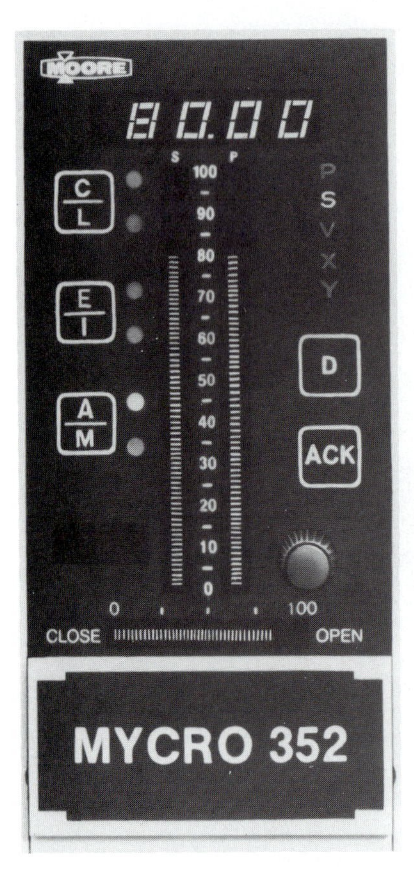

FIGURE B.7
Single-station microprocessor controller. (*Courtesy of Moore Products.*)

FIGURE B.8
Microprocessor control system (TDC 3000). (*Courtesy of Honeywell.*)

FIGURE B.9
Typical control room with computer control. (*Courtesy of Honeywell.*)

FIGURE B.10
Computer control console (CRT display). (*Courtesy of Honeywell.*)

INDEX